PROGRESS IN PHOTOSYNTHESIS RESEARCH

Progress in Photosynthesis Research

Volume 3

*Proceedings of the VIIth International Congress on Photosynthesis
Providence, Rhode Island, USA, August 10–15, 1986*

edited by

J. BIGGINS

*Division of Biology and Medicine, Brown University
Providence, RI 02912, USA*

1987 **MARTINUS NIJHOFF PUBLISHERS**
a member of the KLUWER ACADEMIC PUBLISHERS GROUP
DORDRECHT / BOSTON / LANCASTER

Distributors

for the United States and Canada: Kluwer Academic Publishers, P.O. Box 358, Accord Station, Hingham, MA 02018-0358, USA
for the UK and Ireland: Kluwer Academic Publishers, MTP Press Limited, Falcon House, Queen Square, Lancaster LA1 1RN, UK
for all other countries: Kluwer Academic Publishers Group, Distribution Center, P.O. Box 322, 3300 AH Dordrecht, The Netherlands

ISBN 90-247-3450-9 (vol. I)
ISBN 90-247-3451-7 (vol. II)
ISBN 90-247-3452-5 (vol. III)
ISBN 90-247-3453-3 (vol. IV)
ISBN 90-247-3449-5 (set)

Copyright

PRINTED IN THE NETHERLANDS

GENERAL CONTENTS

VI

CONTENTS TO VOLUME III

4. Metabolite Regulation of Carbon Metabolism

6. Other Chloroplast Enzymes

7. Metabolism of C4 and CAM Plants

8. Integration of Carbon and Nitrogen Metabolism

9. Photorespiration

PREFACE

These Proceedings comprise the majority of the scientific contributions
that were presented at the VIIth International Congress on Photosynthesis.
The Congress was held August 10-15 1986 in Providence, Rhode Island, USA
on the campus of Brown University, and was the first in the series to be
held on the North American continent. Despite the greater average travel
distances involved the Congress was attended by over 1000 active
participants of whom 25% were registered students. This was gratifying
and indicated that photosynthesis will be well served by excellent young
scientists in the future.

As was the case for the VIth International Congress held in Brussels,
articles for these Proceedings were delivered camera ready to expedite
rapid publication. In editing the volumes it was interesting to reflect
on the impact that the recent advances in structure and molecular biology
had in this Congress. It is clear that cognizance of structure and
molecular genetics will be even more necessary in the design of
experiments and the direction of future research.

Shortly after the Brussels Congress in 1983 the photosynthesis community
was grieved to hear of the death of Professor Warren I Butler. Warren was
very enthusiastic about the prospect of holding the VIIth International
Congress in the USA and was anxious not only to participate in the
scientific program, but to welcome and host colleagues from overseas in
his country. A special issue of Photosynthesis Research will be published
shortly containing articles closely related to his field of study. Other
outstanding scientists who also passed away during this time period
include Drs. A Shlyk (USSR), E Roux (France), A Faludi-Dániel (Hungary)
and G Akoyunoglou (Greece). All were recognised during the Congress at
Brown University in symposia dedicated to their memory.

The development of the scientific program and planning of the VIIth
International Congress was the responsibility of the US Organising
Committee and the International Photosynthesis Committee, and their
assistance is gratefully acknowledged. At the local level I wish to thank
my colleagues at Brown University for their support and, in particular,
the outstanding effort provided by Professor Sam I Beale. Special thanks
are also due to Professor Frank Rothman, Dean of Biology, for extensive
logistical support and encouragement, and to Kathryn Holden, the Congress
Secretary. Finally it is a pleasure to acknowledge the long-term
assistance of Ir Adrian C Plaizier of Martinus-Nijhoff publishers for
providing guidance in the production of these Proceedings, and for
bringing the publication to fruition.

ACKNOWLEDGMENTS

The organizers and Congress participants wish to express their appreciation for the financial support received from the following agencies and companies:

United States National Science Foundation
United States Department of Energy
United States Department of Agriculture

VIth International Congress, residual fund

CIBA-GEIGY Corporation
Campbell Soup Company
*E.I. du Pont de Nemours and Company
Monsanto Company
*Pepsico Incorporated
Pfizer Incorporated
Philip Morris
Proctor and Gamble
*Rohm and Haas Company
Shell Development Company
Weyerhaeuser Company

* Benefactor contributor

Commercial products were gratefully received from the following companies:

R.C. Bigelow Incorporated
Coca Cola Bottling Company of Northampton
E.I. du Pont de Nemours and Company
R.T. French Company
Frito-Lay, Incorporated
J and J Corrugated Box Corporation
Nyman Manufacturing Company
Rhode Island Lithograph Company

INTERNATIONAL PHOTOSYNTHESIS COMMITTEE

J.M. Anderson (Australia)
C.J. Arntzen (U.S.A.)
M. Baltscheffsky (Sweden)
J. Barber (U.K.)
J. Biggins (U.S.A.), Chairman
R. Douce (France)
G. Forti (Italy)

H. Heldt (F.R.G.)
R. Malkin (U.S.A.)
N. Murata (Japan)
N. Nelson (Israel)
V.A. Shuvalov (U.S.S.R.)
C. Sybesma (Belgium)
R.H. Vallejos (Argentina)

US ORGANIZING COMMITTEE

C.J. Arntzen
J. Biggins (Chairman)
N.I. Bishop
L. Bogorad
A.L. Christy
A.R. Crofts
P.L. Dutton

M. Gibbs
G. Hind
A.T. Jagendorf
R.E. McCarty
K. Sauer
I. Zelitch

The committee is especially grateful to Dr. E. Romanoff of the National Science Foundation and Dr. R. Rabson of the Department of Energy for valuable advice and support during organization of the Congress.

CONGRESS SECRETARIAT: J. Biggins, Section of Biochemistry, Division of Biology and Medicine, Brown University, Providence, Rhode Island 02912

Localization of the nucleotide binding sites on subunits of CF_1 and TF_1 using the photoreactive nucleotide analog 3-O-(4-benzoyl) benzoyl ATP

Noun Shavit
Biology Department, Ben Gurion University of the Negev, Beer Sheva, Israel

Introduction

The F_1F_O proton translocating ATP synthetases[1] from mitochondrial, chloroplast thylakoids and bacterial plasma membranes are the principal catalyst in the transduction of energy derived from electron transport (1-3). These enzyme complexes are generally composed of an integral membrane sector, F_O, which mediates H^+ translocation, and a peripheral protein sector, F_1, which catalyzes the terminal step of ATP synthesis from ADP and P_i. F_1 is composed of five different polypeptide chains designated $\alpha-\epsilon$ in order of decreasing M_r. The molecular weight of F_1 is about 380,000 and the subunit stoichiometry is $\alpha_3\beta_3\gamma\delta\epsilon$. The binding sites for ADP and ATP are located on the α and β subunits and the catalytic sites are thought to be on the β subunits. The F_1-ATPases from mitochondria (MF_1) and *E. coli* (EF_1) were shown to contain six nucletide binding sites (1, 2). The F_1-ATPase from the thermophilic bacterium, PS3 (TF_1), contains at least four nucleotide binding sites (4) and CF_1, from chloroplast thylakoid membranes, contains three nucleotide binding sites (5,6). TF_1 has three noncatalytic nucleotide binding sites putatively in the α subunits and at least one catalytic site on the β subunit (4). Soluble CF_1 has two sites where ADP can bind, and in the presence of Mg^{2+} and ATP, another single site for ATP was described (6). Tight binding of ADP and ATP to the membrane bound CF_1 was also described (7-10). Although the role of these nucleotide binding sites has been studied extensively, it is difficult to distinguish between nucleotides bound at catalytic and noncatalytic sites. Several lines of evidence amply documented by Boyer and coworkers (7) support the participation of tightly bound nucleotides as catalytic intermediates in ATP synthesis and hydrolysis. However, such a role for tightly bound ATP in photophosphorylation is questionable in view of its relatively low apparent rate of formation. Moreover, during steady-state phosphorylation of various ADP analogs, relatively high rates are observed without the appearance of significant levels of bound products (11).The level of adenine nucleotides tightly bound to membranal CF_1 correlates well with the loss of its Pi-ATP exchange, ATPase and ATP synthetase activities on one hand, and with the reciprocal of energy input on the other hand. It was therefore suggested that energy induced activation of the ATP synthetase with the concomitant release of bound nucleotides may serve to modulate the enzymic activity. In this alternative the role of the nucleotide binding sites is strictly regulatory. The enzyme is active

---

[1]Abbreviations used: MF_1, EF_1, TF_1 and CF_1, for the F_1-ATPase from mitochondria, *E. coli* plasma membranes, plasma membranes of the thermophilic bacterium PS3 and chloroplast thylakoids, respectively. FSBA, 5'-p-fluorosulfonylbenzoyladenosine; BzADP (BzATP), 3-O-(4-benzoyl) benzoyl - ADP and -ATP, respectively; ATPase, CF_1 E.C. 3.6.1.3; SDS-PAGE, polyacrylamide gel electrophoresis in the presence of sodium dodecyl sulphate.

Biggins, J. (ed.), Progress in Photosynthesis Research, Vol. III. ISBN 90 247 3452 5
© *1987 Martinus Nijhoff Publishers, Dordrecht. Printed in the Netherlands.*

when the bound ADP is replaced by ATP. The inactivation process involves binding of ADP to the deenergized form of the enzyme (3, 8-10). Therefore, tightness or looseness of binding are not useful criteria to distinguish between catalytic sites and noncatalytic sites.

Covalent chemical modification and in particular photoaffinity labeling is a useful tool to determine the specific regions on a subunit polypeptide chain where nucleotides bind. Photoreactive nucleotide analogs of the nitrene type, 2-azido-ADP as well as the 2-azido-ATP newly synthesized on the enzyme, were shown to bind to the β subunit of the membrane bound CF_1 (12). 3-O-(4-benzoyl) benzoyl ADP (BzADP), a carbene type of photoreactive adenine nucleotide analog, was also shown to bind to the nucleotide binding sites on subunits of CF_1 and TF_1. BzADP is not a substrate for photophosphorylation, but acts as a strong competitive inhibitor with respect to adenine nucleotides in ATP synthesis or hydrolysis and Pi-ATP exchange catalyzed by membrane bound CF_1. BzADP binds competitively with respect to ADP to the tight binding site(s) and protects against inactivation of the enzyme induced by tight binding of ADP. Modification of CF_1 *in situ* (in the membrane) results in the covalent incorporation of BzADP to both α and β subunits (13). BzADP also inhibits the Ca^{2+}-ATPase (14) and Mg^{2+}- ATPase (15) activities of soluble CF_1 and TF_1, respectively. Binding of 2-3 mol of BzADP per mol enzyme results in the complete loss of enzymic activity. In contrast with the pattern of incorporation observed with the membrane bound CF_1, BzADP is incorporated almost exclusively into the β subunit of soluble CF_1 and TF_1 (14, 15).

Proteolytic digestion of photolabeled β subunits obtained by interaction of soluble CF_1 with 2-azido ADP, 2-azido ATP (16) and with BzADP (M.S. Abbott et al., unpublished), demonstrated that these nucleotide analogs are incorporated to the same polypeptide region of the β subunit. However, interaction of the membrane bound enzyme with BzADP followed by proteolytic digestion of the β subunits showed a different pattern of labeling of the β subunit. It was therefore suggested that the differences in labeling patterns obtained with different photoaffinity analogs and different type of enzyme preparations (soluble versus membrane bound) reflect differences in the conformation of the nucleotide binding sites on CF_1. In order to address these questions, we have studied the labeling patterns of membrane bound CF_1, soluble CF_1 and TF_1 with [^{14}C]BzATP. We describe here the polypeptide labeling patterns of the labeled β subunit. After purification the labeled polypeptides will be sequenced in order to establish the amino acid(s) to which the analog binds.

Materials   and Methods

BzATP was prepared following the procedure of Williams and Coleman (17) as modified by Bar-Zvi and Shavit (13). [γ-^{32}P]BzATP was prepared with [γ-^{32}P]ATP (14) and [^{14}C]BzATP with [1-^{14}C-carboxy]-4-benzoylbenzoic acid (18), a generous gift from Dr. R. G. Yount, Washington State University, Pullman, Washington. ^{32}P$_i$ was obtained from the Nuclear Research Center-Negev, Israel.

Enzyme preparation and assays
CF_1 and TF_1 were prepared and stored as described (13, 15). Latent CF_1 (5-10 μg) was activated by heating at 60 °C for 4 min in an activation buffer containing, in a final volume of 40 μl, as follows: 40 mM Tricine-NaOH (pH 8.0) / 2 mM EDTA / 25 mM ATP and 5 mM dithiothreitol. Ca^{2+}- ATPase activity was determined as described earlier (14). The

reaction mix (1 ml) for TF_1 - ATPase activity contained as follows: 50 mM glycylglycine buffer (pH 8.6) / 5 mM ATP (containing $4 \cdot 10^7$ cpm / nmol $[\gamma - {}^{32}P]ATP$) / 5mM $MgCl_2$ and 3-5 µg TF_1. P_i released was determined as described (19).

Binding of nucleotides

Reversible noncovalent binding of BzATP to thylakoid membranes, CF_1 and TF_1 was done as previously described (8,9,15). Covalent binding of labeled BzATP to these preparations was as described (14,15). Quantitative estimation of label incorporated was done as follows: an ice cold perchloric acid solution was added (final concentration, 5%) to samples irradiated in the presence of $[\gamma-{}^{32}P]$- or $[^{14}C]$- BzATP. After centrifugation for 2 min at 12 000 x g the pellets were washed thrice with 0.2 ml of 5% perchloric acid. Washed pellets were resuspended in 5% sodium dodecyl sulphate and radioactivity determined.

Subunit distribution of labeled BzATP

The acid washed preparations after covalent binding of $[\gamma-{}^{32}P]BzATP$ or $[^{14}C]BzATP$, were resuspended in the sample buffer containing 5 % sodium dodecyl sulphate / 10 mM Tris-HCl (pH 8.0) / 0.01 % bromophenol blue / 5 % glycerol. The pH was adjusted to 7 with 0.5 M Tris-HCl (pH 8.0), and the samples were incubated for 30 min at room temperature. Electrophoresis of proteins was carried out in the presence of sodium dodecyl sulphate, in .8-mm thick, 11 % polyacrylamide slab gels according to Laemmli (20). Radioactivity and protein content of bands in the gel were determined as described (14).

Partial Proteolytic digestion and electrophoresis of the β subunit

The β subunit of CF_1 and TF_1 was purified from the photolabeled, soluble enzymes or from photolabeled thylakoid membranes by SDS-PAGE as described above. The section of each lane containing the β subunit was excised, and the gel slices were equilibrated in buffer containing 125 mM Tris-HCl (pH 6.8), 0.1 % SDS, and 1 mM EDTA, and stored at -20°C. Protease V8 from *Staphylococcus aureus* (Miles Laboratories Inc.) was dissolved and diluted in the above buffer containing 10 % glycerol (v / v). Digestion of gel bands with protease was carried out essentially according to Cleveland *et al.* (21) on a 1.5-mm thick, 10 -20 % polyacrylamide slab gel. Both the reservoir buffers and the gel solutions contained 1 mM EDTA. Gel slices, containing the photolabeled β subunit, were inserted into the 6-mm wide wells of the stacking gel. Buffer solutions, containing various amounts of protease, were layered into the wells, and electrophoresis at room temperature was carried out as described (22). Low molecular weight standards (Pharmacia Inc.) were used throughout.

Autoradiography and fluorography

Gels containing proteins labeled with $[\gamma - {}^{32}P]BzATP$ were stained for 2 - 4 hours and destained in 10 % acetic acid and 10 % ethanol. The destained gels were dried between two sheets of uncoated cellophane and subjected to radioautography at -80 °C using X-omat AR film (Kodak) (21). Gels containing proteins labeled with $[^{14}C]BzATP$ were stained and destained as above. Then the gels were prepared for fluorography (22) and dried between one cellophane and a polyethylene sheet. After peeling of the plastic sheet, gels were exposed to film for about a month as described above.

Results

BzATP is not a substrate for ATP hydrolysis and Pi-ATP exchange reactions catalyzed by the membrane bound CF_1 nor in ATP hydrolysis by soluble CF_1 and soluble TF_1. Without photoactivation, BzATP inhibits the hydrolysis of ATP by both soluble CF_1 and TF_1 (see Fig. 1), and the Pi-ATP exchange by the membrane bound CF_1, in a competitive manner with respect to ATP. Ki(s) values of about 15 μM (Fig. 1) for the inhibition of TF_1-ATPase were calculated. Thus BzATP resembles the behaviour of BzADP (13, 14) which is not a substrate for phosphorylation but inhibits strongly phosphorylation and hydrolysis in chloroplasts, and the hydrolysis of ATP by the soluble ATPases. As with BzADP, covalent incorporation of BzATP and inactivation of either enzyme was prevented by the presence of either ADP or ATP. The degree of inhibition was only slightly increased if preincubation and irradiation were done at high ionic strength (not shown). The simplest interpretation would be that both the ADP and ATP analogs bind to the same site(s) on the ATPase, where covalent incorporation of either noncovalently bound analog results in the inactivation of the enzyme. Alternatively, there could be distinct nucleotide binding sites on the enzyme to which each analog could bind independently. We have recently prepared two types of radioactive BzATP analogs, labeled with [^{32}P] in the γ-phosphate of the ATP portion or with either ^{3}H or ^{14}C in the benzoylbenzoic acid moiety. The latter compounds when activated react with the protein to form covalent adducts that are stable and do not release the label during the isolation and analysis procedures. These analogs should be useful to determine the distribution of nucleotides bound to proteolytic fragments of labeled subunits and to establish the amino acid which was modified. As shown in Fig. 2 the covalent binding of [^{14}C]BzATP to TF_1 correlates well with the loss of enzymic activity.

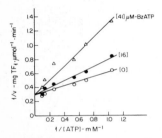
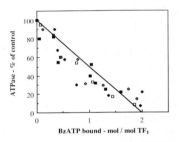

Fig. 1.Inhibition of ATP hydrolysis by noncovalent binding of BzATP. ATP hydrolysis was assayed as described in Materials and Methods. In the absence of BzATP (O), with 16 μM (●) and with 40 μM (△)BzATP.

Fig. 2. Covalent binding of [^{14}C]BzATP and inactivation of the ATPase activity. Irradiation conditions: 115 μM [^{14}C]BzATP and 0.64 μM TF_1 with either 5mM $MgCl_2$ (◇,□) or with 1 mM EDTA (◆,■) for 60 min. Molecular weight of TF_1 used, 380 000

Complete inactivation of the enzyme is attained after incorporation of 1.96 mol BzATP / mol TF_1, independent of the Mg^{2+} concentration during irradiation. The extent of incorporation required for complete inhibition is similar to that obtained with BzADP (14), and somewhat lower than that reported for CF_1 modified with the 2-azido nucleotide analogs (16) [γ- ^{32}P] BzATP is covalently incorporated almost exclusively (>95%) into the β subunit of both the soluble CF_1 and TF_1, Figs. 3A and B, respectively. The pattern of labeling was not changed when photoactivation was carried out with or without Mg^{2+} and for longer periods, except that the extent of labeling increased with Mg^{2+} or time of irradiation (14, 15). However, when BzATP reacted with the membrane bound CF_1, *in situ*, both the α and β subunits of the CF_1 isolated thereafter were labeled (Fig. 4). Labeling of both the α and β subunits was also observed by electrophoressing dissolved labeled thylakoid membranes, thus ruling out a transfer of label during the manipulation of the samples.

Fig. 3. Photolabeling of soluble CF$_1$ and TF$_1$ by [γ- ^{32}P]BzATP. A) Purified CF$_1$ (3.3 μM) was incubated at room temperature for 30 min in the presence of 50 μM [γ- ^{32}P]BzATP. Samples were irradiated with and without 5 mM MgCl$_2$ for the time indicated and the photolabeled protein (35 μg) was analyzed by gel electrophoresis and autoradiography as described in Materials ans Methods. Lane 1 - Coomassie blue stained, photolabeled CF$_1$; autoradiograms: lane 2- 30 min; lane 3- 60 min and lane 4 with Mg^{2+}- 60 min.

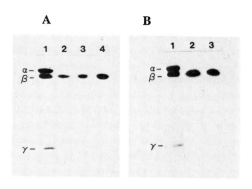

B) Purified TF$_1$ (2.6 μM) was incubated for 60 min in the presence of 37 μM [γ-^{32}P]BzATP at room temperature. Irradiation with and without 5 mM MgCl$_2$ for 60 min was as described and the photolabeled protein (30 μg) was analyzed as in A.

Fig. 4. Photolabeling of membrane bound CF$_1$ by [γ- ^{32}P]BzATP. Thrice washed chloroplasts were prepared, activated and irradiated as described (13). Chloroplast thylakoids (200 μg chlorophyll) with 4.5 μmol of [γ- ^{32}P]BzATP were irradiated with UV light (360 nm) for 30 min in the presence of 30 μM FCCP. An aliqout was taken for extraction of CF$_1$, the rest was solubilized with 5 %SDS buffer. Gel electrophoresis and radioautography was as described.Lane 1 and 3 - Coomassie blue stained, photolabeled CF$_1$ and thylakoid proteins,respectively. Lanes 2 and 4 - autoradiograms prepared from these gels.

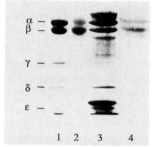

The protease V8 from *S. Aureus* was used to digest the TF$_1$ β subunits labeled with [^{14}C]BzATP. This protease, which hydrolyzes peptide bonds adjacent to aspartate and glutamate residues, produces a specific and reproducible digestion pattern. Lanes 1, 3 and 5 of Fig. 5 show the protein stained electrophoretic pattern of digestion of isolated TF$_1$ β subunits with increasing amounts of protease V8. Two major polypeptides (apparent M$_r$ 47 and 39 kDa) are produced after digestion with 10 ng of protease (lane 3). Incorporation of photoaffinity nucleotide analogs into the β subunit does not affect the electrophoretic pattern of proteolytic fragments obtained (not shown). With higher amounts of protease three more polypeptides are obtained (29, 23 and 15 kDa). From the intensity of the stained bands it is probable that the higher molecular weight polypeptides and the β subunit are further degraded to produce the smaller polypeptides. The [^{14}C]BzATP labeled polypeptide distribution (lanes 2, 4 and 6 of Fig. 5) indicates that all five polypeptides are labeled. A 10-11 kDa polypeptide which migrates immediately behind the dye front is also produced at high enough protease concentration but this band is very close to self digestion products of the protease and is almost certainly contaminated with protease. However it is not labeled with radioactive nucleotide. Therefore, comparing with the labeling pattern obtained with higher protease concentrations the two smaller fragments (23 and 15 kDa) appear to be enriched in radioactivity as proteolysis proceeds and the 14-15 kDa polypeptide is the only labeled product left, after extensive proteolysis, as shown in lane 3 of Fig. 6.

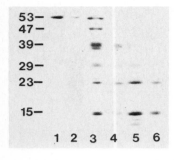 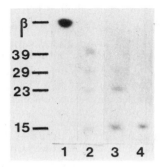

Fig. 5. Partial proteolysis of the β subunit of photolabeled TF_1. Soluble TF_1 photolabeled with [^{14}C]BzATP (5.7 μmol / mg TF_1) in the presence of 5 mM $MgCl_2$ and the subunits were separated by SDS-PAGE. V8 protease (0, 10, 20, 40 ng to lanes 1 - 6, respectively) was added to the β subunits and proteolysis, eletrophoresis and fluorography was as described in Materials and Methods. Lanes 1, 3 and 5 are Coomassie blue stained gels; lanes 2, 4 and 6 are autoradiograms of the proteolysis pattern of these gels.

Fig. 6. Partial proteolysis of the β subunit of photolabeled TF_1. β subunits from TF_1 photolabeled in the presence of $MgCl_2$ (6.8 μmol [^{14}C]BzATP / mg TF_1) were digested with V8 protease 30, 100 and 300 ng (lanes 1-3, respectively) for one hour at room temperature. Electrophoresis and fluorography were as described in Materials and Methods.

Discussion

The data presented clearly show that interaction of BzATP with the membrane bound CF_1 results in the incorporation of label into both the α and β subunits. In contrast, interaction of BzATP with either soluble CF_1 or TF_1 results in the photolabeling of the β subunit only. Previous studies have shown similar patterns of photolabeling with BzADP (13-15). The labeling pattern of membrane bound CF_1 obtained with these photoaffinity nucleotide analogs indicates that the adenine nucleotide binding sites are probably located at the junction of an α and β subunit. The pattern of labeling does not depend on the conditions of photolabeling or on the treatment of CF_1 for gel electrophoresis. The fact that these nucleotide analogs bind to the two subunits only when the protein is attached to the membrane but not to CF_1 in solution supports the view that the conformation of the protein *in situ* differs from that in solution. The existence of conformational changes in CF_1 as thylakoid membranes are energized and deenergized, or on soluble CF_1 is well supported (see Ref. 3). These differences in the pattern of labeling may be due to the existence of two different interconvertible forms of CF_1 in the membrane, one adapted for ATP synthesis and the other adapted for ATP hydrolysis (24, 25). These results are also in agreement with previous studies which have shown that under deenergized conditions, membrane bound CF_1 binds both ADP and ATP (8-10). Upon energization, the ADP present on the membrane bound CF_1 is rapidly released. Both ADP and ATP can rebind to this site during or after energization (23, 24).

Two rather different photoreactive adenine nucleotide analogs, the nitrene (12) and the carbene (14,17) type, react similarly with CF_1 in solution (12,14). Complete inactivation of the CF_1-ATPase activity is obtained upon incorporation of approximately 3 mol analog per mol enzyme. Similarly for complete inactivation of the TF_1-ATPase activity, incorporation of 2-3 mol of BzADP (15) or 2 mol of BzATP (Fig. 2) were also required. The nucleotide binding sites assumed to be located on the α subunits of TF_1 (4) are not labeled by BzADP or BzATP (15, Fig. 3), while the site on the β subunit that is labeled was suggested to be a catalytic site (4). However, if the sites on the β subunits are indeed catalytic sites, the fact that all three sites on the β subunit must be modified for complete inactivation is difficult to reconcile with mechanisms for ATP hydrolysis as postulated by Boyer and coworkers (7). Complete inactivation of MF_1 with FSBA also occurred only after all three β subunits were modified (25). It has also been proposed that fewer than three copies of the β subunit in MF_1 contain catalytic sites (26). Only one β subunit may have an active site which is modulated by ligands bound to the "latent catalytic sites" on the other two β subunits. Alternatively, it was suggested that two β subunits are catalytically active and that the third β subunit contains a regulatory site which modulates the two catalytic sites. To distinguish between these possible models it is important to determine which amino acid residue(s) on the β subunit polypeptide chain in each conformation of the protein reacts with the photoaffinity nucleotide analog.

The labeling pattern of proteolytic fragments obtained with TF_1 is very different from that reported for CF_1 (16). TF_1 contains more aspartic and glutamic acid residues than CF_1 and the number of appropriately sized peptide fragments produced is consequently higher than that with CF_1. Nevertheless, a peptide with an apparent M_r of 14.5 - 15 kDa appears to participate in the binding of the labeled nucleotide analog as is the case with CF_1 (16). We cannot at this stage acertain if the 15 kDa peptide is a degradation product of the 23 kDa peptide or these two peptides are distinct products of proteolysis of the β subunit. However, it is clear that the 15 kDa peptide is the only labeled peptide detected after extensive digestion of the β subunit with the V8 protease. This observation is rather interesting since the 15 kDa peptide was a product of the proteolytic digestion of labeled β subunits isolated from both CF_1 and TF_1 in solution but not from CF_1 labeled *in situ* (M.S. Abbott, et al., unpublished). This suggests that, within the limitations of this technique, the ribose-binding portion of the nucleotide binding site on the β subunit of the two proteins in solution is structurally similar. If there is only one binding site on the β subunit and complete inactivation of the ATPase requires binding of nucleotides to all three subunits, the site on the β subunits may indeed be the catalytic sites. If there are two adenine nucleotide sites on the β subunit of TF_1 the site modified by BzADP and BzATP need not be a catalytic site. These possibilities could be tested by determining the amino acid residues to which the nucleotide analogs bind. Such experiments are currently underway.

Acknowledgments

This work was supported in part by a grant from I. B. Black, Bournemouth, U.K. We wish to thank Dr. R. G. Yount for his generous gift of [1-^{14}C]benzoyl benzoic acid.

References

1. Futai, M. and Kanazawa, H. (1983) Microb. Rev. 47, 285 - 312
2. Senior, A.E. and Wise, J.G. (1983) J. Memb. Biol. 73, 105 - 124
3. Shavit, N. (1980) Ann. Rev. Biochem. 49, 111 - 138
4. Yoshida, M. and Allison, W.S. (1986) J. Biol. Chem. 261, 5714 - 5721
5. Shoshan, V. Chipman, D.M. and Shavit, N. (1978) Biochim. Biophys. Acta 504, 108 - 122
6. Bruist, M.F. and Hammes, G.G. (1981) Biochemistry 20, 6298 - 6305
7. Rosen,G., Gresser, M., Vinkler, C. and Boyer, P.D. (1979) J. Biol. Chem. 254, 10654 - 10661
8. Schumann, J. and Strotmann, H. (1980) Proc. Int. Congr. Photosynth., 5th, pp.789 - 800
9. Bar-Zvi, D. and Shavit, N. (1980) FEBS Lett. 119, 68 - 72
10. Dunham, K.R. and Selman, B.R. (1981) J. Biol. Chem. 256, 212 - 218
11. Aflalo, C. and Shavit, N. (1982) Eur. J. Biochem. 126, 61 - 68
12. Czarnecki, J.J., Abbott, M.S. and Selman, B.R. (1983) Eur. J. Biochem. 136, 19 - 24
13. Bar-Zvi, D., Tiefert, M.A. and Shavit, N. (1983) FEBS Lett. 160, 233 - 238
14. Bar-Zvi, D. and Shavit, N. (1984) Biochim. Biophys. Acta 765, 340 - 346
15. Bar-Zvi, D., Yoshida, M. and Shavit, N. (1985) Biochim. Biophys. Acta 807, 293 -299
16. Abbott, M.S., Czarnecki, J.J. and Selman, B.R. (1984) J. Biol. Chem. 259, 12271 - 2278
17. Willams, N. and Coleman, P.S. (1982) J. Biol. Chem. 257, 2834 - 2841
18. Nakamaye, K.L. and Yount, R.G. (1985) J. Labeled. Compd.& Radiopharm. 12 607 - 613
19. Lindberg, O. and Ernster, L. (1956) Methods Biochem. Anal. 3, 1 - 24
20. Laemmli, U.K. (1970) Nature 227, 680 - 685
21. Cleveland, D.W., Fisher, S.G., Kirschner, M.W. and Laemmli, U.K. (1977) J. Biol. Chem. 252, 1102 - 1106
22. Bonner, W.M. and Laskey, R.A. (1974) Eur. J. Biochem. 46, 83 - 88
23. Shavit, N. Bar-Zvi, D., Aflalo, C. and Tiefert, A.T. (1984) Proc. Int. Congr. Photosynth. 6th, Vol. II, pp.493 - 500
24. Schumann, J. (1984) Biochim. Biophys. Acta 766, 334 - 342
25. Bullough, D.A. and Allison, W.S. (1986) J. Biol. Chem. 261, 5722 - 5730
26. Soong, K.S. and Wang, J.H. (1984) Biochemistry 23, 136 - 141

NUCLEOTIDE BINDING TO THE ISOLATED CHLOROPLAST COUPLING FACTOR ONE

JÜRGEN SCHUMANN   Botanisches Institut der Universität Düsseldorf
Universitätsstraße 1   D-4ooo Düsseldorf 1 Federal Republic of Germany

INTRODUCTION
On the isolated chloroplast coupling factor CF_1, three nucleotide binding
sites were found (1): site 1 binds an ADP molecule which remains bound
even after isolation; this ADP is exchanged during incubation of the
isolated enzyme with ADP or ATP. Site 2 binds a MgATP molecule; CF_1 stored
under ammonium sulfate contains no nucleotide on this site. Upon addition
of ATP and magnesium ions, this site is filled with MgATP which remains
bound even after prolonged incubation with unlabeled nucleotides and/or
EDTA. Site 3 binds ADP or ATP in a reversible manner; the dissociation
constant is about 2 to 5 μM.
The exchange of bound ADP or MgATP during ATP hydrolysis was found to be
slower than hydrolysis (1); it was assumed that only site 3 is involved in
the catalytic ATP cleavage. Recent results from another group, however,
indicate a faster exchange of bound ADP which enables this site to be involved in an alternating binding change mechanism of ATP hydrolysis (2).
In this paper, the kinetics of nucleotide binding was investigated in
further detail; the results do not exclude a direct participation of ADP
bound to site 1 in ATP hydrolysis. From our data it seems obvious that
tight binding of ATP or ADP to site 1 requires binding of one medium
nucleotide to a site with an affinity of 3 to 5 μM and subsequent release
of bound ADP. Thereafter, a second nucleotide from the medium binds to
the enzyme, and one of these nucleotides is converted into a tightly
bound ADP.

MATERIALS AND METHODS
CF_1 from market or greenhouse spinach was isolated, stored and desalted
as described elsewhere (3). (8-^{14}C)ADP or -ATP was incubated with the
enzyme (o.4 mg/ml) at room temperature. At the times indicated, the protein was separated from free nucleotides by centrifugation of the sample
through syringes with Sephadex G-5o fine (4). Radioactivity of eluted
samples were counted in scintillation cocktail, and protein was determined
according to Lowry et al. (5).
Activation of CF_1 was done by incubation with dithiothreitol (5o mM, 2 h)
(3). ATP hydrolysis was determined by thin-layer chromatography of (^{14}C)-
labeled nucleotides on PEI-cellulose plates with o.75 M NaH_2PO_4.

RESULTS AND DISCUSSION
Upon addition of (^{14}C)labeled ADP or -ATP (1o μM) to the isolated coupling
factor (1 μM), the nucleotides were incorporated into tight binding sites
on the enzyme at different rates and to different extents (Fig. 1). A
maximum stoichiometry of one nucleotide per CF_1 was obtained after about
4 hours when ADP or ATP were added without magnesium ions. Binding is
completed after 15 to 3o minutes if $MgCl_2$ is supplied with the nucleotide;
the maximum stoichiometry is 1 for ADP and 2 for ATP (Fig. 1).
This result is in agreement with the data of Bruist and Hammes (1); they
described an exchange of bound ADP (on site 1 of the isolated enzyme) for
ATP or ADP in the absence or presence of Mg^{2+}. An additional binding site

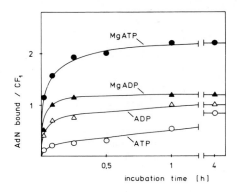

Fig. 1
Time course of nucleotide binding
to isolated latent CF_1.
CF_1 (1 µM) was incubated with
(^{14}C)ADP or (^{14}C)ATP (1o µM)
(+/- 5 mM $MgCl_2$) for the times
indicated.

is occupied after incubation with MgATP. Binding to site 2 was found to
be fast, but binding of nucleotides to site 1 was described to be much
slower (half time about 4 hours) than in the experiment shown in Fig. 1.
Binding of MgATP to site 2 might be explained by the conversion of one
reversible site on the isolated enzyme to a tight site; the exchange of
bound MgATP for medium nucleotides is nearly impossible (1).
Binding of nucleotides to site 1, however, is at least a two-step mechanism because the isolated enzyme contains one ADP bound to this site which
is not released in the absence of medium nucleotides. Release of bound
ADP is therefore induced by binding of medium nucleotides to an unknown
site followed by binding of another nucleotide to the empty site 1. Another possible explanation is that the unknown nucleotide site (occupied
with a loosely bound adenine nucleotide) is converted into a tight ADP
site (with site-1 characteristics).
The rate of nucleotide binding to site 1 was studied using different concentrations of labeled nucleotides in the medium. In a parallel series,
the experiment was carried out in the presence of $CaCl_2$ (5 mM) because
Mg ions would lead to additional binding of MgATP to site 2. The incubation was terminated after 1o seconds by centrifugation through Sephadex
columns (Fig. 2).
If calcium is absent, about o.15 nucleotides were bound per CF_1 in 1o s.
Half maximum binding rates were obtained with about 4 to 5 µM nucleotides
in the absence and in the presence of $CaCl_2$. Since the binding rate is

Fig. 2
Concentration dependence
of nucleotide binding to
latent CF_1.
CF_1 (1 µM) was incubated
with (^{14}C)ADP or (^{14}C)ATP
(+/- 5 mM $CaCl_2$) for 1o
seconds.

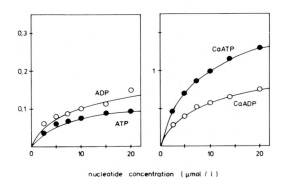

much higher in the presence of Ca, the total number of bound nucleotides reaches 1 per CF_1 in 1o s. Therefore, the obtained dissociation constant might not reflect the affinity of the unknown binding site for nucleotides in the presence of calcium.

Nevertheless, saturation of binding at low concentrations of adenine nucleotides (in the absence of Ca) indicates the participation of a binding site with an apparent dissociation constant of about 5 µM. This site might well be the reversible nucleotide site (site 3) described by Bruist and Hammes (1).

In order to determine the apparent order of the binding reaction, the concentration of the enzyme as well as of the ligand was varied. Increased concentrations of CF_1 (1 to 4 µM at ADP concentrations of 1 to 5 µM) do not increase the rate of binding (data not shown); this is an indication that the reaction depends on a conformational change of the enzyme which is slower than the association of nucleotides with the unknown binding site.

Binding of CaATP to site 1 was studied in the same way; a complex pattern of ATP binding was found (Fig. 3) The results are best explained by the

Fig. 3
Binding of CaATP to latent CF_1 as a function of ATP and CF_1 concentrations.
CF_1 was incubated with $(^{14}C)ATP$ and 5 mM $CaCl_2$ for 45 seconds.
The simulation (right side) shows the equilibrium concentration of $E(ATP)_2$ in
$$E + 2\ ATP = E(ATP)_2$$
with $K_d(1) = K_d(2) = 0.3$ µM.

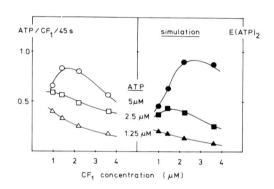

assumption that the extent of bound nucleotides does not depend on the total amount of ATP in the reaction mixture but on the concentration of <u>free</u> ATP. With similar concentrations of enzyme and ligand, nearly all medium nucleotides are bound to the unknown binding site. After the induced release of bound ADP, a second nucleotide molecule is assumed to bind to the empty site 1 and becomes tightly bound. After re-occupation of site 1, the loosely bound nucleotide is released from the unknown site. According to the alternating site hypothesis for ATP hydrolysis by the membrane-bound as well as the isolated coupling factor, the bound ADP on site 1 is an intermediate of ATP cleavage (2). Since conflicting results were reported, we tried to measure the nucleotide turnover on site 1 in comparison with ATP hydrolysis by the activated coupling factor. CF_1 was activated in 5o mM dithiothreitol for 2 hours and then incubated with $(^{14}C)ATP$ (2.5 µM) and $CaCl_2$ (5 mM) at room temperature. Parts of the assay were analyzed for bound nucleotides, and other parts were deproteinized for determination of the amounts of ATP and ADP by thin layer chromatography. The result is shown in Fig. 4.

Under the experimental conditions, binding of labeled nucleotides is even faster (o.53 nucleotides bound per CF_1 in 1o s) than hydrolysis of added ATP (o.36 ATP hydrolyzed per CF_1 in 1o s). Addition of a competitive

Fig. 4
ATP binding and hydrolysis
by DTT-activated CF_1
Activated CF_1 (1 μM) was
incubated with 2.5 μM
(^{14}C)ATP and 5 mM $CaCl_2$;
reactions were terminated
as described in the text.
$\triangle$, $\blacktriangle$: Reactive Red 4
(C.I. 18105) was added to
the reaction mixture
(50 μM).

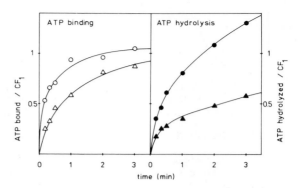

inhibitor of ATP hydrolysis (Reactive Red 4; C.I. 18105) decreased the
rate of binding as well as the rate of ATP hydrolysis. It seems that
hydrolysis is more affected by the inhibitor than binding: within the
first 3 minutes, about 1 ATP is bound per enzyme, while only about 0.5
ATP are hydrolyzed. However, additional experiments are necessary to determine the fate of bound ATP in the absence and presence of the
inhibitor.
Although the results do not exclude that bound ADP might be an intermediate during ATP hydrolysis, one has to explain the differences in the
dissociation constant for nucleotide binding (3 to 5 μM; see Fig. 2) and
the Michaelis and inhibitor constants during ATP hydrolysis (K_m for CaATP
150 μM; K_i for ATP 400 μM, ADP 160 μM, CaADP 50 μM). Preliminary experiments revealed no linear, but a sigmoidal dependence of the reaction rate
on low concentrations of CaATP (0.5 to 10 μM); a similar effect was described by Nelson (6) when ATP is hydrolyzed in the presence of ADP.
The sigmoidicity is more pronounced in the presence of the inhibitor
Reactive Red 4 (data not shown). It might therefore be that binding of
the first ATP molecule to an unknown site with high affinity for nucleotides opens a hidden site with lower affinity, or decreases the affinity
of another site to the values found during ATP hydrolysis.

ACKNOWLEDGEMENTS
Excellent technical assistance by M.Gokus, K.Kiefer, H.Schüll and B.Weyers
is gratefully acknowledged. Part of this work was supported by the
Deutsche Forschungsgemeinschaft.

REFERENCES
1 Bruist,M.F., Hammes,G.G. (1981) Biochem. 20, 6298-6305
2 Feldman,R.I., Boyer,P.D. (1985) J.Biol.Chem. 260, 13088-13094
3 Schumann,J., Richter,M.L., McCarty,R.E. (1985)
      J.Biol.Chem. 260, 11817-11823
4 Penefsky,H.S. (1977) J.Biol.Chem. 252, 2891-2899
5 Lowry,O.H., Rosebrough,N.J., Farr,A.L., Randall,R.J. (1951)
      J.Biol.Chem. 193, 265-275
6 Nelson,N. (1976) Biochim.Biophys.Acta 456, 314-338

EFFECTS OF ORGANIC SOLVENTS ON THE ENZYME-BOUND ATP SYNTHESIS BY ISOLATED CF_1

HIDEHIRO SAKURAI and TORU HISABORI
        Dept. of Biology, Sch. of Education, Waseda Univ.,
        Nishiwaseda, Shinjuku, Tokyo 160, Japan

1. INTRODUCTION

Soluble chloroplast coupling factor 1 (CF_1) catalyzes enzyme-bound ATP formation (single-turnover ATP synthesis) of Feldman and Sigman (1). Because the reaction did not depend on medium ADP, they suggested that the tightly-bound ADP was the ADP substrate. Our previous studies have identified at least four kinds of ADP binding sites on CF_1: a tightly binding site (site A), a slowly exchangeable high-affinity site (site B, Kd=0.021 µM), another slowly exchangeble high-affinity site (site C, Kd=1.6 µM) and several low affinity sites. The sites A, B, and C are not catalytic sites, and the binding of nucleotides to site B or C induces red shifts of the nucleotide UV spectra (2-5). As the tightly bound ADP on site A does not seem to participate in catalytic turnover, we tried to analyze these discrepancies by studying the effects of organic solvents which enhance the ATPase activity of isolated CF_1 (6-10).

2. MATERIALS and METHODS

CF_1 was extracted, purified by Sepharose 6B column chromatography, and stored at 4°C in 50% $(NH_4)_2SO_4$(5). Unless otherwise indicated, CF_1 precipitate was dissolved in 10 mM Tricine-KOH (pH 8.0) and the solution was passed through a Sephadex G-50 centrifuge column equilibrated with 2 mM EDTA-100 mM Tris-maleate (pH 6.0), then twice through the columns equilibrated with 10 mM Tricine. The reaction was started by adding CF1 to the reaction mixture, and terminated by adding 0.1 volume of 3.5 M perchloric acid. ADP had been treated with agarose-bound hexokinase. The reaction time was 40 min at 25°C unless otherwise indicated. The mixture was neutralized by Tricine-KOH, and the ATP concentration in the supernatant was analyzed by the luciferase method. AMP and ADP, when necessary, were converted to ATP by incubation with myokinase, pyruvate kinase and phosphoenolpyruvate (see 5).

3. RESULTS AND DISCUSSION

In harmony with Feldman and Sigman (1), the ADP added did not significantly increase the ATP formed by isolated CF_1 in a medium containing no methanol (Table 1). However, in the presence of 30% (v/v) methanol, medium ADP significantly increaed the ATP formation. Although the CF_1 contained 1.09 adenine nucleotides (mol/mol), the enhancing effect of methanol, if any, was very small in the absence of medium ADP.

Biggens, J. (ed.), Progress in Photosynthesis Research, Vol. III. ISBN 90 247 3452 5
© *1987 Martinus Nijhoff Publishers, Dordrecht. Printed in the Netherlands.*

Table 1.   Effects of ADP and methanol on ATP formation by CF_1

Additions		ATP formation
30% methanol	ADP	(mol/mol CF_1)
−	−	0.07
−	+	0.09
+	−	0.09
+	+	0.23

CF_1 from the first centrifuge column was passed through a Sephadex G-50 column (1.5x20 cm) equilibrated with 10 mM Tricine. CF_1 (per mol) contained 1.09 mol adenine nucleotides, of which 0.04 mol was ATP.   This value was subtracted in the above Table.   Reaction mixture: 50 mM MES-KOH (pH 5.8), 40 mM Pi (pH 5.8), 2 mM $MgCl_2$, 1 µM diadenosine pentaphosphate (A_2P_5, Sigma), and 0.93 µM CF_1.   Where indicated, the mixture also contained 30% (v/v) methanol and/or 20 µM ADP.

The effects of ADP on the single-turnover ATP synthesis in a medium containing 30% methanol is shown in Fig. 1.   ADP greatly enhanced the ATP formation.   The concentration of ADP required for the half maximal ATP formation varied from experiment to experiment: sometimes less than 1 µM, but usually 3-10 µM.   Under optimal conditions, as high as 0.65 mol ATP was synthesized per mol of CF_1.
      Some organic solvents enhanced ATP formation (Fig. 2). The optimal concentration was: about 30% (v/v) for methanol, 25% for ethanol, and 25% for acetone.   n-Propanol was also effective in enhancing ATP formation (data not shown).   These organic solvents also greatly enhanced Mg^{2+}-ATPase activity of isolated CF_1 (6).   n-Butanol which slightly enhanced $Mg^2{\pm}$ATPase

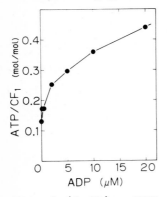
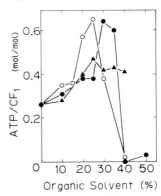

Fig. 1 (left).   Effects of ADP.   Reaction mixture: 25 mM MES-KOH (pH 6.0), 40 mM Pi, 2 mM $MgCl_2$, ADP as indicated, 2.5 µM A_2P_5, 30% methanol, 0.49 µM CF_1.
Fig. 2 (right).   Effects of organic solvents.   Reaction mixture: 25 mM MES-KOH (pH 6.0), 40 mM Pi, 2 mM $MgCl_2$, 50 µM ADP, 2.5 µM A_2P_5, organic solvents as indicated, 0.45 µM CF_1.   (●) Methanol, (○) ethanol, (▲) acetone.

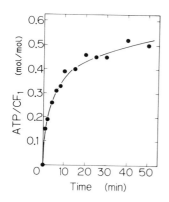
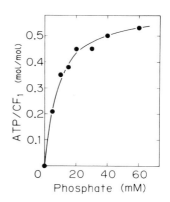

Fig. 3 (left). Time course of ATP formation. Reaction mixture: 25 mM MES-KOH (pH 6.0), 40 mM Pi, 2 mM $MgCl_2$, 50 μM ADP, 2.5 μM A_2P_5, 30% methanol, 0.26 μM CF_1.
Fig. 4 (right). Effects of Pi on ATP formation. Reaction mixture: 50 mM MES-KOH (pH 6.0), Pi as indicated 2 mM $MgCl_2$, 40 μM ADP, 1 μM A_2P_5, 35% propane-1,2-diol, 0.38 μM CF_1.

of CF_1 was not an effective activator of ATP formation. On the other hand, ethyleneglycol, propane-1,2-diol, and propane-1,3-diol which did not significantly enahnce Mg^{2+}-ATPase activity of CF_1 also enhanced ATP formation. Glycerol and dimethylsulfoxide were ineffective.

   Fig. 3 shows the time course of ATP formation. The half time, usually being 2-15 min, was about 3.5 min in this experiment. Feldman and Sigman reported a half-time of 20-60 sec under their experimental conditions (1).

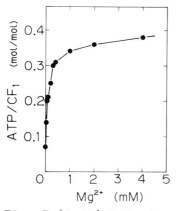
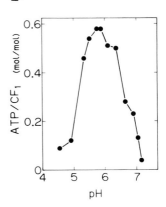

Fig. 5 (left) Effects of $MgCl_2$. Reaction Mixture: 25 mM MES-KOH (pH 6.0), 40 mM Pi, $MgCl_2$ as indicated, 50 μM ADP, 2.5 μM A_2P_5, 30% methanol, 0.49 μM CF_1.
Fig. 6 (right) Effects of pH. Reaction mixture: 100 mM MES-KOH or MOPS-KOH, 40 mM Pi, 2 mM $MgCl_2$, 50 μM ADP, 2.5 μM A_2P_5, 30% methanol, 0.26 μM CF_1.

Fig. 4 shows the effects of Pi concentration on ATP formation. The half optimal concentration of Pi was 8.5 mM in this experiment. Note that no ATP formation occurred unless Pi was added, which argues against the suspicion of activity of myokinase contaminating CF_1. In addition, when the reaction mixture was passed through a Sephadex G-50 centrifuge column, about 80% of ATP in the reaction mixture was recovered in the flow-through fraction containing CF_1.

Fig. 5 shows the effects of Mg^{2+} concentration on ATP formation. The half optimal concentration of Mg^{2+} was about 0.1 mM.

The optimum pH was about 5.7-6.0, which was similar to that of Feldman and Sigman ([1]). In contrast with their results in which ATP formation at pH 7 was still half as much as the maximal value, ATP formation in the presence of propanediol declined sharply towards higher pH's, and at pH 7, less than one-fourth of the maximum ATP formation occurred (Fig. 6).

We have characterized at least four kinds of nucleotide binding sites on CF_1: site A, a tightly binding and noncatalytic site, from which ADP is hardly removable; site I, a tightly binding and catalytic site, from which ADP is slowly removable by gel-permeation chromatography; sites B and C, high affinity, exchangeable and non-catalytic sites, on which binding of nucleotides induces UV spectral changes. There are less well characterized low affinity sites ([2]). Our conclusion is "CF_1 has two kinds of tightly nucleotide binding sites".

References
1 Feldman, R.I. and Sigman, D.S. (1982) J. Biol. Chem. 257, 1676-1683
2 Hisabori, T. and Sakurai, H. (1984) Plant Cell Physiol. 25, 483-493
3 Tanaka, M. and Sakurai, H. (1980) Plant Cell Physiol. 21, 1585-1593
4 Chiba, T., Suzuki, H. Hisabori, T. and Sakurai, H. (1981) Plant Cell Physiol. 22, 5510-560
5 Hisabori, T. and Sakurai, H. (1985) Plant Cell Physiol. 26, 505-514
6 Sakurai, H., Shinohara, K., Hisabori, T. and Shinohara, K. (1981) J. Biochem. 90, 95-102
7 Sakurai, H. (1976) Tampakushitsu Kakusan Koso, special issue 2, 76: 76-80
8 Tiefert, M.A. (1981) in Photosynthesis(ed. Akoyunoglou, G.) vol.2, pp.893-902, Balaban International Science, Philadelphia
9 Mal'yan, A.N. (1981) Photosynthetica 15, 474-483
10 Anthon, G.E. and Jagendorf, A.T. (1986) Biochim. Biophys. Acta 848, 92-98

(Work supported by Grants from the Ministry of ESC, Japan)

ATPase-CATALYZED PHOSPHATE EXCHANGE BETWEEN ATP AND ADP

SUSANNE BICKEL-SANDKÖTTER AND HEIKE SCHÜLL, BOTANISCHES INSTITUT DER UNIVERSITÄT DÜSSELDORF, UNIVERSITÄTSSTRASSE 1, D-4000 DÜSSELDORF

1. INTRODUCTION
    The chloroplast ATPase (CF_1-CF_0-complex) catalyzes photophosphorylation of ADP as well as hydrolysis of ATP, depending on the thermodynamic conditions and the activation state of the enzyme. Three nucleotide binding sites are located on the CF_1 part of the CF_1-CF_0-complex (1) which are responsible for catalysis and regulation. These sites have been identified on the three β-subunits of CF_1 (2), they are postulated to form a triangle with almost similar distances (3).

    The present paper reports on a light-dependent γ-phosphate exchange between ATP and ADP. This reaction is catalyzed by membrane bound chloroplast ATPase and can be observed under condition of phosphate deficiency. The results are discussed in context with the structural arrangement of the cooperating nucleotide binding sites.

2. MATERIAL AND METHODS
    Spinach chloroplasts were isolated and washed four times in order to remove adenylate kinase as described earlier (4). Measurements of phosphate exchange were carried out in a medium containing 25 mM Tricine pH 8, 50 mM NaCl, 5 mM $MgCl_2$, 50 μM PMS and the indicated concentrations of either $[8-^{14}C]$ADP and ATP or $[8-^{14}C]$ATP and ADP. Illumination of the chloroplasts was carried out with white light (400 $W \cdot m^{-2}$). After stop with 0.5 M $HCLO_4$ and neutralization, nucleotides were analyzed by TLC (5) or column chromatography (6).

3. RESULTS AND DISCUSSION
    In the absence of inorganic phosphate isolated thylakoids show a light-dependent phosphate exchange activity between ATP and ADP which is Mg^{2+}-dependent like photophosphorylation:

$$(^{14}C)ADP + ATP \underset{}{\overset{Mg^{2+}, \text{ light}}{\rightleftharpoons}} (^{14}C)ATP + ADP$$

Fig. 1 shows the kinetics of the increase of (^{14}C)ATP in the light or dark and the concomitant decrease of (^{14}C)ADP. The slow increase of (^{14}C)ATP in the dark is due to residual adenylate kinase. Addition of 200μM diadenosylpentaphosphate suppresses the dark reaction.
The ATPase inhibitor phlorizin inhibits the exchange reaction to the same extent as photophosphorylation (I_{50} = 0.75mM). This result and the obvious energy dependence indicate that the exchange is catalyzed by chloroplast ATPase and is not due to a nucleosidediphosphate kinase reaction.
    Kahn and Jagendorf (7) found a chloroplast enzyme catalyzing a phosphate exchange between ATP and ADP. This reaction which is independent of membrane energization is not identical with the here reported exchange.

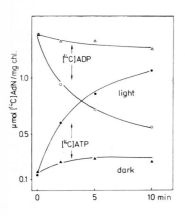

FIGURE 1. Kinetics of formation of (^{14}C)ATP and concomitant decrease of (^{14}C)ADP in the light and in the dark. The reaction medium contained 200 µM (^{14}C)ADP and 200 µM ATP. Chlorophyll was 200 µg/ml.

Isolated thylakoids contain a low level of inorganic phosphate, even after extensive washing. This P_i could give rise to ATP formation in the light under the conditions employed in Fig.1 and lead to misinterpretation. Therefore the reaction was carried out with the substrate couples ATP / (^{14}C)ADP and (^{14}C)ATP / ADP. The kinetics are compared in Fig. 2. Actually a difference between the two modes of measurements is observed which may be explained by ADP phosphorylation via endogenous P_i. As to be expected, phosphorylation is terminated after 30 seconds while the exchange reaction continues.

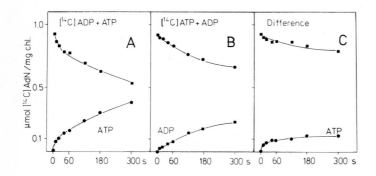

FIGURE 2. Kinetics of the phosphate exchange between (^{14}C)ADP/ATP (A) and (^{14}C)ATP/ADP (B). The difference between the two reactions is plotted in C. Nucleotide concentrations: 200 µM, chlorophyll-content 200 µg/ml.

A substantial rate of (^{14}C)ATP formation from (^{14}C)ADP is also observed when the reaction mixture is preilluminated with the unlabeled substrates and the carrier-free (^{14}C)ADP is added after 5 minutes. Under these conditions the available endogenous P_i should have been consumed during the pretreatment.

K_m (ADP) for the exchange reaction is 47 µM, which is comparable to K_m for photophosphorylation under the employed conditions (8), while the K_m-value for ATP in this reaction is 510 µM. V_{max} varies between 12 and 20 µmol/mg chl·h. If ATP was replaced by the non-hydrolyzable analog AMP-P-C-P, no exchange of γ-phosphate could be observed, indicating again that the formation of (^{14}C)ATP from (^{14}C)ADP needs the γ-phosphate of ATP (Fig.3).

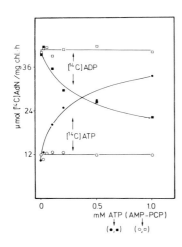

FIGURE 3. ATP/ADP-phosphate exchange as a function of ATP concentration (■,●) or AMP-P-C-P concentration (□,○). (^{14}C)ADP concentration was 100 μM, chlorophyll concentration 200 μg/ml, illumination time 30 seconds.

Table 1 shows that the rate of ATP hydrolysis in the light is much lower than the rate of exchange measured under the same conditions. The rate of ATP hydrolysis in the light is not increased by omitting ADP. Hence, the exchange cannot be interpreted as the result of ATP hydrolysis and re-phosphoryation of ADP by liberated P_i. Moreover, the high K_m (P_i) (0.5 mM) for phosphorylation would make P_i rebinding even more unlikely.

TABLE I. Hydrolysis of ATP and ATP/ADP-phosphate exchange measured under identical experimental conditions. The reaction medium contained additional 5 mM DTT. Nucleotide concentrations were 200 μM (^{14}C)ADP and ATP. Rates of ATP hydrolysis were determined by following the formation of P_i (9).

conditions	hydrolysis (μmol P_i/mg chl·h)	ATP/ADP exchange (μmol (^{14}C)ATP/mg chl·h)
light	1.0	7.8
dark	2.0	1.5
dark after 2 min. pre-illumination	10.8	1.1

Dark ATP-P_i exchange is induced by pre-illumination of thylakoids in the presence of DTT (10). It has been concluded that this exchange reaction results from rephosphorylation of medium ADP formed by ATP hydrolysis and driven by ATP-induced $\Delta\mu_H$+ (10). K_m(ADP) for phosphorylation is at least one order of magnitude lower than K_m(P_i). No ATP/ADP-phosphate exchange is observed with light-triggered chloroplasts (Table I). Under the employed conditions (200 μM ADP and 200 μM ATP) a part of the ATPase molecules are inactivated by tight binding of ADP (11). This explains the relatively low rate of ATP hydrolysis (20% of the control rate in the absence of ADP). As a consequence, a low ΔpH will be established, which is probably insufficient to drive the exchange reaction.

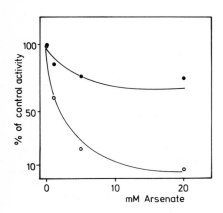

FIGURE 4. ATP/ADP exchange (●) and photophosphorylation (o) as a function of arsenate concentrations, measured under comperable conditions. For measurement of photophosphorylation (o) 100 μM P_i were employed instead of 200 μM ATP (●). Control activities: Photophosphorylation: 138 μmol (^{14}C)ATP/ mg chl·h, Exchange: 15.6 μmol (^{14}C)ATP/ mg chl·h

Arsenate is known to inhibit photophosphorylation by competition with inorganic phosphate. As shown in Fig. 4, the ATP/ADP-phosphate exchange is inhibited to 25% by 20 mM arsenate, while photophosphorylation is decreased to 7% of the control rate. The lower effectiveness of arsenate can be explained by assuming that γ-phosphate exchange takes place in a condition of the enzyme where the reactive domain of the protein is largely screened from the medium.

The results suggest that at least two of the cooperating catalytic sites may approach each other during the catalytic process under certain conditions, so that a transfer of phosphate from ATP which is hydrolyzed on one site to ADP on the second site occurs without intermediate P_i release into the medium. The question is open, whether the energy is needed to re-synthezise ATP or to facilitate approach of the sites or release of the substrates.

REFERENCES
1 Strotmann, H. and Bickel-Sandkötter, S. (1984) Ann. Rev. Plant Physiol. 35, 97–120
2 Abbott, M.S., Czarnecki,J.J. and Selman, B.(1984)J.Biol. Chem. 259,12271–12278
3 Cerione,R.A. and Hammes,G.G.(1982) Biochemistry 21, 745–752
4 Strotmann,H.,Bickel,S. and Huchzermeyer,B.(1976) FEBS Lett.61,194–198
5 Goody,R.S. and Eckstein,F.(1971)J.Am.Chem.Soc.93, 6252–6257
6 Shavit,N. and Strotmann,H.(1980)Methods Enzymol.69,321–326
7 Kahn,J.S. and Jagendorf,A.T.(1961)J.Biol.Chem.236, 940–943
8 Bickel-Sandkötter,S. and Strotmann,H,(1981) FEBS Lett.125,188–192
9 Strotmann,H.(1972) in Proceedings of the 2nd Int.Congress on Photosynthesis Research(Forti,G.,Avron,M. and Melandri.A.,eds)Vol.II.1319–1328, Junk, N.V. Publishers, The Hague
10 Davenport,J.W. and Mc Carty,R.E.(1981) J.Biol.Chem. 256, 8947–8954
11 Schumann,J. and Strotmann,H.(1981) in Photosynthesis II. Photosynthetic Electron Transport and Photophosphorylation(Akoyunoglou,G.,ed.)223–230, Balaban International Science Services, Philadelphia,PA

ON THE MECHANISM OF REGULATION OF CATALYTIC ACTIVITY OF CHLOROPLAST COUPLING FACTOR 1 ATPase

A.N.MALYAN AND O.I.VITSEVA
INSTITUTE OF SOIL SCIENCE AND PHOTOSYNTHESIS, USSR ACADEMY OF SCIENCES, PUSHCHINO, MOSCOW REGION, 142292, USSR

INTRODUCTION

Chloroplast coupling factor one ATP ase (CF_1-ATPase) after its addition to the reaction mixture containing ADP and Mg^{2+} ions undergoes a slow reversible transition from a metastable active state to a stable steady state with low activity [1]. A similar transition occurs with the membrane-bound CF_1-ATPase [2]. The low level of the steady state CF_1-ATPase activity was shown to be due to the tight binding of ADP to a nucleotides binding site of the enzyme [1-4]. In this work we demonstrate that inactivation of CF_1-ATPase is accompanied by the gradual increase in the affinity of one nucleotide binding site to ADP and simultaneous decrease in the affinity of another site to ATP. The binding of ATP to the last site gives rise to the reactivation of CF_1-ATPase.

PROCEDURE

CF_1-ATPase was isolated, purified, activated and desalted as described [1]. Mg^{2+} ATPase activity was determined by coupling the generation of ADP to the oxidation of NADH [5]. The reaction mixture (0.5 ml) at $20^{\circ}C$ contained 50 mM KCl, 20 mM tris-HCl (pH 6.5), 10 mM $MgCl_2$, 0.015-2.0 mM ATP, 2 mM phosphoenolpyruvate, 0.15 mM NADP, 8 units lactate dehydrogenase, 4-40 units pyruvate kinase, 10-25 ug CF_1. Spectrophotometer tracings were initiated within 3s of CF_1 addition (0.025-0.1 ml) and mixing. ADP content was calculated according to [4], using the rate constant of pyruvate kinase reaction (167 ml·min^{-1}·mg^{-1}) determined under the conditions similar to those in experiments. Protein concentration was determined by the method of Bradford [6].

RESULTS AND DISCUSSION

After its addition to the reaction mixture the activity of CF_1-ATPase falls rapidly, approaching in 1-2 min a rather low steady state level (Fig.1). The level rises with the increase in the ATP content. To clarify the effect of inactivation on the ability of the enzyme to interact with the nucleotides we have studied the dependence of the reaction rate on ATP and ADP concentrations. The variations in the ADP concentrations have been carried out by the change in the pyruvate kinase content. Figure 2 shows that Lineweaver-Berk plots obtained at the 4 s after initiation of the reaction, are linear. The calculated values of the apparent Michaelis constant and the constant of competitive inhibition are 19 uM and 24 uM, respectively. The time-dependent decrease in the reaction rate is accompanied with the change in the pattern of the plots: on the range of high ATP concentrations they become curvilinear (Fig. 3,4).

The values of the maximal reaction rate, however, remain practically unchanged. The curvilinear pattern of the Lineweaver–Berk plots is usually interpreted as a result of participation in the reaction of 2 nucleotide binding sites, with high and low affinity to ATP. According to this interpretation in the low ATP concentration range the site with a high affinity to ATP ($K_{M,1}$ about 14 μM) is involved in the reaction. With the increase in the ATP concentration its interaction with the second site becomes essential, resulting in a significant acceleration of the reaction rate. The calculation of $K_{M,2}$ value in the high ATP concentration range shows that inactivation is accompanied with the decrease in the affinity of the site to ATP (Fig.5). In the low ATP concentration range (loss than 50 μM) the pattern of ATP hydrolysis inhibition by ADP under steady state conditions, is noncompetitive (Fig.4). The values of the inhibition constants decrease in the time–dependent manner from 24 to 1 uM (Fig.5). Rather close values of the reaction rates obtained at a variable ADP content as well as a pronounced curvity of the plots in the high ATP concentration range do not allow to calculate precisely the affinity of the second type of nucleotide binding sites to ADP (Fig.4). The data are however sufficient to suggest that in this concentration range the inhibition by ADP is competitive and comparatively weak.

The $K_{M,2}$ value obtained under steady state conditions at high ATP concentrations is of the order of effective concentrations of oxianions which activate Mg^{2+}–dependent ATP hydrolysis /7/. To determine the specificity of the activating effect of ATP we have studied the dependence of oxianion activation on ATP concentrations. Figure 6 shows that the effect of oxianions is completely inhibited by ATP.

The dependence of the steady state reaction rate on ATP concentrations indicates the existance of 2 sites of ATP binding: one with a high affinity ($K_{M,1}$ about 14 μM), and another with a low affinity ($K_{M,2}$ about 13 μM). ATP binding to the first site leads to the formation of the $CF_1 \cdot ATP$ complex with low activity ($V_{max} \sim$ 0.1 $\mu mol \cdot min^{-1} \cdot mg^{-1}$). The noncompetitive inhibition of the reaction by ADP in the low ATP concentration range (under steady state conditions) as well as the reversible inactivation of the enzyme by ADP (under presteady state conditions) support the suggestion that in the presence of ADP the formation of inactive complex $ADP \cdot CF_1 \cdot ATP$ occurs. The saturation of the second nucleotide binding site at a high ATP concentration causes the reduction in the affinity of the CF_1–ATPase to ADP and the change in the pattern of inhibition from noncompetitive to a competitive one (Fig.4). The data imply that the increase in the ATP concentration induces dissociation of ADP, and, consequently transition of the enzyme from the state with a low activity to the active state. The necessity of ADP dissociation for activation of CF_1–ATPase, and the capability of ADP binding site to change its regulatory function for a catalytic one were shown earlier /8,9/. Since CF_1 possesses several nucleotide binding sites it seems probable that the ADP dissociation is induced by the binding of the second ATP molecule resulting in the formation of a transient intermediate complex with 3 nucleotide molecules:

$$ADP \cdot CF_1 + ATP \Longleftrightarrow ADP \cdot CF_1 {\overset{.ATP}{.ATP}} \underset{}{\overset{ADP}{\Longleftrightarrow}} CF_1 {\overset{.ATP}{.ATP}}$$

It is worth noting that the proposed mechanism, in contrast to the mechanism of activation of mitochondrial F_1–ATPase /9/, does

not require ATP hydrolysis for the dissociation of ADP at another
site. Several compounds incapable of hydrolysis, such as AMPPCP
 4 , a number of oxianions which competitively interact with the site of loose ATP binding, can induce the reduction of ADP binding
by CF_1 (Fig.6). This may imply that the second enzyme bound ATP
molecule has a regulatory rather than substrate function promoting
ADP dissociation and maintaining the active state of the enzyme.
     The question arises concerning the interrelation between the
initial active state and steady staty of the enzyme. It is reasonable
to suggest that in case of the active state the linear pattern of
Lineweaver–Berk plot is due to the similarity of the properties of 3
nucleotide binding sites. This suggestion is supported by the competitive inhibition of the reaction by ADP as well as by the closeness
of the Michaelis constant and inhibition constant values. Such kinetic pattern could be observed for the binding change mechanism proposed by Boyer et al 10 . During inactivation, differentiation of the
properties and functions of nucleotide binding sites occurs: the affinity of one site to ADP increases whereas the affinity of another
site to ATP simultaneously decreases.

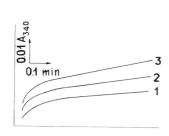

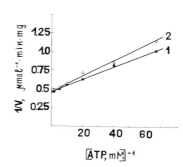

FIGURE 1. Inactivation of CF_1-ATPase. ATP concentrations: 1 –
          0,025 mM, 2 – 0.05 mM; 3 – 0.2 mM; CF_1– 50 ug/ml, py-
          ruvate kinase 250 ug/ml.

FIGURE 2. Lineweaver–Berk plot obtained at 4s after addition of
          CF_1-ATPase to reaction mixture. Pyruvate kinase con-
          centrations: 1 – 250 ug/ml, 2 – 25 ug/ml; CF_1– 50 ug/ml.

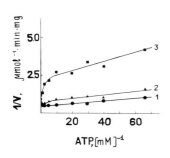

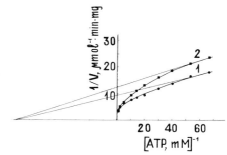

FIGURE 3. The change in the pattern of Lineweaver–Berk plots du-
          ring inactivation. 1– after addition of CF_1-ATPase, 2– 10s,
          3 – 22s.

FIGURE 4. Inhibition of the steady state rate of ATP hydrolysis by ADP. Concentrations pyruvate kinase: 1 – 250 ug/ml, 2 – 25 ug/ml.

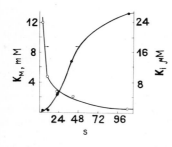 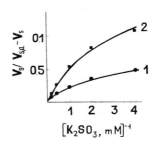

FIGURE 5. The dependences of Michaelis constant and ADP inhibition constant values on the time of incubation of the enzyme in the reaction mixture.

FIGURE 6. Suppression of activating effect of sulfite by ATP. 1 – 0.4 mM ATP, 2 – 2 mM ATP. Concentrations of pyruvate kinase 90 ug/ml, CF_1– 20 ug/ml. V_s and $V_{s,a}$ are the reaction rates in the absence and in the presence of sulfite, respectively.

REFERENCES

1 Malyan,A.N. and Makarov,A.D. (1976) Biokhimiya (USSR) 41, 1087–1093
2 Malyan,A.N. (1981) Photosynthetica 15, 474–483
3 Dunham,K.R. and Selman,B.R. (1981) J.Biol.Chem. 256, 212–218
4 Malyan,A.N. and Vitseva,O.I. (1983) Biokhimiya (USSR) 48,718–724
5 Pullman,M.E., Peneffsky,H.S., Datts,A. and Racker (1960) J.Biol. Chem. 235, 3322–3339
6 Bradford,M.M. (1976) Anal.Biochem. 72, 248–254
7 Malyan,A.N. and Akulova,E.A. (1978) Biokhimiya (USSR) 43, 1206–1211
8 Malyan,A.N., Kuzmin,A.N. and Vitseva,O.I. (1982) Biochem.Intern. 5, 325–328
9 Vinogradov,A.D. (1984) Biokhimiya (USSR) 49, 1220–1238
10 Kayalar,C., Rosing,J. and Boyer,P.D. (1977) J.Biol.Chem, 252, 2486–2491

SUBUNIT–SUBUNIT–INTERACTIONS IN F_oF_1 AS REVEALED BY LIGAND BINDING AND INTRINSIC FLUORESCENCE

Matthias RÖGNER and Peter GRÄBER

Max-Volmer-Institut für Biophysikalische und Physikalische Chemie, Technische Universität Berlin, Strasse des 17. Juni 135, 1000 Berlin 12, FRG

1. INTRODUCTION
    For a better understanding of the molecular mechanism of the H^+-ATP-synthase, a prerequisite is to know more about the correlation between structural and functional aspects. One possibility to study this correlation is to compare the properties of the subunits of this enzyme complex with the changes that occur upon reconstitution of the bigger complexes. Because of its extreme stability, the ATP-synthase of the thermophilic bacterium PS3 is especially well suited for these kinds of experiments (1,2). Using this enzyme we have compared structural properties, ligand binding and ATP hydrolysis of the holoenzyme with those of the isolated subunits. For experiments concerning ligand binding and functional properties, the adenine-nucleotide analogue naphthoyl-ATP (N-ATP) has been used (3,4); the advantages of this ligand are: (1) being fluorescent, (2) having a high affinity for the ATP-binding sites, (3) affecting the intrinsic fluorescence of TF_1, (4) being a potent inhibitor of the ATPase reaction. Preparation procedures and the methods have been described in detail elsewhere (5-7).

2. RESULTS

2.1 Structural properties
    Examination of the intrinsic fluorescence of the subunits and aggregates of subunits (TF_o, TF_1 and TF_oF_1) yields information about their contents of Tyr- and Trp-residues and the respective microenvironment of these amino acids. Fig. 1 shows the intrinsic fluorescence spectra of TF_o, TF_1, and TF_oF_1 in native state. TF_1 shows a typical Tyr-fluorescence (maximum 305-310 nm) as do all its subunits (6). Trp-fluorescence obviously is quenched in the native state of these proteins

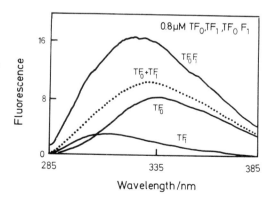

FIGURE 1. Fluorescence emission spectra of TF_1, TF_o and TF_oF_1 for λ_{ex} = 275 nm; the dotted curve is the calculated sum of the spectra of TF_o and TF_1.

and can be seen only after denaturation with 8 M guanidine-HCl in TF_1 and α-subunit (strong), δ- and ε-subunit (weak fluorescence)(6). In contrast to this, native TF_o and TF_oF_1 show strongly blue-shifted Trp-fluorescence, indicating a very hydrophobic environment of these residues. As only the c-subunit of TF_o contains no Trp (8), these hydrophobic residues are located on a- and b-subunit. Presumably, most of them belong to the a-subunit, which is known to be very hydrophobic. As all spectra of Fig. 1 are recorded with the same sensitivity, it is obvious that the fluorescence of TF_oF_1 is dominated by (hydrophobic) Trp-residues of TF_o, while the contribution of the TF_1-fluorescence is negligible. Furthermore, the calculated addition of TF_o- and TF_1-spectrum is depicted by the dotted curve in Fig. 1. Obviously, the natural TF_oF_1 molecule shows a much stronger Trp-fluorescence than the sum of single TF_o and single TF_1 fluorescence. Presumably, binding of TF_1 to TF_o results in a more hydrophobic micro-environment of some Trp-residues leading to an enhanced Trp fluorescence. According to the intrinsic fluorescence spectra, these Trp-residues may be located on the b-subunit of TF_o and on the δ-/ε-subunit of TF_1. Thus, intrinsic fluorescence shows that these Trp-residues may be used as sensitive indicators of conformational changes in the TF_o-TF_1-binding area and also in the α-subunit.

2.2 Ligand binding properties

Fig. 2 shows the intrinsic protein fluorescence of TF_1, α- and β-subunit with and without ligand (N-ATP). Upon binding of N-ATP, the Tyr-fluorescence of all three proteins decreases. Additionally, in case of TF_1, there is a strong fluorescence increase at about 365 nm. Model experiments show (5) that this peak is a second N-ATP peak which obviously reflects the dielectric constant of this environment; therefore, it can be concluded that the N-ATP binding sites on TF_1 are more hydrophobic than those on the isolated subunits. In case of TF_1, the fluorescence decrease with N-ATP can be used to determine the number of binding

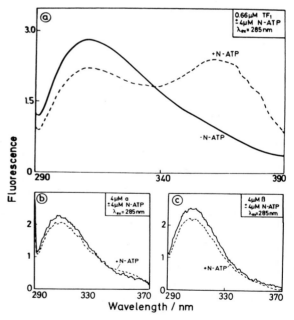

FIGURE 2. Intrinsic fluorescence of TF_1, α- and β-subunit with (dashed lines) and without 1-naphthoyl-ATP (N-ATP)(solid lines)

sites, n, and the dissociation constant, K_D. For both, TF_1 and the isolated α- and β-subunits, the fluorescence anisotropy of N-ATP increases upon binding to these proteins, and this effect can be used as well for a K_D- and n-determination. The results are shown in Table 1. There is strong evidence for 2 N-ATP binding sites on TF_1, which is surprising because both isolated α- and β-subunits can bind one molecule of N-ATP. The result of 2 instead of the expected 6 binding sites on TF_1 ($3\alpha + 3\beta$) points to subunit-subunit or subunit-ligand interactions within TF_1. A similar conclusion can be drawn from the dissociation constants for N-ATP. Table 1 shows that the K_D-values for the isolated subunits are about a factor of 10 higher than the one for TF_1.

TABLE 1. Thermodynamic and kinetic constants of N-ATP binding to TF_1 and isolated α- and β-subunits

	TF_1	α	β
Dissociation constant, $K_D/\mu M$	0.25	4.6	1.9
Number of binding sites, n	1.8	1.0	0.7
Rate constant of binding, $k_1/10^3 \ M^{-1}s^{-1}$	6.6	50	50
Rate constant of release, $k_{-1}(A)/s^{-1}$	$3 \cdot 10^{-3}$	0.14	0.14
Rate constant of release, $k_{-2}(A)/s^{-1}$	$2 \cdot 10^{-4}$	-	-

2.3 Functional properties

Using the change of fluorescence anisotropy of N-ATP, the kinetics of N-ATP binding to and the dissociation from TF_1 can be determined as shown in Fig. 3. In this measurement, the amount of N-ATP was chosen twice as high as TF_1 because of the two N-ATP-binding sites. The displacement of bound N-ATP by natural ligands like ATP requires about 100 times more ATP than N-ATP to see a displacement kinetics. Analysis of the kinetics shows that N-ATP binding is a second order, monophasic process. In contrast to this, the release of

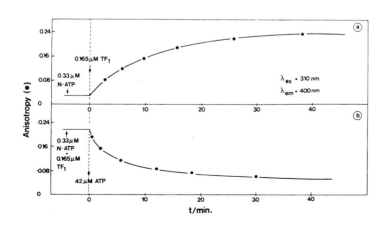

FIGURE 3. Kinetics of N-ATP binding to (a) and release from TF_1 on addition of ATP (b). Fluorescence anisotropy of N-ATP was measured at 400 nm.

N-ATP in the presence of ATP is a biphasic process of pseudo-first-order (5). All rate constants are shown in Table 1. It is obvious that the rate constants of N-ATP binding as well as the ones of N-ATP release in the presence of ATP,k(A), are much higher for the isolated α- and β-subunit than for TF_1. Since the release of N-ATP in the presence of ATP is biphasic, ATP obviously induces a kind of negative cooperativity. In order to check whether the F_o-part (or the surrounding membrane) influences the kinetics of ATP hydrolysis catalyzed by both TF_1 and reconstituted TF_oF_1, experiments with N-ATP as an inhibitor have been performed (7). The inhibitory constant, K_i(N-ATP), for TF_1 (K_i = 150 µM) and TF_oF_1 (K_i = 3 µM) show that N-ATP is a much stronger inhibitor for TF_oF_1 vesicles than for TF_1. As in case of TF_oF_1 vesicles, the initial rate of ATP hydrolysis has been measured, i.e., in a state when the membrane potential has not yet been built up, the difference between TF_1 and reconstituted TF_oF_1 presumably is due to the F_o-part and not due to the membrane energization.

3. DISCUSSION
   All these results taken together indicate the following:
(1) Subunit-subunit-interactions in the reconstituted ATPase change the properties of the isolated subunits.
(2) Specific interactions of the natural substrates (AT(D)P) with the ATPase induce a negative cooperativity of the binding sites. The binding of N-ATP does not cause such an effect. A suitable explanation for the reduction of the number of binding sites in TF_1 and an alteration of their affinities could be that the binding sites of the ATPase complex are located between the α- and the β-subunits. This is corroborated by (a) an increasing affinity for ligands (by interaction of at least two binding sites); (b) a more hydrophobic environment of the binding sites on TF_1 in comparison to isolated α- and β-subunits; (c) a diminished rate constant (factor of at least 7) for N-ATP binding to TF_1 (due to less free accessibility in comparison to the free subunits).

4. ACKNOWLEDGEMENTS
   We are grateful to Prof. Kagawa, Dr. Ohta, Dr. Sone, Dr. Yoshida (Japan) and Prof. Schäfer, Dr. Lücken, Dr. Tiedge and J. Weber (Lübeck) for good cooperation. The financial support of the DFG (Sfb 9) and the DAAD (M.R.) is gratefully acknowledged.

REFERENCES
1 Kagawa, Y. (1980) J. Membr. Biol. 55, 1-8
2 Yoshida, M., Sone, N. et al. (1977) J. Biol. Chem. 252, 3480-3485
3 Onur, G., Schäfer, G. & Strotmann, H. (1983) Z. Naturforsch. 38c,49-59
4 Tiedge, H. et al. (1982) Eur. J. Biochem. 127, 291-299
5 Rögner, M. et al. (1986) Biochim. Biophys. Acta 849, 121-130
6 Rögner, M. and Gräber, P. (1986) J. Biochem. 99, 993-1003
7 Rögner, M. and Gräber, P. (1986) Eur. J. Biochem. (in press)
8 Hoppe, J. and Sebald, W. (1984) Biochim. Biophys. Acta 768, 1-27

REGULATION OF CHLOROPLAST ATPase BY NUCLEOTIDES, P_i, and $\Delta\mu_H^+$

HEINRICH STROTMANN, SIGRID NIGGEMEYER AND ABDEL-RAHMAN MANSY
Botanisches Institut II, Universität Düsseldorf, Universitätsstraße 1,
D-4000 Düsseldorf, FRG

1. INTRODUCTION
     Regulation of the chloroplast H^+ATPase is an important prerequisite to
optimize energy conservation under the conditions of changing energy supply
caused by variations of light intensity in the natural habitat of plants. Depending on the external conditions, the individual ATPase molecules may be
active or inactive, chemically modified or even altered in their structural
relationship to the membrane (1). The inactive enzyme is characterized by
the presence of one "tightly bound" ADP molecule in one of the three β subunits of CF_1. Activation requires a transmembrane $\Delta\mu_H^+$ and includes a change
of protein conformation related with release (or exchange) of the tightly
bound nucleotide molecule. In the activated state a particular disulfide bond
in γ subunit is exposed which may be reduced to a dithiol group ("thiol modulation") by a suitable reductant (2). In broken chloroplasts this can be
achieved by addition of DTT (3). The natural reductant in the intact chloroplast is thioredoxin which is reduced by the photosynthetic electron transport chain through ferredoxin-thioredoxin reductase (4). The activated oxidized ATPase looses its activity with deenergization of the thylakoids,
whereas the reduced ATPase remains active after relaxation of $\Delta\mu_H^+$ under certain conditions. Hence the oxidized form is essentially capable of catalyzing
ATP synthesis in the light, while the thiol-modulated form can catalyze the
reverse reaction, too. Both forms re-incorporate ADP or ATP added after membrane deenergization. Binding of ADP is retarded by inorganic phosphate (5,6).
The activated modulated ATPase is rapidly deactivated upon re-binding of ADP
(5-7) but stabilized by incorporation of ATP (8,9). Binding of either ADP or
ATP does not change the catalytic properties of the already inactive oxidized
enzyme.
     In vivo it may occur that a chloroplast exposed to full light is sudden-
ly subjected to low light or (in extreme) darkness. At the time of transition, the ATPase would be in its thiol-modulated form because of the high
reduction state of the thioredoxin system in strong light and this form
might be maintained for some time in the dark since reoxidation is a relatively slow process (10). Since the chloroplast stroma contains ATP, ADP
and P_i, multiple interactions of these compounds with the active ATPase may
occur. In order to understand the complex co-operation of these reactions,
the mutual interference between catalytic and regulatory processes as well as
energetic parameters are studied in the present paper. For better manipulation
of the system, preactivated and thiol-modulated isolated thylakoids are
employed.

2. EXPERIMENTAL
     Thylakoids from spinach leaves were isolated as in (11). Thiol-mdoula-
tion, measurements of ATP synthesis and hydrolysis were carried out as in
(12). The technique employed for nucleotide binding was described in (13)
and measurements of ΔpH by the 9-aminoacridine method were carried out as
in (14).

3. RESULTS AND DISCUSSION

Two extreme modes of reaction of chloroplasts at light-dark transition are shown in Fig. 1. Thiol-modulated thylakoids are supplied with ADP and P_i and allowed to form ATP in the light until a steady state is attained. In one of the samples ADP, in the other one P_i is the limiting substrate (50 μM).

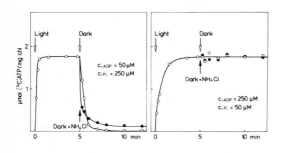

FIGURE 1. ATP synthesis and hydrolysis by thiol-modulated chloroplasts in a light-dark change. Thylakoids were pre-illuminated for 2 min at 250 Wm^{-2} in a medium containing 25 mM Tricine, pH 8, 50 mM NaCl, 5 mM $MgCl_2$, 10 mM DTT and 50 μM PMS before the substrates (^{14}C)ADP and P_i were added at the indicated concentrations. NH_4Cl was added at a concentration of 1 mM. Nucleotide analysis was carried out by TLC (14).

In the two samples the steady state concentrations of ATP reached in the light are quite similar. Moreover the calculated steady state phosphate potentials $(ATP)/(ADP) \cdot (P_i)$ are also the same ($\Delta G_p = 59$ kJmol^{-1}) due to identity of the product $(ADP) \cdot (P_i)$. When the light is turned off, hydrolysis of ATP takes place as expected for thermodynamic reasons, but only in the sample containing a low light steady state concentration of ADP and excess P_i. In contrast, no ATP hydrolysis is observed when in the light the steady state concentration of ADP is high and that of phosphate is low. Obviously under these conditions the reaction is kinetically inhibited by inactivation of the ATPase. The uncoupler NH_4Cl is without effect, but accelerates ATP hydrolysis in the non-inhibited sample. ATP hydrolysis generates a $\Delta \mu_H^+$ which on the other hand controls the rate of ATP hydrolysis. This control is released by uncoupling. Under conditions of excess ADP present at light-dark transition, no ΔpH is maintained in the dark (not shown). This result also shows that the ATPase is inactive. Deactivation of the ATPase upon deenergization can be explained by tight incorporation of ADP. At a 4-fold excess of ADP (200 μM) over ATP (50 μM) virtually all of the ATPase molecules incorporate ADP in less than 1 s, since the rate constants for ADP binding (0.02 s^{-1}μM^{-1}) and ATP binding (0.0025 s^{-1}) differ by one order of magnitude (8). However, in the sample where ADP is the limiting substrate, less than 5 μM ATP is present. Hence at least half of the ATPase molecules would incorporate ATP instead of ADP. Moreover tight binding of ADP would be further decreased by phosphate (see below).

While ATP is hydrolyzed, ADP accumulating in the medium should progressively deactivate the ATPase. In order to investigate the process or self-inhibition, the activity of the ATPase is measured during the course of hydrolysis of 50 μM ATP. The kinetics of ATP hydrolysis is shown in Fig. 2. For determination of ATPase activity at a chosen time, a pulse of PEP and a high concentration of pyruvate kinase together with carrier-free (^{32}P)ATP is employed and the activity is determined by following the formation of $(^{32}P)P_i$. The pyruvate kinase system provides instantaneous reconversion of ADP to ATP, so that hydrolysis of (^{32}P)ATP starts at maximum substrate concentration. In addition, the pulse mix contains NH_4Cl as an uncoupler in

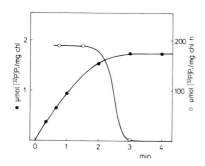

FIGURE 2. Time course of hydrolysis of 50 μM (^{32}P)ATP by activated thiol-modulated thylakoids (●) and ATPase activity (o). For determination of ATPase activity parallel samples were allowed to hydrolyze unlabeled ATP. At the indicated times carrier-free (^{32}P)ATP with phosphoglycerate kinase (30 U/ml), PEP (1 mM) and NH$_4$Cl (1 mM) was added. Release of (^{32}P)P$_i$ was measured as in (14).

order to measure the total capacity of the ATPase. Fig. 2 shows that the enzyme activity is constantly high throughout the progressing reaction. However, when the rate of the reaction is decreasing, the activity of the enzyme also decays to zero within about 1 min. The sudden deactivation is caused by two different effects which amplify each other. As long as enough $\Delta\mu_H^+$ is maintained by ATP hydrolysis, ADP-dependent deactivation can be reverted due to the fact that membrane energization promotes release of tightly bound ADP (15). With increasing exhaustion of ATP ΔpH may drop below a critical value necessary for release. As a consequence more ATPase molecules cease activity which results in acceleration of the decrease of ΔpH and so forth. Actually a breakdown of ΔpH with a similar kinetics as the decrease of ATPase activity has been observed by the 9-aminoacridine technique (not shown). This interpretation can explain mechanistically why ATP hydrolysis in coupled chloroplasts proceeds linearly with time at sufficient substrate concentration, but declines after initial stimulation in the presence of an uncoupler (16).

TABLE 1. Effect of P$_i$ on ATP-induced ΔpH

ΔpH was measured by the 9-aminoacridine method as in (14). ATP (100 μM) was added 45 s after activation (see Fig. 1).

P$_i$ conc (mM)	0	0.21	0.25	0.5	1	2.5	4
ΔpH	2.96	2.91	2.87	2.83	2.74	2.70	2.69

ATP hydrolysis catalyzed by light-activated and thiol-modulated chloroplast thylakoids is virtually not affected by inorganic phosphate at concentrations up to 5 mM. At higher concentrations some inhibition is observed. 50% inhibition is obtained at concentrations as high as 20 mM. In contrast, ATP-induced formation of a proton gradient is decreased by low P$_i$ concentrations (Table 1). This effect is saturated at about 2.5 mM and half-maximal at about 0.5 mM P$_i$. The extent of decrease of ΔpH is about 0.25 units. Inhibition of ΔpH has been convincingly explained by Davenport and McCarty (17) to be related with ATP/P$_i$ exchange. Since ATP/P$_i$ exchange is the result of ATP hydrolysis and rephosphorylation of ADP driven by the ATP-induced proton gradient, the gradient is diminished as in light-dependent phosphorylation. The K$_m$(P$_i$) for ATP/P$_i$ exchange as determined by us is 0.5 mM which corresponds to K$_m$(P$_i$) in photophosphorylation. The fact that P$_i$ in this concentration range does not affect ATP hydrolysis, suggests, that catalytic P_i binding does not interfere with catalytic ATP binding.

Inorganic phosphate has been reported to stabilize the activated state of light-triggered ATPase (18). This effect is referred to as retardation of tight ADP binding (5,6). Inhibition of ADP binding by P_i was reported to show complex kinetics (6). Those measurements, however, were carried out with inadequate time resolution. If actual initial rates are determined as a function of ADP concentration starting from nucleotide-free membranes, the inhibitory effect of P_i is clearly non-competitive (Fig. 3). ADP binding is regarded as a two step process including a loosely bound ADP as an intermediate form (13). In deenergized membranes the formation of the loose complex may be a reversible reaction whereas the subsequent transition to the tight complex is irreversible. By applying the theory of non-competitive inhibition P_i might interact with both, the non-occupied form and the one containing loosely bound ADP, thus prohibiting the transfer to the tight enzyme-ADP complex. The $K_i(P_i)$ for inhibition of tight ADP binding is 0.6 mM, which is very close to $K_m(P_i)$ in photophosphorylation. Therefore phosphate might interact with a catalytic site while inhibiting tight ADP binding.

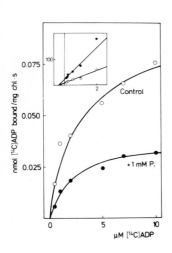

FIGURE 3. Initial rates of tight $(^{14}C)ADP$ binding by activated membranes in the dark as a function of concentration in the absence and presence of P_i (1 mM). The reaction was quenched (13) after 2 s incubation with the labeled nucleotide.

In an experiment shown in Table 2, activated thylakoids were allowed to react with 10 µM ADP in the presence or absence of P_i for 10 s before substrate ATP was added. The levels of tightly bound nucleotides as well as the rates of ATP hydrolysis at the time of substrate addition were determined. The results show that the activity of the ATPase is correlated with the number of tight nucleotide binding sites which are not occupied by ADP. As expected, the presence of P_i restricts incorporation of ADP and thus preserves activity. If preincubation is carried out with 10 µM ATP instead of ADP, the pattern of tightly bound nucleotides after 10 s is 60% ATP and 40% ADP. Again inhibition of ATPase activity correlates with the amount of bound ADP. It seems that the individual enzyme molecule incorporates either ADP or ATP, because the maximum level of tightly bound nucleotides always is 1 per 1 CF_1 molecule, irrespective of the nucleotide added. ATPase molecules containing a tightly bound ATP, retain catalytic activity. During ATP hydrolysis tigthly bound ATP is exchanged. The rate of exchange is, however, by a factor of 100 lower than the overall rate of ATP hydrolysis. Hence we may also conclude that ATP can be hydrolyzed with a tightly bound ATP sticking on the enzyme.

TABLE 2. Relationship between tight nucleotide binding and inactivation of ATPase. 30 s after activation (see Fig. 1) the indicated substrate concentrations were added. For measurement of nucleotide binding, ADP and ATP were ^{14}C-labeled. After 10 s a combined ADP/hexokinase quench (14) was employed in order to discriminate between bound ADP and ATP. For measurement of ATPase activity, 0.2 mM (^{32}P)ATP instead of the quench was added after 10 s exposition with the unlabeled nucleotides and P_i, respectively.

	bound nucleotides		%sites occupied by ADP	ATPase activity	
	nmol/mg chl			μmol P_i/	% in-
	ADP	ATP		mg chl·h	hibition
total binding	0.611	—			
control activity				52.3	
10 μM ADP	0.484	—	79.2	12.7	75.7
10 μM ADP + 0.2 mM P_i	0.118	—	19.3	40.8	22.0
10 μM ATP	0.141	0.216	23.1	42.0	19.7

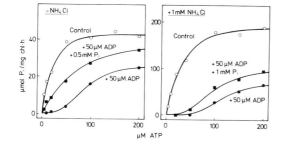

FIGURE 4. ATP hydrolysis as a function of ATP concentration in the absence and presence of 50 μM ADP or 50 μM ADP + 0.5 mM P_i. ATP was ^{32}P-labeled.

Fig. 4 shows rates of ATP hydrolysis as a function of ATP concentration in the absence and presence of ADP and ADP + P_i, respectively, both under coupled and uncoupled conditions. Normal Michaelis-Menten type kinetics are observed in the controls. In the presence of 50 μM ADP, the hyperbolic curves are converted to sigmoidal curves. The apparent cooperativity may be explained by irreversible ADP-dependent enzyme deactivation at low substrate ATP superimposed by ATP-dependent enzyme stabilization and increase of $\Delta\mu_H^+$ with increasing the ATP concentration. When P_i is also added, the sigmoidal curve is reconverted to a hyperbolic curve, but only in the absence of uncoupler. The effect of P_i can be explained by abolition of ADP deactivation as discussed above. The kinetics obtained in the presence of ADP and P_i shows a higher apparent K_m (ATP) but levels off at the same V_{max} as the

control. Since P_i at the employed concentration does not affect ATP hydrolysis itself (Table 1), the increase of apparent K_m probably indicates the
pure competitive effect of ADP on the catalytic reaction. The calculated
K_i(ADP) in this experiment is 12 µM while K_m(ATP) of the control is 25 µM.
As in photophosphorylation (19) K_m(ATP) in ATP hydrolysis seems to increase
with higher efficiency of the reaction. Similarly K_i(ADP) also increases, so
that the ratio K_m(ATP)/K_i(ADP) is always about 2.

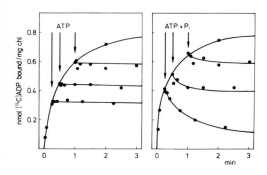

FIGURE 5. Time course of
tight ([14C])ADP binding as
affected by addition of ATP
(0.2 mM) or ATP + P_i (1 mM).
The ([14C])ADP concentration
was 5 µM.

In Fig. 5 kinetics of tight ADP binding are shown. At different times during
the course of the reaction, excess ATP or ATP + P_i was added to initiate ATP
hydrolysis. Addition of ATP leads to an immediate stop of further ADP binding and the level of already bound ADP remains constant. However, when ATP
is given together with P_i, bound ADP is released, provided that the level of
bound ADP was not too high at the time of the chase. Re-release of ADP is
completely abolished by simultaneous addition of an uncoupler (not shown).
Under similar conditions rates of ATP hydrolysis were measured (Table 3).

TABLE 3. Reactivation of the ADP-deactivated ATPase by phosphate and $\Delta\mu_{H}+$

At the indicated times after activation (see Fig. 1) the indicated compounds
were added at the following concentrations: ADP 5 µM, ATP 200 µM, P_i 2 mM,
Nigericin 0.5 µM. ATP was [32P]-labeled.

| Additions at t = | | | ATP hydrolysis | |
15 s	30 s	45 s	(µmol P_i/mg chl·h)	
ADP, ATP			48.6	(100 %)
ADP	ATP		14.4	(30 %)
ADP		ATP	2.7	(6 %)
ADP, ATP, P_i			54.6	(100 %)
ADP	ATP, P_i		40.8	(75 %)
ADP	ATP, P_i, Nigericin		4.8	(9 %)
ADP		ATP, P_i	5.1	(9 %)

After 15 s preincubation with ADP, 7O % of the original ATPase activity disappears, but the enzyme can be reactivated by P_i added together with the substrate. No reactivation occurs under uncoupled conditions. If 94 % of the enzyme molecules are inactive after 3O s reaction with ADP, no reactivation is observed by P_i even in the absence of uncoupler. Accordingly P_i-dependent reactivation by removal of tightly bound ADP in addition to P_i needs a proton gradient, which can not be generated to a sufficient extent, when the ratio of active to inactive ATPases decreases below a critical value.

4. CONCLUSION

On transition of chloroplasts with activated thiol-modulated ATPase from light to dark in a medium containing ATP, ADP and P_i the following reactions may occur: (i) ATP hydrolysis accompanied with generation of a transmembrane proton gradient (ii) rephosphorylation of ADP at the expense of the proton gradient (ATP/P_i exchange) (iii) competitive inhibition of ATP hydrolysis by ADP with a K_i which is about half of the K_m(ATP), (iv) deactivation of ATPase by ADP (v) competitive inhibition of ADP-deactivation and stabilization of activity by ATP (vi) non-competitive inhibition of ADP-deactivation by P_i and (vii) reactivation of ADP-deactivated enzymes by $\Delta\mu_H+$. The rates of the individual reactions which depend on the substrate concentrations and on $\Delta\mu_H+$ change during ATP hydrolysis. At latest after exhaustion of ATP all of the ATPase molecules will be deactivated by tight binding of ADP.

The ATPase most probably contains 3 catalytic sites in the 3 β subunits of CF_1 which are assumed to work cooperatively (2O). It has been suggested from photoaffinity labeling studies with 2-azido ADP, that the site where ATP is formed in photophosphorylation and the site which binds ADP or ATP tightly, are basically identical in peptide structure (21). This means that catalysis and regulation would occur in the same domain of the protein. Furthermore P_i-binding which takes place in the catalytic domain most probably interferes with tight ADP binding. Tight binding of ADP occurs only in the absence of both P_i and of a sufficiently high $\Delta\mu_H+$. Transition of a loosely bound to a tigthly bound ADP includes a conformational change leading to enclosure of the occupied site within the protein sturcture. In a cooperative manner the unoccupied sites seem to get inavailable, too, since no further catalytic or tight binding by the membrane-associated ATPase is observed after one ADP was incorporated (19,22). In contrast, the empty sites remain accessible to medium substrates when the tightly bound nucleotide is ATP. The fact that ATP can occupy one of the sites without inhibiting ATP hydrolysis and without being exchanged at a catalytically competent rate seems to indicate that less than three sites participate in catalysis of ATP cleavage.

ACKNOWLEDGEMENTS

This work was supported by Deutsche Forschungsgemeinschaft. A.R. Mansy was recipient of a scholarship from Deutscher Akademischer Austauschdienst. The authors thank Mrs. Karin Kiefer and Miss Heike Schüll for competent technical assistance and Mrs. Rita Reidegeld for the careful typing of this paper.

REFERENCES

1 Strotmann, H. and Bickel-Sandkötter, S. (1984) Ann. Rev. Plant Physiol. 35, 97-120
2 Ketcham, S.R., Davenport, J.W., Warncke, K. and McCarty, R.E. (1984) J. Biol. Chem. 259, 7286-7293
3 McCarty, R.E. and Racker, E. (1968) J. Biol. Chem. 243, 129-137
4 Mills, J.D. and Mitchell, P. (1982) Biochim. Biophys. Acta 679, 75-83
5 Schumann, J. and Strotmann, H. (1981) in Photosynthesis II. Electron Transport and Photophosphorylation (Akoyunoglou, G., ed.) pp. 881-892, Balaban Int. Sci. Serv., Philadelphia, Pa.
6 Dunham, K.R. and Selman, B.R. (1981) J. Biol. Chem. 256, 10044-10049
7 Bar-Zvi, D. and Shavit, N. (1982) Biochim. Biophys. Acta 681, 451-458
8 Schumann, J. (1981) in Energy Coupling in Photosynthesis (Selman, B.R. and Selman-Reimer, S., eds.) pp. 223-230, Elsevier/North Holland, New York, Amsterdam, Oxford
9 Bar-Zvi, D. and Shavit, N. (1982) J. Bioenerg. Biomembr. 14, 467-478
10 Shahak, Y. (1985) J. Biol. Chem. 260, 1459-1464
11 Strotmann, H. and Bickel-Sandkötter, S. (1977) Biochim. Biophys. Acta 460, 126-135
12 Strotmann, H., Kiefer, K. and Altvater-Mackensen, R. (1986) Biochim. Biophys. Acta 850, 90-96
13 Strotmann , H., Bickel-Sandkötter, S. and Shoshan, V. (1979) FEBS Lett. 101, 316-320
14 Shigalowa, T., Lehmann, U., Krevet, M. and Strotmann, H. (1985) Biochim. Biophys. Acta 809, 57-65
15 Strotmann, H., Bickel, S. and Huchzermeyer, B. (1976) FEBS Lett. 61, 194-198
16 Bakker-Grunwald , T. (1974) Biochim- Biophys. Acta 368, 386-392
17 Davenport, J.W. and McCarty, R.E. (1981) J. Biol. Chem. 256, 8947-8954
18 Carmeli, C. and Lifshitz, Y. (1972) Biochim. Biophys. Acta 267, 86-95
19 Bickel-Sandkötter, S. and Strotmann, H. (1981) FEBS Lett. 125, 188-192
20 Boyer, P.D. and Kohlbrenner, W.E. (1981) see Ref. 8, pp. 231-240
21 Abbott, M.S., Czarnecki, J.J. and Selman, B.R. (1984) J. Biol. Chem. 259, 12271-12278
22 Bickel-Sandkötter, S. (1983) Biochim- Biophys. Acta 723, 71-77

EFFECT OF PREILLUMINATION OF SPINACH LEAVES ON THE COUPLING EFFICIENCY OF CHLOROPLASTS

J.M.Wei, Y.K.Shen, D.Y.Li and H.P.Dai

Shanghai Institute of Plant Physiology, Academia Sinica

1. INTRODUCTION

The stoichiometry of noncyclic photophosphorylation is still a question in debate, i.e. whether the P/O ratio equals 1, 1.3 or 2 (Hall 1984). In previous papers, we have reported that the degree of coupling between the electron transport and photophosphorylation was not always complete in many chloroplast preparations and P/O ratio might be increased by treatment with a number of chemicals which we called coupling efficiency improvers (Shen 1985). Recently we have found that the preillumination of spinach leaves could also increase the coupling efficiency of chloroplasts isolated from them.

2. MATERIALS AND METHODS

Chloroplast preparation: Chloroplasts were rapidly isolated from spinach leaves maintained in darkness or under preilluminated conditions (20 °C for 90 s, at 500 W incandescent lamp through 10 cm water layer). Before isolation, spinach leaves of control and the preilluminated were both immediately put into cold suspension solution at about 0 °C. Assays of chlorophyll, electron transport and photophosphorylation were carried out as previously described (Wei et al. 1983).

3. RESULTS AND DISCUSSION

3.1. Effect of preillumination of spinach leaves on the coupling efficiency of chloroplasts

Morita et al.(1982) and Vallejos et al.(1983) both reported that chloroplasts isolated from preilluminated spinach leaves showed higher rate of noncyclic and cyclic photophosphorylation. We found that preillumination of spinach leaves could not only increase the rate of photophosphorylation, but also the P/O ratio (Table 1).

TABLE 1 Activaties of photophosphorylation (PSP) and P/O ratio of chloroplasts prepared from preilluminated spinach leaves

	Control				preilluminated			
	1	2	3	4	1	2	3	4
Endogenous-PSP	13	11	19	-	31	23	20	-
Cyclic-PSP	556	291	353	1065	643	610	379	1024
Non-cyclic-PSP	250	238	164	226	303	343	284	396
Electron transport	453	407	367	648	460	518	554	749
P/O ratio	1.10	1.17	0.89	0.89	1.32	1.32	1.03	1.06

In our previous paper, we reported that the phosphate concentration in the reaction mixture could affect the P/O ratio (Wei et al.1983). The stimulation by preillumination on P/O ratio was more prominent in the presence of higher concentration of phosphate.

3.2. Effect of preillumination of spinach leaves on the high energy state of chloroplasts
We measured the accumulation of high energy state of the chloroplasts with two stage photophosphorylation method (Shen and Shen 1962). The results (Table 2) showed that the accumulation of high energy state of chloroplasts was increased by preillumination treatment.

TABLE 2 Accumulation of high energy state of chloroplasts prepared from preilluminated spinach leaves

	Photophosphorylation (μmolesATP/mgchl.h.)	High energy state (nmolesATP/mgchl.)	P/O
Control	262	27	1.03
Preilluminated	321	37	1.16

3.3. Effect of polymyxin on photophosphorylation and P/O ratio
Polymyxin (one of the coupling efficiency improvers found by our laboratory) can induce the conformational change of CF_1 and reduce the non-productive energy loss through CF_1 during photophosphorylation (Yin et al.1979). We found that Polymyxin showed no or little effect on the rate of photophosphorylation and P/O ratio of preilluminated chloroplasts.

3.4. Effect of quercetin and antibody against CF_1 on photophosphorylation
Quercetin can interact with the catalytic center of α- and β- subunits of CF_1 and inhibit the activity of ATPase (Deters and Racker 1978). Antibody against CF_1 acts as energy transfer inhibitor (Shen et al.1984). They all showed higher inhibitory effect on photophosphorylation of the chloroplasts prepared from preilluminated spinach leaves than that of control (Only the data for quercetin were shown in Fig 1).

3.5. Effect of the chloroplasts treated with NEM on photophosphorylation
Chloroplasts treated in light with NEM, which can combine with the sulfhydryl groups of γ-subunit in CF_1 , showed lower activity of photophosphorylation (McCarty et al. 1972, Magnuso et al. 1975). When the chloroplasts prepared from preilluminated spinach leaves were treated with NEM in light, the inhibitory effect of NEM on photophosphorylation was reduced (Fig 2).

3.6. Effects of DCCD and pyridine on photophosphorylation

DCCD was known to react with CF_0 and hence block the proton channel of CF_0 (Bouthyette and Jagendorf 1982). Pyridine can also inhibit the light-stimulated opening of proton channel of CF_0-CF_1 complex (Ho and Wang 1981). Experiments with DCCD or pyridine showed that the chloroplasts isolated from preilluminated leaves were less sensitive to DCCD and pyridine inhibition (Only the data for DCCD were shown in Table 3).

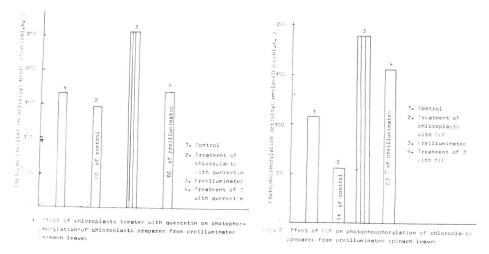

1 Effect of chloroplasts treated with quercetin on photophosphorylation of chloroplasts prepared from preilluminated spinach leaves

Fig. 2 Effect of DCCD on photophosphorylation of chloroplasts prepared from preilluminated spinach leaves

TABLE 3. Effect of DCCD on photophosphorylation of chloroplasts prepared from preilluminated spinach leaves

| | Photophosphorylation (μmoles ATP/mgchl.h.) | | P/O ratio | |
	1 %	2 %	1	2
Control	357 100	172 100	1.18	0.72
+DCCD 5×10^{-5} M	105 51	94 55	-	-
Freilluminated	385 100	219 100	1.27	0.86
+DCCD 5×10^{-5} M	235 61	146 67	-	-

From these results it is suggested that preillumination treatment can induce the sulfhydryl groups in γ-subunit of CF_1 to become less exposed, the catalytic center in α-and β-subunits of CF_1 more easily attacked by chemical modifiers, and the opening of proton channel of CF_1 increased.
It is argued that the P/O ratio in vivo may be higher than that determined conventionally with isolated chloroplasts without preillumination.

REFERENCES

Bouthette,P.Y., Jagendorf , A.T. (1982) Plant Physiol. 69: 888-896

Deters, D.W., Racker, E. (1975) J. Biol. Chem. 250:1041-1047

Hall, D.O., Rao, K.K. (1981) In Energy Coupling in Photosynthesis (Selman and Selman Reimer eds.) 241-248

Magnuson, R.P.,McCarty, R.E. (1975) J. Biol. Chem. 250:2593-2598

McCarty, R.E., Pagan J. (1973) Biochemistry 12:1503-1507

Morita, S., Itoh, S., Nishimura, M. (1982) Biochim. Biophys. Acta 679:125-130

Shen, Y.K., Shen, G.M. (1962) Acta Biochimica et Biophysica Sinica 2:58-64

Shen, Y.K., Xu, C.H., Wei, J.M., Li, D.Y., Feng, Y. and Huang, Z. H. (1984) In Advances in Photosynthesis Research (C. Sybesma ed.) Martinus Nijhooff/Dr. W.Junk Publisher The Hague/Boston/Lancaster 2:11.5 371-378

Shen, Y.K. (1985) Physiol. Vẽg. 23(5):725-729

Vallejos, R.H., Arana, J.L., Ravizzni, R.A. (1983) J. Biol. Chem. 258:7317-7322

Wei, J.M., Shen, Y.K. and Li, D.Y. (1983) Acta Phytophysiologia Sinica 9:231-240

FACTORS INVOLVED IN LIGHT ACTIVATION OF THE CHLOROPLAST H^+-ATPase

YOSEPHA SHAHAK, BIOCHEMISTRY DEPARTMENT, WEIZMANN INSTITUTE OF SCIENCE, REHOVOT 76100, ISRAEL.

1. INTRODUCTION

The chloroplast H^+-ATPase is highly regulated by light. Under Physiological conditions it becomes activated in the light, while in the dark it decays back to a latent state. Several factors are thought to mediate the activation of the enzyme by light: (i) The proton electrochemical gradient which is formed coupled to electron transport in the light, induces conformational changes in the enzyme (reviewed in ref. (1)). (ii) Reduction/oxidation of a dithiol regulatory site (on the γ subunit, ref. 2) is involved in the activation/deactivation, respectively (3,4). Both the physiological reductant (thioredoxin) and oxidant (peroxide?) are products of the electron transport process in the light. However, the interaction of the enzyme with each of them is selective and depends on light (namely on $\Delta \mu H^+$). In the light it interacts only with the reductant, while in the dark only with the oxidant (5). (iii) Dissociation of tightly bound nucleotides from CF_1 in the light and reassociation in the dark, correlate with activation and deactivation (6). (iv) Displacement of ϵ subunit in the light probably also contributes to activation of the ATPase (7). We further suggest the involvement of Mg^{2+} ions in the regulation of the enzyme, in two levels: The thioredoxin system (ferredoxin, Fd-Th reductase and thioredoxin) forms a loose complex which is bound to the stromal thylakoid surface in a Mg^{2+} dependent mode (8). In addition, we have recently found the Mg^{2+} modifies the sensitivity of the ATPase to uncouplers (9). Mg^{2+} effect is opposite in the light vs. dark. Its presence during light activation or absence during a dark period given between light activation and ATP hydrolysis assay, yield "normal" activity which is stimulated by uncouplers. However, the asbsence of Mg^{2+} from the activation step or its presence in the dark interval, yield an "abnormal" enzyme which is then inhibited by uncouplers in the assay. We suggest that Mg^{2+} is involved in the light/dark changes in the interactions between CF_1 and CF_0 subunits, which modify the affinity of a regulatory site on CF_0 to protons.

In this communication we report on further studies on three of the factors involved in regulation of the ATPase: thiol oxidation, modulation of the uncoupler sensitivity of the enzyme and the contribution of $\Delta \mu H^+$ to the activation.

2. MATERIALS AND METHODS

Isolation of chloroplasts (type A and C) from spinach leaves and assays for ATP hydrolysis and synthesis were described before (3, 8). It should be emphasized that the leaves were picked up from dark adapted plants, and the chloroplasts were isolated and kept in darkness until use to ensure that the ATPase is fully oxidized. Otherwise a significant fraction of the enzyme population was thiol modulated to begin with.

3. RESULTS AND DISCUSSION
3.1. **Differential effect of thiol oxidants**

Based on our finding that thiol-oxidants deactivate the ATPase in the dark, but not in the light, we have suggested that in the presence of $\Delta \mu H^+$ the reduced dithiol regulatory site becomes inaccessible. $\Delta \mu H^+$ can also be formed in the dark, coupled to ATP hydrolysis. It was thus predicted that thiol-oxidants would not deactivate the ATPase if added during hydrolysis. Fig. 1 shows that this indeed was

Biggens, J. (ed.), Progress in Photosynthesis Research, Vol. III. ISBN 90 247 3452 5
© *1987 Martinus Nijhoff Publishers, Dordrecht. Printed in the Netherlands.*

the case. No effect of o-iodosobenzoate was observed, if added to the ATPase assay mix under coupled conditions. However, in the presence of an uncoupler (which by itself stimulated ATP hydrolysis, as expected), oxidant-induced deactivation of the enzyme occurred within less than 1 min. Similar results were obtained with ferricyanide. The results give further support to the model of thiol activation which involves two kinds of conformational changes in the γ subunit: changes induced by $\Delta\mu H^+$ and changes induced by oxidation-reduction of the dithiol site (5).

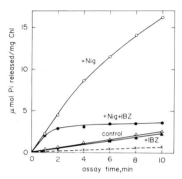

FIGURE 1: **Effect of thiol-oxidant on coupled and uncoupled ATPase activity.** Intact chloroplasts were light activated for 3 min, then osmotically shocked into the ATPase assay mix. Where indicate, 0.4 μM nigericin (Nig) and/or 1 mM o-iodosobenzoate (IBZ) were present during assay.

3.2. Modulation of the uncoupler sensitivity of the ATPase

We have previously reported that Mg^{2+} ions affect dark deactivation of the ATPase in intact chloroplasts which are osmotically shocked after light activation. If Mg^{2+} is present in the shocking medium, the apparent deactivation is fast, while in its absence, the ATPase remains active for 3 min in the dark, and then drops down (10). It was later found that Mg^{2+} does not induce deactivation, but rather modulates the sensitivity of the ATPase to uncouplers. In chloroplasts which were shocked into a Mg^{2+}-containing medium 90 sec before assay, ATPase activity was inhibited rather than stimulated by uncouplers; while in chloroplasts shocked in the absence of Mg^{2+}, "normal" stimulation by uncouplers was obtained (9). Here we report that the "abnormal" uncoupler sensitivity develops also in the absence of Mg^{2+}, but it takes longer. As shown in Fig. 2, the stimulation of ATP hydrolysis by NH_4Cl was maximal if assayed shortly after the dark osmotic shock. Between 3 to 4 min dark interval, there was a sharp decrease inthe uncoupler-stimulation, and then uncoupler-inhibition was observed. The coupled rate of ATP hydrolysis measured after 1-6 min dark intervals was essentially constant. Upon reillumination (given after the dark osmotic shock) the capacity of the ATPase to be stimulated by uncouplers was restored (not shown).

Mills and Mitchell have suggested a model of dual pH optimum requirement for activation (11). One can visualize two regulatory sites, one sensing the stromal pH and the other - the lumen pH, to be involved in ΔpH-activation. Modulation of the uncoupler-sensitivity of the ATP hydrolyzing activity is suggested to reflect changes in the affinity of the lumen-regulatory-site to proton. These changes might result from changes in subunit interactions, which occur slowly in the dark (Fig. 2), and can be stimulated by Mg^{2+} ions (9). Changes in the ATPase uncoupler sensitivity after light activation were also reported by Komatsu-Takaki for type C chloroplasts (11). It should be emphasized that in this session as well as the following one, we relate to the involvement of ΔpH in the activation, rather than in the catalytic mechanism.

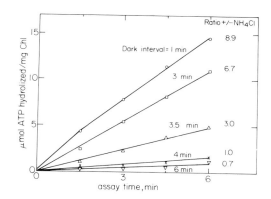

FIGURE 2. **Dependence of NH$_4$Cl-stimulation of ATPase activity on the dark interval between preillumination and assay.** Intact chloroplasts were light activated, then osmotically shocked by 10-fold dilution in 20 mM Na-Tricine pH 8 and kept in darkness for the indicated interval, before ATP hydrolysis assay, which contained 2 mM NH$_4$Cl. On the right, ratios of ATPase rate with and without NH$_4$Cl are indicated.

3.3. Can pure $\Delta \mu H^+$ activate the ATPase?

It is well known that photophosphorylation does not require thiol modulation. However in the recent years it became clear that ATP synthesis also responds to the activated state of the enzyme. The major factor in activation was suggested to be $\Delta \mu H^+$, while thiol reduction only reduces the size of the proton gradient required for activation (2, 11, 13). If the same mechanism activates the enzyme to work in both directions, then $\Delta \mu H^+$ is expected to activate the hydrolytic activity of the enzyme even without thiol reduction. In the experiment designed to answer this question, several points were taken care of: (i) We ensured that the enzyme is fully oxidized before illumination (see section 2). In some reports this was not done, yielding mis-leading results. (ii) Uncpouler was added to the assay mix to reduce the thermody-namic inhibition by $\Delta \mu H^+$ (iii) ATP regenerating system was added, and Pi omitted from the hydrolysis assay to prevent photophosphorylation.

As summarized in Table 1A, DTT was found to be obligatory for activation of ATP hydrolysis. No activation by pure $\Delta \mu H^+$ could be observed. Hydrolysis was assayed either in the dark, or in the light under reduced $\Delta \mu H^+$. Photophosphoryla-tion was measured under similar conditions, for comparison (Table 1B). The phos-phorylation results are compatible with (11): thiol modulation improves the rate of phosphorylation, especially under uncoupler-reduced $\Delta \mu H^+$ conditions.

Therefore, there seems to be an apparent difference in the activation of the forward and the backward reactions of the enzyme. The former, but not the latter reaction can be activated by pure $\Delta \mu H^+$. At present the origin of this difference is not clear and requires further investigation.

Acknowledgements

This work was supported in part by the US-Israel Binational Science Founda-tion. The technical assistance of Mrs. Drora Nadav is highly appreciated. Y.S. is an incumbant of the Mark Stanley Shriro Career Development Chair.

TABLE 1. Can pure $\Delta\mu H^+$ activate ATP hydrolysis by the H^+-ATPase?

	Preactivation conditions	Assay in	0	0.3	NH$_4$Cl in assay 1	3mM
A.					ATP hydrolysis	
I	Light+DTT	Light	9	54	198	335
	-DTT		0	2	0	0
II	Light+DTT	Dark	117	250	271	234
	-DTT		4	2	3	0
III	Dark+DTT	Light	0	5	22	24
	-DTT		4	4	0	0
B.					ATP synthesis	
I	Light+DTT	Light	582	464	279	67
	-DTT		441	326	83	10
III	Dark+DTT	Light	722	639	177	34
	-DTT		741	515	149	25

Type C chloroplasts were preilluminated by saturating white light in the presence of pyocyanine+MgCl$_2$ + or - DTT, then diluted 10-fold in assay mix. Hydrolysis assay mix contained pyocyanine, MgCl$_2$ [γ-^{32}P]ATP, PEP and pyruvate kinase. Synthesis assay mix contained pyocyanine, MgCl$_2$, ADP, ^{32}Pi, glucose and hexokinase.

REFERENCES

1. Jagendorf, A.T. (1981) in Photosynthesis: Proc. 4th Int. Congr. Photosynthesis (Akoyunoglou, G., ed.) Vol. 2, pp. 719-730, Balaban Int. Sci., Philadelphia.
2. Ketcham, S.R., Davenport, J.W., Warncke, K. and McCarty, R.E. (1984) J. Biol. Chem. 259, 7286-7293.
3. Shahak, Y. (1982) Plant Physiol. 70, 87-91.
4. Mills, J.D. and Mitchell, P. (1982) Biochim. Biophys. Acta 679, 75-83.
5. Shahak, Y. (1985) J. Biol. Chem. 260, 1459-1464.
6. Shavit, N. (1980) Ann. Rev. Biochem. 49, 111-138.
7. Finel, M., Rubinstein, M. and Pick, U. (1984) FEBS Lett. 166, 85-89.
8. Shahak, Y. (1982) FEBS Lett. 145, 223-229.
9. Shahak, Y. (1986) Eur. J. Biochem. 154, 179-185.
10. Shahak, Y. (1984) in Advances in Photosynthesis Research. Proc. 6th Int. Congr. Photosynthesis (Sybesma, C. ed.) Vol. 2 pp 6527-6530. M. Nijhoff/W. Junk, The Hague.
11. Mills, J.D. and Mitchell, P. (1984) Biochim. Biophys. Acta 764, 93-104
12. Kamatsu-Takaki, M. (1984), FEBS Lett. 175, 433-439.
13. Graber, P., Schlodder, E. and Witt, H.T. (1984) in: H^+-ATPase (ATP synthase): Structure, Function, Biogenesis (Papa, S., Altendorf, K., Ernster, L. and Packer, L., eds.) pp. 431-440, Adriatica Editrice, Bari.

PARTICIPATION OF THREE DISTINCT ACTIVE STATES OF CHLOROPLAST
ATPase COMPLEX CF_0CF_1 IN THE ACTIVATION BY LIGHT AND DTT

MIZUHO KOMATSU-TAKAKI, DEPARTMENT OF CHEMISTRY, FACULTY OF PHARMACEUTICAL
SCIENCES, TEIKYO UNIVERSITY, SAGAMIKO, KANAGAWA 199-01, JAPAN

1. INTRODUCTION

CF_0CF_1 in isolated thylakoids exhibits its ability to hydrolyze ATP only
after activation by energization in the presence or absence of a thiol compound. Recently, it has been shown that ATP hydrolysis activated by illumination in the presence of dithiothreitol (DTT) differs from that activated
by illumination in the absence of a thiol compound in two respects. The
former is relatively insensitive to pH and sensitive to stimulation by an
uncoupler (NH_4Cl). The latter is relatively sensitive to pH and insensitive
to stimulation by NH_4Cl (1). Here, a new method of activating CF_0CF_1 in
thylakoids is reported. CF_0CF_1 in the preilluminated thylakoids is activated in the dark by addition of both DTT and Pi. The DTT/Pi-activated ATP
hydrolysis is highly sensitive to pH. The pH dependence of the light/DTT-
activated ATP hydrolysis approaches that of the DTT/Pi-activated ATP hydrolysis in the dark. The pH dependence of the ATP hydrolysis activated by illumination in the absence of a thiol compound is quite different from that
of the DTT/Pi-activated ATP hydrolysis. The possible correlation among the
light/DTT-activated state, the light-activated state, and the DTT/Pi-acti-
vated state is discussed.

2. MATERIALS AND METHODS

Spinach chloroplast thylakoids were prepared as described (1).

Light/DTT-activated ATP hydrolysis: 0.1 ml of thylakoids was added to
1.4 ml of medium A containing 50 mM KCl, 25 mM Tricine-KOH (pH 8.3), 5 mM
$MgCl_2$, 30 mM DTT, and 0.1 mM methyl viologen. The mixture was illuminated
for 60 s. At 10 s after turning the light off (at dark 10 s), 0.5 ml of the
activated mixture was added to 0.5 ml of medium B containing 50 mM KCl, 25
mM Tricine-KOH (pH 8.3), 5 mM $MgCl_2$, pyruvate kinase (20 units/ml), 1 mM
ATP, 10 μM Ap_5A, and 3 mM phosphoenolpyruvate (PEP). After incubation for
40 s, the reaction was terminated by addition of 0.25 ml of 15 % perchloric
acid. The content of pyruvate plus ADP was assayed as described (1).

DTT/Pi-activated ATP hydrolysis: 50 μl of thylakoids were added to 0.95
ml of medium A which lacked DTT but contained, additionally, pyruvate
kinase (10 units/ml). The mixture was illuminated for 60 s, and at dark 10
min, 5 mM Pi and 30 mM DTT were added. After incubation for 10 min in the
dark, 0.5 mM ATP, 5 μM Ap_5A, and 1.5 mM PEP were added to the activated
mixture and incubated for 30 s.

pH dependence of the light/DTT-activated ATP hydrolysis: 0.1 ml of thylakoids was added to 1.4 ml of medium C containing 50 mM KCl, 5 mM Tricine-
KOH (pH 8.3), 5 mM $MgCl_2$, 0.1 mM methyl viologen, and 30 mM DTT. The mixture was illuminated for 60 s, and at dark 10 s, 0.5 ml of the activated
mixture was added to 0.5 ml of medium D (pH 7.0-8.7) containing 50 mM KCl,
5 mM $MgCl_2$, 50 mM Tricine-KOH, 50 mM Hepes-KOH, pyruvate kinase (20 units/
ml), 1 mM ATP, 10 μM Ap_5A, and 3 mM PEP. The mixture was incubated for 40 s.

3. RESULTS AND DISCUSSION

Preincubation of thylakoids in the dark with 30 mM DTT remarkably accel-

Biggens, J. (ed.), Progress in Photosynthesis Research, Vol. III. ISBN 90 247 3452 5
© *1987 Martinus Nijhoff Publishers, Dordrecht. Printed in the Netherlands.*

erated the subsequent light activation of CF_0CF_1 (Fig.1). Preincubation in the absence of DTT did not accelerate the light activation. This result confirms the previous report that CF_1 in thylakoids undergoes reduction by DTT in the dark (2). Preincubation of thylakoids in the dark with DTT caused no activation of $CF_0 \cdot CF_1$. Therefore, it seems that, unlike soluble CF_1 (2), CF_0CF_1 in thylakoids must experience further activation, in addition to the reduction, to exhibit its activity.

In the dark after illumination, addition of both DTT and Pi activated CF_0CF_1 in thylakoids (Fig.2). The activity increased with increasing time interval between the addition of DTT and Pi and the addition of ATP. The dependence on the incubation time of the DTT/Pi activation is similar to that of the dark reduction of CF_0CF_1 in thylakoids (Figs.1,2). Therefore, the reduction of CF_0CF_1 may be a central part of the DTT/Pi activation. With increasing time interval between the end of illumination and the addition of DTT and Pi, the ability of DTT and Pi to activate CF_0CF_1 decreased (Fig.3). In the dark without preillumination, incubation of thylakoids with both DTT and Pi caused no activation of CF_0CF_1. Therefore, it is likely that in the post-illumination dark, CF_0CF_1 converts its state from the one (E_1) which is inactive but activatable by the reduction alone to the one (E) which is inactive and is not activated by the reduction alone. As shown in Fig.4, the

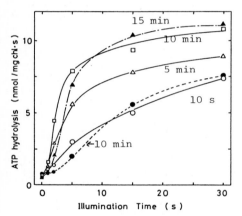

coexistence of ADP with DTT and Pi in the post-illumination dark inhibited the activation of CF_0CF_1. ADP may accelerate the conversion from E_1 to E and Pi inhibits this acceleration.

Fig.1. <u>Effect of Preincubation with DTT on the rate of light activation.</u> Before illumination, 50 µl of thylakoids was added to 0.7 ml of medium A or to medium A which lacked DTT (●). The mixture was incubated for the time indicated before illumination. (●), 30 mM DTT was added just before illumination.

Fig.2. <u>Activation of CF_0CF_1 in thylakoids by DTT and Pi in the post-illumination dark.</u> 0.2 ml of thylakoids were added to 4.3 ml of medium A which lacked DTT but contained pyruvate kinase. The mixture was illuminated for 60 s and at dark 10 min, Pi and DTT were added. After further incubation for the time indicated, DTT, Pi ATP, Ap5A and PEP were added to make their final concentrations, 30 mM, 5 mM, 0.25 mM, 2.5 µM and 1.5 mM, respectively.

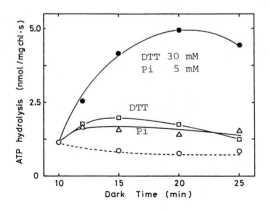

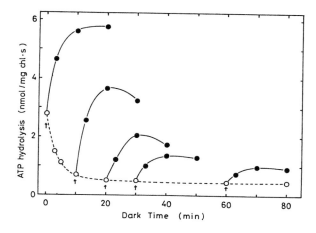

Fig.3. <u>Decrease in the DTT/Pi activation in the post-illumination dark.</u> 30 mM DTT and 5 mM Pi were added at the time indicated by the arrows.

Fig.5. <u>Change in pH dependence of light/DTT activated ATP hydrolysis in the post-illumination dark.</u> Light/DTT activated mixture was mixed with medium D at dark 10 s (O), 1 min (△), 3 min (□), or 5 min (●). ATP hydrolysis activated by light in the absence of DTT (■). DTT/Pi-activated ATP hydrolysis(▲).

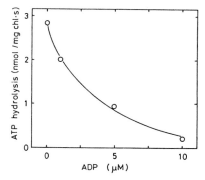

Fig.4. <u>Effect of ADP on the DTT/Pi activation.</u> ADP was added with DTT and Pi at dark 10 min.

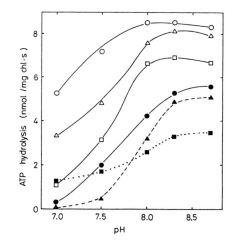

As shown in Fig.5, the DTT/Pi-activated ATP hydrolysis was highly sensitive to pH. The light/DTT-activated ATP hydrolysis increased its sensitivity to pH with increasing time interval between the end of illumination and the addition of ATP, and its pH dependence approached that of the DTT/Pi-activated ATP hydrolysis. This suggests that in the dark, the light/DTT-activated, relatively pH-insensitive state (E_2^*) converts to the state (E_2i^*) which is the same as (or similar to) the state formed by the DTT/Pi activation. The pH dependence of ATP hydrolysis activated by illumination in the absence of a thiol compound was different from those of both the light/DTT-activated ATP hydrolysis assayed after 10 s of dark incubation and the DTT/Pi-activated ATP hydrolysis.

NH$_4$Cl stimulates the light/DTT-activated ATP hydrolysis and the stimulating effect decreases with increasing time interval between the end of illumination and the addition of ATP. ATP hydrolysis activated by illumination in the absence of a thiol compound has been shown not to be stimulated

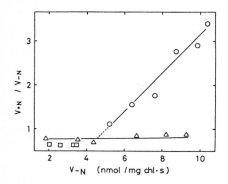

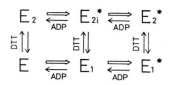

Fig.7. <u>Model for the activation of</u> <u>CF_0CF_1 in thylakoids by light and</u> <u>DTT</u>. A possible correlation among three distinct active states.

Fig.6. <u>Dependence of the stimulating effect of NH_4Cl on the rate of ATP hy-</u> <u>drolysis</u>. The light/DTT-activated ATP hydrolysis in the absence of NH_4Cl (V_{-N}) was measured by addition of 0.5 ml of the activated mixture to 0.5 ml of medium B with ATP at 20 µM to 2.0 mM, at dark 10 s (**O**) or at dark 3 min (**△**). (**□**), ATP hydrolysis activated by illumination in the absence of DTT. For the measurement of the ATP hydrolysis in the presence of NH_4Cl (V_{+N}), 4 mM NH_4Cl was added to medium B.

by NH_4Cl (1). To evaluate the stimulating effect of an uncoupler, however, we must take the possibility into account that the stimulating effect of an uncoupler depends on the rate of ATP hydrolysis. This possibility was exam- ined by measuring the stimulating effect of NH_4Cl on the light/DTT-activ- ated ATP hydrolysis at varied rates (Fig.6). The rate of ATP hydrolysis was varied by addition of varied concentrations of ATP at 10 s after turning the light off. The stimulating effect of NH_4Cl decreased with the decrease in the rate of ATP hydrolysis and at the rate of 5.2 nmol/mg of chl (the limiting rate), NH_4Cl exhibited little stimulation. The rate of ATP hydrol- ysis activated by illumination in the absence of a thiol compound is usual- ly near or below this value. Therefore, it is impossible to evaluate its sensitivity to NH_4Cl through the stimulating effect of NH_4Cl. As described, the stimulating effect of NH_4Cl on the light/DTT-activated ATP hydrolysis decreases with increasing time in the dark. ATP hydrolysis assayed after 3 min dark incubation was not stimulated by NH_4Cl at all, even at the rate higher than the above mentioned limiting rate. This result shows that in the post-illumination dark, the light/DTT-activated CF_0CF_1 converts its state from the uncoupler sensitive one to the uncoupler insensitive one.

Here, a possible model which contains three distinct active states is proposed (Fig.7). The presence of DTT in the dark converts CF_0CF_1 from E to a reduced form E_2. Illumination converts E_2 to active E_2^* through E_{2i}^*. Il- lumination of thylakoids in the absence of a thiol compound converts E to active E_1^* through E_1. ATP hydrolysis catalyzed by E_2^*, E_1^*, or E_{2i}^* is rel- atively insensitive, relatively sensitive, or highly sensitive to pH, res- pectively. ATP hydrolysis catalyzed by E_2^* is sensitive to stimulation by an uncoupler. ATP hydrolysis catalyzed by E_{2i}^* is insensitive to stimulation by **an** uncoupler. E_{2i}^* is likely under a certain uncoupled condition.

The full paper of this summary is ref.3.

REFERENCES
1. Komatsu-Takaki, M. (1986) J. Biol. Chem. <u>261</u>, 1116-1119
2. McCarty, R.E., and Racker, E. (1968) J. Biol. Chem. <u>243</u>, 129-137
3. Komatsu-Takaki, M. (1986) J. Biol. Chem. in press

THIOL MODULATION OF THE THYLAKOID ATPase IN THE GREEN ALGA UNDERLINE{DUNALIELLA}

GRAHAM D. NOCTOR AND JOHN D. MILLS, DEPARTMENT OF BIOLOGICAL SCIENCES, UNIVERSITY OF KEELE, KEELE, STAFFORDSHIRE, ST5 5BG, U.K.

1. INTRODUCTION

Studies dating back twenty years have shown that illumination of photosynthetic systems increases the activity of various subsequently assayed stromal enzymes (1). This light-induced enzyme activation has been explained in terms of light-induced increases in stromal Mg^{2+} concentration and pH and generation of reductants, including thioredoxin, by electron transport (1). The process of light activation of enzymes carries important implications regarding the induction period of CO_2 fixation. Upon illumination, photosynthetic systems capable of high rates of CO_2 fixation display a lag or induction period before the steady-state rate of CO_2 fixation is reached. This phenomenon has been explained both in terms of a requirement for build-up of stromal metabolite levels and a requirement for light-mediated activation of key enzymes (2).

The light-induced activation of the thylakoid protonmotive ATPase (CF_0-CF_1) is interesting as it has been shown to be activated by both pH and thiol-reductants (3). Under the appropriate conditions, CF_0-CF_1 will synthesize ATP immediately on illumination. However, a preillumination period is necessary if high rates of ATP hydrolysis are to be observed. It has been proposed that the anisotropic CF_0-CF_1 complex has a dual pH optimum requirement, a condition which is met on illumination when ΔpH is generated across the thylakoid membrane (4). This 'activated' form of the enzyme can then be thiol-reduced to produce a thiol-modulated enzyme that requires a smaller ΔpH for activation and is therefore capable of high rates of ATP hydrolysis (4). Thus, the rate of ATP hydrolysis observed is proportional to the degree of thiol-modulation of the enzyme.

In order to assess the contribution of enzyme activation mediated by thiol-reductants to the lag in CO_2 fixation, we have compared the kinetics of onset of ATP hydrolysis and CO_2 fixation in the green alga UNDERLINE{Dunaliella} tertiolecta. Dunaliella has a glycocalyx coat but no cell wall and therefore can be lysed comparatively easily: thus the state of activation of enzymes may be measured immediately upon release from the in vivo environment.

2. MATERIALS AND METHODS

Algae were grown at 25°C in 1.5 M NaCl supplemented with an artificial medium adapted from McLachlan (5). Cells were harvested by centrifuging at 480*g for 3 minutes. CO_2 fixation was measured at 25°C in a medium of 0.5 M NaCl supplemented with 2 mM sodium bicarbonate. Rates of ATP hydrolysis were assayed at 25°C using a two-stage procedure. The preillumination stage consisted of algae suspended in 0.33 M sorbitol, 30 mM tricine (pH 8.0), 20 mM NaCl, and 5 mM $MgCl_2$ at a concentration of 150-200 mg chlorophyll ml^{-1}. CO_2 fixation was assayed in this medium and no difference in either rates or kinetics was found relative to the NaCl

Biggens, J. (ed.), Progress in Photosynthesis Research, Vol. III. ISBN 90 247 3452 5
© 1987 Martinus Nijhoff Publishers, Dordrecht. Printed in the Netherlands.

medium. Experimental conditions, including light intensity (red light, obtained using a Corning 2-62 filter) and illumination time were varied in the preillumination stage. Aliquots containing 20 μg chlorophyll were sampled into an assay medium (30 mM Tricine (pH 8.0), 2.5 mM ATP, 2.5 mM $MgCl_2$, 0.5 mM NH_4Cl) in which experimental conditions were constant. The resulting 1 ml suspension was rapidly passed through a nylon mesh (pore diameter 10 μM) in order to lyse the cells. After 5 minutes the reaction was stopped with 0.5 mls of 5% trichloroacetic acid and ATP hydrolysis was determined by release of inorganic phosphate (6).

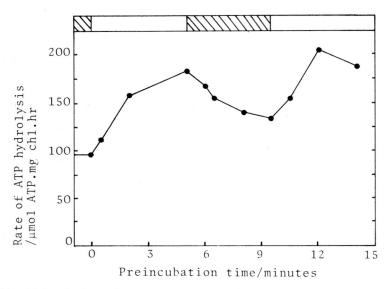

Fig. 1. Light-dark cycle for ATP hydrolysis activity in <u>Dunaliella.</u> Light intensity was 68 $W.m^{-2}$. Hatched areas, darkness; clear areas, illumination.

3. RESULTS AND DISCUSSION

As shown in Fig. 1, preillumination of algae at a light intensity of 68 $W.m^{-2}$ produces a 2-2.5 fold increase in the rate of ATP hydrolysis determined in the subsequent dark period. This activity is completely inducible within 5 minutes illumination and is dark-reversible, total loss of light-induced activity occurring within 10-15 minutes darkness. The light-induced activity can be inhibited by CF_1-specific inhibitors although the light-independent activity cannot (not shown). We conclude that the latter activity is due to cellular ATPases.

10 mM methyl viologen partially inhibits the light-induced activity whereas 100 mM methyl viologen leads to total inhibition (Fig. 2). High concentrations of methyl viologen were necessary, presumably because of low rates of penetration through the glycocalyx coat. Addition of methyl viologen after steady-state ATP hydrolysis rates are reached does not affect ATPase activity. Thus methyl viologen in the preillumination stage causes adjustment of ATPase activity to a new steady-state rate. This has been interpreted as evidence that CF_1 ATPase activity is controlled by a reductant generated by the electron transport chain at a site on the

reducing side of photosystem 1 (4). Such a reductant is probably thioredoxin, which is reduced via its reductase by ferredoxin (1).

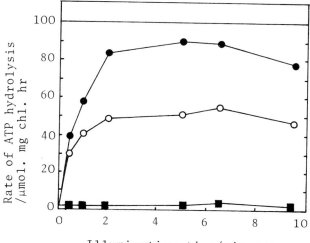

Fig. 2. Effect of methyl viologen on light-induced ATPase activity in Dunaliella. Light intensity was 105 W.m^{-2}. Closed circles, - methyl viologen; Open circles, 10 mM methyl viologen; Squares, 100 mM methyl viologen.

In order to assess the importance of the activation state of stromal enzymes during the induction period of CO_2 fixation, a series of experiments determining the kinetics of induction for both ATP hydrolysis and CO_2 fixation were carried out at different light intensities (Fig. 3). For both ATP hydrolysis and CO_2 fixation the final steady-state rate and the kinetics of induction varied with light intensity. At low light (Fig. 3 (a)) both activities required 5 minutes illumination to reach maximal rate (illumination times of up to 20 minutes were used and did not increase the rate further). At 11 and 42 W.m^{-2}, kinetics of induction of CO_2 fixation and ATP hydrolysis were extremely similar. However the kinetics of induction of both activities changed from linear to hyperbolic with increase in light intensity. Only at high light (Fig. 3 (c)) was the induction of ATPase activity significantly faster than that of CO_2 fixation.

Previous work has provided evidence that thiol-reduction of CF_1 in protoplasts is complete at illumination times and light intensities far lower than those required to saturate CO_2 fixation (6). These results have been interpreted as indicating that enzyme activation is unimportant as a limiting factor during the lag phase of photosynthesis (6). Results shown here obtained at high light intensity support this conclusion. However results obtained at lower light intensities indicate that reductive enzyme activation may play an important role during the induction period of photosynthesis under certain conditions. Similar studies are being carried out on other stromal enzymes in Dunaliella as a means of further assessing the contribution of reductive activation to the induction period of photosynthesis.

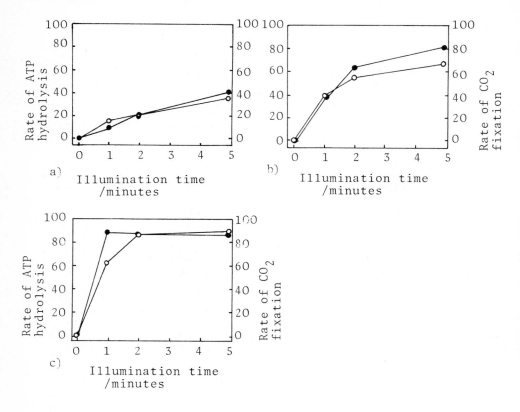

Fig. 3. Effect of light intensity on induction of CO2 fixation and ATP hydrolysis in <u>Dunaliella</u>. Results are normalised as a percentage of the highest rate observed. Closed circles, ATP hydrolysis; Open circles, CO_2 fixation. a) Light intensity = 11 $W.m^{-2}$. b) Light intensity = 42 $W.m^{-2}$ c) Light intensity = 600 $W.m^{-2}$.

REFERENCES
1 Buchanan, B.B. (1980) Ann. Rev. Plant Physiol. 31, 341-374
2 Leegood, R. and Walker, D. (1980) Plant Cell Environ. 4, 59-66
3 Bakker-Grunwald, T. (1974) Biochim. Biophys. Acta 368, 386-392
4 Mills, J.D., Mitchell, P. and Schurmann, P. (1981) in Photosynthesis 2. Electron Transport and Photophosphorylation (Akoyunoglou, G., ed.) pp. 839-848, Balaban International Science Services, Philadelphia
5 McLachlan, J. (1960) Can. J. Microbiol. 66, 367-379
6 Mills, J.D. and Quick, W.P. (1986) Biochim. Soc. Trans. 14, 20-22

ACTIVATION OF THE MAIZE CHLOROPLAST COUPLING FACTOR ATPase

William J. Patrie and Donald Miles, Division of Biological Sciences, University of Missouri, Columbia, Missouri 65211 U.S.A.

1. INTRODUCTION

The chloroplast coupling factor (CF_1) is part of the proton-translocating ATPase (H+-ATPase) complex which phosphorylates ADP in the process of photophosphorylation. CF_1 consists of 5 different kinds of subunits, denoted $\alpha - \varepsilon$ in order of decreasing molecular weight.

In broken spinach chloroplast membranes CF_1 does not normally hydrolyze ATP. However, after activation by illumination in the presence of thiol reducing agents such as dithiothreitol (DTT), the enzyme can catalyze Mg^{2+} specific ATP hydrolysis that is coupled to proton uptake. This reduced enzyme is likewise activated for ATP synthesis, and this form of the enzyme is the predominant physiological form in illuminated leaf tissue (1,2). After purification, soluble CF_1 has little ATPase activity unless activated by one of several rather diverse means: protease treatment, heat treatment, DTT treatment, or treatment with one of several detergents or aliphatic alcohols. Activation by DTT correlates with a reduction of a disulfide in the subunit. In addition, the thiol activated enzyme is the only activated form that is competent for photophosphorylation; the other activated forms are proton leaky, and result in the uncoupling of ATP synthesis.

Relatively little work has been published concerning CF_1 from C4 monocot plants. One of the better studied is CF_1 from maize. Work by Cohen (3) suggested that the CF_1CF_0 complex of maize controlled proton efflux in a manner different from pea and spinach chloroplasts. In addition, little stimulation of a Mg^{2+}-ATPase activity by light and DTT treatment of maize thylakoids was observed, quite in contrast to the membranes of spinach and pea (W.S. Cohen, personal communication). This latter observation in particular suggested significant differences in the activation and regulation of the maize enzyme and prompted our present study of the activation of the soluble maize CF_1.

2. METHODS

Maize CF_1 was prepared using the small scale method described by Cerione (4) for spinach CF_1, with the modification that the CF_1 fraction was eluted from the DEAE-cellulose column directly using buffer containing 0.4M NaCl without the 0.2M MaCl washing step.

Preparations of maize and spinach CF_1 were activated

Biggens, J. (ed.), Progress in Photosynthesis Research, Vol. III. ISBN 90 247 3452 5
© *1987 Martinus Nijhoff Publishers, Dordrecht. Printed in the Netherlands.*

by the indicated methods and the activities under different reaction conditions were compared. Control CF_1: Latent, untreated enzyme. DTT CF_1: CF_1 treated in 50mM Tris HCl pH 8.0, 2mM EDTA and 1mM ATP (TEA) containing 50mM DTT for 4 hours. Heat CF_1: CF_1 heated at 62 C for 5 min. in TE containing 40mM ATP. Heat DTT CF_1: CF_1 heat activated in the presence of 10mM DTT. Trypsin CF_1: CF_1 (100ug) treated with lug trypsin for 5 min. in TEA. Trypsin DTT CF_1: DTT activated enzyme was subsequently trypsin treated. Ca^{2+}-ATPase: ATPase activity in 50mM Tris HCl pH 8.0, 5mM $CaCl_2$, 5mM ATP. Mg^{2+}-ATPase: ATPase activity in 50mM Tris HCl pH 8.0, 1.5mM $MgCl_2$, 4 mM ATP. $OGMg^{2+}$-ATPase: Mg^{2+}-ATPase activity as above in the presence of 30mM octylglucoside. MeOH Mg^{2+}-ATPase: ATPase activity in 50mM Tris HCl pH 8.8, 5mM $MgCl_2$, 5mM ATP, 34% MeOH. SO_3^{2-} Mg^{2+}-ATPase: ATPase activity in 50mM Tris HCl pH 8.0, 80mM Na_2SO_3, 8mM $MgCl_2$, 8mM ATP. All assays were at 37 C.

CF_1 preparations were labelled with the fluorogenic sulfhydryl reagent, anilinonaphthylmaleimide as described in reference 5.

3. RESULTS AND DISCUSSION
3.1 Activation of maize CF_1

Table 1 compares the activites of maize and spinach CF_1 preparations treated by various procedures known to activate spinach CF_1, as described in "Methods". Notable differences include: 1) A lack of significant Ca^{2+}-ATPase activities for maize CF_1 with the exception of trypsin-treated DTT-CF_1. 2) The presence of $OGMg^{2+}$-ATPase activities only for otherwise activated maize CF_1 preparations; spinach CF_1, in contrast, has an $OGMg^{2+}$-ATPase activity independent of various activation

TABLE 1. Comparison of ATPase activities of maize and spinach CF_1 preparations

	Ca^{2+}		Mg^{2+}		OG Mg^{2+}		MeOH Mg^{2+}		SO_3^{2-} Mg^{2+}	
	c	s	c	s	c	s	c	s	c	s
CONTROL	1.0	6.0	0.8	2.2	2.2	26	12	20	1.4	7.1
DTT	2.0	25	1.4	3.5	25	27	23	20	6.7	33
HEAT	1.2	17	1.0	3.9	1.9	24	5.0	20	1.1	20
H DTT	1.7	31	1.7	5.3	19	24	17	20	6.7	44
TRYP	4.4	21	1.7	2.6	6.5	28	17	22	8.1	29
TRYP DTT	32	37	4.0	5.6	49	28	37	19	39	54

procedures. A similar activation dependent pattern is

observed for both maize and spinach CF_1 preparations for the SO_3^{2-} dependent Mg^{2+}-ATPase activity. A similar, although less notable pattern is observed for the maize CF_1 MeOH Mg^{2+}-ATPase activity. 3) There is no significant activation of maize CF_1 by heat treatment under any assay condition. Since heat activation, as well as activation by MeOH and octylglucoside of spinach $CF1$ all appear related to the interactions of ε subunit with the complex, the decreased activity observed for the maize enzyme would suggest a different, perhaps stronger interaction than with the spinach enzyme.

3.2 In vivo activation and deactivation of maize CF_1
Figure 1 demonstrates the in vivo activation of the $OGMg^{2+}$-ATPase in maize leaves. Plants (dark adapted for 12

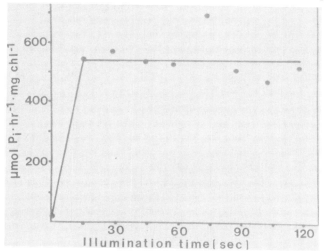

Figure 1. Light activation of maize CF_1 OG Mg^{2+}-ATPase

hours) were illuminated and 2-3 leaves from individual plants were rapidly removed and frozen in liquid nitrogen at the indicated times. The leaves were subsequently homogenized in 5ml TEA and filtered. Aliquots (50ul) of the filtrates were assayed for $OGMg^{2+}$-ATPase activity, an activity dependent upon the prior activation of the maize enzyme, in this case presumably reduction. Activation is quite rapid and is complete within 15 seconds. The dark inactivation (not shown) is much slower with a $t_{1/2}$ of about 5 min.

3.4 The sulfhydryl content of maize CF_1 subunits
Figure 2 compares the incorporation of the fluorescent sulfhydryl reagent anilinonaphthylmaleimide (ANM) into the γ subunit of latent and DTT activated CF_1 preparations

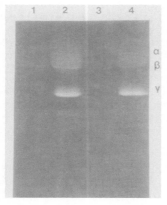 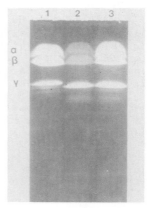

Figure 2.(Left) Reduction of a disulfide by DTT activation of maize CF_1

Figure 3.(Right) Sulfhydryl content of maize CF1

analyzed using SDS-PAGE. In lanes 1 and 2 CF_1 was pretreated with N-ethylmaleimide to block accessible sulfhydryl groups. Lanes 1 and 3 contain latent CF_1, lanes 2 and 4 contain DTT-activated CF_1. Very little ANM is incorporated into the latent enzyme, whereas the γ subunit of the activated enzyme incorporates a good deal more label. Incorporation of the label into disulfide sulfhydryls also results in a shift to higher mobility of the γ subunit. This is more clearly demonstrated in Figure 3. In lane 1, ANM incorporation into SDS denatured latent CF_1 is illustrated, showing much greater incorporation of label into α, β and γ subunits compared to Figure 2. No incorporation is observed in the presumed ε subunit, in contrast to spinach CF_1. In lane 2, latent CF_1 was first blocked with NEM, then DTT activated and labelled with ANM.

Similar amounts of ANM incorporation are observed in lanes 1 and 2, but a shift in mobility is observed. In lane 3, latent CF_1 that was not previously blocked was reduced and labelled with ANM, showing still greater incorporation into γ subunit, but maintaining the higher mobility. Assuming only 1 disulfide in maize γ subunit as there is in spinach CF_1, there would likewise appear to be 2 free sulfhydryls in the latent enzyme.

REFERENCES
1 Ketcham, S.R., Davenport, J.W., Warncke, K. and McCarty, R.E. (1984) J. Biol. Chem. 259, 7286-7293
2 Marchant, R.H. (1981) Proc. Int. Congr. Photosyn. Res 2, 999-1008
3 Cole, R.M., Macpeek, W.A., and Cohen, W.S. (1981) Plant Physiol. 68, 610-615
4 Cerione, R.A., McCarty, R.E. and Hammes, G.G. (1983) Biochemistry 22, 769-776
5 Nalin, C.M. and McCarty, R.E. (1984) J. Biol. Chem. 259, 7275-7280

COMPARISON OF METHODS OF ACTIVATION OF THE ATPase ACTIVITY OF THE CHLOROPLAST COUPLING FACTOR 1

Mark L.Richter, James D.Lampton and Richard E.McCarty, Section of Biochemistry, Molecular and Cell Biology, Cornell University, Ithaca, NY 14853 USA.

1. INTRODUCTION

The isolated chloroplast coupling factor 1 (CF_1) can be activated to hydrolyse ATP in the presence of Ca^{2+} ions by reduction of the disulfide bond of the gamma subunit or by removal of the epsilon subunit. Each method results in partial activation, whereas a combination of both methods results in an enzyme with approximately twice the rate of ATP hydrolysis observed with either method alone [1]. A similar increase in activation can be obtained by a brief exposure of either the reduced or ε-deficient enzymes to trypsin. Under these conditions the γ subunit is rapidly cleaved [2]. CF_1 can also be induced to hydrolyze Mg^{2+}-ATP by assaying in the presence of methanol, ethanol or certain detergents, or by including oxyanions such as sulfite in the assay mixture following pretreatment of the enzyme by reducing agents or ε removal. These conditions are apparently necessary for overcoming an inhibition caused by free Mg^{2+} ions (see 3 for review).

We have examined the temperature dependence for ATP hydrolysis of CF_1 following different activation pretreatments. The results indicate that CF_1 can exist in two distinct states of activation differing by about 5 Kcal per mol in activation energy. The same two states exist for either Ca^{2+}- or Mg^{2+}-dependent ATPase activity, and are shown to be very similar to different active states induced in the membrane-bound enzyme.

2. PROCEDURES

2.1 Preparations:

Thylakoid membranes and CF_1 were prepared from market spinach [2]. $CF_1(-\varepsilon)$, dithiothreitol-reduced $CF_1(-\varepsilon)$ and trypsin-treated $CF_1(-\varepsilon)$ were prepared as described elsewhere [1]. Reduction of the γ disulfide bond, and limited tryptic proteolysis of membrane-bound CF_1 were achieved as in [2].

2.2 Assays:

Ca^{2+}-ATPase reaction mixtures contained 50 mM Tris-HCl (pH 8), 5 mM ATP and 5 mM $CaCl_2$. Mg^{2+}-ATPase reaction mixtures contained 50 mM Tris-HCl (pH 8), 4 mM ATP, 2 mM $MgCl_2$ and Na_2SO_3 as indicated. 2-8 µg of enzyme preparation were added to start the reaction and P_i formation was measured colorimetrically after 2-4 min at 37°C (unless otherwise indicated).

3. RESULTS

Treatment of $CF_1(-\varepsilon)$ with either dithiothreitol or trypsin resulted

Biggens, J. (ed.), Progress in Photosynthesis Research, Vol. III. ISBN 90 247 3452 5

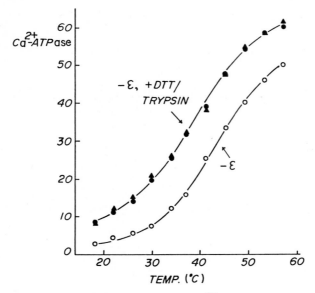

<u>Figure 1</u> *Temperature Dependence for Ca^{2+}-ATP Hydrolysis. $CF_1(-\varepsilon)$, ○—○;*
$CF_1(-\varepsilon)$ preincubated with 10 mM dithiothreitol for 15 min at room temp.
●—●; and $CF_1(-\varepsilon)$ preincubated with trypsin (1 µg per 50 µg $CF_1(-\varepsilon)$, 4 min
at room temp. followed by 20 µg trypsin inhibitor), ▲—▲ , were assayed
for Ca^{2+}-ATPase activity (µmol per min per mg) at the temperatures
indicated.

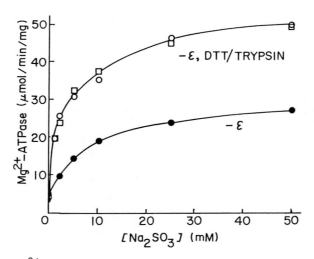

<u>Figure 2</u>   *Mg^{2+}-ATP Hydrolysis as a Function of Na_2SO_3 Concentration.*
$CF_1(-\varepsilon)$, ●—●; $CF_1(-\varepsilon)$ treated with dithiothreitol, ○—○; $CF_1(-\varepsilon)$ treated
with trypsin, □—□ .   (See Fig.1).

in an essentially identical change in the temperature dependence curve for Ca^{2+}-ATP hydrolysis (Figure 1). The activation energies for the reaction, calculated from these data, were 17.1 Kcal per mol for $CF_1(-\varepsilon)$ and 12.3 Kcal per mol following either of the additional treatments. A very similar effect was observed for Mg^{2+}-ATP hydrolysis when the three different forms of the enzyme were assayed as a function of the Na_2SO_3 concentration present during assay (Figure 2). Activation energies for the different forms, determined from the temperature dependence in the presence of 40 mM Na_2SO_3, were 23.6 Kcal per mol for $CF_1(-\varepsilon)$ and 18.5 Kcal per mol after further activation, again a difference of about 5 Kcal per mol.

Treatment of membrane-bound CF_1 with either dithiothreitol or trypsin during thylakoid illumination results in a permanent activation of the enzyme for dark, Na_2SO_3-dependent Mg^{2+}-ATPase activity (see 4 and Larson & Jagendorf in this volume). Interestingly, the reduced and trypsin-treated enzymes showed a strong resemblance to the partially and fully activated states respectively of soluble CF_1 in terms of their Na_2SO_3 dependencies (Figure 3).

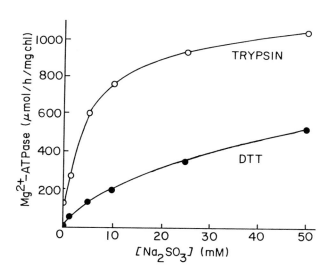

<u>Figure 3</u>   *Na_2SO_3 Dependent Mg^{2+}ATPase Activity of Membrane-Bound CF_1. Thylakoid membranes (equivalent to 100 µg chlorophyll) were preincubated in the presence of 5 mM dithiothreitol or 7 µg of trypsin in 50 mM Tricine-NaOH (pH 8), 25 µM pyocyanin and 4 mM $MgCl_2$ for 5 min AT 35°C during continuous illumination. The membranes were stored in the dark on ice for at least 20 min before dilution into the assay medium. 50 µg of trypsin inhibitor were added immediately after trypsin treatment.*

4. DISCUSSION

Most of the known methods of activation of soluble CF_1 can be described in terms of their effects on either the γ or ε subunits, both of which appear to be intimately involved in the activation process (1). The results described here indicate that the enzyme may be induced into one of two discrete active conformations irrespective of the nature of the activating pretreatments. This was apparent for both Ca^{2+}-ATPase and sulfite-induced Mg^{2+}-ATPase activities. There are some indications also (5,6) that similar states of activation may exist for Mg^{2+}-ATPase induced by the presence of alcohols or detergents during assay, but this remains to be clearly established

Activation of the Mg^{2+}ATPase activity of membrane-bound CF_1 normally requires both reduction of the γ disulfide bond, and maintenance of a low transmembrane proton gradient (7). It was observed recently (4), however, that the reduced enzyme is capable of Mg^{2+}-ATP hydrolysis when provided with sulfite ions in the complete absence of a transmembrane proton gradient. In this case, reduction of the disulfide bond appears to induce a partially activated state of the enzyme very similar to that of the partially activated soluble enzyme. That this is only a partial activation is clearly evident from the fact that cleavage of the γ subunit during illumination results in an additional activation, similar to the fully activated state of the soluble enzyme (Figure 3).

It will be particularly interesting to determine the relationship, if any, between these different forms of the membrane-bound enzyme and those previously characterized in terms of their nucleotide-binding properties (8).

REFERENCES

1. Richter, M.L., Snyder, B., McCarty, R.E. and Hammes, G.G. (1985) Biochemistry 24, 5755-5763
2. Schumann, J., Richter, M.L. and McCarty, R.E. (1985) J. Biol. Chem. 259,11121-11128
3. McCarty, R.E. and Moroney,, J.V. (1985) in *The Enzymes of Biological Membranes* (Martinosi, A.,Ed.) 2nd ed., pp383-413, Plenum Press, New York
4. Larson, E.M. and Jagendorf, A.T. (1986) Ann. Meet. Amer. Soc. Plant Physiol. *abstract*
5. Anthon, G.E. and Jagendorf, A.T. (1986) Biochim. Biophys. Acta 848, 92-98
6. Yu, F. and McCarty, R.E. (1985) Arch. Biochem. Biophys. 238, 61-68
7. McCarty, R.E. (1979) Annu. Rev. Plant Physiol. 30, 79-104
8. Schumann, J. (1984) Biochim. Biophys. Acta 766, 334-342

DTT Stimulation of ATP Synthesis by Chloroplast Thylakoids
Correlates with Dark Decay of ATP-P$_i$ Activity

J.W.Davenport and R.E.McCarty[*]

Department of Human Genetics, Yale Medical School, New Haven, CT
06510 and * Department of Biochemistry, Cornell University,
Ithaca, NY 14850

Introduction

Illumination of chloroplast thylakoids in the presence of
DTT activates ATP hydrolysis assayed in subsequent darkness. It
also activates ATP synthesis in that the ∆pH required to support
a given rate of photophosphorylation is reduced by approximately
0.3 pH units (1;2;figure 2).

ATP-P$_i$ exchange differs from photophosphorylation only in
that the proton gradient which drives it is generated by
hydrolysis instead of electron flow (3). The dependence of the
rate of ATP-P$_i$ exchange upon ∆pH is similar to that of activated
photophosphorylation when the latter is assayed using a protocol
very similar to that for exchange (3; figure 1). However, the
reduction in the ∆pH required to support a given rate of
phosphate esterification (the "shift")is smaller than that seen
for photophosphorylation as it is usually assayed (compare
figures 1 and 2). We present further data on the variable nature
of this shift.

Experimental Procedure

All experiments were performed as described in referece 3.

Results

In a number of experiments we have seen little or no shift
in the dependence of exchange upon ∆pH (figure 2). In these
experiments dark decay of ATP-P$_i$ exchange activity was negligible
(insert figure 2). The activation of photophosphorylation was
normal.

In several experiments we have found that the shift in the
dependence of exchange upon ∆pH increases in parallel with
increasing dark decay (figure 3).

Biggens, J. (ed.), Progress in Photosynthesis Research, Vol. III. ISBN 90 247 3452 5
© *1987 Martinus Nijhoff Publishers, Dordrecht. Printed in the Netherlands.*

Conclusions

The shift in the dependence of the rate of ATP-P_i exchange upon ΔpH correlates with the extent of dark inactivation. The variable rate of dark inactivation probably reflects variable amounts of ADP carried through the thylakoid preparation.

Activation of photophosphorylation is usually assayed with thylakoids which have been stored in the dark with 2 mM ADP after the activating illumination and before assay; they should be completely inactivated and the shift reproducibly maximal. In contrast, by its very nature, ATP-P_i exchange is observed with thylakoids the dark decay of which is incomplete, so the shift is less than maximal. When photophosphorylation is assayed with thylakoids which have not completed thier dark decay, the shift of photposphorylation is reduced to that of exchange (figure 1).

ATPase activity was not measured in these experiments. Its decay is presumably parallel to that of ATP-P_i exchange. We speculate that hydrolytic activity is enhanced immediately upon reduction of CF_1, but that the stimulation of the synthetic reactions seen as the shift requires subsequent binding of ADP and inactivation. This is puzzling in view of the Haldane relationship and the evidence that bound ADP is released from CF_1 active in photophosphorylation.

References

1. Mills,J.D. and Mitchell,P. FEBS Lett (1982) <u>144</u> 63-67
2. Ketcham,S.J.,Davenport,J.W.,Warncke,K. and McCarty,R.E. JBC (1984) <u>259</u> 7286-7293.
3. Davenport,J.W. and McCarty,R.E. JBC (1981) <u>256</u> 8947-8954.

CHARACTERIZATION OF THE CATALYTIC SITE ON THE β SUBUNIT OF THE H^+-ATP SYNTHASE-ATPase COMPLEX AND DEMONSTRATION OF FUNCTIONAL HOMOLOGY OF β SUBUNITS ISOLATED FROM DIFFERENT SOURCES

Z. GROMET-ELHANAN, DEPARTMENT OF BIOCHEMISTRY, THE WEIZMANN INSTITUTE OF SCIENCE, REHOVOT 76100, ISRAEL

1. INTRODUCTION

The molecular mechanism of ATP synthesis and hydrolysis carried out by the $F_0 \cdot F_1$ enzyme complex is still unknown. A large number of studies using different approaches point to the F_1-β subunit as the one that contains the catalytic site and is involved, together with the α subunit, in substrate binding (1-4). A detailed characterization of substrate binding sites on F_1 and their relation to its catalytic site is, however, very difficult because of the complexity of the F_1 structure. A promising approach to the elucidation of this problem is the study of isolated, functionally active, α and β subunits.

Reconstitutively active β subunits were obtained from three different respiratory bacteria (5-7), from a photosynthetic bacterium (8), and recently also from spinach chloroplasts (9). Experiments with respiratory bacteria started with F_1 complexes which were dissociated into their individual subunits. Upon reassembly of the isolated, purified α, β, and γ subunits a soluble ATPase activity was obtained, indicating the functional activity of these subunits (5-7). A similar dissociation of a δ- and ϵ-less CF_1-ATPase enabled Richter et al. (9) to isolate and purify the CF_1-β subunit ($CF_1\beta$), but its functional activity could not be assayed directly due to the absence of isolated α and γ subunits. We have earlier developed a method for the selective extraction of the β subunit from the membrane-bound $F_0 \cdot F_1$ complex of the photosynthetic bacterium *Rhodospirillum rubrum*, leaving a β-less chromatophore preparation that lost all its ATP synthesis and hydrolysis activities (8). The extracted *R. rubrum* β subunit ($Rr\beta$) was purified to homogeneity (10) and optimal conditions for its reconstitution into the β-less *R. rubrum* chromatophores were determined (8, 10, 11). The purified $Rr\beta$ has no catalytic activity by itself (8), but its native functional state is retained, since its reconstitution into the β-less chromatophores leads to full restoration of their lost ATP-linked activities (8, 10, 11).

This *R. rubrum* system provides a unique opportunity to investigate both the isolated, reconstitutively active, $Rr\beta$ and the resulting membrane-bound β-less $RrF_0 \cdot F_1$ complex. On the purified $Rr\beta$ substrate binding sites have been identified (12-14) and their relationship to the catalytic site of the $F_0 \cdot F_1$ enzyme complex has been clarified (4). The β-less chromatophores have been found to form also active hybrid $F_0 \cdot F_1$ complexes with β subunits isolated from different sources, such as the *Escherichia coli* β subunit ($Ec\beta$) and the $CF_1\beta$ (15, 9). The main results of these studies will be presented in this lecture together with new information, which demonstrates that the catalytic activity of each hybrid $F_0 \cdot F_1$ complex is determined by the properties of its specific β subunit. These results have identified one of the substrate binding sites on $Rr\beta$ with the catalytic site of the $RrF_0 \cdot F_1$ and have clearly established the functional homology of β

Biggens, J. (ed.), Progress in Photosynthesis Research, Vol. III. ISBN 90 247 3452 5
© *1987 Martinus Nijhoff Publishers, Dordrecht. Printed in the Netherlands.*

subunits isolated from various prokaryotic and eukaryotic, photosynthetic and respiratory sources.

2. PROCEDURE

Isolation and purification of Rrβ, its reconstitution into β-less *R. rubrum* chromatophores and assays of their restored ATP synthesis and hydrolysis are described in (16). Chemical modifications of Rrβ are outlined in (4), and the substrate binding assays in (12-14). Isolation of CF$_1\beta$ and Ecβ and their reconstitution into β-less chromatophores are described in (9) and (15), respectively. In all cases reconstitution was carried out at pH 8.0, and the reconstituted chromatophores were centrifuged. In this set-up all the remaining unbound β subunit was removed from the chromatophores before they were assayed for activity.

3. RESULTS AND DISCUSSION

3.1. *Identification of the catalytic substrate binding site on Rrβ.*

Since Rrβ has no catalytic activity by itself we could measure directly its capacity to bind labeled ATP, ADP, and Pi. Such direct binding studies have revealed the presence of two binding sites for ATP and ADP and one for Pi on the isolated, native Rrβ (Table 1). One of the nucleotide binding sites is a Mg-independent high-affinity site with Kd values ranging between 5 and 9 μM (12, 13). The second is a Mg-dependent low-affinity site with Kd values of 200 μM for ATP (12) and 90 μM for ADP (13). The Pi binding site is a Mg-dependent low-affinity one with a Kd value of 270 μM (14).

TABLE 1.   Binding stoichiometry and reconstitutive activity of native and chemically modified Rrβ

Type of assay	Assay carried out in the absence (−) or presence (+) of MgCl$_2$ with			
	Native Rrβ		WRK-Rrβ	
	(−)	(+)	(−)	(+)
Binding of [a]:				
^{32}Pi	0	0.93	0	0.01
(^{3}H)ATP	0.97	1.90	0.85	0.86
(^{3}H)ADP	0.83	1.88	0.90	1.88
ATP synthesis[b]				
Restored to β-less chromatophores	−	798	−	17

[a]Binding stoichiometry in mol/mol β.
[b]ATP synthesis in μmol per h per mg bacteriochlorophyll.

As is illustrated in Table 1, full occupation of both nucleotide binding sites as well as of the Pi binding site on native Rrβ, can be demonstrated only in the presence of MgCl$_2$. This native Rrβ is functionally active since its reconstitution into β-less chromatophores restores completely their ATP synthesis activity. Table 1 also illustrates the effect of the carboxyl group reagent Woodward's reagent K(WRK) on the substrate binding properties as well as the reconstitutive activity of Rrβ. The WRK modified Rrβ has lost its capacity to bind Pi and ATP to their Mg-dependent site, but could still bind ATP to its Mg-independent site and ADP to both binding sites. These results confirm our earlier data (14), which have indicated that Pi binds at the site occupied by the γ-phosphate group of ATP in its Mg-dependent site, a location that remains empty during binding of ADP. The WRK-Rrβ lacking this binding location can still rebind to β-less chromatophores (4) but loses completely its ability to restore their photophosphorylation (Table 1). Similar results have also been obtained with Rrβ modified by the histidine reagent diethyl pyrocarbonate (4). These data led us to conclude that this binding location, which has essential histidine and carboxyl residues, is an integral part of the catalytic site of the RrF$_0$.F$_1$-ATP synthase.

3.2. *Demonstration of the functional homology of β subunits isolated from diffrent sources.*

The above described studies have confirmed earlier suggestions (1-3) that the β subunit is the catalytic subunit of the F$_0$.F$_1$ enzyme complex. Assuming a similar mechanism of action for all F$_0$.F$_1$ complexes one would expect their catalytic β subunits to exhibit a close structural and functional homology. A high degree of structural homology has been suggested from a comparison of amino-acid sequences of F$_1$-β subunits from various photosynthetic or respiratory eukaryotic and prokaryotic sources (17). The functional homology of various isolated β subunits has recently been investigated by employing the β-less *R. rubrum* chromatophores as a sensitive test system (9, 15). Reconstitution of these chromatophores with either CF$_1\beta$ (9) or Ecβ (15) has been found to result in restoration of their capacity to catalyze ATP-linked reactions, indicating the formation of active hybrid F$_0$.F$_1$ complexes between the β-less *R. rubrum* chromatophores and the isolated CF$_1\beta$ or Ecβ.

The restored activity of these hybrid F$_0$.F$_1$ complexes could not be due to activation of any reisdual Rrβ, that might have remained in the extracted β-less chromatophores. Because, as we have earlier shown by a number of different experimental approaches (8, 16, 18), all the Rrβ has been removed from these chromatophores. Furthermore, as is shown in Fig. 1, the hybrid and homologous systems exhibit different properties under identical conditions. Increasing amounts of the various β subunits during reconstitution cause a different degree of restoration of the Mg^{2+}-ATPase activity. With Rrβ and CF$_1\beta$ the restoration follows a simple saturation curve, but the level of the restored activity reached with CF$_1\beta$ is only about 50% of the level obtained with Rrβ. With Ecβ, on the other hand, the restored activity increases linearly with increasing concentrations, reaching even higher rates than those observed with saturating concentrations of Rrβ (Fig. 1).

The hybrid and homologous systems differ also in their sensitivity to various effectors (Table 2). Sulfite has been reported to stimulate by several fold the Mg-dependent ATPase activity of F$_1$ enzymes from various sources, including *R. rubrum* (19) and chloroplasts (20), but by less than 20% the *E. coli* F$_1$-ATPase (21). As can be seen in Table 2 the Mg^{2+}-ATPase activity of β-less chromatophores reconstituted with Rrβ and CF$_1\beta$ is stimulated by sulfite between 2 to 4 fold, whereas with Ecβ there is only 20% stimulation. A completely different pattern has been observed with tentoxin, which is a highly specific inhibitor of ATP-linked reactions in

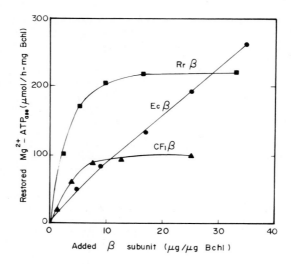

FIGURE 1.   Effect of the amount of various β subunits added during reconstitution of β-less *R. rubrum* chromatophores on the degree of restoration of their Mg^{2+}-ATPase activity.

thylakoids (22). Incubation of the reconstituted chromatophores with tentoxin results indeed in >95% inhibition of the sulfite stimulated ATPase activity of the $CF_1\beta$ hybrid but has no effect on the $Ec\beta$ hybrid or on the homologous $Rr\beta$ system (Table 2). Such a marked difference in the sensitivity to tentoxin of the hybrid $CF_1\beta$ and the homologous $Rr\beta$ systems has also been observed in the absence of sulfite, when ATP synthesis as well as hydrolysis activites could be tested (9). These results indicate that the restored activity of the homologous or hybrid $F_0 \cdot F_1$ complexes reflects the properties of their specific β subunits.

TABLE 2.   The effect of sulfite and tentoxin on the Mg^{2+}-ATPase activity of β-less *R. rubrum* chromatophores reconstituted with various β subunits[a]

Source of β-subunit	ATPase activity assayed in presence of		
	No addition	Sulfite	Sulfite and tentoxin
R. rubrum	404	703	662
E. coli	212	254	250
Chloroplasts	129	441	28

[a]Reconstitutions were performed with equal amounts of the respective β subunits at a ratio of 10 μg/μg bacteriochlorophyll.

The homologous Rrβ has always restored ATP synthesis and hydrolysis activities to the same extent (8, 10, 11), whereas the heterologous CF$_1\beta$ and Ecβ restored ATP hydrolysis much more efficiently than ATP synthesis (9, 15). Thus, when the reconstitution was carried out at pH 8.0 using a ratio of 10μg β/1μg of bacteriochlorophyll, Rrβ restored about 90% of both activities (10, 11). But CF$_1\beta$ restored 50% of the ATPase activity and only 10% of the ATP synthesis activity (9), and Ecβ restored 50-90% of the first (23, and Table 2) and only 6-10% of the second activity (15). The low activity in ATP synthesis could not be explained by poor binding of the heterologous β subunits to the β-less R. rubrum chromatophores, since CF$_1\beta$ was shown to compete rather effectively with the binding of Rrβ (9). Also, the high activity in ATP hydrolysis as compared to synthesis could not be explained by a general uncoupling effect of CF$_1\beta$ or Ecβ. Because (a) ATP synthesis in the hybrid F$_0$.F$_1$ enzyme complex was as sensitive to gramicidin as in the native RrF$_0$.F$_1$ (23) and (b) an identical light-induced quenching of quinacrine fluorescence was obtained with the hybrid and homologous chromatophores (15).

The low ATP synthesis activity of the hybrid F$_0$.F$_1$ enzyme is consistent with a loose coupling of the hybrid F$_1$ complex to proton translocation through the RrF$_0$ channel during catalysis. This loose coupling could be due to a less efficient interaction of Ecβ or CF$_1\beta$ with the other subunits of the F$_0$.F$_1$ enzyme complex. A schematic presentation of this loose coupling is given in Fig. 2.

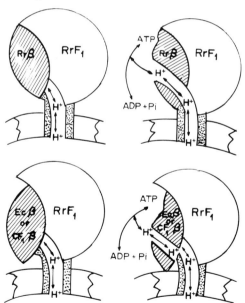

FIGURE 2.   A schematic presentation of the interaction of the homologous and heterologous β subunits with the β-less R. rubrum chromatophores.

The system presented here enabled us to identify the catalytic substrate binding site on the isolated β subunit and to demonstrate the functional homology of β subunits isolated from different prokaryotic and eukaryotic, respiratory and photosynthetic sources. It also provides an ideal system for future investigations on the interaction of the β subunit with other $F_0 \cdot F_1$ subunits during its reassembly into an active $F_0 \cdot F_1$ complex.

ACKNOWLEDGMENTS
    This research is supported by a grant from the United States-Israel Binational Science Foundation (BSF), Jerusalem, Israel, and by the Minerva Foundation, Munich, Germany. The material presented in this lecture is the result of a productive collaboration with S. Philosoph, D. Khananshvili and S. Weiss from the Weizmann Institute, H. Kanazawa and M. Futai from Okayama University, and M.L. Richter and R.E. McCarty from Cornell University.

REFERENCES
1. Futai, M. and Kanazawa, H. (1980) Curr. Top. Bioenerg. 10, 181-215
2. Nelson, N. (1981) Curr. Top. Bioenerg. 11, 1-33
3. Futai, M. and Kanazawa, H. (1983) Microbiol. Rev. 47, 285-312
4. Khananshvili, D. and Gromet-Elhanan, Z. (1985) Proc. Natl. Acad. Sci. USA 82, 1886-1890
5. Yoshida, M., Sone, N., Hirata, H. and Kagawa, Y. (1977) J. Biol. Chem. 252, 3480-3485
6. Dunn, S.D. and Futai, M. (1980) J. Biol. Chem. 255, 113-118
7. Hsu, S.Y., Senda, M., Kanazawa, H., Tsuchiya, T. and Futai, M. (1984) Biochemistry 23, 988-993
8. Philosoph, S., Binder, A. and Gromet-Elhanan Z. (1977) J. Biol. Chem. 252, 8747-8752
9. Richter, M.L., Gromet-Elhanan, Z. and McCarty, R.E. (1986) J. Biol. Chem. in press
10. Khananshvili, D. and Gromet-Elhanan, Z. (1982) J. Biol. Chem. 257, 11377-11383
11. Gromet-Elhanan, Z., Khananshvili, D. and Philosoph, S. (1981) in Energy Coupling in Photosynthesis (Selman, B.R. and Selman-Reimer, S. eds.) pp. 323-331, Elsevier/North-Holland, New York
12. Gromet-Elhanan, Z. and Khananshvili, D. (1984) Biochemistry 23, 1022-1028
13. Khananshvili, D. and Gromet-Elhanan, Z. (1984) FEBS Lett. 178, 10-14
14. Khananshvili, D. and Gromet-Elhanan, Z. (1985) Biochemistry 24, 2482-2487
15. Gromet-Elhanan, Z., Khananshvili, D., Weiss, S., Kanazawa, H. and Futai, M. (1985) J. Biol. Chem. 26, 12635-12640
16. Gromet-Elhanan, Z. and Khanashvili, D. (1986) Methods in Enzymol. 126, 528-538
17. Falk, G., Hampe, A and Walker, J.E. (1985) Biochem. J. 228, 391-407
18. Philosoph, S. and Gromet-Elhanan, Z. (1981) Eur. J. Biochem. 119, 107-113
19. Webster, G.D., Edwards, P.A. and Jackson, J.B. (1977) FEBS Lett. 76, 29-35
20. Anthon, G.E. and Jagendorf, A.T. (1983) Biochim. Biophys. Acta 723, 358-365
21. Takeda, K., Hirano, M., Kanazawa, H., Nukiwa, N., Kagawa, Y. and Futai, M. (1982) J. Biochem. 91, 695-701
22. Selman, B.R. and Durbin, R.D. (1978) Biochim. Biophys. Acta 502, 29-37

23. Gromet-Elhanan, Z. (1986) in Perspectives of Biological Energy Transduction (Mukohata, S. et al., eds.) Academic Press, in press

PROTON SPIN-ECHO SPECTRA OF THE ISOLATED BETA SUBUNIT OF THE CF_0-CF_1 ATP SYNTHASE

INGRID J. APEL[*][#], ROBERT R. SHARP[*] AND WAYNE D. FRASCH[#]
DEPARTMENT OF CHEMISTRY[*] AND DEPARTMENT OF BIOLOGY[#]
THE UNIVERSITY OF MICHIGAN, ANN ARBOR MI 48109

1. INTRODUCTION

The application of spin-echo methods (1) in NMR provides a means of selecting resonances on the basis of spin-spin relaxation , which in turn is related to the mobility of residues or molecules involved. Spin-echo spectroscopy has been applied previously in this laboratory to the selective observation of mobile amino acid sidechains on CF_1, the proton translocating ATPase of the chloroplast thylakoid membrane (2,3). These experiments have demonstrated the existence of a small subset of amino acid sidechains, approximately 25 in number, which extend into the aqueous phase and exhibit mobility much higher that that of the protein as a whole. Two of the sidechains, one aspartate and one glutamate, are immobilized in a highly selective manner by nucleotide binding to the active site. The immobilized groups lie at a distance of approximately 15Å from the metal cofactor bound at the active site.

Proton spin-echo NMR is used here to examine highly mobile amino acids associated with the isolated, purified β subunit of CF_1 (Apel et al., these Proceedings). Previous covalent modification and photoaffinity labeling experiments have indicated that the site of ATPase activity resides on the β subunit. The present experiments have characterized the 10 ms spin-echo spectrum of the purified β subunit in three forms: (1) in the latent form; (2) following depletion of nucleotides; (3) following addition of MgADP to the depleted β subunit.

2. MATERIALS AND METHODS

The β subunit of the CF_0-CF_1 ATP synthase was isolated and purified as described (Apel et al., these Proceedings). Just prior to use, the latent protein was passed through a BioGel P30 column (0.7 x 30 cm) equilibrated in 40 mM Borate·D_2SO_4, pH 8.0. Nucleotide-depleted β was prepared by column chromatography of the purified protein using two successive BioGel P30 columns (0.7 x 30 cm) equilibrated in 20 mM $MgCl_2$ and 40 mM Tricine, pH 8.0, then followed by the Borate·D_2O column. Further washing and deuteration was performed as described (2) using an Amicon PM30 membrane. For experiments observing the response of readdition of nucleotides to the SE spectrum of the β, MgADP was added directly to the stripped β in the NMR tube.

High resolution ^{1}H-NMR spectra were obtained at 45 ± 1°C

Biggens, J. (ed.), Progress in Photosynthesis Research, Vol. III. ISBN 90 247 3452 5
© *1987 Martinus Nijhoff Publishers. Dordrecht. Printed in the Netherlands.*

using a Bruker WM 360 spectrometer. The residual DHO peak
from the solvent was suppressed by decoupling at all times
except during accumulation of the FIDs. The spin-echo
sequence 90-τ-180-τ-accumulate) was used; 5000 transients per
spectrum of 16384 data points were collected with a bandwidth
of 5000 Hz. The total accumulation time was 3 hours per pulse
spacing.

3. RESULTS AND DISCUSSION

The upfield region of the 360 MHz ^1H spin-echo NMR
spectrum of the latent β subunit of the CF_0-CF_1 ATP synthase
is shown in Fig. 1A. Tentative assignments of the resolved
peaks (Table I) have been made on the basis of chemical shift
information and multiplet structure. For this purpose,
chemical shifts and spin-spin coupling constants tabulated for
amino acid side chains in short random coil aqueous peptides
by (4) have been used.

Figure 1. Upfield region of the 10ms proton spin-echo
spectrum of the purified β subunit before depletion of
nucleotides (A), after depletion of nucleotides (B) and after
addition of 0.4mM MgADP to the nucleotide-depleted protein
(C). Vertical gain of (C) is half that of (A) and (B).

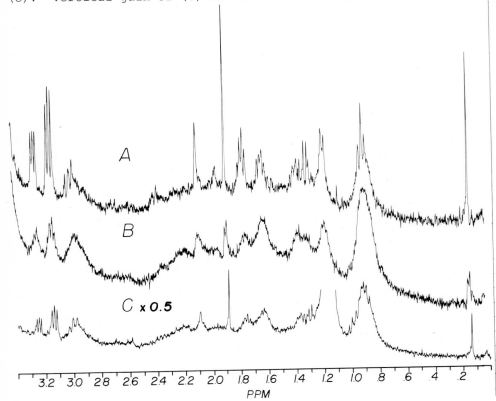

Table I: Assignment of resolved peaks in the 10 ms spin-echo
spectrum of the β subunit.

Resonances	δ, ppm	^3J, Hz
ARG δ-CH$_2$	3.25	5.3
(ARG or LYS CH$_2$)	3.12	6.2
LYS ε-CH$_2$	3.01	8.6
MET CH$_3$	1.88	---
LYS δ-CH$_2$	1.78	---
ARG γ-CH$_2$	1.65	---
THR ε-CH$_3$	1.35	7.6
ILE ε-CH$_3$	0.88	8.2

The three resolved triplets at 3.25 ppm (J=5.3 Hz), 3.12 ppm (J=6.2 Hz) and 3.01 ppm (J=8.6 Hz) arise from three highly mobile, charged sidechains of either lysine ε-CH$_2$ protons or arginine δ-CH$_2$ protons (these protons occur at 3.02 ppm and 3.30 ppm, respectively, in aqueous peptides). The long T_2's and near first order structure of these resonances show that the methylene protons are in virtually unhindered segmental motion in the aqueous phase and are nearly chemically equivalent. The chemical shift values and coupling constants indicate that one arginine (3.25 ppm) and one lysine (3.01 ppm) are involved; the assignment of the triplet at 3.12 is less certain. Resolved structure from other CH$_2$ protons on these sidechains is also present in the region 1.40-2.20 ppm; in particular, multiplets near 1.65 ppm and 1.78 ppm are assigned to lysine δ-CH$_2$ and arginine γ-CH$_2$ resonances, respectively.

Three highly resolved methyl peaks are evident in the region from 1.9 to 0.1 ppm. The narrow singlet at 1.88 ppm arises from a methionine methyl (2.13 ppm in random coil aqueous peptides), and the doublet at 1.35 ppm is tentatively assigned to threonine ε-CH$_3$. Assignments of other resonances from sidechains that give less distinctive multiplets require systematic decoupling experiments for definitive assignment.

The spin-echo spectrum changes dramatically when the β is depleted of bound nucleotides (Fig. 1B). A decrease in the mobilities of all of the sidechains that contribute resolved structure to Fig. 1A is evident both in the shortened T_2's and in the loss of the simple first order coupling patterns observed for the untreated enzyme. All resolved resonances are affected, suggesting that these amino acids may lie on a common segment of the peptide chain that exhibits high mobility only in the presence of bound nucleotides.

The addition of one equivalent of MgADP to the nucleotide-depleted enzyme reverses the loss of mobility that was induced by removal of nucleotide (Fig. 1C). The most pronounced effects occur in the methylene triplets of lysine and arginine (3.0-3.3 ppm), which regain their first order structure and exhibit lengthened T_2's (note the decrease in vertical gain of Fig. 1C), and in the lengthened T_2 of the methionine ϵ-CH_3 singlet at 1.88 ppm.

The spin-echo experiments reported here have shown that the isolated β subunit of CF_1, like intact CF_1 itself, contains a small number of highly mobile amino acid sidechains. This mobility does not seem to be a general property of surface-exposed, charged sidechains. Of the 21 lysyl and 29 arginyl residues in the β subunit (5), only 3 contribute resolved structure in the 10 ms spin-echo spectrum. Furthermore, the highly mobile amino acids include certain nonpolar sidechains (one Met-CH_3 out of 16 and one Ile-CH_3 out of 32 total in β). The small number of sidechains that contribute to the spin-echo spectrum as well as the participation of both nonpolar and charged sidechains in the the NMR-visible subset suggests that the spin-echo experiment is monitoring a short unstructured segment of the peptide chain. The mobility of this segment is altered in a reversible manner by the depletion and readdition of tightly bound ADP. The reversibility of these changes in mobility upon readdition of bound nucleotides suggests that the changes are not the result of irreversible and/or random denaturation of the protein.

REFERENCES
1 Campbell, I.D., Dobson, C.M., Williams, R.J.P. and Wright, P.E. (1975) FEBS Lett. 57, 96-99
2 Sharp, R.R. and Frasch, W.D. (1985) Biochem. 24, 5449-5454
3 Frasch, W.D. and Sharp, R.R. (1985) Biochem. 24, 5454-5458
4 Bundi,A.and Wuthrich, K. (1979) Biopolymers 18, 85-297
5 Zurawski, G., Bottomley, W. and Whitfield, P.R.(1982) Proc. Natl. Acad. Sci. USA 79, 6260-6264

ACKNOWLEDGEMENTS
This study was supported by the USDA (83-CRCR-1-1339). We also acknowledge the technical assistance of Matthew A. Sanders.

PURIFICATION OF THE β SUBUNIT OF THE CHLOROPLAST
H$^+$-TRANSLOCATING ATPase OF SPINACH THYLAKOIDS

INGRID J. APEL, ALFONSO MEJIA AND WAYNE D. FRASCH
DEPARTMENT OF BIOLOGY, THE UNIVERSITY OF MICHIGAN, ANN ARBOR MI

1. INTRODUCTION
    A significant portion of the catalytic site of CF_0-CF_1 is
located on the β subunit (for a review, see (1)). This enzyme
contains three copies of the β subunit which suggests that there
are three active sites. Experiments that involve isotope
exchange (2), substrate trapping (3) and binding of Mn (4)
suggest that at least two of the active sites are coupled.
Soluble CF_1 is a latent ATPase that can be activated by a
variety of procedures (5), some of which are believed to require
the exchange of disulfide bonds of the γ subunit (6).
    Recently, it has been possible to dissociate the β subunit
from membranes of *E. coli*, *R. rubrum* and the thermophilic
bacterium PS3 and reconstitute ATPase activity of these depleted
membranes upon readdition of the β subunit (7-10). The purified
β from *R. rubrum* has two bin- ding sites (11,12) for ADP with
dissociation constants of 6.7 and 80 μM. The low affinity site
requires Mg(II) to bind the nucleotide. The β subunit from *R.
rubrum* also exhibits ATPase activity at a level of about 0.1% of
the intact enzyme (13).
    We report the purification of the β subunit of the CF_0-CF_1
ATPsynthase from spinach thylakoids with 1 M LiCl in a form that
can bind up to two moles of MgADP and retains ATPase activity.

2. MATERIALS AND METHODS
    Thylakoids were isolated from spinach leaves using a grind
buffer that contained 0.4 M sucrose, 10 mM NaCl, 10 mM $MgCl_2$ and
50 mM Tricine, pH 7.3. This suspension was then centrifuged 5
min at 3500 x g; the pellet washed once with the same buffer,
centrifuged 30 s at 100 x g, and the supernatant centrifuged 5
min at 3500 x g. The pellet was resuspended in incubation
buffer (.25 M sucrose, 4 mM $MgCl_2$, 4 mM ATP, 50 mM Tricine, pH
8.0) and stirred at 4°C for 1-2 hrs. After centrifugation for 5
min at 3500 x g, the thylakoids were diluted to 0.3 mg Chl/ml in
incubation buffer with 1 M LiCl added, which eluted the β from
the membrane. This suspension was centrifuged for several hours
at 32000 x g. The β subunit in the supernatant was concentrated
by pressure dialysis, diluted four-fold in elution buffer (4mM
ATP, 4mM $MgCl_2$, 50 mM Tricine, pH 8.0) and concentrated again.
Final purification was obtained by chromatography with a 1.4 x
78 cm BioGel P150 column. To prevent proteolysis, PMSF and NaN_3
were included in all buffers used.
    ATPase activity was determined using [γ^{32}P]-ATP and measur-
ing the formation of $^{32}P_i$. Calculation of the concentrations of
substrate, inhibitor and cofactor required to yield final concentrations of MgATP and MgADP was done by successive approxima-

Biggens, J. (ed.), Progress in Photosynthesis Research, Vol. III. ISBN 90 247 3452 5

tion using the stability constants as in (14). Assays were run for 5 min at 37°C and were begun with the addition of 22 mg of purified β to the reaction mixture which had been pre-equilibrated for 3 min at the same temperature. The reaction was quenched with $HClO_4$ and the $^{32}P_i$ extracted as in (15). Kinetic constants were determined using the Fortran programs as in (16).

The β subunit was depleted of bound nucleotides by chromatography on three successive 0.7 x 30 cm Sephadex G50 columns. Equilibrium binding studies of ^3H-ADP were performed as in (12) by incubating 5 μM β subunit with ^3H-ADP at a specific activity of 8.4 x 10^6 cpm/μmol ADP for 90 min in the presence of 20 mM $MgCl_2$ and 50 mM Tricine, pH 8.0. The unbound nucleotides were removed by centrifugation chromatography with Sephadex G50 equilibrated in 20 mM $MgCl_2$ and 50 mM Tricine, pH 8.0 as in (12). Analysis of the protein purity was determined by SDS-PAGE on 12% gels as in (17) and were silver stained as in (18).

3. RESULTS AND DISCUSSION

Extrinsic membrane proteins which were solubilized from thylakoids with 1 M LiCl were purified by chromatography on BioGel P150. The proteins that eluted from the column were analyzed by SDS-PAGE as shown in Fig. 1. Purified CF_1 was run as a standard in Lane 5. The first protein-containing fraction from the column (Lane 1) contained the β subunit in the highest purity. In subsequent fractions, the β was present in greatest abundance although increasing amounts of smaller molecular mass proteins become apparent, one of which has a molecular mass similar to the γ subunit of CF_1. Washing thylakoids in 1 M LiCl was also found to remove significant amounts of the ε subunit as well. The ε subunit was found to copurify with the β unless the crude extract was diluted with elution buffer. This treatment decreases the concentration of sucrose which apparently effects the binding of the ε to the β subunit.

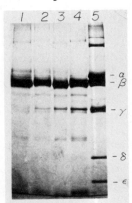

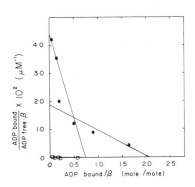

FIGURE 1. Electrophoretic profile of fractions eluted from the P150 column. Lane 1: the first fraction to contain protein; Lane 2-4: successive fractions; Lane 5: CF_1.

Figure 2. Scatchard plot of ADP binding to the purified β in the presence (circles) and absence (squares) of Mg(II).

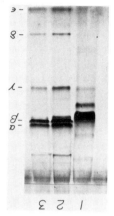

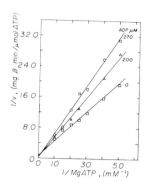

Figure 3. Comparison of the relative abundance of subunits of CF_1 in the purified β preparation. Lane 1: 15.0 nmol purified β; Lane 2 and 3: 15 nmol and 0.15 nmol β in soluble CF_1, respectively.

Figure 4. Inhibition kinetics of purified β by MgADP.

TABLE I. Kinetic Constants of the MgATPase of the Purified β by MgADP.

Treatment	V^*	V/K	$K (mM^{-1})$
Latent β	1.858	269	6.90
20% EtOH	0.068	83	0.82
50mM DTT	0.653	248	2.63
60°C, 5mM DTT	0.942	159	5.90
Octyl glucoside	3.365	391	8.60

*μmol ATP hydrolyzed/mg β.min

A Scatchard plot of the binding of ADP to the β subunit is shown in Fig. 2. Two MgADP binding sites with different affinities are present on the protein with dissociation constants of 15.3 μM and 202 μM (circles). These values are comparable to the dissociation constants of ADP observed for the purified β from R. rubrum (8,9). However, unlike the β subunit from R. rubrum, Mg(II) was required for the binding of nucleotides to both the high and low affinity sites on the β from spinach.

The purified β subunit exhibited measurable rates of ATPase activity. The kinetic constants of the MgATPase activity are shown in Table I. Calcium was not an effective cofactor for the ATPase activity even when the β was heat-activated. The MgATPase did not require activation. Only activation by octyl glucoside was found to cause a 2-fold increase in both V and V/K. The V was 25% of CF_1 when expressed on a per mg protein basis and about 10% of CF_1 when expressed on a per mole of β subunit (assuming 3 β per mole CF_1).

The possibility that the ATPase activity observed in the β subunit preparation may have resulted from a 10% contamination with soluble CF_1 was dismissed as a result of the experiment of Fig. 3. Lane 1 of the SDS-PAGE gel of Fig. 3 contains 15.0 µmol of purified β subunit. The amount of soluble CF_1 in Lanes 2 and 3 contained 15 nmol and 0.15 nmol of β subunit, respectively. If the ATPase activity catalyzed by the preparation of purified β subunit resulted from a residual contamination of soluble CF_1, the amount of the α, γ, δ and ε subunits observed in Lane 3 should be apparent in Lane 1. The absence of these subunits in Lane 1 indicates that the observed activity does not result from contamination with intact CF_1.

High concentrations of ADP inhibit the MgATPase activity of the β subunit. The kinetics of this inhibition (Fig. 4) show that MgADP appears to be a simple competitive inhibitor with a K_{is} of 690 µM. Concentrations of MgADP in excess of 150µM cause soluble CF_1 to exhibit a sigmoidal dependence on substrate, which suggests a cooperative interaction between active sites. With the purified β subunit, the substrate dependence appears linear even in the presence of 270 µM MgADP. This suggests that either the β subunits are completely dissociated or that these concentrations of MgADP can no longer induce cooperation between the active sites.

REFERENCES
1 Amzel,S. and Pedersen,P.(1985) Ann.Rev.Biochem. 52,801-824
2 Kohlbrenner,W. and Boyer,P.(1983) J.Biol.Chem. 258,10881-10886
3 Grubmeyer,C. and Penefsky,H.(1981) J.Biol.Chem. 256,3728-3734
4 Hiller,R. and Carmeli,C.(1985) J.Biol.Chem. 260,1614-1617
5 Nelson,N.(1976) Biochem.Biophys.Acta 45,314-338
6 Nalin,C. and McCarty,R.(1984) J.Biol.Chem. 259,7275-7280
7 Futai,M.(1977) Biochem.Biophys.Res.Comm. 79,1231-1237
8 Philosoph,S.,Binder,A. and Gromet-Elhanan,Z.(1977)
   J.Biol.Chem. 252,8747-8752
9 Khananshvilli,D. and Gromet-Elhanan,Z.(1982) J.Biol.Chem.
   257,11377-11383
10 Yoshida,M.,Sone,N.,Hirata,H. and Kagawa,Y.(1977) J.Biol.Chem.
   252,3480-3485.
11 Khananshvilli,D. and Gromet-Elhanan,Z.(1984) FEBS Lett.
   178,10-14
12 Gromet-Elhanan,Z. and Khananshvilli,D.(1984) Biochemistry
   23,1022-1028
13 Harris,D.,Boork,J. and Baltscheffsky,M.(1985) Biochemistry
   24,3876-3883
14 Pecoraro,V.,Hermes,J. and Cleland,W.(1984) Biochemistry 23,
   5262-5271
15 Frasch,W. and Selman,B.(1982) Biochemistry 21,3636-3642
16 Cleland, W.(1967) Advances Enzymol. 29,1-36
17 Chua, N-H.(1980) Methods in Enzymol. 69,434-446
18 Sammons,W.,Adams,L. and Nishizawa, E.(1981) Electrophoresis
   2, 135-141

ACKNOWLEDGEMENTS
    This study was supported by the USDA (83-CRCR-1-1334). We also acknowledge the technical assistance of Matthew A. Sanders.

ESSENTIAL AMINO ACIDS AND REGULATION OF THE H⁺-ATPases

VALLEJOS, R.H., ARANA, J.L., RAVIZZINI, R.A. and VIALE, A.M.

CENTRO DE ESTUDIOS FOTOSINTETICOS Y BIOQUIMICOS,CEFOBI(CONICET, F.M. LILLO, U.N.ROSARIO) SUIPACHA 531, 2000 ROSARIO, ARGENTINA.

1. INTRODUCTION

F1 isolated from very different sources like chloroplasts, mitochondria, photosynthetic bacteria, E.coli and termophilic bacteria have similar structural and catalytic properties (1,2). It is generally accepted that the catalytic site resides in the β subunit of the enzyme which shows a high degree of homology in the different F1's (2).

Chemical modification of F1's with group-specific reagents has shown that the activities catalyzed by the F1-ATPases are inhibited by modification of arginine, lysine, tyrosine and carboxyl groups (see (3) and references therein). The chloroplast enzyme (Fo-CF1) has been found to possess a latent ATPase activity that can be unmasked by incubation of thylakoids with DTE in the light (4). Upon this treatment ATP is readily hydrolyzed by Fo-CF1, although the activated state decays in the subsequent dark period (5). The activation of the ATPase complex occurs when an electrochemical potential is developed across the thylakoids in the presence of a reducing dithiol compound. In intact chloroplasts (6,7) and leaves (8-10) CF1 may be activated to a similar state by light in the absence of added reductant. Probably, thioredoxin provides reducing equivalents for the light triggering of the ATPase in physiological conditions (11). The γ subunit contains four cysteine residues (12) involved in ATPase activation (13). These results were confirmed recently (14,15). A similar activation mechanism involving a redox change of a disulfide bond was observed for the membrane bound enzyme in whole leaves (10). Deactivation of the membrane bound ATPase in the dark may be due to the dissipation of proton gradient, thiol oxidation, or both (10,16). Using different thiol oxidants the activation process in broken chloroplasts can be reversed (17,18).

The purpose of this work is to summarize the information obtained with chemical modification studies comparing amino acid sequences in the β subunit of different F1's around the residues chemically modified, and to discuss the contribution of the proton gradient and thiol redox changes on ATPase deactivation in physiological conditions.

2. RESULTS AND DISCUSSION
2.1. <u>Functional amino acids in the F1-ATPases as detected by chemical modification studies</u>

The following residues have been shown to be modified by group-specific reagents concomitantly with an inhibition of the activities catalyzed by the F1-ATPases :

<u>Arginine residues:</u> essential arginines have been described in CF1, MF1, TF1 and RF1 (3,19-21). In the chloroplast enzyme, we have identified Arg 312 to be modified by phenylglyoxal during ATPase inactivation (21). As shown in Fig. 1, this arginine residue is present in all the β subunits from different sources, in a highly conserved region. Moreover, this region has been proposed to participate in adenine nucleotide binding in the β subunit

Biggens, J. (ed.), Progress in Photosynthesis Research, Vol. III. ISBN 90 247 3452 5
© *1987 Martinus Nijhoff Publishers, Dordrecht. Printed in the Netherlands.*

of E.coli (2). The Arg residue is very close to residues that have been shown in MF1 to be modified by 8-azido ATP (22) and to the Tyr residue modified by 7-chloro-4-nitrobenzofurazan (Nbf) (see below) (23,24). This evidence, and the postulated role for arginine residues in proteins in the recognition, binding and orientation of anionic ligands (25), led us to postulate that Arg 312 participates in the binding of adenine nucleotides (21). It is likely that Arg 312 is not the only arginine residue participating in the binding of adenine nucleotides or Pi to the catalytic or regulatory sites as suggested by chemical modification studies (3,20).

Carboxyl groups: essential carboxyl groups were reported in CF1, MF1, RF1, TF1 and EF1 (3,26). In CF1 (3) and RF1 (26), two types of essential carboxyl groups, located in the β subunit, were described by using N,N'-dicyclohexylcarbodiimide (DCCD) and Woodward's reagent K (WRK) as chemical modifiers. The Glu residues modified by DCCD in the β subunit from MF1, EF1 and TF1 have been identified (27-29). Two conserved regions contain the essential Glu (see Fig. 1). One of them is found in TF1 and the other (10 residues removed from the first) in MF1 and EF1. Neither the residues modified by WRK nor the ones modified by DCCD in other F1's have been identified yet. It is interesting to note that the latter is protected against the reagent by divalent cations (3,27-29) and it has been postulated to function in the metal binding or in the binding of the metal moieties of the adenine nucleotide-metal complexes (3). On the other hand, only adenine nucleotides protected against modification by WRK (3,26). Thus, it would be very interesting to identify the carboxyl groups modified by this reagent.

Tyrosine residues: the presence of essential Tyr residues has been shown by two kind of reagents: a) Nbf: an adenine analogue, inactivates CF1, MF1, EF1, TF1 and RF1 (3,30). In MF1 and TF1 the modified Tyr residues in the β subunit have been identified (23,24,31). This residue is present also in all the β subunits in a highly conserved region (see Fig. 1) and this region has been postulated to interact with nucleotides (2). This Tyr is close to the Arg modified by phenylglyoxal in CF1 (21) and to the amino acids modified by 8-azido-ATP after photolysis in MF1 (22). The role of this Tyr residue is uncertain. It has been postulated to participate in the propagation of conformational signals between adjacent subunits rather than in catalysis (23), responding to the binding of adenine nucleotides. It is interesting to note that the Tyr is protected against Nbf modification by Pi and not by adenine nucleotides (30,32) suggesting that the Tyr residue is located in a regulatory site for Pi that is different from the catalytic one. This site has been postulated to play a role in the exit or entrance of Pi during catalysis (30). b) FSBA: this adenine nucleotide analogue has been shown to inactivate MF1 and CF1 (3). The Tyr and His residues modified in MF1 β subunit have been identified (23,34). The Tyr residue is present in all the β subunits, as shown in Fig. 1, also a conserved region. This residue has been postulated to participate in the binding of adenine nucleotides to catalytic or regulatory sites (34).

Lys residues: inactivation of F1's by lysine reagents like PLP has been described in CF1 and MF1 (3), although no residues have been identified yet. Lys residues have been shown to be modified in the β subunit of MF1 and TF1 when Nbf group bound to Tyr migrates to a Lys (35,36). The so-labelled Lys residue lies in a conserved region (see Fig. 1). This region is known in several proteins to bind adenine nucleotides (2,35,36). Therefore, it is likely that this Lys participates in the binding of adenine nucleotides to the active site. Moreover, this Lys residue is close to the Glu residues modified by DCCD (see Fig. 1). Recently, Yoshida and Allison (37) provided strong evidence for the existence of the active site in the β subunit. It has been reported that modifications of MF1 and TF1 with some chemical modifiers abolishes ATP hydrolysis without affecting phosphorylation of enzyme-bound ADP (37,38). These results are in line with the fact that all of the identified essential residues are located in this subunit in F1. However, some questions remain to be answer about the role of these residues : Which of them participate in the binding of substrates and which in catalysis?, How do they do that? In which conformational state of the β subunit do they participate? Do they have the same function at the same time in all the β subunits? How does the conformational state of the β subunit affect their participation? Answering these questions will contribute to elucidate the molecular mechanism of catalysis by proton ATPases.

2.2. Physiological deactivation of the chloroplast ATPase

Chloroplasts rapidly prepared from preilluminated spinach leaves (light chloroplasts) show a Mg-ATPase activity at 0ºC that was stable for 40 min at least and was 20 times higher than the activity of chloroplasts obtained from leaves kept in the dark (dark chloroplasts) (10). This type of preparation was employed for studying the participation of proton gradient and thiol modulation in the mechanism of deactivation in vivo and in vitro of the chloroplast ATPase.

Incubation of light chloroplasts at 25ºC during 20 min diminished their ATPase activity to 7 % of the control (Table 1). Surprisingly, the activity of soluble CF1 isolated from the temperature inactivated light chloroplasts, was as high as the control.

Alquilation with iodoacetamide of the accessible sulfhydryl groups of CF1 activated by DTE did not affect the activity of the enzyme and prevented inhibition by oxidants (13). Similarly, inactivation by ferricyanide of light chloroplasts and of CF1 obtained thereof was prevented by the iodoacetamide pretreatment. On the other hand, iodoacetamide did not prevent the temperature dependent decay of Mg-ATPase (Table 1). These results clearly indicate that the temperature dependent deactivation process of light chloroplasts does not require dithiol oxidation.

Gramicidin induced a rapid inactivation of the Mg-ATPase activity of light chloroplasts incubated at 0ºC (Fig. 2). However, CF1 isolated from chloroplasts treated with gramicidin was active. This result suggests that the in vitro deactivation is mainly due to the dissipation of the proton gradient.

Inactivation at 20ºC of the proton ATPase of light chloroplasts was strongly dependent on the presence of cations in the incubation medium (Table 2). Mg was more effective than Ca and Mn. The rate of decay increased with the concentration of the divalent cation. Monovalent cations were effective only at higher concentrations.

In shocked preilluminated intact chloroplasts the presence of MgCl2, accelerated the deactivation process (48). It was suggested that Mg affect the ATPase conformation by binding to a regulatory site of the enzyme. On the other hand, cations may help to the dissipation of the proton gradient across the thylakoid membrane.

The striking stability of the ATPase activity at 0ºC (Fig.3, ref.10) and the slow inactivation at 20ºC (10) seems to involve a stable or slowly dissipating electrochemical proton gradient that may be related to the high buffering special domains described by others (49-51). The dissipation of the proton gradient at 25ºC related to the active conformation of the ATPase has a time constant of several minutes, similar to that affecting photosystem II (49). Gramicidin accelerates both the equilibration of the special domains

```
                        175                  200                 309                                                    381
CF1 (spinach)          GVG K TVL...GGVG E RTREGNDLYM E MKE...LQE R ITSTKEGSITSIQAV Y VPA...GEEH Y ETA
CF1 (tobacco)          GVG K TVL...GGVG E RTREGNDLYM E MKE...LQE R ITSTKEGSITSIQAV Y VPA...GEEH Y ETA
CF1 (maize)            GVG K TVL...GGVG E RTREGNDLYM E MKE...LQE R ITSTKKGSITSIQAV Y VPA...GNEH Y ETA
CF1 (barley)           GVG K TVL...GGVG E RTREGNDLYM E MKE...LQE R IASTKKGSITSIQAV Y VPA...GNEH Y ETA
CF1 (wheat)            GVG K TVL...GGVG E RTREGNDLYM E MKE...LQE R IASTKKGSITSIQAV Y VPA...GNEH Y ETA
                            *                                                                         *
MF1 (beef heart)       GVG K TVF...AGVG E RTREGNDLYH E MIE...MQE R ITTTKKGSITSVQAI Y VPA...GSEH Y DVA
                                         *             *         *      *      *                  *
MF1 (Nicotiana pl.)    GVG K TVL...AGVG E RTREGNDLYR E MIE...LQE R ITTTKKGSITSVQAI Y VPA...GEDH Y NTA
MF1 (yeast)            G K TVF...AGVG E RTREGNDLYH E MED...LQE R ITTTKKGSVTSVQAV Y VPA...GQEH Y DVA
EF1 (E. coli)          GVG K TVN...AGVG E RTREGNKFYH E MTD...LQE R ITSTKTGSITSVQAV Y VPA...GQEH Y DTA
                            *                              *                            *
TF1 (PS3)              GVG K TVL...AGVG E RTREGNDLYH E M  .......·  ............I Y VPA.........·  ...·
                                         *
RF1 (R. rubrum)        GVG K TVL...AGVG E RTREGNDLYH E MID...LQE R ITSTKKGSITSVQAI Y VPA...GEEH Y KVA
RpsF1 (Rps. blastica)  GVG K TVL...AGVG E RTREGNDLYH E FIE...LQE R TTSTKAGSITSVQAI Y VPA...GEEH Y NTA
                            Lys          Glu           Glu        Arg               Tyr           Tyr
```

FIGURE 1 : Alignment of F1 β sequences around chemically modified amino acid side chains..The asterisks indicate the labelled residues. The sequences were taken from refs. 2,22,39-47. The numbers correspond to amino acid residues on the β subunit of CF_1 (39).

with the bulk phase and the ATPase decay.

TABLE 1 : Effect of iodoacetamide on ATPase inactivation. Chloroplasts were prepared from preilluminated leaves by a rapid procedure described earlier (10) and were incubated twice at 0.2 mg chl./ml during 20 min as stated. After each incubation, chloroplasts were centrifugated and resuspended at 0ºC. CF1 was isolated by adding aliquots of 40 mg of chlorophyll to a final volume of 0.5 ml of 0.75 mM EDTA and incubating for 5 min at 20ªC Then they were centrifugated 4 min at 10,000 x g and the ATPase activity of supernatants (0.4 ml) were assayed as previously described (10). IA, 10 mM iodoacetamide; Fecy, 5 mM ferricyanide.

Thylakoids treatments		Membrane bound Mg-ATPase (μmol Pi/mg chl.h)	Soluble Ca-ATPase μmol Pi/mg prot.min)
1st.	2nd.		
None	0ºC	496	16.2
None	25ºC	35	14.6
Fecy	0ºC	210	6.3
IA, Fecy	0ºC	374	16.8
IA	0ºC	510	17.0
IA	25ºC	40	15.4

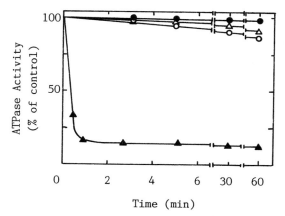

FIGURE 2 : Inhibition of chloroplasts proton ATPase by gramicidin. Light chloroplasts (0.5 mg chl./ml) were incubated at 0ºC in darkness in the absence (O , ●) or in the presence (Δ , ▲) of 50 nM gramicidin D. At the times stated membrane bound Mg-ATPase (● , ▲) and Ca-ATPase solubilized thereof (O ,Δ) were measured as described in Table 1. The activities of the controls were 523 μmol Pi/mg chl. h for the Mg-ATPase and 12.8 μmol Pi/mg prot. min for the Ca-ATPase.

TABLE 2 : Effect of cations on the dark deactivation kinetics of Mg-ATPase of light chloroplasts. Chloroplasts prepared as described in Table 1 were centrifugated, resuspended at 0°C in a medium containing 2 mM Tris-Tricine pH 8, 200 mM sucrose and the salt stated and incubated in the dark at 20°C. Half time of deactivation (t 0.5) is reported. Activity of the control was 310 μ mol Pi/mg chl. h.

Additions (mM)	t 0.5 (min)
None	26
MgCl₂ (1)	11
MgCl₂ (5)	4.5
MgCl₂ (10)	2.3
CaCl₂ (1)	14
CaCl₂ (5)	7.1
MnCl₂ (5)	14
NaCl (100)	6.0
KCl (100)	5.5

While incubation in the dark at 25°C of the light chloroplasts inactivated the membrane bound ATPase and had no effect on the soluble ATPase (Table 1), both activities diminished with similar rates upon dark incubation of intact leaves (Fig. 3).

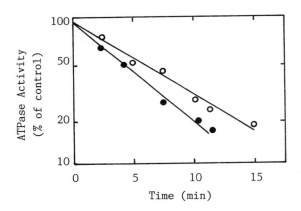

FIGURE 3 : Physiological deactivation of the chloroplast ATPase. Spinach plants were illuminated during 90 sec and their leaves were kept at 25°C in darkness. At the time stated, chloroplasts and CF1 were prepared as described in Table 1. (●-●) Mg-ATPase, (O-O) Ca-ATPase. The activities of the controls were 747 μ mol Pi/mg chl. h for the Mg-ATPase and 14.4 μ mol Pi/mg prot. min for the Ca-ATPase.

An apparent deactivation half time of 4 min was obtained for the membrane bound ATPase. In the same experimental conditions the soluble

ATPase showed also a fast inactivation, with an apparent deactivation half time of 5.7 min. These values are very closed to the half time of 4.7 min found in the in vitro dark inactivation of the membrane bound enzyme at the same temperature (10).

The data suggest that the dark in vitro deactivation of light chloroplasts is only associated with the dissipation of the proton gradient while the physiological deactivation also involves a change in CF1, which probably is oxidation of the vicinal dithiol of the γ subunit (10). The rate of this process indicates that the dithiol oxidation is an important factor for in vivo deactivation of the ATPase.

REFERENCES

1 Cross, R.L. (1981) Annu. Rev. Biochem. 50, 681-714.
2 Futai, M. and Kanazawa, H. (1983) Microbiol. Rev. 47, 285-312
3 Vallejos, R.H. (1981) In Energy Coupling in Photosynthesis (Selman, B. and Selman-Reimer, S. eds.) pp 129-139, Elsevier, Amsterdam
4 Carmeli and Avron (1967) Eur. J. Biochem. 2, 318-326
5 Carmeli, C. (1969) Biochim. Biophys. Acta 189, 256-266
6 Mills, J.D. and Mitchell, P. (1982) Biochim. Biophys. Acta 679, 75-83
7 Shahak, Y. (1982) Plant Physiol. 70, 87-91
8 Marchant, R.H. (1981) Proc. Int. Congr. Phosynth. Res. 2, 999-1008
9 Morita, S., Itoh, S. and Nishimura, M. (1982) Biochim. Biophys. Acta 679, 125-130
10 Vallejos, R.H., Arana, J.L. and Ravizzini, R.A. (1983) J. Biol. Chem. 258, 7317-7321
11 Mills,J.P., Mitchell,P. and Schürmann,P. (1980) FEBS Lett. 112, 173-177
12 Ravizzini, R.S., Andreo, C.S. and Vallejos, R.H. (1980) Biochim. Biophys. Acta 591, 135-141
13 Arana, J.L. and Vallejos, R.H. (1982) J. Biol. Chem. 257, 1125-1127
14 Moroney, J.V., Fullmer, C.S. and McCarty, R.E. (1984) J. Biol. Chem. 259, 7281-7285
15 Nalin, C.M. and McCarty, R.E. (1984) J. Biol. Chem. 259, 7275-7280
16 Vallejos, R.H. and Ravizzini, R.A. (1983) Proc. Int. Congr. Photosynth.Res. 2, 519-522
17 Andreo,C.S. and Vallejos,R.H. (1976) Biochim. Biophys. Acta 423, 590-601
18 Shahak, Y. (1985) J. Biol. Chem. 260, 1459-1464
19 Takabe, T., De Benedetti, E. and Jagendorf, A.T. (1982) Biochim. Biophys. Acta 682, 11-20
20 Viale, A.M., Andreo, C.S. and Vallejos, R.H. (1982) Biochim. Biophys. Acta 682, 135-144
21 Viale, A.M. and Vallejos, R.H. (1985) J. Biol. Chem. 260, 4958-4962
22 Hollemans, M., Runswick, M.J., Fearnley, I.M. and Walker, J.B. (1983) J. Biol. Chem. 258, 9307-9313
23 Andrews, W.W., Hill, F.C. and Allison, W.S. (1984) J. Biol. Chem. 259, 8219-8225
24 Sutton, R. and Ferguson, S.J. (1985) Eur. J. Biochem. 148, 551-554
25 Riordan, J.F. (1979) Mol. Cell. Biochem. 26, 71-92
26 Ceccarelli,E. and Vallejos,R.H. (1983) Arch.Biochem.Biophys.224, 382-388
27 Esch, F.S., Böhlen, P., Otsuka, A.S., Yoshida, M. and Allison, W.S. (1981) J. Biol. Chem. 256, 9084-9089
28 Yoshida, M., Allison W.S., Esch, F.S. and Futai, A. (1982) J. Biol. Chem. 257, 10033-10037
29 Yoshida, M., Posher, J.W., Allison, W.S. and Esch, F.S. (1981) J. Biol. Chem. 256, 148-153

30 Cortez, N., Lucero, H.A. and Vallejos, R.H. (1983) Biochim. Biophys. Acta 724, 396-403

31 Verburg, J.G., Yoshida, M. and Allison, W.S. (1986) Arch. Biochem. Biophys. 245, 8-13

32 Perez, J.A., Greenfield, A.J., Sutton, R. and Ferguson, S.J. (1986) FEBS Lett. 198, 113-118

33 Esch, F.S. and Allison, W.S. (1978) J. Biol. Chem. 253, 6100-6106

34 Bullough, D.A. and Allison, W.S. (1986) J. Biol. Chem. 261, 5722-5730

35 Andrews, W.A., Hill, F.C. and Allison, W.S. (1984) J. Biol. Chem. 259, 14378-14386

36 Andrews, W., Yoshida, M., Hill, F.C. and Allison W.S. (1984) Biochem. Biophys. Rev. Res. Comm. 123, 1040-1046

37 Yoshida, M. and Allison, W.S. (1986) J. Biol. Chem. 261, 5714-5721

38 Sakamoto, J. and Tonomura, Y. (1982) J. Biochem. (Tokyo) 93, 265-275

39 Zurawski, G., Bottomley, W. and Whitfeld, P.R. (1982) Proc.Natl.Acad.Sci. USA 79, 6260-6264

40 Shinozaki,K., Deno,H., Kato,A. and Sugiura,M. (1983) Gene 24, 147-155

41 Krebbers, E.T., Larrinna,Z.M., McIntosh,L. and Bogorad,L. (1982) Nucleic Acid Res. 10, 4985-5002

42 Zurawski, G. and Clegg,M.T. (1984) Nucleic Acid Res. 12, 2549-2559

43 Boutry,M. and Chua N-H (1985) EMBO J. 4, 2159-2165

44 Saltzgaber-Müller, J., Kunapuli, S.P. and Douglas, M.G. (1983) J. Biol. Chem. 258, 11465-11470

45 Falk, G., Hampe,A. and Walker, J.E. (1985) Biochem.J. 228, 391-407

46 Tybulewicz, V.L.J., Falk, G. and Walker, J.E. (1984) J. Mol. Biol. 179, 185-214

47 Howe, C.J., Fearnley,I.M., Walker, J.E., Dyer, T.S. and Gray,J.C. (1985) Plant Mol. Biol. 4, 333-345

48 Shahak, Y. (1983) Proc. Int. Congr. Photosynth. Res. 2,527-530

49 Theg, S.M., Johnson,J.D. and Homann,P.H. (1982) FEBS Lett. 145, 25-29

50 Theg, S.M. and Junge, W. (1983) Biochim. Biophys. Acta 723, 294-307

51 Laszlo, J.A., Baker, G.M. and Dilley, R.A. (1984) Biochim. Biophys. Acta 764, 160-169

INHIBITION OF Spirulina maxima PHOTOPHOSPHORYLATION BY N-ETHYLMALEIMIDE
CENTRO DE INVESTIGACION Y ESTUDIOS AVANZADOS DEL I.P.N.

C. GOMEZ-LOJERO AND CLAUDIA LERMA. DEPTO. DE BIOQUIMICA. APARTADO POSTAL
14-740, 07000 MEXICO CITY, MEXICO.

INTRODUCTION
   After the important work of McCarty (1), a great deal of attention has
been brought to the sulfhydryl groups of the H^+ adenosine triphosphate synthase (2,3). Monofunctional maleimides, which react with sulfhydryl groups,
cause a light dependent inhibition of various reactions of the chloroplast
H^+ATPase. These reactions are photophosphorylation, the ATP driven proton
pump and the ATPase activated by trypsin, dithiothreitol or methanol (1).
It has been observed that the H^+ATPase of cyanobacteria is activated by dithiothreitol (DTT) (4). However, more direct studies concerning the sulfhydryl groups of cyanobacterial enzyme have not been reported. The present
work deals with the effect of N-ethylmaleimide (NEM) on photophosphorylation and related partial reactions of highly coupled cyanobacterial membrane vesicles. The vesicles were obtained from the mesophilic cyanobacterium Spirulina maxima.

METHODS
   Spirulina maxima membrane vesicles were prepared by an already described
technique (5). To study the effect of NEM on the H^+ATPase, membranes were
preincubated for two minutes at 25°C with or without the maleimide in different conditions and their photophosphorylation, proton pump or ATPase activities were then measured in a second incubation. The SMT buffer used for
the membranes preincubation contained 0.5M sucrose, 50mM NaCl, 5mM $MgCl_2$,
10^{-4}M phenazine methosulfate (PMS) and 50mM tricine-NaOH pH 8.0. In the
case of the photophosphorylation experiments, different samples were preincubated as described in Table 1 and photophosphorylation was assayed as
described earlier (5). When the proton pump or ATPase experiments were performed, the Spirulina membrane sample was divided in two aliquots. One aliquot was preincubated with 5mM NEM, 2mM phosphate and 3mM ADP in 40 ml of
SMT buffer. The other aliquot was used as a control and preincubated in the
sample buffer. In both cases, membranes corresponding to 10µg Chl/ml were
preincubated in the light. The 6mM DTT was added to the membrane samples
and they were centrifuged at 20,000 x g for 30 min at 4°C and resuspended
in one ml of SMT buffer to assay their H^+ pump. This pump was measured by
means of a corning model 12, pHmeter. Membranes (20µg Ch/ml) were added to
2.5 ml of a medium containing 0.25M KCl, 5mM $MgCl_2$, 15mM NaCl, 0.1% w/v albumin, 10^{-4}M PMS, pH 6.0 to 6.4. The reaction was started with light. For
the measurement of ATPase activity after the preincubation, membranes were
added (25µg Chl/ml) to 2.5 ml of reaction medium containing 25mM NaCl, 5mM
ATP, 2.5mM $MgCl_2$, 35% methanol, and 25mM Tris-HCl (pH 8.8). ATPase was assayed as described before (4). For the assay of Mehler reaction (electron
transport), the reaction medium and method have already been described (5).
When cyclic photophosphorylation of spinach chloroplast was assayed by the
method of Nishimura (6), the medium contained 0.1M KCl, 5mM $MgCl_2$, 3mM
KH_2PO_4, 1mM ADP and 10^{-4}M PMS, pH 8.0 to 8.4. Light intensity used in all
our experiments was 4.4×10^5 erg/cm^2 per s.

Biggens, J. (ed.), Progress in Photosynthesis Research, Vol. III. ISBN 90 247 3452 5
© 1987 Martinus Nijhoff Publishers, Dordrecht. Printed in the Netherlands.

TABLE I: CONDITIONS FOR N-ETHYLMALEIMIDE INHIBITION OF PHOTOPHOSPHORYLA-
TION IN SPIRULINA AND SPINACH THYLAKOIDS

In both cases, thylakoids (20µg Chl/ml) were first incubated in 0.5ml buffer. When indicated, 10mM NEM was added before the first incubation in the light or dark. After the first incubation, 6mM DTT was added, as well as the required amounts of the reagents needed to assay photophosphoryla-tion in a 2nd incubation. In this step no more NEM was added.

CONDITIONS DURING FIRST INCUBATION	PHOTOPHOSPHORYLATION ASSAYED IN 2nd INCU̱BATION moles ATP/mg Chl per h	
	Spinach chloroplasts	Spirulina membranes
Light	381 ± 20	502 ± 73
Light + NEM	101 ± 1	30 ± 2
Light + NEM..DTT added after incubation	180 ± 10	648 ± 109
Dark	492 ± 40	864 ± 213
Dark + NEM	317 ± 60	31 ± 5
Dark + NEM..DTT added after incubation	461 ± 15	974 ± 121

RESULTS

The results shown in Table 1 can be summarized as follows. No preincu-bation step with the maleimide in light or darkness before the enzyme as-say was necessary for NEM to inhibit photophosphorylation of Spirulina maxima membranes. In contrast, as described by McCarty (1), a preincuba-tion of chloroplasts with NEM in the light was essential to inhibit phos-phorylation, since ADP present during the phosphorylation assay apparently protect the enzyme against this effect (1). Moreover, in the experiments with Spirulina vesicles, the presence of ADP was essential for the light dependent inhibition of photophosphorylation by NEM, almost 100% inhibi-tion was observed when 10mM was present during the cyanobacterial photo-phosphorylation assay (Table 1). Hence, the effect of NEM was instantane-ous.

The last results are consistent with the time dependent effect shown in Figure 1-Panel A. In this experiment, the addition of 5mM NEM was enough to inhibit completely and immediately the photophosphorylation of Spirulina membranes. According to McCarty's interpretation about ADP protection a-gainst NEM-inhibition (1), the addition of the maleimide during chloro-plast phosphorylation, only affected the reaction slightly (Figure 1-Panel B). The concentration dependence of NEM-inhibition upon cyanobacterial ATP-synthesis is shown in Figure 2. Fifty percent inhibition was attained at 2mM of NEM, but surprisingly, almost 100% inhibition was reached at higher concentrations. In contrast, in chloroplasts a 50% inhibition of photophos-phorylation was obtained with lower concentrations of NEM (0.5mM) and con-centrations higher than 1mM of the maleimide, caused only a 70% (maximal) inhibition. Since NEM-inhibition was instantaneous and ADP did not protect cyanobacterial photophosphorylation against this inhibition, we were able to explore the effect of the maleimide on the coupled Mehler reaction on Spirulina membranes. Panel 1 of Figure 3 shows that 4mM NEM did not impair basal electron transport, and it did not inhibit the gramicidin stimulated Mehler reaction either (Figure 3-Panel II). This figure also shows the transition caused by the addition of the channel former which is consistent with the stimulation effect of ADP (Panel II). In Panel II, inhibition of Spirulina Mehler's reaction by NEM under coupling conditions (ADP and Pi present) is also shown. Panel III shows that NEM's inhibitory effect of Mehler reaction was eliminated by the addition of the uncoupler. These re-

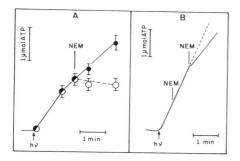

FIGURE 1. Effect of N-ethylmaleimide on photophosphorylation of spinach chloroplasts and of a membrane preparation of <u>Spirulina maxima</u>.Panel A. Spirulina membranes photophosphorylation. ATP synthesis assayed by incorporation of ^{32}P to ATP.Reaction medium and assay conditions described in methods, control without NEM. Experiment with 5mM NEM added when indicated.Panel B. Chloroplast photophosphorylation. Chloroplasts added to get 10µg Chl/ml. NEM was added to get final concentrations of 1mM (first addition) and 5mM (second addition).(hv)=Light.

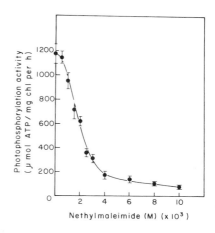

Figure 2. Quantity of N-ethylmaleimide necessary for the inhibition of Spirulina maxima's photophosphorylation. ATP synthesis was assayed as in Figure 1. Panel A.

sults suggest that NEM inhibits cyanobacterial phosphorylation acting as an energy transfer inhibitor and not as an uncoupler. This is also the case concerning monofunctional maleimide inhibition in chloroplasts(1).The following result validates the former observation: The maximal activity of H^+ pump (with the PMS shuttle) of Spirulina membrane vesicles was recorded. This activity varied from 350 to 668 nmoles H^+/mg Chl in different preparations and was completely abolishe by the addition of 25µM gramicidin. When Spirulina membranes were preincubated under cyclic photophosphorylation assay conditions in the presence of NEM. This treatment did not inhibit the light driven proton pump. If maleimide were an energy transfer inhibitor, NEM treatment causing inhibition of photophosphorylation of Spirulina membranes, would also inhibit other reactions of the ATPase. Methanol stimulated ATPase of Spirulina membranes was assayed (134±17µmol Pi/mg per h).When chloroplasts were illuminated in the presence of NEM, subsequent methanol induced ATPase activity was reduced by about 50% (7). However, no inhibitory effect of methanol activated Spirulina membrane ATPase was observed after preincubation cyanobacterial vesicles with NEM, ADP and Pi in the light.

DISCUSSION

NEM inhibition of cyclic photophosphorylation in Spirulina membranes was immediate in conditions in which the absence of the maleimide. Would otherwise elicit ATP synthesis. In contrast to chloroplasts in which ADP or ATP confer partial protection from NEM inhibition. Moreover, preincubation in the absence of ADP with NEM in the light was necessary for inhibition (1). McCarty has shown that two cysteine residues present in the chloroplast of CF_1 gamma subunit are preferential targets for NEM, when this inhibitor is added at mM concentrations (1). However, these two residues are not involved in ATPase activity stimulated by DTT (2). Nevertheless, chloroplast CF_1 has other cysteine residues present: one in α (8), one in β (9), two more in γ (10) and one in ε(9). All the cysteine residues of γ of the spinach chloroplast ATPase react with the thiol reagent under different condi-

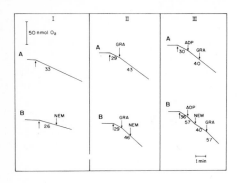

FIGURE 3. Effect of N-ethylmaleimide
on Mehler reaction of Spirulina vesicles. Membranes equivalent to 5μg Chl/
/ml were added to the buffer described
in Methods (30°C). Additions: 3mM ADP,
25μM gramicidin (GRA) and 4mM N-ethylmaleimide (NEM). Panel I: A) Basal photosynthetic electron transport. B) Addition of NEM during Mehler reaction.
Panel II: A) Stimulation of electron
transport stimulated by gramicidin,
panel III: A) Addition of gramicidin
did not increase the electron transport already stimulated by ADP. B) Inhibitory effect of NEM on electron
transport stimulated by ADP and reversibility of this inhibition by the
addition of gramicidin. The values
shown are μmol O_2/mg Chl h, ↑ Light on.

tions. One reacts with NEM, when chloroplasts are incubated in the dark.
Another is exposed by energization of the thylakoids. Recently McCarty
showed that the other two cysteines of the subunit are bonded in a disulfide bridge. These cysteine residues react with NEM only after a DTT treatment of the thylakoids and appear to be involved in the enzyme activation
by DTT (2). In Spirulina membranes, ATPase activity was triggered by a
light plus DTT treatment (4). Unfortunately, we do not know anything about
the aminoacid composition of the different subunits of the cyanobacterial
coupling factor. Further work in this interesting organism is required for
a proper interpretation.

REFERENCES

1. McCarty R Pittman P and Tsuchiya Y (1972) J Biol Chem 247:3048-3051.
2. Ketcham S Daveport J Warnicke K and McCarty R (1984) J Biol Chem
   259:7286-7293.
3. Ravizzini R Andreo C and Vallejos R (1980) Biochem Biophys Acta 591:
   135-141.
4. Lerma C and Gomez-Lojero C Photosynthesis Research (In Press).
5. Lerma C and Gomez-Lojero C (1982) Biochem Biophys Acta 680:181-186.
6. Nishimura M Ito J and Chance B (1982) Biochem Biophys Acta 59:177-
   183.
7. Anthon G and Jagendorf A (1983) Biochem Biophys Acta 723:358-365.
8. Zurawski G Bottomly W and Whitteld P (1982) Proc Natl Acad Sci (USA)
   79:6260-6264.
9. Deno H Shinizaki K and Sugiura M (1983) Nucleic Acid Research 11:2185.
   2191.
10. Moroney J Fullmer C and McCarty R (1984) J Biol Chem 259:7281-7285.

THE RATE OF ATP SYNTHESIS AND ATP HYDROLYSIS CATALYZED BY
RECONSTITUTED CF_oF_1 LIPOSOMES

Günter SCHMIDT and Peter GRÄBER

Max-Volmer-Institut für Biophysikalische und Physikalische
Chemie, Technische Universität Berlin, Strasse des 17. Juni
135, 1000 Berlin 12, FRG

1. INTRODUCTION

The membrane-bound chloroplast ATP-synthase (ATPase),
CF_oF_1, catalyzes reversibly proton transport-coupled ATP synthesis and hydrolysis. For ATP synthesis a maximum turnover
of about 400 ATP $(CF_1 \cdot s)^{-1}$ and for ATP hydrolysis a maximum
turnover of about 90 ATP $(CF_1 \cdot s)^{-1}$ has been found (1,2).
CF_oF_1 has been isolated, purified and reconstituted into
liposomes (3). However, the rates of ATP synthesis and ATP
hydrolysis catalyzed by the reconstituted CF_oF_1 were usually
very low; i.e., almost 1% of the turnover found in chloroplasts (3-8). Recently, we have shown that reconstituted
CF_oF_1 catalyzes ATP synthesis with a rate of about 200 ATP
$(CF_1 \cdot s)^{-1}$, when the membrane was energized artificially by an
acid/base transition and a K^+/valinomycin diffusion potential
(9). This rate was measured with a rapid mixing quenched flow
apparatus (1). In this work we reported on the functional
dependence of the rate of ATP synthesis on ΔpH and $\Delta \Psi$ and
we optimized the conditions for measuring high rates of ATP
hydrolysis (about 20 ATP $(CF_1 \cdot s)^{-1}$).

2. RESULTS
2.1 The rate of ATP synthesis

CF_oF_1 was isolated and reconstituted into asolectin liposomes as described earlier (9). The ATP yield after genera-
tion of ΔpH and $\Delta \Psi$ was measured as a function of the reaction
time and the rate was determined as in refs. 1 and 9. Fig. 1
shows the ATP yield as a function of the reaction time. The
pH_{out} was 8.35, the K^+ concentration inside was 6 mM and outside 63 mM. The linear increase of the yield demonstrates
that under these conditions the transmembrane ΔpH and $\Delta \Psi$ do
not change significantly. Thus, the slope of the curves
directly gives the rate of ATP synthesis at constant ΔpH and
$\Delta \Psi$. Increase of the transmembrane ΔpH also increased the
rate.

Fig. 2 shows the rate of ATP synthesis as a function of
ΔpH when different K^+/valinomycin diffusion potentials are
generated simultaneously with ΔpH. The data are from Fig. 1
and similar sets of measurements at other diffusion potentials.
The dependence of the rate on ΔpH is sigmoidal with a maximum of about 200 s^{-1}.

At constant ΔpH an additional diffusion potential (the
parameter given in Fig. 2 is the ratio between inner and
outer K^+ concentration) increases the rate unless the maximum

Biggens, J. (ed.), Progress in Photosynthesis Research, Vol. III. ISBN 90 247 3452 5
© *1987 Martinus Nijhoff Publishers, Dordrecht. Printed in the Netherlands.*

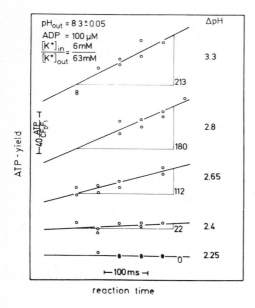

FIGURE 1. ATP yield as a function of the reaction time. The slope of these curves gives the rate of ATP synthesis, the numbers give the rate in ATP(CF_oF_1 s)$^{-1}$.

has been reached. If we assume the same permeability coefficients as in chloroplasts, diffusion potentials of 68 mV, 47 mV and 0 mV are calculated for the ratios of K^+_{in}/K^+_{out}: 0.5 mM/60 mM, 6 mM/63 mM and 120 mM/120 mM at ΔpH = 3.2 from the Goldmann-Hodgkin-Katz equation. Using these potentials, the sum of ΔpH and $\Delta \psi$ does not determine the rate of ATP synthesis as found for chloroplasts (1). This is probably due to different permeability coefficients for asolectin and the thylakoid membranes.

The half-maximal rate at $\Delta \psi$ = 0 mV is obtained at ΔpH = 3.25. This is nearly identical with the result from chloroplasts at ΔpH = 3.35(1). The maximal rate observed here was 200 s^{-1}. This is almost half the rate observed with chloroplasts under the same conditions (1).

2.2 The rate of ATP hydrolysis

CF_oF_1 can exist in - at least - four different states:

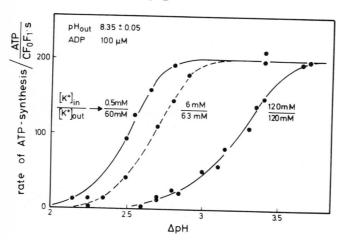

FIGURE 2. The rate of ATP synthesis as a function of ΔpH and $\Delta \psi$. Data are from Fig. 1 and similar sets of experiments. The parameter given at the curves is the ratio of inner and outer K^+ concentration.

an inactive, oxidized state, E_i^{ox}, an inactive, reduced state, E_i^{red}, and both species can be brought into an active state by energization of the membrane (10). Functionally, the inact-

ive states cannot catalyze ATP hydrolysis and ATP synthesis
(or they can catalyze only a very low rate of ATP hydrolysis).
The rate of ATP hydrolysis of the isolated CF_oF_1 before and
after reconstitution is about $1 \ s^{-1}$. When, after reconstitution, CF_oF_1 is reduced (incubation with 50 mM dithiothreitol
for 2 h) the rate is practically the same (see Fig. 3). Activation of CF_oF_1 by a preceding acid/base transition gives a
rate of ATP hydrolysis of
$12 \ s^{-1}$ (+DTT + Δ pH)(The
$\gamma - ^{32}P$-ATP was added together with 10 mM NH_4Cl 15 s
after the acid/base transition.) Oxidation of CF_oF_1
(incubation with 2 mM iodosobenzoate for 20 min) followed by activation with
ΔpH gives a rate of $2 \ s^{-1}$
(+IBZ + ΔpH). If the oxidation by IBZ is reversed by a
subsequent incubation with
DTT after activation again a
rate of $14 \ s^{-1}$ is observed;
i.e., practically the same
as before IBZ treatment.
    When different prepara-
tions of reconstituted CF_oF_1
were used, the following
rates of ATP hydrolysis were
found for E_a^{ox} 0-2 s^{-1}, for
E_i^{ox} 0-2 s^{-1}, for E_i^{red} 0-2 s^{-1}
and for E_a^{red} 12-25 s^{-1} (11).

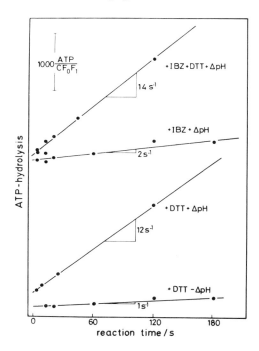

1000 $\frac{ATP}{CF_oF_1}$

+IBZ+DTT+ΔpH

14 s^{-1}

+IBZ + ΔpH

2 s^{-1}

ATP-hydrolysis

+DTT + ΔpH

12 s^{-1}

+DTT -ΔpH

1 s^{-1}

0   60   120   180

reaction time / s

FIGURE 3. Release of phosphate
as a function of the reaction
time after different conditions.
Details see text.

3. DISCUSSION
    In comparison with earli-
er work on the reconstituted
CF_oF_1 high rates of ATP synthesis and ATP hydrolysis
have been obtained here.
Aside from optimization of
the reconstitution and assay
procedures, we assume that
for ATP synthesis this is due to the high energization obtained by ΔpH and $\Delta\psi$. For ATP hydrolysis this is due to the
fact that the enzyme is brought into the form E_a^{red} before the
measurement of the rate.
    Compared to the results obtained with chloroplasts, the ATP
synthesis rate catalyzed by reconstituted CF_oF_1 is a factor
two smaller (400 s^{-1} versus 200 s^{-1}); whereas, the rate of ATP
hydrolysis is by a factor 3-4 smaller (90 s^{-1} vs. 25 s^{-1}).
Since some denaturation of CF_oF_1 may occur during isolation
and reconstitution, the rate per correctly reconstituted,
active CF_oF_1 is actually higher. Therefore, we conclude that
with respect to ATP synthesis and ATP hydrolysis CF_oF_1 in the
thylakoid membrane and in the asolectin membrane are functionally not different.

One might ask whether the inactive enzyme (both forms, the oxidized and the reduced one) can catalyze ATP hydrolysis. Our results in this respect are not clear yet. Rates between $0-2$ s^{-1} were found for the oxidized and the reduced species, most preparations show rates of about 1 s^{-1}. This activity might be either an intrinsic property of the species E_i^{ox} and E_i^{red} or it is an artifact of the preparation procedure: (1) During the isolation of the ATPase proteases might partially digest the ATPase and this might lead to an ATP hydrolysis activity; (2) octylglucoside in the presence of DTT is used to extract the ATPase from the membrane. When part of CF_1 is disconnected from CF_o, the octylglucoside can stimulate ATP hydrolysis of CF_1^o (12); (3) CF_oF_1 is reconstituted into the liposomes by dialysis (5 h, 30° C). This heat treatment might also elucidate some hydrolysis activity.

Since several of our preparations showed no ATP hydrolysis when treated with IBZ and when treated with DTT, we assume at present that the intact, reconstituted ATPase in the form E_i^{ox} and E_i^{red} shows no ATP hydrolysis.

ACKNOWLEDGEMENTS
This work has been supported by a grant from the Deutsche Forschungsgemeinschaft.

REFERENCES
1 Gräber, P., Junesch, U. and Schatz, G.H. (1984) Ber. Bunsenges. Phys. Chem. 88, 599-608
2 Junesch, U. and Gräber, P. (1986) in: VII. International Congr. on Photosynthesis
3 Pick, U. and Racker, E. (1979) J. Biol. Chem. 245, 2793-2799
4 Hauska, G., Samoray, D., Orlich, G. and Nelson, N. (1980) Eur. J. Biochem. 111, 535-543
5 Dewey, T.G. and Hammes, G.G. (1981) J. Biol. Chem. 256, 8941-8946
6 van Walraven, H.S., Lubberding, M.J., Marvin, M.J.P. and Kraayenhof, R. (1983) Eur. J. Biochem. 137, 101-106
7 Shahak, Y. and Pick, U. (1983) Arch. Biochem. Biophys. 223, 393-406
8 Pick, U., Gounaris, K., Admon, A. and Barber, J. (1984) Biochim. Biophys. Acta 765, 12-20
9 Schmidt, G. and Gräber, P. (1985) Biochim. Biophys. Acta 880, 46-51
10 Junesch, U. and Gräber, P. (1985) Biochim. Biophys. Acta 809, 429-434
11 Schmidt, G. and Gräber, P. (1986) Eur. J. Biochem. in press
12 Pick, U. and Bassilian, S. (1982) Biochemistry 21, 6144-6152

PROTEOLIPOSPOMES FROM RHODOSPIRILLUM RUBRUM F_0F_1 AND SOYBEAN ASOLECTIN: ATP-HYDROLYSIS AND 9-AMINO-ACRIDINE FLUORESCENCE QUENCHING.

L. SLOOTEN AND S. VANDENBRANDEN, BIOPHYSICS LABORATORY, VRIJE UNIVERSITEIT BRUSSEL, PLEINLAAN 2, B-1050 BRUSSLES, BELGIUM

1. INTRODUCTION

F_0F_1-ATPases incorporated into liposomes continue to constitute interesting model systems for investigation of various aspects of energy coupling. However, when ATP-synthesis is reconstituted in such systems, the rates obtained are generally rather low (e.g. [1,2]), and this might be taken occasionally as a potential objection against the significance of the results so obtained. We felt that the low rates may be due in part at least, to the fact that often these systems are somewhat uncoupled; this is evident from relatively low stimulation of ATP-hydrolysis by uncoupler (e.g. [1,3]). For this reason we set out to devise a method for obtaining well-coupled proteoliposomes, at least as judged by ATP-hydrolysis. Here we present such a procedure.

ATPase induced membrane energization in our preparation was evidenced by 9-amino-acridin fluorescence changes. Fluorescence increases as well as decreases were observed. We present some experiments pertaining to the factors responsible for these variable outcomes.

2. METHODS

R. rubrum was grown, and chromatophores were prepared as described [4,5]. F_0F_1-ATPase was prepared as follows. Chromatophores (1 mM bacteriochlorophyll) were incubated for 15 min at 32°C with 15 mM octylglucoside, 2.4 mM deoxycholate, and 10 µM PMSF in 43 mM KCl, 175 mM NaCl, 4 mM HEPES, pH 7.9. The suspension was centrifuged for 1 h at 100,000 x g (without brake). The supernatant was immediately layered onto a linear, stepwise sucrose gradient (0.16-0.52 M) in 10 mM HEPES, pH 7.9, and centrifuged for 17 h at 60,000 x g in a swing-out rotor. The ATPase containing fractions (lower half of the tubes) were pooled and concentrated by means of dry Sephadex G-50. The concentrate (0.3-0.5 mg/ml protein) was stored at 0°C and could be used for at least 4 weeks.

Soybean asolectin (Sigma, type IV-S) was partially purified [6] and stored at -20°C. Liposomes were prepared by adding phospholipid to an aqueous solution and vortexing under nitrogen until a milky suspension was obtained. If necessary, the pH was readjusted at this stage. 2 ml of this suspension was sonicated to clarity with a probe (MSE, 125 W) operated at 3 µ amplitude.

Proteoliposomes were prepared by mixing, in a final volume of 0.5 ml, approx. 0.1 mg/ml F_0F_1, liposomes corresponding to 9.8 mg/ml lipid, and 8.8 mg/ml cholate in 133 mM sucrose, 0.2 mM EDTA, 10.2 mM $MgCl_2$, and 75 mM KCl (final conc.), pH 7.9. After 30 min incubation at 0°C the suspension was centrifuged [7] through 5-ml columns containing Sephadex G-50 equilibrated with 133 mM sucrose, 0.2 mM EDTA, 3.2 mM $MgCl_2$, 10 mM HEPES, 70 mM KCl, pH 7.9. ATPase was measured as in [8].

3. RESULTS

Table 1 shows the yield and specific activity at various stages of purification of F_0F_1-ATP-ase. The activities were measured in the presence of sulfite because this abolishes the Mg^{2+}-inhibition of uncoupled ATPase [9]. Thus, activities obtained with intact chromatophores are directly

Table 1. Preparation of R.rubrum F_0F_1.

Fraction	ATPase°)	Protein	Specific activity	yield
	μmol/min	mg	μmol/min mg pr.	%
Chromatophores	894+)	312	2.87	100
Detergent extract	483	35.6	13.6	55
Sucrose grad. pool	322	ND°°)	ND	36
Concentrate	262	4.96	55.8*)	29.3

°)Medium: 35 mM K_2SO_3-133 mM sucrose-3.2 mM $MgCl_2$-0.2 mM EDTA-10 mM HEPES-pH 7.9. Reaction started by addition of 3 mM ATP. Reaction time, 2 min.; +)Plus 50 μM CCCP; °°)not determined; *)89.5% inhibited by 6 μg/ml oligomycin.

comparable with those obtained with isolated F_0F_1. Table 1 shows that the yield of F_0F_1-ATPase was about 30%, and the specific activity was increased about 20-fold after gradient centrifugation and concentration. ATP-hydrolysis by purified F_0F_1 was strongly inhibited by oligomycin.

The uncoupler-dependence of ATP-hydrolysis in proteoliposomes is shown in fig. 1. During assays with sulfite, CCCP generally caused a 3.5 to 5-fold (3.85-fold in fig.1, left) increase in ATPase activity. This stimulation was much higher in assays with chloride salts: When liposomes were prepared in a medium with KCl, CCCP caused a 17.2 to 17.9-fold increase in activity (fig. 1, middle panel). When liposomes were prepared and assayed in a medium with ammonium chloride, valinomycin caused a 14.3-fold increase in activity (fig. 1, right). This, incidentally, indicates that the permeability of the liposome membrane for NH_4^+ is very low compared with the permeability for NH_3 (c.f. [10]).The middle panel also shows that, as in chromatophores [9,11], the activity in the presence of uncoupler is very sensitive to Mg^{2+}-inhibition in the absence of sulfite.

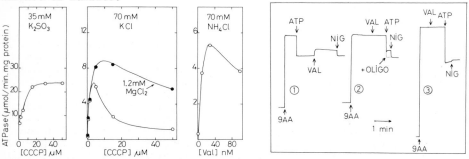

Fig.1. Uncoupler dependence of ATP-hydrolysis in proteoliposomes. Conditions: Left, as in Table 1. Middle, right: Same, but K_2SO_3 replaced by 70 mM of the indicated salt, and, where indicated, 1.2 mM $MgCl_2$; Lipid, 0.26 mg/ml. Protein 2.9 μg/ml.

Fig.2. ATP-induced 9-amino-acridin fluorescence changes in proteoliposomes. Lipid, 0.24 mg/ml. Protein, 5.1 μg/ml. 9-amino-acridin 1 μM (traces 1,2) or 6 μM (trace 3). 3 mM ATP, 0.1 μM valinomycin, 0.15 μM nigericin, 6μg/ml oligomycin added where indicated. Other conditions as in Table 1.

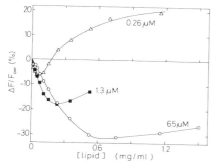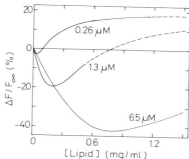

Fig.3. 9-amino-acridin fluorescence changes induced by an acid-base jump. Liposomes were prepared at 30 mg/ml in 30 mM K-phosphate, 0.1 mM EDTA, 60 mM K_2SO_4, pH 5.0. Samples were added to 2.5 ml of the same medium but of higher pH, such that the final pH was 8.0 and the 9-amino-acridin concentration was as indicated. F_∞ was obtained by addition of 0.1 µM nigericin. Data corrected for light scattering by liposomes and for changes in fluorescence yield due to alkalinization.

Fig.4. Simulation of the data shown in Fig.3. Assumptions: <u>Tight-binding sites</u>: 2 nmoles/mg lipid; K_d, 0.1 µM; the fluorescence of 9-<u>amino-acridin</u> bound to these sites increases by 50%. <u>Weak-binding sites</u>: 10 nmoles/mg lipid; K_d, 2 µM; the fluorescence of 9-amino-acridin <u>bound</u> to these sites is quenched 100%. Portions of the curves with lipid concentrations outside the measured ranges are dashed.

ATPase-induced 9-amino-acridin fluorescence changes are shown in fig. 2. At the high concentrations used, ATP decreased the fluorescence yield of 9-amino-acridin by direct interaction with the probe. Apart from that, ATP induced a fluorescence change which was largely dependent on valinomycin (trace 1), and which was abolished by oligomycin (trace 2), or by nigericin (traces 1-3). Hence, the changes correspond qualitatively with the establishment of a trans-membrane ΔpH, negative inside, as usual [12]. However, the direction of the change depended on the probe concentration: At 6 µM (trace 3) we found a (small) fluorescence quenching, but at 1 µM 9-amino-acridin a fluorescence increase was observed.

The concentration-dependence of 9-amino-acridin fluorescence changes was studied in more detail in experiments involving acid-base jumps in liposomes without F_0F_1. Fig. 3 shows results obtained with a jump from pH 5 (inside) to 8 (outside). The lipid- dependence of the magnitude of the fluorescence change (in %) was complex and depended also on the probe concentration. The data could be explained reasonably well with a model in which the acid-base jump exposes two kinds of probe-binding sites: A small population of high-affinity sites where the fluorescence of bound 9-amino-acridin is increased; and a larger population of low-affinity binding sites where the fluorescence of bound 9-amino-acridin is quenched. The results of a simulation are shown in fig. 4.

4. DISCUSSION

Fig. 1 shows that the ATPase activity of proteoliposomes is strongly enhanced by uncouplers. The lesser enhancement with sulfite may be accounted for in part with the assumption that not all F_0F_1 has been incorporated into liposomes. Free F_0F_1 has virtually no Mg-ATPase activity in a chloride medium; however, in a sulfite medium its activity is probably as high as in completely uncoupled liposomes. Apart from this, the reconstituted ATPase is well-coupled, as shown in the middle and right-hand panel of fig. 1. The occurrence of ATPase-induced 9-amino-acridin fluorescence changes (fig. 2) supports this conclusion. Mg^{2+}-inhibition of the ATPase activity, counteracted by membrane energization (fig. 1, middle panel) occurs as in intact chromatophores [9,11].

The fact that 9-amino-acridin responds to ATPase activity (fig. 2) or pH-jumps (fig. 3) with a fluorescence increase is, in retrospect, not wholly surprising. In early experiments with egg lecithin, bacteriochlorophyll was included in the liposomes because, without it, enhancement rather than quenching of atebrin fluorescence was observed after pH-jumps. The inclusion of bacteriochlorophyll also enhanced quenching of 9-amino-acridin fluorescence after pH-jumps [13]. Also, at low probe concentrations, 9-amino-acridin fluorescence quenching decreased rather than increased, when the chloroplast or liposome concentration was raised above a certain value [14]. We speculate that fluorescence enhancement occurs when the probe binds to a hydrophobic site, or is adsorbed into a hydrophobic environment, somewhere in the membrane interior. Fluorescence quenching, on the other hand, is probably due to binding of the probe to more exposed, hydrophilic sites [15]. In the experiments of Deamer et al. [13], it may be supposed that bacteriochlorophyll occupied the hydrophobic sites, leaving the hydrophilic sites for probe binding. We stress that our simulations show that even if the proportion of strong binding hydrophobic sites is so low that fluorescence increases are not observed, this may still cause serious deviations from the behavior expected if only one type of probe-binding site is present (c.f. [14]). The occurrence of two types of probe binding site poses the additional problem that the relative proportions in which these sites are exposed, vary with the circumstances. For example, the data shown in fig. 2 were obtained with 0.24 mg/ml lipid. The 9-amino-acridin dependence of the procentual fluorescence change was different in these experiments than in the experiments shown in fig. 3, at that lipid concentration. This observation is interesting in its own right and deserves further exploration. However, it poses additional serious problems for the use of 9-amino-acridin as a probe for ΔpH.

REFERENCES

1 Dewey, T.G. and Hammes, G.G. (1982) J.Biol.Chem. 256 8941-8946
2 Oren, R., Weiss, S., Garty, H., Caplan, R.S. and Gromet-Elhanan, Z. (1980) Arch.Biochem.Biophys. 205 503-509
3 Pick, U. and Racker, E. (1979) J.Biol.Chem. 254 2793-2799
4 Slooten, L. and Branders, C. (1979) Biochim.Biophys.Acta 547 79-90
5 Slooten, L. and Nuyten, A. (1983) Biochim.Biophys.Acta 725 49-59
6 Kagawa, Y. and Racker, E. (1971) J.Biol.Chem. 245 5477-5487
7 Penefsky, H.S. (1979) Meth.Enzymol. 56 527-530
8 Slooten, L. and Nuyten, A. (1981) Biochim.Biophys.Acta 638 305-312
9 Webster, G.D. and Jackson, J.B. (1977) FEBS-Lett. 76 29-35
10 Briller, S. and Gromet-Elhanan, Z. (1967) Biochim.Biophys.Acta 205 263-272
11 Slooten, L. and Nuyten, A. (1981) Biochim.Biophys.Acta 638 313-326
12 Schuldiner, S., Rottenberg, H. and Avron, M. (1972) Eur.J.Biochem. 25 64-70
13 Deamer, D.W., Prince, R.C. and Crofts, A.R. (1972) Biochim.Biophys.Acta 274 323-335
14 Fiolet, J.W.T., Bakker, E.P.B. and Van Dam, K. (1974) Biochim.Biophys. Acta 368 432-445
15 Kraayenhof, R. and Arents, J.C. (1977) in Electric Phenomena at the Biological Membrane Level (Roux, E. ed.) Elsevier Scient. Publ. 493-504

PARTIAL PROTEIN SEQUENCE OF THE SUBUNIT DELTA FROM SPINACH AND
MAIZE CF_1 AND TOPOGRAPHICAL STUDIES ON THE BINDING REGION
BETWEEN CF_1 AND CF_0

BERZBORN, R.J., FINKE, W., OTTO, J., VÖLKER, M., MEYER, H.,
NIER, W., OWORAH-NKRUMA, R., *BLOCK, J.
Biochemistry of Plants, Dep. of Biology, Ruhr-Universität,
D-4630 Bochum, FRG
*Max-Planck-Inst. f. Ernährungsphysiologie, D-4600 Dortmund

It is well accepted that within the photosynthetic ATPsynthase CF_1CF_0 the
site of ATP synthesis is located in the peripheral CF_1, on the beta subunit, whereas the integral CF_0 is conducting H^+ down the $\Delta \mu H^+$-gradient.
According to current theories two energy transforming reactions take
place: The electrochemical gradient of H^+ brings about a conformational
change that in turn enables CF_1 to produce ATP against the thermodynamic
equilibrium.
Thus one central question is: How is the power of H^+ efflux transformed to
a conformational change, and exactly where within the complex is this
happening? Obviously there are two possibilities; the H^+ efflux could be
guided across the connecting structures between CF_0 and CF_1 and only CF_1
could move, or the H^+ flux could lead to a conformational change already
in the CF_0, which would be transduced into CF_1. Therefore the biochemical
details of the specific binding regions between CF_0 and CF_1 have to be
known.

Statement 1: One can quantitatively determine that subunit
             delta binds to CF_1 as well as to CF_0.
Conditions for specific quantitative microdetermination by electro immuno
diffusion, also called rocket electrophoresis, for free sub. delta are as
described for the determination of CF_1 (ref. 1); for determination of
delta in situ, i.e. in the thylakoid or unresolved CF_1CF_0 complex, a previous dissociation by desoxycholate or octylglucoside is essential. Upon
EDTA treatment delta does not dissociate from CF_1 (same proportion of
resolved CF_1 and delta; and anti delta precipitates complete CF_1, analyzed
by polyacrylamide gel electrophoresis, PAGE). Isolated delta binds specifically to CF_0 (rebinding titrates to 100%, if excess of protein is added,
ref. 2), but sub. delta is not essential for CF_1 binding (ref. 2,3). Sub.
delta is essential, however, for reconstitution of photophosphorylation
capacity, both in EDTA and NaBr treated thylakoids (ref. 2-4). Delta does
not by itself lead to structural reconstitution (ref. 2,3). Titration of
residual delta is only done by us.

Statement 2: Need for excess delta in reconstitution experi-
             ments may be an artefact of proteolysis.
Sub. delta, 21 kD, is easily degraded to a 20 kD delta', as shown by SDS-
PAGE and Western blot with anti delta serum. But also sub. beta during
isolation and handling of CF_1 often is digested to a 19 kD delta'' (positive Western blot of some anti beta, e.g. serum 249, with 19 kD delta'';
positive Western blot of some anti delta, e.g. serum 120, with beta,
which means that this serum is not monospecific and crossreacts with beta,
probably because the immunogen delta was contaminated with some proteo-

lysis product from beta). Sequence analysis from isolated 19 k D polypeptide (automated determination of PTH derivatives) gave: A V Y V P A D D L T which is identical to a sequence in beta: A_{326} V Y V P A D D L T. Thus the delta'' may comprise to 173 C-terminal amino acids of beta, starting with A_{326}. Homogeneity of preparations of isolated delta are difficult to prove by routine PAGE!

Statement 3: A 25 kD polypeptide of maize CF_1CF_0 complex can be identified to be the delta subunit.

Isolated EDTA-CF_1 from Zea mays shows on SDS-PAGE a 25 kD polypeptide, but no 21 kD. Reconstitution with this CF_1 is possible, also hybrid reconstitution with EDTA thylakoids and EDTA CF_1 between spinach and maize (ref. 5). Anti 25 kD precipitates the complete CF_1 from maize, as does anti 21 kD precipitate CF_1 from spinach. There is no immunochemical crossreactivity in precipitation, however, between spinach and maize delta, except for very little reaction in Western blot, i.e. after SDS treatment. The comparison of N-terminal amino acid sequences, as shown under statement 4, futher supports statement 3.

Statement 4: Sequence homologies between N-terminal amino acids of the delta subunits of spinach and maize CF_1, reported here, and four respective polypeptides from the literature, are obvious. The region with high homologies is situated in an amphipathic alpha helix.

The 21 kD delta from spinach and the 25 kD delta from maize were separated from other polypeptides on PAGE, extracted, checked for purity, and the N-terminal sequences determined (automated analysis of PTH derivatives).

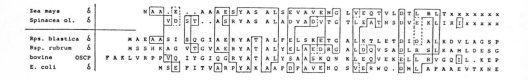

Tab. 1 shows the achieved sequences and the proposed alignment with the respective polypeptides from other species (ref. 6-8). Sequence homologies are obvious, up to 35% in the stretch reportet.
We had our computer calculate and plot the secondary structure predictions according to the algorithms of Scheraga, Chou Fasman, Nagano, Robson and Wittmann-Liebold. Different algorithms did not lead to the same structure. But with some probability the initial 18 (spinach) and 20 (maize) amino acids form an alpha helix, which then has a strong amphipathic character (fig. 1,2). Note the position of tyr_8 and in 5 cases, the close vicinity of at least 2 acidic side chains in pos. 14,17,18. Since the complete amino acid sequence of any higher plant delta is still lacking, we are not able to extend our analysis further. From the comparison of the 4 known sequences a pattern of homologous domains and some common features of secondary structure are already emerging (J. Block, R.J. Berzborn, unpublished).

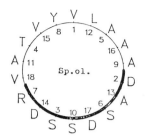 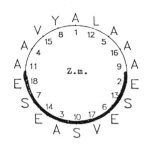

fig. 1 and 2: Amphipathic character of N-terminal amino acid sequences of spinach and maize subunit delta of CF_1, as determined by us.

Statement 5: Subunit delta may be hidden within CF_1CF_0 complex

For immunochemical determination, i.e. accessibility of delta, a dissociation is necessary (c.p. statement 1). Anti delta does not agglutinate thylakoid membrane suspensions as fast and strong as expected from the precipitation titer of the antiserum.

No monoclonal antibodies against delta could be selected yet, which would react with delta in the undissociated CF_1CF_0 complex. Monoclonal antibodies could be produced against spinach delta, reacting with delta in isolated CF_1, i.e. the binding surface towards CF_0, and others reacting with dissociated free delta only, i.e. the binding surface towards CF_1 (W. Finke, R.J. Berzborn, unpublished).

Statement 6: CF_0 subunits II and III are accessible and exposed at the thylakoid in presence of CF_1.

Antisera 169 and 187, produced with CF_1CF_0 and absorbed with CF_1, contained anti II and anti I (Western blot), and inhibited photophosphorylation to about 80%, serum 171, an anti III and anti epsilon, inhibited about 20% (ref. 9).

Newly produced antiserum 240 (anti III) and 242 (anti II) are monospecific (Western blot) and inhibit cyclic photophosphorylation, 70 and 55% of the control with 50 µl/10 µg chl., resp.

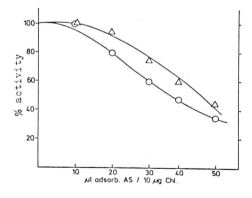

fig. 3: Inhibition of cyclic photophosphorylation with PMS by antiserum 240 (anti III) o , and 242 (anti II) △ .

Take home: Models of CF_1CF_0 complex should resemble more a flower pot than an ice cream cone.

References:
1. Roos, P., Berzborn, R.J. (1983) Z. Naturforsch. 38c, 799-805
2. Roos, P., Berzborn, R.J. (1982) Sec. EBEC, Lyon, CNRS ed. 99-100
3. Andreo, C.S., Patrie, W.J., McCarty, R.E. (1982) J. Biol. Chem. 257, 9968-9975
4. Nelson, N., Karny, O. (1976) FEBS Lett. 70, 249-253
5. Pucheu, N.L., Berzborn, R.J. (1983) in Adv. in Photosynthesis Res. (C. Sybesma, ed.), pp. II.6.571-574, Martinus Nijhoff/Dr. W. Junk Publishers, The Hague
6. Tybulewicz, V.L.J., Falk, G., Walker, J.E. (1984) J. Mol. Biol. 179, 185-214
7. Ovchinnikov, Y.A., Modyanov, N.N., Grinkevich, V.A., Aldanova, N.A., Trubetskaya, O.E., Nazimov, I.V., Hundal. T., Ernster, L. (1984) FEBS Lett. 166, 19-22
8. Walker, J.E., Runswick, M.J., Saraste, M. (1982) FEBS Lett. 146, 393-396
9. Klein-Hitpaß, L., Berzborn, R.J. (1983) in Adv. in Photosynthesis Res. (C. Sybesma, ed.), pp. II.6.563-566, Martinus Nijhoff/Dr. W. Junk Publishers, The Hague

THE RELATIONSHIP BETWEEN THE FUNCTION OF THE COUPLING FACTOR
AND ITS COMBINATION WITH THE THYLAKOID
Y.K.Shen, H.P.Dai, J.M.Wei and Y.Q.Qian

Shanghai Institute of Plant Physiology, Academia Sinica

1. INTRODUCTION

The coupling factor (CF_1) is a protein complex on the thylakoid membrane that catalyzes ATP formation in photophosphorylation. When CF_1 either on the membrane or detached therefrom is activated through some treatment, it shows ATP hydrolysis activity. Though the mechanism of its action has been studied intensively in a number of laboratories, it is still unclear in many aspects. One of them is that there is nearly no information concerning the effect of combination of CF_1 with the membrane on its function. Recently we have carried out some studies on this problem.

2. METERIALS AND METHODS

Plants of spinach (Spinacea oleracea,L.) were grown in the phytotron and field. Chloroplasts were isolated from spinach leaves as described previously (Wei et al. 1980). CF_1 were prepared according to Strotmann et al.(1973). CF_0-CF_1 complex were extracted and partially purified with Pick and Bassilian's method (1982). ATPase activities were measured according to Anthon and Jagendorf (1983), and photophosphorylation activities were measured with methods used before (Shen and Shen, 1962).

3. RESULTS AND DISCUSSION

3.1. The effect of combination of CF_1 with the membrane on its Mg^{++}-ATPase activated by methanol

Methanol can activate CF_1 to show Mg^{++}-ATPase activity whether it is combined with the membrane or detached from it (Anthon and Jagendorf,1983). This fact may be used to study the effect of membrane on the function of CF_1. Our results show that the concentration of methanol required for activation of CF_1 to its maximum activity is higher when CF_1 is combined with the membrane and its maximum activity is still lower than that in the isolated state (Fig 1). Furthermore, the Arrhenius plot of Mg^{++}-ATPase is a straight line, while that of membrane bound CF_1 shows a break point around 22 °C (Fig 2). The CF_1 is bound to CF_0 in the thylakoid membrane. The properties of the Mg^{++}-ATPase of CF_1 in the isolated CF_0-CF_1 complex are similar to those combined with the membrane (Table 1). These results indicate that the effect of membrane on Mg^{++}-ATPase of CF_1 is mainly produced by CF_1 and its tightly bound lipid, and the other part

Biggens, J. (ed.), Progress in Photosynthesis Research, Vol. III. ISBN 90 247 3452 5
© *1987 Martinus Nijhoff Publishers, Dordrecht. Printed in the Netherlands.*

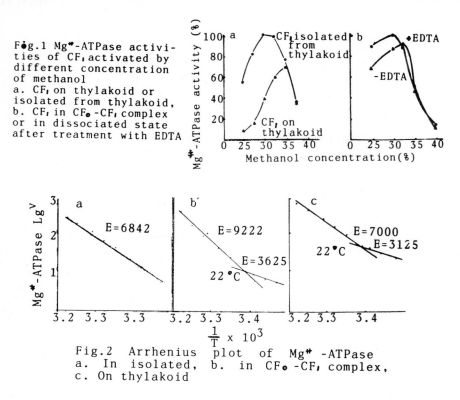

Fig.1 Mg*-ATPase activities of CF, activated by different concentration of methanol
a. CF, on thylakoid or isolated from thylakoid,
b. CF, in CF₀-CF, complex or in dissociated state after treatment with EDTA

Fig.2 Arrhenius plot of Mg*-ATPase
a. In isolated, b. in CF₀-CF, complex,
c. On thylakoid

TABLE 1. Effects of methanol on ATPase activities of CF, at different state

	ATPase activity (relative)	
	Mg*-ATPase	Ca*-ATPase
CF, on thylakoid	100	6.4
CF, isolated from thylakoid	100	38.3
CF, in CF₀-CF, complex	100	7.4

of the membrane may enhance the effect.

3.2. The dependence of CF, Mg*-ATPase activities on their structural changes

Isolated CF, may be activated by methanol to show both Mg*-ATPase and Ca*-ATPase activities, but when CF, is bound to the membrane or in the CF₀-CF, complex, it only shows Mg*-ATPase activity. As the thylakoids have been treated with NEM to react with the sulfhydryl group of the γ-subunit in CF, under illuminated conditions, the Mg*-ATPase activity of CF, is significantly lowered no matter whether it is bound to or isolated from the treated thylakoid (Fig 3). However, the Ca* -ATPase activity of this isolated CF, is not affected. After the thylakoids have been treated with trypsin

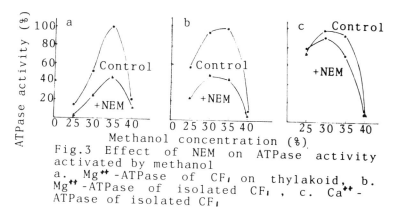

Fig.3 Effect of NEM on ATPase activity activated by methanol
a. Mg^{++}-ATPase of CF$_1$ on thylakoid, b. Mg^{++}-ATPase of isolated CF$_1$, c. Ca^{++}-ATPase of isolated CF$_1$

to break some of the peptide bonds of CF$_1$ under the illuminated condition, the CF$_1$ isolated from these treated thylakoids shows Ca^+-ATPase activity immediately, and adding methanol does not increase, but decreases its activity. However, this isolated CF$_1$ still requires methanol to activate its Mg^+-ATPase (Fig 4). From these

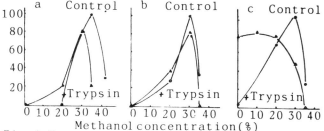

Fig.4 Treatment of thylakoid with trypsin on its ATPase activities of CF$_1$ activated by methanol
a. Mg^+-ATPase of CF$_1$ on thylakoid, b. Mg^+-ATPase of CF$_1$ isolated from thylakoid, c. Ca^+-ATPase of CF$_1$ isolated from thylakoid

results,it may be seen that when the CF$_1$ is modified by various treatments,it shows different effects on Mg^+-ATPase and Ca^{++}-ATPase activities. Whenever the CF$_1$ is bound to the membrane or in the CF$_0$-CF$_1$ complex, it imposes strong limitations on its expression of Ca^+-ATPase activity. As for the Mg^+-ATPase, it is more movement of some

closely correlated to the turn-over of the thylakoids of the groups in CF$_1$. The illumination of the thylakoids to bring it to the energized state or treatment with methanol is in favor of such movements.

3.3 The relationship between the function of CF$_1$ and its location on the thylakoid

The CF$_1$ and the protein complexes of photosystem I are mainly distributed in the stroma thylakoids and the exposed part of the grana thylakoids, while those of photosystem II are predominantly found in stacked areas of grana thylakoids. What is the effect of location on the function of CF$_1$ and its coordination with the energy conservation sites along the electron transport chain is an important problem. Our results show that CF$_1$ on the stacked thylakoids is more liable to react with antibody against CF_I than that on unstacked thylakoids, and after antibody treatment the Mg^+-ATPase of the CF_I in the former case is inhibited more severely (Fig 5). The cyclic photophosphorylation

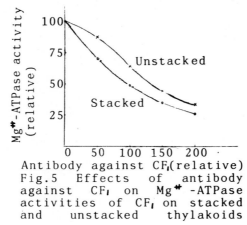

Fig.5 Effects of antibody against CF_I on Mg^+-ATPase activities of CF_I on stacked and unstacked thylakoids

(mainly related to photosystem I) and noncyclic photophosphorylation (related to both photosystems I and II) activities are lowered to similar degrees after the antibody treatment in stacked thylakoids, but the cyclic photophosphorylation is decreased to a much greater extent after the antibody treatment in unstacked thylakoids (Table 2). These results indicated that the location of CF_I on the thylakoid does influence its reaction, and the mechanism deserves further study.

TABLE 2. Effects of antibody against CF_I on photophosphorylation of stacked and unstacked thylakoids

		Photophosporylation (% of inhibition)			
		Expt. No.1		Expt. No.2	
		Cyclic	Noncyclic	Cyclic	Noncyclic
Stacked:	Control	100	100	100	100
	+antibody	67	70	79	84
Unstacked:	Control	100	100	100	100
	+antibody	51	58	78	94

REFERENCES

1. Wei J.M., Shen Y.K., Li D.Y. and Zhang Y.X. (1980) Acta Phytophysiologica Sinica 6, 393
2. Strotmann, H. et al.,(1973) Biochim. Biophys. Acta 314, 202
3. Pick, U. and Bassilian, S. (1982) Biochemistry 21, 6144
4. Anthon, G.E. and Jagendorf, T.J. (1983) Biochim. Biophys. Acta 723, 258
5. Shen Y.K. and Shen G.M. (1962) Scientia Sinica 11, 1097

SOLUBILIZATION AND RECONSTITUTION OF ATP SYNTHASE ACTIVITY FROM
Chlamydomonas reinhardtii CHLOROPLAST THYLAKOID MEMBRANES

Lloyd M. Yu and Bruce R. Selman, Department of Biochemistry, University
of Wisconsin-Madison, Madison, Wisconsin, 53706, USA

1.   INTRODUCTION
    The chloroplast thylakoid ATP synthase (CF_0-CF_1) has been isolated
from several sources (1) and appears to contain at least 8 different
subunits. Unlike the enzyme complex from $\underline{E}$. $\underline{coli}$ (2), the subunit
stoichiometry of the ATP synthase from thylakoids has not been
established. Since the F_0-F_1 type enzymes most likely share a common
mechanism for ATP synthesis, it is pertinent to determine the subunit
identity and stoichiometry of the enzyme complex from chloroplast
thylakoids, particularly because the composition of other eukaryotic ATP
synthases appears to be different from that found in the prokaryote $\underline{E}$.
$\underline{coli}$ (3). We have chosen the green alga $\underline{Chlamydomonas}$ $\underline{reinhardtii}$ as
our source of thylakoid membranes and CF_0-CF_1 because this organism is
suitable for mass culture, mutagenesis, and genetic analysis (4) and
serves as a model photosynthetic system for vascular plants.
    Prior to establishing the subunit identity and stoichiometry for the
$\underline{C}$. $\underline{reinhardtii}$ chloroplast ATP synthase, it is essential to first purify
an active enzyme complex. In this paper we summarize our progress
towards this goal. We have had success fractionating detergent extracts
of crude thylakoid membranes with polyethylene glycol (PEG) and
reconstituting certain fractions with bacteriorhodopsin and soybean
phospholipids to form liposomes that catalyze light dependent ATP
synthesis. The activity is sensitive to energy transfer inhibitors,
uncouplers, and anti-CF_1 serum. The larger subunits of CF_1 are clearly
visible in SDS-polyacrylamide gels of active PEG fractions. We conclude
that our solubilization procedure yields a partially purified and active
CF_0-CF_1.

2. PROCEDURE
2.1 Methods and Materials
2.1.1 Solubilization of ATP synthase activity. The solubilization
mixture contained washed, crude thylakoid membranes at 2 mg Chl/ml, 30
mM Na-Tricine (pH 8.5), 5 mM $MgCl_2$, 10 mM DTT, 1 M KCl, and 0.4% Na-
deoxycholate. The mixture was agitated for one hour and then
centrifuged for one hour (w^2=2.36x10^7). The yellowish green supernatant
was precipitated twice with PEG 3550 to remove some protein and most of
the pigment and once again to concentrate ATP synthase activity. All
steps were performed at 0-2°C. The final PEG precipitated pellet was
suspended in resuspension buffer containing 30 mM Na-Tricine (pH 8.5), 1
mM $MgCl_2$, 50 mM KCl, 10 mM DTT, 1 M KCl, 0.5 mM ATP, and 0.4% Na-
deoxycholate and stored at -80°C.

2.1.2 Liposome preparation. Liposomes were prepared in two stages. In
the first stage, purple membranes and phospholipids were sonicated for
15 min (in a bath type sonicator) in a suspension containing 5 mM Na-
Tricine (pH 8.0), 62.5 mg/ml soybean phospholipids, 0.1 M sucrose, and

Biggens, J. (ed.), Progress in Photosynthesis Research, Vol. III. ISBN 90 247 3452 5
© *1987 Martinus Nijhoff Publishers, Dordrecht. Printed in the Netherlands.*

1.25 mg/ml purple membrane (bacteriorhodopsin) to produce a clear, dark purple suspension. In the second stage, PEG fractions in resuspension buffer were added and the mixtures were frozen in a dry ice/acetone bath. After thawing at room temperature, the liposome suspensions were sonicated for 20 sec.

2.1.3 ATP synthase assay. Reaction mixtures in 200 μl final volume contained 20 μl of the liposome suspension, 30 mM Na-Tricine (pH 8.5), 2 mM ADP, 2 mM (^{32}P)-P$_i$ (ca. 4 to 7x10^6 cpm/ml), 0.5 mM DTT, 10 mM glucose, 10 U/ml hexokinase, and 1 mg/ml defatted BSA. Reactions were normally terminated after 15 min and esterified phosphate was determined as previously described (5). Light dependent P$_i$ esterification was linear for at least 50 min at 25°C.

2.1.4 Miscellaneous. Chlamydomonas reinhardtii (strains CW15, 21 gr, and 137) was cultured and crude thylakoid membranes were prepared as previously described (6), except that a nitrogen cavitation bomb or a glass bead beater was used to break the cells. Halobacterium halobium was cultured and purple membrane prepared essentially as described by Oesterhelt and Stoeckinius (7). Sodium dodecylsulphate polyacrylamide gel electrophoresis (SDS-PAGE) was performed essentially as described by Laemmli (8).

3. RESULTS AND DISCUSSION

3.1 Reconstituted ATP synthase assay. Table 1 summarizes the results of a reconstitution experiment in which liposomes, containing bacteriorhodopsin, were prepared either without (panel A) or with (panel B) protein from a 10-20% PEG fraction derived from detergent solubilized C. reinhardtii thylakoid membranes. Clearly only those liposomes that contained bacteriorhodopsin as well as the 10-20% PEG fraction were capable of light dependent phosphate esterification. That ATP synthase activity had indeed been reconstituted was shown by a product analysis using thin-layer chromatography (data not presented), in which the light dependent reaction product migrated with Glucose-6-Phosphate, and, in the absence of hexokinase, with ATP.

TABLE 1. LIGHT DEPENDENT PHOSPHATE ESTERIFICATION REQUIRES PROTEIN FROM Chlamydomonas reinhardtii

10-20% PEG Fraction	Light	CPM	Average CPM	nmol P$_i$ esterified / mg C. reinhardtii protein·min
(A) —	+	99 131 128	119	----
—	—	127 123 116	122	----
(B) +	+	4152 3634 4355	4047	22.8
+	—	140 136 150	142	0.80

The specific activity of the most active PEG fractions was somewhat variable and ranged from 10-30 nmol P_i esterified/mg/min. The highest activities were obtained when the resuspension buffer contained a high concentration of KCl (1 M), which apparently facilitated the incorporation of the ATP synthase into the liposomes (data not shown).

3.2 <u>Sensitivity of the ATP synthase activity to various inhibitors</u>. As shown in Fig. 1, the ATP synthase activity reconstituted from a 10-20% PEG fraction is sensitive to dicyclohexylcarbodiimide (DCCD) and carbonyl cyanide m-chlorophenyl hydrazone (CCCP), classic inhibitors of ATP synthesis for F_0-F_1 type enzymes. A monospecific rabbit antiserum directed against the CrCF$_1$ also inhibited the ATP synthase activity of this fraction (Fig. 1). The same concentration ranges of these inhibitors were similarly effective in inhibiting photophosphorylation catalyzed by <u>C</u>. <u>reinhardtii</u> thylakoid membranes (data not shown).

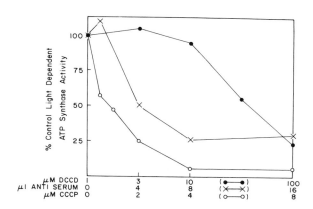

FIG. 1. SENSITIVITY OF ATP SYNTHASE ACTIVITY TO DICYCLOHEXYLCARBODIIMIDE (DCCD), ANTI-SERUM DIRECTED AGAINST THE C. reinhardtii CF$_1$, AND CARBONYL CYANIDE m-CHLOROPHENYL HYDRAZONE (CCCP).

3.3 <u>SDS-PAGE of the partially purified ATP synthase</u>. Fig. 2 shows the SDS-polyacrylamide gel patterns of various fractions of the partial purification. The larger subunits of CF$_1$, α and β, were clearly depleted from the high-speed pellet (lane D) that resulted after the centrifugation of (solubilized) membranes (lane C). The staining intensities of the α and β subunits were greatest in the 10-20% PEG fraction of the high-speed supernatant (lane H)--precisely where the most ATP synthase activity (88%) and the highest specific activity (22.2 nmol P_1 esterified/mg/min) was found--and least in the higher PEG fractions where little protein was precipitated (lanes I and J). The normalized specific activities of the 3.33-6.66%, the 6.66-10%, and the 10-20% PEG fractions were, respectively, 1, 19, and 117. However, the staining intensities of the α and β subunits for these PEG fractions clearly did not increase as rapidly as did the respective specific activities (lanes F, G, and H). Several initial experiments (data not shown) suggest the existence of a factor(s) in the lower percentage PEG fractions that inhibits the expression of ATP synthase activity by disrupting the normal incorporation of CF$_0$-CF$_1$ into liposomes.

FIG. 2. SDS-POLYACRYLAMIDE GEL ELECTROPHORESIS OF FRACTIONS FROM THE
PARTIAL PURIFICATION OF ATP SYNTHASE ACTIVITY.

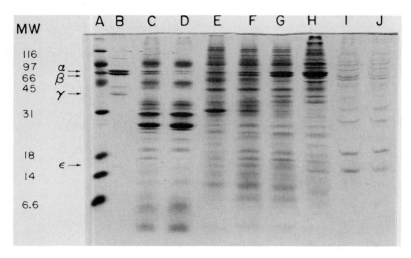

LEGEND: Lane A--Molecular weight standards with approximate kilodalton
values tabulated in the far left column; Lane B--Purified C. reinhardtii
CF$_1$, with subunits identified to the left of Lane A; Lane C--
Unsolubilized thylakoid membranes; Lane D--The high-speed pellet of
solubilized membranes; Lanes E-F are PEG fractions of the high speed
supernatant of solubilized membranes; Lane E--0-3.33% PEG, Lane F--3.33-
6.66% PEG; Lane G--6.66-10% PEG; Lane H--10.0-20.0% PEG; Lane I--20.0-
30.0% PEG; Lane J--30.0-40.0% PEG.

ACKNOWLEDGEMENTS. We thank Kathleen McGiness of W. Stoeckinius'
laboratory for H. h. strain Et1001 and culture procedures. This
research was supported in part by grants from the College of
Agricultural and Life Sciences, University of Wisconsin-Madison, and the
National Institutes of Health (GM 31384).

4. REFERENCES.

1. Pick, U. (1982) in Methods in Chloroplast Molecular Biology
   (Edelman, Hallick and Chua, eds.) pp. 863-872, Elsevier Biomedical
   Press, Amsterdam.
2. Foster, D.L. and Fillingame, R.H. (1972) J. Biol. Chem. 257, 2009-
   2015.
3. Amzel, L.M. and Pederson, P.L. (1983) Ann. Rev. Biochem. 52, 801-824.
4. Sagu, R. (1972) in Cytoplasmic Genes and Organelles pp. 49-101,
   Academic Press, New York.
5. Selman, B.R. (1976) J. Bioenerg. Biomemb. 8, 143-156.
6. Selman-Reimer, S., Merchant, S. and Selman, B.R. (1981) Biochemistry
   20, 5476-5482.
7. Oesterhelt, D. and Stoeckinius, W. (1973) in Methods in Enzymology
   (Van Vunakis and Langone, eds.) 31, 667-678, Academic Press, New
   York.
8. Laemmli, U. (1970) Nature (London) 227, 680-685.

A COMPARISON OF CHLOROPLAST COUPLING FACTOR ONE, CF_1, USING POLYCLONAL CHICKEN ANTIBODIES

Susanne Selman-Reimer and Bruce R. Selman, Department of Biochemistry, University of Wisconsin-Madison, Madison, WI 53706, USA

1. INTRODUCTION

Antibodies directed against the chloroplast thylakoid coupling factor 1, CF_1, the extrinsic membrane-bound sector of the reversible energy transducing H^+-ATP synthase complex, have been widely used to study both the enzyme structure (1-3) and subunit function and accessibility (4,5). Most antibody preparations that have been employed have been monospecific polyclonal antisera or IgG enriched fractions usually isolated from rabbits (6) although more recently some studies employing mouse hybridoma monoclonal antibodies have been reported (e.g. ref. 3). Although it is relatively easy to raise antibodies in rabbits, it usually takes several months until the animals achieve peak titre, and both the immunization and subsequent bleeding processes can be rather traumatic (7). The failure of more wide spread application of monoclonal antibodies is testament to the enormous effort required to obtain clones and subsequent tumors, a process that can require a year.

A great deal has been learned using polyclonal antisera directed against CF_1 (6). For example, there is a high degree of species cross-reactivity of polyclonal antibodies directed against holo coupling factors, first documented by Rott and Nelson (8). This is not surprising in light of the high degree of conservation in the primary amino acid sequence of the β-subunit (9,10). Apparently this is not the case for the other subunits (8,11), with the possible exception of subunit III of the membrane sector portion of the energy transducing complex (12,13), and has been used as an argument for the location of at least a portion of the active site being on the β-subunit (14). Thus, the rule has been that inhibitory antibodies directed against coupling factors almost always inhibit the catalytic activity of coupling factors from closely related species (e.g., ref. 6).

In this communication, we compare the immunological cross-reactivity of coupling factors isolated from a variety of plant species. We have raised our antibodies in chickens, which have provided a number of advantages over rabbits. We have been able to accumulate a large collection of antibody preparations in a relatively short period of time. Our collection contains antibody preparations obtained from different animals and preparations that vary temporally with respect to the immunization schedule for individual animals. Some of these preparations display some unusual immunological properties with respect to their cross-reactivities and abilities to inhibit in vitro ATPase activity and should prove to be extremely valuable tools in studies on both the structure and function of the enzyme and its subunits without necessitating the preparation of monoclonal antibodies.

2. PROCEDURE
2.1 Methods and Materials

Biggens, J. (ed.), Progress in Photosynthesis Research, Vol. III. ISBN 90 247 3452 5
© *1987 Martinus Nijhoff Publishers, Dordrecht. Printed in the Netherlands.*

2.1.1 Isolation of proteins. Four subunit, δ-less, algal chloroplast coupling factors (CF$_1$) were isolated and purified from <u>Dunaliella</u> <u>bardawil</u> (DbCF$_1$) (15), <u>Dunaliella</u> <u>salina</u> (DsCF$_1$) (15), and <u>Chlamydomonas</u> <u>reinhardtii</u> (CrCF$_1$) (16) as previously described. Five subunit vascular plant coupling factors were isolated from spinach and lettuce (SpCF$_1$ and LeCF$_1$, respectively) (17-19). Ribulose bisphosphate carboxylase/oxygenase (Rubisco) was isolated and purified from cell extracts of <u>Dunaliella</u> <u>salina</u> using the method described by Hall and Tolbert (20) for spinach Rubisco.

2.1.2 Preparation of IgG. Highly enriched IgG fractions were prepared from egg yolks as described by Polson et al. (21). The following notation has been adopted to identify the different antibody preparations:

Inoculating antigen(animal identification number/inclusive day(s) of egg production following the primary inoculation)

2.1.3 Analytical assays. Double diffusion tests were performed as described by Ouchterlony$_2$ (22). The algal coupling factors were assayed as ethanol-activated, Mg^{2+}-dependent ATPases using incubation conditions previously detailed (17). Spinach and lettuce CF$_1$ were assayed either as heat-activated, Ca^{2+}-dependent (23) or octylglucoside-activated, Mg^{2+}-dependent (24) ATPases. Enzyme Linked Immunosorbent Assays (ELISA) were performed essentially as described by Engvall (25). Sodium dodecylsulfate denaturing polyacrylamide gel electrophoresis (SDS-PAGE) was performed essentially as described by Laemmli (26). Protein transblotting of polypeptides separated by SDS-PAGE was performed essentially as described by Towbin et al. (27). Protein was determined according to the method described by Bradford (28).

2.1.4 Preparation of agarose-linked spinach and <u>C</u>. <u>reinhardtii</u> CF$_1$ and the enrichment of inhibitory antibodies directed against the DbCF$_1$. SpCF$_1$ and CrCF$_1$ were covalently linked to cyanogen bromide activated cross-linked agarose essentially as described in Pharmacia Fine Chemical AB, "Affinity Chromatography: Principles and methods". Details on the methods used for affinity adsorption chromatography will be presented elsewhere (Selman-Reimer and Selman, manuscript submitted). The affinity purified DbCF$_1$ antibody preparations were denoted as follows:

Ap(Sp)Db(360/25-39): Db(360/25-39) purified on SpCF$_1$-linked agarose
Ap(Cr)Db(360/25-39): Db(360/25-39) purified on CrCF$_1$-linked agarose

2.1.5 Miscellaneous. [γ-^{32}P]ATP was prepared as previously described (29). Spinach, <u>D</u>. <u>bardawil</u>, and <u>C</u>. <u>reinhardtii</u> CF$_1$ were treated with dicyclohexylcarbodiimide (DCCD) as previously described (30).

3. RESULTS AND DISCUSSION
3.1 Specificity of hen IgG's. An IgG preparation [Sp(432/16)] from a hen immunized with SpCF$_1$ demonstrates the high specificity of these antibodies. Whereas this preparation did not precipitate any of the algal coupling factors, including DsCF$_1$, DbCF$_1$, and CrCF$_1$ (Table I), it did precipitate coupling factors isolated from other vascular plants (e.g., lettuce, not shown). [Note that it did not cross-react with

TABLE I
SUMMARY OF OUCHTERLONY TESTS

	Test Antigen				
Antibody Preparation	DbCF$_1$	DsCF$_1$	SpCF$_1$	CrCF$_1$	Rubisco
Db(360/14-23)	+	+	+	+	-
Db(360/25-39)	+	+	+	+	-
Sp(432/16)	-	-	+	-	-
Ap(Sp)Db(360/25-39)[a]	+	ND[b]	-	ND	ND
Ap(Cr)Db(360/25-39)	+	ND	-	-	ND
Rubisco(428/19-38)	-	+	-	-	+

[a]Ap ≡ affinity purified   [b]ND = not determined

Rubisco (Table I).] As shown in Fig. 1, Sp(432/16) inhibited the detergent-activated, Mg^{2+}-dependent SpCF$_1$ and LeCF$_1$ ATPase activities but not the solvent-induced, Mg^{2+}-dependent DsCF$_1$ ATPase (nor any of the other algal ATPases tested, not shown). These results suggest that this preparation is at best only weakly cross-reactive with the algal coupling factors, a suggestion confirmed by the ELISA assay shown in Fig. 2.

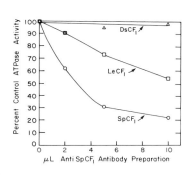

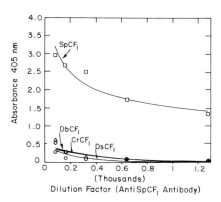

Figure 1 (Left): **Ability of a chicken IgG preparation directed against the SpCF$_1$ to inhibit the ATPase activity of various coupling factors.** Antibodies were prepared from the egg yolks of a chicken immunized with SpCF$_1$ and tested for their ability to inhibit the octylglucoside-induced, Mg^{2+}-dependent SpCF$_1$ (o-o) and LeCF$_1$ (□-□) and ethanol-induced, Mg^{2+}-dependent DsCF$_1$ (Δ-Δ) ATPase activities.

Figure 2 (Right): **Cross-reactivity as determined by ELISA assays of various coupling factors to an antibody preparation directed against the SpCF$_1$.**

A different pattern was found for two other IgG preparations [Db(360/14-23) and Db(360/25-39)] isolated from a hen immunized with DbCF$_1$. Both of these preparations precipitated all other coupling factors tested (Table I), showed strong positive signals in ELISA assays when tested against SpCF$_1$ and CrCF$_1$ (Fig. 3), a somewhat less positive signal with DsCF$_1$ (Fig. 3), but no signal with Rubisco (not shown). Both preparations were effective inhibitors of the solvent-induced, Mg^{2+}-dependent DbCF$_1$ and DsCF$_1$ ATPase activities, but neither preparation inhibited the detergent-induced, Mg^{2+}-dependent SpCF$_1$ ATPase or the solvent-induced, Mg^{2+}-dependent CrCF$_1$ ATPase (not shown).

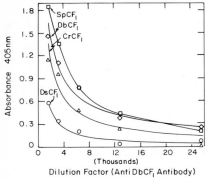

Dilution Factor (AntiDbCF$_1$ Antibody)

Figure 3: **Cross-reactivity as determined by ELISA assays of various coupling factors to an antibody preparation directed against the DbCF$_1$.**

The above results suggest that there are different classes of antigenic sites recognized by the antibodies in the IgG preparations directed against the DbCF$_1$. Clearly those sites recognized by the preparations that react with SpCF$_1$ and CrCF$_1$ are not functionally significant for those two ATPases. In order to determine if those same antigenic determinants are functionally important for the _Dunaliella_ coupling factors, the following adsorption-type experiment was performed. DbCF$_1$, SpCF$_1$, and CrCF$_1$ were chemically modified with dicyclohexylcarbodiimide (DCCD) under conditions that led to a greater than 95% irreversible loss of ATPase activity (30). Prior to assaying the DbCF$_1$ ATPase, the reaction mixtures (in the absence of untreated ATPase) containing either pre-immune IgG or immune IgG (sufficient to inhibit the DbCF$_1$ ATPase activity 50 to 80%) were pre-incubated at room temperature for 4 min in the presence of variable amounts of the DCCD-inactivated coupling factors. Thereafter, DbCF$_1$ was added and the ATPase activity assayed as usual.

Fig. 4A shows that the pre-incubation of the immune IgG preparation with DCCD-treated DbCF$_1$ (at an IgG/DCCD-treated DbCF$_1$ ratio equivalent to the final IgG/untreated DbCF$_1$ ratio; 27 μg IgG/5 μg DbCF$_1$) resulted in a greater than 90% prevention of subsequent inhibition of the DbCF$_1$ ATPase activity. These results indicate that the treatment of the (DbCF$_1$) coupling factor with DCCD does not alter its affinity for inhibitory IgG's. Fig. 4B shows the results for the same experiment in which the immune IgG preparation was pre-incubated with DCCD-inactivated SpCF$_1$. In this case the inhibition of the DbCF$_1$ ATPase activity by the IgG preparation was not relieved even upon the addition of a two fold excess of DCCD-inactivated SpCF$_1$. Similar results were found for the DCCD-inactivated CrCF$_1$ (not shown). Assuming that the treatment of

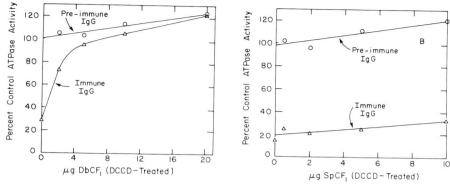

Figure 4: **Ability of DCCD-treated CF$_1$ to prevent an antibody preparation directed against the DbCF$_1$ from inhibiting the DbCF$_1$ ATPase activity.** DbCF$_1$ (Fig. 4A) and SpCF$_1$ (Fig. 4B) were treated at pH 6.5 with 2.0 mM DCCD for 20 min at 37°C as previously described (30). ATPase reaction mixtures contained either pre-immune IgG (o-o) [Db(371/1)] or immune IgG (Δ-Δ) [Db(360/25-39)] at 270 μg/mL. Increasing amounts of DCCD-treated CF$_1$ were added and the mixtures incubated for 4 min at room temperature. Thereafter untreated DbCF$_1$ was added, and the ATPase activity was determined.

either SpCF$_1$ or CrCF$_1$ does not alter the affinity of the proteins for antibodies, those antibodies that remain after absorption of the preparation with SpCF$_1$ or CrCF$_1$ appear to be specifically directed against the DbCF$_1$, and at least a fraction of those antibodies recognize unique antigenic determinants on the DbCF$_1$ that when complexed to antibodies can have a critical effect on the activity of the enzyme.

3.2 Specificity of the chicken IgG's with respect to the subunits of CF$_1$. The results described above for the specificity of the chicken antibody preparations Db(360/25-39) and Sp(432/12-27) strongly suggest that there is only partial immunological cross-reactivity between the algal and vascular plant coupling factors. This is more clearly documented in the composite results shown in Fig. 5. Fig. 5A is the silver stained gel of the subunits of the five coupling factors separated on 18% SDS-PAGE prior to transblotting onto nitrocellulose paper. (Note that in order to insure a good separation between the algal α and β-subunits, the ε-subunits were lost from the gel.) Figures 5B and 5C show immunodecorated transblots using as the primary antibody Sp(432/12-27) and Db(360/25-39), respectively. Qualitatively, the three algal coupling factors, DbCF$_1$, DsCF$_1$, and CrCF$_1$ (lanes 1, 2, and 3, respectively) and the two vascular plant coupling factors, SpCF$_1$ and LeCF$_1$ (lanes 4 and 5, respectively) appear to be immunologically very similar although some differences can be seen. For example, a comparison of the two vascular plant coupling factors shows that the spinach antibody preparation Sp(432/12-27) cross-reacts very strongly with the LeCF$_1$ α and β-subunits, but very weakly, if at all, with the LeCF$_1$ γ or δ-subunits (Fig. 5B). With respect to the algal coupling factors, the antibody preparation Db(360/25-39) cross-reacts with all

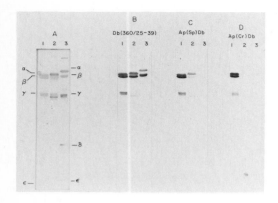

Figure 5: **Separation of CF$_1$ subunits by SDS-PAGE and immunodecoration of protein transblots after binding various chicken IgG antibody preparations.** Lanes 1, DbCF$_1$; Lanes 2, DsCF$_1$; Lanes 3, CrCF$_1$; Lanes 4, SpCF$_1$; Lanes 5, LeCF$_1$.

three of the largest subunits, α, β, and γ (as well as the smallest subunit, ϵ; see Fig. 6, lanes 1 and 2), although the cross-reactivity with the CrCF$_1$ γ-subunit appears to be weaker than with either the α, β, or ϵ-subunits (Figures 5C and 6).

More interesting, however, is the comparison between the algal and vascular plant coupling factors. Whereas the antibody preparation Sp(432/12-27) recognizes determinants on both of the two largest vascular plant coupling factor subunits, only the β-subunits of the algal coupling factors bind antibodies from this preparation (Fig. 5B). And, reciprocally, whereas the preparation Db(360/25-39) strongly cross-reacts with the three largest subunits of the algal coupling factors, only the β-subunits of the vascular plant enzymes strongly cross-react, the cross-reactivity with the α-subunits being much weaker (Fig. 5C).

These results confirm the previous findings of Rott and Nelson (9), who clearly demonstrated immunological cross-reactivity of β-subunits from coupling factors associated with a wide variety of energy transducing membranes, as would be anticipated from the observation that there is a great deal of homology in the primary amino acid sequence of β-subunits (9,10).

3.3 Characterization of affinity purified Db(360/25-39). In an attempt to more accurately define those regions on the DbCF$_1$ that, when complexed with antibodies, result in a loss of ATPase catalytic activity, antibodies in the preparation Db(360/25-39) that cross-reacted with either SpCF$_1$ or CrCF$_1$ were removed by affinity adsorption chromatography. SpCF$_1$ and CrCF$_1$ were covalently linked to CNBr activated cross-linked agarose, and the antibody preparation Db(360/25-39) was passed through columns in order to adsorb out of the preparation the non-inhibitory SpCF$_1$ and CrCF$_1$ cross-reacting antibodies. These were designated Ap(Sp)Db(360/25-39) and Ap(Cr)Db(360/25-39) for the antibody preparations after passage through the SpCF$_1$- and CrCF$_1$-linked agarose columns, respectively.

Table I shows that, after passage of the antibody preparation Db(360/25-39) through the CF$_1$-linked agarose columns, the resultant IgG solutions still retained precipitating antibodies directed against the

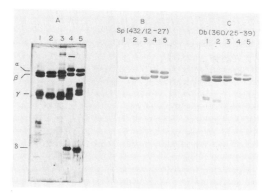

Figure 6: Comparison of the immunodecorated protein transblots of the CF_1 subunits using a chicken antibody directed against the $DbCF_1$ before and after affinity adsorption chromatography. Lanes 1, $DbCF_1$; Lanes 2, $CrCF_1$; Lanes 3, $SpCF_1$.

$DbCF_1$ but not the $SpCF_1$ or $CrCF_1$. Furthermore, not only were the cross-reacting precipitating antibodies removed from Db(360/25-39), but, as revealed in ELISA assays, virtually all of the $SpCF_1$ cross-reacting antibodies were removed from Ap(Sp)Db(360/25-39) and $CrCF_1$ cross-reacting antibodies from Ap(Cr)Db(360/25-39) (data not shown). Whereas Ap(Sp)Db(360/25-39) still retained some cross-reactivity with $CrCF_1$, Ap(Cr)Db(360/25-39) was virtually unreactive with $SpCF_1$ as detected at this level of sensitivity. As expected, these preparations still inhibited the $DbCF_1$ and $DsCF_1$ ATPase activities (not shown).

Figures 6B, 6C, and 6D respectively compare the immunodecorated protein transblots for the subunits of $DbCF_1$ (lanes 1), $CrCF_1$ (lanes 2), and $SpCF_1$ (lanes 3) using Db(360/25-39), Ap(Sp)Db(360/25-39), and Ap(Cr)Db(360/25-39) as the primary antibodies. Whereas Db(360/25-39) cross-reacts with the $SpCF_1$ α and β-subunits, virtually all of the antibodies directed against the $SpCF_1$ α-subunit and almost all of the antibodies directed against the $SpCF_1$ β-subunit have been removed from Ap(Sp)Db(360/25-39). When compared to the cross-reactivity with the $CrCF_1$ subunits, Ap(Sp)Db(360/25-39) is substantially enriched for antibodies that cross-react with the $CrCF_1$ α-subunit. Apparently there is far more immunological homology between the $CrCF_1$ and $SpCF_1$ β-subunits than between the $DbCF_1$ and $CrCF_1$ β-subunits. Similarly, there appears to be more immunological homology between the $CrCF_1$ and $DbCF_1$ α-subunits than between the algal and $SpCF_1$ α-subunits. Ap(Sp)Db(360/25-39) still contains antibodies directed against the $DbCF_1$ α, β, γ, and ϵ-subunits.

A comparison of the cross-reactivity of Db(360/25-39), Ap(Sp)Db(360/25-39), and Ap(Cr)Db(360/25-39) with the algal and vascular plant coupling factors leads to the following conclusions: (i) Not surprisingly, there is a great deal of immunological homology amongst the CF_1 β-subunits. (ii) The β-subunits of the $SpCF_1$ and $CrCF_1$ appear more similar than the $CrCF_1$ and $DbCF_1$ β-subunits. (iii) The $DbCF_1$ and $CrCF_1$ α and ϵ-subunits are more immunologically homologous than the $CrCF_1$ and $SpCF_1$ α and ϵ-subunits. (iv) There is no immunological similarity between the $DbCF_1$ and $SpCF_1$ γ-subunits, and only partial immunological similarity between the $CrCF_1$ and $DbCF_1$ γ-subunits.

Acknowledgements

This research was supported in part by grants from the College of Agricultural and Life Sciences, University of Wisconsin-Madison, and the National Institutes of Health (GM 31384).

REFERENCES

1. Lien, S., Berzborn, R.J., and Racker, E. (1972) J. Biol. Chem. <u>247</u>, 3520-3524.
2. Suss, K.-H. and Manteuffel, R. (1983) FEBS Lett. <u>153</u>, 134-140.
3. Tiedge, H., Lunsdorf, H., Schafer, G., and Schairer, H.U. (1985) Proc. Natl. Acad. Sci. <u>82</u>, 7874-7878.
4. Nelson, N., Deters, D.W., Nelson, H., and Racker, E. (1973) J. Biol. Chem. <u>248</u>, 2049-2055.
5. Berzborn, R.J., Kopp, F., and Mühlethaler, K. (1974) Z. Naturforsch. <u>29c</u>, 694-699.
6. Berzborn, R.J. and Lockau, W. (1977) In: Encyclopedia of Plant Physiology: New Series (A. Trebst, and M. Avron, eds.) pp.283-296, Springer-Verlag, Berlin-Heidelberg.
7. Hurn, B.A.L. and Chantler, S.M. (1980) In: Methods in Enzymology (H. van Vunakis, and J.J. Langone, eds.) Vol. 70, pp.104-142, Academic Press, New York.
8. Rott, R. and Nelson, N. (1981) J. Biol. Chem. <u>256</u>, 9224-9228.
9. Walker, J.E., Saraste, M., and Gay, N.J. (1984) Biochim. Biophys. Acta <u>768</u>, 164-200.
10. Zurawski, G., Bottomley, W., and Whitfield, P.R. (1982) Proc. Natl. Acad. Sci. USA <u>79</u>, 6260-6264.
11. Perlin, D.S. and Senior, A.E. (1985) Arch. Biochem. Biophys. <u>236</u>, 603-611.
12. Sebald, W., Graf, T., and Ludins, H.B. (1979) Eur. J. Biochem. <u>93</u>, 587-599.
13. Hoppe, J. and Sebald, W. (1980) Eur. J. Biochem. <u>107</u>, 57-65.
14. Senior, A.E. (1985) Current Topics in Membranes and Transport <u>23</u>, 135-151.
15. Selman-Reimer, S., Finel, M., Pick, U., and Selman, B.R. (1984) Biochim. Biophys. Acta <u>764</u>, 138-147.
16. Selman-Reimer, S., Merchant, S., and Selman, B.R. (1981) Biochemistry <u>20</u>, 5476-5482.
17. Strotmann, H., Hesse, H., and Edelmann, K. (1973) Biochim. Biophys. Acta <u>314</u>, 202-210.
18. Hesse, H., Jank-Ladwig, R., and Strotmann, H. (1976) Z. Naturforsch. <u>31c</u>, 445-451.
19. Binder, A., Jagendorf, A., and Ngo, E. (1978) J. Biol. Chem. <u>253</u>, 3094-3100.
20. Hall, N.P. and Tolbert, N.E. (1978) FEBS Lett. <u>96</u>, 167-169.
21. Polson, A., von Wechmar, M.B., and van Regenmortel, M.H.V. (1980) Immunol. Comm. <u>9</u>, 475-493.
22. Ouchterlony, O. (1962) Prog. Allergy <u>6</u>, 30-154.
23. Vambutas, V.K. and Racker, E. (1965) J. Biol. Chem. <u>240</u>, 2660-2667.
24. Pick, U. and Bassilian, S. (1982) Biochemistry <u>21</u>, 6144-6152.
25. Engvall, E. (1980) In: Methods in Enzymology (H. van Vunakis, and J.J. Langone, eds.) Vol. 70, pp.419-439, Academic Press, New York.
26. Laemmli, U. (1970) Nature (London) <u>227</u>, 680-685.
27. Towbin, H., Staehelin, T., and Gordon, J. (1979) Proc. Natl. Acad. Sci. USA <u>76</u>, 4350-4354.
28. Bradford, M.M. (1976) Anal. Biochem. <u>72</u>, 248-254.
29. Magnusson, R.P., Portis, A.R., Jr., and McCarty, R.E. (1976) Anal. Biochem. <u>72</u>, 653-657.
30. Shoshan, V. and Selman, B.R. (1980) J. Biol. Chem. <u>255</u>, 384-389.

CHARACTERIZATION OF THE MN(II) BINDING SITE OF CHLOROPLAST COUPLING FACTOR ONE: PROTON MAGNETIC RELAXATION FIELD DEPENDENCE

ALICE E. HADDY,* ROBERT R. SHARP* AND WAYNE D. FRASCH**
DEPARTMENTS OF CHEMISTRY* AND BIOLOGY,** THE UNIVERSITY OF
MICHIGAN, ANN ARBOR, MICHIGAN 48109

1. INTRODUCTION
     The metal binding sites of chloroplast coupling factor one have been studied using paramagnetic Mn(II), an effective ATPase cofactor which can be monitored by magnetic resonance techniques (1,2). The environment of bound Mn(II) can be characterized by the magnetic field dependence of the paramagnetic contribution to the nuclear magnetic resonance relaxation rate of solvent nuclei. Values for the hydration number of the bound ion, the distance between the water protons and the ion, and the correlation times for the water-ion interaction have been deduced for several metal-enzyme complexes (3). We have examined the magnetic field dependence of proton spin-lattice relaxation induced by Mn(II) bound to latent CF_1 in order to characterize the Mn(II) binding sites on CF_1 in the absence and presence of added ADP.

2. MATERIALS AND METHODS
     Spinach CF_1 was prepared and quantitated, assuming a molecular weight of 400 kD, as described previously (2). For experiments in the presence of ADP, the nucleotide was added to a concentration equal to that of the CF_1. Proton spin-lattice relaxation rates, $R_1 = 1/T_1$, were measured at Larmor frequencies between 5 and 60 MHz (2).
     The concentration of bound Mn(II) in each CF_1-$MnSO_4$ solution was calculated from the relaxation rate at 20.7 MHz using the method of relaxation enhancement (2). In this method, the fraction of bound Mn(II) is related to the observed bulk enhancement through the enhancement of the bound Mn(II). For the high-affinity Mn(II) sites of CF_1, enhancements of 9.01 (2) and 11.8 (4) in the absence and presence of ADP, respectively, were employed. The enhancement of the low-affinity sites was similar to that of the high-affinity sites (2) and, at the Mn(II) to CF_1 binding ratios used here, contributed negligibly to the observed relaxation rates.
     The frequency dependence of the observed paramagnetic nmr spin-lattice relaxation rate, $1/T_{1p}$, was analyzed according to the Solomon-Bloembergen-Morgan (SBM) equations (5). A proton-Mn(II) distance, r, of 2.77 Å was assumed, and the rotational correlation time, τ_r, for CF_1 was taken to be the Stokes-Einstein value of 0.1 μs. The electron spin relaxation rate, $1/\tau_s$, was assumed to consist of two terms representing distortion of the inner coordination sphere, one a frequency dependent term $1/\tau_{s,d}$, characterized by the distortion time,

Biggens, J. (ed.), Progress in Photosynthesis Research, Vol. III. ISBN 90 247 3452 5
© *1987 Martinus Nijhoff Publishers, Dordrecht. Printed in the Netherlands.*

τ_d, and the other a frequency independent term, $1/\tau_{s,v}$, representing vibrational modulation of the zero field splitting (4). $1/T_{1p}$ was calculated from each observed rate by subtracting the diamagnetic contribution (measured using the apoenzyme) and the contribution from free Mn(II) (calculated using the SBM equations). For the CF_1-Mn(II)-ADP solution, a correction for the contribution from CF_1-Mn(II) was also made (4).

Proton and deuteron nmr data, including experimental field dispersion profiles, were analyzed using a nonlinear least-squares fitting procedure to find four parameters: the hydration number of bound Mn(II), q; the water residence time, τ_m; and the zero field electron spin relaxation times, $\tau_{s,d}^0$ and $\tau_{s,v}^0$ (= $\tau_{s,v}$ for these field strengths). The experimental frequency range did not accurately define the ligand field distortion time, τ_d, but a minimum value of 50 ps was evident from the profile characteristics at low field. The fitting procedure was then carried out using values of τ_d between 50 and 200 ps. Values calculated for q, τ_m and $\tau_{s,v}$ were independent of τ_d. Deuteron relaxation data were used to provide an independent check on the parameters obtained from the proton data (4), resulting in a unique fit.

3. RESULTS AND DISCUSSION
3.1. Field dispersion profiles for varying Mn_b/CF_1 ratios:
The magnetic field dependence of the paramagnetic spin-lattice relaxation rate, $1/T_{1p}$, due to the binding of Mn(II) to latent CF_1 was examined for five samples with ratios of Mn(II) bound per CF_1 (Mn_b/CF_1) ranging from 0.33 to 1.8 (Figure 1). Nonlinear least-squares fits to the data, represented by the solid curves of Figure 1, gave well-defined values for the hydration number, q, and the residence time of water, τ_m, with standard deviations of 6% or less. However, the value of the zero field electron spin relaxation time due to inner sphere distortion, $\tau_{s,d}^0$, depended strongly on the inner sphere distortion time, τ_d, which was poorly defined for the frequency range of the data.

The parameters obtained from the computer fits showed distinct trends with changes in Mn_b/CF_1 (Table 1). The apparent residence time, τ_m, increased from about 120 to 360 ns as Mn_b/CF_1 increased from 0.33 to 1.8. Likewise, the apparent hydration number, q, increased from about 1.0 to 1.9 over the same range. These variations reflect an alteration in the environment of high-affinity bound Mn(II) as Mn_b/CF_1 increased. The trend in τ_m suggests that water molecules were bound more securely within the inner coordination sphere (i.e. became less labile) as the Mn(II) to CF_1 binding ratio increased.

The hydration number, q, increased from one at the lowest ratio of Mn_b/CF_1 (0.3), which probably represents the singly-bound CF_1-Mn(II) complex, to a higher value of 1.9 when the Mn_b/CF_1 ratio increased to 1.8. The observed trends probably reflect either successive occupation of two types of Mn(II) binding sites or a structural alteration of the site itself. The first option would be inconsistent with the positive cooperativity of Mn(II) binding (1,2), according to current

III.1.**121**

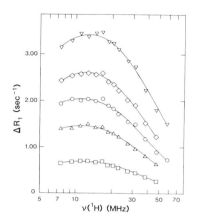

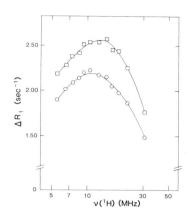

FIGURE 1. Magnetic field dependence of $1/T_{1p}$ for various
ratios of Mn_b/CF_1. Measurements were made at 24°C on
solutions of 26 or 27 μM
latent CF_1. The ratios of
Mn_b/CF_1 were: □, 0.33; Δ,
0.70; O, 1.0; ◇, 1.3; ▽, 1.8.

FIGURE 2. Magnetic field
dependence of $1/T_{1p}$ for
CF_1-Mn(II) with and without
ADP. Measurements were
made at 26°C on solutions
of 31 μM CF_1 with 35 μM
$MnSO_4$: O, no ADP; □, 31 μM
ADP.

theory. Thus the observed increase in q suggests that an
alteration in the Mn(II) binding sites took place as a result
of cooperative binding such that the hydration number
increased from one in the singly-bound Mn(II)-enzyme to two in
the multiply-bound Mn(II)-enzyme.

3.2. Field dispersion profiles in the presence of ADP:
    The field dispersion profile of $1/T_{1p}$ due to Mn(II) bind-
ing to CF_1 in the absence of added nucleotide was compared to
an identically prepared sample containing one molar equivalent
of ADP (Figure 2). Only a small shift in the $1/T_{1p}$ maximum
was observed, indicating little change in the electron spin
relaxation processes. The presence of ADP augmented the
relaxation rate at all frequencies. Nonlinear least-squares
fits revealed an apparent increase in the hydration number, q,
of about 30% in the presence of ADP (Table 1). This is
surprising since, if anything, one would expect a decrease in
q due to the displacement of exchangeable water. This
increase in q is reflected in the increase in the enhancement
at 20.7 MHz from 9.01 (2) to 11.8 (4). It should be noted
that the calculated hydration number, q, is an apparent value
based on an assumption of constant distance, r, from Mn(II) to
water protons (2.77 Å). Variations in r could account for some
change in the apparent value of q. However it is clear from
the physically reasonable range of r that ADP did not displace

TABLE 1. Nonlinear least-squares calculations of hydration numbers, q, and residence times, τ_m, for CF_1-Mn(II) samples

Mn_b/CF_1	ADP	τ_m	q
0.33	−	124 (±6%)	0.99 (±3%)
0.70	−	194	1.26
1.0	−	232	1.41
1.3	−	234	1.43
1.8	−	357	1.88
0.88	−	192 (±6%)	1.28 (±3%)
0.85	+	192 (±7%)	1.67 (±4%)

For all fits the values of τ_d and $(1/\tau_{s,v} + 1/\tau_r)$ were 50 ps and 10^8 sec^{-1}, respectively. Errors represent standard deviations generated by the least squares fitting procedure.

exchangeable water molecules at the high-affinity Mn(II) binding site.

The observation that ADP binding to CF_1-Mn(II) did not displace exchangeable waters from the coordination sphere of the Mn(II) cofactor suggests that the metal ion did not directly participate in nucleotide binding. It is possible that ADP coordinated the metal by displacing non-exchangeable water or liganding groups from the protein. In this case, however, we would have expected a much larger perturbation of the magnetic interaction at the Mn(II) site, particularly in the correlation times characterizing electron spin relaxation and in the residence time of exchangeable water. Thus the nmr data are not easily reconciled with binding models involving direct coordination of Mn(II) to exchangeable ADP.

4. ACKNOWLEDGEMENTS

This work was supported by grants from the U.S. Department of Agriculture (83-CRCR-1-1339) to W.D.F. and R.R.S. and from the Phoenix Memorial Research Foundation to W.D.F.

REFERENCES
1 Hiller, R. and Carmeli, C. (1985) J. Biol. Chem. 260, 1614-1617.
2 Haddy, A.E., Frasch, W.D. and Sharp, R.R. (1985) Biochemistry 24, 7926-7930
3 Dwek, R.A. (1973) in Nuclear Magnetic Resonance (NMR) in Biochemistry, pp. 247-284, Clarendon Press, Oxford
4 Haddy, A.E., Frasch, W.D. and Sharp, R.R., submitted
5 Bloembergen, N. and Morgan, L.O. (1961) J. Chem. Phys. 34, 842-850.

KINETICS OF Mn^{2+} BINDING TO CF_1
C. Carmeli, R. Hiller[1], P. D. Boyer[2].

1. INTRODUCTION

The uncertainty in the identification of the three active sites in CF_1 (1) required the use of divalent metal ions as probes to the active sites. Divalent metal ions are essential for catalysis and are also involved in nucleotide binding (1). Divalent metal-ATP complex probably is the true substrate for ATPase activity in CF_1 (2,3). Various divalent metal-ATP complexes were found to serve as substrates among which MnATP was as effective as MgATP which is the in vivo substrate. Using Mn^{2+} as probe we have shown that CF_1 contains six binding sites: three loose noninteracting sites and three tight sites with strong positive cooperative interaction (4). The three tight sites were suggested to be in the active sites of the enzyme as covalent modification of an arginine by naphthylglyoxal which caused inhibiton of catalysis (5) also inhibited the cooperative interaction among the Mn^{2+} binding sites. Nucleotides decreased the Ka of Mn^{2+} binding and anions such as carbonate or maleate which are effectors of the activity, similarly decreased the Ka for binding and the Km for MnATP as substrate (3). The Ka for binding to the tight sites was also similar to the Ki of free Mn^{2+} as inhibitor of ATPase activity.

The positive cooperative interaction in Mn^{2+} binding is expressed in a 20 fold decrease in the apparent rate constant for the preseady state acceleration of ATPase activity (6). The acceleration of ATPase activity is probably a result of negative cooperative interaction among the three active sites of the enzyme induced by hydrolysis of ATP. Cooperative interaction was suggested to be essencial for a sequential mechanism of catalysis (7). In this work we measured the kinetcs of Mn^{2+} binding and release from the three tight binding sites. The cooperative interaction among the binding sites was expressed in the kinetics of binding and release of the cations and was found to be accelerated following the activation of the enzyme.

2. PROCEDURE

2.1. Enzyme preparation: CF_1 was isolated either from lettuce chloroplasts (4) or from spinach chloroplasts (8). Stored CF_1 (5,) was passed through a centrifuged G-50 sephadex column equilibrated with 40 mM,HEPES-NaOH,pH8 and 1mMATP. Following incubation for 1 hr the enzyme was passed twice through centrifuged G-50 sephadex column equilibrated with 40 mM HEPES-NaOH,pH8. ATPase activity was determined (3) following heat activation (9). Protein concentration was determined by UV absorption (9).

2.2 Assay of $^{54}Mn^{2+}$ binding: CF_1,15uM was incubated in a medium containing 1mM $^{54}MnCl_2$ (containing 10^6cpm/ml), 40mM HEPES-NaOH,pH8 for time periods as indicated at $20^{\circ}C$. Aliquots of 100ul were passed through 1ml centrifuge G-50 sephadex column and radioactivity was counted in the effluent.

2.3 Rapid Kinetic Measurement of Mn^{2+} binding: Time dependent changes of the concentration of the free Mn^{2+} were monitored by Varian E-109 EPR spectrometer interphased with Nicolet 1280 computer. The absorption of the free Mn^{2+} was recorded with the field set at 3419 G,

----

[1]Department of Biochemistry, Tel-Aviv University, Tel-Aviv, Israel.
[2]Department of Chemistry and biochemistry, University of California, Los Angeles, CA 90024, USA.

modulation amplitude of 25 and microwave power of 10mV. The free Mn^{2+} gives a sharp absorption spectrum while the CF_1 bound Mn^{2+} gives an extremely broad spectrum at the X-band frequency region. Solution of approximately 40uM CF_1 and 40mM HEPES-NaOH,pH8 was mixed in a rapid mixing device (10) with a solution of 100uM $MnCl_2$ (or as indicated) and 40mM HEPES-NaOH,pH8.

3. RESULTS AND DISCUSSION

The kinetics of substrate binding and the thermodynamic dissociation constants are important aspects of the mechanism of catalysis. In a multibinding site enzyme such as CF_1 the knowledge of these parameters is essential for the determination of the involvement of a given site in catalysis or in regulation. In this work we measured the kinetics of binding and release of divalent metal ions. The metal ions are good probes as they bind with ATP to form the true substrate for ATPase (3). In steady state measurements only small changes in binding during time periods longer than a few minutes after mixing of CF_1 with Mn^{2+} were observed (4). Different results were obtained when binding of $^{54}Mn^{2+}$ was measured by column centrifugation. One Mn^{2+} per CF_1 binds within a minute and a second ion binds during the next 30 minutes (Fig. 1). while the steady state method measures

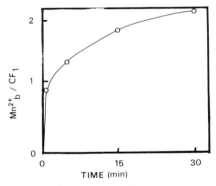

FIG.1. Binding of slowly released Mn^{2+} to CF_1. Time course of $^{54}Mn^{2+}$ which remained bound (M_b) following the passage of CF_1 through centrifuged G-50 sephadex column.

all the binding sites, the column centrifugation technique determines only binding sites from which Mn^{2+} is slowly released. It seems that Mn^{2+} binds quickly to all 6 sites of CF_1. The bound Mn^{2+} induces time dependent conformational changes in the enzyme that cause slower release of the ions from some of the binding sites. Evaluation of the rate of Mn^{2+} release was done by immediately passing the eluted enzyme through a second centrifuged sephadex column. Only residual $^{54}Mn^{2+}$ was left on the enzyme following the second column. Most of the bound

$^{54}Mn^{2+}$ was released when the enzyme was passed through a column preequilibrated with Mn^{2+} instead of HEPES; indicating that the exchange was faster then the release of the bound Mn^{2+}. In similar experiments it was found that up to two Mg^{2+} ions bind to CF_1 (11).

Measurements of the fast binding and release of Mn^{2+} required the use of a technique with a better time resolution. Time resolution down

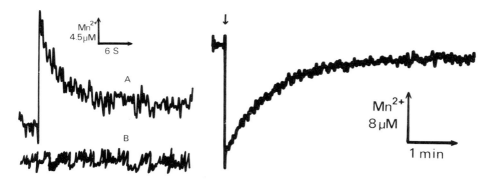

FIG.2. Time course of Mn^{2+} binding to CF_1. Latent (A) or heat activated (B) CF_1 were rapidly mixed with $MnCl_2$. The change in absorption was monitored by EPR spectrometer.

FIG.3. Time course of the release of bound Mn^{2+} from CF_1. CF_1 which was preincubated with $MnCl_2$ was mixed with $MgCl_2$ and the absorption measured.

to 8mS was obtained by the recording of the changes in the concentration of free Mn^{2+} by EPR spectrometer following rapid mixing. A fast increase in the concentration of the free Mn^{2+} which occurs during the mixing of the latent CF_1 with $MnCl_2$ (upward deflection) is followed by a time resolved binding of the cation (Fig. 2A). Measurements of the amount of bound Mn^{2+} at the end of the run indicated that part of the cation binds faster than the response time of the instrument and therefore is not recorded. Indeed when heat activated CF_1 is rapidly mixed with $MnCl_2$ no transiet change is recorded (Fig. l.B) Similar binding capacity was determined for both latent and the activated CF_1. Therefore, the lack of recorded changes is an indication that the kinetics of binding of Mn^{2+} to the heat activated CF_1 is faster than 8 mS which is the time resolution of the measuring system. It is also faster than the slow phase but could be similar to the kinetics of the faster phase of Mn^{2+} binding to the latent enzyme.

The kinetics of the release of Mn^{2+} was measured by rapidly mixing $MgCl_2$ with CF_1 to which Mn^{2+} was bound by preincubation with $MnCl_2$. The Mg^{2+} ions do not give a signal in the EPR. Therefore, the exchange of bound Mn^{2+} by Mg^{2+} ions was recorded as an increase in the absorption. The rate of Mn^{2+} release was slower than the rate of its binding to the latent enzyme (Fig. 2). In the presence of 3.3 fold excess of Mg^{2+} to Mn^{2+} a time course of the release of Mn^{2+} from 1.5 sites was recorded. Calculations indicate that practically all the released Mn^{2+} was recorded.

It was reasonable to assume that the observed slow kinetics of Mn^{2+} binding to the latent enzyme and the slower kinetics of Mn^{2+} release are a result of conformational changes in the enzyme which cause changes in the binding properties. These could be the changes that induce the cooperative interation among the three tight Mn^{2+} binding sites in CF_1 (4). A mechanism is suggested to which the data were fitted. The mechanism assumes that the metal free CF_1 (F) could bind Mn^{2+} (M) to one site exciting a reversible confromational change (E*M) which is expressed in a decrease in the dissociation constants

of the metal binding to all three sites. In the metal free enzyme one
of the sites is in a different conformation than the others. In this
conformation it has the capacity for faster rate of excitation of the
structural change on binding of metal than the two other sites. This
difference causes a preference in the sequence of binding between the
first and the rest of the sites. The second and the third sites
undergo additional decrease in their dissociation constants on binding
of metal ions. These changes occur however independently.

$$E+M \underset{K_{-1}}{\overset{K_1}{\rightleftharpoons}} EM \underset{K_{x-1}}{\overset{K_{x1}}{\rightleftharpoons}} EM^*+M \underset{K_{-2}}{\overset{K_2}{\rightleftharpoons}} EM_2 \underset{K_{x-2}}{\overset{K_{x2}}{\rightleftharpoons}} EM_2^{**}+M \underset{K_{-3}}{\overset{K_3}{\rightleftharpoons}} EM_3^{**} \underset{K_{x-3}}{\overset{K_{x3}}{\rightleftharpoons}} EM_3^{***}$$

K_1, K_2, K_3 and K_{-1}, K_{-2} and K_{-3} are the respective kinetic association
and dissociation constant of the three binding sites. K_{x1}, K_{x2}, K_{x3} and
K_{x-1}, K_{x-2} and K_{x-3} are the respective excitation and relaxation constants of the changes in the dissociation constants for each of the
three sites. The fit of the preliminary data indicated that the suggested mechanism can account for the observed data. The calculated
thermodinamic dissociation constants were in harmony with the measured
apparent dissociation constants for the three cooperative sites in
CF_1 (4).

The approximate kinetic association constants were in the order
of diffusion control reaction. K_{x1}, K_{x2}, K_{x3}, were 10^{3i} sec^{-1} and
3 sec $^{-1}$ for the first and the two others, respectively while K_{x-1}, K_{x-2},
and K_{x-3} were $5x10^{-1}$ sec^{-1} and $2x10^{-1}$ sec^{-1} for the first and the two
others, respectively. The kinetics of Mn^{2+} release was also similar to
the kinetics of the presteady state acceleration of the rate of ATPase
activity induced by ATP in metal bound CF_1 (6).

Analysis of the kinetics of Mn^{2+} binding yielded information on
the kinetics of the conformational changes transmitted among the three
binding sites. The kinetics of this cooperative interaction, which
probably occurs among the three active sites of the enzyme, gives a
detailed analysis of an important control mechanism and possibly also
of some aspects of the mechanism of catalysis (6).

REFERENCES
1. McCarty, R. E., and Carmeli, C (1982) in Photosynthesis: Energy
   Conversion by Plants and Bacteria (Govindjee, ed.) Vol. 1, pp.
   647-695, Academic Press, New York.
2. Hochman, Y., and Carmeli, C. (1981) Biochemistry 20, 6278-6292.
3. Hochman, Y., and Carmeli, C. (1981) Biochemistry 20, 6293-6297.
4. Hiller, R., and Carmeli, C. (1985) J. Biol. Chem. 260, 1614-1617.
5. Takabe, T., Debenedetti, E., and Jagendorf, A. T. (1982) Biochim.
   Biophys. Acta 682, 11-20.
6. Carmeli, C., Lifshitz, Y. and Gutman, M. (1981) Biochem. 20,
   3940.
7. Smith, L. T.., Rosen, G. and Boyer, P. D. (1983) J. Biol. Chem.
   258, 10887-10894.
8. Binder, A., Jagendorf, A. T., and Ngo, E. (1978) J. Biol. Chem.
   253, 3094-3100.
9. Farron, F., Racker, E. (1970) Biochem. 9, 3829.
10. Califso, D. S. and Habbell, W. L. (1982) Biophys. J. 39, 263-272.
11. Girault, G. and Galmiche, J. M., Lemaire, C. and Stulzafz, O.
    (1982). Eur. J. Biochem. 128, 405-411.

GENERATIVE AND DISSIPATIVE PATHWAYS OF THE PROTONMOTIVE FORCE IN PHOTO-
TROPHIC BACTERIA

J B JACKSON, J F MYATT, M A TAYLOR and N P J COTTON
DEPARTMENT OF BIOCHEMISTRY, UNIVERSITY OF BIRMINGHAM, PO BOX 363,
BIRMINGHAM B15 2TT, UK

1. INTRODUCTION
    In Rhodopseudomonas capsulata growing phototrophically, the major
chemiosmotic proton circuit across the cytoplasmic membrane is comprised
of a Δp generator, the cyclic electron transport chain and a Δp consumer,
the ATP synthase [1]. Tight controls over the generative and dissipative
ionic currents are essential to the energy economy of the cell. In this
report we show that the diodic nature of the cytoplasmic membrane,
formerly revealed by light intensity reduction [1,2], can also be demonstrated by titration with electron transport inhibitors. This property
leads to an exaggerated differential effect of antimycin and myxothiazol
on the maximum value of $\Delta\psi$ observed in steady-state light [3,4,5]. It is
also shown that the conductance properties of the cytoplasmic membrane
change with the specific growth rate of the bacterial culture while under
anabolic substrate limitation. Evidence is presented for a futile H^+
cycle across the cytoplasmic membrane whose function might be to
dissipate excess energy whilst permitting rapid turnover of photosynthetic electron transport.

2. PROCEDURES
    Rps. capsulata strain N22 was grown phototrophically either in batch
culture as described [1,2] or in continuous culture in a chemostat. In
the latter malate was made the growth limiting substrate by reducing its
concentration in the reservoir to 5 mM from the medium described [6]. It
was ensured that the incident light intensity was saturating and rigid
precautions were taken to ensure strict anaerobiosis.
    Intact cells were prepared from batch culture or from continuous
cultures as described [2]. Chromatophores were isolated as in [1].
Analysis of the dependence of the membrane ionic current (J_{dis}) on $\Delta\psi$
was carried out as in [7]. Briefly, intact cells or chromatophores in
anaerobic suspension were illuminated until $\Delta\psi$ indicated by the
carotenoid electrochromic absorption change reached a constant value
(0.1- 4s). J_{dis} was taken to be proportional to the initial rate
of decay of the absorption change upon darkening the suspension. The
value of $\Delta\psi$ was decreased in separate experiments either by reducing the
actinic light intensity or by the addition of electron transport
inhibitor. Experiments were performed in conditions in which $\Delta\psi$ was the
sole contributor to Δp.

3. RESULTS AND DISCUSSION
3.1 Inter-relations between electron transport rate, H^+/e ratio, J_{dis}
    and $\Delta\psi$ revealed by inhibitor titration
    Fig. 1 shows the dependence of J_{dis} on $\Delta\psi$ in chromatophores as
    revealed by titration with either myxothiazol or antimycin. The
    diodic relationship is not influenced by the nature of the inhibition

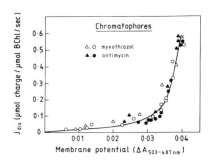

FIGURE 1. The dependence of membrane ionic current on Δψ as revealed by titration with myxothiazol and antimycin in chromatophores.

The symbol sets represent duplicate experiments. Each symbol is the result of a separate experiment. The suspension medium was 10% sucrose, 50 mM KCl, 8 mM $MgCl_2$, 50 mM tricine-KOH (pH 7.6), 0.5 mM NADH, 0.5 mM K^+ succinate. Bchl = 10 μM. J_{dis} was calibrated as described [13].

and is similar to that observed when Δψ is progressively reduced with light intensity [1,2]. Thus in these conditions there is no evidence for interaction between the primary ion pump and the dissipative current pathways across the membrane other than through Δp.

It is evident from Fig. 1 that lower values of Δψ are reached with myxothiazol than with antimycin. This is better illustrated in Fig.2 where Δψ is plotted against inhibitor concentration. With both

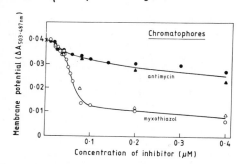

FIGURE 2. The inhibition of light-induced Δψ by myxothiazol and antimycin. See Fig. 1.

reagents inhibition is sigmoidal and a component of inhibition is reached approximately at equivalence with the reaction centre concentration (0.1 μM). However in confirmation of the results Remennikov and Samuilov [3,4] and Kotova et al. [5] who used tetraphenylboron uptake as an indicator, myxothiazol lowers Δψ more effectively than does antimycin.

Because measurements were performed in steady state where J_{dis} = J_E x n, data from the experiments described in Figs. 1 and 2 can also be used to illustrate the dependence of the product of electron transport rate (J_E) and H^+/e^- stoichiometry (n) on the concentration of inhibitor (Fig. 3). Again inhibition is sigmoidal [cf 8] and again myxothiazol is the more potent inhibitor. However the critical difference between Figs. 2 and 3 is that inhibition of protonmotive

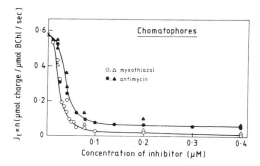

FIGURE 3. The inhibition of J_E x n by myxothiazol and antimycin. See Fig. 1.

activity (J_E x n) by both myxothiazol and antimycin is much more pronounced than inhibition of $\Delta\psi$.

This difference can be explained by an appreciation of the diodic nature of the membrane conductance (Fig. 1) and by recognising that the cytochrome b/c_1 complex is capable of limited rates of electron transport even in the presence of saturating titres of antimycin. Following Rich's suggestion [9] for the operation of the cytochrome b_6/f complex in thylakoids, we propose that when the UQ reductase site of the chromatophore b/c_1 complex is blocked with antimycin [10] the UQH_2 oxidase can turn over slowly by directing both reducing equivalents through the Rieske FeS centre. This curtailed pathway would operate with an H^+/e^- ratio of one in contrast to the stoichiometry of two in uninhibited electron transport by a modified Q cycle [11]. On this basis we calculate (Fig. 3) that the electron transport rate is reduced from 0.29 to 0.06 µmol/µmol BChl/sec by antimycin. This limited rate still includes the electrogenic reactions of the photosynthetic reaction centre and therefore still contributes to the formation of $\Delta\psi$. Because the membrane is diodic even this low rate of electrogenic electron transport is capable of maintaining a substantial $\Delta\psi$. In contrast inhibition of UQH_2 oxidase by myothiazol more completely inhibits cyclic electron transport and the generation of $\Delta\psi$ is minimal.

3.2 <u>Evidence for the operation of a gated ionophore in the cytoplasmic membranes of bacteria grown with limiting concentrations of anabolic substrate</u>

It has long been recognised that the energy-yielding reactions in growing bacteria might not be tightly coupled to the needs of anabolism. The concepts of uncoupled growth, energy spillage, overflow metabolism have been propounded (e.g. [12]). A number of mechanisms have been proposed to account for such phenomena including the possibility of futile cycles. In the experiments described below evidence has been sought for the operation of futile cycles of H^+-translocation. During phototrophic growth of <u>Rps</u>. <u>capsulata</u> malate can fullfil a strictly anabolic role since photosynthetic electron transport is cyclic and proceeds independently of carbon metabolism (contrast chemoheterotrophic growth on malate). Conditions were established to ensure that the concentration of malate in the chemostat was growth-limiting and that light intensity was saturating. It was then reasoned that reduction of the chemostat dilution rate (equals specific growth rate in the steady state) would lead to increased activity of existing metabolic overflows.

Fig. 4 shows the results of experiments in which $J_{dis}/\Delta\Psi$ curves were monitored in cells which had been harvested from the steady state chemostat at high and low dilution rate.

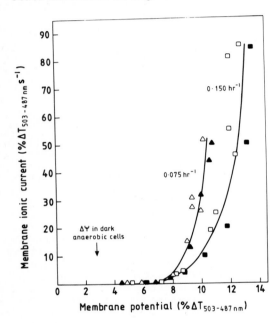

Figure 4. The dependence of J_{dis} on $\Delta\Psi$ in intact cells harvested from a steady-state chemostat operating at a dilution rate of either 0.075 hr⁻¹ or 0.15 hr⁻¹

The two symbol sets represent the results of duplicate experiments from separate runs of the chemostat. Cells were harvested, washed and resuspended in 10 mM Na-phosphate pH 7.0 to give BChl = 10 µM. Light intensity was progressively reduced to give the experimental profiles.

In separate experiments (not shown) it was found that the calibration factors relating the amplitude of the electrochomic absorption change to $\Delta\Psi$ and the rate of change of the absorption change to membrane current were similar to chromatophores prepared from the high growth rate and low growth rate cells. Therefore it can be concluded that the differences between high and low growth rate cells in Fig. 4 reflect real differences in the conductance properties of the membrane and not trivial differences in the electrochromic response of endogenous pigments.

Fig. 5 shows the effect of the ATP synthase inhibitor venturicidin on the $J_{dis}/\Delta\Psi$ profiles of the same high and low growth rate cells analysed in Fig. 4. It was established (not shown) that the venturicidin concentration was sufficient to inhibit completely light-induced changes in the ATP levels in these cells. The pronounced decrease in J_{dis} at constant $\Delta\Psi$ resulting from venturicidin treatment in both sets of cells shows that a substantial fraction of

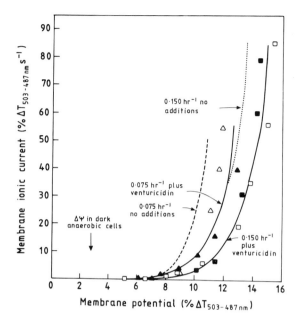

FIGURE 5. The effect of venturicidin on the $J_{dis}/\Delta\Psi$ profiles of intact cells from a chemostat operating at a dilution rate of either 0.075 hr^{-1} or 0.15 hr^{-1}.

Conditions as in Fig. 4 except that venturicidin was present at 5 μg/ml. The dashed and dotted lines are transcribed from Fig. 4.

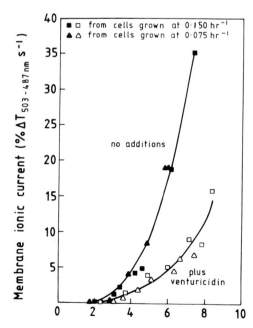

FIGURE 6. $J_{dis}/\Delta\Psi$ profiles in the presence and absence of venturicidin in chromatophores prepared from cells grown at 0.075 hr^{-1} or 0.15 hr^{-1} in a chemostat.

Conditions as Fig. 1 except that chromatophores were prepared from cells harvested from a chemostat at the growth rate shown. Where indicated venturicidin was present at 0.1 μg/ml. The profiles were obtained through progressive reduction in light intensity. Open and closed symbols give duplicate experimental data from separate chromatophore preparations.

the membrane ionic current had been proceeding through the ATP synthase. However even in the presence of venturicidin, high values of J_{dis} were observed: the threshold region for increased current flow was apparently shifted to higher $\Delta\psi$. Moreover the differences in the $J_{dis}/\Delta\psi$ curves between the high and low growth rate cells were preserved in the presence of venturicidin.

These results are interpreted to mean that another major pathway for the dissipation of $\Delta\psi$ becomes dominant when the F_0 channels of the ATP synthase are blocked. The threshold for this pathway, a gated protonophore, is slightly higher than that required to activate the ATP synthase. In cells whose growth rate is heavily restricted by the supply of anabolic substrate the threshold for the gated protonophore is lowered to permit more rapid dissipation of $\Delta\psi$ and allow continued turnover of photosynthetic electron transport.

In chromatophores isolated from bacteria taken from high and low growth rate cultures the venturicidin-sensitive component of J_{dis} is still in evidence but the venturicidin-insensitive current carrying component is not detectable (Fig. 6). Moreover the conductance properties of the two sets of chromatophores are similar. It seems that a soluble component responsible for activation of the gated protonophore is lost on chromatophore preparation.

ACKNOWLEDGEMENTS

This work was supported by a grant from the Science and Engineering Research Council.

REFERENCES
1   Clark, A.J., Cotton, N.P.J. and Jackson, J.B. (1983) Biochim. Biophys. Acta 723. 440-453
2   Jackson, J.B. (1982) FEBS Lett. 139, 139-143
3   Remennikov, V.G. and Samuilov, V.D. 1979) Biochim. Biophys. Acta 548, 216-223
4   Remennikov, V.G. and Samuilov, V.D. (1980) Arch. Microbiol. 125, 271-275
5   Kotova, E.A., Oleskin, A.V. and Samuilov, V.D. (1983) Photobiochem. and Photobiophys. 6, 211-221
6   Aiking, H and Sojka, G. (1979) J. Bact. 139, 530-536
7   Jackson, J.B. and Nicholls, D.G. (1986) Methods in Enzymol. 127 (Packer, L., ed.) pp. 557-577, Academic Press, Orlando, Florida
8   Kroger, A. and Klingenberg, M. (1973) Eur. J. Biochem. 39, 313-323
9   Rich, P.R. (1984) Biochim. Biophys. Acta, 768, 53-79
10 Van den Berg, W.H., Prince, R.C., Bashford, C.L., Takamiya, K., Bonner, W.D. and Dutton, P.L. (1979) J. Biol. Chem., 254, 8594-8604
11 Crofts, A.R., Meinhardt, S.W., Jones, K.R. and Snozzi, M. (1983) Biochim. Biophys. Acta 723, 202-218
12 Tempest, D.W. and Neijssel, O.M. (1984) Ann. Rev. Microbiol. 38, 459-486
13 Cotton, N.P.J., Clark, A.J. and Jackson, J.B. (1984) Eur. J. Biochem. 142, 193-198

COMPLETE TRACKING OF PROTON FLOW MEDIATED BY CFO-CF1 AND BY CFO

WOLFGANG JUNGE, GERALD SCHOENKNECHT AND HOLGER LILL
Biophysik, Fachbereich Biologie/Chemie, Universitaet Osnabrueck, Postfach
4469, D-4500 Osnabrueck, F.R.Germany

INTRODUCTION

We investigated proton flow through the integral chloroplast ATP synthase, CFO-CF1, and also through its channel portion, CFO, alone. Proton pumps of thylakoids were pulse stimulated by single-turnover flashes of light and the backflow of protons was completely tracked by measurements in three subcompartments, in the lumen (pH-transients via neutral red), in the membrane (electric transients via electrochromism) and in the suspending medium (pH-transients via hydrophilic indicator dyes). It was noteworthy that these three ways to follow proton flow yielded consistent results, in other words, a proton which had left the lumen was detectable as a charge crossing the thylakoid membrane and as a proton appearing in the medium.

By this approach we studied two questions debated in the literature: (A) Do protons released by the pumps (e.g. water oxidation) and taken up by the ATP synthase pass through the lumen? Or are pumps and synthases linked by special proton conducting domains distinct from the lumen (e.g. 1). This question is of particular interest as there are proton buffering domains burried in the thylakoid membrane (e.g.1-3) that can transiently trap pumped protons from water oxidation (4). These domains extend over several protein molecules (5) and they possibly bear on photophosphorylation (6). The other question: (B) Which is the protonic unit conductance of CFO? Published figures for the homologous channels from different bacteria were in the range of some percent of a femto-Siemens (see 7 for references) equivalent to the turnover of in the order of ten protons per second which is much too low to support high rates of proton translocation as observed under photophosphorylation.

MATERIALS AND METHODS

Thylakoids were prepared from two week old pea seedlings according to standard procedures. Flash spectrophotometric experiments were performed in standard equipment. A thylakoid suspension contained in an optical cell was excited with flashing light. pH-transients in the lumen and in the medium were obtained as the difference between transient absorption changes measured with and without addition of an appropriate pH-indicating dye. Neutral red (with BSA added, wavelength 548nm) served as specific and selective indicator of pH-transients at the lumenal interface of thylakoid membranes as previously established (8,9). We detailed the limits of its applicability and dealt with recent criticism elsewhere (10). pH-transients in the suspending medium were measured via either of two hydrophilic pH-indicators, phenol red (pH 7.8, wavelength 559nm) and m-cresol purple (pH 8.3, 575nm). Transients of the transmembrane voltage were measured via electrochromism of intrinsic pigments (522nm) (11, reviewed in Ref.s 12,17). The following conditions differed depending on the application to questions A and B:

(A) Pathway of protons. Thylakoids were suspended at 10 μM chlorophyll in water with the pH adjusted to pH 8.3 by addition of NaOH. 10 μM methyl viologen and 6 U/ml superoxide dismutase were added as electron acceptors. The total proton buffering capacity of the suspension did not exceed 30 μM/pH. If indicated the following components were added: 20-60 μM phosphate, 10-30 μM ADP, 50 μM tentoxin, and 7 μM DCCD (10 min incubation). Samples were excited with one or two groups of short flashes (3ms spacing within each group and 125ms spacing between two groups in a pair) at 0.1 Hz repetition rate. 20 signals were averaged. m-cresol purple served as indicator of pH-transients in the medium.

(B) Unit conductance of CFO. For CF1 extraction thylakoids were incubated in low salt media with EDTA. Aliquods of concentrated thylakoids (typ. 0.5 mg/ml chlorophyll) were incubated with 1mM EDTA,$pH 7.8 for 10 min on ice and then centrifuged for 10 min at 20,000xg ("1st EDTA wash"). In some cases this incubation was repeated once ("2nd EDTA wash"). The degree of CF1-depletion was determined by rocket-immunoelectrophoresis both for the pellet and the supernatant. Phenol red served as indicator of pH-transients in the medium, 200 μM hexacyanoferrate was the electron acceptor and the pH was 7.3.

RESULTS AND DISCUSSION

(A) The pathway of protons from water oxidation into CFO-CF1.

CHARGE-AND PROTON FLOW WITH P,(60μM) AND ADP (20μM)

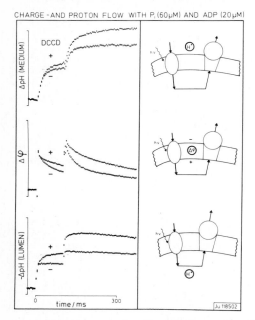

FIGURE 1: The right hand side of Fig.1 illustrates the thylakoid membrane with light driven (wavy arrow) proton pump, ATP synthase, proton flow (straight arrows) and three observables (encircled), namely, protons in the medium (top), transmembrane voltage (middle) and protons in the lumen (bottom). The left hand side shows the transient changes of these observables under excitation with two groups of three closely spaced flashes and in the presence of ADP and P. The traces with greater extent were obtained with CFO blocked by covalent modification with 7 μM DCCD (14).

The traces in the bottom revealed the following: Upon firing of the first group of three flashes water oxidation and photosystem II turned over three times while photosystem I, due to the close spacing between flashes, turned over approximately once. In Fig.1 (bottom) this was reflected in a rapid three-stepped ejection of protons by water oxidation (see Fig. 4 for higher time resolution), and by lesser extent of slow proton release by plastoquinone oxidation (see Ref.15 for kinetics of proton release). Similar behaviour was observed after firing of the second group. In the absence of the DCCD-block to CFO the slow component of the lumenal acidification was diminished. This was interpreted as a compensation of slow proton release by slow backflux of protons mediated by CFO-CF1 in the presence of ADP and P (see Fig. 4) and mainly driven by electric force (see 19,20). The correctness of this interpretation could be infered from inspection of the middle and top traces in Fig.1. The middle traces showing the generation and decay of the transmembrane voltage revealed extra charge flow if CFO was not blocked by DCCD. The top trace showing the alkalinization of the medium revealed lesser extent of alkalinization, if DCCD was absent, which was again compatible with partial backflux of protons into the medium during the slow rise of the alkalinization (see Refs. 13, 19 and 20 for the origin of the slow rise.) (We should mention that the pH-transients in the medium were only measurable because of the very low buffering capacity contributed by ADP and P. This was due to the choice of alkaline pH, 8.3, and to an unexpectedly low apparent dissociation constant between P and CF1, namely 23 μM, allowing to use low concentrations.)

Extra proton flow in the presence of ADP and P was tracked in Fig. 1. Its relation to CFO was suggested by its sensitivity to the CFO-blocker DCCD. Fig.2 shows that the extra flow was also sensitive to a specific antagonist of the catalytic portion, tentoxin (16). Together with the ADP-sensitivity (see Fig. 4) it corroborated the attribution of proton flux to the integral enzyme, CFO-CF1.

FIGURE 2: Transients of the transmembrane voltage (top, left) and of the lumenal pH (bottom, left) in the presence of ADP and P and with and without tentoxin. The right hand side side shows the difference of the traces in the left, with and without tentoxin. The difference for the lumenal pH was plotted over the one for the voltage (see text).

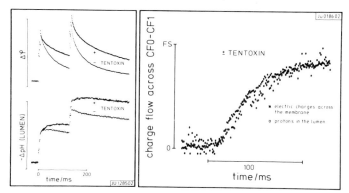

To a fair approximation the number of protons (and charges, respectively) which had left the lumen (or crossed the membrane) via the CFO-CF1-related pathway was represented by the difference between each pair of traces in the left of Fig. 2. These differences were plotted in the right hand side of Fig. 2, both, for protons leaving the lumen and for charges crossing the membrane and superimposed to each other. (Their extent was normalized under consideration of the known charge and proton stoichiometries under flashing light (reviewed in Ref. 17)). Proton efflux from the lumen was approximately synchronous to charge passage across the membrane, in other words, the extra flow of protons was completely tracked in a quantitative sense. We mention without proof that the extra flow of protons per group of three flashes was approximately 0.6 protons per 1000 chlorophyll molecules or, in other terms, 0.6 protons per CF1 molecule.

We used fairly high measuring light intensities (100-200 $\mu W/cm^2$) which established a subthreshold pH-difference. Extra proton flow was not initiated unless the protonmotive force exceeded the threshold for the activation of CF1 (22) due to a surplus of the flash light induced voltage. We asked for the nature of the rate limiting step of the electrically driven proton flow through CFO-CF1. Fig. 3 shows the extra charge flow across the membrane under variation of the ADP- and the P-concentration (top) and upon isotopic substitution of deuterium for hydrogen in the suspending medium (bottom). While the two transients obtained at different substrate concentrations coincided (top) the rate of extra charge flow was halved if deuterium was substituted for hydrogen. This proved rate limitation by a particular protolytic reaction. Studies on the nature of this rate limiting reaction are under way.

FIGURE 3: CFO-CF1-mediated extra charge flow, obtained as difference between two voltage transients with and without DCCD (as in Fig.2). Variation of the ADP,P-concentration (top) and substitution in the medium of hydrogen by deuterium (bottom).

FIGURE 4: pH-transient in the lumen with (trace D) and without (C) 15 μM ADP and 45 μM P_i added; with ADP and P_i but after incubation with 7 μM DCCD (A) and in the presence of 0.3 nM gramicidin (trace B).

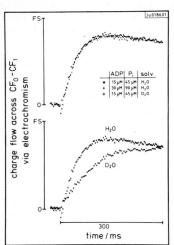

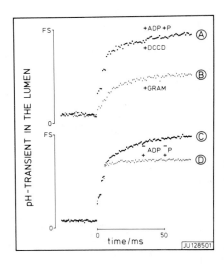

Figs. 1-3 served to establish the complete tracking of proton flow through CFO-CF1 and to demonstrate its relation to the functional enzyme via its substrate-, DCCD- and tentoxin-sensitivity. Fig. 4 brings us back to the initial question, namely, are protons produced by water oxidation first released into the lumen before they are taken up by CFO-CF1? The two traces at the bottom of Fig. 4 gave a positive answer. Following three flashes at 3 ms intervals three rapid acidification steps were clearly visible, which are characteristic for water oxidation (see Ref. 15). The extra flow of protons out of the lumen which was induced by ADP and P followed at a much longer time scale (typical half-rise time 35ms). With the sound physicochemical evidence for the selectivity of neutral red (BSA added) for pH-transients at the lumenal surface of the thylakoid membrane (8-10) the bottom traces of Fig. 4 showed to us that protons were first released into the lumen before they interacted with CF0-CF1. One or the other collegue in the bioenergetic community may mistrust "sound physicochemical evidence" in such complicated matter. Those who do may wish to inspect the upper two transients in Fig. 4, in particular trace B. Previous work in Dilley's (1,2), Homann's (3) and our laboratory (4,5) showed that very low concentrations of gramicidin at alkaline pH can activate proton storage domains, that can transiently trap protons from water oxidation (4). This was reproduced in trace B of Fig. 4. Thus the message resulting from Fig. 4 was the following: Yes, proton trapping domains can be activated (see Fig. 4B) but they are not active under conditions of ADP,P-induced proton flow via CF0-CF1. This result is at variance from several hypotheses on localized coupling which are popular in the bioenergetic literature (see Ref. 23 for a survey).

(B) The unit conductance of CF0.

By similar techniques we studied the unit conductance of CF0. Thylakoids were CF1-depleted by EDTA-treatment. They were excited with single flashes and hexacyanoferrate (III) was used as electron acceptor. At 200 µM concentration it acted only on photosystem I (20).

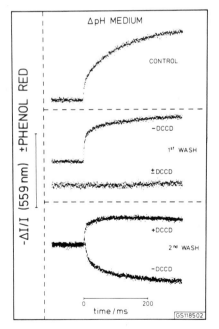

FIGURE 5: pH-transients in the suspending medium of thylakoids which had undergone no (top), one (middle) and two washes with EDTA (bottom). The degree of CF1-depletion of these membranes was 12% of total after the first wash and further 20% after the second one.

The control in Fig. 5 showed a slowly rising and long lasting alkalinization of the suspending medium (relaxation in 14s, not shown). After the first EDTA-wash the rise of the alkalinization was drastically accelerated and the extent diminished. Both effects are well understood (20) and they don't matter in this context. More importantly, there was neither rapid proton efflux from the lumen nor any DCCD-sensitivity. Both appeared only after the second wash with EDTA. The negative directed trace in the bottom of Fig. 5 indicated rapid proton efflux from the lumen which was prevented by DCCD (positive directed trace). The proportion of the alkalinization (with DCCD) and of the net acidification (without DCCD) was approximately 1:1 as expected with hexacyanoferrate (III) as electron acceptor. (Under these conditions one flash of light caused the uptake of one proton from the medium at photosystem II but the release of two protons, one at water the other one at plastoquinol oxidation. The net result was an acidification by one proton (17)). We checked the attribution of this rapid proton flux to CFO by using further known FO-blockers as tributyltin and venturicidin, all of which worked as did DCCD.

The message contained in Fig. 5 was twofold: 1.) A certain fraction of CF1 can be removed from thylakoids without that exposed CFO conducts protons. This behaviour may be discussed in the context of evidence that "mild extraction" of CF1 leaves the delta-subunit of CF1 behind where it acts like a plug to the proton channel, CFO (24 and Lill et al., these proceedings). 2.) Under higher degrees of extraction proton channels appear which are sensitive to known modifiers of FO-channels (review in Ref.25).

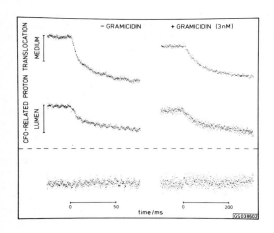

FIGURE 6: Proton translocation via CFO as viewed from the medium (top) and from the lumen (middle). The difference plotted at the bottom (extent normalized). Left: without gramicidin, Right: with 3 nM gramicidin added.

We determined the proton conductance of these channels. The difference between the two traces in the bottom of Fig. 5 reflects the time course of proton influx into the medium as discussed in the context of Fig. 2. This difference was plotted in the upper left of Fig. 6 and likewise the time course of proton efflux from the lumem (middle). After normalization of the extent both transients were identical as shown in the bottom of this figure. Again proton flux, here via CFO, was completely tracked. Protons disappearing from the lumen synchronously appeared in the medium. The right hand traces in Fig. 6 were measured with the transmembrane voltage shunted by gramicidin. This slowed the efflux drastically, in terms of the relaxation time from 7 ms (rapid phase in the left of Fig. 6) to 85 ms. Consequently, the rapid phase (Fig. 6, left) was attributable to electric relaxation with protons as charge carrier. Previously we measured the electric relaxation in thylakoids after CF1-extraction (28). It is noteworthy that here, for the first time, this relaxation was measured via proton flow, which allowed to calculate the protonic unit conductance of CFO. The specific conductance of the thylakoid membrane, g, followed from the capacitor equation, $g = c/t$, with c the specific capacitance, $1 \, \mu F/cm^2$, as usual, t the relaxation time, 7ms. The data implied a specific conductance of $1.4 \times 10^{-4} \, S/cm^2$. This was related to the surface density of exposed CFO. Taking accepted figures for the frequency of CFO in terms of CF1/chl, appr. 1:1000, and for the surface density of chlorophyll-molecules, $(2.2 \text{ nm}^2)^{-1}$, into account we calculated a unit conductance of 10 fS, if any CFO molecule out of the 30%-fraction of exposed ones was actually proton conducting (see Ref. 7 for details). This was by orders of magnitude larger than published figures for any FO-channel. At an electric driving force of 100 mV it meant the turnover of 6200 protons per second, which was already satisfactory in the light of the highest turnover rates for photophosphorylation, approximately 300 ATP (22) or 1000 protons per second at a stoichiometric factor of 3 H^+/ATP.

This figure for the unit conductance of "exposed CFO" certainly underestimated the true "time-averaged" unit conductance of "active CFO". We learned from Fig. 5 that CF1 extraction may lead to exposure of CFO without concomitant proton conduction.

SUMMARY

Proton pumps in thylakoids were stimulated by flashing light. Backflux of protons was completely tracked. Disappearence of protons in the lumen, charge passage across the membrane and appearence of protons in the medium were measured and found to occur in synchrony.

We studied ADP,P-induced extra flow of protons via CFO-CF1. The apparent K_m for P_i was surprisingly low, 23 µM. Extra proton flow was to be blocked at the channel portion (Fig.1) or at the catalytic portion (Fig.2) of CFO-CF1. Its velocity was steered by a protolytic reaction (isotope effect in Fig.3). Uptake of protons from the lumen was preceeded by deposition of protons from water oxidation into the lumen (Fig.4). The respective half-rise times were 35 ms and appr. 1ms. In the presence of very low concentrations of gramicidin, 0.3 nM or GRAM/chl=1/33333, on the other hand, protons from water oxidations were transiently trapped without appearing in the lumen (Fig.4 and Ref.4). These two classes of phenomena were observed in parallel experiments (Fig.4), they implied: Intramembrane domains which can transiently trap and retain protons exist (1-5, this article), but they are not neccessarily active under conditions of ADP,P-related proton translocation via CFO-CF1. Although this statement is confined to the repetitive excitation conditions of this article (see 6 for others) it weakens the proposed relevance for photophosphorylation of these undoubtedly existing domains.

We studied the protonic unit conductance of CFO after its exposure by CF1-extraction. For this we measured the relaxation of the pH-difference across the thylakoid membrane under electric driving force (Fig.s 5 and 6). The calculated lower limit, 10 fS, was equivalent to the passage of 6200 H^+/s at 100 mV driving force. This lower limit was by orders of magnitude greater than any published figure for FO-type proton channels (see 7 for references). It was obtained under the assumption that every exposed CFO molecule was proton conducting. This was untrue (see Fig.5 and Ref.s 7,26). Our further studies showed that only a few percent of exposed CFO were proton conducting (26), one possible reason being the plugging action of subunit delta of CF1 (see Ref.s 24, 27 and Lill et al., these proceedings). Thus the true unit conductance of CFO may be as high as 170 fS (see Ref.26).

ACKNOWLEDGEMENTS

We are very grateful to our colleques Dr.Siggi Engelbrecht for essential biochemical input and to Peter Jahns, Hella Kenneweg and Karin Schuermann for expert technical assistance. This work was financially supported by the DFG (Sonderforschungsbereich 171, B3).

REFERENCES

1 Baker,G.M., Bhatnagar,D., and Dilley,R.A.(1981) Biochem.20,2307-2315
2 Dilley,R.A., Prochaska,L.J., Baker,G.M., Tandy,N.E., and Millner,P.A.
   (1982) Curr.Top.Membr.Transp. 16,345-369
3 Theg,S.M., and Homann,P.H. (1982) Biochim.Biophys.Acta 679,221-234
4 Theg.S.M., and Junge,W. (1983) Biochim.Biophys.Acta 723,294-307
5 Polle,A., and Junge,W. (1986) FEBS Lett. 198,263-267
6 Dilley,R.A., and Schreiber,U. (1984) J.Bioenerg.Biomembr. 16,173-193
7 Schoenknecht,G., Junge,W., Lill,H., and Engelbrecht,S. (1976) FEBS Lett.
   202,23-28
8 Junge,W., Auslaender,W., McGeer,A.J., and Runge,T. (1979) Biochim.
   Biophys.Acta 546,121-141
9 Hong,Y.Q., and Junge,W. (1983) Biochim.Biophys.Acta 722,197-208
10 Junge,W., Schoenknecht,G., and Foerster,V. (1986) Biochim.Biophys.Acta
   -in press-
11 Junge,W., and Witt,H.T. (1968) Z.Naturforsch. 23b,244-254
12 Witt,H.T. (1971) Quart.Rev.Biophys. 4,365-477
13 Polle,A., and Junge,W. (1986) Biochim.Biophys.Acta 848,274-278
14 Linnett,P.E., and Beechey,R.B. (1979) Meth.Enzymol. 25,472-518
15 Foerster,V., and Junge,W. (1985) Photochem.Photobiol. 41,183-190
16 Steele,J.A., Uchytil,T.F., Durbin,R.D., Bhatnagar,P., and Rich,D.H.
   (1976) Proc.Natl.Acad.Sci. USA 73,2245-2248
17 Junge,W. (1982) Curr.Top.Membr.Transp. 16(1982)431-465
18 Junge,W., Rumberg,B., and Schroeder,H. (1970) Eur.J.Biochem. 14,575-581
19 Graeber,P., Schlodder,E., and Witt,H.T. (1977) Biochim.Biophys.Acta
   461,426-440
20 Polle,A., and Junge,W. (1986) Biochim.Biophys.Acta 848,257-264
21 Junge,W., and Polle,A. (1986) Biochim.Biophys.Acta 848,265-273
22 Junesch,U., and Graeber,P. (1985) Biochim.Biophys.Acta 809,429-434
23 Westerhoff,H.V. (1983) Dissertation, Universiteit van Amsterdam
24 Junge,W., Hong,Y.Q., Qiang,L.P., and Viale,A. (1984) Proc.Natl.Acad.Sci.
   USA 81,3078-3082
25 Vignais,P.V., and Satre,M. (1984) Molec.Cell.Biochem. 60,33-70
26 Lill,H., Engelbrecht,S., Schoenknecht,G., and Junge,W. (1986) Eur.J.
   Biochem. -in press-
27 Engelbrecht,S., Lill,H., and Junge,W. (1986) Eur.J.Biochem.-in press-
28 Schmid,R., Shavit,N., and Junge,W.(1976)Biochim.Biophys.Acta 430,145-153

Delta subunit of chloroplast coupling factor 1 (CF_1) inhibits proton leakage through coupling factor 0 (CF_0)

Holger Lill, Siegfried Engelbrecht, and Wolfgang Junge

Biophysik, Fachbereich Biologie/Chemie
Universität Osnabrück, Postfach 4469
D-4500 Osnabrück, F.R. Germany

Introduction

ATP synthases of chloroplasts, mitochondria, and bacteria are composed of two parts: F_1, extrinsic to the membrane, contains the binding sites for nucleotides, and F_0, a membrane spanning complex, is considered to act as proton conducting device. CF_1, the ATPase of chloroplasts, is composed of five different polypeptides, named alpha, beta, gamma, delta, and epsilon in order of decreasing molecular mass. The large subunits, alpha and beta, form the nucleotide binding sites. The gamma subunit might regulate proton flow through CF_0CF_1. The epsilon subunit of CF_1 binds to gamma and acts as an inhibitor of ATP-hydrolysis (for a recent review on CF_1, see [1]).

Junge et al. have found the content of delta in CF_1 extracted by EDTA-treatment of thylakoids to vary with the extraction conditions [2]. CF_1-depleted membranes were leaky when delta appeared with solubilized CF_1, they where tight if delta was lacking. This led to the suggestion of delta acting as a stopcock for the proton channel through CF_0.

In this work, we compared the effects of extraction of delta-free versus delta-containing CF_1 on the leakiness to protons and on the rates of ATP synthesis. The total degree of CF_1-extraction was kept constant while the subunit composition of solubilized CF_1 was altered by variation of incubation times and of NaCl-content of the EDTA-extraction buffer. In a reverse approach, we measured reconstitution of photophosphorylation by CF_1 and CF_1 (-delta) on EDTA-depleted thylakoids.

Results

Traces in Fig.1 represent proton efflux into the medium from thylakoids after single flash excitation. The three samples were subjected to different extraction protocols (see Tab.1). For comparison, a baseline as observed with control thylakoids is indicated by a broken line ("0"). The insert of the figure shows immunoprecipitation lines which illustrate the degree of CF_1-extraction. In experiment "a", despite 13% loss of total CF_1, almost no proton efflux occured. In experiment "b" , the extraction degree again was 13%, but proton efflux was much greater. The extraction conditions in experiment "c" resulted in 55% CF_1-extraction and in greater extent and higher rate of proton efflux .

Biggens, J. (ed.), Progress in Photosynthesis Research, Vol. III. ISBN 90 247 3452 5
© *1987 Martinus Nijhoff Publishers, Dordrecht. Printed in the Netherlands.*

Fig.1: Time course of
proton efflux from
thylakoids into the
outer suspending
medium under differ-
ent protocols for
CF_1-depletion as
measured after single
flash excitation with
phenol red as indica-
tor for transient pH-
changes
Insert: CF_1-content
of extracted thyla-
koids as detected
by rocket-immuno-
electrodiffusion

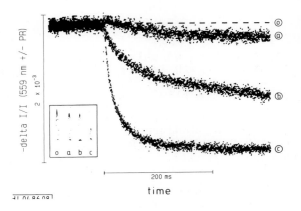

In order to examine the subunit composition of solubilized CF_1, extraction supernatants were concentrated by
ultrafiltration and subjected to HPLC on TSK DEAE 650 (S).

Fig.2: Purification
of solubilized CF_1
and analysis of
subunit composition
by HPLC

Insert: SDS-PAGE of
samples from peak
a and b

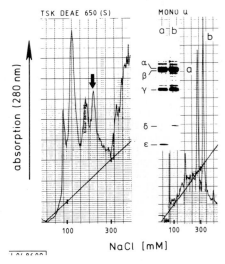

Fig.2 shows the elution profiles of a typical HPLC run
with the supernatant of experiment c. The TSK DEAE 650 (S)
elution profile (left), showed several peaks, but only one
contained ATPase-activity (marked by an arrow). This peak was
analyzed further by HPLC on Mono Q (right profile). The
amounts of protein separated by this run were too small for
further analysis by gel electrophoresis. Therefore, we
subjected 1 mg of pea CF_1 from a large scale preparation to
Mono Q-HPLC. Two peaks were eluted at the same two NaCl
concentrations as in Fig.2, and the protein yield was high

enough for further analysis by gel electrophoresis. As documented in Fig.2, the peak at 278 mM NaCl contained CF_1 (-delta) and the peak at 320 mM NaCl CF_1 (- epsilon). Peak integration of the profile given in Fig.2 revealed that the peak at 278 mM comprised 46% of total CF_1, i.e. the ratio CF_1 (-delta):CF_1 (+delta) was about 1:1 in this particular extraction experiment. The supernatants produced by the two other extraction protocols were also subjected to this analysis. Table 1 summarizes the ATP-synthesis rates, the degree of CF_1-extraction (cf. Fig.1,insert), and the ratio of delta-free versus delta-containing enzyme for control thylakoids and for the three extracted samples. Comparison of experiments a and b revealed that a higher amount of delta-containing CF_1 in the supernatant (at constant extraction degree) was accompanied by higher proton leakage (cf. Fig.1), and by a greater decrease of the photophosphorylation rate. As the decrease of the latter was higher than the decrease of CF_1 content, it was mainly caused by dissipation of protonmotive force (in agreement with measured proton efflux, cf. Fig.1.) and only to lesser extent by loss of catalytic activity.

Table 1: ATP-synthesis after different extraction protocols

Extraction procedure	Curve in Fig. 1	ATP-synthesis (%)	residual CF_1 (%)	% delta content
control	0	100	100	--
5min,20 mM NaCl	a	88	87	18
1.5min,5 mM NaCl	b	78	87	36
10min,1 mM NaCl	c	19	45	54

In a reverse experiment, we examined the reconstitution of ATP-synthesis in CF_1-depleted thylakoids by CF_1 and CF_1 (-delta) and either species in DCCD-inhibited form. CF_1 and CF_1 (- delta) were prepared via Mono Q-HPLC as in Fig.2 with a large scale preparation of spinach CF_1. CF_1-extraction from pea thylakoids was as under "c" in Table 1. Table 2 shows that CF_1, CF_1 (-delta), and DCCD-inhibited CF_1 did reconstitute ATP-synthesis, while DCCD-inhibited CF_1 (-delta) did not.

Table 2:Reconstitution of ATP-synthesis by various CF_1-samples

Sample	ATP-synthesis (%)
control thylakoids	100
CF_1-depleted thylakoids	37
depleted thyl. + CF_1	60
depleted thyl. + CF_1 (-delta)	50
depleted thyl. + DCCD - CF_1	49
depleted thyl. + DCCD - CF_1 (-delta)	36

Discussion

From the results given above, a function of delta as a "stopcock" to CF_0 seems highly probable: If delta remained bound to CF_0 no dissipative proton efflux occured after removal of the four other subunits of CF_1. A loss of delta during the extraction and concentration procedure was unlikely, since delta containing CF_1 in considerable amounts was detected under other extraction conditions.

In reconstitution experiments with uninhibited CF_1 and CF_1 (-delta) and DCCD-modified, inhibited CF_1 and CF_1 (-delta), among the inhibited varieties only integral CF_1 enhanced phosphorylation rates. Whereas the uninhibited enzyme could restore higher synthesis rates in two ways, namely by adding catalytic capacity and by blocking proton leaks, inhibited CF_1 could act only by plugging proton leaks. As it did restore photophosphorylation, this confirmed the above conclusion that subunit delta can block proton leakage through open CF_0. A failure of rebinding of DCCD-CF_1 (-delta) was unlikely because enhanced synthesis rates (due to addition of catalytic capacity) were obtained with uninhibited CF_1 (-delta). Our own experiments gave no information about the role of subunit delta in rebinding of CF_1 to CF_0: We assumed that CF_1 (-delta) rebound to CF_0 (+delta).

Although subunit delta of CF_1 obviously acts as an inhibitor of proton flow through exposed CF_0, its function in phosphorylating CF_0CF_1 must go beyond this merely passive role. If direct participation of protons in ATP-synthesis is assumed after their passage through CF_0 into CF_1 during the catalytic cycle [3], the block of proton flow by delta must be relieved. A possible candidate for this control function on delta is the gamma-subunit of CF_1, which was proposed to be a proton gate. On the other hand, if one assumes more indirect, conformational coupling between proton flow and ATP-synthesis [4], delta could transmit conformational changes to CF_1 resulting from proton flow through CF_0, perhaps by analogy to the rod and the piston in a combustion engine. Transmission of information from CF_0 (binding of DCCD) into CF_1 (altered affinities of nucleotide binding sites) has been demonstrated recently for the mitochondrial F_0F_1 [5].

References

1 Vignais, P.V. & Satre, M. (1984) Molec. and Cell Biochem. 60, 33-70
2 Junge,W., Hong,Y.Q., Quian,L.P. & Viale,A. (1984) Proc. Natl. Acad. Sci. USA 81, 3078-3082
3 Mitchell, P. & Moyle, J. (1974) Biochem. Soc. Spec. Publ. 4, 91-111
4 Boyer, P.D. (1979) in: Membrane Bioenergetics, eds. Lee et al. (Addison - Wesley, Reading, M.A.), pp. 461 - 479
5 Penefsky, H.S. (1985) Proc. Natl. Acad. Sci. USA 82, 1589-1593

Acknowledgements

We would like to thank Mrs. K. Schürmann for highly skilled and dedicated assistance, and Mrs. H. Kenneweg for graphs and photographs. Financial support by the DFG (SFB 171-84/B2/B3) is gratefully acknowledged.

FLASH-INDUCED PROTON RELEASE FROM SPHEROPLASTS OF RHODOPSEUDOMONAS SPHAEROIDES

HIROYUKI ARATA, DEPARTMENT OF BIOLOGY, FACULTY OF SCIENCE, KYUSHU UNIVERSITY 33, FUKUOKA 812, JAPAN

1. INTRODUCTION

Vectorial proton translocation coupled with the electron transfer is one of the important subject in the field of the bioenergetics. The advantage of the cyclic electron transfer system of the purple photosynthetic bacteria in the study of the coupling mechanism is that we can make rapid stoichiometric excitation using short flashes.

Kinetics of the flash-induced proton uptake on the cytoplasmic side of the chromatophore membrane has been studied in detail (1-6). Proton release on the periplasmic side, on the other hand, has not been studied extensively. Petty and Dutton (4) measured proton release to the inside of the chromatophores using pH indicators trapped in the vesicles. Taylor and Jackson (7) measured flash-induced rapid proton release from intact cells of Rhodopseudomonas capsulata. Use of the intact cells has, however, some disadvantages: first, we can not control the redox potential of the system; second, the cell wall and/or the invaginated membrane system could be a barrier to the proton movement; third, the high turbidity makes the spectrophotometric measurements difficult.

In the present paper, we measured the flash-induced proton release using right-side-out membrane vesicles prepared by digestion of the cell wall by lysozyme followed by hypotonic treatment. Proton release under the continuous excitation has been measured with the right-side-out vesicles (8), but no kinetic analysis with flash excitations has been reported.

2. MATERIALS AND METHODS

Cells of Rhodopseudomonas sphaeroides green mutant were cultured anaerobically in the slightly modified medium of Cohen-Bazire et al. (9). Spheroplasts were prepared by the method described by Witholt et al. (10) with some modifications. Cells were suspended in 50 mM Tris-HCl (pH 8)/2.5 mM Na-EDTA/0.5 M sucrose at a concentration of 50-100 mg wet weight/ml. Lysozyme was added at a final concentration of 0.1 mg/ml. The cell suspension was diluted twice by water, and incubated at 30 °C for 40 min. The spheroplasts formed were collected by centrifugation at 5,000×g for 20 min and hypotonic treatment was made by suspending them in water. 1 M KCl solution, in a volume equal to the original solution, was added and the suspension was centrifuged at 10,000×g for 30 min. The pellet was resuspended in and washed with 1 M KCl, and was finally suspended in 100 mM KCl. The obtained preparation showed no photooxidation of cytochrome c_2, which was washed out during the preparation.

Absorption change measurements and redox potentiometry was done as described in (11). The pH of the medium was monitored by a glass electrode inserted into the cuvette and adjusted by adding NaOH or HCl. The samples were excited by a xenon flash (Sugawara Laboratories, MF-1500-U3), which gave single turnover saturating flashes (10 µs duration). pH change was measured by absorption change at 586 nm in the presence of 20 uM bromcresol purple. It was calibrated by adding a known amount of HCl. Less than 10 % of bromcresol purple was bound to the membrane: More than 90 % of the added dye was found in the supernatant after centrifugation of the reaction mixture. For the measurements of oxidation-reduction of the cytochromes and the carotenoid band shift, absorption changes at two wavelengths were measured successively and the differences were recorded.

3. RESULTS AND DISCUSSIONS

Fig. 1 shows the flash-induced pH changes (Traces A-D) and the absorbance changes due to the oxidation-reduction of the primary electron donor (Trace E), oxidation-reduction of cytochrome c (Trace F) and carotenoid shift (Trace G). Rapid increase in the absorbance at 586 nm was observed in the presence of bromcresol purple. The origin of this absorbance increase was not clear. It was not due to the increase of pH of the medium, because 10 mM phosphate buffer did not affect it (Traces A and C). After the initial rise, an absorbance decrease followed in the absence of the buffer, showing acidification of the medium (Traces B and D).

The rate of the proton release was increased by the addition of horse heart cytochrome c (Trace D), which mediate the electron transfer from the

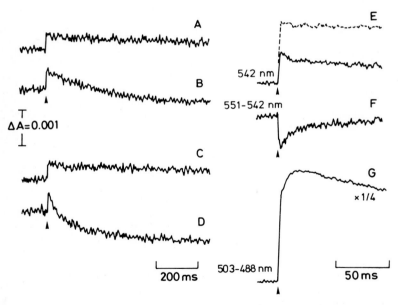

FIGURE 1. pH change (A-D) and absorbance change (E-G) by the flash excitation of spheroplasts. The upward deflection corresponds to the absorbance increase. The reaction mixture contained 100 mM KCl, 20 µM DAD and 10 µM pyocyanine. Further additions; 5 µM cytochrome c for C, D, E, F and G; 20 µM bromcresol purple for A, B, C and D; 10 mM Pi buffer for A, C, E, F and G. E_h=150 mV, pH=6.3. The dotted line in Trace E shows measurement with only 100 mM KCl and represents the total primary donor oxidized. A-D, averages of 16 recordings; E-G, averages of 4 recordings.

cytochrome $b-c_1$ complex to the reaction center. The rereduction of the photo-oxidized primary donor and the oxidation of cytochrome c_1 were slow in the absence of exogenous cytochrome c, as observed by Wood (12). In the presence of cytochrome c, the rate still appeared to be slower (about 70 ms in the half time) than the rereduction of cytochrome c (Trace F) or that of the electrical field formation (Trace G).

The number of protons released was 1.42 ± 0.23 H^+ per reaction center (average of three measurements with different preparations). About 30-40 % of the oxidized primary electron donor was not rereduced by cytochrome c (Trace E). Therefore, the number of protons released per rereduced primary donor was about 2. It was constant in E_h range of 0-160 mV. The value agrees with the H^+/e^- ratio obtained with chromatophores (15) and with the mitochondrial cytochrome $b-c_1$ complexes (13, 14). It is not clear why 30-40 % of the reaction center did not react

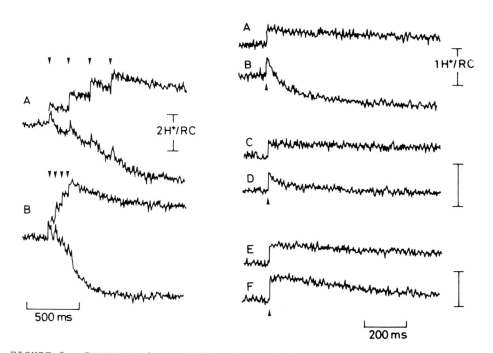

FIGURE 2. Proton release following multiple flashes. The spheroplasts were excited by four flashes with frequencies of 5 Hz (A) and 16 Hz (B). The experimental conditions were as in Traces A-D of Fig. 1. Upper traces, both in (A) and (B), represent the measurement in the presence of 10 mM Pi buffer. Averages of 8 recordings.

FIGURE 3. Effects of inhibitors of cytochrome $b-c_1$ complex on the proton release. The experimental conditions were as in Traces A-D of Fig. 1. A and B, no further addition; C and D, 20 μM antimycin A; E and F, 20 μM myxothiazol. Traces A, C and E represent the measurements in the presence of 10 mM Pi buffer. Averages of 16 recordings.

with added cytochrome $\underline{c}$, while all cytochrome $\underline{c}_2$ could be washed out. Possibly, part of the periplasmic face of the exposed membrane folded again to form inside-out or multi-layered vesicles.

In trace A of Fig. 2, the sample was excited by four flashes with 200 ms intervals. The intervals were long enough so that the pH changed to the maximal extent before the next excitation, and each flash released approximately the same amount of protons. In trace B, the intervals were shorter (62.5 ms) than the apparent half time of the proton release, but the same amount of protons as with the longer intervals were finally released after the fourth flash. This indicated that the proton release itself was rapid, but its delocalization was slower.

The released proton could be localized in a "sink" within the membrane, as suggested by Taylor and Jackson (7). Alternative possibility is that the membrane vesicles aggregated and formed small aqueous domain between membranes. The cytochrome $\underline{c}$ added would diffuse into the domain slowly but the released proton would not diffuse out quickly.

Fig. 3 shows effects of inhibitors of cytochrome $\underline{b}$-$\underline{c}_1$ complexes on the proton release. Antimycin A, which inhibits the reoxidation of cytochrome $\underline{b}$, decreased the amount of released proton (Trace D). In the presence of the inhibitor, quinol still reduce cytochrome $\underline{b}$ and Rieske iron sulfur center at the Q_z-site. If all protons of the quinol were released to the periplasmic aqueous phase, two protons per reduced cytochrome $\underline{b}$ would be observed. Alternatively, the reduced cytochrome $\underline{b}$ could retain one of the protons, because the midpoint redox potential of cytochrome $\underline{b}$ depends on pH in the pH range used here (4). The number of protons released per cytochrome $\underline{b}$ reduced was smaller than 2, but the conclusive measurement was not possible because of the low signal to noise ratio.

Myxothiazol inhibits the oxidation of quinol at the Q_z-site (15). It inhibited the proton release almost completely (Trace F), as expected.

REFERENCES
1 Jackson, J.B. and Crofts, A.R. (1969) Eur. J. Biochem. 10, 226-237
2 Chance, B., Crofts, A.R., Nishimura, M. and Price, B. (1970) Eur. J. Biochem. 13, 364-374
3 Cogdell, R.J. Jackson, J.B. and Crofts, A.R. (1972) J. Bioenergetics 4, 413-429
4 Petty, K.M. and Dutton, P.L. (1976) Arch. Biochem. Biophys. 172, 335-345
5 Petty, K.M., Jackson, J.B. and Dutton, P.L. (1977) FEBS Lett. 84, 299-303
6 Wraight, C.A. (1979) Biochim. Biophys. Acta 548, 309-327
7 Taylor, M.A. and Jackson, J.B. (1985) FEBS Lett. 180, 145-149
8 Hochman, A., Fridberg, I. and Carmeli, C. (1975) Eur. J. Biochem. 58, 65-72
9 Cohen-Bazire, G., Sistrom, W.R. and Stanier, R.Y. (1975) J. Cell Comp. Physiol. 49, 25-68
10 Witholt, B., Boebhout, M., Brock, M., Kingma, J., Van Heerikhuizen, H. and De Leij, L. (1976) Anal. Biochem. 74, 160-170
11 Takamiya, K. and Hanada, H. (1980) Plant Cell Physiol. 21, 979-988
12 Wood, P.H. (1980) Biochem. J. 192, 761-764
13 Mitchell, P. (1976) J. Theor. Biol. 62, 327-367
14 Rich, P. (1984) Biochim. Biophys. Acta 768, 53-79
15 Glaser, E.G. and Crofts, A.R. (1984) Biochim. Biophys. Acta 766, 322-333

EFFECTS OF 2-(N-HEPTYL)- AND 2-(N-NONYL)-4-HYDROXYQUINOLINE N-OXIDE (HQNO, NQNO) ON FLASH-INDUCED PROTON TRANSLOCATIONS IN THYLAKOIDS

Y. HONG AND A. B. HOPE[*], INSTITUTE OF PLANT PHYSIOLOGY, ACADEMIA SINICA, SHANGHAI, CHINA, AND [*]SCHOOL OF BIOLOGICAL SCIENCES, FLINDERS UNIVERSITY, BEDFORD PARK, SOUTH AUSTRALIA 5042 ([*]Address for correspondence)

1.   INTRODUCTION

In previous papers it has been shown that proton uptake stimulated by the addition of valinomycin (1,2) and the slow phase of the electrochromic shift (3-6) are consistent with the operation of a Q-cycle in thylakoids, under some conditions.   Continuous cycling was shown to depend on preventing the increase of an electric potential difference that otherwise made energetically unfavourable the passage of an electron across the membrane in the b/f complexes.   An estimated voltage of c. 100 mV was shown to half-inhibit both the slow phase and valinomycin-induced proton uptake (2,5).

As a test of the involvement of the cytochrome b563s in the process of proton uptake under conditions suitable for Q-cycling we have studied the effects of HQNO and NQNO, reputed inhibitors of cytochrome b536 oxidation. Proton uptake, proton deposition, and electron transport were measured, in both oxidizing conditions (electrons from water to ferricyanide, methyl viologen, or methyl purple) or (proton deposition) reducing conditions (electrons from duroquinol to methyl viologen, +DCMU, PS1 only).

2.   PROCEDURE
2.1   Materials and methods

2.1.1. Chloroplasts: These were prepared from 10-14 day old pea leaves by blending, straining, centrifuging and resuspending as described earlier (7), sometimes with an additional centrifugation through Percoll (8) after the standard steps mentioned.

2.1.2. Chlorophyll: Chlorophyll was determined using the wavelengths and extinction coefficients recently adopted by Graan and Ort (9) as offering improved accuracy.

2.1.3 Proton uptake: The dye phenol red at 20 μM was added to chloroplast suspensions and the change in transmittance at 560 nm was taken as proportional to the extent of proton uptake (10,11).   Also in the medium were ferricyanide 0.3 mM, sorbitol 330, KCl 10 and $MgCl_2$ 3 mM, pH 7.8.   The chlorophyll was 10 μM.

2.1.4. Proton deposition: The dye neutral red, 20 μM, was used as previously described (12,11).

2.1.5. Light Sources: Single-turnover (<0.5 us), saturating (20-40 mJ incident at the cuvette face) laser flashes were delivered from a Phase-R (N.H., USA) type DL-32 dye laser with the dyes Rhodamine 640 and Oxazine 720 mixed to give a wavelength of 675 nm.   For flashing at 5 Hz a "Stroboslave" xenon flashlamp was used, λ > 630 nm, <3 μs duration at half-height, and energy that was 84% saturating for 10 μM chlorophyll.   The measuring beam derived from an Oriel (Stamford USA) double monochromator with regulated tungsten lamp.

Biggens, J. (ed.), Progress in Photosynthesis Research, Vol. III. ISBN 90 247 3452 5
© *1987 Martinus Nijhoff Publishers, Dordrecht. Printed in the Netherlands.*

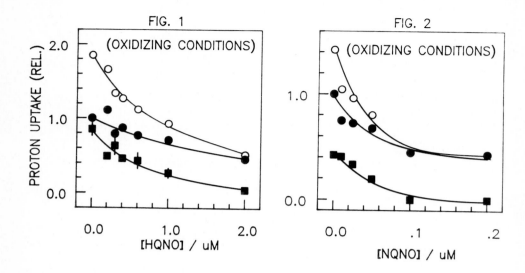

Figure 1. The effect of HQNO on proton uptake in the absence (•) and presence (o) of 15 nM valinomycin. Proton uptake is represented as relative to that minus HQNO and minus val. The relative stimulation due to valinomycin (■) was estimated as described in the text. The uptake was that following 21 flashes at 10 or 20 Hz; other details in the text. Error bars where given are SEM of 5–18 measurements. HQNO (or NQNO, Fig. 2) were left in contact with thylakoids for at least 120 s to obtain the full effect.

Figure 2. The effect of NQNO on proton uptake, similar conditions as for Fig. 1 except a flash series of 11 at 5 Hz and chl now 20 uM; means of 3 expts; the stimulation by val is less than in Fig. 1 because of the lower flash frequency (2).

3. RESULTS AND DISCUSSION

3.1. Proton uptake in oxidizing conditions

Figs 1–2 show the extent of proton uptake due to a train of flashes as affected by HQNO, and NQNO. The data is normalized to zero concentration of inhibitor. In each experiment a run was recorded without inhibitor, followed by its addition, a delay of 120 s and a further train of flashes. In other experiments of the same format, inhibitor was added at the beginning and the second series of flashes was after the addition of valinomycin at a concentration (10–15 nM) that was found maximally to stimulate proton uptake by a factor of 1.6–1.8 (2). The fractional extent of stimulation due to valinomycin, $\{\Delta H_o^+(+val) - \Delta H_o^+(-val)\}/\Delta H_o^+(-val)$, has been plotted against the concentration of inhibitor in the same Figures. The stimulation by valinomycin was more sensitive to either HQNO (or NQNO) than was the control (−valinomycin) proton uptake, with a $c_{1/2}$ of 0.6 μM (0.04). However, the control uptake was unexpectedly also inhibited, with $c_{1/2}$ of 1.7 μM (0.1). The fast electrogenic signal representing turnovers at reaction centres was unaffected by up to 0.3 μM NQNO.

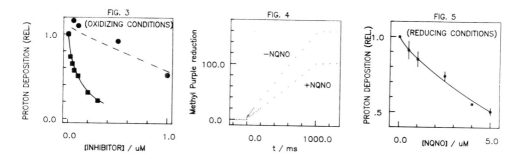

Figure 3. The effects of HQNO and NQNO on proton deposition, without val.; for HQNO, 20 flashes averaged at 0.3 Hz; for NQNO, 11 at 5 Hz.

Figure 4. The effect of 100 nM NQNO on the transmittance change at 590 nm proportional to the reduction of methyl purple in a 1 s exposure to shuttered actinic light ,λ>660 nm; methyl purple 10 μM, chl 20 μM; the control rate of electron transport was 60/(s.P700).

Figure 5. The effect of NQNO on proton deposition, minus valinomycin; 11 flashes at 5 Hz; means of 2-7 expts, SEM shown if n $\geqslant$ 3.

3.2. Proton deposition in oxidizing conditions

In Figure 3, the effect is shown of HQNO and NQNO on proton deposition in oxidizing conditions, in the absence of valinomycin. The sensitivity to the inhibitors was similar to that of proton uptake. A kinetic separation into the signals representing the protons from water oxidation and those from plastoquinol oxidation was not attempted (12,11). The inhibiting effect of HQNO and NQNO on proton uptake or deposition in the absence of valinomycin is unexpected if these substances bind only to the Qc sites on cytochrome b/f complexes (13). We therefore queried their effect on electron transport. The study of Barton et al. (14) showed that at least for HQNO there is likely to be a binding site near PS2 since electron transport from water to PS1 acceptors or to PS2 acceptors was inhibited with a $c_{1/2}$ of 0.5-2 μM when FeCN was the electron acceptor, depending on the order of addition. (The oxidation of PQH_2 at Qz sites was apparently inhibited by HQNO (15,16), but at concentrations of the order of 50 μM, ten times greater than those considered here). The data in Figs. 1 and 3 are thus understandable, the $c_{1/2}$ for inhibition of proton translocation in the absence of valinomycin being similar to those reported earlier (14,15). Comparable data has not appeared for NQNO; our observations showed a $c_{1/2}$ of about 1 μM for electron transport to methyl viologen or ferricyanide, measured with an oxygen electrode, and a similar sensitivity of photophosphorylation. Only when the electron transport in the first few turnovers was observed with the aid of methyl purple (17,1) was it clear that there was a temporary inhibition by as little as 100 nM NQNO (Fig.4), for about 6 turnovers, after which the rate gradually increased. NQNO is surely inhibiting at Qz sites, and as the concentration   of competing PQH_2 rises as flashing continues, the inhibition is relieved.

3.3   Proton deposition in reducing conditions

Fig. 5 shows that proton deposition in reducing conditions, minus valinomycin, is half-inhibited by $5 \mu M$ NQNO.   The slow electrogenic phase was much more sensitive (50 nM) but the fast phase was unaffected (Y. Hong and A. B. Hope, unpublished).   The much higher concentration of NQNO needed to half-inhibit proton deposition under these conditions is also interpreted as an effect at Qz sites, but since the concentration of PQH_2 is so much higher than in oxidizing conditions much more NQNO is needed.

4.   CONCLUSIONS

With both HQNO and NQNO the valinomycin-stimulated proton uptake, which has been attributed to Q-cycle activity (1,2), was more sensitive to the inhibitors than the control (-val) uptake, with $c_{1/2}$ of 0.6 μM and c.0.04 μM respectively.   These results strongly support the Q-cycle model, as these substances inhibit the re-oxidation of cytochrome b563 by binding to Qc sites (13).   The inhibition of control proton translocations is readily explained in terms of weaker binding of these inhibitors to QB or Qz sites. The specificity at the Qc sites for NQNO exceeds that for HQNO by a factor of about 10, while at Qz sites the specificity ratio may be much higher.

5.   ACKNOWLEDGMENTS

The project was supported by the Australian Research Grants Scheme and Flinders University, for which we are grateful.   Thanks go to West Hiscock, Ian Modistach and Peter Bond for essential assistance.   The NQNO was a generous gift of Dr John Whitmarsh.   YH is grateful for a short-term Research Fellowship from Flinders University.

REFERENCES

1 Graan, T. and Ort, D.R. (1983) J. Biol. Chem. 258, 2381-6
2 Hope, A.B., Handley, L. and Matthews, D.B. (1985) Aust. J.   Plant Physiol. 12, 387-94
3 Selak, M.A. and Whitmarsh, J. (1982) FEBS Lett. 150, 286-92
4 Jones, R.W. and Whitmarsh, J. (1985) Photobiochem.   Photobiophys. 9, 119-127
5 Hope, A.B. and Matthews, D.B.(1986a) Aust. J. Plant Physiol. (submitted)
6 Hope, A.B. and Matthews, D.B.(1986b) Aust. J. Plant Physiol. (submitted)
7 Hope, A.B. and Matthews, D.B. (1983) Aust. J. Plant Physiol. 10, 363-72
8 Robinson, S.P. (1982) Plant Physiol. 70, 1032-6
9 Graan, T. and Ort, D.R. (1984) J. Biol. Chem. 259, 14003-10
10 Junge, W. and Auslander, W. (1974) Biochim. Biophys. Acta 333, 59-70
11 Hope, A.B. and Morland, A. (1979) Aust. J. Plant Physiol. 6, 289-304
12 Auslander, W. and Junge, W. (1975) FEBS Lett. 59, 310-5
13 Hauska, G., Hurt, E. Gabellini, N. and Lockau, W. (1983) Biochim. Biophys. Acta 726, 97-133
14 Barton, J.R., MacPeek, W.A. and Cohen, W.S. (1983) J.   Bioenerg. Biomembr. 15, 93-104
15 Barton, J.R. and Cohen, W.S. (1983) Plant Sci. Lett. 32, 109-13
16 Houchins, J.P. and Hind, G. (1983) Biochim. Biophys. Acta 725, 138-45
17 Hill, R., Crofts, A., Prince, R., Evans, E., Good, N. and Walker, D. (1976) New Phyt. 77, 1-9

PATHWAYS FOR DARK PROTON EFFLUX FROM THYLAKOIDS

Mordechay Schonfeld, Bente Sloth Kopeliovitch and Hedva Schickler
The Hebrew University of Jerusalem, Faculty of Agriculture, Department of
Agricultural Botany, P.O. Box 12, Rehovot 76100, Israel.

INTRODUCTION

Light induced electron transport in chloroplasts is coupled to proton pumping into the thylakoid lumen and generates a proton gradient across its membrane. Proton efflux down this gradient and through an H^+ - ATPase is coupled to ATP synthesis (1). Chloroplasts can carry out considerable rates of electron transport, even in the absence of ADP and Pi, indicating that protons can leak out of thylakoids at proportional rates without driving ATP synthesis. Proton efflux in such experiments is monitored indirectly by assuming a fixed stoichiometry of $2H^+/e$ (2,3). The partial inhibition of non-phosphorylating electron transport by energy transfer inhibitors is an evidence for part of the proton leakage being associated with the ATPase proton channel (4,3). The residual proton leakage, which can still be substantial, is usually described as "passive" or "non-specific".

We have previously used parallel measurements of electron transport, and of internal and external proton concentrations ($[H^+]i$ and $[H^+]o$ respectively), to resolve proton efflux in the light into three different components (5): 1. Proton efflux associated with the ATPase and characterized by an exponential dependence on $[H^+]i/[H^+]o$; 2. an additional leakage pathway characterized by a linear dependence on $[H^+]i/[H^+]o$; 3. efflux by diffusion characterized by a linear dependence on $[H^+]i$. These results (5, and see also 3,6) can be summarized by writing proton efflux in the light (J_H^1) as a sum of three terms corresponding to these three leakage pathways.

$$ J_H^1 = P_H * [H^+]i + k_1 \frac{[H^+]i}{[H^+]o} + k_2 \left(\frac{[H^+]i}{[H^+]o}\right)^b \qquad (1) $$

where P_H, K_1, k_2 and b are constants.

Instead of measuring proton efflux indirectly in steady state illumination, one can follow <u>dark</u> proton efflux directly by monitoring the decay of a light induced proton gradient. This decay was reported to be a process with first order kinetics (7), in apparent contradiction with the complex kinetics of proton efflux in the light (eq 1). In the present study we re-investigated the kinetics of dark proton efflux from thylakoids and its dependence on $[H^+]i$ and $[H^+]o$. A short report of part of this study was recently published (8).

MATERIALS AND METHODS

Envelope-free chloroplasts were isolated from lettuce leaves as previously described (9). Measurements of $[H^+]i$ via flourescence changes of 9-aminoacridine were conducted according to Schuldiner et al (10), with a Jasco FP-500 spectrofluorometer. The magnetically stirred cuvette was equipped with a glass combination electrode and an oxygen electrode and permitted simultaneous measurements of $[H^+]i$, $[H^+]o$ and electron transport. A projector-lamp supplied a maximum of 1000 uE $* m^{-2} * s^{-1}$ of red actinic light (> 600 nm) at the position of the cuvette.

RESULTS AND DISCUSSION

The initial rate of dark proton efflux from thylakoids ($J_H{}^d$), at a given pH, was found to be proportional to the internal proton concentration (Fig. 1). Proton gradients were generated by preilluminating chloroplasts under non-phosphorylating conditions at various light intensities with methyl viologen as acceptor.

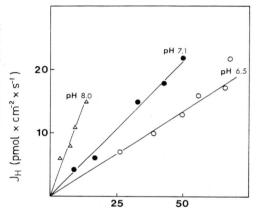

FIGURE 1. Dependance of dark proton efflux ($J_H{}^d$) on $[H^+]i$ at different pHo levels. The reaction mixture contained in a volume of 3ml: 50mM KCl, 0.2mM methyl viologen, 1mM NaN_3, 2uM 9-aminoacridine and chloroplasts equivalent to 24 ug/ml. The pH changes were calibrated by addition of aliquotes of standard HCl solutions. A value of 2.2 nm^2 was assumed for the membrane area occupied by one Chl molecule, in order to obtain fluxes and permeabilities in conventional units (8). See Materials and Methods for further details.

The linear relationship between $J_H{}^d$ and $[H^+]i$ was observed between pH 6.0 and 8.5 (Fig. 1). The nature of the dependence of $J_H{}^d$ on $[H^+]o$ is clarified in Figure 2 where $J_H{}^d/[H^+]i$ is plotted _vs_ $1/[H^+]o$ and a straight line is obtained. Dark proton efflux is accordingly given by:

$$J_H{}^d = P_H * [H^+]i + k_1 * \frac{[H^+]i}{[H^+]o} \qquad (2)$$

A comparison of eq 2 and eq 1 seems to indicate that proton efflux in the dark proceeds in two parallel pathways, vs the three obsrved in the light (5). The first term can be ascribed to efflux by diffusion; considering the absence of

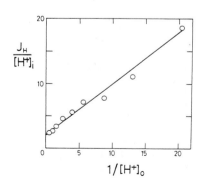

FIGURE 2. Dependence of "the apparent proton permeability"-$J_H{}^d/[H^+]i$ on $[H^+]o^{-1}$. Numbers in the ordinate are cm/s $* 10^5$, while numbers in the abcissa are ml/mol $* 10^{-10}$. Conditions as in Figure 1.

membrane potential and that $[H^+]i >> [H^+]o$. P_H— the proton permeability of the thylakoid membrane, is obtained from the intercept at the ordinate in Figure 2. Values obtained in the dark for P_H were usually between $2*10^{-5}$ and $3*10^{-5}$ cm/s i.e. similar to those obtained in the light (5). The second term in eq 2 is interpreted as indicating the operation of an additional proton leakage pathway. As observed in the light (5 and see also 3), this "special pathway" is characterized by a linear dependence on the proton concentration ratio. At a given $[H^+]o$, eq 2 can be simplified to give:

$$J_H^d = P_H' * [H^+]i \qquad (3)$$

accounting for the linear relationship between dark proton efflux and the internal proton concentration illustrated in figure 1. P_H' "the apparent proton permeability" of the membrane at a given pH, is defined by:

$$P_H' = P_H + k_1/[H^+]o.$$

Figure 3 illustrates the pH dependence of J_H^d, and of its components. Proton gradients were generated at saturating light intensities. J_H^d exhibited a broad optimum around neutral pH, declining gradually at higher pH levels. Efflux by diffusion predominated at low pH and declined with increase of pH above 7 (curve 1). Efflux through the special pathway was minimal at low pH, and peaked at pH 8 where it was the predominant pathway (curve 2).

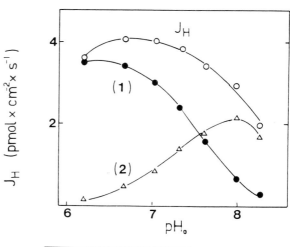

FIGURE 3. The effect of pH on dark proton efflux (J_H^d) and on its two components: 1. $P_H * [H^+]i$ (efflux by diffusion). 2. $k_1 * [H^+]i/[H^+]o$ ("Special pathway"). Experimental conditions as in Figure 1.

FIGURE 4. Effect of DCCD on light (J_H^l) and dark (J_H^d) proton efflux. Experimental conditions as in Figure 1.

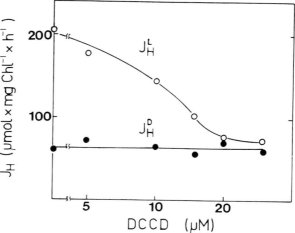

Proton efflux through the H^+ - ATPase was not evident in the dark. 'Dark' efflux was insensitive to DCCD while 'light' efflux, which was significantly higher, was inhibited by DCCD to the level of the dark process (figure 4). 'Light' proton efflux was calculated from rates of electron transport, which were measured immediately before the light was turned off for 'dark' efflux measurements. Similar results (not shown) were obtained with ATP substituting for DCCD. Efflux through the H^+ - ATPase proton channels is apparently deactivated under these experimental conditions, within a short time after cessation of illumination.

These conclusions are in an apparent contradiction with the demonstration of postillumination ATP synthesis (11). Dark proton efflux through CF-channels is probably made possible in this case by the lower pH threshold for proton efflux under phosphorylating conditions (9), and possibly also by higher initial ΔpH levels achieved with a mediator of cyclic electron transport instead of methyl viologen used here. Our results differ from those of McCarty et al (12) who reported inhibition of dark proton efflux by ATP. The reason for this difference is not clear.

Early analyses of proton uptake and release from thylakoids were based on the assumption that 'dark' and 'light' proton efflux processes are identical (7,13). More recent studies (14,15) have concluded that proton efflux in the light is greater than in the dark. Proton efflux in the light was measured by reducing the light intensity so as to allow a previously established proton gradient to relax to a lower steady state level. It was not evident however that such a relaxation portrays proton efflux in the light quantitatively, as proton efflux through the ATPase is continuously deactivated and the rate of proton **influx** is continuously attenuated during such relaxations.

REFERENCES

1 Mitchell, P. (1966) Biol. Rev. 41, 445-502
2 Junge, W. and Auslander, W. (1974) Biochim. Biophys. Acta 333, 59-70
3 Davenport, J.W. and McCarty, R.E. (1984) Biochim. Biophys. Acta 766, 363-374
4 Portis, A.R. and McCarty, R.E. (1976) J. Biol. Chem. 251, 1610-1617
5 Schonfeld, M. and Schickler, H. (1984) FEBS Letters 167, 231-234
6 Graber, P. and Witt, H.T. (1976) Biochim. Biophys. Acta 423, 141-163
7 Karlish, S.J.D. and Avron, M. (1968) Biochim. Biophys. Acta 153, 878-888
8 Schonfeld, M. and Kopeliovitch, B.S. (1985) FEBS Letters 193, 79-82
9 Schonfeld, M. and Neumann, J. (1977) FEBS Letters 73, 51-54
10 Schuldiner, S., Rottenberg, H. and Avron, M. (1972) Eur. J. Biochem 25, 64-70
11 Jagendorf, A.T. an Uribe, E.G. (1966) Brookhaven Symp. Biol. 19, 215-246
12 McCarty, R.E., Fuhrman, J.S. and Tsuchiya, Y. (1971) Proc. Natl. Acad. Sci. USA 68, 2522-2526
13 Schwartz, M. (1968) Nature 219, 915-919
14 Ho, Y.K., Linn, C.J., Saunders, D.R. and Wang, J.H. (1979) Biochim. Biophys. Acta. 547, 149-160
15 Abbott, M.S. and Dilley, R.A. (1983) Archiv. Biochem. Biophys. 222, 95-104

BINDING OF 2´(3´)-O-(2,4,6 TRINITROPHENYL)-ADP OPENS THE
PATHWAY FOR PROTONS THROUGH THE CHLOROPLAST ATPase COMPLEX

Richard Wagner[1], Gudrun Ponse[2] and Heinrich Strotmann

[1] Universitaet Osnabrueck, Fachbereich Biologie/Chemie-Biophysik
Postfach 4469, 4500 Osnabrueck (FRG) [2] Botanisches Institut (Lehrstuhl
II, Biochemische Pflanzenphysiologie) Universitaet Duesseldorf,
Universitaetsstrasse 1, D-4000 Duesseldorf (FRG)

INTRODUCTION
Photosynthetic ATP synthesis in green plants is catalyzed by the
oligomeric ATPase complex which acts as a reversible ATP dependent proton
as proposed by Mitchell (3). Its molecular mechanism is a matter of
intensive research. Although basically the "energy linked binding change
mechanism" is becoming widely accpeted (1,2,6), there are open problems
concerning the detailed mechanism. One of them is, how the protons of
the electrochemical gradient are transferred to the catalytic site(s) or,
if they are not involved in the chemical reaction directly, how the
energy stored in the electrochemical gradient is conducted to the active
site(s) and how the proton conductance or the input of energy is
regulated. In this paper we report on experiments which show, that in
the absence of phosphate the nucleotide analogue TNP-ADP increases proton
permeability of the thylakoid membrane in a similar way as ADP plus P_i.
Further results suggest, that this is achieved by screening of the P_i
binding site by the trinitro- phenyl substituent in a way that a
conformation of the active site is induced which resembles the one
induced by binding of ADP and P_i. The results suggest that proton
conductance through the CF_0-CF_1 complex is triggered by simultaneous
binding of ADP and P_i.

MATERIALS AND METHODS
Broken chloroplasts from spinach or peas were prepared as in (10) or
(11), respectively. Photophosphorylation and electron transport were
measured as described in reference 12. Measurement of flash induced
absorbance changes in chloroplast suspensions, namely, the pH_{in} and pH_{out}
indicating absorbance changes of neutral red and cresol red as well as
the elcetrochromic absorbance changes were measured repetitively as
described in detail elsewhere (13,14,15). TNP-ADP was synthesized from
ADP and 2,4,6-trinitrobenzosulfonic- acid by a procedure given in Ref.16.

RESULTS AND DISCUSSION
Rates of phosphorylation as a function of ADP concentration in the
absence and presence of TNP-ADP are shown in Fig.1. Lineweaver-Burk
plots reveal a competitive type of inhibition by this ADP derivative.
(K_i values between 0.7 and 2 µM) In contrast to ADP, TNP-ADP in the
absence of P_i accelerates basal electron transport (Fig.2). Average
stimulation in different experiments was about 30% at TNP-ADP saturation
(half maximal effect at about 2 µM). The stimulative effect was
abolished by simultaneous addition of ADP (Fig.2). At low TNP-ADP/ADP
ratios the ADP-dependent inhibition on basal electron transport is
apparent whereas with increasing TNP-ADP/ADP ratio the stimulativ effect
by TNP-ADP is becoming more pronounced.

Biggens, J. (ed.), Progress in Photosynthesis Research, Vol. III. ISBN 90 247 3452 5
© *1987 Martinus Nijhoff Publishers, Dordrecht. Printed in the Netherlands.*

Figure 1

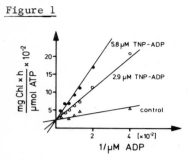

Figure 2.

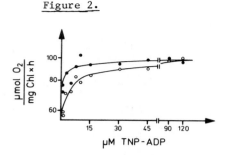

Figure 1: Double reciprocal plot of photophosphorylation with ADP as variable substrate in the prescence of TNP–ADP. Figure 2: Rate of electron transport from H_2O to MV in the absence ($\bullet$) and in the presence of 5 μM ADP (o) at variable TNP–ADPconcentration. Figure 3: Rate of electron transport from H_2O to MV in the presence of 30 μM TNP–ADP and variable concentration of P_i. Dashed line: rate of electron transport in the absence of P_i. The stimulation of basal electron transport by TNP–ADP is also abolished by the addition of P_i as shown in Fig.3. The half maximal effect is obtained at P_i concentration between 20–50 μM. TNP–ADP-dependent stimulation of electron transport is completely suppressed if chloroplasts are preincubated with 20 μM DCCD for 10 min (not shown). H^+-permeability of the thylakoid membrane was monitored by the changes of proton concentration in the medium as well as in the thylakoid lumen, employing flash-induced absorption changes of the pH indicator dyes cresol red (CR) and neutral red (NR), respectively (13,14).

Figure 3

Figure 4

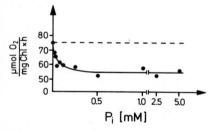

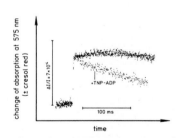

Figure 4: Time course of the pH_{out} indicating absorption changes of cresol red at 575 nm. The sample, 3 ml, contained chloroplasts equivalent to 20 μg chlorophyll; pyocyanin, 20 μm; cresol red, 15 μM; KCl, 50 mM; $MgCl_2$, 2.5 mM. The pH was adjusted to pH 7.9 before measurement. Time resolution was 200 μs/point; the shown signals were averaged over 15 single events. TNP–ADP when added was 20 μM. Figure 5: Time course of the pH_{in} indicating absorption changes of neutral red at 548 nm. The pH_{in} indicating absorption change resulted from the difference of the signals obtained in the presence of neutral red and in the absence of the dye. The sample 3 ml, contained; chloroplasts equivalent to 20 μg chlorophyll/ml; pyocyanin, 20 μM; KCl, 25 $MgCl_2$, 2.5 mM; BSA, 1.3 mg/ml; $\pm$ neutral red, 13 μM. Before measurement the pH was adjusted to 7.4. Time resolution was 500 μs/point; the shown signals were averaged over 20 single events.

upper trace: control chloroplasts (top) and chloroplasts in presence of
        20 µM TNP-ADP.
lower trace: chloroplasts treated for 10 min with 20 µM DCCD (20°C)
        before the addition of TNP.
Figure 4 shows the time course of absorption changes of the
non-permeative dye cresol red at 575 nm in the absence (upper trace) and
presence of 20 µM TNP-ADP (lower trace). In the shown time range there
is virtually no decrease of absorption in the control. In experiments
conducted at a different time scale, proton efflux of the control
proceeded with a half time of 12 s (data not shown). A fast decay of the
signal is observed (half time 150 ms) with chloroplasts incubated in the
presence of TNP-ADP, indicating an increase of the membrane permeability
for protons (ADP 50µM added after TNP-ADP incubation completly reversed
the effect,not shown). Figure. 5 shows the time course of the pH_{in}
indicating absorption change of neutral red at 548 nm. The lower one
shows the time course obtained for a chloroplast suspension containing
TNP-ADP (20 µM) after pretreatment with 20 µM DCCD (10 min). With the NR
method the increase of the rate of proton efflux by TNP-ADP is likewise
evident 9half time 250 ms compared with 10 s in the control). DCCD
treatment largely abolished the effect of TNP-ADP, indicating that the
nucleotide analogue affects proton flux through the ATPase complex rather
than increaes the proton permeability of the membrane in an unspecific
way.
Figure 5   Figure 6

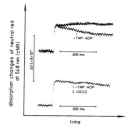

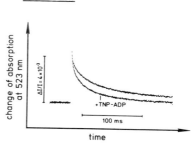

Figure 6 shows the effect of 20 uM TNP-ADP on the electrochromic
absorption change at 523 nm after a single turnover flash. TNP-ADP
accelerates the decay of the signal and this acceleration is again
reveresed by ADP or by preincubation of the thylakoids with DCCD (not
shown). The difference of the decay kinetics [+]TNP-ADP reveals ahalf time
of 200 ms. This time is in good agreement with the half times obtained
with the pH_{out} or pH_{in} indicating dyes, suggesting that the
counterflowing charges which lead to the observed acceleration of the
electrochromic absorption change are protons. As demonstrated in Figure
1, TNP-ADP, a trinitrophenylether of ADP, is an effective competitive
inhibitor of photophosphorylation. TNP-ADP is poorly phosphorylated to
TNP-ADP by chloroplasts (1-2% of ADP phosphorylation), thus fulfilling
another important property of a competitive inhibitor. The inhibitory
power of TNP-ADP in ADP phosphorylation decreases with increasing
phosphate concentration. The apparent K_i values at 1 and 5 mM P_i were
determined as 0.8 and 1.4 µM, respectively. These results suggest that
TNP-ADP not only covers the ADP binding site but occupies part of the P_i
binding site at the catalytic center. Hence, lack of phosphorylation of
TNP-ADP may be explained by prohibiting the access of inorganic

phosphate. ADP and ATP not only serve as substrates but also regulate chloroplast ATPase in multiple ways. ADP in the absence of P_i (and also ATP) inhibits proton efflux through the ATPase complex (7,8). As a consequence, basal electron transport is also inhibited by feed back control (7). The nucleotide concentrations required as well as the nucleotide specificities suggest that tight nucleotide binding to CF_1 is responsible for the observed effect. TNP-ADP-binding in the absence of P_i obviously effects increase in proton conductance of the ATPase complex above the one obtained with the nucleotide-free enzyme. Hence, TNP-ADP is recognized like ADP plus phosphate rather than ADP alone which decreases proton conductance. Provided that the trinitrophenyl residue not only screens the P_i binding site but also interacts with the site in a way similar to the substrate P_i, this interaction may trigger the opening of the proton channel. The finding that the effect of TNP-ADP is abolished by ADP as well as by P_i alone additionally support this view. The opening of the proton channel at the moment when both substrate molecules are ready for reaction would be a highly economical way of energy transduction.

REFERENCES

1. Amzel, L.M. and Pedersen, P.L. (1983) Annu. Rev. Biochem. 52, 801-821
2. Strotmann, H. and Bickel-Sandkoetter, S.B. (1984) Ann. Rev. Physiol. 35, 97-120
3. Mitchell, P. (1966) Chemiosmotic coupling in oxidative and photophos phorylation, pp. 1-192, Glynn Res. Lab. Bodmin, England
4. Senior, A.E. and Wise, G.J. (1983) J. Membr. Biol. 73, 105-124
5. Futai, M. and Kunazawa, H. (1983) Microbiol. Rev. 47, 285-312
6. Boyer, P.D. and Kohlbrenner, W.E. (1981) In: Inergy coupling in photosynthesis
7. McCarty, R.E., Fuhrmann, J.S. and Tsuchiya, Y. (1971) Proc. Natl. Acad.Sic. USA, 68, 2522-2526
8. Graeber, P., Burmeister, M. and Hortsch, M. (1981) FEBS Lett. 136, 25-31
9. Loehr, A., Willms, I. and Huchzermeyer, B. (1985) Arch. Biochem.
[A   Biophys. 236, 832-840
10. Strotmann, H. and Bickel-Sandkoetter, S.B. (1977) Biochim. Biophys. Acta 460, 126-135
11. Foerster, V., Hong, Y.Q. and Junge, W. (1981) Biochim. Biphys. Acta 638, 141-152
12. Avron, M. (1960) Biochim. Biophys. Acta 40, 257-272
13. Auslaender, W. and Junge, W. (1975) FEBS Lett. 59, 310-315
14. Junge, W. and Auslaender, W. (1973) Biochim. Biophys. Acta 333, 59-70
15. Junge, W. In: Chemistry and Biochemitry of Plant Pigments, Godwin, ed. 2nd ed., Academic Press, London, San Francisco, New York, pp. 233-332
16. Hiratsuka, T. and Uchida, K. (1973) Biochim. Biophys. Acta 320, 635-647

ACKNOWLEDGEMENTS
Financial support of the Deutsche Forschungsgemeinschaft is gratefully acknowledged (SFB 171/B2). Valuable discussion with Dr. Junge and Dr. J. Schumann as well as skillful preparation of the figures by Mrs H. Kenneweg are very much appreciated.

PROTONS CONTAINED IN THE THYLAKOID SEQUESTERED DOMAINS ARE UTILIZED FOR
ENERGIZING ATP SYNTHESIS

STEVEN M. THEG AND RICHARD A. DILLEY   Department of Biological Sciences,
Purdue University, West Lafayette, IN, 47907, U.S.A.

Evidence has accumulated in recent years indicating that protons can
be sequestered in membrane-associated domains in non-energized
thylakoids. These protons form a metastable pool and remain out of
equilibrium with the bulk aqueous phases for periods as long as an hour
unless uncouplers are added [1].
Dilley and Schreiber [2] demonstrated that the uncouplers, CCCP and
desaspidin, could be added to thylakoid membranes and then completely removed by a subsequent addition of BSA; apparently BSA binds the uncoupler
and functionally removes it from solution. They used this technique to
empty the domains of protons, and then assayed the chloroplasts for ATP
synthesis in flashing light in the presence of K^+ and valinomycin. They
found that after reversible uncoupling in alkaline medium, i.e., when the
domains were depleted of protons, an additional 10 flashes were required
to reach a ΔpH sufficient to overcome the thermodynamic threshold for ATP
synthesis.   In a second flash train applied to the same sample, the lag
for ATP synthesis was returned to that of the control. Those data
indicated that the domains must be full of protons in order for ATP
formation to begin, and that electron-proton transport refills the
domains with protons after a reversible uncoupling treatment.   The
hypothesis associated with Dilley and Schreiber's work was that the
domains form a localized pathway for protons to follow between their
sites of release from electron transport components to the chloroplast
coupling factors.   Their results, however, did not rule out the possibilities either that the domains form a "blind alley" buffering space
unrelated to photophosphorylation which must be filled before protons
can be deposited into the lumen, or that protons simply pass through the
domains before entering the lumen *en route* to the CF_0-CF_1 complexes.
An experiment to test whether protons in the domains enter the
coupling factors directly is to let the flash-induced $\Delta\psi$ be the predominant contributor to the protonmotive force (Δp) under conditions in
which the buried domains are either protonated at the beginning, or
emptied of protons prior to the excitation flash sequence.   In this case,
protons would play a more passive role in ATP synthesis.   Rather than
contributing significantly to the threshold Δp, they would be electrophoresed through the coupling factors and participate mainly in the enzymatic phosphorylation of ADP.   It is expected that those protons with the
easiest access to the coupling factors would be the ones used during $\Delta\psi$-
mediated ATP synthesis.   When the preincubation period is long enough to
allow equilibration of the pH in the inner and outer aqueous phases, the
number of flashes to the onset of ATP synthesis should be independent of
the protonation state of the domains if protons in the lumen are poised
to pass next through the coupling factors upon membrane energization.
On the other hand, if domain protons are those used during $\Delta\psi$-driven

Biggens, J. (ed.), Progress in Photosynthesis Research, Vol. III. ISBN 90 247 3452 5
© *1987 Martinus Nijhoff Publishers, Dordrecht. Printed in the Netherlands.*

phosphorylation, then an additional number of flashes should be required
before ATP synthesis begins after a reversible uncoupling treatment.
The experiments reported below are consistent with the latter prediction.

MATERIALS AND METHODS

Chloroplasts were isolated as described in [3], except 5 mM DTT was
present in the grinding and resuspension media. ATP synthesized in
flashing light was monitored by following luminescence from luciferin-
luciferase [2]. The ATP onset lags were determined as the flash at which
the luciferin luminescence first rose above the baseline (first number),
or by the flash to which the steady-state ATP flash yield extrapolated
back to the baseline (second number). The two numbers are reported in the
figures separated by a colon. All experiments were performed at 10° C.

RESULTS AND DISCUSSION

Fig. 1 shows that the ATP formation onset lag, when $\Delta\psi$ contributes
significantly to Δp, is sensitive to the protonation state of the sequestered domains. In dark-adapted samples (left side), curves a and b were
obtained with samples having full domains and display lags of 4:18 and
8:21 flashes, respectively. Emptying the domains of protons by reversible uncoupling with CCCP and BSA resulted in an extension of the onset
lag to 16:33 flashes (trace c). Complete withdrawl of the CCCP by BSA is
proven by the high ATP flash yields obtained in traces b and c. That $\Delta\psi$
was a major contributor to the Δp at the onset of ATP synthesis is
demonstrated by the long lag obtained in the presence of valinomycin
(trace d).

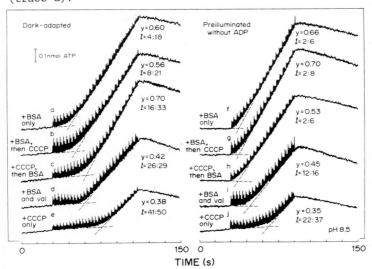

Figure 1. The effect of reversible uncoupling on the flash lag for ATP synthesis. Traces show ATP-dependent luciferin luminescence. Time (t) = 0, chloroplasts were added to an alkaline phosphorylation medium containing the indicated compounds and lacking ADP; t = 30 s, 1 mg BSA/ml or 400 nM CCCP added for the second or third traces; t = 1 min, 10 s preillumination of samples on the right; t = 3.5 min, + ADP; t = 4 min, + luciferinluciferase; t = 5 min, start flashes at 1 Hz. val = valinomycin; y = nmol ATP formed per mg Chl per flash; l = ATP onset lag.

If 10 s of illumination were given to comparable samples <u>after</u> the
CCCP and BSA treatments (right side), there was no difference in the

onset lags for the top three samples. This is consistent with the notion that the illumination refilled the protonatable domains, and in that state those protons were "in line" to be driven into the CF_o-CF_1 in energy linked proton flow. Four minutes of dark adaptation intervened between the 10 s illumination and the start of the flash sequence, and ADP was withheld until after the preillumination to avoid any ATP formation and subsequent proton pumping during the dark period.

Figure 2 shows the result of an experiment similar to that in the left side of Fig. 1, but performed at pH 7.0. It can be seen that the reversible uncoupling treatment was essentially without effect on the ATP onset lags at this pH. This behavior would be predicted if the initial pH in the domains before the flash sequence was close to the external pH, in which case no proton movement would be induced by the uncoupler. A similar pH dependence, titrating with a pK between 7.2 and 7.8, has been found of every phenomenon so far purported to be governed by domain protons (c.f. [4]).

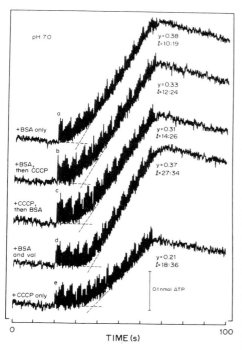

Figure 2. Effect of reversible uncoupling on the ATP onset flash lags at neutral pH. Chloroplasts were assayed in pH 7.0 buffer. Reversible uncoupling was completed by t = 30 s; flash sequence at 2 Hz initiated at t = 3 min. val, y and l defined as in Fig. 1.

One possible role for the protons in the domains could be to provide an activating proton flux through the coupling factors [5], rather than act to in the enzymatic formation of ATP. We tested this possibility by examining the effect of DDT activation of the coupling factors on the ability of the reversible uncoupling treatment to lengthen the ATP onset lags. Traces a-c in Fig. 3 show that considerable ATPase activity could be demonstrated by adding ATP (at t = 8 min) to thylakoids that had received a prior two minute preillumination (at t = 1-3 min) in the presence of 5 mM DTT. It can be seen that the reversible uncoupling treatment (at t = 6 min) had no effect on the activation of the coupling factors. When thylakoids similarly DTT activated were assayed for flash-driven ATP synthesis activity (at t = 8 min), the reversibly uncoupled sample still displayed an extension of the ATP onset lag (traces d and e). This demonstrates that protons in the domains were not required simply to convert the latent coupling factors to an activated state analogous to that achieved by illumination in the presence of DTT.

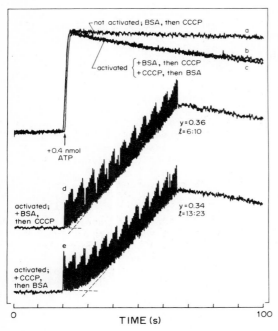

Figure 3. Effects of DTT
activation of the coupling
factor on the extension of the
ATP onset flash lag by reversible uncoupling. Protocol:
t = 0, chloroplasts added to
pH 8.5 assay medium containing
neither BSA nor CCCP (DTT
present); t = 1 min, 2 min
steady-state preillumination
with red light started; t = 6
min, first compound added (BSA
or CCCP); t = 6.5 min, second
compound added (CCCP or BSA);
t = 6.75 min, ADP added; t = 7
min, luciferin-luciferase
added; t = 8 min, 0.4 nmol ATP
added (traces a-c) or sequence
of flashes at 2 Hz intitated
(traces d and e). The sample
not activated (trace a) was
not preilluminated between 1
and 3 min. y and l defined as
in Fig. 1.

Other experiments not shown indicate that our results cannot be
explained as either an incomplete equilibration of protons in the lumen
with those in the external phase before the flash sequence was initiated,
or as an incomplete removal of the CCCP by BSA before the Δp threshold
for ATP synthesis was reached.

The experiments reported here establish that protons in the domains
are the ones electrophoresed through the coupling factor during Δψ-driven
photophosphorylation. This suggests that the easiest route through which
protons gain access to the CF$_o$ portion of the coupling factors is through
the domains, not through the lumen. As discussed in recent reviews
[4,6], these data, taken with others showing that protons do not pass
through the lumen when traveling between their sites of release and the
coupling factors, make it reasonable to propose that the domains provide
a localized pathway for proton movement in energized thylakoid membranes.

1. Laszlo, J.A., Baker, G.M. and Dilley, R.A. (1984) J. Bioenerg.
   Biomembr. 16, 37-51.
2. Dilley , R.A. and Schreiber, U. (1984) J. Bioenerg. Biomembr. 16, 173-
   193.
3. Theg, S.M., Belanger, K.M. and Dilley, R.A. (1986) J. Bioenerg.
   Biomembr., in press.
4. Theg, S.M. and Dilley, R.A. (1987) Ann. Rev. Plant Physiol., in
   preparation.
5. Abbot, M.S. and Dilley, R.A. (1983) Arch. Biochem. Biophys. 222, 95-
   104.
6. Jagendorf, A.T. and Anthon, G.E. (1985) in: Molecular Biology of the
   Photosynthetic Apparatus (Steinback, K.E., et al., eds.) Cold Spring
   Harbor.

FURTHER EVIDENCE THAT KCL-THYLAKOID ISOLATION INDUCED A LOCALIZED PROTON GRADIENT ENERGY COUPLING TO BECOME BULK PHASE DELOCALIZED

W.A. Beard and R.A. Dilley

Dept. of Biological Sciences, Purdue University,
West Lafayette, Indiana, 47907, U.S.A.

INTRODUCTION

Thylakoid ATP formation can be driven by a bulk phase delocalized protonmotive force, Δp, such as in the acid-base experiment with succinate as an internal H^+ reservoir (1) or in the post-illumination phosphorylation (PIP) protocol with pyridine as the internally accumulated H^+ reservoir (2). Other experiments were not consistent with bulk phase delocalized coupling and a localized Δp has been postulated (3). The anomalies in the results demand resolution. Following a suggestion by Sigalat et al. (4), we recently demonstrated that thylakoids isolated in the presence of KCl exhibited pyridine-dependent increases in the length of the lag for the onset of ATP formation and of PIP, while in thylakoids isolated in a resuspension buffer lacking KCl, these parameters were pyridine-insensitive when the external pH was 8 (5). To verify that pyridine was acting as an internal pH buffer, we demonstrate here that at an external pH of 7, the pyridine-dependent effects were as expected from the predictions based on theoritical considerations. These results support and extend our earlier report (5) that KCl induces a shift from localized to bulk phase delocalized proton gradient energy coupling.

MATERIALS AND METHODS

Chloroplast thylakoids were isolated from growth chamber-grown spinach as described by Ort and Izawa (6). The thylakoids were washed once and resuspended in 200 mM sucrose, 5 mM Hepes (pH 7.5), 2 mM $MgCl_2$, and 0.5 mg/ml BSA or in a resuspension buffer where the sucrose was replaced with 100 mM KCl.

Saturating single-turnover flashes were used to initiate electron transfer and luciferin-luciferase luminescence was used to follow ATP formation as described earlier (5).

RESULTS AND DISCUSSION

Our assay utilized single-turnover flashes and the luciferin-luciferase ATP detection method (5). The flash excitation allows control over the magnitude of energization and permits easy detection of the lag for the onset of ATP formation. The sensitive luciferin-luciferase system permits direct observation of PIP after the last flash in a flash sequence. The effect of permeable buffers on extending the flash lag for the onset of ATP formation in the presence of valinomycin-K^+ and on changing the PIP yield are used as indicators of bulk phase delocalized proton gradient energization (2,7,8). The absence of an effect by a permeable buffer on the ATP onset lag (3) and PIP ATP yield (5) are taken as indicative of localized proton gradient coupling.

Biggens, J. (ed.), Progress in Photosynthesis Research, Vol. III. ISBN 90 247 3452 5
© *1987 Martinus Nijhoff Publishers, Dordrecht. Printed in the Netherlands.*

The effect of high-KCl storage treatment on the onset of ATP formation at pH 7 was quite dramatic (fig. 1). Compared to the usual resuspension buffer, the high-KCl treatment caused the lag to increase from 31/38 to 66/95 flashes (see legend to fig. 1 for explanation of flash parameters). Table 1 gives a compilation from several assays documenting the point illustrated in fig. 1 (a, - KCl; c, + KCl).

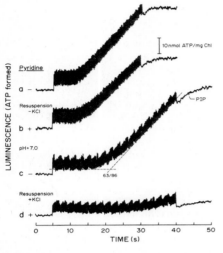

FIGURE 1. The effect of pyridine on single-turnover flash initiated phosphorylation with thylakoids resuspended in the absence or presence of 100 mM KCl. Flashes (125 in a,b; 175 in c,d) were delivered at a rate of 5 Hz to thylakoids with 15 μM chlorophyll in a reaction medium containing 50 mM MOPS-KOH (pH 7.0), 10 mM sorbitol, 3 mM $MgCl_2$, 1 mM KH_2PO_4, 5 mM DTT, 0.1 mM ADP, 0.1 mM methyl viologen, 400 nM valinomycin and 5 μM diadenosine pentaphosphate. The flash lag for the onset of ATP formation was determined with the aid of two criteria: 1) the first detectable rise in luminescence (63 in c) and 2) the back extrapolation of the steady rise in the flash-induced luminescence increase to the x-axis (86 in c). The increase in luminescence after the last flash was due to post-illumination phosphorylation (PIP).

TABLE 1. Effect of a KCl resuspension treatment and pyridine on the onset lag for ATP formation at pH 7.0. Conditions were as in fig. 1. The lags for the onset of ATP formation were determined as described in fig. 1 and represent the actual/extrapolated lags. The ATP yield per flash was determined from the linear rise in luminescence (see fig. 1). The results are the means of 4 observations ± SE.

Conditions	Number of flashes to the onset of ATP formation	ATP yield per flash [nmol ATP (mg Chl flash)$^{-1}$]
- KCl storage		
- pyridine	31 ± 1/ 38 ± 2	0.71 ± 0.09
+ pyridine	36 ± 3/ 45 ± 1	0.62 ± 0.06
+ KCl storage		
- pyridine	66 ± 3/ 95 ± 6	0.41 ± 0.03
+ pyridine	86 ± 6/109 ± 7	0.09 ± 0.02

When the external pH is 7, 85% of the pyridine will become protonated before a ΔpH of 2.3 units is built up to reach the energetic threshold, therefore delaying the onset of ATP formation to a greater extent than when the external pH is 8. Clearly fig. 1 illustrates the sensitivity of the lag for the onset of ATP formation toward 5 mM pyridine (pK_a = 5.4) when thylakoids were isolated in the presence of high KCl. However, thylakoids resuspended in a buffer lacking KCl exhibited lags which were nearly insensitive toward pyridine. These

points are documented in table 1. Thus, thylakoids which were isolated
in the absence of KCl exhibit a localized energy coupling behavior, while
high-KCl isolation induces a bulk phase delocalized coupling.

The luciferin-luciferase technique used in these assays permitted
direct detection of PIP, as shown in fig. 1 as the increase in
luminescence after the last flash. Thus, the effects of permeable
buffers on PIP allow an additional test for distinguishing localized from
bulk phase delocalized energy coupling. In those experiments, pyridine
had no effect on the PIP yields in the control thylakoids (- KCl), but
significantly inhibited the PIP yield in the high KCl-stored sample (fig.
2 and table 2). When the external pH was 8, pyridine stimulated the PIP
yield with thylakoids isolated in the presence of high KCl, as expected
(5). Lowering the external pH to 7 alters the sensitivity to pyridine so
as to observe an inhibition with high-KCl isolated thylakoids. At an
external pH of 8, pyridine can serve as a large reservoir of protons
after the energetic threshold is reached. However, at pH 7, very little
pyridine is available (15%) after the energetic threshold has been
reached. Since unprotonated pyridine can cross the thylakoid as
internally located pyridine becomes protonated (2), even longer
illumination times are required to observe a stimulation of the PIP ATP
yield at pH 7.0 (8). Kinetic analysis of the luminescent signals in fig.
2 indicate that ATP formation decays as a single exponential with $t_{1/2}$ =
1.1 s [-KCl ($\pm$ pyridine)] and a $t_{1/2}$ = 3.9 s [+KCl ($\pm$ pyridine)]. This
may indicate that the buffering groups contributing to the PIP ATP yield
are of different origin. The two fold increase in the extent of the PIP
ATP yield observed with KCl-isolated thylakoids is consistent with this
view. The sensitivity of the PIP ATP yield toward pyridine indicates
that the slower decaying ATP formation is of lumenal origin. In contrast
to pH 8 where the decay of the PIP is longer in the presence of pyridine
for KCl-isolated thylakoids [fig. 2 of (5)], pyridine has no effect on
the decay at pH 7 due to lumenally-located protonated pyridine's
inability to serve as a reservoir of energetically competent protons.

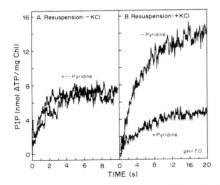

FIGURE 2. *The effect of pyridine on
the luminescence resulting from post-
illumination phosphorylation at pH 7.0
with thylakoids resuspended in the
absence (A) or presence (B) of 100 mM
KCl. The PIP portions of the
luminescent signals in fig. 1 are
presented.*

The substitution of 100 mM KCl for 200 mM sucrose in the thylakoid
resuspension buffer induced a sensitivity, of both the lag for the onset
of ATP formation and the PIP ATP yields, toward pyridine in a manner
consistent with bulk phase delocalized coupling. In previous work which
indicated that permeable buffers could increase the lag for the onset of
ATP formation (7,8) the conditions for the phosphorylation assays
included high concentrations of KCl, whereas the experimental conditions

TABLE 2. *Effect of a KCl resuspension treatment and pyridine on post-illumination ATP formation at pH 7.0. Conditions were as in fig. 1. PIP ATP yield was determined from the increase in signal after the last flash in a flash sequence, while the remaining rise in signal from the onset of ATP formation to the beginning of PIP represents the total ATP yield from the flash train. The results are means of 4 observations ± SE.*

Conditions	Post-illumination ATP yield [nmol ATP (mg Chl)$^{-1}$]	Total ATP yield during flash train
- KCl storage		
- pyridine	7.7 ± 1.0	63 ± 6
+ pyridine	7.9 ± 0.8	51 ± 5
+ KCl storage		
- pyridine	11.5 ± 0.9	34 ± 4
+ pyridine	5.4 ± 1.6	7 ± 1

for the first demonstration of a lack of an effect of such buffers on the lag for the onset of ATP formation (3) had only enough KCl to collapse the membrane potential with the aid of valinomycin. It should be emphasized that all our assays were performed in the same phosphorylation medium which lacked KCl but had enough MOPS-KOH present to collapse the membrane potential via valinomycin-dependent K^+ flux.

The increase in the lag for the onset of ATP formation induced by pyridine for high-KCl stored thylakoids is about that calculated for the increase in buffering capacity of the lumen by pyridine under the assumption that a bulk phase delocalized proton gradient could protonate pyridine. If the energetic threshold ΔpH for ATP formation is 2.3 pH units (9), then the calculated extension of the onset of ATP formation by pyridine at pH 7.0 is 19 flashes compared to the measured 20 flash extension (table 1).

The KCl effect reported here agrees with the suggestion of Sigalat et al. (4) and extends the findings observed at pH 8 (5). It appears that the earlier contradictory results concerning localized or bulk phase delocalized proton gradients (3,7,8) may now be resolved. However, studies to determine the mechanism behind the high KCl inducing a shift from localized to bulk phase delocalized energy coupling remain to be done.

REFERENCES
1 Jagendorf, A.T. and Uribe, E. (1966) Proc. Natl. Acad. Sci. U.S.A. 55, 170-177
2 Nelson, N., Nelson, H., Naim, Y. and Neumann, J. (1971) Arch. Biochem. Biophys. 145, 263-267
3 Ort, D.R., Dilley, R.A. and Good, N.E. (1976) Biochim. Biophys. Acta 449, 108-124
4 Sigalat, C., Haraux, F., de Kouchkovsky, F., Hung, S.P.N. and de Kouchkovsky, Y. (1985) Biochim. Biophys. Acta 809, 403-413
5 Beard, W.A. and Dilley, R.A. (1986) FEBS Lett. 210, 57-62
6 Ort, D.R. and Izawa, S. (1973) Plant Physiol. 52, 595-600
7 Davenport, J.W. and McCarty, R.E. (1980) Biochim. Biophys. Acta 589, 353-357
8 Vinkler, C., Avron, M. and Boyer, P.D. (1980) J. Biol. Chem. 255, 2263-2266
9 Graeber, P., Junesch, U. and Schatz, G.H. (1984) Ber. Bunsenges. Phys. Chem. 88, 599-608

DELOCALIZATION OF ENERGY COUPLING IN THYLAKOIDS BY AMINES

Y. DE KOUCHKOVSKY, C. SIGALAT AND F. HARAUX. Laboratoire de Photosynthèse, C.N.R.S., B.P. 1, F 91190 Gif-sur-Yvette, France.

SUMMARY. The previously proposed microchemiosmotic scheme, in which a resistive proton pathway is separated from the bulk medium by diffusion barriers, implies that their lowering should delocalize the coupling between redox chains and F_0F_1-ATPases. This was already obtained by various treatments (osmotic, ionic) and now also by the use of permeant amines. The latter point explains the discrepancy of results found in literature.

INTRODUCTION
    Although chemiosmotic principles offer the best ground for understanding the energetic coupling of biomembranes, their classical formulation faces several theoretical and experimental challenges (review [1]). Among them are the flux-force relationships, where flux is for instance the rate of ATP formation (V_p) and force is the proton electrochemical potential $\Delta\mu H$. If the coupling device comprises only two aqueous compartments, separated by a membrane, in each of which protons are delocalized, these relationships should be at steady-state independent on the way each term is varied. Since this is generally not the case, several alternatives were recently proposed, some [2] close to 'direct coupling', other [3] bringing a new life to 'chemical' or 'conformational' hypotheses. Actually, we think that chemiosmotic theory must take into account additional facts (see [4]): 1/ a membrane is separated from the medium by an interface (cf. [5]); 2/ the membrane organization is heterogeneous, and the proton 'producers' (redox chains) and 'consumers' (coupling factors) are spatially separated, especially with system II [6]; 3/ because H^+ are translocated first via the membrane, they may move in part along the interface — and even on or in the membrane — in addition of travelling across the delocalizing bulk phase of free water (which should barely exist in a lumen of usual thickness); 4/ this lateral pathway may be sufficiently resistant to cause a small lateral pH drop, making that $\Delta\mu H$ is higher at the level of redox pumps and lower at that of coupling factors; and 5/ the balance between localized and delocalized pathways (which must be separated by some diffusion barriers) thus the pH lateral drop, should strongly depend on the membrane topography and on the physico-chemical characteristics of the medium. By reducing the lateral resistance and rupturing the diffusion barriers, via osmotic and ionic effects [7], one could shift from localized to delocalized behavior. The latter was consistently observed with thylakoids by McCarty's group [8], which led us to extend our investigation on the flexibility of coupling. This was done here by using buffers to favor a random distribution of H^+. Among them, permeant amines [9] were good candidates, especially since a stimulation of ATP synthesis in their presence was reported by Giersch and coworkers [10,11].

METHODOLOGY
    Thylakoids were prepared from lettuce as in [7], which also describes the experimental protocol: ΔpH was measured with 9-aminoacridine (4 μM) and ATP by the luciferase reaction. The medium at pH 7.8 (in an open, stirred and thermostated at 20°C, spectrophotometric cuvette) contained in mM: sorbitol 200 + Tricine 10 + Hepes 10 + KCl 10 + $MgCl_2$ 6 + K_2HPO_4 2 + ADP 0.5;

valinomycin (50 nM) was added to supress $\Delta\Psi$ and di-adenosine pentaphosphate (10 μM) to inhibit possible kinases. Chlorophyll was 10 μM and maximum red light was 1.3 kW m^{-2} but was varied with neutral filters (up to < 1%). Cyclic System I chain was induced by pyocyanin (PYO), linear System II chain by dimethylquinone (DMQ). All experiments were at steady-state.

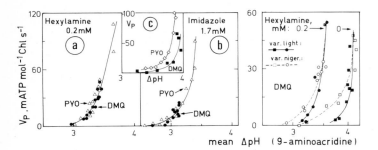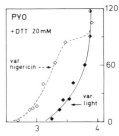

mean ΔpH (9-aminoacridine)

FIGURE 1. Delocalizing effect of amines (a: hexylamine 0.2 mM, b: imidazole 1.7 mM) on the coupling of phosphorylation rate (V_P) to proton gradient (restricted here to ΔpH: valinomycin present). Insert (c): typical behavior of SI (pyocyanin PYO) and SII (dimethylquinone, DMQ) chains without amines. Conditions: see Methodology, Curves traced by varying light.

FIGURE 2. Effect of hexylamine (0.2 mM) on the response of phosphorylation rate (V_P) to the adjustment of ΔpH by a light decrease (—) or by up to 40 nM nigericin addition at maximum light (- -). Conditions: see Methodology; dimethylquinone SII-chain.

FIGURE 3. Absence of dithiothreitol (DTT) effect on the differential sensitivity of phosphorylation rate (V_P) to ΔpH adjustment by light decrease (—) or by up to 800 nM nigericin addition at maximum light (- -). Conditions: see Methodology; pyocyanin chain. Before experiment, F_1 was reduced by 3 mn illumination in presence of 20 mM DTT, then the system relaxed 2 mn; after experiment, F_1 was inhibited by 5 μM tri-N-butyltin to prevent dark hydrolysis of ATP. (The control situation — no DTT modulation — gives similar curves, as Fig. 8c in [7].

RESULTS

One amine which deserves special attention is hexylamine, since it is frequently used when radioactively labelled for measuring ΔpH [8]. The authors of this reference saw no difference between SI and SII chains in their phosphorylating ability, at variance with us. We have already shown that the medium composition plays a key role in this type of experiments: high KCl in absence of neutral osmoticum (sorbitol), as in [8], may indeed reduce the discrimination between SI and SII: see merged curves with open symbols, Fig. 9 right, in [7]. The other difference between our two groups is the nature of ΔpH probe. We are not disputing here their pros and their cons — we could confirm with glass electrode the validity of comparative studies made with 9-aminoacridine [7] — but we wish to point out to the fact that hexylamine is used at concentration high enough to possibly affect proton conduction in the lumen. Giersch [11] had already noticed a stimulating effect of low amount of methylamine, analogous to hexylamine, on ATP synthesis. Actually, the response concentration curves of phosphorylation are complex, but at high enough concentration, the main effect is proton delocalization. This is shown Fig. 1: whereas in absence of amine, the V_P vs. ΔpH curves with pyocyarin (SI) and dimethylquinone (SII) are distinct, they are superimposed, even in our normal osmotic conditions, if enough hexylamine (a) or imidazole (b) is present. We have also observed that 50 µM hexylamine in even suffi- cient to render comparable the SI and SII ability for ATP synthesis at given ΔpH in a medium of high KCl, without sorbitol, all conditions met in [8].

We always noticed [7] that a medium which physico-chemical characteris- tics favor such a SI-SII equality is still unable to fully suppress the dif- ference of behavior observed when ΔpH is set at a given value by acting on H$^+$ efflux (nigericin) or influx (light). We investigated how active are ami- nes in this respect and Fig. 2 illustrates the case of hexylamine on SII chain. Indeed, the nigericin curve (dashed) becomes then less distant from the light one (continuous) considering the ΔpH factor; the actual magnitude of this effect depends on the type of chain, the nature and concentration of amine, and the medium. A second point is apparent on this Fig. 2: a shift of the curves on ΔpH scale, due to an overall ΔpH lowering at the same time ATP synthesis is relatively stimulated; this will be discussed below. How- ever, a possible explanation of the higher phosphorylation in presence of nigericin could be an increased electron flow which would help to activate coupling factors. It is indeed known that reduction of a S-S bridge in F_1 enhances its sensitivity to ΔpH [12]. Although the thioredoxin system is un- likely to operate in washed thylakoids, we examined the impact, on the nige- ricin vs. light adjustment of ΔpH, of a preliminary reduction of this enzyme by dithiothreitol. Fig. 3 (no amine present) demonstrates that this treat- ment does not change the picture. (A comparison of SI and SII chains is not possible because dimethylquinone gets chemically reduced).

TABLE 1. Summary of amine effects near optimum concentration (see text)

Mean values (3-5 samples)	1.7 mM Imidazole		50 µM Hexylamine	
	PYO	DMQ	PYO	DMQ
ΔpH control	3.85	3.90	4.06	3.96
ΔpH amine	3.71	3.50	3.87	3.74
Vp control	72.4	45.8	81.1	47.4
Vp amine	100.5	49.4	96.9	51.3

DISCUSSION

Table 1 summarizes mean effects, at steady-state, of amines at concen- trations generally used in this work. These are a compromise between the be- ginning of uncoupling (significant ΔpH lowering and decrease of phosphoryla- tion) and the maximum of stimulation (highest phosphorylation for about un- changed ΔpH). That phosphorylation may be stimulated whereas $\Delta\mu$H is unchan- ged or even decreased was interpreted by Giersch as due to a 'direct type of coupling' between redox chains and coupling factors, not impaired by uncou- pler-induced H$^+$ leakage [10]. In fact, this would explain an insensitivity of phosphorylation to uncoupler, not a stimulation, and the additional idea of removing a 'kinetic barrier' [11], needs to be elaborated. Thus, in agreement with our previous conclusions drawn on numerous experiments (see [4], we think that this stimulation is due to proton delocalization. Indeed amines may carry back H$^+$ across the membrane if they are not fully imper- meant when charged, but moreover they may shuttle these H$^+$ along the membra- ne and the interface (not mentioning the lumen). That is the diffusion bar- riers between membrane/interface and bulk domains are short-circuited and the lateral resistance is shunted. Consequently, the actual $\Delta\mu H$ across F_1F_0 is raised with respect to the average $\Delta\mu H$, the only which is measurable and is expressed, as ΔpH, on the abscissa of the figures. At high enough amine concentration, the system becomes completely delocalized, hence identical phosphorylating ΔpH-yields of SI and SII (Fig. 1).

The nigericin concentration required to scan the V_p vs. ΔpH curve is one order of magnitude higher for SI than for SII (see Figs 2 and 3). This may result from a faster SI than SII turnover, hence a more intense H^+ influx able to compensate the nigericin leakage. Finally, the fact that even in the presence of these amines, the nigericin vs. light curves remain distinct, though less than in the control, could simply reflect that at low ΔpH induced by nigericin, the internal concentration of amine, which pumping is ΔpH-dependent [9], is then insufficient. Thus, without ruling out more complex pictures, one may explain the present data simply by the existence of two phases, separated by amine-(and nigericin-)sensitive barriers: the membrane and interface domain, somewhat resistant, and the bulk delocalizing medium. One do not need separate theories, as mentioned in the Introduction, to account for localized and delocalized situations. Also, one easily understands the reason of divergent results as in [8].

ACKNOWLEDGMENT
    This work was supported by a CNRS-ATP contract.

REFERENCES
 1 Haraux, F. (1985) Physiol. Vég. 23, 397-410
 2 Westerhoff, H.V., Melandri, B.A., Venturoli, G., Azzone, G.F. and Kell, D.B. (1984) FEBS Lett. 165, 1-5
 3 Slater, E.C., Berden, J.A. and Herweijer, M.A. (1985) Biochim. Biophys. Acta 811, 217-231
 4 de Kouchkovsky, Y., Sigalat, C., Haraux, F. and Phung Nhu Hung, S. (1986) in Ion Interactions in Energy Transfer Biomembranes (Papageorgiou, G.C., Barber, J. and Papa, S., eds), pp. 119-131, Plenum Press, New York
 5 Kell, D.B. (1979) Biochim. Biophys. Acta 549, 55-99
 6 Andersson, B. and Anderson, J.M. (1980) Biochim. Biophys. Acta 593, 427-440
 7 Sigalat, C., Haraux, F., de Kouchkovsky, F., Phung Nhu Hung, S. and de Kouchkovsky, Y. (1985) Biochim. Biophys. Acta 809, 403-413
 8 Davenport, J.W. and McCarty, R.E. (1984) Biochim. Biophys. Acta 766, 363-374
 9 Pick, U. and Avron, M. (1976) Eur. J. Biochem. 70, 569-576
10 Giersch, C. (1983) Biochim. Biophys. Acta 725, 309-319
11 Giersch, C. and Meyer, M. (1984) Bioelectrochem. Bioenerg. 12, 63-71
12 Mills, J.D. and Mitchell, P. (1982) Biochim. Biophys. Acta 679, 75-83

THE ACTIVATION OF THE REDUCED CHLOROPLAST ATP-SYNTHASE BY ΔpH

Ulrike JUNESCH and Peter GRÄBER

Max-Volmer-Institut für Biophysikalische und Physikalische Chemie, Technische Universität Berlin, Strasse des 17. Juni 135, 1000 Berlin 12, FRG

1. INTRODUCTION

The chloroplast ATP-synthase, CF_oF_1, can exist in - at least - four different states: an inactive, oxidized state, E_i^{ox}, an active, oxidized state, E_a^{ox}, an inactive, reduced state, E_i^{red}, and an active, reduced state, E_a^{red} (1). The change of the redox state of CF_oF_1 can be effected by mediators, e.g., dithiothreitol, DTT, iodosobenzoate, IBZ, etc. The change of the activation state can be effected by membrane energization, e.g., by ΔpH and $\Delta\Psi$. In this work we investigate the functional dependence of the activation of the reduced CF_oF_1 on ΔpH. The principle of our experiment is as follows: chloroplasts are illuminated in the presence of DTT; thereby CF_oF_1 is activated and reduced; i.e., the form E_a^{red} is generated. Rebinding of ADP then leads to inactivation, to the form E_i^{red}. When an acid/base transition is now carried out with this form, a part of the enzyme is transformed into E_a^{red}, the magnitude of this part being dependent on ΔpH. When all enzymes are transformed into E_a^{red} the maximum rate of ATP hydrolysis should be observed. Therefore, the fraction of CF_oF_1 transformed into E_a^{red} can be measured by the relative rate of ATP hydrolysis.

2. RESULTS

Class-II-chloroplasts were reduced by DTT treatment (2), ATP hydrolysis was measured by the release of ^{32}P; energization of the membrane by an acid/base transition was carried out as in ref. 3 and high time resolution was obtained by using a rapid-mixing quenched flow system (3).

2.1 The rate of ATP hydrolysis

Fig. 1 shows the P_i released as a function of the reaction time in the presence of 3 mM NH_4Cl and in its absence, when the CF_oF_1 was in the form E_a^{red}. The slope of these curves gives the rate of ATP hydrolysis which was 150 mM ATP(M Chl $\cdot$ s)$^{-1}$ in the presence of the uncoupler. In the absence of NH_4Cl the rate was a factor 7.5 lower. The amount of Chl per CF_4 was 600 as determined by rocket-immunelectrophoresis. This gives a rate of 90 ATP $(CF_1 \cdot s)^{-1}$. This is about a factor two higher than reported in earlier work. Whereas the curve in the presence of NH_4Cl extrapolates to zero when the reaction time becomes zero, the extrapolation in the absence of NH_4Cl gives a burst of P_i, indicating a fast phase of ATP hydrolysis shortly after initiating the reaction. With the

Biggens, J. (ed.), Progress in Photosynthesis Research, Vol. III. ISBN 90 247 3452 5

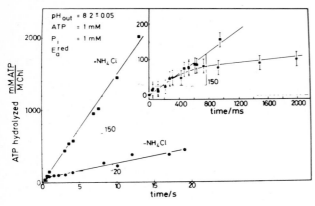

FIGURE 1.
ATP hydrolysis
catalyzed by CF_oF_1
in the form E_a^{red} in
the presence and ab-
sence of 3 mM NH_4Cl.
Insert: initial part
of the curve with
high time resolution.

rapid mixing quenched flow technique we investigated this
initial phase (Fig. 1, insert). It can be seen that in the
first 300 ms the rates of ATP hydrolysis in the presence and
absence of NH_4Cl are indentical and give a rate of 150 mM ATP
(M Chl s)$^{-1}$. In the presence of NH_4Cl this rate is constant up to about 20 s; whereas, in the absence of NH_4Cl after
300 ms the rate decreases and after about 1s the rate of 20 mM
ATP (M Chl s)$^{-1}$ is observed which is constant for at least
30 s. This result shows that during the first 300 ms the internal proton concentration is so low that the rate of ATP
hydrolysis is not limited by the deprotonation reaction at the
inside. Due to the hydrolysis-coupled influx of H^+, the proton concentration increases and after about 1 s the rate is
limited by the deprotonation reaction and a steady state is
reached where the hydrolysis-coupled H^+ influx is balanced by
the efflux of H^+ due to the basal permeability of the membrane.
Since the same rate is observed in the first 300 ms and in the
presence of NH_4Cl, it can be concluded that the NH_4Cl concentration is high enough so that ATP hydrolysis is not limited
by the deprotonation inside and that, on the other hand, the
NH_4Cl concentration is low enough not to decrease the activity
of the enzyme. We conclude, therefore, that the observed rate
(90 s^{-1}) is the maximum turnover of the enzyme under these
conditions (pH = 8.2, 1 mM ATP, 20° C).

2.2 The reversibility of the activation/deactivation of E^{red}
When the activation of the reduced ATP-synthase is to be
measured first, all CF_oF_1 must be brought into the form E_i^{red}.
This was done as follows: class II-chloroplasts were illuminated in the presence of DTT giving the species E_a^{red}. The deactivation process (due to ADP rebinding) was followed by
measuring the rate of ATP hydrolysis at different times after
the illumination. Fig. 2 shows the result: at the top the P_i
released as a function of the reaction time is shown, the
slopes represent the rate of ATP hydrolysis. With increasing
storage time after illumination the rate decreases. On the
left, the results in the absence of P_i are shown; on the right,
in the presence of 2 mM P_i. At the bottom, the rate of ATP

hydrolysis as a function of storage time is shown. In the absence of P_i after 10 min and in its presence after about 30 min no ATP hydrolysis is observed. The reversibility of this deactivation is demonstrated by energizing the membrane with ΔpH and $\Delta\psi$ before measuring the rate of ATP hydrolysis. It can be seen in Fig. 2, bottom, that at any storage time the inactivation can be completely reversed by membrane energization ($+\Delta\tilde\mu_H+$) in the presence and absence of P_i. This implies that the activation/inactivation process is completely reversible.

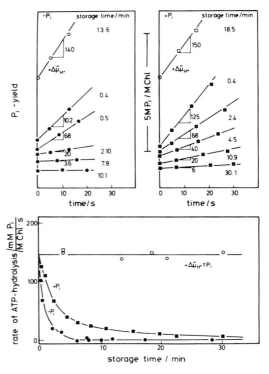

FIGURE 2. ATP hydrolysis as a function of the storage time after preillumination. Rates are measured in the absence ($-P_i$) and in the presence of 2 mM P_i ($+P_i$).

2.3 The activation of E^{red} as a function of ΔpH

Based on this result the procedure of the activation of E_i^{red} was as follows: the enzyme was brought in its inactive, reduced state, E_i^{red}. With these chloroplasts an acid/base transition was carried out under conditions where $pH_{out} = 8.2$, $\Delta\psi = 0$ mV and $P_i = 1$ mM. Five seconds after the acid/base transition 3 mM NH_4Cl and $\gamma-^{32}P$-ATP (1 mM) was added and the release of ^{32}P was measured as a function of the reaction time. This is shown in Fig. 3. It can be seen that the preceding pH jump greatly increases the rate of ATP hydrolysis.

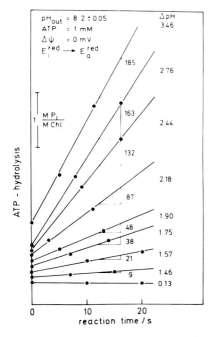

FIGURE 3. ATP hydrolysis as a function of ΔpH when CF_oF_1 was brought into the form E_i^{red} before starting the experiment. See text.

In Fig. 4 the rate of ATP hydrolysis is depicted as a function of ΔpH in the activating ΔpH jump. (The rate is measured always in the presence of NH_4Cl.) A sigmoidal dependence can be seen with a maximum rate of 180 mM ATP (M Chl s)$^{-1}$. When all CF_oF_1 was brought into the reduced state (i.e., $E^{red}(max) = E_t$), the relative rate of ATP hydrolysis gave the fraction of reduced, active ATP-synthases; i.e., $v/v(max) = E_a^{red}/E_t$ (E_t = total amount of CF_oF_1).

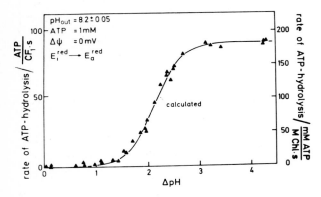

FIGURE 4.
Rate of ATP hydrolysis as a function of ΔpH in the preceding activation step-Data from Fig. 3 and similar measurements. See text for details.

3. DISCUSSION

We have shown earlier that the activation of the oxidized form requires a higher energization than the catalytic reaction (4). Therefore, when the ATP-synthase is in the form E_i^{ox}, the rate of ATP synthesis as a function of ΔpH reflects the ΔpH dependence of the activation of E_i^{ox} (3). If we assume an equilibrium between the inactive and active form, i.e., $E_i + bH_{in}^+ \rightleftharpoons E_a + bH_{out}^+$, the fraction of active ATPases, E_a/E_t, as a function of ΔpH can be calculated. For the oxidized species it resulted for the equilibrium constant $pk_E^{ox} = 5.9$, for the reduced species $pk_E^{red} = 3.7$, and for the number of H^+ it resulted b = 1.7 in both cases. The pH for half-maximal activation (i.e., the pk value for protonation from the inside) is for the oxidized species $pk^{ox} = 4.7$ and for the reduced species $pk^{red} = 6.0$. This implies that the activation of the reduced species occurs at a lower ΔpH (1.3 units lower) than that of the oxidized species. This result is similar to that obtained by Rumberg and Becher (5).

REFERENCES
1 Junesch, U. and Gräber, P. (1985) Biochim. Biophys. Acta 809, 429-434
2 Junesch, U. and Gräber, P. (1986) Biochim. Biophys. Acta in press
3 Gräber, P., Junesch, U. and Schatz, G.H. (1984) Ber. Bunsenges. physik. Chem. 88, 599-608
4 Gräber, P., Schlodder, E. and Witt, H.T. (1977) Biochim. Biophys. Acta 461, 426-440
5 Rumberg, B. and Becher, U. (1984) in: H^+-ATPase (Papa, S. et al., eds.), pp. 421-430, Adriatica Editrice, Bari, Italy

THE CHLOROPLAST ATP-SYNTHASE: THE RATE OF THE CATALYTIC REACTION

Peter GRÄBER, Ulrike JUNESCH and Gerlinda THULKE

Max-Volmer-Institut für Biophysikalische und Physikalische Chemie, Technische Universität Berlin, Strasse des 17. Juni 135, 1000 Berlin 12, FRG

1. INTRODUCTION

The membrane-bound chloroplast ATP-synthase, CF_oF_1, catalyzes reversibly proton transport-coupled ATP-synthesis and ATP-hydrolysis. The chloroplast ATP-synthase can exist in - at least - four different states, an inactive, oxidized state, E_i^{ox}, an inactive, reduced state, E_i^{red}, and both species can be brought into an active state by membrane energization. The relation between these different states is shown in Fig. 1.

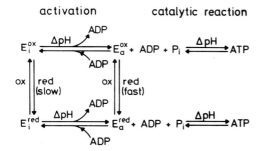

activation   **catalytic reaction**

FIGURE 1.
Simplified scheme for activation, redox reaction and catalytic reaction of the chloroplast ATP-synthase

In isolated class-II-chloroplasts the ATP-synthase is usually in the state, E_i^{ox}. Energization of the membrane (e.g., by illumination) leads to the release of tightly bound ADP and to the activation of the enzyme, E_a^{ox}. If dithiothreitol, DTT, is present, the ATP-synthase is reduced giving the species E_a^{red}. Binding of ADP to E_a^{red} then leads to the reduced, inactive form, E_i^{red}. Functionally, both inactive forms cannot catalyze ATP synthesis or hydrolysis. Both active forms can catalyze ATP synthesis/hydrolysis.

In Fig. 2 the scheme from Fig. 1 is formulated with more detail: The activation of the enzyme occurs upon proton binding from the inner aqueous phase and leads to the release of tightly bound ADP. This process takes place with the oxidized as well as with the reduced form. The redox process, $E^{ox} \rightleftharpoons E^{red}$, can occur with the non-protonated as well as with the protonated species. However, for clarity this reaction is depicted only for one species. It is well known that the rate of the reduction by dithiothreitol is much faster when the

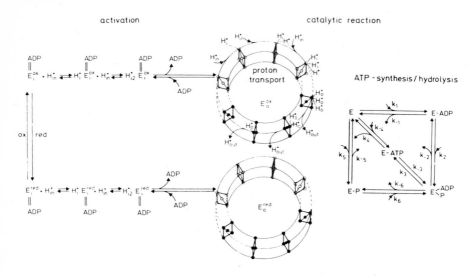

FIGURE 2. Detailed reaction scheme for activation, proton
transport and ATP synthesis/hydrolysis

enzyme is in the protonated form. The protons, necessary for
activation, presumably are not used in the catalytic reaction.
We assume that upon deactivation they are released into the
internal phase. After activation the enzyme is able to bind
protons from the inside for the catalytic reaction (the
"activating" protons being still bound to the enzyme). The
simplest mechanism for ATP synthesis/hydrolysis formulated for
the reaction at one site is depicted on the right side of
Fig. 2. The catalytic side can bind ADP, then P_i (or P_i first
and then ADP); then ATP is formed and at last ATP is released
from the enzyme. Until now it is not clear in which of these
steps the binding of the catalytic protons occurs. Therefore,
this 5-species scheme is depicted in a simplified version in
the center of Fig. 2.

Schematically it is shown that the enzyme species, E, is
protonated three times from the inside. A conformational
change (dotted lines)   then makes the protons accessible
to the outside and they are released. A conformational
change (dotted lines) then brings the enzyme back into its
original conformation. At present it is not clear which
enzyme species is protonated. Therefore, the scheme includes
all possibilities. For example, we can start with the species
E, then bind two H_{in}^+, then P_i, then the third H_{in}^+, and then
ADP. This leads to the formation of ATP at the enzyme. The
conformational change then allows release of ATP and of 3 H_{out}^+
and lastly the enzyme reacts back into its initial state and
a second cycle can start. The way in which the protolytic
reactions are coupled with ATP synthesis/hydrolysis is one

question of our current research interest.

2. RESULTS

2.1 Activation of CF_oF_1 and catalytic reaction as function of ΔpH

The reaction described above can be catalyzed by the oxidized as well as by the reduced enzyme; however, because of the short life time of the species E_a^{ox}, the reactions can be investigated practically only with the species E_a^{red}. This seems to be also the most important species as the catalytically active species in vivo is presumably the reduced form (1,2). The scheme in Fig. 2 predicts that these four different reactions of the ATP-synthase might depend on ΔpH: the activation of the oxidized species, the activation of the reduced species and the catalytic reaction catalyzed by the oxidized and the reduced species.

When CF_oF_1 is in its oxidized form, energization of the membrane leads to E_a^{ox}. From parallel measurements of the activation (by release of ^{14}C-ADP) and ATP synthesis as a function of energization it was concluded that the activation of the oxidized form requires a higher energization than ATP synthesis itself - at least at low phosphate potentials (3,4). Correspondingly, ATP hydrolysis is hardly observed in class II-chloroplasts, since at low energization the ATP-synthase is inactive and ATP synthesis occurs at high energization. A consequence of the high energization necessary for activation is that the rate of ATP synthesis as a function of ΔpH does not reflect properties of the catalytic reaction but gives the dependence of the fraction of active, oxidized CF_oF_1 on ΔpH. In Fig. 3 such measurements are depicted (4). The measurement of the activation of the reduced CF_oF_1 has been reported at this congress (5) and the results are also depicted in Fig. 3. Additionally, ATP synthesis catalyzed by the reduced CF_oF_1, E_a^{red}, is shown in Fig. 3 (6). We assume that this curve reflects ATP synthesis as a function of ΔpH for the catalytic reaction. It is not yet clear whether the catalytic reaction catalyzed by E_a^{ox} has the same ΔpH depend-

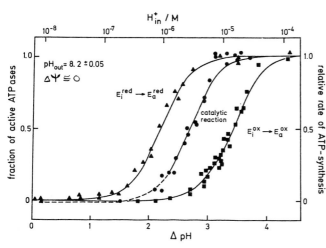

FIGURE 3. Fraction of active ATPases and the relative rate of the catalytic reaction as a function of ΔpH and internal proton concentration at $pH_{out}=$ 8.2

ence as that catalyzed by E_a^{red}, since we do not know any possibility to measure this rate without involvement of the activation process.

2.2 The rate of the catalytic reaction as a function of ΔpH and $\Delta\Psi$

We have shown earlier that for the activation of the oxidized enzyme ΔpH and $\Delta\Psi$ are kinetically equivalent (4). This means that an increase of ΔpH increases the rate by the same factor as a corresponding increase of $\Delta\Psi$. The question remains whether this is also true for the catalytic reaction. Therefore, we have measured the rate of ATP synthesis as a function of ΔpH and $\Delta\Psi$ when the ATP-synthase is in the form E_a^{red}. Fig. 4 shows the result.

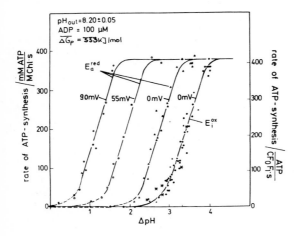

FIGURE 4.
Rate of ATP synthesis as a function of ΔpH when K^+/valinomycin diffusion potentials of 0 mV, 55 mV and 90 mV are generated

The rate of ATP synthesis was measured with a rapid mixing quenched flow apparatus as described in (7). For comparison, the rate catalyzed by E_i^{ox} at $\Delta\Psi = 0$ mV is depicted, too. It can be seen that an additional $\Delta\Psi$ increases the rate at constant ΔpH. In Fig. 5 these data are replotted as a function of the electrochemical potential difference of protons, $\Delta\tilde{\mu}_H+$. The diffusion potentials have been calculated from the Goldmann-Hodgkin-Katz equation as described earlier (4). Under these conditions the data of Fig. 4 can be described - within error limits - by a single curve; i.e., the rate is determined by $\Delta\tilde{\mu}_H+$ irrespective of the relative contributions of ΔpH and $\Delta\Psi$. This result implies that for two different reactions, the activation of the ATP-synthase and the catalytic reaction, CF_oF_1 is able to use energy derived from the concentration difference of protons and from an electric potential difference. It is not yet known in which way this is realized by the ATP-synthase. However, a simple explanation is that the internal proton concentration is in equilibrium with the "proton concentration on the enzyme" and that CF_o is a proton selective channel. Under these conditions CF_o acts as a "proton well" transforming a $\Delta\Psi$ in a proton concentration

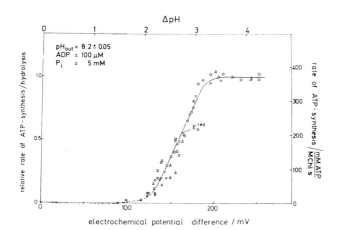

FIGURE 5.
Rate of ATP syn-
thesis catalyzed
by E_a^{red} as a
function of the
electrochemical
potential differ-
ence of protons.
Data from Fig. 4.

difference (8). This would explain the experimentally observed kinetic equivalence of ΔpH and $\Delta\Psi$. However, it should
be mentioned that other mechanisms can also lead to the same
conclusion.

2.3 The rate of ATP synthesis and ATP hydrolysis as a function
    of Δ pH
      The rate of ATP synthesis in our previous work was measur-
ed under conditions where the product concentration (ATP) was
very small ($<$ 1 µM). Under these conditions the back reaction
is practically not observed. Therefore, we have investigated
ATP synthesis and ATP hydrolysis under conditions where both
reactions can occur. We added ^{32}P to the medium to measure
only the synthesis direction and for the hydrolysis direction
we added $\gamma - ^{32}$P-ATP. The measurement was carried out with a
rapid-mixing quenched flow apparatus as described earlier (4).
Under these conditions all substrate and product concentrations as well as Δ pH and $\Delta\Psi$ remain constant (for about
300 ms) at their initial value. Fig. 6 shows the rate of ATP
synthesis and ATP hydrolysis as a function of Δ pH (at $\Delta\Psi =$
0 mV) under the same conditions; i.e., ADP = 50 µM, ATP =
50 µM and P_i = 1 mM. The dotted curve is the calculated net
reaction, i.e., $v_{synth} - v_{hydr}$. The maximum rate of ATP
synthesis is 120 ATP (CF$_1$ s)$^{-1}$ (at Δ pH $>$ 3.2); the maximum
rate of ATP hydrolysis is 30 ATP (CF$_1$ s)$^{-1}$ (at Δ pH $<$ 1.2).
The net rate is zero at Δ pH = 2.0. With ΔG_p^0 = 33.9 kJ/mol
(9) it results ΔG_p = 51 kJ/mol for our experimental conditions. At zero net rate, i.e., at equilibrium, we obtain for

$$n = \frac{\Delta G_p}{\Delta \tilde{\mu}_{H^+}} = 4.5$$

This ratio is considerably higher than what has generally
been accepted up to now (n = 3). If we accept n = 3 as the
true value, we have to conclude that ΔG_p^0 = 16.6 kJ/mol, a
conclusion which was made by Rumberg and Becher (10).

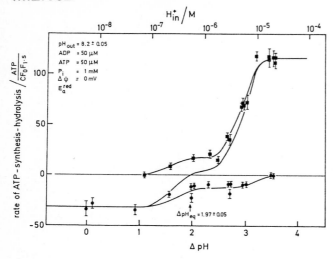

FIGURE 6.
Rate of ATP syn-
thesis and ATP
hydrolysis as a
function of ΔpH
measured in an acid
base jump with a
rapid-mixing
quenched flow
apparatus

2.4 The rate of ATP synthesis as a function of ADP concentration

The rate of ATP synthesis shown in the previous figures was measured at a constant ADP concentration. We now investigate the rate of ATP synthesis catalyzed by E_a^{red} at a constant ΔpH = 3.2 as a function of the ADP concentration. The result is shown in Fig. 7.

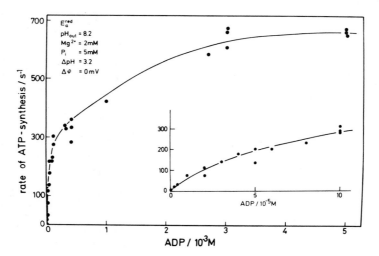

FIGURE 7. Rate of ATP synthesis catalyzed by E_a^{red} at ΔpH = 3.2, $\Delta\Psi$ = 0 mV, pH_{out} = 8.2 as a function of the ADP concentration. The rate is given in ATP $(CF_1 \cdot s)^{-1}$.

In contrast to other experiments (11), the functional dependence under these conditions is not a typical Michaelis-Menten kinetics; the shape of the curve seems to reflect some negative cooperativity between ADP-binding sites. The maximum rate observed here is 700 s^{-1}. This implies a rate of 2100-2800 H^+ $(CF_1 \cdot s)^{-1} \approx 2100-2800$ H^+ $(e\text{-chain} \cdot s)^{-1}$ (estimated for n = 3-4 and 1000 Chl/CF_1). Such high rates cannot be obtained by light-induced proton transport, they can only be observed with the artificial generation of ΔpH and $\Delta \Psi$ used in our work.

At low ADP concentrations the rate is presumably diffusion controlled. This can be discussed as follows: The rate is given in this case by $v = k_1$ $[E]$ $[ADP]$. If the free enzyme concentration, E, is approximately identical with the total enzyme concentration, E_t, we can estimate from Fig. 7 at the ADP concentration 10^{-5} M

$$\frac{v}{[E_t] [ADP]} \approx k_1 \approx 6 \cdot 10^6 \text{ M}^{-1} \text{ s}^{-1}$$

If the diffusion coefficient of the membrane-bound enzyme can be neglected it results from the simple collision theory for k_1 with $r_E = 50$ Å radius of the enzyme; $r_{ADP} = 5$ Å, radius of ADP, and an estimated diffusion coefficient $D_{ADP} \approx 5 \cdot 10^{-6}$ cm^2 s^{-1}:

$$k_1' = 4 \pi (r_E + r_{ADP}) \cdot D_{ADP} \cdot N_L \approx 2 \cdot 10^{10} \text{ M}^{-1} s^{-1}$$

The active binding site for ADP is only a small fraction of the enzyme surface. In the simplest model the active area has a magnitude of the area of ADP and it results for the fraction of active enzyme surface

$$f_E = \frac{\pi r_{ADP}^2}{4 \pi r_E^2} \approx 2.5 \cdot 10^{-3}$$

Also, not every orientation of the ADP leads to a binding at the enzyme. A reasonable estimate of the fraction of active substrate surface is $f_{ADP} \approx 0.2$. Then it results for the rate constant

$$k_1 = k_1' \cdot f_E \cdot f_{ADP} \approx 1 \cdot 10^7 \text{ M}^{-1} s^{-1}$$

Since this is the maximum expected rate constant for a diffusion-controlled reaction, it is in agreement with the measured rate constant which is a factor of about two smaller. Preliminary experiments show that the activation energy of this process is indeed very small as expected for a diffusion-controlled step. Also, different experimental evidence had led to the conclusion that ADP binding is diffusion-controlled (12).

3. DISCUSSION

It may be asked whether the regulation of the redox state has physiological relevance in vivo. In the dark the ATP-synthase is usually in the oxidized form, i.e. without energization the enzyme is inactive and therefore, ATP-hydrolysis does not take place and a high ATP-level is maintained (13). Upon illumination the membrane is energized,the enzyme is activated and ATP is synthesized with a small rate. When enough redox equivalents become available thioredoxin is reduced (via ferredoxin) and CF_oF_1 is reduced by thioredoxin to the form E_a^{red} (1,2). This implies according to Fig. 3 that at constant ΔpH the rate of ATP-synthesis increases considerably, leading to an increased ATP-supply of the chloroplast. When the membrane is deenergized in the dark the species E_a^{red} hydrolyzes ATP with high rates. However, reoxidation by oxygen leads to the species E_a^{ox} which is rapidly inactivated to E_i^{ox} and therefore, ATP-hydrolysis is prevented.

The results reported here reflect the high complexity of the ATP-synthase and its reactions: 1) contact with two aqueous phases, 2) response to ΔpH and $\Delta\Psi$, 3) regulation of its redox state, 4) regulation of its activity by ΔpH and $\Delta\Psi$, 5) coupling of single-site catalysis with protolytic reactions, 6) interactions between different catalytic sites and coupling with protolytic reactions.

ACKNOWLEDGEMENTS

This work has been supported by grants from the Deutsche Forschungsgemeinschaft and the Fonds der Chemischen Industrie. We thank Prof. B. Rumberg for stimulating discussions during the course of this work.

REFERENCES
1 Mills, J.D., Mitchell, P. and Schürmann, P. (1980) FEBS Lett. 144, 63-67
2 Shahak, Y. (1982) Plant Physiol. 70, 87-91
3 Gräber, P., Schlodder, E. and Witt, H.T. (1977) Biochim. Biophys. Acta 461, 426-440
4 Gräber, P., Junesch, U. and Schatz, G.H. (1984) Ber. Bunsenges. Phys. Chem. 88, 599-608
5 Junesch, U. and Gräber, P. (1986) in: Proc. of VII. Congress of Photosynthesis
6 Junesch, U. and Gräber, P. (1985) Biochim. Biophys. Acta 809, 429-434
7 Junesch, U. and Gräber, P. (1987) BBA submitted
8 Mitchell, P. (1968) Chemiosmotic Coupling and Energy Transduction, Glynn Res. Ltd., Bodmin, UK
9 Rosing, J. and Slater, E.C. (1972) Biochim. Biophys. Acta 267, 275-286
10 Rumberg, B. and Becher, U. (1984) in: H^+-ATPase (Papa et al., eds.), pp. 421-430, Adriatica Editrice, Bari, Italy
11 Bickel-Sandkötter, S. and Strotmann, H. (1981) FEBS Lett. 125, 188-193
12 Alfalo, C. and Shavit, N. (1984) in: Advances in Photosynthesis Research, C. Sybesma, ed., pp. II,6.559-563, M. Nijhoff Publ., The Hague
13 Giersch, C. et al. (1980) Biochim.Biophys.Acta,590,59-73

COUPLING MECHANISM BETWEEN PROTON TRANSPORT AND ATP SYNTHESIS
IN CHLOROPLASTS

T. Tran-Anh and B. Rumberg
Max-Volmer-Institut, Technische Universität Berlin,
D-1000 Berlin 12, Germany

1. INTRODUCTION

ATP synthesis from ADP and phosphate is a process in need of
energy which is made feasible by
coupling to transmembrane H^+
translocation down the H^+ gradient, the coupling membrane in
chloroplasts being the thylakoid
membrane and the coupling device
being a membrane protein called
ATP-synthase. Direction of H^+
translocation to drive ATP synthesis is from the inside to the
outside water phase of the thylakoid vesicles (see Fig. 1). The
aim of our work is to get insight
into the coupling process at the

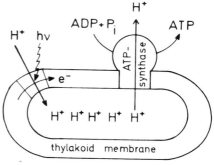

Fig. 1 Scheme of H^+ transport
across the thylakoid membrane.

ATP-synthase by analysis of the kinetical pattern as controlled
by transmembrane ΔpH.

2. EXPERIMENTAL PROCEDURE

Experiments were performed
with suspensions of enve-
lope-free chloroplasts isolated from spinach. The
suspension was stirred and
pH changes were continuously monitored by means of a
glass electrode. The rate
of ATP synthesis/hydrolysis
was obtained from the rate
of the coupled pH changes
(0.936 H^+ consumed per ATP
produced at pH 8.0). Internal acidification was calculated from the pH jump
after a pulse of 100 µM

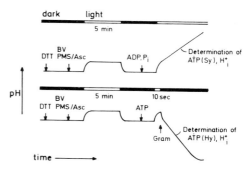

Fig. 2 Schematic diagram of the experimental procedure to obtain data
of ATP synthesis (above) and ATP
hydrolysis (below).

imidazole (for further details see [1]). Recently it was recognized that control of the rate by ΔpH is due to the ΔpH dependence of the catalytic and of an activation process. The con-

trol due to activation, however, is eliminated if the ATP-synthase is preactivated in the presence of the thiol reducing agent dithiothreitol (see [1]). Therefore the measurements have been performed under such circumstances only (see Fig. 2).

3. RESULTS

3.1 ATP synthesis

We measured the rate of ATP synthesis in dependence on the concentration of phosphate, internal and external H^+. Variation of the internal acidification was obtained by addition of the H^+ permeability inducing agent gramicidin D. With respect to phosphate a Michaelis-Menten relationship is found which is, however, dependent on H_{in}^+ and H_{out}^+, namely the Michaelis constant of phosphate is direct proportional to H_{out}^+ and reciprocal proportional to H_{in}^+ squared (this is true for $[P] \leq 1$ mM). The results are shown in Fig. 3 and Fig. 4.

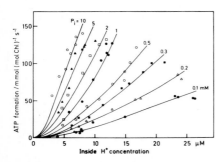

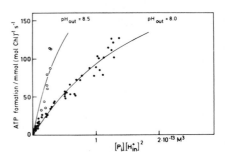

Fig. 3 Rate of ATP synthesis as a function of the internal H^+ concentration. $[ADP] = 0.5$ mM, $pH_{out} = 8.0$.

Fig. 4 Data of Fig. 3 replotted as a function of $[P_i] \cdot [H_{in}^+]^2$. In addition data obtained at $pH_{out} = 8.5$.

The results are interpreted as follows:

1. Dependence of ATP synthesis on $[P]$. $[H_{in}^+]^2$ indicates that the transfer of two H_{in}^+ to the reaction centre takes place during the rate-limiting step of the reaction cycle.

2. Dependence of ATP synthesis on $[H_{out}^+]^{-1}$ indicates that the binding of phosphate is preceded by the release of H^+ from the substrate-binding-centre to the outer phase ($K_H \sim 10^{-9}$ M).

3.2 ATP hydrolysis

The chloroplast ATP-synthase which under normal conditions catalyzes the synthesis of ATP from ADP and phosphate may be forced to operate in reverse direction. Under such circumstances ATP hydrolysis is coupled to H^+ pumping from the outside to the inside water phase of the thylakoid vesicles. Prerequisite of this artificially induced reaction is to push the ATP-synthase in its active state by short preillumination (see Fig. 2). We measured the rate of ATP hydrolysis in dependence on the con-

centrations of ATP, H^+_{in} and H^+_{out} in order to get insight into
the molecular coupling mechanism from the analysis of the data.
The internal H^+ concentration during steady state conditions is
controlled by both the H^+ pump activity and the H^+ permeability
of the thylakoid membrane. In order to be able to adjust H^+_{in}
values independent of the ATP concentration we changed the H^+
permeability by addition of the H^+ permeability inducing agent
gramicidin D.

The results are shown in Fig. 5 and Fig. 6 and may be described
as follows:
1. With respect to H^+_{in}, inhibition is found independent of
the ATP concentration, which is linearly dependent on H^+_{in}.
This gives evidence that the release of one H^+ into the inner
phase takes place during the rate limiting step of the reaction
cycle.

2. With respect to ATP, a Michaelis-Menten relatonship independent of H^+_{in} is found. The Michaelis constant, however, is
inversely related to H^+_{out}. This gives evidence that the binding of ATP is preceded by the binding of H^+ from the outer
phase to the substrate-binding-centre ($K_M \sim 10^{-7}$ M).

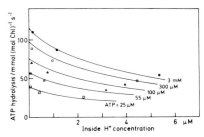

Fig. 5 Rate of ATP hydrolysis
as a function of internal H^+
concentration. [ADP] = 30 µM,
$[P_i]$ = 30 µM, pH_{out} = 8.0.

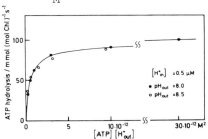

Fig. 6 Data of Fig. 5 replotted
as a function of $[ATP] \cdot [H^+_{out}]$.
In addition data obtained at
pH_{out} = 8.5.

4. DISCUSSION

The results of the ATP synthesis measurements are most easily
interpreted by the following reaction sequence (compare Fig. 7):

1. Release of H^+ from the substrate-reaction-centre E to the
outer water phase ($E^{2+} \to E^+$).

2. Binding of HPO_4^{2-} and $ADPMg^-$ from the outer phase to E^+.

3. Binding of H^+ from the inner water phase to binding sites
B located at the F_o-F_1 interface ($B \to B^+$).

4. Inward movement of E^+ with the ligands $ADPMg^-$ and HPO_4^{2-}
towards B^+ (by rotation or conformational change of the α, β-
sub-units and favoured by electrostatic attraction between the
E-ligand-complex and B^+) and twofold H^+-transfer from B^+ to the
E-ligand-complex to produce $ATPMg^{2-}$ bound to E^{2+} and H_2O. In
the case of low H^+_{in} concentration this reaction step will be
the rate-limiting one, in the case of high H^+_{in} concentration

the release of H^+_{out} preceding phosphate binding and/or the release of ATPMg^{2-} will become rate limiting.
Analysis of the kinetical pattern of ATP-hydrolysis gives evidence for the reaction steps which complete the reaction cycle:

5.   Further H^+ transfer from B^+ to E-ligand-complex and outward movement of E^{3+} with the ligand ATPMg^{2-} away from B^+ to expose bound ATPMg^{2-} to the outer water phase (by rotation or conformational change and favoured by electrostatic repulsion between the E-ligand-complex and B^+).

6.   After release of ATPMg^{2-} and H^+ into the outer water phase E^{2+} is left behind.
In the case of ATP-hydrolysis the reaction cycle takes place in reversed direction and reversed step No. 5 will be the rate-limiting one.

The reaction sequence results in a stoichiometry of 3 H^+ translocated from inside to outside for the synthesis of one ATP molecule. This agrees with the direct determination of the H^+/ATP stoichiometry [2]. The results rule out the hypothesis of indirect H^+ coupling by conformational changes put forward by Boyer [3]. They rather prove principally right Mitchell's hypothesis of direct H^+ coupling [4].

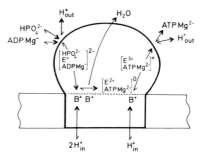

Fig. 7 Path of reaction for ATP synthesis/hydrolysis at the ATP-synthase.

ACKNOWLEDGEMENT

The technical assistence of Marianne Cetin and financial support of the Deutsche Forschungsgemeinschaft are gratefully acknowledged.

REFERENCES

[1]   Rumberg, B. and Becher, U. (1984) in: H^+-ATPase: Structure, Function, Biogenesis (Papa, S. et. al., eds.) pp. 421-430, Adriatica Editrice, Bari.

[2]   Rathenow, M. and Rumberg, B. (1980) Ber. Bunsenges. Phys. Chem. 84, 1059-1062.

[3]   Boyer, P.D. (1983) in: Biochemistry of Metabolic Processes (Lennon, D.F. et. al., eds.) pp. 465-477, Elsevier, New York.

[4]   Mitchell, P. (1985), FEBS Lett. 182, 1-7.

QUANTITATIVE RELATIONSHIP BETWEEN 9-AMINOACRIDINE FLUORESCENCE QUENCHING
AND INTERNAL pH IN BROKEN CHLOROPLASTS

ThanhVu Van, Thomas Heinze, J. Buchholz and Bernd Rumberg
Max-Volmer-Institut, Technische Universität Berlin,
D-1000 Berlin 12, Germany

1. INTRODUCTION

Illuminated thylakoids create a transmembrane proton gradient by translocating protons from the outside phase into the inside one. Measurement of this
inside proton concentration $[H^+]_{in}$ is not a simple problem because the thylakoids with their tiny sizes are not accessible even to the smallest proton
sonde. A versatile method to evaluate pH_i based on the distribution of amines
was introduced by Rottenberg et. al. (Eur.J.Biochem. 25/1972/54,64,71) assuming that only the uncharged species of amines is freely permeable across
the membrane. The distribution of amines upon membrane energization could
be determined by: 1. radioactive labeling,
                   2. fluorescence quenching,
                   3. proton uptake.
Although these methods are widely applied, there still exist some discrepan-
ces, mainly from the fluorescence quenching method due to possible complications arising from dimerization, membrane binding, uncoupling effect, and
eventually from a not satisfying theory of this light induced fluorescence
quenching. In this respect we examined the pH_{in} determination for the two
cases, namely by measuring the additional proton uptake ΔH^+_{add} in the presence of imidazole as a standard procedure and the fluorescence quenching
$\Delta F/F$ of 9-aminoacridine (9-AA).

2. PROCEDURE
2.1 Theory

On the basis of the assumption introduced by Rottenberg et.al. (1972) which
is mentioned above, the following formulae were derived:
a) A monoamine generally distributes between two phases as follows:

$$\frac{(A)_{t,in}}{(A)_{t,out}} = \frac{K + (H^+)_{in}}{K + (H^+)_{out}} \tag{1}$$

b) For the case of proton uptake $(K \overset{\sim}{=} [H^+]_{out})$

$$\Delta H^+_{add} = (A)^o_{t,out} \cdot K \cdot V_{in} \cdot \frac{\left((H^+)_{in} - (H^+)_{out} \right)}{(K + (H^+)_{out})^2 + (K + (H^+)_{out}) \cdot (K + (H^+)_{in}) \cdot (Chl) \, V_{in}} \tag{2}$$

$(A)^o_{t,out}$: initial total amine concentration

V_{in} : internal volume of the thylakoids in $dm^3/molChl$

c) For the case of fluorescence quenching $(K << [H^+]_{out} << [H^+]_{in})$

$$\frac{\Delta F}{F} = \frac{(H^+)_{in} / (H^+)_{out} - 1}{1/ V_{in}(Chl) + 1} = \frac{(H^+)_{in}}{(H^+)_{out}} V_{in} (Chl) \tag{3}$$

Biggens, J. (ed.), Progress in Photosynthesis Research, Vol. III. ISBN 90 247 3452 5
© *1987 Martinus Nijhoff Publishers, Dordrecht. Printed in the Netherlands.*

2.2 Experimental

Broken chloroplasts were extracted from spinach loaves as described by Winget et. al. (Biochem.Biophys.Res.Commun. 70/1976/1283).
We measured at 20 °C $\Delta F/F$ of 9-AA and ΔH^+_{add} by injection of imidazole
under steady state illumination with red actinic light.
The reaction medium contained 50 mM KCl, 0.5 mM tricine, 0.1 mM benzyl
viologen, 0.6 or 3µM 9-AA, 0.1 mM imidazole.
V_{in} was varied with sucrose and $[H^+]_{in}$ was varied with the uncoupler
gramicidin-D.

3. RESULTS
3.1 Light induced additional proton uptake with imidazole

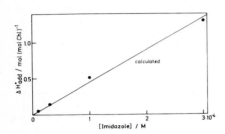

Fig.1a: Additional H^+-uptake as
a function of amine concentration

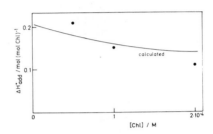

Fig.1c: Additional H^+-uptake as a
function of chlorophyll concentration

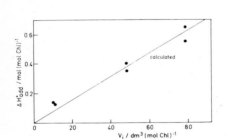

Fig.1b: Additional H^+-uptake as
a function of internal volume

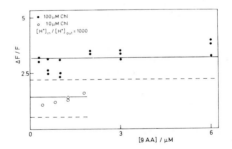

Fig.2: Fluorescence quenching as
a function of amine concentration

The solid lines in the figs. 1a to 1c were calculated with eq. 2.

These results show that imidazole, the light-induced uptake of which was
reflected by pH increase in the outside phase, exhibits neither aggregation nor binding to the membrane and its uptake is quantitatively described
on the basis of the Rottenberg's model. It does not uncouple (not shown
here).

3.2 Light induced fluorescence quenching of 9-AA

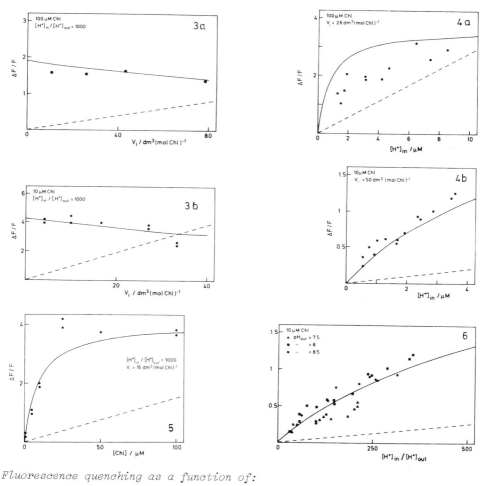

Fluorescence quenching as a function of:

fig. 3a,b: *internal volume*
fig. 4a,b: *inside H+-concentration*   } *at 2 different Chl-*
fig. 5   :   *chlorophyll concentration*   *concentrations*
fig. 6   :   *outside H+-concentration*

In the figs. 2 to 6:   - - -   calculated with eq. 3
                        ———   calculated with eq. 6   for α = 550
                                                               β = 0.825

As can be seen in the above figures, equation 3 derived from Rottenberg's model cannot describe the experimental results which indicate that 9-AA:

a) is bound to the inside membrane surface; the internal volume V_{in} had only a very weak influence (fig.3),

b) still exhibits fluorescence from the inside phase (saturation took place in fig. 5).

These results can, however, be described rather satisfactorily by a new model which is based on the fact that binding take place and this phenomenon is responsible for the fluorescence quenching.

4. MODEL

Due to its very low dissociation constant ($pK \simeq 10$), 9-AA appears at pH below 6 in the inside phase almost only as the protonated species AH^+.
New assumtions:

a) The binding of AH^+ to the membrane can be described by a Langmuir-isotherm (linear region only is considered)

$$n_{AH^+, bound} \simeq \alpha \cdot [AH^+]_{in, bulk} \tag{4}$$

α contains the binding constant and the number of binding sites.

b) 9-AA molecules, once being bound to the membrane, can still exhibit fluorescence, but only with a considerably reduced intensity.

$$\Delta F = const. \; \beta \cdot n_{AH^+, bound} \tag{5}$$

β : quenching factor $0 < \beta < 1$

Combining these equations, the fluorescence quenching can now be expressed as follows:

$$\left(\frac{\Delta F}{F} \right) = \left(\frac{\Delta F}{F} \right)_\infty \cdot \frac{(Chl) \cdot (H^+)_{in} / (H^+)_{out}}{(Chl) \cdot (H^+)_{in} / (H^+)_{out} + \left(\frac{\Delta F}{F} \right)_\infty \frac{1}{\alpha \beta}} \tag{6}$$

$$\text{with} \quad \left(\frac{\Delta F}{F} \right)_\infty \equiv \frac{\beta}{1 - \beta + V_{in}/\alpha}$$

The solid curves in the figures 2 to 6 which were calculated with this equation assuming $\alpha = 550$ and $\beta = 0.825$ can describe rather satisfactorily the experimental data. The pH_{in} values obtained with this model also agree with those obtained by the quite different method of Rumberg and Siggel (Naturwiss. 56/1969/130).

ACKNOWLEDGEMENT

The authors gratefully acknowledge Mrs. Marianne Cetin for her extensive technical assistance.

SINGLE AND DOUBLE INHIBITOR TITRATIONS OF BACTERIAL PHOTOPHOSPHORYLATION

M.Virgili[o], D.Pietrobon[+], G.Venturoli[o], and B.A. Melandri[o], Institute of Botany, University of Bologna[o] and CNR Center for the Study of the Physiology of Mitochondria, Padua[+], Italy

1. INTRODUCTION

In the minimal scheme of chemiosmotic coupling the primary and secondary proton pumps are coupled through a delocalized pool of common intermediates, the high-energy protons located at one side of the coupling membrane. This mechanism is therefore only consistent with a fully delocalized interaction of the primary and secondary proton pumps, since the model assumes complete electrochemical equilibrium for protons in the bulk phases at both sides of the membranes, and also complete protonic equilibrium with the protolytic sites of both types of pump. Titration of the rate of phosphorylation with inhibitors of the ATP synthetase and of the electron transfer chain, or with uncouplers, are traditionally important tools for the study of the mechanism of energy transduction and for establishing its degree of delocalization. Indeed, using this experimental approach results have been obtained which have been interpreted as non consistent with a fully delocalized coupling mechanism. These data indicated the absence of a univocal correlation in the flow-force relationship for the ATP synthetase, when the system was perturbed with different agents. These results relied heavily on the accuracy in the quantitative estimation of the transmembrane potential and pH difference (1,2).

The study of the combined effect of two inhibitors, one of the electron transfer chain, and the other of the ATP synthetase, on the rate of ATP synthesis has been proposed as a criterium for testing the degree of delocalization of coupling completely independent of any evaluation of $\Delta \mu_H^+$. The intuitive rationale underlined in this type of experiments was that, if the two pumps are coupled through a delocalized pool of intermediates, inhibition of one pump (e.g. the primary pump) should affect the inhibitor response of the other pump, since it will alter the control of the latter on the rate of the overall processes (3,4). The results of double inhibitor titration experiments, on the contrary, indicated that the relative inhibition of ATP synthesis by an inhibitor of the ATPase was not altered by partial inhibition of the electron flow. These results have been considered as unequivocally inconsistent with a delocalized mechanism of coupling (4).

This simple intuitive reasoning has been however criticized when a more quantitative analysis of the double inhibitor response was performed using models based on linear or non-linear flow-force relationships for the primary and secondary pumps, and when non-negligible leaks were introduced in the proton flow balance. It has been pointed out that the existence of significant leaks complicates the response and renders dubious a simple intuitive espectation (5). The response of a double

Biggens, J. (ed.), Progress in Photosynthesis Research, Vol. III. ISBN 90 247 3452 5
© *1987 Martinus Nijhoff Publishers, Dordrecht. Printed in the Netherlands.*

inhibitor titration is also not obvious when the flow-force relation for either pump is not linear, i.e. when the change of the reaction rate on changing $\Delta\mu_{H}^{+}$ varies with the value of $\Delta\mu_{H}^{+}$ (6).

In a recent lucid analysis of single and double inhibitor titration, performed utilizing a linear model of chemiosmotic coupling, Pietrobon and Caplan (7) have shown that the relative inhibition of the rate of ATP synthesis by an inhibitor of the ATP synthetase or of the electron transfer chain, is univocally related to the ratio of the conductances of the two pumps, i.e. to the ratio of the slopes of the dependance of the flow on $\Delta\mu_{H}^{+}$. This conclusion valid only with linear relations between flows and and $\Delta\mu_{H}^{+}$ and in the presence of a negligible leak is the basis for the validity of the rule of thumb commonly used in analyzing the results of double inhibitor experiments. The presence of non-negligible leaks causes the attenuation of the differential response of a delocalized vs. a localized coupling mechanism, and, therefore, a decrease in the validity of the double inhibitor approach as a mean of discriminating between these two models. The introduction of non-linear flow-force relationship for either pump (8), moreover determines a complex dependence on $\Delta\mu_{H}^{+}$ of the conductance ratio, which is constant in a linear model. In this case the results of inhibitor titrations are highly dependent on the value of $\Delta\mu_{H}^{+}$ and on the kind of dependence of the flows on $\Delta\mu_{H}^{+}$ in the experimental conditions under study.

It is clear therefore that in order to draw an unequivocal conclusion from a double inhibitor titration, a complete set of parameters must be evaluated, including the relations between flows and $\Delta\mu_{H}^{+}$ for the primary and secondary pumps and for the leaks. This approach is therefore not completely independent of any evaluation of $\Delta\mu_{H}^{+}$, thus losing its main claimed advantage as a mean of discriminating between different energy coupling mechanisms. There is however another aspect of the inhibitor titrations which might allow such a discrimination without the need of knowing the flow-force relationships. It consists in the comparison of the relative inhibition of the rate of ATP synthesis when a certain fraction of primary and secondary pumps are inhibited. A relative inhibition of the rate of ATP synthesis equal to both the fraction of inhibited primary and secondary pumps is never expected within a delocalized model in the presence of a small leak (7). We have performed this kind of experiment in chromatophores, measuring the rate of photophosphorylation as a function of the fraction of inhibited ATPsynthetases and as a function of the fraction of inhibited bc_1 complexes. The results indicate that, under the experimental conditions used (pH = 8.5, 50 mM KCl, and nigericin in continuous light) the coupling mechanism can be delocalised.

2. MATERIALS AND METHODS

Chromatophores were prepared by French-press disruption and differential centrifugation of cells of Rhodobacter sphaeroides GA, grown photoheterotrophically in a malate synthetic medium (9).

Photophosphorylation was measured at 30°C and at pH = 8.5 in a medium containing: 50 mM glycylglycine, 5 mM Mg acetate, 50 mM KCl, 0.2 mM sodium succinate, 1 mM KCN, 1 mg/ml of bovine serum albumine, 1 μM nigericin, 2 mM ADP, 10 mM phosphate, and chromatophores corresponding to about 30 μM bacteriochlorophyll. Samples were illuminated for 1 to 3 minutes, and quenched with 4% w/v of trichloroacetic acid. The concentration of ATP in the supernatant was measured with luciferine-luciferase as specified in ref. (9). Linearity versus time was controlled under all experimental conditions. The free-energy change for ATP synthesis in the initial conditions of the assay was about 39 KJ moles^{-1}; this value increased to 43 KJ moles^{-1} after 3 minutes of illumination.

ATPase activity was assayed colorimetrically under the same conditions as described in (9). The rate of formation of ΔpH induced by ATP hydrolysis was measured by the quenching of the fluorescence of 9 amino-acridine, in the presence of 2 µM valinomycin; the calibration of the 9 amino-acridine response was performed as described in (10).

The electrochromic signal of carotenoids, induced by light was monitored at 503 nm in a single beam spectrophotometer or at 503-486 nm in a double wavelength spectrophotometer. Calibration of the ionic current and of the value of the membrane potential was obtained by comparison of the electrochromic signal induced by one single-turnover flash in the same sample inhibited with 20 µM antimycin (11). The concentration of reaction center was measured in the same sample after addition of valinomycin (9).

The kinetics of reaction center rereduction after a train of flashes was followed at 542 nm in uncoupled chromatophores (9).

3. RESULTS

3.1. The fraction of ATP synthetases inhibited by DCCD

Crucial to a correct interpretation of an inhibitor titration is the measure of the fraction of active pumps in partially inhibited chromatophores. In the present experiments the ATP synthetase has been inhibited irreversibly by incubation for one hour at 0°C with DCCD (70 to 560 nmoles per mg bacteriochlorophyll); after this treatment the degree of inhibition was kept constant by storing the chromatophores (suspended in a 60% glycerol containing buffer) at -18°C (9). In principle, the fraction of enzyme inhibited can be evaluated by measuring the hydrolyzing activity in uncoupled chromatophores. This approach is however complicated by the observation that about 20-25 per cent of the ATPase, measured in the dark, is insensitive to oligomycin, and that complete uncoupling causes total inactivation of the oligomycin-sensitive activity (12). We have therefore measured the ATPase activity in the presence of only nigericin, a condition which stimulates the activity close to the maximum level observed at optimal concentration of uncouplers. We have moreover chosen to consider as indicative of the active ATPase fractions only the oligomycin sensitive activity, since all energy requiring processes are fully sensitive to this antibiotic. In support to this assumption is the result of Figure 1 showing a proportional relation between the fraction of inhibited ATPase, derived from a titration with DCCD of the uncoupled ATP hydrolysis,and the relative inhibition by DCCD of the initial rate of ΔpH formation induced by ATP hydrolysis in the dark. Since the ATP-induced formation of ΔpH is related to the protonic flow by the differential buffering capacity of chromatophores - a constant parameter in the initial conditions of the assay both in control and in DCCD-treated membranes - these results indicate that the proton pumping activity of the ATP synthetases can be evaluated from the oligomycin sensitive ATPase activity, measured in the presence of nigericin.

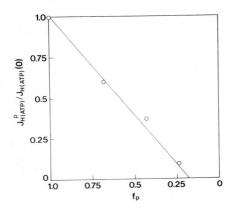 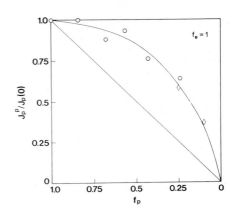

FIGURE 1. Linear correlation between the relative inhibition by DCCD of oligomycin sensitive ATPase (f_p) and of the rate of proton translocation induced by ATP hydrolysis in the dark ($J^P_{H(ATP)}/J_{H(ATP)}(0)$).

$$f_p = (v_{ATPase}(DCCD) - v_{ATPase}(oligo))/(v_{AT\ Pase}(0) - v_{ATPase}(oligo))$$

$J^P_{H(ATP)}/J_{H(ATP)}(0)$ was evaluated from the initial rate of ΔpH formation in DCCD treated chromatophores relative to control.

FIGURE 2. Lack of linear correlation between the relative inhibition by DCCD of oligomycine- sensitive ATPase (f_p) and the rate of photophosphorylation ($J^P_p/J_p(0)$). Different symbols refer to different sets of experiments.

3.2. Effect of DCCD on the rate of phosphorylation in continuous light.

Preincubation with DCCD resulted in a progressive inhibition of ATP synthesis (9). Figure 2 shows that the relative inhibition of the synthetic activity was always significantly smaller than the fraction of inhibited ATPases. These results are indicative of a marked delocalized effect, by which the fraction of active ATPases turns over more rapidly and partially compensates for the decreased number of active enzymes (7). The conclusion drawn from these results differs therefore from that offered in ref.(9). This is due to the different interpretation of the ATPase activity: in the previous paper in fact the entire ATPase activity, and not the oligomycin sensitive part, was considered as indicative of the number of active ATP synthetases (9).

3.3 The fraction of electron transfer chain inhibited by antimycin

Antimycin blocks the cyclic electron transfer of bacterial photosynthesis binding very effectively at the Q_c site of the bc_1 complex (13). The effect of the inhibitor concentration on the fraction of active bc_1 complexes was evaluated from the kinetics of rereduction of reaction centers in uncoupled chromatophores following activation by a train of closely spaced (20 ms) single turnover flashes (9). The extent of the fast relaxing phase of reduction was taken as a measure of the fraction of bc_1 still active after partial inhibition with antimycin (compare ref. 9). The reasonable assumption implicit in these measurements is that antimycin binding is the same in energized and uncoupled chromatophores. It is also postulated implicitely that the number of primary proton pumps coincides with that of active bc_1 complexes. This is perfectly reasonable given the essentiality of this complex for cyclic electron transfer.

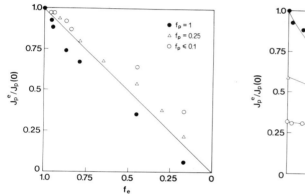 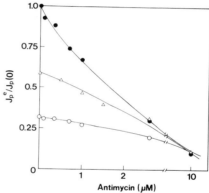

FIGURE 3. Correlation between the relative inhibition of the photosynthetic chains by antimycin (f_e) and the relative inhibition of photophosphorylation ($J_p^e/J_p(0)$). For the evaluation of f_e see text. Closed symbols, control chromatophores; open symbols, DCCD-treated chromatophores.

FIGURE 4. The response to antimycin inhibition of the rate of photophosphorylation in controls (closed circles) and DCCD-treated chromatophores (open symbols). The rate of photophosphorylation in controls ($J_p(0)$) was 240 μmoles h^{-1} mg Bchl^{-1}. Symbols and f_p values as in Fig.3.

3.4. The effect of antimycin on photophosphorylation

The relative inhibition of photophosphorylation by antimycin is plotted in Figure 3 as a function of the fraction of active primary pumps (f_e). Opposite to what observed with DCCD, in this case the inhibition of ATP synthesis was nearly proportional (Figure 3, full circles) or, particularly at higher concentrations of the inhibitor, even larger than the relative inhibition of the primary pumps. These results are in agreement with previous data from this laboratory (14), and confirm that in bacterial chromatophores the primary redox pumps control entirely the rate of phosphorylation in the absence of uncouplers.

3.5. The combined effect of antimycin and DCCD on photophosphorylation

The effect of antimycin was also studied in chromatophores treated with subsaturating concentrations of DCCD (Figure 4). In this case the relative inhibition by antymycin was decreased, as compared to that of DCCD-untreated chromatophores. The same experimental results are replotted in Figure 3 as a function of the fraction of active primary redox pumps. When the ATP synthetase is 75% inhibited ($f_p = 0.25$, open triangles), the inhibition of photophosphorylation, relative to the same DCCD-inhibited chromatophores but in the absence of antimycin, falls systematically above the diagonal (corresponding to perfect linearity between the relative inhibition of ATP synthesis and f_e). This pattern is confirmed and enhanced if the ATPase is more that 90% inhibited ($f_p \leq 0.1$, open circles). In this latter case the delocalized role of the primary pumps in the overall coupling mechanism becomes evident; this effect is however observable only when a very large fraction of the ATP synthetases has been inactivated by DCCD.

4. DISCUSSION

The interpretation of a single- or of a double-inhibitor titration in terms of delocalized or localized coupling mechanism depends critically on the evaluation of the fraction of active primary or secondary pumps (f_e and f_p). Indeed the main reason for the different interpretation of the results relative to the inhibition by DCCD offered in the present paper, as compared to (9), is in the definition of f_p, there taken as the fraction of the total ATPase activity, and not of only the oligomycin sensitive part. The present inhibitor titration data are not inconsistent with, and actually suggest, a delocalized mechanism of coupling. These results do not, however, remove all difficulties in interpreting the steady-state kinetics of photophosphorylation (1,2,14), since the absence of a single-valued flow-force relationship during uncoupler- vs. electron transfer inhibitor-titrations was confirmed also under the present experimental conditions (not shown). In addition, much of the evidence in favour of localized coupling phenomena is based on non-steady-state experiments (9,16). In particular, it should be stressed that the constant response to DCCD inhibition of the ATP yield per flash during trains of flashes fired at different frequencies (9), an experiment often quoted as an example of double-inhibitor titration (5,7), cannot be properly analyzed with a steady-state kinetic approach.

Passive leaks interfere with the results of inhibitor titrations (5-8) and decrease the efficacy of these experiments as a validity test for delocalized coupling. In chromatophores the leaks can be very significant: judging from the dissipative current evaluated from the carotenoid electrochromic signal in the presence of oligomycin, they can amount to about 30% of the ionic current in phosphorylating conditions

(300 μmoles h^{-1} mg^{-1} BChl. vs. 720 μmoles h^{-1} mg^{-1} BChl for the phosphorylating proton current, taken as three fold the rate of ATP synthesis). In the presence of such important leaks the titration with an inhibitor of the primary pump can be hyperbolic also in delocalized model ($J_p^e/J_p(0) < f_e$ (7)). This behaviour is indeed confirmed in the single-inhibitor titration with antimycin (Figure 3). In spite of the leaks, however, in the titration with DCCD, the relative inhibition of ATP synthesis is always significantly lower than f_p. In a delocalized model this behaviour can be simulated when the conductance ratio of the secondary to the primary pump is higher than 1, i.e. when the primary pump controls the overall coupled process more than the secondary pump. This can be true if the dependence of the rate of ATP synthesis on $\Delta\mu_{H^+}$ is very steep, and simultaneously the control of $\Delta\mu_{H^+}$ on the electron flow is very poor. This latter condition can be real in the present experimental conditions (0.2 mM succinate, 1 mM KCN, aerobiosis), in which the ambient redox potential can be far from the optimal one for a fast electron transfer (17,18). As an alternative interpretation a marked upward deviation from a linear behaviour in the $J_p^P/J_p(0)$ vs. f_p plot, can be simulated in a delocalized model by assuming that a considerable leak occurs through uncoupled ATPases (Pietrobon, D., unpublished). Experimental evidence that a large part of the leak current in chromatophores is through uncoupled ATPases is provided by the larger dissipative ionic current measured in non phosphorylating conditions in the absence of ATPase inhibitor than in its presence (19). This type of leaks will be titrated by DCCD in parallel with the active ATP synthetases and the decrease in dissipative flow will partially compensate for the decrease in the number of active enzymes.

ACKNOWLEDGEMENTS

This work has been supported by Consiglio Nazionale delle Ricerche and by Ministero per la Pubblica Istruzione of Italy.

REFERENCES

1 Westerhoff, H.V., Melandri, B.A., Venturoli, G., Azzone, G.F. and Kell, D.B. (1984) Biochim. Biophys. Acta 768, 257-292
2 Ferguson, S.J. (1985) Biochim. Biophys. Acta 811, 47-97
3 Baum, H., Hall, G.S., Nalder, J. and Beechey, R. (1971) in Energy Transduction in Respiration and Photosynthesis (Quagliariello et al., eds.), pp. 747-755, Adriatica Editrice, Bari
4 Hitchens, G.D. and Kell, D.B. (1982) Biochem. J. 206, 351-357
5 Westerhoff, H.V. (1983) Mosaic Non-Equilibrium Thermodynamics and (the Control of) Biological Free Energy Transduction, Ph.D. Thesis, Univ. of Amsterdam
6 Davenport, J.W. (1985) Biochim. Biophys. Acta 807, 300-307
7 Pietrobon, D. and Caplan, S.R. (1986a) Biochemistry, in press
8 Pietrobon, D. and Caplan, S.R. (1986b) Biochemistry, in press
9 Venturoli, G. and Melandri, B.A. (1982) Biochim. Biophys. Acta 680, 8-16
10 Casadio, R. and Melandri, B.A. (1985) Arch. Biochem. Biophys. 238, 219-228

11 Cotton, N.P.J., Clark, A.J. and Jackson, J.B. (1984) Eur. J. Biochem. 142, 193-198

12 Casadio, R. and Melandri, B.A. (1984) in Proton-ATPase (ATP synthase): Structure, Function, Biogenesis (Papa, S. et al., eds.), pp. 411-420, Adriatica Editrice, Bari

13 Van den Berg, W.H., Prince, R.C., Bashford, C.L., Takamiya, K., Bonner, W.D. and Dutton, P.L. (1979) J. Biol. Chem. 254, 8594-8604

14 Baccarini Melandri, A., Casadio, R. and Melandri, B.A. (1977) Eur. J. Biochem. 78, 389-402

15 Casadio, R., Baccarini Melandri, A. and Melandri, B.A. (1978) FEBS Lett. 87, 323-328

16 Melandri, B.A., Venturoli, G., De Santis, A. and Baccarini Melandri, A. (1980) Biochim. Biophys. Acta 592, 38-52

17 Prince, R.C. and Dutton, P.L. (1977) Biochim. Biophys. Acta 462, 731-747

18 Crofts, A.R., Meinhardt, S.W., Jones, K.R. and Snozzi, M. (1983) Biochim. Biophys. Acta 723, 202-218

19 Clark,A.J., Cotton,N.P.J. and Jackson,J.B. (1983) Biochim. Biophys. Acta 723, 440-453

CHANGES IN THE APPARENT Km FOR ADP BY CF_0-CF_1 UNDER DIFFERENT CONDITIONS OF LIGHT INTENSITY AND UNCOUPLER CONCENTRATION.

W. PAUL QUICK AND JOHN D. MILLS, DEPARTMENT OF BIOLOGICAL SCIENCES, UNIVERSITY OF KEELE, KEELE, STAFFORDSHIRE, ST5 5BG, U.K.

1.   INTRODUCTION

There have been several reports that the apparent Michaelis - Menton constant (Km) of the chloroplastic ATPase (CF_0-CF_1) for its substrate ADP varies according to the experimental conditions (1-3); this has also been observed in liver submitochondrial particles (4,5). Reduction of the rate of ATP synthesis by lowering the light intensity (1-3) or by adding an electron transport inhibitor (2) was shown to decrease the apparent Km for ADP. However, reduction of the rate of ATP synthesis using an uncoupler elevated the apparent Km for ADP (1,3). These results have been interpreted in several ways and used as evidence to support models of ATP synthesis involving a direct activation of CF_0-CF_1 by $\Delta\mu H^+$ (2,4,5) or by electron transport (1,3).

Previous workers have failed to measure the ΔpH during their experiments and have thus not included another potential variable which is uncontrolled during the experiment. We have repeated the experiments of Vinkler (1) but have included simultaneous measurements of the ΔpH. The results are in broad agreement with other reports (1-5) in that decreasing the light intensity decreases the apparent Km for ADP whereas increasing the uncoupler concentration increases the apparent Km for ADP. However, by taking account of ΔpH changes that occur during the Km analysis the results can be explained by a simple delocalised chemiosmotic hypothesis. Data obtained from a mathematical model of ATP synthesis, described previously (6), based on chemiosmotic principles show results in close agreement with those measured.

2.   MATERIALS AND METHODS

Intact chloroplasts were isolated from Pisum sativum (var. Feltham First) as described previously (7) except the chloroplasts were lysed in a medium containing 5mM $MgCl_2$, 5mM ascorbate and 20mM MES-KOH buffer (pH 6.5) and then recentrifuged before the final resuspension. The rate of ATP synthesis and the magnitude of the trans-thylakoid ΔpH were measured simultaneously in a medium containing 20mM KCl, 5mM $MgCl_2$, 1mM Pi, 0.1mM MeV, 15mM Glucose, 10 units/ml hexokinase, 2.5 µg/ml diadenosine pentaphosphate, 20µM 9-amino acridine, 1000 units/ml catalase, 0.4µCi ^{32}Pi (Amersham International) and 30mM Tricine-KOH (pH 8.1) in an oxygen electrode at 20°C. Illumination was provided by a Halogen lamp filtered through an RG 665 red cut-off filter and the intensity was varied using neutral density filters. 9-amino acridine fluorescence was measured as described previously (8). The reaction was initiated by illumination and terminated after 2 mins by the addition of 200µl of 20% trichloroacetic acid per 1ml of assay medium containing 15µg chlorophyll and the

Biggens, J. (ed.), Progress in Photosynthesis Research, Vol. III. ISBN 90 247 3452 5
© *1987 Martinus Nijhoff Publishers, Dordrecht. Printed in the Netherlands.*

indicated ADP concentration. The rate of ATP synthesis was determined by ^{32}Pi incorporation into glucose-6-phosphate as described previously (9).

3. RESULTS AND DISCUSSION

CF$_0$-CF$_1$ is a complicated enzyme with three potential substrates (ADP, Pi and Mg^{2+}) and $\Delta\mu$H$^+$ could be regarded as a fourth. Nevertheless in the presence of saturating Pi and Mg^{2+} concentrations (approx. 1mM) the enzyme shows a linear Michaelis-Menton plot with respect to ADP allowing the determination of an apparent Km and Vmax (1-3). However, previous reports have ignored how the ΔpH varied during their Km analysis and how this variation was effected under different experimental conditions.

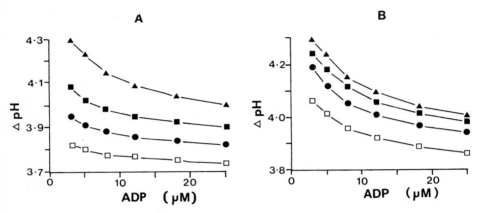

Fig. 1. ΔpH values determined by 9-amino acridine fluorescence. Thylakoid membranes undergoing ATP synthesis were incubated with various concentrations of ADP. In (A) the rate of ATP synthesis was reduced by adding the uncoupler nigericin (▲-0nM, ■-30nM, ●-60nM, □-90nm). In (B) the rate of ATP synthesis was reduced by lowering the light intensity (▲-240, ■-110, ●-53, □-29 W.m^2).

Fig. 1. shows how the ΔpH determined by 9-amino acridine fluorescence measured simultaneously to the rate of ATP synthesis varied in the presence of differing concentrations of ADP. The results show that as the ADP is increased the magnitude of the steady-state transthylakoid pH gradient diminishes. This is a predictable result since the ΔpH is a result of a balance between protons pumped by electron transport and protons dissipated across the membrane by either the ATPase or membrane leakage; thus increased ATP synthesis will lead to an increased rate of proton dissipation. In Fig. 1(A) the effect of ADP concentration on ΔpH is shown in the presence of several concentrations of the uncoupler nigericin. As the concentration of nigericin is increased the ΔpH is lowered but the change in ΔpH due to ADP is decreased; in the absence of nigericin the ΔpH decreases by approximately 0.25 pH units for a change in the ADP concentration from 3 to 25µM, in the presence of 90nM nigericin a similar ADP concentration change results in a change of less than 0.1 pH units. In Fig. 1(B) the ΔpH was reduced by lowering the light intensity and it can be seen that at all light intensities used a similar large change in ΔpH was induced by increasing the ADP concentration from 3 to 25 µM.

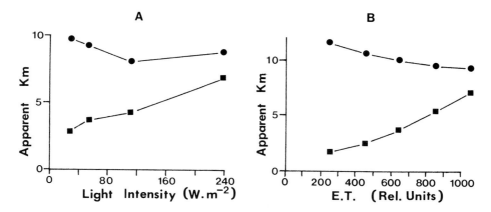

Fig. 2. Dependence of the apparent Km of CF_0-CF_1 (ADP) on the light intensity (A) in the absence (■) and presence (●) of 90nm nigericin. The calculated apparent Km values determined by a mathematical model (see text for details) are shown in (B) for various rates of electron transport with either a low (■) or high (●) membrane permeability constant for protons.

The apparent Kms of CF_0-CF_1 (ADP) were determined by Michaelis-Menton double reciprocal plots and Fig. 2(A) shows how the apparent Km varied with increasing light intensity in the presence and absence of uncoupler. As the light intensity was increased the apparent Km for ADP increased. For any given light intensity the presence of an uncoupler increased the apparent Km and this was more marked as the light intensity decreased. In the presence of uncoupler increasing the light intensity caused a slight decrease in the apparent Km. These results are in broad agreement with other reports (1-5).

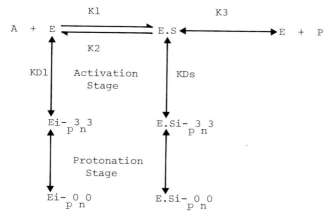

Scheme 1. An outline of a proposed mechanism of ATP synthesis utilised in the mathematical model. Details of the model and values for the rate constants will be presented elsewhere.

We have attempted to simulate these results using a simple mathematical model based on a fully delocalised chemiosmotic coupling mechanism as described previously (6) but with slight modifications to allow for changes in the substrate (ADP) concentration. The model assumes simple Michaelis-Menton kinetics for CF_0-CF_1 with respect to its substrate ADP but allows for the effect of ΔpH as either a thermodynamic driving force or an activation phenomenon (Scheme 1). The results of this model are shown in Fig. 2(B) and it can be seen to be qualitatively identical to the experimental results shown in Fig. 2(A). Thus our analysis of these data suggest that the complex behaviour of Km(ADP) can be a consequence of a fully delocalised chemiosmotic system and that therefore there is no need to invoke alternative hypotheses involving direct links between electron transport and CF_0-CF_1. The model will be discussed further and more details will be presented elsewhere.

REFERENCES

1 Vinkler, C. (1981) Biochem. Biophys. Res. Comm. 99, 1095-1100
2 Bickel-Sandkotter, S. and Strotmann, H. (1981) FEBS Lett. 125, 188-192
3 Loehr, A., Wilms, I. and Huchzermeyer, B. (1985) Arch. Biochem. Biophys. 236, 832-840
4 Kayalar, C., Rosing, J. and Boyer, P.D. (1976) Biochem. Biophys. Res. Comm. 72, 1153-1159
5 Hatefi, Y., Yagi, T., Phelps, D., Wong, S., Vik, B. and Galante, Y. (1982) Proc. Nat. Acad. Sci. USA 79, 1756-1760
6 Mills, J.D. and Mitchell, P. (1983) in Proc. VI. Int. Cong. Phot., ed. Symbesma, C., Nijhoff/Junk Publ., p523-526
7 Mills, J.D., Mitchell, P. and Schurmann, P. (1980) FEBS Lett. 112, 173-177
8 Mills, J.D. and Mitchell, P. (1982) Biochim. Biophys. Acta 679, 75-83
9 Mills, J.D. and Mitchell, P. (1982) FEBS Lett. 144, 63-67

THE ROLE OF COUPLING FACTOR ACTIVATION IN REGULATING THE INITIATION OF ATP FORMATION

ROGER HANGARTER[†], PATRICIA GRANDONI AND DONALD R. ORT, DEPARTMENT OF PLANT BIOLOGY, USDA/ARS, UNIVERSITY OF ILLINOIS, URBANA, ILLINOIS 61801

INTRODUCTION

The activity of the chloroplast coupling factor complex is highly regulated. The light-driven electron transfer reactions in thylakoid membranes generate an electrochemical potential difference consisting of a transmembrane imbalance of electrical charge ($\Delta\Psi$) and of hydrogen ion activity (ΔpH). It is this electrochemical potential that is used to drive the various coupling factor catalyzed reactions. In washed dark-adapted thylakoid membranes, ATP synthesis begins after just a few milliseconds of illumination and maximum efficiency (ATP/e_2) is achieved several milliseconds later (1). When electron transfer is driven by saturating single-turnover flashes under conditions in which a membrane potential is prevented from forming, only about 15 flashes are required to build up the requisite ΔpH for the onset of ATP formation and maximum ATP/e_2 is attained shortly thereafter (2).

The efficiency and rapid induction of ATP synthesis by the coupling factor complex is in contrast to what is observed for the behavior of the reverse reaction of ATP hydrolysis (3). Under non-energized conditions washed chloroplast thylakoid membranes are unable to catalyze the thermodynamically favorable hydrolysis of ATP. Seconds, rather than milliseconds, of illumination are required before maximal rates of ATP hydrolysis can be observed. However, the highest rates of ATP hydrolysis require preillumination of thylakoid membranes in the presence of sulfhydryl reagents, such as dithiothreitol (DTT), but even this activation requires a prolonged illumination.

Until recently the relevance of DTT-modulation of CF_1 to studies of ATP synthesis has been largely overlooked since, as mentioned above, rapid and efficient ATP synthesis occurs in isolated thylakoid membranes in the absence of such thiol reagents. However recent reports have indicated that, at least under conditions of limited energization, the rate of ATP synthesis can be significantly influenced by pretreatment with DTT. In this communication we report on the effects of DTT-modulation of thylakoid membranes on the initiation of ATP synthesis and other coupling factor activities. We found that flash-induced ATPase, ATP synthesis, and release of tightly bound [¹⁴C]ADP were initiated at lower flash numbers in the DTT-modified membranes than in untreated membranes, or membranes in which the DTT-modification was reversed by ferricyanide. The yield per flash for all of these CF_1 reactions, once initiated, increased more rapidly in the DTT-treated membranes so that maximum levels were reached up to 30 flashes earlier than in the controls. DTT-treatment increased the efficiency of ATP synthesis (ATP/e_2) over the controls but only when the level of

[†]Present Address: Botany Department, Ohio State University, Columbus, Ohio 43210

Biggens, J. (ed.), Progress in Photosynthesis Research, Vol. III. ISBN 90 247 3452 5
© *1987 Martinus Nijhoff Publishers, Dordrecht. Printed in the Netherlands.*

energization was suboptimal. The effect of the ΔG_{ATP} on the onset of ATP synthesis indicates that a significant energy barrier must still be over-come for activation of the DTT-modified membranes, but this energy barrier was lower than in unmodified membranes.

METHODS AND MATERIALS
Chloroplast thylakoid membranes were isolated from commercial spinach (Spinacea oleracea L.) as described elsewhere (2). Reactions were carried out in a water-jacketed reaction chamber thermostatted at the indicated temperature. Light flashes, of 6 µs duration at half peak height, were provided by two xenon lamps (Model FX-200 flash tube, EG&G, Salem, MA) discharged simultaneously on opposite sides of the sample at a frequency of 10 Hz. For the DTT-treatment thylakoid membranes were illuminated for 5 min at 20°C with 10 mM DTT. Demodulated membranes were prepared by treating the DTT-modified membranes with 0.5 mM ferricyanide. The amount of electron transport occurring per flash was determined by measuring the flash-induced oxygen uptake associated with electron transfer from H_2O to methylviologen. ATPase activity was assayed at 18°C by measuring the net acidification of the medium associated with ATP hydrolysis. In some experiments thylakoid membranes were prelabeled with [^{14}C]ADP under the reaction conditions used for the DTT-treatment except that 5 µM [^{14}C]ADP was added to the reaction medium. The ATP formation resulting from a series of light flashes at 4°C was measured as the incorporation of ^{32}Pi into ATP as described previously (2).

RESULTS AND DISCUSSION
When the formation of a membrane potential is prevented by the pre-sence of nonactin and potassium the number of flashes required to energize the membrane depends quantitatively on the number of electron transport chains that turn over per flash. This is because the threshold proton pool required to achieve a sufficiently large ΔpH for the initiation of ATP formation is filled by the cooperative action of at least several hundred electron transfer complexes (2). This property of the chloroplast system makes it possible to use the number of flashes required for the onset of coupling factor activity as a sensitive monitor of the energy transduction process.

The DTT-treatment of thylakoid membranes is known to reduce a specific disulfide in the γ subunit of CF_1 (4). The conditions for the treatment, however, potentially could alter other components of the energy trans-ducing system. For example if the DTT-treatment were to protect electron transfer complexes from photoinhibition during the 5 min preillumination the filling of the threshold proton pool might then require fewer flashes in the DTT-treated membranes than in the preilluminated controls. Fortun-ately the protocol we followed for the DTT-modulation and demodulation of the thylakoid membranes had no effect on the electron transfer flash yields (Table I).

An additional complicating factor associated with the preillumination used for the DTT-treatment is that it results in the generation of a transmem-brane pH difference. During the following dark period the pH difference decays as ions move across the membrane to reach equilibrium concentra-tions but the permeability properties and the buffer capacity of thylakoid membranes are such that it can take several minutes for the light-induced ΔpH to fully relax. In fact, Hangarter and Good (5) found that preillumi-nated thylakoid membranes can retain the capacity to synthesize ATP for more than 12 min in the dark. For this reason care must be taken not to be mislead by the effects of any residual ΔpH that can be present follow-ing the preillumination. Junesch and Gräber recently reported (6) that they had to use their DTT-modified membranes within five minutes of the preillumination treatment to see an effect of the treatment on phosphoryl-ation. Mills and Mitchell (7,8) used their DTT-treated membranes within 30 min after the preillumination and, in addition, their controls were not preilluminated. Thus it is possible that residual membrane energization may have had some effect on their results. In our experiments the pre-illuminated membranes were stored in the dark for at least one hour prior to use to ensure the relaxation of any ΔpH. We also used as controls membranes that were exposed to identical or nearly identical preillumina-tion conditions.

Table I. Lack of effect of the DTT-modulation and demodulation treatments on flash-induced electron transfer. Electron transfer was determined at 18°C by measuring the flash-induced uptake of oxygen associated with elec-tron transfer from H_2O to methylviologen.

Membranes	Electron Transfer (meq/mol Chl/flash)
native	1.94 ± 0.16
DTT-modulated	1.93 ± 0.10
demodulated	2.03 ± 0.21

Although we are confident that the effects of DTT-modification of CF_1 are not due to the presence of a residual ΔpH or to an effect on the electron transfer system there remained the possibility that the DTT may have altered the buffer capacity of the thylakoid lumen. It is probably not possible to measure directly small changes in the buffer capacity of the lumen but, in earlier work, we found that the pseudosteady state amount of flash-induced proton uptake was the same in DTT-treated and untreated mem-branes (9). Since the flash yields for electron transfer were the same for both membrane preparations (Table I) this suggests that the proton binding capacity was not altered by the DTT-treatment. In addition, the DTT-treatment had no effect on the subsequent dark-efflux of the protons that had accumulated in the thylakoid membrane vesicles during a flash series (4,9). Thus, we believe that the effects described below reflect the ability of the thylakoid membranes to use the flash-induced ΔpH rather than their ability to generate or maintain a ΔpH.

Dark-adapted thylakoid membranes are unable to hydrolyze ATP until a transmembrane electrochemical potential of sufficient magnitude is generated (3), but even after prolonged illumination the ATPase activity remains low. On the other hand, illumination of thylakoid membranes in the presence of DTT increases the level of ATPase activity several fold (3). Figure 1 shows these effects when electron transfer was driven by saturating single-turnover flashes. Membranes that had been pretreated with DTT were able to hydrolyze ATP after about 10 flashes and maximum rates were achieved after a series of less than 50 flashes. In contrast, when the disulfide groups that had been reduced by the DTT treatment were reoxidized, by washing with ferricyanide, several more flashes were required for the onset of ATPase activity and the maximum rate achieved was only about 20% of the rate reached by the DTT-modulated membranes.

Fig. 1 **The flash-number dependence of the onset of ATPase activity in DTT-modulated and demodulated thylakoid membranes.** DTT-modulation, demodulation and ATPase measurements, were carried out as described under methods and materials. In dark-adapted DTT-modulated membranes ATPase activity began after about 10 saturating single-turnover flashes. The activity reached the maximum rate after a series of less than 50 flashes. In demodulated membranes more flashes were required to initiate ATPase activity and the maximum rate of ATP hydrolysis was only about 20% of the rate achieved with the modulated membranes.

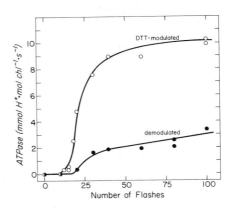

Fig. 2 **Flash-induced release of tightly-bound [¹⁴C]ADP from DTT-modulated and demodulated thylakoid membranes.** In DTT-modulated membranes flash-induced release of the tightly-bound [¹⁴C]ADP was initiated by as few as 6 flashes and the total amount that had been bound was released by about 50 flashes. However, if the disulfide groups that had been reduced by the DTT are reoxidized by washing the membranes with ferricyanide the initiation of the release of tightly-bound ADP is delayed until about the twelfth flash and the subsequent flash-dependence is more gradual. These data suggest that the threshold ΔpH required for initiating the activity of the coupling factor is significantly lower when the enzyme is in the reduced form.

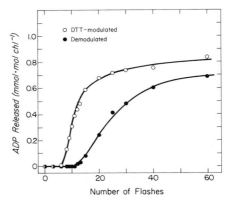

In dark-adapted thylakoid membranes the coupling factor has on average one molecule of ADP tightly-bound to it and the energy-induced release of this tightly-bound ADP is associated with the activation of the coupling factor (10). In Figure 2 the flash-induced release of tightly-bound [¹⁴C]ADP from DTT-modulated and demodulated membranes is shown. The [¹⁴C]ADP was incorporated into the tight binding site during the preillumination with the DTT. The release of the tightly-bound [¹⁴C]ADP from the DTT-modulated membranes could be observed after as few as 6 flashes and complete release was achieved by about 50 flashes. Oxidation of the DTT-modulated membranes delayed the initiation of the release of the [¹⁴C]ADP until about the twelfth flash and the post-threshold flash-dependence was more gradual. The total amount of tightly-bound [¹⁴C]ADP was the same for both the DTT-modulated and demodulated membranes and in the experiment shown in Fig. 2 was 0.86 mmol·mol Chl^{-1}. This is equivalent to about one tightly-bound [¹⁴C]ADP/CF$_1$ (9).

The threshold number of flashes required to initiate ATP formation is also sensitive to DTT-modulation (Fig. 3). The onset of ATP synthesis in the DTT-modulated membranes was observed after about 10 flashes but in the native, untreated membranes up to 15 flashes were required. When the results are expressed as ATP yield per flash it can be seen that the capacity to synthesize ATP increases more steeply in the DTT-modulated membranes than in the native membranes. This correlates well with the onset of coupling factor activation depicted by the ATPase activity (Fig. 1) and the release of tightly-bound [^{14}C]ADP (Fig. 2). However, the release of the tightly-bound [^{14}C]ADP could always be observed a few flashes before the onset of ATPase activity or ATP synthesis.

Fig. 3 Dependence of ATP synthesis in DTT-modulated and native thylakoid membranes on the number of single turnover flashes. The upper panel shows the amount of ATP synthesized by flash series of increasing number of flashes. ATP synthesis can be observed by the tenth flash in membranes that had been pretreated with DTT whereas unreduced membranes require up to 15 flashes. The lower panel shows the results expressed as ATP yield per flash. The flash yields were estimated by calculating the average yields between consecutive data points from the experiment shown in the upper panel. The solid lines in the lower figure were calculated from the lines drawn through the data in the upper figure. These results show that the onset of ATP synthesis follows closely the release of tightly bound [^{14}C]ADP (Fig. 2).

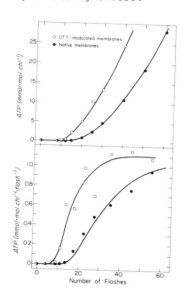

Since the electron transfer yields are not affected by the DTT-treatment the data in the lower panel of Figure 3 indicate that the efficiency of ATP synthesis is higher for DTT-modulated membranes during the transition from the onset of ATP synthesis to when maximum flash yields are reached. The increased efficiency observed during the transition period suggested that the DTT-modification might increase the ATP/e$_2$ under conditions of suboptimal energization. Table II shows the effect of DTT-modulation on the efficiency of ATP synthesis when electron transfer was inhibited by about 90% and the ATP/e$_2$ was correspondingly reduced by about 60% in demodulated and in native membranes. The DTT-modulation caused about a 40% increase in the ATP/e$_2$ ratio under the conditions of limited electron transfer but had no effect on the efficiency when saturating flashes were used.

The results presented above indicate that the DTT-modulated thylakoid membranes are able to use a lower ΔpH to initiate the activity of the coupling factor complex. We examined the effect of the phosphate potential on the initiation of the release of tightly-bound [^{14}C]ADP and on the onset of photophosphorylation in order to determine the changes in the energetic requirements that appear to be induced by the DTT-treatment.

The ΔG_{ATP} was varied by changing the concentrations of ATP and ADP in the reaction mixture and the values were calculated according to the method described by Lemasters (11).

Table II. **The effect of DTT-modulation on the efficiency of flash-induced photophosphorylation.** The ATP/e$_2$ ratios were estimated from the oxygen uptake associated with electron transfer to methylviologen from 400 single-turnover flashes and the flash yields for ATP synthesis averaged between 200 and 500 flashes. Electron transfer was 2.02 meg·mol Chl⁻1 flash⁻¹ with saturating flashes and 0.24 meq·mol Chl·flash⁻¹ with the attenuated flashes.

| treatment | expt. | ATP/e$_2$ | |
		saturating flash	attenuated flash
DTT-modulated	1	1.10	0.77
	2	1.09	0.88
demodulated	1	1.05	0.57
	2	1.06	0.60

Fig. 4 **Lack of effect of ΔG_{ATP} on the flash-induced release of [¹⁴C]ADP in DTT-modulated membranes.** Changing the ΔG_{ATP} from 38.5 kJ·mol⁻¹ (50 µM ADP, 10 µM ATP) to 56.9 kJ·mol⁻¹ (50 µM ADP, 1.5 mM ATP) had no effect on the initiation of the release of tightly-bound ADP even though the onset of photophosphorylation is delayed by high ΔG_{ATP} (Fig. 5).

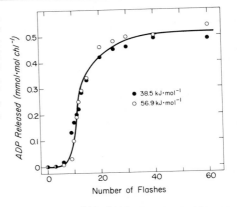

The initiation of the release of tightly-bound [¹⁴C]ADP was totally un-effected in DTT-modulated membranes when the ΔG_{ATP} was increased from 38.5 kJ·mol⁻¹ to as high as 56.9 kJ·mol⁻¹ (Fig. 4). This was also true for native and demodulated membranes (data not shown) although the threshold flash number was about 12 for these membranes in which the coupling factor was in its oxidized state (see Fig. 2). This is in contrast to the effect of the ΔG_{ATP} on the onset of ATP formation (Fig. 5). With saturating flashes the onset of ATP synthesis in native membranes was delayed by ΔG_{ATP} values below about 51 kJ·mol⁻¹ (Fig. 5A).

Because the threshold number of flashes for the onset of ATP formation in the DTT-modulated membranes was considerably lower than in the native or demodulated membranes it was more difficult to determine the effect of the ΔG_{ATP} with saturating flashes since the lags induced by high ΔG_{ATP} values are proportionately shorter due to the extensive pooling of protons (2). Thus we repeated the experiments shown in Figure 5A with attenuated flashes so that only 40% of the reaction centers turned over per flash. The reduced electron transfer increased the number of flashes needed to fill the threshold pool and caused a proportional increase in the lags induced by increasing phosphate potentials (Fig. 5B). Although the measurements made with the attenuated flashes have a lower signal to noise ratio the data show clearly that ATP synthesis in DTT-modulated membranes is sensitive to lower phosphate potentials than are the native membranes. For both the DTT-modulated and control membranes the threshold flash number remained constant until the ΔG_{ATP} reached a critical level. This was true with saturating and attenuated flashes. Thus the DTT-modulated membranes must still overcome a threshold energy requirement for activation of the coupling factor prior to the initiation of ATP synthesis in spite of the "preactivation" by DTT. That is, even in DTT-pretreated membranes, at ΔG_{ATP} values below 45 kJ·mol^{-1} the Δp becomes energetically adequate to phosphorylate ADP prior to becoming adequate to activate reduced CF$_1$. The energy required, however, is substantially lower than that required by the demodulated or untreated membranes. The ΔG_{ATP} at which the threshold flash number is effected should indicate the minimum energy required for the activation of the coupling factor. Assuming 3H$^+$·ATP^{-1} we estimate these threshold energy levels to be about 17 kJ·mol^{-1} and about 15 kJ·mol^{-1} multiplied by the number of charges involved in the activation process for native membranes and for DTT-treated membranes. These values are equivalent to ΔpH values of 2.9 for native membranes and 2.6 for DTT-reduced membranes.

Fig. 5 The onset of photophosphoryl-ation from single-turnover flashes at various phosphate potentials in DTT-modulated and native thylakoid membranes. A. The effect of increasing ΔG_{ATP} on the flash number threshold for the onset of ATP formation in native membranes. B. The effect of increasing ΔG_{ATP} is shown for both DTT-modulated and unmodulated membranes for a flash intensity that caused only 40% of the reaction centers to turnover per flash. The reduced electron transfer increased the number of flashes needed to fill the threshold proton pool and thus caused a proportional increase in the lags induced by the higher phosphate potentials.

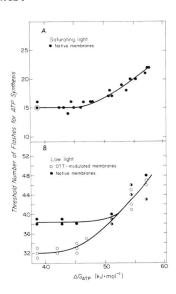

The effects of DTT-modification of thylakoid membranes on the onset of flash-induced ATP synthesis, ATP hydrolysis, and release of tightly bound [^{14}C]ADP strongly support the notion that the reduced form of the coupling factor complex requires less energy for activation. In all cases we found that the coupling factor activity of the DTT-modified membranes was initiated by fewer flashes than any of the control treatments. In preliminary experiments we found that the effect of the DTT-modification was stable for up to 4 hr in the dark. From the considerations discussed above it seems likely that the DTT-treatment results in a change in the energy transduction system of the thylakoid membrane at the level of the coupling factor complex, a notion that supports observations made by others (4,7,8). In all likelihood the change is due to the well characterized modification of the disulfide in the γ subunit of CF_1 (4). A kinetic model has been proposed to explain the effects of ΔpH on the rates of ATP synthesis and hydrolysis (6). According to the model there is an energy-dependent equilibrium between active and inactive states of CF_1 and at a given ΔpH there is a greater proportion of reduced CF_1 in the active state than for oxidized CF_1.

The effects of DTT-modification of thylakoid membranes on flash-induced coupling factor activity are quite apparent during the transition from dark to light and when electron transfer is severely limited. However, once a pseudosteady state is reached with reasonable levels of electron transfer (greater than 50% of chains turning over) both the oxidized and reduced forms of the membrane synthesize comparable amounts of ATP. Since in saturating continuous light maximum rates of ATP synthesis are achieved in milliseconds (1) it seems likely that the oxidation state of CF_1 plays its most significant role in regulating ATP synthesis under conditions of limited energization where the reduction of CF_1 could allow higher rates and efficiencies of ATP synthesis (4,). Furthermore, oxidation of the reduced form may play an important role in deactivating the ATPase upon light to dark transitions to prevent wasteful ATP hydrolysis (12).

REFERENCES
1.   Graan, T. and Ort, D.R. (1981) Biochim. Biophys. Acta 637, 447-456.
2.   Hangarter, R. and Ort, D.R. (1985) Eur. J. Biochem. 149, 503-510.
3.   Bakker-Grunwald, T. (1977) In Encyclopedia of Plant Physiology - Photosynthesis I (Trebst, A. and Avron, M., eds.), Vol. 5, pp. 369-373, Springer-Verlag, Berlin.
4.   Ketcham, S.R., Davenport, J.W., Warncke, K. and McCarty, R.E. (1984) J. Biol. Chem. 259, 7286-7293.
5.   Hangarter, R.P. and Good, N.E. (1984) Biochemistry 23, 122-130.
6.   Junesch, U. and Gräber, P. (1985) Biochim. Biophys. Acta 809, 429-434.
7.   Mills, J.D. and Mitchell, P. (1982) FEBS Lett. 144, 63-67.
8.   Mills, J.D. and Mitchell, P. (1984) Biochim. Biophys. Acta 764, 93-104.
9.   Hangarter, R.P. and Ort, D.R. (1986) Eur. J. Biochem. In Press.
10. Strotmann, H., Bickel-Sandkotter, S. and Shoshan, V. (1979) FEBS Lett. 101, 316-320.
11. Lemasters, J.J. (1984) J. Biol. Chem. 259, 13123-13130.
12. Shahak, Y. (1982) Plant Physiol. 70, 87-91.

ROLE OF THE SUPEROXIDE ANION (O_2^-) AND HYDROXYL RADICAL ($OH^\cdot$) IN ATP SYNTHESIS OBTAINED WITH SPINACH CHLOROPLASTS IN DARKNESS

E. TYSZKIEWICZ AND E. ROUX*, Service de Biophysique, CEN Saclay, 91191 GIF-SUR-YVETTE cedex, France

INTRODUCTION

It is well know since the work of M. Avron and N. Sharon (1) on photophosphorylation performed with oxygen 18, that it is orthophosphate, but not ADP which loses one of its oxygens before its fixation to ADP, to form ATP. On the other hand, we know that chloroplasts synthesize ATP in the light, when they are in presence of orthophosphate, ADP and of an electron mediator such as phenazine methosulfate (PMS) or pyocyanine (Py). In previous work (2) we have reported that chloroplasts can synthesize ATP in darkness, at constant pH 8, only if reduced PMS is added to the medium and if the reaction is performed under oxygen.

According to J. Davis and P.J. Thornaley (3) the mechanism of the autoxidation of reduced PMS (PMSH) by oxygen leads to the formation of the superoxide anion and $PMSH^+$ radical, then in a further step to a hydrogen peroxide and hydroxyl radical, following the reactions :

$$PMSH + O_2 \rightleftharpoons PMSH^+ + O_2^- \xrightarrow{+ H^+} PMSH^+ + H_2O_2 \qquad \text{1a}$$

$$PMSH + H_2O_2 + H^+ \longrightarrow PMSH^+ + OH^\cdot + H_2O \qquad \text{1b}$$

Our purpose is to study the effect of these radicals on the activation of the orthophosphate ion i.e. its dehydroxylation, preceeding its fixation on ADP.

In this work we examine the influence of the presence in the phosphorylating medium of superoxide anions (O_2^-) and of hydroxyl radicals ($OH^\cdot$) on ATP synthesis performed in darkness at constant pH with spinach chloroplasts (class B) as prepared in (2).

Superoxide anion (O_2^-)

To check the influence of the superoxide anion we have studied :
1) The ability of reduced pyocyanine (by NADH) to induce ATP synthesis in darkness. Indeed, according to the above mentioned authors (3), the reoxidation of reduced pyocyanine generates superoxide anion and $PyH^\cdot$ radical, but no hydroxyl radical, nor hydrogen peroxide are formed, following the reactions :

$$PyH_2 + O_2 \rightleftharpoons PyH^\cdot + O_2^- + H^+ \qquad \text{2a}$$

$$PyH^\cdot + O_2 \rightleftharpoons Py + O_2^- + H^+ \qquad \text{2b}$$

2) The action of the superoxide dismutase (SOD) (catalizing the dismutation of superoxide anion into hydrogen peroxide and oxygen), on ATP

*E. Roux died on August 21, 1985, but this work was initiated by him.

Biggens, J. (ed.), Progress in Photosynthesis Research, Vol. III. ISBN 90 247 3452 5
© *1987 Martinus Nijhoff Publishers, Dordrecht. Printed in the Netherlands.*

synthesis in darkness, in presence of chemically reduced PMS or
chemically reduced pyocyanine.
3) The action of xanthine oxidase, (producing superoxide anion in presence
of xanthine), on ATP synthesis in darkness in the absence of electron
mediator.

EXPERIMENTAL RESULTS
1) Reduced pyocyanine induces ATP synthesis in darkness.
2) Superoxide dismutase inhibits the ATP synthesis in darkness in the
presence of reduced PMS or reduced pyocyanine (Table 1).

Table 1 - ATP syntheis in darkness, in the presence or absence, of
superoxide dismutase

Addition to the medium	SOD (units)	ATP $nmol.h^{-1}$ mg^{-1} Chl.	% inhibition
Py	0	40	
Py + NADH	0	483	0
Py + NADH	100	441	8,5
Py + NADH	200	438	9
Py + NADH	300	273	43,5
PMS	0	42	
PMS + DTT	0	606	0
PMS + DTT	98	510	15,9
PMS + DTT	142	462	23,7
PMS + DTT	195	384	36,4

Phosphorylations in darkness were performed as described in (2), PMS
and DTT (dithiotreitol) : 10^{-2} M ; chloroplasts : 0.2 mg chlorophyll ;
reaction time : 2 minutes.

3) The presence of xanthine and xanthine oxidase in the phosphorylating
medium, increases the ATP synthesis obtained in darkness (Table 2).

Table 2 - ATP synthesis in darkness, in the presence or absence of
xanthine and xanthine oxidase.

Addition to the medium	ATP $nmol.h^{-1}$ mg^{-1} Chl.	Yield (%)
no addition	90	100
xanthine + xanth-ox	139	154
xanthine + xanth-ox boiled	95	105

Phosphorylations in darkness were performed as described in (2) except
for xanthine : 5.10^{-4} M ; xanthine oxidase : 0.25 unit ; chloroplasts :
0.2 mg chlorophyll ; reaction time : 2 minutes.

Hydroxyl radical OH$^{\cdot}$

To check the role of hydroxyl radicals, we have studied the influence of

their presence in the phosphorylating medium.
According to W.A. Pryor and R.H. Tang (4) methional is a good scavenger of the hydroxyl radical following the reaction :

$$CH_3-S-CH_2-CH_2-CHO + OH^\bullet \longrightarrow CH_2=CH_2 + HCOOH + \frac{1}{2}CH_3SSCH_3 \qquad 3a$$

Ethylene produced can be determined by capillary gaz chromatography. By this means we have checked that ethylene is formed in a phosphorylating medium containing methional and reduced PMS in presence of oxygen. That is to say that $OH^\bullet$ radicals are formed in the phosphorylating medium, when reduced PMS is reoxidized by oxygen.
To check the hydroxyl radicals action we have studied :
1) The influence of the concentration of orthophosphate ions on the production of ethylene i.e. the production of $OH^\bullet$ radicals in the phosphorylating medium.
2) The influence of the presence of methional in the phosphorylating medium on the ATP synthesis obtained with chloroplasts maintained in darkness or in the light.

EXPERIMENTAL RESULTS
1) The increase of orthophosphate ions concentration in the phosphorylating medium increases ethylene production, others conditions being maintained (Table 3).

Table 3 – Ethylene formation dependence on orthophosphate concentration

Orthophosphate mM	ethylene formed $nmol.ml^{-1}$ of the reaction medium
0	0.32
0.1	0.55
1.0	2.9
10.0	8.6

Reaction medium : pH8 ; orthophosphate as indicated ; $MgCl_2$: $10^{-3}M$; EDTA : 10^{-4} M; PMS $5.10^{-5}M$; DTT : 10^{-2} M ; methional : $0.72.10^{-3}$ M ; reaction time : 10 minutes ; temperature : 26°C ; ethylene is measured by capillary gaz chromatography : Varian, model 3700.

2) Addition of methional to the phosphorylating medium increases phosphorylations performed in darkness, up to 254 % and in the light up to 126 %, when PMS is used as a mediator(Table 4).

Table 4 – ATP synthesis in the presence or absence of methional

Addition to the medium	Methional mM	ATP $\mu mol.h^{-1}$ $mg^{-1}.Chl.$	Yield %
		in darkness	
PMS + DTT	0	288.10^{-3}	100
PMS + DTT	1,7	732.10^{-3}	254
PMS + DTT	4,2	375.10^{-3}	130
		in the light	
PMS	0	113	100
PMS	1.1	143	126
PMS	2,75	124	109

Phosphorylations in darkness were performed as described in (2) ;

photophosphorylations as described in (5) ; PMS : 5.10^{-5} M ; chloroplasts : 0.2 mg chlorophyll ; reaction time : 5 minutes.

DISCUSSION

Chloroplasts synthesize ATP in darkness in presence of orthophosphate and ADP, when the phosphorylating medium induces a reduction of molecular oxygen, i.e. in presence of superoxide anions. This is obtained by reoxidation of reduced PMS or reduced pyocyanine, or by the action of xanthine-oxidase on xanthine, in presence of oxygen.

When superoxide anions are scavenged by the action of superoxide dismutase, an inhibition of the ATP synthesis in darkness is observed. This may be due to the elimination of superoxide anions from the medium, and/or to the hydrogen peroxide formation (by the action of SOD) which reoxidize PMSH giving rise to the formation of hydroxyl radicals (by reaction 1b) which have an inhibitory effect on the phosphorylations. This result suggest that superoxide anion formation is a necessary step in the activation of the orthophosphate ion, preceeding its fixation on the ADP. On the other hand, the scavenging of hydroxyl radicals from the phosphorylating medium leads to an important increase of the ATP synthesis, both for the phosphorylations performed in darkness as well as for those performed in the light. Their continuous elimination may increase the rate of reactions 1b, then 1a, which are necessary steps for the orthophosphate activation, as suggested above.

This result eliminates the possibility of a direct participation of hydroxyl radicals in the orthophosphate activation. Hydroxyl radical formation is a consequence but not the cause of its activation.

Those results suggest two hypotheses : 1) superoxide anion activates directly the orthophosphate ion, producing metaphosphate and hydroxyl radical, or 2) superoxide anion acts producing hydrogen peroxide, reaction which activates orthophosphate.

ACKNOWLEDGEMENTS

The first author wish to express her gratitude to Drs Galmiche J.M., Girault G. and Berger G. for fruitfull discussions.

REFERENCES

1. Avron, M. and Sharon, N. (1960) Biochem. Biophys. Res. Comm. 2, 236-339
2. Tyszkiewicz, E., Bottin, H. and Roux, E. (1982) Bioelectrochem. Bioenerg. 141, 157-166
3. Davis, J. and Thornaley, P.J. (1983) Biochim. Biophys. Acta 724, 456-464
4. Pryor, W.A. and Tang, R.H. (1978) Biochem. Biophys. Res. Comm. 81, 498-464
5. Tyszkiewicz, E., Nicolic, D., Popovic, R. and Saric, M. (1979) Phys. Plant. 46, 324-329

ANALYSIS OF THE INDUCTION PHENOMENA IN INITIAL PSP

Y.Z.Li,Z.Y.Du,J.Wei,B.J.Guo,Y.Q.Hong and B.X.Tong

Shanghai Institute of Plant Physiology, Academia Sinica

1. INTRODUCTION
There are different views on the initial lag in photo-phosphorylation (1-6). Some experiments showed that PSP initiates within a few milli-seconds after the begining of illumination. The existence of the time lag indicates that a critical threshold potential is necessary for activating ATP synthase to evoke the energy-dependent conformational change of the CF_0-CF_1 complex. But some other experiments showed no detectable lag. These results raise the question on the relationship between PSP after begining of illumination and the formation of the electrochemical potential difference of protons across the thylakoid membrane.

2. MATERIALS AND METHODS
Chloroplasts were prepared from spinach leaves grown in phytotrone. ΔpH was measured by 9-aminoacridne fluorescence quenching under illumination (7). Electrochromic absorption change ($\Delta\phi$) and H_{in}^+ were measured by μs-flash spectroscopy device at 515nm and by the absorption change of neutral red (8). ATP formation was measured by ^{32}Pi esterification.

3. RESULTS AND DISCUSSION
3.1.Analysis of time lag in initial PSP
In our experiments we found that when the energy input is low (4×10^4 ergs.cm^{-2}.s^{-1}), a significant time lag in the initiation of ATP synthesis is observed. If the energy input is higher, that is in saturating light (4×10^6 ergs.cm^{-2}.s^{-1}) no lag is detected(Fig.1). We had also measured the ATP formation in saturating single ms or μs flash (not shown). Since we could not end the PSP reaction by TCA within μs or ms, so we had not actually measured the ATP formation under μs or ms illumination in these experiments. However the rapid responding electrochromic absorption change was fast enough for monitoring $\Delta\phi$. PSP causes an accelerated decay in the carotenoid shift and no induction phase can be observed in this accelerated decay (Fig.2).
Therefore we may conclude that so long as the input energy is enough to satisfy the ΔG of ATP formation, the onset of PSP reaction is not limited by the initiation rate of conformational change of CF_0-CF_1 complex which is fast enough byitself (see 9).

3.2.The relationship between initial ATP synthesis and $\Delta\phi$, H_{in}^+ formation.
It is well known that ATP formation is driven by ΔpH

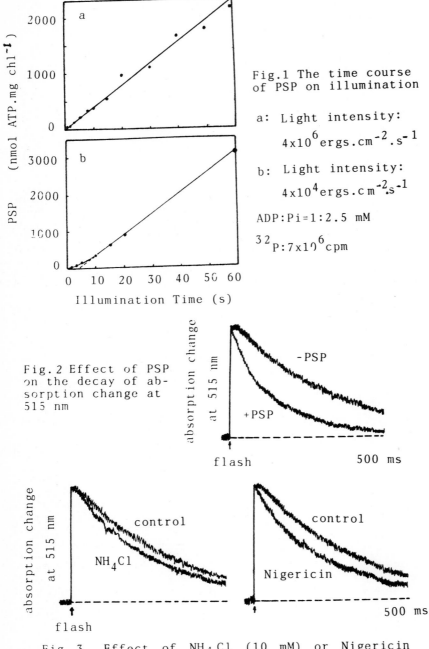

Fig.1 The time course of PSP on illumination

a: Light intensity:
$4 \times 10^6 \text{ergs.cm}^{-2}.\text{s}^{-1}$

b: Light intensity:
$4 \times 10^4 \text{ergs.cm}^{-2}.\text{s}^{-1}$

ADP:Pi=1:2.5 mM
$^{32}\text{P}:7 \times 10^6 \text{cpm}$

Fig.2 Effect of PSP on the decay of absorption change at 515 nm

Fig 3. Effect of NH_4Cl (10 mM) or Nigericin (0.1 μM) the absorption change at 515 nm

and $\Delta\phi$ across the thylakoid membrane. ΔpH is the major contributor for PSP in steady state, whereas the main force for initial PSP is $\Delta\phi$ (10). However the initial ATP formation driven by $\Delta\phi$ is also due to the H_{in}^{+} efflux through the proton channel of CF_O-CF_I (9,11). No matter that the driving force is ΔpH or $\Delta\phi$, protonefflux plays a key role in both steady state and initial PSP. We have further observed that NH_4Cl or Nigericin can accelerate 515nm decay (Fig.3). It implies that H^{+} flux is involved in the decay of electrochromic absorption change at 515nm. We also have measured the formation of H_{in}^{+} by neutral red with one single flash (Fig.4b). It seems likely that there is a direct close relation between ATP and H_{in}^{+} formation under us flashes. However in the initial stage the H_{in}^{+} efflux is markedly delayed by simultaneous PSP and accumulation of H_{in}^{+} is instantly increased (Fig.4a). We have further studied this phenomenon. Valinomycin accelerates the 515nm decay, but delays H_{in}^{+} efflux (Fig.5). It means H_{in}^{+} efflux may be driven by $\Delta\phi$, and when $\Delta\phi$ is eliminated by Valinomycin, H_{in}^{+} efflux is delayed. Therefore, When $\Delta\phi$ is eliminated by simultaneous PSP in the initial stage, H_{in}^{+} efflux is immediately delayed.

From these results, we may conclude that ATP formation in us flashes is dependent on the membrane potential ($\Delta\phi$)

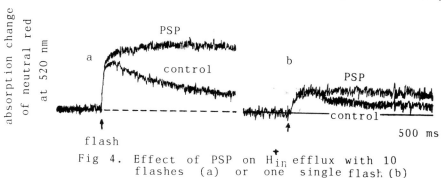

Fig 4. Effect of PSP on H_{in}^{+} efflux with 10 flashes (a) or one single flash (b)

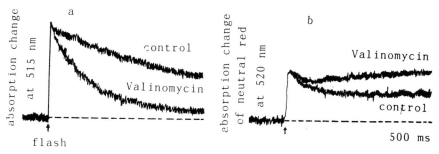

Fig 5. Effect of Vainomycin (10 nM) on the absorption change at 515nm (a) or the absorption change of neutral red at 520nm (b)

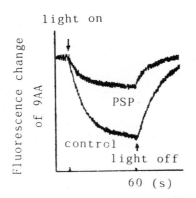

Fig.6 Effect of PSP on 9AA fluorescence change in steady state

and needs no H^+ participation. It is not same as in steady state where H^+_{in} is consumed significantly by PSP (Fig.6). According to our results, no H^+/ATP ratio can be calculated in initial stage. Lemaire et al (11) compared the yield of ATP synthesized by flash group and the absorbance changes at 515nm to determine the concomitant proton flux through the ATP synthase and got H^+/ATP ratio . But he did not examined the effect of simultaneous PSP on H^+_{in} flux directly. Probably, the H^+_{in} efflux is not driven by membrane potential through the proton channel (CF_0) to activate the ATP synthase in initial PSP and the $\Delta\phi$ may control the activation of H^+-ATPase directly . It is possible that in initial steps of PSP the proton chan nel of CF_0-CF_1 may serve in the transference of electric charge.

REFERENCES
1.   Graan,T. and Ort,D. (1981) Proc. 5th Intern. Photosyn. Congr. Vol. II p. 935-944
2.   Boork,J. and Baltscheffsky,M. (1983) In Photosynthetic Prokaryotes: Cell Differentiation and Function, ed by G.C.Papageorgion and L.Packer, pp 389-399
3.   Davenport,J. and McCarty,R. (1980) Biochim. Biophys. Acta, 589:353-357
4.   Melandri,B. et al (1980) Biochim. Biophys. Acta, 592: 38-52
5.   Vinkler,C. et al (1980) J.Biol.Chem. 255:2263-2266
6.   Keister,C. and Minton,N. (1969) Proc.Natl.Sci.U.S.A. 63:489-495
7.   Z.Y.Du, and Y.Z.Li (1985) Acta Biochimica et Biophysca Sinica, 17:363-372
8.   Y.Hong et al (1986) KEXUE TONGBAO 4:312-315
9.   Gräber,P. et al (1984) Ber. Bunsenges Phys. Chem. 88:599-608
10. Avron,M. (1978) FEBS Lett. 96:225-235
11. Lemaire,C. et al (1985) Biochim.Biophys.Acta, 807:285-292

ACKNOWLEDGEMENT
     This work was supported by grants from the Deutsche Stiftung Volkswagenwelk.

QUANTUM THERMODYNAMICS APPROACH TO PHOTOPHOSPHORYLATION, PHOTOSYNTHESIS
AND PHOTOTROPHIC GROWTH EFFICIENCIES

V.D. TRAN AND M. TABI, SERVICE DE RECHERCHE EN SOLS, MAPAQ, 2700 EINSTEIN,
STE-FOY, QUEBEC, CANADA, G1P 3W8

Photosynthesis may be regarded as the transformation of carbon dioxide
(CO_2) into the photosynthate ($CH_\gamma O_2$) by reaction of the former with
protons and electrons:

$$CO_2 + \gamma H^+ + \gamma e^- \longrightarrow CH_\gamma O_2$$

Both the protons and the electrons in the above reaction would come from
the decomposition of water in the thylakoid:

$$\frac{\gamma}{2} H_2O \longrightarrow \gamma H^+ + \gamma e^- + \frac{\gamma}{4} O_2$$

They are transported into the stroma space of the chloroplast (where
carbon fixation takes place) with the help of light energy. In the
stroma space, the electrons are carried by NADPH. For reason of symmetry, one may assume that the protons are carried by ATP. The fixation
of carbon may therefore be written as

$$CO_2 + \gamma ATP^{4-} + \frac{\gamma}{2} NADPH \longrightarrow CH_\gamma O_2 + \gamma ADP^{3-} + \gamma Pi^{2-} + \frac{\gamma}{2} NADP^+ + \frac{\gamma}{2} H^+$$

The photosynthetic process may be divided thus into two steps. The first
one, which may be called, for lack of a better designation, photophosphorylation, consists in the production of ATP and NADPH using light
energy. It is an irreversible step. The other step consists in the
biosynthesis of the photosynthate from CO_2, ATP and NADPH. This step
is assumed to be reversible.

In a steady state photosynthesis, protons and electrons are transported
at the same rate. Since two photons are known to be required for the
transport of one electron {1} plus, of necessity, one proton, one may
invoke the principle of equipartition of energy and postulate that each
photon is capable of transporting either one electron or one proton.
This is to say that the chemical potentials of the transported electron
and the transported proton are the same. In other words, one NADPH is
equivalent to two ATP. Thus, there is a one to one stoichiometry between the photon, the proton, the electron and the ATP.

Consequently, on the sole basis of stoichiometric considerations, the
quantum requirement for the fixation of one C-atom at the level of reduction γ_M is given by

$$R_\phi = 2\gamma_M \tag{1}$$

On the other hand, the quantized loss of energy during the photophosphorylation may be expressed as

$$^uE_{ATP} = E_\phi - {^r}E_{ATP} \tag{2}$$

Biggens, J. (ed.), Progress in Photosynthesis Research, Vol. III. ISBN 90 247 3452 5
© 1987 Martinus Nijhoff Publishers, Dordrecht. Printed in the Netherlands.

where E_{ATP} is the heat of hydrolysis of ATP in the chloroplast, E_ϕ the average energy content of the photons and u and r are integers designating respectively the number of ATP equivalents wasted and conserved, per photon. The energy efficiency of the photophosphorylation is by definition

$$\eta_{pp} = \frac{r}{u+r} \tag{3}$$

It is identical to the energy efficiency of the whole photosynthetic process (η_{ps}) because the biosynthesis step is assumed to be reversible. Thus

$$\eta_{pp} = \frac{r}{u+r} = \eta_{ps} = \frac{\gamma_M E_0}{2X_\phi E_\phi} \tag{4}$$

where E_0 is the heat of reduction of oxygen and X_ϕ the thermodynamic quantum requirement for the fixation of one gram atom of carbon. The number of photons consumed per C-atom fixed (Z_ϕ) may be calculated from the expression of energy conservation:

$$Z_\phi E_\phi = X_\phi E_\phi - \frac{\gamma_M E_0}{2} \tag{5}$$

or

$$Z_\phi = \frac{u}{r} \cdot \frac{\gamma_M}{2} \cdot \frac{E_0}{E_\phi} \tag{6}$$

The number of ATP equivalents used in the biosynthesis of one gram-atom of carbon is consequently:

$$n_{ATP} = rZ_\phi = \frac{u\gamma_M}{2} \cdot \frac{E_0}{E_\phi} \tag{7}$$

If one defines the anabolic substrate as the state of carbon to which ATP and NADPH energy should be added in order to bring it to the level of the photosynthate, then, one may write:

$$\frac{\gamma_A E_0}{2} + n_{ATP} E_{ATP} = \frac{\gamma_M E_0}{2} \tag{8}$$

where γ_A is the degree or reduction of the anabolic substrate. Eliminating n_{ATP} in the last two equations, one obtains

$$\frac{\gamma_M - \gamma_A}{\gamma_M} = \frac{u E_{ATP}}{E_\phi} = \frac{u}{u+r} \tag{9}$$

An obvious solution of the above equation is

$$\gamma_M = u+r \tag{10}$$

$$\gamma_A = r$$

$$\gamma_M - \gamma_A = u$$

The energy efficiency of the photosynthesis may now be expressed as

$$\eta_{ps} = \frac{r}{u+r} = \frac{r}{\gamma_M} = \frac{\gamma_M E_0}{2X_\phi E_\phi} \tag{11}$$

Hence

$$X_\phi = \frac{\gamma^2_M}{2r} \cdot \frac{E_0}{E_\phi} \tag{12}$$

Since both E_ϕ and E_0 are comprised between the limits of about 45 Kcal (for red light and the oxidation of elemental carbon) and about 70 Kcal (for violet light and the oxidation of dihydrogen) one may assume that their mean values are the same. Thus

$$X_\phi = \frac{\gamma^2 M}{2r} \qquad (13)$$

In view of the 1:1 stoichiometry between the photon and ATP ($r = 1$) one finally has

$$X_\phi = \frac{\gamma^2 M}{2} \qquad (14)$$

Table 1. Effect of the degree of reduction of the photosynthate on energy efficiencies

Photosynthate (Examples)	Degree of reduction	Quantum requirement	Energy efficiency		
			Photosyn- thetic	Hetero- trophic[a]	Autotro- phic[b]
Formic acid Glyoxylic acid	2	2	$\frac{1}{2}$	$\frac{1}{5}$	$\frac{1}{10}$
Formaldehyde Carbohydrate	4	8	$\frac{1}{4}$	$\frac{3}{5}$	$\frac{3}{20}$
Alkanol Alkene	6	18	$\frac{1}{6}$	$\frac{3}{5}$	$\frac{1}{10}$
Methane	8	32	$\frac{1}{8}$	$\frac{2}{5}$	$\frac{1}{20}$
Hydrogen	∞	∞	0	0	0

a) Permissible values for aerobic heterotrophic growth efficiency are 0, 1/5, 2/5 and 3/5. Actual values depend on the degree of reduction of the photosynthate. (Tran and Tabi (1986) In press)

b) Autotrophic efficiency = Photosynthetic efficiency X Heterotrophic efficiency.

It can be seen that photosynthates with a degree of reduction $\gamma_M = 4$ will satisfy both quantum requirements for the steady flows of matter and energy ($R_\phi = X_\phi$). Such photosynthates will also give rise to the highest autotrophic energy efficiency (Table 1).

Equation (13) shows that there cannot be a steady state production of photosynthates with odd degrees of reduction, except for $r = 0$. Thus biosynthesis of glycolic acid or malic acid (both with $\gamma_M = 3$) can only reach a steady state at the so-called compensation points, where the heat of hydrolysis of ATP would predictably be 18 Kcal/mole, as compared to a value of 13.5 Kcal/mole for the steady state photosynthesis of carbohydrate ($\gamma_M = 4$), assuming $E_\phi = E_0 = 54$ Kcal. In the latter case, the simultaneous operation of the glycolate pathway or the malate pathway affects the net rate of photosynthesis but does

not alter the time-independent energy efficiency. For each captured
photon, both C-3 and C-4 plants will convert one fourth of its energy
into chemical energy.

The flows of matter and energy during photosynthesis of carbonhydrate
is represented in Figure 1. Of the eight ATP-equivalents of energy
made available by eight photons, six are used directly in the fixation
of one C-atom ($n_{ATP} = Z_\phi = 6$) while two are used in "futile" cycles
(including the RuDP cycle)— all of which amounts to the evolvement of
two photon equivalents (E_ϕ) of heat, while four photon-equivalents are
dissipated as fluorescence. Since there is a decrease of the system's
entropy equivalent to 1.5 E_ϕ/T (T: absolute temperature) during biosynthesis, the net production of entropy per C-atom fixed, is consequently
equal to 4.5 E_ϕ/T if the fluorescence is allowed to dissipate as heat
e.g., in a black-wall chamber. In a mirrored-wall chamber, all the
fluorescence would be recaptured and the overall energy efficiency would
be 1/2; the apparent ATP/photon ratio, 2; the apparent quantum requirement, 4, and the net production of entropy, 0.5 E_ϕ/T per C-atom.

The anabolic substrate in photosynthesis is the bicarbonate ion which
by virtue of substrate-level phosphorylation (generation of one proton)
has actually a degree of reduction equal to 1 ($\gamma_A = 1$ instead of zero).
The uncertainty associated with the use of the number of available
electrons per C-atom as a measure of the energy potential of a given material is a characteristic feature of the present quantum thermodynamics.

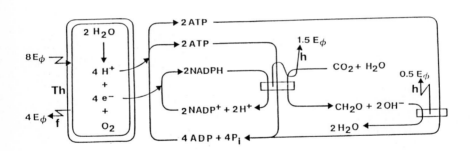

Figure 1. Flows of matter and energy in the steady state photosynthesis
of carbohydrate. Th, thylakoid; f, fluorescence; h, heat
evolvement.

REFERENCE

1. Govindjee (1975) in Bioenergetics of photosynthesis (Govindjee, ed.)
   pp 2-43. Academic Press, New York.

PYROPHOSPHATASE OF RESPIRATORY MEMBRANES FROM Rhodospirillum rubrum.

Irma Romero, Juan Quinto and Heliodoro Celis
Instituto de Fisiología Celular, Universidad Nacional Autónoma de México
Apdo. Postal 70-600, 04510, México, D.F. MEXICO.

1. INTRODUCTION
    The bacteria Rhodospirillum rubrum is member of the Rhodospirillaceae
family, it can grow chemotrophically in the dark and phototrophically in
the light. Indeed, these organisms are remarkably versatile for obtaining
growth energy through alternative mechanisms.
    It has been reported (1) that the photosynthetic membranes (chroma-
tophores) contain a membrane-bound Pyrophosphatase which is coupled to
proton gradient (2,3). This enzyme coexist in chromatophores with H^+ATPase.
Melandri et al (4) have demonstrated that the H^+ATPase of chromatophores
is the same molecular entity that present in Respiratory membranes.
However, it has not been described if a membrane-bound PPiase exist in
Respiratory membranes. In this work, we report the existence and characteristics of a membrane-bound Pyrophosphatase in Respiratory Membranes of
Rs. rubrum.

2. PROCEDURE
2.1.Materials and Methods.
    2.1.1.Growth Conditions.
        Rs. rubrum were grown in the medium of Cohen-Bazire et al (5).
The aerobic culture was carried out in 1 litre. Erlenmeyer flasks which
contained 200 ml of growth medium and were incubated in the dark at 30°C
while intense aeration was provided by giratory shaking. Cells harvested
from those aerobic cultures contained 0.03 µg bacteriochlorophyll per mg
of cell protein. Phototrophic cultures were performed under anaerobic
conditions with 40 watts tungsten lamps.
    2.1.2.Membrane Preparations.
        Chromatophores were prepared as described (6), Respiratory
membranes were obtained as in (7).
    2.1.3 Assays.
        Protein was measured by the method of Lowry et al (8).
Bacteriochlorophyll content was determined as Clayton (9). Hydrolysis of
pyrophosphate was determined in the dark in the conditions described under
Results. The reaction was arrested by 6% trichloroacetic acid (final
concentration). Phosphate was determined in the supernatant as described
in Ref. (10). The rate of oxygen uptake was measured polarographically in
the conditions described in Fig. 2.

3. RESULTS AND DISCUSSION.
    As can be seen in Table I, the PPiase specific activity refered to
protein is 30-50% to that found in chromatophores. However, when the
activity is refered to bacteriochlorophyll, a very high specific activity
is found due to a decrease of the bacteriochlorophyll content in the
preparation. This suggest that the PPiase activity of Respiratory membranes
is not due to a chromatophore contamination.

Biggens, J. (ed.), Progress in Photosynthesis Research, Vol. III. ISBN 90 247 3452 5
© *1987 Martinus Nijhoff Publishers, Dordrecht. Printed in the Netherlands.*

TABLE I

| | μmol Pi/h/mg Protein | | μmol Pi/h/mg Bchl | |
	PPiase	ATPase	PPiase	ATPase
Chromatophores	6.872	2.430	119.0	42.18
Respiratory Membranes	3.774	4.198	50320.0	55983.00

The media contained: For PPiase, 50 mM Tris-HCl pH 8.0, PPi-Na 2 mM, $MgCl_2$ 5 mM and 1 mg protein of Chromatophores or Respiratory membranes. Temperature 30°C. For ATPase, Tris-acetate 50 mM pH 7.5, ATP 3 mM, $MgCl_2$ 5mM, PEP 1.45 mM and Pyruvate Kinase 7 μg.

The PPiase activity is inhibited by DCCD, NaF, IDP and MDP, but not by Oligomycin an LiCl, this pattern of sensibility to inhibitors is the same that the found in chromatophores (1,3,11-14). This indicates that the PPiase of Respiratory membranes is very similar to that of chromatophores.

In order to assay the specifity of the enzyme, several phosphate compounds were proved (β-glycerophosphate,α-β glycerophosphate, D-glucose-1-phosphate, D-glucose-6-phosphate, AMP, ADP and ρ-nitrophenylphosphate. None of them were hydrolyzed by the enzyme. Thus the PPiase of Respiratory membranes is highly specific for Pyrophosphate as substrate. For the first time it is demonstrated the presence of a Pyrophosphatase activity in Respiratory membranes.

To compare the PPiase of Respiratory membranes with the one of chromatophores, two important characteristics were studied: the pH dependence and the divalent cations requirement.

The PPiase activity of Respiratory membranes has a pH optimum of 6.5 and the same is for the activity of Chromatophores (15).

In respect of the divalent cations requirement to form the metal-PPi complex (which is the real substrate) the Respiratory membranes PPiase can hydrolyze the complex Mg-PPi and Zn-PPi as well, in minor proportion, the complex Co-PPi is also hydrolyzed and the complex Ca-PPi is not hydrolyzed. This selectivity is the same to that found in chromatophores (15). It is interesting to note that at high concentrations, either Zn^{2+} and Mg^{2+} inhibit the hydrolytic activity, this is most likely due to an inhibitory effect of these cations when are in its free form (15).

All the characteristics studied of the Respiratory membranes PPiase, taking together and compared with those in chromatophores, strongly suggest that the two enzymes are the same.

In order to demonstrate that the PPiase activity of Respiratory membranes is due to a membrane-bound enzyme, the next experiments were made.

First, it has been demonstrated (16) that 1-butanol is a selective inhibitor of the membrane-bound PPiase but not of the soluble cytoplasmic PPiase or H^+ATPase. In this work, we show that the PPiase activity of Respiratory membranes is also inhibited by 1-butanol.

In order to corroborate that the PPiase in Respiratory membranes is membrane-bound, the preparation was treated with Triton X-100. More than the 80% of the activity was released to the supernatant (Fig. 1). This solubilized enzyme was stimulated 30-40% by the addition of phospholipids.

We can conclude that the PPiase of Respiratory membranes is an integral membrane protein.

An important question is if the membrane-bound PPiase of Respiratory membranes is coupled to the proton gradient produced by the oxidation of substrates. In Fig. 2., it is shown that the addition of phosphate produce an increment in the oxygen uptake, with a Respiratory control of 1.7. This is inhibited by NaF to 1.2. This suggest that there is a PPi synthesis which is coupled to the oxidation of succinate.

In the future we will study the kinetic characteristics of Synthesis, PPi-Pi exchange and the proton gradient produced by the hydrolysis of PPi in Respiratory membranes.

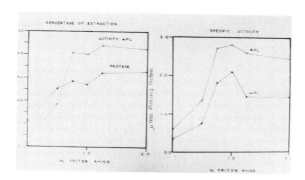

Fig. 1) SOLUBILIZATION OF THE MEMBRANE-BOUND PPiase FROM RESPIRATORY MEMBRANES WITH TRITON X-100.

The experiment was made as follow: 7 mg protein of Respiratory membranes were resuspended in a buffer containing 50 mM Tris-HCl, pH 8.15, 25% Ethyleneglycol, 600 mM $MgCl_2$ and the indicated concentration of Triton X-100; after gentle stirring on ice for 20 min, the suspension was centrifuged at 110 000 x g for 2 h., the supernatants were analized for PPiase activity, in the following media: 50 mM Tris-maleate pH 6.5, 2 mM PPi-Na, 5 mM $MgCl_2$. In case phospholipids were present, they were prepared as follows: 60 mg/ml of L-α-lecithin from soybean, commercial grade (Sigma Chem. CO.) were suspended in 10 mM Tris-HCl, pH 7.5 and sonicated until clarity, the final concentration in the assay tubes were 3 mg.

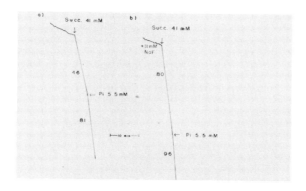

Fig. 2) RESPIRATORY CONTROL BY Pi IN RESPIRATORY MEMBRANES.

The assay media contained: 20 mM Tris-HCl pH 8.0, 250 mM sucrose, 5 mM $MgCl_2$, 16.6 μM of reduced cytochrome C from horse heart and 2.7 mg/ml of protein of respiratory membranes. The membranes were prepared in the presence of 5% of bovine albumin (Fraction V). Respiratory control in trace a) is 1.7 and in trace b) plus 11 mM NaF is 1.2.

Partially supported by CONACyT (PCSABNA-022637) MEXICO.

REFERENCES.
1) Baltscheffsky, H., Von Stedingk, L.V., Heldt, H.W. and Klingenberg, M. (1966) Science. 153, 1120-1124.
2) Moyle, J., Mitchell, R. and Mitchell, P. (1972) FEBS Lett. 23, 233-236
3) Guillory, R.J. and Fisher, R.R. (1972) Biochem. J. 129, 471-481.
4) Melandri, B.A., Baccarini-Melandri, A., San Pietro, A. and Gest. H. (1971) Science 174, 514-516.
5) Cohen-Bazire, G., Sistrom, W.B., and Stainer, R.Y. (1957) J. Cell Comp. Physiol. 49.
6) Scholes, P., Mitchell, P. and Moyle, J. (1969) Eur. J. Biochem. 8, 450-454.
7) Fenoll, C. and Ramírez, J.M. (1984) Arch. Microbiol. 137, 42-46.
8) Lowry, O.H., Rosebrough, N.J. Farr, A.L. and Randal, R.J. (1951) J. Biol. Chem. 193, 265-275.
9) Clayton, R.W. (1963) Biochim. Biophys. Acta. 75, 312-323.
10) Sumner, J.B. (1944) Science (Washington D.C.) 100, 413-415.
11) Baltscheffsky, M., Baltscheffsky, H. and Boork, J. (1982) in Electron Transport and Photophosphorylation (Barber, J. ed) Top. Photosynth. Vol. 4, pp 249-272. Elsevier, Amsterdam, New York.
12) Keister, D.L. and Minton, N.J. (1971) Biochem. Biophys Res. Commun. 42, 932-939.
13) Nyrén, P. and Baltscheffsky, M. (1983) FEBS Lett. 155, 125-130.
14) Keister, D.L. and Minton, N.J. (1971) Arch. Biochem. Biophys 147, 330-338.
15) Celis, H., and Romero, I. (1986) submited.
16) Baltscheffsky, M. (1968) in Regulatory Functions of Biological Membranes (Jarnefelt, J. ed) BBA Library Vol. 11 277-286. Elsevier, Amsterdam, New York.

PHOSPHOTRANSACETYLASE AND ACETATE KINASE FROM *RHODOPSEUDOMONAS PALUSTRIS*

KARL KNOBLOCH, HELMUT VIGENSCHOW and HANS-MARTIN SCHWARM
Institut für Botanik und Pharmazeutische Biologie der Universität
Erlangen-Nürnberg, Staudtstr. 5, D-8520 Erlangen, Fed.Rep. of Germany

1. INTRODUCTION
    Energy-linked NAD^+ reduction and energy-dependent $NADP^+$ transhydrogenation in cell-free extracts from *Rhodopseudomonas palustris* are catalyzed by the supernatant fraction S-144 (1,2). Reduction of pyridine nucleotide is mediated by an ATP-driven reversal of electron transfer including cytochrome *c* oxidation and concomitantly occurring NAD^+ reduction (1). The ATP-dependent reversed electron flow and the ATP-mediated $NADP^+$ transhydrogenase reaction are catalyzed as well in case ATP is replaced by AcP, or by Ac-CoA in the presence of inorganic *o*-phosphate (2). The corresponding enzymatic systems, phosphate acetyl transferase (acetyl-CoA:orthophosphate acetyltransferase) (EC 2.3.1.8) and acetate kinase (ATP:acetate phosphotransferase) (EC 2.7.2.1), have been purified and characterized. So far, neither a phosphotransacetylase nor an acetate kinase have been described from photosynthetic bacteria.

$$Ac{-}CoA + P_i \; \underset{PTA}{\overset{}{\rightleftharpoons}} \; CoA + AcP \underset{}{\overset{AK}{\rightleftharpoons}} \; ATP + AcO^-$$
$$+ ADP$$

2. MATERIALS AND METHODS
2.1. Cell Growth. *Rhodopseudomonas palustris* (ATCC 11168) was grown photosynthetically on thiosulfate or acetate in the medium modified after (2).

2.2. Preparation of the cell-free fraction S-144, polyacrylamide gel electrophoresis and high performance liquid chromatography were performed according to (3). Gel chromatography, preparative isoelectric focusing and determination of the relative molecular mass were carried out as described elsewhere (4).

2.3. Assays of enzymatic activities. Ac-CoA formation was determined after (5). Generated AcP was assayed after (6) and (7) with some modifications (4). ATP formed by the acetate kinase in the presence of AcP and ADP was determined after (8) in a special luminometer (9) as described in (4).

2.4. Purifications of the phosphotransacetylase and acetate kinase were performed after (4).

2.5. For graphic determination of K_m, Lineweaver-Burk plots were applied. Arrhenius plots were employed to determine the activation energies of the phosphotransacetylase reaction and the acetyl phosphate formation.

Abbreviations: Ac-CoA, acetyl-coenzyme A; AcO^-, acetate; AcP, acetyl phosphate; AK, acetate kinase; Me^+, Me^{2+}, monovalent or divalent metal ion; PTA, phosphotransacetylase; SDS, sodium dodecyl sulfate.

Table 1. Purification procedures for a phosphotransacetylase (PTA) and an acetate kinase (AK) from *R. palustris*

Fraction	Total protein [mg]		Total activity [μmol·min⁻¹]		Specific activity [μmol·min⁻¹·mg⁻¹]		Recovery [%]		Purification [fold]	
	PTA	AK	PTA	AK	PTA	AK	PTA	AK	PTA	AK
Supernatant S-144	725	473	87	22	0.12	0.047	100	100	–	–
Eluate from DEAE-Sephadex A-50	39	74	84	13	2.17	0.18	97	59	18	3.8
Eluate from Phenyl-Sepharose CL-4B	3.0	6.8	56	6.9	18.7	1.01	64	31	156	21
Eluate from DEAE-Trisacryl M	0.56	3.4	27	2.9	48.2	0.85	31	13	402	18
Eluate from Sephacryl S-300	0.25	0.69	7.9	2.2	31.6	3.16	9.1	10	263	67

Table 2. Characterization of the phosphotransacetylase and acetate kinase from *R. palustris*

	M_r	M_r (after treatment with SDS)	pI	pH optimum	temperature optimum [°C]	activation energy [kJ/mol]
PTA	55 500	52 500	6.3	7.8	38	23 (up to 15°C) 16 (15 to 30°C)
AK	47 500	45 000	4.9	6.5 to 7.1	60	21 (5 to 60°C)

	activated by	K_m(AcP) [mM]	K_m(CoA) [mM]	K_m(AcO⁻) [mM]	K_m(MgATP2⁻) [mM]	K_m(MgADP2⁻) [mM]
PTA	Me⁺	4.7	0.15	–	–	–
AK	Me²⁺	0.0026	–	40.0	1.1	0.087

3. RESULTS AND DISCUSSION

3.1. Phosphotransacetylase

The phosphotransacetylase turned out to be labile if purification procedures were carried out without the additions of $(NH_4)_2SO_4$. Avoiding time consuming and inactivating purifications, such as dialysis or concentration steps, an enzyme protein could be separated which appeared homogenous as judged by polyacrylamide gel electrophoresis. A 260-fold increase in specific activity and a recovery of 9% could be achieved (Table 1). The fraction obtained earlier, after DEAE-Trisacryl M chromatography, showed three minor protein bands due to impurities, however revealed a 400-fold specific activity with 30% in recovery compared to the crude cell-free extract S-144 (Table 1). For this reason the experiments were performed with the fraction obtained after the DEAE-Trisacryl M separation step.

The enzyme was characterized by a relative molecular mass of 54 500, an isoelectric point of 6.3 and the absence of dissociable subunits. Maximum reaction rates were measured at pH 7.8 and 38°C. The enzyme appeared very labile at elevated temperatures or in diluted solutions. The stability could be increased distinctly in case sulfate or ammonium ions were added. The activity also was influenced by inorganic salts. Potassium and ammonium ions activated the enzymatic reaction. Sulfate ions revealed an inhibitory influence. A strong substrate inhibition was found with coenzyme A as substrate. The Arrhenius plot revealed a discontinuity at 15°C which most likely corresponds to a conformational change of the enzyme protein.

The specific phosphotransacetylase activity determined in the cell-free supernatant fraction S-144 equals about 0.15 $\mu mol \cdot min^{-1} \cdot mg$ protein^{-1}. This value appears to be low if compared to specific activities of the corresponding enzymes found in other sources (5,10,11).

The enzyme apparently does not contain dissociable subunits. The corresponding enzyme of *Veillonella alcalescens* and *Clostridium thermoaceticum* were determined to be dimers (M_r= 75 000) (11) and tetramers (M_r= 88 000) (12), respectively. For the *Lactobacillus fermenti* phosphotransacetylase a relative molecular mass of 68 000 was reported (10), and a number of 90 000 for the corresponding enzyme from *Bacillus subtilis* (13), however, informations on subunit compositions were not given.

3.2. Acetate kinase

The specific activity of an acetate kinase in the crude cell-free supernatant fraction (S-144) was low (Table 1) compared to other origin (7,14,15). No increase in specific activity was observed after photosynthetic growth on acetate as the only source of carbon.

The acetate kinase was purified to apparent homogeneity by high resolving chromatography steps (Table 1). The isoelectric point was determined to be 4.9. The enzyme was characterized by a relative molecular mass of 46 500 with no dissociable subunits detectable. Likewise, the *Escherichia coli* acetate kinase was described to be a monomeric enzyme (15). In contrast, the corresponding enzymes from *Bacillus stearothermophilus* and *Veillonella alcalescens* were reported to contain 4 (15) and 2 (16) subunits, respectively.

The purified enzyme was found to be thermolabile, but remained stable for some days at 0 to 4°C. The maximum reaction rate occurred at pH 6.5 to 7.1 and 60°C.

There was an absolute requirement for divalent metal ions in the enzymatic reaction. It was activated most strongly by Mg^{2+} cations, and it is concluded that the $(Mg-ADP,-ATP)^{2-}$complex acts as the true substrate (4). High concentrations of Mg^{2+} had an inhibitory effect. The apparent K_m values for Co^{2+}, Zn^{2+} and Ca^{2+} appeared lower than those for Mn^{2+} and Mg^{2+}. However, the formation of acetyl phosphate is clearly favoured in the presence of Mg^{2+} (4). The apparent K_m(ATP) was not influenced by the $[Mg^{2+}]/[ATP]$ ratio. There seems to be no regulation by an interaction of Mg^{2+} and an allosteric site of the enzyme. The inhibition of the acetate kinase activity at high concentrations of Mg^{2+} may be explained by an interaction of free Mg^{2+} with the active center of the enzyme (17).

The acetate kinase can serve for the activation of acetate and for ATP formation as well. For acetate kinase from other origin, K_m(ATP) values were reported to be within the same range when compared to the K_m(ADP) numbers (7,18). The ratio $K_m(AcO^-)/K_m(AcP)$ was reported to range from 50 (16) to 400 (18). The *R. palustris* enzyme is characterized by a K_m(ATP)/K_m(ADP) ratio of more than 10 and a $K_m(AcO^-)/K_m(AcP)$ quotient of 15 000 (Table 2). The thermodynamic equilibrium $[AcP]\cdot[ADP]/[ATP]\cdot[AcO^-]$ has been calculated to be in the range of 0.01 (7), i.e., the formation of ATP is favoured thermodynamically.

REFERENCES

1 Knobloch, K., Eley, J.H. and Aleem, M.I.H. (1971) Biochem. Biophys. Res. Commun. 43, 834–839
2 Knobloch, K. (1975) Z. Naturforsch. 30c, 771–776
3 Schwarm, H.-M., Vigenschow, H. and Knobloch, K. (1986) Biol. Chem. Hoppe-Seyler 367, 119–126
4 Vigenschow, H., Schwarm, H.-M. and Knobloch, K. (1986) Biol. Chem. Hoppe-Seyler 367 (in press)
5 Bermeyer, H.U., Holz, G., Klotzsch, H. and Lang, G. (1963) Biochem. Z. 338, 114–121
6 Lipmann, F. and Tuttle, L.C. (1945) J. Biol. Chem. 159, 21–28
7 Rose, I.A., Grunberg-Manago, M., Korey, S.R. and Ochoa, S. (1954) J. Biol. Chem. 211, 737–756
8 Lundin, A., Richardsson, A. and Thore, A. (1976) Anal. Biochem. 75, 611–620
9 Müller, H., Neufang, H. and Knobloch, K. (1982) Eur. J. Biochem. 127, 559–566
10 Nojiri, T., Tanaka, F. and Nakayama, I. (1971) J. Biochem. 69, 789–801
11 Whiteley, H.R. and Pelroy, R.A. (1972) J. Biol. Chem. 247, 1911–1917
12 Drake, H.L., Hu, S.-I. and Wood, G.H. (1981) J. Biol. Chem. 256, 11137–11144
13 Rado, T.A. and Hoch, J.A. (1973) Biochim. Biophys. Acta 321, 114–125
14 Sagers, R.D., Benziman, M. and Gunsalus, I.C. (1961) J. Bacteriol. 82, 233–238
15 Nakajiama, H., Suzuki, K. and Imahari, K. (1978) J. Biochem. 84, 193–203
16 Bowman, C.M., Valdez, R.O. and Nishimura, J.S. (1976) J. Biol. Chem. 251 3117–3121
17 Morrison, J.F. (1978) Methods Enzymol. 63, 257–283
18 Vahane, I. and Muhlrad, A. (1979) J. Bacteriol. 137, 764–772

REGULATION OF PHOTOSYNTHETIC ELECTRON FLOW BY LIGHT-DEPENDENT METABOLISM

RENATE SCHEIBE, LEHRSTUHL PFLANZENPHYSIOLOGIE, UNIVERSITÄT BAYREUTH, UNIVERSITÄTSSTR 30, 8580 BAYREUTH, WEST GERMANY.

INTRODUCTION
Green plants are equipped with the photosynthetic apparatus which generates ATP and NADPH needed for the production of organic carbon compounds. Thus illuminated green cells have the potential to cover their needs solely from light energy and in addition store or export reduced carbon as a source of energy for heterotrophic conditions (in the dark or in non-green tissues). This sophisticated machinery needs to be balanced in order to be protected from over-reduction or depletion of intermediate pools under changing environmental and metabolic conditions. The regulatory mechanisms involved both, to achieve the maximal benefit and to avoid the detremental effects of light energy are acting at various levels:
1. Protein biosynthesis, 2. Post-translational modification, 3. Influence of metabolite levels on modification, 4. Influence of metabolite levels on active enzymes. These four levels of regulation will be discussed in the following.

1. LIGHT-INDUCED EXPRESSION OF PHOTOSYNTHETIC ENZYMES.
In greening pea seedlings light serves as a signal to increase the biosynthesis of those proteins that are exclusively needed for photoautotrophic growth, as e.g. the Calvin-cycle enzymes (Table 1). There is a ten-fold increase in the amount of Rubisco, and 5- to 10-fold increases in the enzyme capacities of phosphoribulokinase (PRK), NADP-glyceraldehyde dehydrogenase (NADP-GAPDH), and chloroplast fructose-1,6-bisphosphatase (FBPase), each measured subsequently to full activation by DTT, thus showing the increase in potentially active enzyme molecules. This change in the enzyme composition adjusts the metabolic properties of the cells to photoautotrophic life on a long term when light-generated energy replaces largely that derived from glycolytic pathways.
In contrast, the synthesis of two chloroplast enzymes which are also subject to regulation by light at the post-translational level, namely NADP-malate dehydrogenase (NADP-MDH) and glucose-6-phosphatedehydrogenase (G6PDH), as well as glutathione reductase (GSH-reductase) appears to be independent of the light stimulus. These enzymes involved in the generation of reducing equivalents either by producing directly NADPH (chloroplastic G6PDH) or the hydrogen transporter malate (NADP-MDH) are present in etioplasts and may therefore have a more general function in metabolism. The in vivo activation of NADP-MDH (and probably inactivation of G6PDH), mediated by thioredoxin, which is also shown to be present at high levels in etiolated seedlings (Table 1), possibly occurs via a NADPH-dependent system (1).

Biggens, J. (ed.), Progress in Photosynthesis Research, Vol. III. ISBN 90 247 3452 5
© *1987 Martinus Nijhoff Publishers, Dordrecht. Printed in the Netherlands.*

Table 1. Development of etiolated and of greening pea seedlings[a]

| | eight days dark grown | | | |
| | + 80 h dark | | + 80 h light | -fold increase |
	(per mg protein)			
Chlorophyll	0.0	µg	59.0 µg	∞
Rubisco	20.0	µg	190.0 µg	9.5
FBPase (chloroplast)	0.008	U	0.08 U	10.0
NADP-GAPDH	0.06	U	0.54 U	9.5
PRK	0.06	U	0.32 U	5.2
NAD-MDH	4.8	U	4.0 U	0.8
NADP-MDH	0.15	U	0.21 U	1.4
NADP-MDH	4.0	µg	5.5 µg	1.4
G6PDH (total)	0.04	U	0.04 U	1.1
G6PDH (chloroplast)	0.008	U	0.008 U	1.0
GSH-reductase	0.05	U	0.06 U	1.2

[a] R. Schiffelholz and R. Scheibe, manuscript in preparation

2.REDUCTIVE ACTIVATION/INACTIVATION OF ENZYMES.

Several key enzymes in the chloroplast can be converted from potentially active enzyme molecules into actually active enzymes, by being linked to the light/dark modulation system. This covalent modification occurs within minutes or seconds and is mediated by the ferredoxin/thioredoxin-system (1) or by the LEM-system (2). It is most likely that the nature of this process is reductive in vivo, since the redox-state of the mediator thioredoxin changes upon light-dark transitions (3). However, up to now there is clear evidence for the reversible formation of thiol-groups upon activation of NADP-MDH and FBPase only in reconstituted systems where electrons are provided by the reductant DTT (4, 5, 6). Clearly this process depends on the redox potential and can be driven equally well by dithiols and by monothiols as GSH (7). This could occur in vivo, e.g. under anaerobic conditions even in the dark, as has been shown for NADP-MDH (8). In a reconstituted system complete G6PDH inactivation is also achieved with GSH, which is kept in the 100 % reduced state by NADPH/GSH-reductase (K. Fickenscher and R. Scheibe, unpublished).

3.REGULATION OF LIGHT-MODULATION BY METABOLITES.

While the above described light-modulation of several chloroplast enzymes is linked inevitably to photosynthetic electron flow via the mediator system(s), thus providing an on/off-switch, additional "tuning" of the rate and the extent of light-modulation is required. Several effects of metabolites upon the reductive activation of Calvin-cycle enzymes have been described:
Fructose-1,6-bisphosphate enhances the light- and DTT-reduced thioredoxin-mediated activation of chloroplast FBPase (9). Similarly sedoheptulose-1,7-bisphosphate and thioredoxin exhibit this type of concerted action upon SBPase activation

(10).Inorganic phosphate and 1,3-diphosphoglycerate enhance the reductive activation of NADP-GAPDH (11). The activation of PRK by DTT is largely prevented by ATP (12). NADP inhibits light- as well as DTT-mediated activation of NADP-MDH (13). NADPH in- creases the rate of inactivation by DTT and DTT-reduced thio- redoxin of the chloroplast G6PDH (Fig. 1, K. Fickenscher and R. Scheibe, unpublished).

FIGURE 1. Effect of NADPH on the inactivation of chloro- plastic G6PDH by DTT and DTT-reduced thioredoxin. Partially purified G6PDH was incubated at room temperature at pH 8.0 under nitrogen in the presence of a: 1 mM NADPH (control), b: 5mM DTT, c: 5 mM DTT and 1 mM NADPH, d: 5 mM DTT and thioredoxin m and e: 5 mM DTT, thioredoxin m and 1 mM NADPH, respectively. 100 % activity corresponds to 0.01 units per assayed sample of 0.05 ml.

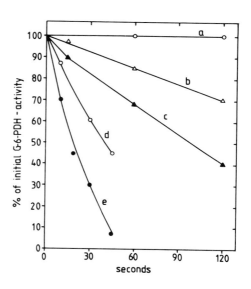

Thus it appears that metabolites can act as mediators which by reflecting the actual metabolic situation can adjust the fluxes through certain key enzymes. The steady state levels of their active forms achieved by a varying light modulation respond to the needs of the system in a dynamic process. Such an effect of metabolites on the modification of various chloroplast enzymes could indeed provide an effective system resembling the "mono- cyclic cascade" described by Chock et al. (14), where the pos- sible range of adjustment is wider than that which could be achieved by product inhibition of a constant amount of enzyme. In this context it is of importance to stress the fact that light-modulation in the presence of oxygen requires continuous electron flow on to the enzyme, but that the short half-time of the light-modulated state (10 sec) is the prerequisit for the high flexibility of this system.
The rapid response of this system to metabolic changes during the light phase (addition of NADP to broken chloroplasts or of bicarbonate to intact illuminated chloroplasts) is shown in Fig.2. NADP induces only a transient change in NADP-MDH acti- vity, since it is used up due to photoreduction under these conditions. On the other hand, it has been demonstrated that even at low light-intensities (10W /m^2) NADP-MDH remains in its

activated state, which demonstrates that changes in electron
flow do not directly influence rate and extent of light activation (R. Scheibe, unpublished).

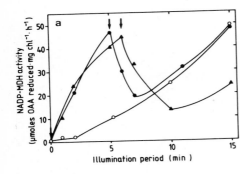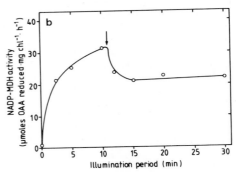

FIGURE 2. Light activation of NADP-MDH in osmotically shocked
(a) and intact (b) pea chloroplasts.
a) Broken pea chloroplasts (1 mg chlorophyll/ml) were
illuminated in hypotonic buffer pH 7.4 at 25 ° C under nitrogen
atmosphere. The arrows indicate the addition of 2.5 and 5.0
μmoles NADP per mg chlorophyll, respectively, after 5 min
illumination (samples 1 and 2). In 3 addition of 2.5 μmoles
NADP was prior to illumination. b) Isolated intact pea chloroplasts (0.6 mg chlorophyll/ml) were illuminated in isotonic
CO_2-free resuspension medium (pH 7.6) at 26°C under nitrogen
atmosphere. The arrow indicates the addition of 10 μmoles
sodium bicarbonate (pH 8.0) to a final concentration of 10 mM.

The physiological relevance of such a highly flexible, rapidly
responding system will be now discussed for some well defined
conditions posed upon isolated intact chloroplasts. The first
preference for a functioning chloroplast appears to be to direct light energy primarily into CO_2-fixation, and, in fact,
apart from a transient lag there is no competition by other
electron acceptors, with nitrite as an exception, a case where
additional regulatory mechanisms might be involved in vivo
(15). The Calvin-cycle has to be protected from depletion or
accumulation of its intermediates. Increase in fructose-1,6-
bisphosphate, sedoheptulose-1,7-bisphosphate and inorganic
phosphate signalize turnover in the Calvin-cycle and at the
same time increase the flux through FBPase, SBPase and GAPDH by
enhancing their activation. Such a tight regulation of FBPase
has also been postulated from the finding that in intact leaves
the fructose-1,6- bisphosphate/fructose-6-phosphate-ratio remains constant over a wide range of CO_2-fixation rates (16).
Similarly a decrease in the ATP-level during rapid Calvin-cycle
turnover might relieve the inhibition of light-activation of
PRK and thus increase ribulose-1,5-bisphosphate regeneration.

NADP-MDH, another light-modulated chloroplast enzyme produces malate and NADP from oxaloacetate and NADPH. Malate as opposed to NADPH is readily exported into the cytoplasm via a dicarboxylate translocator system. This pathway of indirect hydrogen transfer however, needs to be under strict control in order to avoid energy dissipation, since the fully active system would channel 100 - 200 µmoles reducing equivalents per mg chlorophyll and hour away from chloroplast metabolism (18).
Actual NADP-MDH activity in the light comes up to 100 % of the potential activity (with DTT) only when no electron acceptor, - not even oxaloacetate - is present as acceptor for reducing equivalents (Table 2). Light activation of NADP-MDH occurs also quite rapidly and to a high extent when the NADPH/NADP ratio is high, signalizing lack of regeneration of NADP as electron acceptor, e.g. during induction. On the other hand, the export of reducing equivalents in the form of malate is largely prevented when active turnover of photoreduced NADPH is indicated by a low NADPH/ NADP-ratio, e.g. under steady-state conditions.

TABLE 2. Light activation of NADP-MDH in isolated intact spinach chloroplasts under various conditions.

Electron acceptor added	NADPH/NADP	NADP-MDH (% of total activity)
none	3.0	100
2.5mM sodium bicarbonate		
during lag phase	1.5	70
during steady state	0.66	50
2.5 mM 3-phosphoglycerate	0.25	20
2.5 mM oxaloacetate	n.d.	20

data taken from (18); n.d. = not determined

FIGURE 3. Modulation of enzyme activities in isolated intact spinach chloroplasts illuminated in the presence of 5 mM sodium bicarbonate. The rate of oxygen evolution is given in parantheses. FBPase was assayed at pH 8.0 with 0.5 mM fructose-1,6-bisphosphate, 5 mM $MgCl_2$ and 14 mM 2-mercaptoethanol in order to only register the activity of the reduced form. Full activation of the enzymes (100 %) was obtained by incubation of chloroplast extract with 20 mM DTT at pH 8.0

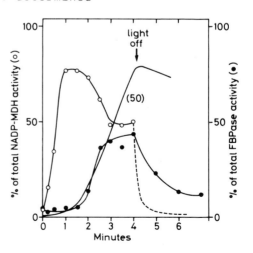

Fig. 3 shows a time-course of the light activation of NADP-MDH during induction of CO_2-fixation reflecting the high electron pressure when no CO_2-acceptor is available during the first minutes. In contrast, the light-activation of FBPase can only occur when the fructose1, 6-bisphosphate pool has been built up. Thus the malate producing and exporting system of the chloroplast has a function as a "valve" for excess electron flow. It allows the export of exactly that portion of reducing equivalents which supports poising of the ATP/NADPH-ratio needed for optimal performance of chloroplast metabolism. Alternatively, it has been shown that electron flow to oxygen with concomitant H_2O_2 production can also serve as an "emergency valve". However, in the intact system photosynthetic oxygen reduction is very limited and has been shown to be in the range of 1 to 10 µmoles H_2O_2 produced per mg chlorophyll and hour (19). These authors have demonstrated that over-reduction of the electron transport chain is more readily relieved by the "malate-valve" (addition of OAA) than by the "oxygen-valve" (H_2O_2-production) (Fig. 4).

FIGURE 4. Poising effect of addition of OAA (2.5 mM) on a unbalanced requirement of ATP and NADPH for photosynthetic CO_2 reduction by intact (90 %) spinach chloroplasts as indicated by H_2O_2- and concurrent O_2 (insert)-evolution. The incubation medium contained $NaHCO_3$ (7.8 mM) and chloroplasts equivalent to 100 µg chl/ml; illumination, saturating red light.

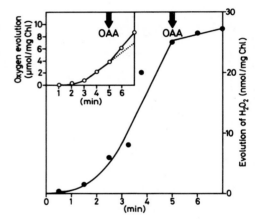

The chloroplast coupling factor (CF_1) might also contribute to relieve pressure by high energy input when it is stimulated to catalyze ATP-hydrolysis under those conditions which regulate removal of the "photosynthetic control" (20). However, it would be expected that the reductive activation of the ATP-hydrolyzing activity which is also mediated by the thioredoxin system (21) would only occur under high ATP/ADP-ratio in the light.

4. DIRECT INFLUENCE OF METABOLITES UPON CATALYTIC PROPERTIES OF THE ENZYMES.

As a clear example of the direct type of regulation of activity the feedback inhibition by NADPH of both isoenzymes of G6PDH should be quoted. While this is the only regulatory mechanism for the cytoplasmic isoenzyme (22), it applies for the chloroplast isoenzyme for its oxidized dark form (Table 3). Furthermore, several effects of metabolites upon the light-activated form of PRK (23,24) and of other chloroplast enzymes have been observed. Although many similar cases have been described, it is not always easy to distinguish this type of regulation from that described under 3.

TABLE 3. Kinetic properties of cytoplasmic and dark-modulated chloroplastic G6PDH from pea leaves.

	cytoplasmic[a] G6PDH	chloroplastic (oxidized)[b]
K_m [G6P]	120 µM	122 µM
K_m [NADP]	14 µM	17 µM
K_i [NADPH]	11 µM	19 µM
type of inhibition	competitive	competitive

[a] data from 22; [b] K. Fickenscher and R. Scheibe, unpublished.

Ideally the conditions during preincubation and in the assay have to be defined in detail and their respective effects both upon the rate of activation and upon the actual activity in the assay (or in situ as in the functioning chloroplast) have to be regarded. When preincubation of the enzyme is taking place in the illuminated intact chloroplast, protoplast or leaf, it would be helpful to always register the metabolite levels, and consider their effect upon the rate and extent of light activation, a reaction which, as shown here, does not occur equally well for all light-activated enzymes.

CONCLUSIONS
1. Plastids of potentially photosynthesizing cells contain several enzymes (NADP-MDH, chloroplast G6PDH, thioredoxin m, GSH-reductase) involved in the regulation of the redox-state even in the absence of light. No significant increase in the content of these enzymes is observed upon greening of etiolated pea seedlings, while those enzymes traditionally linked with greening as the Calvin-cycle enzymes Rubisco, FBPase, NADP-GAPDH and PRK exhibit a 5- to 10-fold increase.
2. The light modulation systems provide the potential to fully light modulate several key enzymes, but metabolites determine the extent of this reaction, thus respecting the metabolic needs of the chloroplast under changing conditions in the light. Examples are given that enzymes involved in the transformation of light-generated chemical energy into carbohydrates, might not primarily be regulated by direct effects upon their catalytic properties, but the flux through them is regulated by changing their amount of active species.
3. Each light-modulated enzyme appears to be adjusted to the required level by the differentiated influence of various metabolites so that the light-generated electron flow can be directed either into the Calvin-cycle, which occurs with first preference, or - if necessary - into energy dissipating systems such as malate production by NADP-MDH. The latter provides a means to export only that portion of reducing equivalents to support poising of the ATP/NADPH ratio, when production and consumption are not balanced.

III.3.**240**

REFERENCES

1 Buchanan, B.B. (1984) Bioscience 6, 378-383
2 Anderson, L.E. (1985) In: Advances in Botanical Research, Vol.12 (J.A. Callow, ed.) pp 1-46, Academic Press, New York
3 Scheibe, R. (1981) FEBS Lett. 133, 301-304
4 Buchanan, B.B. (1981) In "Photosyntesis IV. Regulation of Carbon Metabolism" (G. Akoyunoglou, ed.) pp. 245-256, Balaban International Services, Philadelphia.
5 Scheibe R. (1984) Biochim. Biophys. Acta 788, 241-247
6 Jenkins, C.L.D., Anderson, L.E. and Hatch, M.D. (1986) Plant Sci. 45, 1-7
7 Scheibe, R., Fickenscher, K. and Ashton, A.R. (1986) Biochim. Biophys. Acta 870, 191-197
8 Nakamoto, H. and Edwards, G.E. (1984) Z. Pflanzenphysiol. 114, 315-320
9 Wolosiuk, R.A., Perelmuter, M.E. and Chehebar, C. (1980) FEBS Lett. 109, 289-293
10 Nishizawa, A.N. and Buchanan, B.B. (1981) J. Biol. Chem. 256, 6119-6126
11 Wolosiuk, R.A, Hertig, C.M. and Busconi, L. (1986) Arch. Biochem. Biophys. 246, 1-8
12 Ashton, A.R. (1983) In: "Thioredoxins - Structure and function" (P. Gadal, ed.) pp. 245-250, CNRS, Paris
13 Scheibe, R. and Jacquot, J.-P. (1983) Planta 157, 548-553
14 Chock, P.B, Rhee, S.G. and Stadtman, E.R. (1980) Ann. Rev. Biochem. 49, 813-843
15 Heber, U. (1984) In: Advances in Photosynthesis Research. Vol III (C. Sybesma, ed.) pp. 381-389, Martinus Nijhoff, Dr. W. Junk Publishers, The Hague
16 Dietz, K.-J. and Heber, U. (1986) Biochim. Biophys. Acta 848, 392-401
17 Hatch, M.D., Dröscher, L., Flügge, U.I. and Heldt, H.W. (1984) FEBS Lett. 178, 15-19
18 Scheibe, R., Wagenpfeil, D. and Fischer, J. (1986) J. Plant Physiol. 124, 103-110
19 Steiger, H.-M. and Beck, E. (1981) Plant Cell Physiol. 22, 561-576
20 Strotmann, H., Kiefer, K. and Altvater-Mackensen, R. (1986) Biochim. Biophys. Acta 850, 90-96
21 Mills, J.D., Mitchell, P. and Schürmann, P. (1980) FEBS Lett. 112, 173-177
22 Fickenscher, K. and Scheibe, R. (1986) Arch. Biochem. Biophys. 247, 393-402
23 Gardemann, A., Stitt, M. and Heldt, H.W. (1982) FEBS Lett. 137, 213-216
24 Gardemann, A., Stitt, M. and Heldt, H.W. (1983) Biochim. Biophys. Acta 722, 51-60

ACKNOWLEDGEMENT

I wish to thank Prof. Dr. E. Beck for fruitful discussions and R. Schiffelholz and K. Fickenscher for their contributions of unpublished data. The financial support for parts of this work given by the Deutsche Forschungsgemeinschaft is gratefully acknowledged.

STUDIES ON ENZYME PHOTOACTIVATION BY THE FERREDOXIN/THIOREDOXIN SYSTEM

M. MIGINIAC-MASLOW, M. DROUX, J.-P. JACQUOT
PHYSIOLOGIE VEGETALE MOLECULAIRE, U.A. 1128 CNRS UNIVERSITE DE PARIS-SUD, ORSAY, FRANCE

AND

N.A. CRAWFORD, B.C. YEE AND B.B. BUCHANAN
MOLECULAR PLANT BIOLOGY, UNIVERSITY OF CALIFORNIA, BERKELEY, CALIFORNIA

1. INTRODUCTION
    The ferredoxin/thioredoxin system functions in oxygenic photosynthesis by linking light to the regulation of selected target enzymes Fig. 1 (1).

FIGURE 1. Light activation/dark deactivation of enzymes by the ferredoxin/-thioredoxin system.

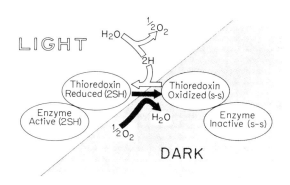

Here, the light signal is converted to reduced thioredoxin by way of ferredoxin, ferredoxin-thioredoxin reductase (FTR) and the photosynthetic electron transport chain, thereby changing the thiol status of target enzymes through transfer of net reducing equivalents. An oxidation of the SH groups formed in the light takes place in the dark in the presence of molecular oxygen.

While the thioredoxin components of the system have been characterized (2), our understanding of the enzyme catalyzing the reduction of thioredoxin by reduced ferredoxin, FTR, is incomplete. We now report some characteristics of FTR, purified to homogeneity from higher plants and cyanobacteria, including evidence for an active disulfide group functional in the reduction of thioredoxin. FTR was found to be synonymous with the protein earlier called ferralterin (3,4).

2. NEW DEVELOPMENTS OF THE FERREDOXIN/THIOREDOXIN SYSTEM
    Affinity chromatography on ferredoxin-sepharose was found successful for the purification of FTR, as was previously the case for ferredoxin-NADP reductase (FNR). FTR showed a relatively strong binding to ferredoxin-Sepharose, requiring 0.20 M NaCl for elution, versus 0.14 M NaCl for FNR. Each FTR examined had a yellowish-brown color and exhibited absorption peaks in the visible (410 nm) and ultraviolet (278 nm) (Fig. 2). Homogeneous FTR contained 4.0 iron and 3.8 sulfide groups per mole enzyme. Similar results were earlier reported for spinach FTR (3). The presence of FTR was also reported recently in Chlamydomonas (Huppe et al., these proceedings) with similar spectroscopic characteristics.

FIGURE 2. Absorption spectra of FTR from chloroplasts (Spinach and Corn) and Cyanobacteria (<u>Nostoc</u> Muscorum).

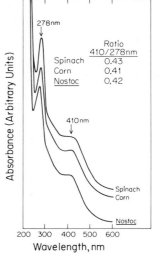

When submitted to SDS-PAGE, each FTR showed two subunits differing in molecular weight as reported previously (3,4). In each case, the molecular weight of the native enzyme was about 30,000. The subunit profile was unique for a particular FTR but each showed a similar subunit with a molecular weight of 13,000. and a second (variable) subunit with a molecular weight > or < 13,000. The similar subunit of the different FTR enzymes also resembled one another immunologically (Droux et al., these proceedings).

An additional common feature of the 13 kDa subunit among the FTR enzymes examined was its capability to react with ^{14}C-iodoacetate following reduction with dithiothreitol (DTT). The results suggest the presence of a dithiol reactive disulfide located on the 13 kDa subunit. The presence of this disulfide was confirmed by an extensive analysis of the number of cyst(e)ines in the spinach protein that revealed 4 free cysteine, 4 cysteine in the iron-sulfur cluster, and one cystine (S-S) (Droux et al., these proceedings).

FTR purified to homogeneity from each of the sources examined was capable of activating both NADP-MDH with thioredoxin m and FBPase with thioredoxin f by way of photoreduced ferredoxin (Table I).

TABLE I. Requirements for Light-dependent Activation of Chloroplast NADP-MDH and FBPase with Pure Components of the Ferredoxin-Thioredoxin System. Thylakoid membranes were washed with NaCl and EDTA.

	Relative Activity*					
	NADP-MDH activation with FTR from			FBPase activation with FTR from		
Treatment	Spinach	Corn	Nostoc	Spinach	Corn	Nostoc
Light						
Complete	100[a]	100[b]	100[c]	100[d]	100[e]	100[f]
minus FTR	4	3	4	2	4	3
" ferredoxin	16	2	2	2	31	4
" thioredoxin	12	5	4	17	8	11
Dark						
Complete	0	0	1	2	6	4

*100% activity corresponds to: (a) 1.5, (b) 3.2, (c) 0.25 μmoles NADPH oxidized per min; (d) 0.16, (e) 0.02, (f) 0.18 μmoles Pi released per min.

These requirements for FTR, which were observed in its original description (5), are in accord with a central role for FTR in facilitating the light-mediated regulation of enzymes of oxygenic photosynthetic organisms (6).

While there is no question as to its role, the means by which FTR utilizes photoreduced ferredoxin for the reduction of thioredoxin is not known. We have, therefore, conducted a study to determine whether the disulfide bridge in FTR is reduced in the light, analogous to earlier findings with NADP-MDH and DTT (7,8). We examined the reduction of FTR, as well as thioredoxin m and NADP-MDH, in a chloroplast system reconstituted from the pure components. The thiol content of the different soluble proteins of the system was determined by following the incorporation of ^{14}C-iodoacetate in the light and in the dark. By measuring the difference between the two treatments, we identified the disulfides in Table II.

TABLE II. Light-Generated Sulfhydryls in Components of the Ferredoxin/Thioredoxin System and in the Target Enzyme, NADP-MDH.

| | SH per mole | |
	Control	+ Urea
Ferredoxin	0	0
FTR	1.6	1.3
Thioredoxin m	1.9	1.5
NADP-Malate dehydrogenase (subunit)	1.6	1.7

The net incorporation of ^{14}C-iodoacetate increased in the light in FTR, thioredoxin, and NADP-MDH, but not for ferredoxin or on the chloroplast thylakoid membranes. The results reflected the reduction of a disulfide bridge in FTR as well as thioredoxin m (9). For NADP-MDH, the appearance of two new SH per subunit agrees with the results previously obtained in the presence of DTT reduced thioredoxin m (7,8). For FTR, the increase in SH seems to be the result of a reduction rather than a conformational change leading to the exposure of internal free SH groups as illustrated by the results obtained in the presence of the denaturing agent, urea (Table II). Significantly, in related experiments, the extent of reduction of NADP-MDH was found to correlate with the activity of the enzyme.

We confirmed the labelling of each component of the ferredoxin/thioredoxin system with ^{14}C-iodoacetate by fluorography/autoradiography following SDS-PAGE (Droux et al., these proceedings). The labelling in FTR seems to be localized in the 13 kDa subunit. It is noteworthy that the requirements for the light-dependent reduction of the components of the ferredoxin/thioredoxin system were identical to those needed for the light activation of NADP-MDH. We have also observed that the system can photoreduce a nonphysiological acceptor, insulin, with identical requirements (Table III), thereby confirming a similarity with the NADP-thioredoxin system (10).

TABLE III. Reduction of Insulin by the Ferredoxin-Thioredoxin System in the Light and Dark.

Treatment	Insulin Reduced SH formed, nmoles	
	Light	Dark
Complete	46.5	3.7
minus Insulin	6.2	3.3
" Thioredoxin m	2.3	1.7
" FTR	2.2	2.1
" Ferredoxin	1.8	1.5

To determine whether the light-generated sulfhydryl group in FTR functions in catalysis, we carried out an experiment where we photoreduced FTR and then added (in the dark) its disulfide acceptor, thioredoxin m, in both the absence and presence of insulin, using as a probe monobromobimane (mBBr). By analyzing the mBBR-labelled protein following separation by SDS-PAGE, we found that the reducing equivalents accumulated on the 13 kDa FTR subunit are transferred in the dark to thioredoxin and insulin (Table IV). The results are consistent with a sulfhydryl-disulfide interchange between reduced FTR, oxidized thioredoxin m and insulin. At this time we have no definitive results on the role of the iron-sulfur cluster in this process.

TABLE IV. Demonstration that Photoreduced Thiol Group in FTR Functions in Catalysis (mBBr as probe). Td = Thioredoxin; Fd = ferredoxin; Thyl = thylakoids.

Phase I (15 min) Light or Dark	Phase II (10 sec) Dark	Relative reduction %		
Treatment	Addition	FTR	Td m	Insulin
Light Thyl + Fd + FTR	Buffer	100	---	---
	Td m	65	34	---
	Td m + Insulin	48	21	30
Dark Thyl + Fd + FTR	Td m + Insulin	0	0	0

3. STUDIES ON ENZYME PHOTOACTIVATION

The light activation process does not consume a large portion of the reducing power of chloroplasts. Light saturation curves show a saturation of the process at 10 to 20 $W.m^2$ under nitrogen and 50 $W.m^2$ under air, whether determined in a reconstituted system or in intact chloroplasts (11). With such a low light requirement, we would expect activation to take place under unfavorable energetic conditions. However, this is not the case. Data obtained as part of an earlier study showed that thylakoids from photosystem II deficient bundle sheath thylakoids from corn, a C_4 plant, were almost unable to photoactivate NADP-MDH or fructose-1,6-bisphosphatase unless supplied with an artificial electron donor to photosystem I (ascorbate/dichlorophenol indophenol). Moreover, NADP photoreduction activity was not completely absent in the thylakoids and represented about 10% of that determined with mesophyll thylakoids (12).

Another type of perturbation condition is brought about after treatment of isolated chloroplasts with high light, a process known as photoinhibition. Exposure of chloroplasts to high light decreases photosystem II activity of thylakoids quite appreciably (13). Under our conditions, an 8 min preillumination of isolated thylakoids decreased their ability to photoreduce NADP about 50%. Unexpectedly, NADP-MDH activation was completely abolished (Table V).

TABLE V. Comparison of Photoinhibition of NADP Photoreduction and NADP-MDH Light Activation

	NADP Photoreduction μmole per mg Chl · h	NADP-MDH Activity Units per mg Chl
Control	96	18.3
Photoinhibited	45	0.5
Photoinhibited + ascorbate/dichlorophenol indophenol	82	20.0

Both activities were restored to their approximate original values by the addition of ascorbate/dichlorophenol indophenol. When the reduction of disulfide groups on protein components of the light activation system was examined, we found that photoinhibition treatment decreased FTR reduction by 40 to 50% and thioredoxin m reduction by 20 to 30%. Full reduction of both proteins was restored by addition of ascorbate/dichlorophenol indophenol (Table VI). It seems therefore, that photoactivation of target enzymes occurs only when thioredoxin is almost completely reduced. This observation agrees with the conclusions of others that the equilibrium of the reaction between dithiothreitol-reduced thioredoxins and target enzymes favors strongly the reduction of thioredoxins (14-16). According to these investigators, the reaction would be shifted towards the target enzyme by continuous electron flow provided by illuminated thylakoids. Our results indicate that this observation applies also to the

TABLE VI. Effect of Photoinhibition on Photoreduction of FTR and Thioredoxin (Td) m

	Light Dependent SH Formed (cpm ^{14}C-iodoacetate)			NADP-MDH Activity Units per mg Chl
	FTR	FTR+Td m	Td m	
Control	944	3757	2813	18
Photoinhibited	418	2705	2287	1
Photoinhibited + ascorbate/dichlorophenol indophenol	902	3646	2755	20

complete light activation system and that even a partial deficiency in photosystem II can strongly affect enzyme light activation. These results likely have physiological significance as a decrease in light activation of enzymes has been reported in crude extracts from leaves of light stressed plants (17).

4. CONCLUSION

FTR--a protein found to be synonymous with ferralterin (4)--was purified from cyanobacteria, C_3 and C_4 plants, and found to be an iron-sulfur protein, containing 4 iron and labile sulfide groups. In each case, the enzyme had a molecular weight of about 30,000 and was composed of 2 nonidentical subunits, one that is similar in the enzyme from the different sources, and another that is variable and characteristic of a particular source. The similar subunit contained a disulfide bridge that is reduced either chemically by a dithiol or photochemically by chloroplasts and ferredoxin. The photochemically generated sulfhydryls reduced the disulfide group of thioredoxin m and, in turn, the target enzyme, NADP-MDH. The results are in accord with the conclusion that FTR is the first member of a thiol regulatory chain leading from the electron transport pathway to selected target enzymes. Furthermore, although the requirement of the system for reducing power is very low, the process is very sensitive to any decrease in photosystem II activity and is inhibited when part of the thioredoxins are in an oxidized form.

ACKNOWLEDGEMENT

Aided by grants from NSF, USDA/CRGO, and CNRS ATP--Pirsem (No. 2248).

REFERENCES
1. Buchanan, B.B. (1984) BioScience 34, 3278-383
2. Maeda, K., Tsugita, A., Dalzoppo, D., Vilvois, F., and Schürmann, P. (1986) Eur. J. Biochem., in press
3. Schürmann, P. (1981) in Proceedings of the 5th Internationals Photosynthesis Congress (Akoyunoglou, ed.) Vol. IV, pp. 273-280, Balaban Press
4. De la Torre, A., Lara, C., Yee, B.C., Malkin, R., and Buchanan, B.B. (1982) Arch. Biochem. Biophys. 213, 545-550
5. Wolosiuk, R.A. and Buchanan, B.B. Nature 266, 565-567
6. Cséke, C. and Buchanan, B.B. (1986) Biochim. Biophys. Acta., Reviews in Bioenergetics, in press
7. Jacquot, J.P., Gadal, P., Nishizawa, A.N., Yee, B.C., Crawford, N.A. and Buchanan, B.B. (1984) Arch. Biochem. Biophys. 228, 170-178
8. Scheibe, R. (1984) Biochim. Biophys. Acta 788, 241-247
9. Tsugita, A., Maeda, M., and Schürmann, P. (1983) Biochem. Biophys. Res. Commun. 115, 1-7
10. Holmgren, A. (1985) Thioredoxin. Annu. Rev. Biochem. 52, 237-272
11. Miginiac-Maslow, M., Jacquot, J.P., and Droux, M. (1985) Photosyn. Res. 6, 201-213
12. Lavergne, D., Droux, M., Jacquot, J.P., Miginiac-Maslow, M., Champigny, M.L., and Gadal, P. (1985) Planta 166, 187-193
13. Cornic, G. and Miginiac-Maslow, M. (1985) Plant Physiol. 78, 724-730
14. Buc, J., Soulié, J.M., Minot, R., and Ricard, J. (1983) In Thioredoxins: Structure et fonctions (P. Gadal, ed.), pp. 159-166, CNRS, Paris

15. Rebeille, F. and Hatch, M.D. (1986) Arch. Biochem. Biophys. 249, 164-170
16. Rebeille, F. and Hatch, M.D. (1986) Arch. Biochem. Biophys. 249, 171-179
17. Powles, S.B., Chapman, K.S.R., and Whatley, F.R. (1982) Plant Physiol. 69, 371-374

FERREDOXIN-THIOREDOXIN REDUCTASE: AN IRON-SULFUR ENZYME LINKING LIGHT TO
ENZYME REGULATION IN CHLOROPLASTS

M. DROUX, J.-P. JACQUOT, M. MIGINIAC-MASLOW, AND P. GADAL
PHYSIOLOGIE VEGETALE MOLECULAIRE, UNIVERSITE DE PARIS-SUD, ORSAY, FRANCE
AND
N.A. CRAWFORD, B.C. YEE AND B.B. BUCHANAN
MOLECULAR PLANT BIOLOGY, UNIVERSITY OF CALIFORNIA, BERKELEY, CALIFORNIA

1. INTRODUCTION
     The ferredoxin/thioredoxin system functions in oxygenic photosynthesis
by linking light to the regulation of selected target enzymes (1). Here,
the light signal is converted to reduced thioredoxin by way of ferredoxin,
ferredoxin-thioredoxin reductase (FTR) and the photosynthetic electron
transport chain, thereby changing the thiol status of target enzymes
through transfer of net reducing equivalents. An oxidation of the SH
groups formed in the light takes place in the dark in the presence of
molecular oxygen.
     While the thioredoxin components of the system have been character-
ized, our understanding of the enzyme catalyzing the reduction of thioredoxin by reduced ferredoxin, FTR, is incomplete. We now report some characteristics of FTR, purified to homogeneity from higher plants and cyanobacteria, including evidence for an active disulfide group functional in
the reduction of thioredoxin.

2. MATERIALS AND METHODS
     FTR was purified from spinach and corn leaves and Nostoc muscorum
cells as described by Droux et al. (2). Previously described methods were
also used for the purification of thioredoxin m and thioredoxin f, (3),
fructose-1,6-bisphosphatase (FBPase) (4) and NADP-malate dehydrogenase
(NADP-MDH) (5). Methods described elsewhere were used in light activation
of NADP-MDH and FBPase with spinach thylakoids (6), for ^{14}C-iodoacetate
experiments (5) and for monobromobimane (mBBR) labelling (7).

3. RESULTS AND DISCUSSION
     Affinity chromatography on ferredoxin-sepha-
rose was found successful for the purification
of FTR, as was previously the case for ferredoxin-NADP reductase (FNR). FTR showed a relatively strong binding to ferredoxin-Sepharose,
requiring 0.20 M NaCl for elution, versus
0.14 M NaCl for FNR. Each FTR examined had a
yellowish-brown color and exhibited absorption
peaks in the visible (410 nm) and ultraviolet
(278 nm). Homogeneous FTR contained 4.0 iron
and 3.8 sulfide groups per mole enzyme.

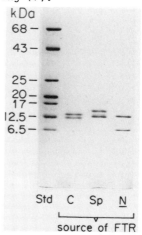

FIGURE 1.   Molecular weight of FTR from
chloroplasts (spinach and corn) and cyanobacteria (Nostoc muscorum) as determined by
SDS-PAGE. In each case 4 µg of FTR was used.

Similar results were earlier reported for spinach FTR (8).

FIGURE 2. Identification
of similar and variable
subunits in spinach, corn
and <u>Nostoc muscorum</u> FTR
with corresponding anti-
bodies (Western blot).

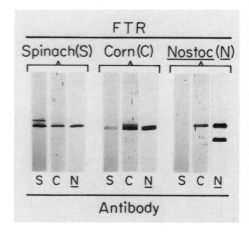

When submitted to SDS-
PAGE, each FTR showed two
subunits differing in molecular weight (Fig. 1). In
each case, the molecular
weight of the native enzyme
was about 30,000. The subunit profile was unique for
a particular FTR but each
showed a similar subunit
with a molecular weight of
13,000. and a second
(variable) subunit with a molecular weight > or < -13,000. The similar
subunit of the different FTR enzymes also resembled one another
immunologically as determined in Western blots (Fig. 2).

An additional common feature of the 13 kDa subunit among the FTR
enzymes examined was its capability to react with ^{14}C-iodoacetate following reduction with dithiothreitol (DTT) (Fig. 3). The results suggest
the presence of a dithiol reactive disulfide located on the 13 kDa subunit. Significantly, monothiols such as 2-mercaptoethanol and reduced
glutathione, did not replace DTT in this reaction, indicating that the
reduction is dithiol-specific. The presence of this disulfide was confirmed by an extensive analysis of the number of cysteine in the spinach
protein that revealed 4 free cysteine, 4 cyst(e)ine in the iron-sulfur
cluster, and one cystine (S-S).

FTR purified to homogeneity
from each of the sources examined
was capable of activating both
NADP-MDH with thioredoxin <u>m</u> and
FBPase with thioredoxin f <u>by</u> way
of photoreduced ferredoxin. The
requirements observed for activation are in accord with a central
role for FTR in facilitating the
light-mediated regulation of enzymes of oxygenic photosynthetic
organisms.

FIGURE 3. Identification of FTR
subunit labelled with ^{14}C-iodo-
acetate after reduction by DTT.
Superimposed densitometric tracings of radioactivity and protein.

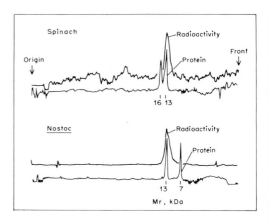

While there is no question as to its role, the means by which FTR utilizes photoreduced ferredoxin for the reduction of thioredoxin is not known. We have, therefore, conducted a study to determine whether the disulfide bridge in FTR is reduced in the light, analogous to earlier findings with NADP-MDH and DTT (5,9). We examined the reduction of FTR, as well as thioredoxin m and NADP-MDH, in a chloroplast system reconstituted from the pure components. The thiol content of the different soluble proteins of the system was determined by following the incorporation of ^{14}C-iodoacetate in the light and in the dark. By measuring the difference between the two treatments, we identified the disulfides in Table 1. The net incorporation of ^{14}C-iodoacetate increased in the light for FTR, thioredoxin, and NADP-MDH, but not for ferredoxin or for possible sulfhydryl groups on the chloroplast thylakoid membranes. The results reflected the reduction of a disulfide bridge, in FTR as well as thioredoxin m (10). For NADP-MDH, the appearance of two new SH per subunit agrees with the results previously obtained in the presence of DTT reduced thioredoxin m (5,9). For FTR, the increase in SH seems to be the result of a reduction rather than a conformational change leading to the exposure of internal free SH groups as illustrated by the results obtained in the presence of the denaturing agent, urea (Table I). Significantly, in related experiments, the extent of reduction of NADP-MDH was found to correlate with the activity of the enzyme.

TABLE 1. Light-generated sulfhydryls in components of the ferredoxin/thioredoxin system and in the target enzyme, NADP-MDH

| | SH per mole | |
	Control	+ Urea
Ferredoxin	0	0
FTR	1.6	1.3
Thioredoxin m	1.9	1.5
NADP-Malate dehydrogenase (subunit)	1.6	1.7

We confirmed the labelling of each component of the ferredoxin/thioredoxin system with ^{14}C-iodoacetate by fluorography/autoradiography following SDS-PAGE (Fig. 4). Staining with Coomassie blue revealed that several polypeptides were released by the thylakoid membranes, but radioactivity was recovered almost entirely in the added soluble proteins, FTR, thioredoxin m and NADP-MDH. The labelling in FTR seems to be

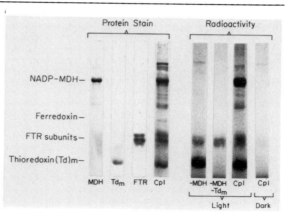

FIGURE 4. Localization of light generated sulfhydryls in components of the ferredoxin/thioredoxin system and the target enzyme, NADP-MDH.

localized in the 13 kDa subunit--a finding predicted by the earlier DTT experiment of Fig. 3. It is noteworthy that the requirements for the light-dependent reduction of the components of the ferredoxin/thioredoxin system were identical to those needed for the light activation of NADP-MDH. We have also observed that the system can photoreduce a nonphysiological acceptor, insulin, thereby confirming a similarity with the NADP-thioredoxin system (11).

To determine whether the light-generated sulfhydryl group in FTR functions in catalysis, we carried out an experiment where we photoreduced FTR and then added (in the dark) its disulfide acceptor, thioredoxin $\underline{m}$, in both the absence and presence of insulin. By analyzing the mBBR-labelled protein following separation by SDS-PAGE, we found that the reducing equivalents accumulated on the 13 kDa FTR subunit are transferred in the dark to thioredoxin and insulin. The results are consistent with a sulfhydryl-disulfide interchange between reduced FTR and oxidized thioredoxin $\underline{m}$ and insulin. At this time we have no definitive results on the role of the iron-sulfur cluster in this process.

4. CONCLUSION

FTR purified from cyanobacteria, C_3 and C_4 plants, is an iron-sulfur protein, containing 4 iron and labile sulfide groups. In each case, the enzyme has a molecular weight of about 30,000 and is composed of 2 nonidentical subunits, one that is similar in the enzyme from the different sources, and another that is variable and characteristic of a particular source. The similar subunit contained a disulfide bridge that is reduced either chemically by a dithiol or photochemically by chloroplasts and ferredoxin. The photochemically generated sulfhydryls reduced the disulfide group of thioredoxin $\underline{m}$ and, in turn, the target enzyme, NADP-MDH. The results are in accord with the conclusion that FTR is the first member of a thiol regulatory chain leading from the electron transport pathway to selected target enzymes.

REFERENCES
1. Buchanan, B.B. (1984) Bioscience 34, 378-383
2. Droux, M., Jacquot, J.P., Miginiac-Maslow, M. Gadal, P., Huet, J.C., Crawford, N.A., Yee, B.C. and Buchanan, B.B., submitted
3. Crawford, N.A., Yee, B.C., Hutcheson, S.W., Wolosiuk, R.A. and Buchanan, B.B. (1986) Arch. Biochem. Biophys. 244, 1-15
4. Nishizawa, A.N. (1982) in Methods in Chloroplast Molecular Biology (Edelman, M. Hallick, R.B. and Chua. N.-H., eds.), pp. 707-714, Elsevier, New York
5. Jacquot, J.P., Gadal, P., Nishizawa, A.N., Yee, B.C., Crawford, N.A. and Buchanan, B.B. (1984) Arch. Biochem. Biophys. 228, 170-178
6. Miginiac-Maslow, M. Jacquot, J.P. and Droux, M. (1985) Photosyn. Res. 6, 201-213
7. Crawford, N.A., Kosower, N.S. and Buchanan, B.B., submitted.
8. Schurmann, P. (1981) in Proceedings of the 5th International Photosynthesis Congress (Akoyunoglou, ed.) Vol. IV, pp. 273-280, Balaban Press
9. Scheibe, R. (1984) Biochim. Biophys. Acta 788, 241-244
10. Tsugita, A., Maeda, M. and Schurmann, P. (1983) Biochim. Biophys. Res. Comm. 115, 1-7
11. Holmgren, A. (1985) Ann. Rev. Biochem. 54, 237-272

(Aided by grants from NSF and USDA/CRGO.)

EFFECT OF LIGHT ON THE THIOL STATUS OF CHLOROPLASTS

NANCY A. CRAWFORD, NECHAMA S. KOSOWER, and BOB B. BUCHANAN
DIVISION OF MOLECULAR PLANT BIOLOGY, HILGARD HALL
UNIVERSITY OF CALIFORNIA, BERKELEY, CA 94720

1. INTRODUCTION

Intact chloroplasts were analyzed for thiol content under light and dark conditions by using the thiol labelling reagent, monobromobimane (mBBr) (1). This reagent, which readily penetrated both the chloroplast envelope and thylakoid membranes, labelled available thiols thereby rendering them fluorescent. The results suggest that light brings about the reduction of the disulfide groups of numerous proteins of chloroplasts, especially those of the soluble protein fraction (chloroplast extract).

2. MATERIALS AND METHODS

Aliquots of freshly prepared intact chloroplasts were either illuminated or kept in the dark for 10 or 15 min. (the "reduction phase" of the reaction). mBBr, purchased from Calbiochem, was added (1.5 mM) and the chloroplast suspensions were held in either light or dark for the indicated time period (the "labelling phase" of the reaction). The labelled intact chloroplasts were then fractionated into soluble and membrane components by lysis and centrifugation. The coupling factor was removed from the membranes by the method of Strotmann et al (2). Proteins were analyzed by SDS-PAGE. Gels were fixed and fluorescent bands were visualized on a U.V. light box and were photographed with Polaroid type 55 positive-negative film. Densitometric scans were made from the 4x5 inch Polaroid negative (1). The non-protein component was isolated and analyzed on a reverse phase HPLC column as described by Newton et al. (3).

3. RESULTS AND DISCUSSION

3.1 Nonprotein SH

The major identifiable nonprotein thiols, glutathione (GSH) and H_2S were calculated to be present at respective average concentrations of 0.2 mM and 0.15 mM. Cysteine was also present but at a much lower concentration (average of 0.013 mM). In each case the concentration ratios of light:dark were close to unity with a rather large deviation from experiment to experiment (cf. 4). A number of unknown non-protein fluorescent products was generated in large quantity when chloroplasts were incubated with mBBr in the light (cf. 5). When the generation of these products was eliminated by denaturing the chloroplast proteins prior to treatment with mBBr, the quantities of known non-protein SH compounds remained similar.

3.2 Chloroplast Extract

Labelling with mBBr was quantitatively assessed by cutting out and weighing the densitometric tracings of mBBr fluorescence profiles. Soluble proteins from light-treated chloroplasts showed an average of 20⁰/o greater total incorporation of mBBr than their dark-treated counterparts (cf. 6,7). Similar results were obtained with [3H]NEM. Such a light dependent increase in thiols was observed in experiments in which reaction time with mBBr (labelling phase), varied from 2 to 30 min.

Biggens, J. (ed.), Progress in Photosynthesis Research, Vol. III. ISBN 90 247 3452 5
© *1987 Martinus Nijhoff Publishers, Dordrecht. Printed in the Netherlands.*

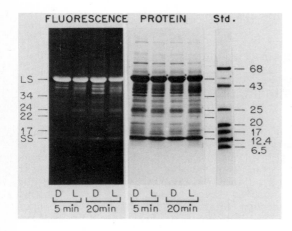

FIGURE 1. Photograph of mBBr-labelled and Coomassie blue-stained soluble chloroplast proteins (chloroplast extract) following separation by SDS-PAGE. Proteins from dark and light-treated chloroplasts are indicated by D and L, respectively. Length of labelling phase is indicated.

When added to intact chloroplasts incubated in the dark, DTT mimicked the effect of light, suggesting that light promotes the reduction of disulfide bonds in some soluble stromal proteins.

Interestingly, with both mBBr and [^{3}H]NEM, the light-dependent increase in the thiol content of soluble chloroplast proteins occurred more as a background hue rather than as changes in discrete protein bands (Fig. 1).This hue is probably due to changes in numerous minor proteins of the stroma which are too low in concentration to be resolved by these methods. Likely included among these proteins are thioredoxins which are reduced by light (8,9).

In contrast, the major soluble proteins from light and dark-treated chloroplasts showed similar labelling. The large subunit (LS) of ribulose 1,5-bisphosphate carboxylase/oxygenase (rubisco) appeared to be labelled by mBBr equally well in the light and dark (Fig. 1). The small rubisco subunit (SS), on the other hand, was labelled by mBBr in a time-dependent

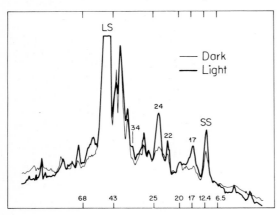

FIGURE 2. Superimposed densitometric tracings illustrating the fluorescence of the major mBBr labelled soluble proteins from light and dark treated chloroplasts (chloroplast extract) shown in Fig. 1. Reaction time with mBBr was 5 min. The profiles were aligned such that the background hue was cancelled.

manner: 80⁰/o more label in the light with a short reaction time (5 min), but a greater amount of label in the dark with a 20 or 30 min reaction time (Fig. 1). Such a change is consistent with the occurrence of a light-induced conformational change that leads to an alteration in the accessibility of the single small subunit SH group (cf. 10).

Of the other major proteins labelled with mBBr, the 34, 24, 22 and 17 kDa species differed in their response to light. Whereas each was consistently more labelled after a 5 min reduction time in the light (Figs. 1 and 2), only the 34, 22 and 17 kDa proteins maintained this pattern with longer reaction times; the 24 kDa protein did not. After 20 or 30 min, the 24 kDa protein became equally labelled in the dark and light. The addition of 1 mM DTT did not alter this pattern. It seems, therefore, that the 24 kDa protein may undergo a conformational change during illumination, or the pKa of its reactive thiol(s) may be altered.

The light induced change was mimicked by 1 mM DTT in the dark only on the 22 kDa protein. The 22 kDa protein, thus, may be an example of a protein that undergoes a net reduction in the light. The response of these various soluble chloroplast proteins to light is summarized in Table I.

TABLE I. EFFECTS OF LIGHT AND DTT ON SULFHYDRYLS OF SEVERAL MAJOR SOLUBLE CHLOROPLAST PROTEINS (mBBr AS PROBE)

Protein	Light effected increase	DTT effected increase	Suggested response to light
Rubisco LS	None	None	None
" SS	Initial	Initial	Nonreductive change*
34 kDa	Sustained	None	" "
24 kDa	Initial	None	" "
22 kDa	Sustained	Sustained	Net reduction
17 kDa	Sustained	None	Nonreductive change*

*i.e., conformational, proteolytic, or pKa change.

It is of interest to note that several soluble proteins were more highly labelled in the dark than in the light, especially when longer reaction times were used. Such behavior was seen, for example, in the cluster of bands in the 36-50 kDa range (illustrated in the D 20 min lane in Fig. 1). Perhaps these proteins either were proteolytically altered or they underwent conformational or pKa changes rendering SH less available to mBBr in chloroplasts in the light. A similar change was seen in experiments with [^{3}H]NEM.

3.3 Thylakoid Membranes

Labelled thylakoid membranes, from which CF$_1$-ATPase had been removed, were examined electrophoretically, and were found to be substantially labelled by mBBr provided that the labelling time was at least 5 min. On average, illumination of parent chloroplasts increased mBBr labelling of thylakoid membranes by 5%, but unlike the soluble proteins (chloroplast extract), 1 mM DTT added in the dark did not mimic the effect of light. Consistent light-dependent differences were apparent with a few bands, migrating at 54, 45, 40, and 19 KDa with longer reaction times. In each case, the protein was found to be more heavily labelled in the dark than in the light. A similar pattern of labelling of these proteins was seen with [^{3}H]-NEM. As noted above for chloroplast extract, it is concluded that the thiols of these membrane proteins are rendered less accessible in the light as a result of conformational change, altered pKa, or changed rate of proteolysis.

In parallel experiments, we observed increased mBBr labelling of the membrane-bound γ subunit of the CF_1-ATPase in the light with short reaction times, e.g., 2 min, but not in longer reaction times, e.g., 20 min. indicating a conformational change. Other investigators have previously reported such a change in CF_1-ATPase after brief exposure to light (11,12).

4. CONCLUSION

In this paper, we have utilized a technique whereby chloroplast thiols can be analyzed under light and dark conditions. We found about 20% greater increase by light in labelling by mBBr and a similar if not larger increase with $[^3H]NEM$. The results indicate that, while an increase in thiol content may occur in numerous minor soluble proteins during illumination, major quantitative changes don't occur in most of the major proteins. Conformational, pKa, and proteolytic changes may account for some of the differences observed between light and dark.

Because some of the effects of light were mimicked by DTT in the dark in mBBr experiments, we assume that a significant portion of the total light induced increase is due to the net generation of thiols. The ferredoxin/thioredoxin system is one mechanism by which light could accomplish this increase in thiols. Such a system would seemingly be independent of GSH which appears not to change its thiol status as a function of illumination (cf. 4,13).

5. ACKNOWLEDGEMENTS

This research was supported by a grant from the National Science Foundation. We gratefully acknowledge the contribution of Professor R.C. Fahey and Mr. G.L. Newton in conducting the analyses for nonprotein thiols.

REFERENCES
1 Kosower, N.S., Kosower, E.M., Zipser, V., Faltin, Z. and Shomrat, R. (1981) Biochim. Biophys. Acta 640, 748-759
2 Strotmann, H., Hesse, H. and Edelmann, K. (1973) Biochim. Biophys. Acta 314, 202-210
3 Newton, G.L., Dorian, R. and Fahey, R.C. (1981) Anal. Biochem. 114, 383-387
4 Halliwell, B. and Foyer, C.H. (1978) Planta 139, 9-17
5 Melis, A., Kosower, N.S., Crawford, N.A., Kirowa-Eisner, E., Schwartz, M. and Kosower, E.M. (1986) Arch. Biochem. Biophys. In press
6 Slovacek, R.E. and Vaughn, S. (1982) Plant Physiol. 70, 978-981
7 Slovacek, R.E. and Monahan, B.C. (1983) Arch. Biochem. Biophys. 224, 310-318
8 Buchanan, B.B. (1980) Annu. Rev. Plant Physiol. 31, 341-374
9 Scheibe, R. (1982) FEBS Lett. 133, 301-304
10 Miziorkio, H.M. and Lorimer, G.H. (1983) Annu. Rev. Biochem. 52, 507-535
11 Ketcham, S.R., Davenport, J.W., Warncke, K. and McCarty, R.E. (1984) J. Biol. Chem. 259, 7286-7293
12 Vallejos, R.H., Arana, J.L. and Ravizzini, R.A. (1983) J. Biol. Chem. 258, 7317-7321
13 Wolosiuk, R.A. and Buchanan, B.B. (1977) Nature 266, 565-567

LIGHT AND DARK ANAEROBIC ACTIVATION OF NADP-MALATE DEHYDROGENASE IN PEA
LEAVES AND CHLOROPLASTS

M. VIVEKANANDAN and GERALD E. EDWARDS
DEPARTMENT OF BOTANY, WASHINGTON STATE UNIVERSITY, PULLMAN, WASHINGTON
99164-4230, USA

1. INTRODUCTION
    Upon illumination, the activities of several enzymes located in the
chloroplast stroma are changed through covalent modification. The chloroplastic NADP-malate dehydrogenase (NADP-MDH) is one such enzyme activated
by light and deactivated in the dark (1,2). In vitro activation of this
enzyme requires reducing conditions and is thought to be mediated be reduction of disulfide groups on the protein (1,2,3,4). In C_4 plants, the in
vivo state of NADP-MDH activation, as mediated by covalent modification,
may reflect the extent of the reductive conditions in the mesophyll chloroplasts of C_4 plants both in dark and light. NADP-MDH in maize has been
shown to be dark activated in vivo under anaerobic conditions (1).
    In this study we evaluated whether NADP-MDH of Pisum sativum L., a C_3
plant, can also be dark activated similar to C_4 plants, and attempted to
identify the possible source of reducing power for dark anaerobic activation. We also examined the influence of various compounds (e.g. reductants
and oxidants) on the activation of NADP-MDH.

2. MATERIALS
2.1. Plant materials and reagents: Pea plants (Pisum sativum) were grown
    in growth chambers under controlled conditions (1). Leaves of the
    first node from the bottom of the plants were used for all experimental
    purposes. Iodoacetol phosphate (IAP) and chloroacetol phosphate (CAP)
    were provided by Dr. F. C. Hartman.
2.2. Isolation and purification of chloroplasts: The procedure for isola-
    tion of chloroplasts was basically the same as described by Scheibe
    and Jacquot (5). Chloroplasts were further purified by centrifuging
    through a 20% (v/v) Percoll solution to minimize contamination by
    catalase. Catalase was omitted in the suspension medium when H_2O_2 was
    included in the experiments with chloroplasts concerning dark activa-
    tion of NADP-MDH.
2.3. Effect of IAP, CAP and H_2O_2 on dark activation of NADP-MDH: The effect
    of pretreatment of chloroplasts with IAP, CAP and H_2O_2 on activation
    of NADP-MDH was carried out in a final volume of 2 ml with approxi-
    mately 200 ug chlorophyll by flushing the glass vials with humidified
    N_2 gas for 2 min and incubating the reaction mixture for 3 min at $25°C$
    in the dark. In separate experiments, dihydroxyacetone-P (DHAP),
    reduced glutathione (GSH) or ascorbate (ASC) were added to test
    whether IAP, CAP and H_2O_2 inhibitions of dark activation of NADP-MDH
    in chloroplasts can be reversed. These compounds were added after cen-
    trifugation at 600 g for 2 min in order to remove the inhibitors from
    the media. The chloroplasts were resuspended and, after incubation for
    3 min with DHAP or reductants, the chloroplasts were again centrifuged
    and resuspended in order to prevent any carry over of these substances

to the enzyme assay medium.
2.4. Effect of IAP and CAP on light activated NADP-MDH: Leaf extracts containing NADP-MDH from pre-illuminated pea leaves were treated with IAP
and CAP in a total volume of 1 ml containing 2-5 ug Chl in the original extract. The possible reversal of IAP or CAP inhibition of the
activated enzyme was studied by initially treating with inhibitors and
subsequently adding 5 mM DTT.

3.   RESULTS AND DISCUSSION
NADP-MDH is normally functional only in the light, since the enzyme is
deactivated in the dark. However, when chloroplasts (Fig. 1) as well as
leaf discs (data not shown), which contained NADP-MDH in the inactive
state, were subjected to anaerobiosis in dark, conditions favorable for
conversion of the enzyme from the inactive to the active form were created.
In addition, activation of the enzyme in isolated pea chloroplasts under
anaerobic treatment in the dark indicates that there is a source of reducing power within the chloroplasts to activate the enzyme under these conditions. The present findings show that dark anaerobic activation of NADP-
MDH can occur in vivo in the C_3 plant pea and are in good agreement with
the previous studies in maize (1). Glycolysis, GSH and ASC are among the
possible sources of reductants available to the chloroplasts in the dark.
To evaluate these possibilities, the effects of several compounds on the
dark anaerobic activation of the enzyme in chloroplasts were studied.
Among these, IAP and CAP, which are potential inhibitors of glycolysis were
examined. These are reactive analogs of DHAP and were originally synthesized as potential active-site reagents for aldolase, triose-P isomerase
and glycerophosphoate dehydrogenase (6). Usuda and Edwards (8) also found
IAP to be a potent inhibitor of NADP-triose-P dehydrogenase (NADP-TPDH).
Both IAP and CAP at 20 uM totally abolished dark anaerobic activation of
NADP-MDH in treated chloroplasts suggesting that the source or reducing
power might come from glycolysis (Table 1). However, from this result it
is uncertain whether prevention of dark activation of the enzyme by IAP and
CAP is due to total inhibition of generation of reducing power from glyco-
lysis or oxidation of sulfhydryl groups on the enzyme, since Hartman (6,7)
reported that both IAP and CAP could also act as sulfhydryl reagents.
In order to determine whether the inhibition of NADP-MDH by IAP and
CAP is reversible, compounds like GSH and ASC were added to the chloroplasts after the removal of the inhibitors. Surprisingly, addition of DHAP
could overcome the inhibition of
IAP and CAP by over 50% (Table 1).
This suggests that DHAP may contribute reducing power through
NADP-TPDH. GSH and ASC also reversed the inhibition by IAP and
CAP suggesting that there is some
potential for these reducing agents
to provide reducing equivalents
directly or indirectly to the enzyme. Further evidence that GSH
can reactivate the inactive NADP-
MDH comes from the fact that GSH
activated NADP-MDH in the crude
leaf extracts which were isolated
from the detached leaves and incubated at 0 to 4°C in pea, barley

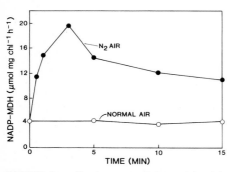

FIGURE 1.   Dark anaerobic activation
of NADP-MDH in chloroplasts
of pea.

Table 1. Reversal of IAP- CAP- and H_2O_2-induced inhibitions of dark activation of NADP-MDH in pea chloroplasts.

Treatment			umol/mg Chl. h
1. Normal air	No addition		0
2. Control (N_2 air)	No addition		21.3
	+ DHAP	(0.1 mM)	20.2
	+ GSH	(1.0 mM)	16.4
	+ ASC	(1.0 mM)	12.2
	+ AT	(1.0 mM)	21.6
3. IAP (20 uM), N_2 air	No addition		0
	+ DHAP	(0.1 mM)	13.8
	+ GSH	(1.0 mM)	7.8
	+ ASC	(1.0 mM)	4.0
4. CAP (20 uM), N_2 air	No addition		0
	+ DHAP	(0.1 mM)	12.3
	+ GSH	(1.0 mM)	9.8
	+ ASC	(1.0 mM)	4.3
5. H_2O_2 (1 mM), N_2 air	No addition		0
	+ AT	(1.0 mM)	0
	+ AT + ASC	(1.0 mM)	14.7
	+ AT + DHAP	(0.1 mM)	13.7
	+ AT + GSH	(1.0 mM)	11.8

Table 2. Effect of IAP and CAP on the activated NADP-MDH in extracts from pre-illuminated pea leaves.

Additions		umol/mg Chl. h
None		174
Assay		
+ IAP	(5 uM)	58
+ CAP	(5 uM)	39
Preincubation		
+ IAP	(5 uM)[a]	19
+ CAP	(5 uM)[a]	24
+ IAP	(5 uM)[b]	36
+ CAP	(5 uM)[b]	42
+ IAP	(5 uM)[c]	97
+ CAP	(5 uM)[c]	85

[a] Preincubated for 15 min
[b] Preincubated simultaneously with either IAP or CAP and 5 mM DTT for 15 min
[c] Preincubated with either IAP or CAP for 5 min and then treated with DTT for 10 min

and wheat (data not shown). GSH (20 mM) but not ASC activated partially purified NADP-MDH in pea in the absence of the regulatory protein, thioredoxin (data not shown). Complete protection against H_2O_2 (2 mM) inhibition of light activated NADP-MDH in chloroplast extract was given by GSH (1 mM) and DTT (1 mM) and not by ascorbate (data not shown). Ascorbate was apparently partially successful in providing reducing equivalents to the chloroplasts only following IAP inhibition of the dark activation of NADP-MDH. However, it should be noted that ascorbate alone in control experiments caused some inhibition of activation (Table 1). The basis for this is uncertain but, perhaps not all of the ascorbate was in the reduced form. Therefore, GSH is a potential source of reducing power for NADP-MDH in chloroplasts in vivo. This is in agreement with GSH acting to stabilize certain enzymes by preventing oxidation of their thiol groups (9) as well as supported by our observation that H_2O_2 (which may oxidize SH groups on enzymes) inhibition of the dark activation of the enzyme is reversed to varying degrees by GSH, ASC and DHAP (Table 1). Aminotriazole (AT), which was added to inhibit any residual catalase and prevent breakdown of H_2O_2, had no effect on the inactivation of NADP-MDH by H_2O_2.

As another part of our study, the effect of IAP and CAP on the light activated NADP-MDH in leaf extracts was examined. Addition of IAP and CAP (5 uM) to leaf extracts significantly decreased the activity of the light activated enzyme (Table 2). Prolonged incubation of the partially deactivated enzyme for 15 min with DTT significantly reversed both IAP and CAP inhibitions by more than 50% further suggesting that these two inhibitors reversibly inhibited enzyme activation by preferentially reacting with SH groups.

In general, from the present experiments we conclude that the reducing power for dark anaerobic activation of NADP-MDH in the chloroplasts is obtained through glycolysis and/or GSH. The possibility that GSH protects the active form of the enzyme or serves as an electron donor to support activation in vivo needs further investigation. The present work indicates that IAP and CAP may be useful for selective oxidation of protein SH groups.

ACKNOWLEDGEMENTS
This research was supported by an Indo-US Fulbright Fellowship to M. V. and NSF Grant DMB-8506197, and we thank Dr. M. S. B. Ku for assistance in reviewing the manuscript.

REFERENCES
1 Edwards, G.E., Nakamoto, H., Burnell, J.N. and Hatch, M.D. (1985) Annu. Rev. Plant Physiol. 36, 255-286
2 Johnson, H.S. and Hatch, M.D. (1970) Biochem. J. 119, 273-280
3 Buchanan, B.B. (1984) Biosci. 34, 378-383
4 Kagawa, T. and Hatch, M.D. (1977) Arch. Biochem. Biophys. 84, 290-297
5 Scheibe, R. and Jacquot, J.-P. (1983) Planta 157, 548-553
6 Hartman, F.C. (1970) Biochemistry 9, 1776-1786
7 Hartman, F.C. (1971) Biochemistry 10, 146-154
8 Usuda, H. and Edwards, G.E. (1981) Plant Physiol. 67, 854-858
9 Jocelyn, P.C. (1972) Biochemistry of the Thiol Groups, Academic Press, New York

IS THERE CHANNELING OF INTERMEDIATES IN THE CHLOROPLAST?
LOUISE E. ANDERSON, IVANO A. MARQUES[1] AND JERZY A. MACIOSZEK

UNIVERSITY OF ILLINOIS AT CHICAGO, DEPARTMENT OF BIOLOGICAL SCIENCES, BOX 4348, CHICAGO IL 60680, USA

1. INTRODUCTION

In experiments designed to examine the relationship between light activation of reductive pentose phosphate cycle enzymes and induction we were surprised to find that the active site concentrations of NADP-linked glyceraldehyde-3-P dehydrogenase and aldolase exceed substrate concentrations. The potential activity of these enzymes is then not sufficient to support observed rates of photosynthetic CO_2 fixation. Since photosynthetic CO_2 fixation does occur either our estimates are in error or some mechanism exists for channeling metabolites between enzymes in the chloroplast. Experiments with broken chloroplast extracts are consistent with transfer of P_2-glycerate between P-glycerate kinase and glyceraldehyde-3-P dehydrogenase without diffusion into the surrounding medium.

2. MATERIALS AND METHODS

CO_2-dependent O_2 Evolution. Chloroplast isolation and measurement of CO_2-dependent O_2 evolution was carried out as described in [1].
Metabolite Levels. Substrate level are based on radioactivity in metabolites separated by high pressure liquid chromatography. (Specific activity was constant after 6 minutes exposure to $^{14}CO_2$.)
Enzyme Assays. Activity was assayed by methods used previously in this laboratory [2,3].

3. RESULTS AND DISCUSSION

Estimated active site concentrations, substrate concentrations and potential activity in the chloroplast for glyceraldehyde-3-P dehydrogenase and aldolase are listed in Table 1.

Direct transfer of P_2-glycerate between P-glycerate kinase and glyceraldehyde-3-P dehydrogenase has been reported for the halibut enzymes [4]. The substrate for the dehydrogenase is apparently P-glycerate-bound P_2-glycerate. P-glycerate enhances the rate of the overall reaction. We find apparent negative cooperativity when P-glycerate is the varied substrate and NADPH is the reducing substrate for the two enzyme kinase, dehydrogenase coupled reaction (Fig. 1), but normal Michaelis-Menten kinetics when NADH is the reducing substrate (data not shown). These kinetics could be indicative of transient complex formation and rate enhancement by P-glycerate. The data can also be analyzed as two enzymes acting on one substrate using the model of [5] (Fig. 2). In either case deviation from normal Michaelis Menten kinetics is greatest when the dehydrogenase is in the dark (less active) form and when NADPH is the

----

[1]Present address: Institute of Biotechnology, Swiss Federal Institute of Technology, Hönggerberg, CH-8093 ZURICH Switzerland.

Biggens, J. (ed.), Progress in Photosynthesis Research, Vol. III. ISBN 90 247 3452 5
© *1987 Martinus Nijhoff Publishers, Dordrecht. Printed in the Netherlands.*

TABLE 1. *Estimated active site concentrations, substrate concentrations and potential activity levels in the photosynthesizing pea chloroplast*

Enzyme	G3PD*	Aldolase	
Substrate	P_2GA	G3P	DHAP
Concentration (mM) in the chloroplast**	0.0015	0.001-0.04	0.35-0.64
Active Site Concentration (mM)	0.03	1.7	
Potential Activity (mmol ml^{-1} min^{-1})	0.03	0.007	
Portion of Required Activity	60%	70%	

*Abbreviations used are: DHAP, dihydroxyacetone-P G3P, glyceraldehyde-3-P G3PD, glyceraldehyde-3-P dehydrogenase F6P, fructose-6-P P_2GA, 1,3-P_2-glycerate.

**P_2-glycerate estimated from measured and published P-glycerate, ATP and ADP concentrations and Keq of the P-glycerate kinase catalyzed reaction. Glyceraldehyde-3-P and dihydroxyacetone-P concentrations are for free keto forms, which are probably the substrate for the enzyme, by analogy to other systems.

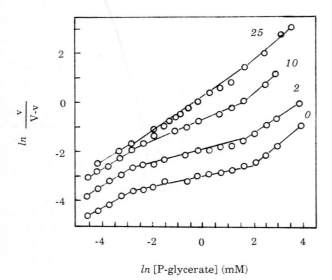

ln [P-glycerate] (mM)

FIGURE 1. *Hill plots for P-glycerate kinase/NADP-linked glyceraldehyde-3-P dehydrogenase activity with P-glycerate as the varied substrate after chloroplasts were illuminated for 0, 2, 10 and 25 minutes.*

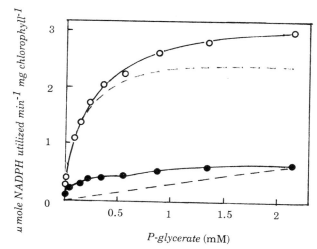

FIGURE 2. *Activity of the two enzyme P-glycerate kinase, NADP-linked glyceraldehyde-3-P dehydrogenase reaction analyzed as two enzymes acting on one substrate. Solid line is sum of activity of both "enzymes". Data points are experimentally obtained values in light (o) or dark (•). Upper dashed line is calculated activity of "Enzyme 1" after chloroplasts have been illuminated 10 minutes. Lower dashed line is calculated activity of "Enzyme 2".*

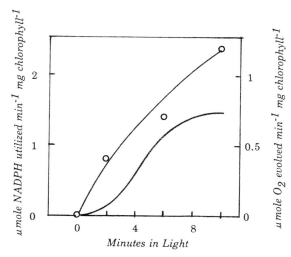

FIGURE 3. *Simulated v₀ for the P-glycerate kinase/NADP-linked glyceraldehyde-3-P dehydrogenase couple when chloroplasts are illuminated. P-glycerate levels were determined experimentally in another experiment, and used in the simulation. Kinetic parameters used in the simulation were estimated after chloroplasts were illuminated for times indicated. Solid line is observed rate of O_2 evolution. The ATP concentrations used were 2.5 mM.*

reducing substrate. The response of the kinase/dehydrogenase couple to illumination is shown in Figs. 1 and 2.

P-glycerate levels in the chloroplast increase during illumination. The effect of light activation and of the increase in P-glycerate levels in the intact chloroplast on the activity of the two enzyme couple is shown in Fig. 3. At *in vivo* levels of ATP (about 1.2 mM [6]) the dehydrogenase activity would be 1.2 μmol min^{-1} ml^{-1}, or sufficient to support observed rates of CO_2 fixation. Direct transfer of P_2-glycerate between P-glycerate kinase and glyceraldehyde-3-P dehydrogenase in the assay cuvette and in the chloroplast then seems possible.

Acknowledgments - This research was supported by US National Science Foundation and Department of Energy. We thank Larry Sykora and staff at the University of Illinois at Chicago Greenhouse for growing the pea plants.

REFERENCES
1 Marques, I.A. and Anderson, L.E. (1985) Plant Physiol. 77, 807-810
2 Pacold, I. and Anderson, L.E. (1975) Plant Physiol. 55, 168-171
3 Anderson, L.E. and Avron, M. (1976) Plant Physiol. 57, 209-213
4 Weber, J.P. and Bernhard, S.A. (1982) Biochemistry 21, 4189-4194
5 Spears, G., Sneyd, J.G.T. and Loten, E.G. (1971) Biochem. J. 125, 1149-1151
6 Heber, U., Takahama, U., Neimanis, S., and Shimizu-Takahama, M. (1982) Biochim. Biophys. Acta 679, 287-299

EFFECT OF LIGHT GROWTH CONDITIONS ON ATP SULPHURYLASE ACTIVITY OF MAIZE LEAVES

ROSSELLA GHISI and CALVINO PASSERA, Inst. of Agricultural Chemistry, University of Padova - Via Gradenigo, 6 - 35131 PADOVA (Italy)

1. INTRODUCTION

Light supplies the ATP and NADPH needed for CO_2 assimilation and regulates the activity of certain enzymes of reductive pentose phosphate cycle, the two additional enzymes of C_4 dicarboxylic acid CO_2 fixation pathway and enzymes of glucose degradation (1). Evidences from a variety of experimental approaches show also that enzymes of nitrogen assimilation require light for their activity (2). On the contrary, experimental evidences showing the effects of light on enzymes of sulphur assimilation are scanty. Therefore, the activity of ATP sulphurylase (EC 2.7.7.4), the first enzyme in the reaction chain of sulphate assimilation (3), was determined in leaves and isolated cells of maize hybrids grown under different light conditions.

2. PROCEDURE

2.1. Materials and methods

Maize (Zea mays L.) hybrids (Dekalb XL 85, Iperon, XL 72 A, Fedro, Rex, XL 75, Baron, Pronto, Lux, Gigas and Sponsor) were grown hydroponically (4) in a growth chamber maintained at 27°C day / 20°C night temperature and 70-80% relative humidity. Daylength was 14 hours. Photosynthetic flux densities at tank height were 320 (high light intensity) or 100 μmol m^{-2} s^{-1}(low light intensity). After 12 days of growth the second leaf of each plant was excised in pieces of tissue between 4-8 cm below the leaf tip, and protein content and the activities of ATP sulphurylase (ATP-s) (5), RuBP carboxylase and PEP carboxylase (6) were determined. Aliquots of tissue pieces (3.0 g) were used to isolate mesophyll (MC) and bundle sheath (BSC) cells (5). Cross contamination was determined measuring RuBP carboxylase and PEP carboxylase activities in isolated cells. Calculations were made according to Edwards et al. (7). Each reported data is the mean of five independent experiments. SD was never higher than 10% of the mean.

3. RESULTS AND DISCUSSION

ATP-s activity in leaves of 11 maize hybrids is given in Tab. 1. Results show that ATP-s activity is dependent on maize genotype. Since the hybrid Lux (with the hybrids Gigas and Sponsor), showed the highest ATP-s activity and the hybrid XL 85 the lowest one, they were selected to determine the effect of light intensity on enzyme activity.

Biggens, J. (ed.), Progress in Photosynthesis Research, Vol. III. ISBN 90 247 3452 5
© *1987 Martinus Nijhoff Publishers, Dordrecht. Printed in the Netherlands.*

TABLE 1. ATP sulphurylase (μmol PO_4^{3-} mg^{-1} protein h^{-1}) activity in leaves of maize hybrids grown under high light intensity.

Hybrid	Enzyme activity
XL 85	3.8
Iperon	5.7
XL 72 A	5.8
Fedro	5.8
Rex	6.0
XL 75	7.1
Baron	7.6
Pronto	8.3
Lux	9.7
Gigas	10.3
Sponsor	10.3

Upon decrease of light intensity the activity of ATP-s decreased in both hybrids: in XL 85 the activity diminished from 3.8 to 1.9 μmol PO_4^{3-} mg^{-1} protein h^{-1}, whereas in hybrid Lux the activity decreased from 9.7 to 4.7 μmol PO_4^{3-} mg^{-1} protein h^{-1}. Protein content also decreased, but the decrease was lower than that of ATP-s activity (Tab. 2).

TABLE 2. Protein content (mg g^{-1} f. w.), ATP-sulphurylase (μmol PO_4^{3-} mg^{-1} protein h^{-1}), RuBP carboxylase (RuBPc) and PEP carboxylase (PEPc) (μmol CO_2 mg^{-1} protein h^{-1}) in leaves of maize hybrids Lux and XL 85 grown under high (HI) and low (LI) light intensity.

	LUX		XL 85	
	HI	LI	HI	LI
Protein	9.5	7.6	9.2	8.0
ATP-s	9.7	4.7	3.8	1.9
RuBPc	20.8	20.7	21.0	21.3
PEPc	82.7	83.0	90.5	93.0

Under the same experimental conditions RuBPc and PEPc did not change significantly (Tab. 2).
Light exercises its regulatory function on enzyme throughout either affecting ion-, protein-mediated mechanisms or promoting de novo synthesis (1). Previous experiments (R. Ghisi, C. Passera unpublished) suggest that the increase in ATP-s activity of dark-grown seedlings of Zea mays L. transferred from dark to light, might be due to enzyme synthesis rather than activation of pre-existing enzyme, since protein inhibitors

blocked the enhancement of ATP-s activity. At the present, our findings seem to indicate that both processes might occur in maize green leaves since there was not a tight correspondence between decreases in protein and ATP-s activity. On the other hand, a mechanism of preferential synthesis or degradation of ATP-s in respect of the other proteins cannot be excluded.

In C_4 plants the enzymes of sulphate reduction mechanism are compartmentalized between bundle sheath and mesophyll cell and the higher load in sulphate assimilation is supported by bundle sheath cells (5, 8). Therefore, the data in Tab. 3, which are corrected for cross contamination, confirm this suggestion since at least 90% of the recovered ATP-s activity was localized in bundle sheath cells. Mesophyll cells contamination by BSC was 6-8%, whereas BSC contamination by MC was 1-2%. In mesophyll cell of hybrid XL 85 ATP-s activity was very low (Tab. 3).

TABLE 3. ATP sulphurylase (μmol PO_4^{3-} mg^{-1} protein h^{-1}), RuBP carboxylase and PEP carboxylase (μol CO_2 mg^{-1} protein h^{-1}) activities in mesophyll (MC) and bundle sheath (BSC) cells of maize hybrids Lux and XL 85 grown under low (LI) and high (HI) light intensity.

Enzyme	Level of light	Lux		XL 85	
		MC	BSC	MC	BSC
ATP-s	LI	2.2	16.1	0.3	10.4
	HI	2.5	26.3	0.3	14.8
RuBPc	LI, HI	1.7	25.6	2.0	26.6
PEPc	LI, HI	108.0	1.5	120.0	2.0

The data of Tab. 3 show also that the ATP-s localized in BSC increased as light intensity enhanced. The same effect was not observed in mesophyll cells. If this response is due to different behaviour of MC ATP-s or to some kind of inactivation during isolation of cells, is unknown. Differential responses of ATP-s from BSC and MC to other factors such as benzylaminopurine (9), N and S deprivation (10, 5) have been previously reported.

REFERENCES
1) Buchanan, B.B. (1980) Ann. Rev. Plant Physiol. 31, 341-374.
2) Naik, M.S., Abrol, Y.P., Nair, T.V.R. and Ramarao, C.S. (1982) Phytochemistry 21, 495-504.
3) Schwenn, J.D. and Trebst, A. (1976) in The Intact Chloroplast (Barber, J., ed.) Vol. 1, pp. 315-334, Elsevier, Amsterdam.
4) Hoagland, D.R. and Arnon, D.L. (1950) Calif. Agr. Sta. Circular 347.

III.3.**268**

5) Passera, C. and Ghisi, R. (1982) J. Exp. Bot. 33, 432–438.
6) Passera, C. and Albuzio, A. (1978) Agrochimica 19, 480–490.
7) Edwards, G.E., Lilley, RMcC., Craig, S. and Hatch, M.D. (1979) Plant Physiol. 63, 821–827.
8) Gerwich, B.C. and Black, C.C. (1979) Plant Physiol. 64, 590–593.
9) Ghisi, R., Anaclerio, F. and Passera, C. (1984) Giornale Botanico Italiano 118, Suppl. 1, 118–119.
10) Ghisi, R., Anaclerio, F. and Passera, C. (1986) Biol. Plantarum 28 (2) 114–119.

Acknowledgements – Research work supported by CNR, Italy. Special grant IPRA – Sub-project 1. Paper N. 944

SULFITE SENSITIVITY OF LIGHT MODULATION OF CHLOROPLAST ENZYME ACTIVITY IS
CONTROLLED BY A CYTOPLASMIC GENE

GYÖRGYI MUSCHINEK[1], RUTH ALSCHER* & LOUISE E. ANDERSON, DEPARTMENT OF
BIOLOGICAL SCIENCES, UNIVERSITY OF ILLINOIS AT CHICAGO, Box 4348, CHICAGO
IL, 60680, USA, and *BOYCE THOMPSON INSTITUTE, CORNELL UNIVERSITY, ITHACA
NY, 14853, USA.

1. INTRODUCTION
Sulfite inhibits light modulation of enzyme activity in higher plants. This
inhibition is thought to be the result of the action of these compounds on
the thylakoid membrane-bound Light Effect Mediator (LEM) [1]. We now report
that the thylakoid-bound component of the light activation system in the
more SO_2-sensitive *Pisum* cultivar "Nugget" [2] is more susceptible to
sulfite poisoning than is the membrane-bound component in the less
SO_2-sensitive cultivar "Progress No. 9". This sensitivity is inherited from
the maternal parent. Experiments with intact chloroplasts indicate that
sulfite has additional sites of action.

2. MATERIALS AND METHODS
*Treatment of Thylakoid Membranes with Sulfite and Preparation of
Reconstituted Broken Chloroplasts.* Chloroplasts were isolated from green
house grown seedlings, lysed, the thylakoids sulfite-treated and
chloroplasts were reconstituted as in [3]. In the experiments with hybrids,
3 or 4 F_1 generation shoots were ground in cold deionized water in a
porcelain mortar, buffer was added to a final concentration of 50 mM
Hepes-NaOH, 1 mM EDTA, 2 mM $MgCl_2$, 10 mM KCl, pH 7.2, the homogenate was
filtered through 2 layers of Lutrasil (a bonded polypropylene fabric) and
centrifuged (27,000g, 15 min). After washing the pellet was used as the
thylakoid fraction. The same proceedure was used to isolate thylakoids from
the parent plants in this set of experiments.
Light Activation. Light activation was assayed as in [3]. Light activation
units are defined here as change in target enzyme activity units min^{-1} mg
Chl^{-1}.
Enzyme Assays. For the NADP-linked malic dehydrogenase assay see [3]. The
fructose-1,6-bisphosphatase (FBPase) activity assay of Charles and Halliwell
[4] was used, but without EDTA and with FBP 0.5 mM, and NADP 0.5 mM. The pH
was 7.8.
CO_2-dependent Oxygen Evolution. Chloroplast isolation and measurement of
CO_2-dependent oxygen evolution were carried out as described in [5] Sulfite
was added 2 min before illumination. In experiments involving
glyceraldehyde-3-P, chloroplasts were preincubated in the dark for 5 min
prior to illumination, and sulfite or arsenite was added 4 min, and
glyceraldehyde-3-P was added 2 min, before illumination.
Reciprocal Crosses. Flowers were destaminated and pollinated.

----

[1]Permanent address: Ministry of Agriculture and Food, Plant Protection and
 Agrochemistry Centre, 1502 Budapest, POB 127, HUNGARY

Biggens, J. (ed.), Progress in Photosynthesis Research, Vol. III. ISBN 90 247 3452 5
© *1987 Martinus Nijhoff Publishers, Dordrecht. Printed in the Netherlands.*

3. RESULTS AND DISCUSSION

Light activation of FBPase and MDH was inhibited in the reconstituted broken chloroplasts when the thylakoids were treated with sulfite (Table 1). Clearly thylakoids from the more SO_2-sensitive variety "Nugget" are more sensitive to sulfite treatment than are thylakoids from the less SO_2-sensitive variety "Progress". The sulfite sensitivity of the light modulation system is therefore an inherited trait.

We found that the extent of sulfite inhibition was the same regardless of whether electrons entered the system from water or from DCPIP-ascorbate and that sulfite at concentrations which affect light modulation had essentially no effect on NADP photoreduction (data not shown). Therefore the sulfite-sensitive site is not directly involved in photosynthetic electron transport to NADP. Our experiments then indicate that one of the

TABLE 1. Inhibition of light activation of MDH and of FBPase in reconstituted chloroplasts after thylakoids were exposed to 1.0 mM sulfite

Cultivar	Inhibition (%)		
	MDH		FBPase
"Nugget"	49 ± 6	(5)	68 ± 4 (5)
"Progress"	28 ± 5	(7)	29 ± 3 (4)
"Nugget" x "Progress"	53 ± 10	(3)	
"Progress" x "Nugget"	24 ± 6	(4)	

Number of experiments in parentheses.

differences in the effects of sulfite in the two cultivars investigated here is related to the susceptibility of a thylakoid membrane-bound factor (presumably the LEM) to sulfite poisoning. This difference in sensitivity could be due to differences in the primary structure of the LEM or to differences in the composition of the thylakoid membrane and hence to the orientation of the LEM in the membrane and the exposure of the sulfite and arsenite-sensitive sites in the two cultivars.

We made reciprocal crosses between "Nugget" and "Progress" and examined thylakoids from the progeny for sensitivity to sulfite. The results indicate that the sulfite sensitivity of the thylakoid component of the light modulation system is controlled by a maternally inherited cytoplasmic gene (Table 1).

In experiments with intact "Little Marvel" chloroplasts Marques and Anderson [6] found that although sulfite inhibits CO_2-dependent O_2 evolution it does not inhibit light activation of fructosebisphosphatase. We repeated this experiment with chloroplasts from the two cultivars used here and obtained similar results (data not shown). The effect of short term exposure of intact chloroplasts to sulfite at these levels is apparently not related to inhibition of light modulation.

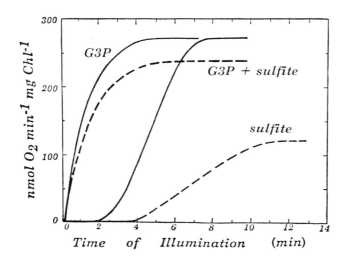

FIGURE 1. Effect of dl-glyceraldehyde-3-P (0.5 mM) on CO_2-dependent oxygen evolution in "Nugget" chloroplasts after sulfite (0.3 mM) poisoning.

Arsenite, like sulfite, inhibits light modulation of enzyme activity in broken chloroplasts [3]. Unlike sulfite, arsenite also inhibits light activation in intact chloroplasts [6]. Sulfite then has effects on photosynthetic CO_2 fixation beyond the effect on light modulation. That one of these effects is enhancement of transport of metabolites out of the chloroplast was suggested by the experiments of Marques and Anderson [6]. Consistent with this result, we found that glyceraldehyde-3-P is effective

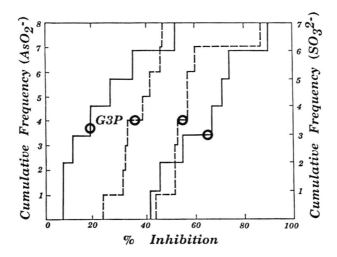

FIGURE 2. Frequency distribution for inhibition of the maximal photosynthetic rate by 0.3 mM sulfite (———) or 0.3 mM arsenite (----) in "Nugget" chloroplasts with and without addition of 0.5 mM dl-glyceraldehyde-3-P.

in relieving sulfite inhibition of steady-state photosynthesis (Fig. 1) and relatively less effective in relieving arsenite inhibition after induction (Fig. 2). Sulfite apparently has more than one site of action in inhibiting photosynthetic CO_2 fixation.

Acknowledgments - This research was supported by US Department of Energy. We thank Larry Sykora and staff at the University of Illinois at Chicago Greenhouse for growing the pea plants and Frank Zimonick, Green Bay Canning Corporation, for some of the "Nugget" seeds used in these experiments. We thank Dennis Nyberg for introducing us to ogives.

REFERENCES
1 Anderson, L.E. (1986) in Advances in Botanical Research (Callow, J.A., ed.), Vol. 12, pp.1-46, Academic Press, New York
2 Alscher, R., Jeske, C.W. and Rogers, A.M. (1983) in Light-Dark Modulation of Plant Enzymes (Scheibe, R., ed.), pp.154-165, University of Bayreuth, Bayreuth
3 Anderson, L.E. and Avron, M. (1976) Plant Physiol. 57, 209-213
4 Charles, S.A. and Halliwell, B. (1960) Biochem. J. 185, 689-693
5 Marques, I.A. and Anderson, L.E. (1985) Plant Physiol. 77, 807-810
6 Marques, I.A. and Anderson, L.E. (1986) Plant Physiol., in press

REGULATION OF CO_2 FIXATION BY THE RIBULOSE 1,5-BISPHOSPHATE CARBOXYLASE IN THE CHLOROPLAST

R. G. JENSEN, D. A. RAYNES, R. E. B. SEFTOR AND S. W. GUSTAFSON,
DEPARTMENTS OF BIOCHEMISTRY AND PLANT SCIENCE, UNIVERSITY OF ARIZONA,
TUCSON, AZ 85721 USA

1. INTRODUCTION
    Photosynthetic CO_2 fixation in the intact leaf is regulated by the activity of the enzyme, ribulose 1,5-bisphosphate carboxylase/oxygenase (Rubisco). The activity depends on the amount of enzyme active sites having formed the activator carbamate in the presence of CO_2 and Mg^{+2}. CO_2 first binds to inactive Rubisco to form the enzyme-activator CO_2 complex (E-C) which is stabilized with Mg^{+2} to give the active enzyme-activator CO_2-Mg^{+2} complex (E-C-M). E-C-M leads to carboxylation or oxygenation by first binding ribulose 1,5-bisphosphate (RuBP) followed by attack by either CO_2 or O_2 (1).

2. ACTIVATION OF RUBISCO
    Light-induced activation changes in Rubisco have been studied from intact leaves by assaying the activity in crude extracts (2,3). In limiting light, the activation of Rubisco in wheat seedlings usually changed proportionally with the net photosynthetic CO_2 exchange rate of the intact plants. Increased rates of photosynthesis, induced by higher light intensities, were accompanied with an increase in Rubisco activation, while RuBP levels remained unchanged and not rate limiting. In darkness and very low light, RuBP levels are low and the photosynthetic rate appeared to be dependent on RuBP availability. The highest levels of RuBP occurred at 25% of light saturation, at levels which were nearly three times the concentration of RuBP binding sites on the enzyme in the stroma (3).

    Experiments have generally confirmed a correlation between the rate of photosynthetic CO_2 exchange and the amount of activated Rubisco in light (2-4). Rubisco does not deactivate to zero in the dark where RuBP availability limits CO_2 fixation. A closer examination of these studies suggest that there are conditions where the correlation does not appear to hold. This may be due to an error inherent in the assay of active Rubisco which, because of its nature, measures both the E-C and E-C-M forms of the enzyme, whereas only the E-C-M form is involved in catalysis (5). Depending on the plant the measurement of Rubisco activity in leaf extracts can pose major problems. Although we have found the techniques to be straight forward with extracts from spinach, wheat and maize, the results can be quite variable with Xanthium, sunflower and sometimes soybean (6).

    Measurements of RuBP pool sizes indicates that there must be some mechanism in chloroplasts of intact leaves which regulates the levels of RuBP in light at concentrations of the RuBP binding sites of Rubisco or higher. The RuBP levels remain fairly constant during steady-state photosynthesis, but rapidly change following a rapid drop of irradiance (7,8). The levels of inorganic phosphate in the chloroplast stroma and

Biggens, J. (ed.), Progress in Photosynthesis Research, Vol. III. ISBN 90 247 3452 5
© *1987 Martinus Nijhoff Publishers, Dordrecht. Printed in the Netherlands.*

cytoplasm most likely play a role in regulating the synthesis of RuBP and, possibly, Rubisco activation. The effect on the carboxylase activity is not likely to be by affecting activation by the Calvin cycle intermediates by competitive binding to Rubisco at the RuBP binding sites. In light, as the RuBP levels in the leaf are usually equal to or greater than the binding site concentration and as the binding affinity of RuBP to the enzyme is considerably greater than the Calvin cycle phosphates, this effectively excludes competitive binding of them versus RuBP on Rubisco (1,3).

3. INHIBITION OF RUBISCO
    Another method used by plants for controlling CO_2 fixation is to regulate the amount of Rubisco enzyme which can be catalytically competent (capable of catalyzing carboxylation or oxygenation). This activity is measured in extracts following total activation with saturating levels of Mg^{+2} and CO_2. Recently it has become very apparent that just because the activator CO_2-carbamate is formed the enzyme may still not exhibit activity. In some plant species, Rubisco total activity was significantly greater following extraction from light-adapted leaves rather than leaves adapted to overnight darkness (9-11). This dark-induced reduction in Rubisco total activity was shown to be the result of an inhibitor of Rubisco catalysis (12,13). The inhibition was not caused by a change in activation but was due to preferential binding of a phosphorylated inhibitor to the activated form of Rubisco at the catalytic site. The inhibitor is formed in the chloroplast and has a maximal concentration by the end of the dark period, just before dawn (11). In light the inhibitor is degraded in the leaf within 10 to 15 min, allowing the already activated Rubisco to again bind RuBP and fix CO_2 or O_2. Both Vu et al (14) and Seemann et al (12) showed a clear distinction between species which exhibit this dark/ light modulation of Rubisco activity and those which lack it. Many of those which show this effect are legumes, but not all. Wheat, maize and spinach are some of the common plants which do not show this light-dependent diurnal effect.

Studies with purified Rubisco have shown that phosphorylated compounds do effect Rubisco activity, most as positive effectors but some as negative effectors (15-19). As pointed out by Miziorko and Lorimer (1), before any physiological significance can be attributed to an effector molecule, two constraints must be filled. a. The concentration of Rubisco active sites within the chloroplast is of the order of 4 mM (20). For any compound to be physiologically effective, it must be present in vivo in amounts approaching or exceeding the concentration of Rubisco active sites. b. The various effectors interact with Rubisco at a single site, the catalytic site for RuBP, and stabilize the activator carbamate, the E-C-M form (15, 16). Thus, an enzyme molecule cannot be simultaneously catalytically competent and activated by an effector as the latter occupies the RuBP binding site. The effector molecule would be expected to be an inhibitor of catalysis. Interestingly, RuBP, itself, is both a potent positive effector and slightly weaker negative effector of the rate of CO_2/Mg^{+2} activation of Rubisco (21). RuBP can bind effectively to both the inactive form of the enzyme, E, and the activated form, E-C-M, and thereby either inhibits binding to or restrains loss of Mg^{+2} and the activator CO_2 from the active site.

The phosphorylated inhibitor described by Seemann et al (12) and Servaites (13) fulfills both constraints of Miziorko and Lorimer (1). It is present in amounts comparable to the concentration of Rubisco active sites and is a very effective inhibitor of catalysis, able to displace bound RuBP. From studies of activity versus fraction of catalytic sites bound by inhibitor (assuming a 1:1 binding of inhibitor to active sites) the inhibitor dissociation constant, K_d, is about 0.1×10^{-6} (12). Although an effective inhibitor of Rubisco activity, the inhibitor does not bind as tight to Rubisco active sites as 2-carboxyarabinitol bisphosphate (CABP), as CABP can effectively replace it. CABP binding with CO_2/Mg^{+2}-activated Rubisco is characterized by a slow process which results in very tight binding (K_d less than 10 pM, 22). Incubation of the enzyme bound with inhibitor but not CABP with alkaline phosphatase could restore Rubisco activity. The rate of activity recovery was significantly reduced when CO_2 was present, which is consistent with the proposal that the inhibitor binds more tightly to activated Rubisco. The mechanism of synthesis and light-enhanced breakdown of this monophosphorylated inhibitor in the leaf are not understood.

4. DIE-OFF OF CO_2 FIXATION WITH ISOLATED CHLOROPLASTS
     Some examples where Rubisco activation does not appear to correlate with photosynthesis are seen with isolated chloroplasts and protoplasts. Most of the experiments on photosynthetic CO_2 fixation with isolated chloroplasts and protoplasts have been conducted with saturating levels of CO_2 (frequently 10 mM bicarbonate). Under this condition the levels of RuBP produced in the chloroplast are usually low and limiting the rate of CO_2 fixation (23). Although it is clear that RuBP levels can limit photosynthesis, studies relating RuBP levels to photosynthetic rates in isolated chloroplasts clearly indicated that factors other than RuBP levels in the chloroplast were involved (24,25). The levels of RuBP became high and remained for 50 min or longer when chloroplasts were illuminated in solutions having CO_2 levels in equilibrium with normal air (26,27). In spite of sufficient RuBP, the rate of fixation was linear for only the first 10 min and then declined to near zero by 40 to 60 min. The enigma was that even if significant levels of CO_2 and RuBP were available to Rubisco in the intact chloroplasts, CO_2 fixation stopped. The decline began with a drop in activity of Rubisco. Next, the light reactions caused gradual leakage of the stroma proteins, including carboxylase, to the suspending medium so that up to 80% of the chloroplast Rubisco leaked out by 90 min. The chloroplast outer envelope appeared to reseal and protect the thylakoids as indicated by the lack of change in apparent intactness as measured with the ferricyanide-dependent Hill reaction. In the dark, under otherwise identical conditions, leakage of stromal proteins does not occur (27). The loss of activity of the chloroplast Rubisco was also not due to the breakdown of the activated E-C complex to the inactive E form. The nature of the enzyme state which the Rubisco was forming is unknown. These studies emphasize that photosynthesis with isolated chloroplasts is the study of a changing, nonsteady-state system. It is small wonder that Robinson et al (28) were unable to correlate CO_2 fixation to Rubisco activation in isolated chloroplasts when RuBP levels were limiting and the plastids were leaking Rubisco.

5. OTHER MECHANISMS OF REGULATION
     Other considerations have recently become important concerning regulation of Rubisco activity. Somerville et al (4) have identified a mutant of Arabidopsis thaliana which appears unable to change Rubisco activation in the light. The mutant requires high levels of CO_2 for growth,

because in CO_2 of normal air, photosynthesis is severely impaired. Rubisco, isolated from both the wild type and the mutant (designated rca), had similar kinetics and respond similarly upon CO_2 and Mg^{+2} activation, indicating that the enzyme was not somehow different in the mutant (29). Soluble polypeptides were isolated from chloroplasts of the wild type and mutant, and compared following 2-D gel electrophoresis with isoelectric focusing in the first dimension and SDS-PAGE in the second. Two poly-peptides of approximately 47 and 50 kD were missing in the mutant stromal extract. Separation of the soluble leaf proteins of the wild type by gel filtration suggested that the two polypeptides combined to form a holo-enzyme. Using a reconstituted chloroplast system, Salvucci et al (29) demonstrated light activation of spinach Rubisco by combining spinach thylakoid membranes with the chloroplast stromal extract containing the unique holoenzyme and with purified Rubisco. This stromal extract was fractioned to give protein between 30 and 300 kD, which excluded low molecular weight compounds and Rubisco. Upon illumination, the reconstituted system increased Rubisco activity over ten fold using chloroplast extracts from spinach or wild-type Arabidopsis, but not from the rca mutant. They postulated that Rubisco activation in vivo is an enzyme mediated process requiring a stromal protein which they have called "Rubisco activase". This protein appears like it could be part of a mechanism by which the light reactions in the chloroplast can participate directly in controlling CO_2 fixation by regulating Rubisco activity. Details how this mechanism operates is currently under study by the Salvucci, Portis and Ogren research groups.

Recently, Sharkey (30) has suggested an interesting approach by expanding on a hypothesis by Walker and Herold (31) that the availability of free inorganic phosphate can limit photosynthesis. The Calvin cycle predicts that carbon going to synthesize starch and sucrose must go by way of triose phosphates at one third the rate of CO_2 fixation. The rate of triose phosphate utilization (TPU) is of key importance as limitation in TPU could reflect short term metabolic interactions, such as between photosynthesis and sink demand. TPU limitation has been studied by feeding six carbon sugars, such as mannose, to leaves so that the resulting hexose monophosphates, which if slowly metabolized such as mannose 6-P, would sequester Pi (32). In this manner it was possible to observe how photosynthesis responded to CO_2 when Pi availability became limiting. Harris et al (33) observed that the photosynthesis of leaf disks after feeding mannose lost sensitivity to O_2 partial pressure seen as the stimulation of CO_2 fixation at 2% versus 21% O_2. Sharkey (34) has suggested that this O_2-insensitive photosynthesis results from physiological limitations to the rate of CO_2 fixation such that it occurs only as fast as triose phosphates are removed from the Calvin cycle by starch and sucrose synthesis. Usually TPU limitation of photosynthesis occurs not by limiting RuBP concentration but by reducing the activity of Rubisco. Obviously when TPU becomes limited, then Rubisco activity and RuBP regeneration rate must be reduced to match the capacity for TPU.

The mechanisms by which TPU capacity and Pi availability regulate the activity of Rubisco remains unclear. Limitations in the availability of Pi could reduce photophosphorylation and the Mg^{+2} concentration in the stroma. Once this is reached, Rubisco could be deactivated by Mg^{+2} loss and CO_2 assimilation would decrease to allow starch and sucrose synthesis to use up the triose phosphate pool and reestablish the Pi pool. What may be contradictory is the supposed large amount of Pi in the vacuole of mesophyll cells making it less likely that a plant grown under sufficient nutrient conditions would be limited because of phosphate nutrition.

Recently we have been studying the effects of CO_2 and O_2 in altering the activity of Rubisco in 7-8 day old wheat and maize seedlings. The seedlings were brought to steady-state photosynthesis in air levels of CO_2 and O_2 with high light, followed by 3 hours of illumination in CO_2-free air having either 1% or no O_2 in N_2. The plants were then restored to normal air levels of CO_2 and O_2 and the recovery to steady-state photosynthesis followed. Leaf samples were taken, extracts made and the amount of activity of Rubisco determined following 10 min incubation with CO_2 and Mg^{+2}. The specific activity of the wheat Rubisco, calculated from measurements at 25°C, ranged from 0.34 to 1.06 μmol CO_2 fixed (mg Rubisco min)$^{-1}$ during and after the 1% O_2 treatment and from 0.16 to 0.66 μmol CO_2 fixed (mg Rubisco min)$^{-1}$ during and after the N_2-only treatment. Maize Rubisco specific activity ranged from 0.15 to 0.36 μmol CO_2 fixed (mg Rubisco min)$^{-1}$ during and after the 1% O_2 treatment and from 0.05 to 0.29 μmol CO_2 fixed (mg Rubisco min)$^{-1}$ during and after the N_2-only treatment. The specific activities of the wheat compared well with previous reported values at 25°C of 1.38 +/- 0.54 μmol (mg Rubisco min)$^{-1}$ (3), but low compared to 3.9 μmol (mg Rubisco min)$^{-1}$ at 30°C based on ^{14}C-CABP determination of the protein (35). Our maize specific activities were especially low compared to 2.12 μmol (mg Rubisco min)$^{-1}$ at 28°C from mature maize using rocket immunoelectrophoresis (36), and to 5.6 μmol (mg Rubisco min)$^{-1}$ at 30°C using ^{14}C-CABP (35). Most likely these differences in specific activity are unique to seedlings which have more enzyme protein per Chl (10.4 mg Rubisco (mg Chl)$^{-1}$ for maize seedlings vs 2.56 mg Rubisco (mg Chl)$^{-1}$ for mature maize (36).

The many fold changes in the activities of the leaf Rubisco during and after the CO_2-free treatment period is a fascinating observation. Apparently these seedlings activate only sufficient Rubisco to sustain the photosynthetic rate with the rest of the Rubisco protein stored to be activated as the leaf matures. This carboxylase is stored in a form which the plant can make catalytically active or inactive upon demand, but which cannot be activated in leaf extracts with high Mg^{+2} and CO_2. The molecular nature of the stored protein and the various mechanisms used by plants for making the enzyme capable of catalyzing carboxylation or oxygenation are unknown. The changes in Rubisco activity in the wheat and maize seedlings are not a result of carbamate formation giving a change in activation. They could be due to preferential binding of a phosphorylated or some other inhibitor to the activated form of Rubisco at the catalytic site. It may be similar in some aspects to the light-modulated inhibitor mechanism of Rubisco catalysis in bean, soybean and tobacco; however, we have been unable to demonstrate that it involves the same stable monophosphate inhibitor. Future experiments will greatly clear up the various mechanisms by which plants regulate Rubisco activity by controlling the amount of enzyme which is catalytically competent.

REFERENCES
1. Miziorko, H.M. and Lorimer, G.H. (1983) Ann. Rev. Biochem. 52, 507-535
2. Maechler, F. and Noesberger, J. (1980) J. Experimental Bot. 31, 1485-1491

3. Perchorowicz, J.T., Raynes, D.A. and Jensen, R.G. (1981) Proc. Natl. Acad. Sci. USA 78, 2985-2989

4. Somerville, C.R., Portis, A.R. and Ogren, W.L. (1982) Plant Physiol. 70, 381-387

5. Seftor, R.E.B., Bahr, J.T. and Jensen, R.G. (1986) Plant Physiol. 80, 599-600

6. Perchorowicz, J.T., Raynes, D.A. and Jensen, R.G. (1982) Plant Physiol. 69, 1165-1168

7. Mott, K.A., Jensen, R.G., O'Leary, J.W. and Berry, J.A. (1984) Plant Physiol. 76, 968-971

8. Perchorowicz, J.T. and Jensen, R.G. (1983) Plant Physiol. 71, 966-960

9. McDermott, D.K., Zeiher, C.A. and Porter, C.A. (1982) Curr. Topics Plant. Biochem. Physiol. 1, 230-238

10. Servaites, J.C., Torisky, R.S. and Chao, S.F. (1984) Plant Sci. Lett. 35, 115-121

11. Vu, C.V., Allen, L.H. and Bowes, G. (1983) Plant Physiol. 73, 729-734

12. Seemann, J.R., Berry, J.A., Freas, S.M. and Krump, M.A. (1985) Proc. Natl. Acad. Sci. 82, 8024-8028

13. Servaites, J.C. (1985) Plant Physiol. 78, 839-843

14. Vu, J.C.V., Allen Jr, L.H. and Bowes, G. (1984) Plant Physiol. 76, 843-845

15. McCurry, S.D., Pierce, J., Tolbert, N.E. and Orme-Johnson, W.H. (1981) J. Biol. Chem. 256, 6623-6628

16. Badger, M.R. and Lorimer, G.H. (1981) Biochemistry 20, 2219-2225

17. Chollet, R. and Anderson, L.L. (1976) Arch. Biochem. Biophys. 176, 344-351

18. Hatch, A.L. and Jensen, R.G. (1980) Arch. Biochem. Biophys. 205, 587-594

19. Jordan, D.B., Chollet, R. and Ogren, W.L. (1983) Biochemistry 22, 3410-3418

20. Jensen, R.G. and Bahr, J.T. (1977) Ann. Rev. Plant Physiol. 28, 379-400

21. Jordan, D.B. and Chollet, R. (1983) J. Biol. Chem. 258, 13752-13758

22. Pierce, J., Tolbert, N.E. and Barker, R. (1980) Biochemistry 19, 934-942

23. Sicher, R.C., Bahr, J.T. and Jensen, R.G. (1979) Plant Physiol. 64, 876-879

24. Sicher, R.C. and Jensen, R.G. (1979) Plant Physiol. 64, 880-883

25. Sicher, R.C., Hatch, A.L., Stumpf, D.K. and Jensen, R.G. (1981) Plant Physiol. 68, 252-255

26. Stumpf, D.K. and Jensen, R.G. (1982) Plant Physiol. 69, 1263-1267

27. Seftor, R.E.B. and Jensen, R.G. (1986) Plant Physiol. 81, 81-85

28. Robinson, S.P., McNeil, P.H. and Walker, D.A. (1979) FEBS Lett. 97, 296-300

29. Salvucci, M.E., Portis Jr, A.R. and Ogren, W.L. (1986) Photosyn. Res. 7, 193-201

30. Sharkey, T.D. (1985) Bot. Rev. 51, 53-105

31. Walker, D.A. and Harold, A. (1977) in Photosynthetic Organelles: Structure and Function (Miyachi, S., Fujita, Y. and Shibata, K., eds.), Special Issue of Plant Physiol., pp.295-310

32. Herold, A. and Lewis, D.H. (1977) New Phytol. 79, 1-40

33. Harris, G.C., Cheesbrough, J.K. and Walker, D.A. (1983) Plant Physiol. 71, 108-111

34. Sharkey, T.D. (1985) Plant Physiol. 78, 71-75

35. Seemann, J.R., Badger, M.D. and Berry, J.A. (1984) Plant Physiol. 74, 791-794
36. Baer, G.R. and Schrader, L.E. (1985) Plant Physiol. 77, 612-616

This research was supported in part by the National Science Foundation under grant DMB-8207687 and the Science and Education Administration under grant 82-CRCR-1-1010.

III.4.**280**

HIGH PERFORMANCE LIQUID CHROMATOGRAPHY OF KEY SUGAR PHOSPHATES INVOLVED
IN PHOTOSYNTHETIC CARBON REDUCTION

ALAN V. SMRCKA and RICHARD G. JENSEN

UNIVERSITY OF ARIZONA, DEPARTMENT OF BIOCHEMISTRY, TUCSON, AZ 85721

Introduction
    We have developed an improved method for the separation and quantita-
tion of phosphorylated sugars by HPLC. The method utilizes a low capacity
strong anion exchange column coupled with a novel detection technique.
Other methods for HPLC analysis of sugar phosphate esters have used anion
exchange chromatography with detection based on radioactivity (1) or inorganic phosphate released after hydrolysis of the resolved sugar phosphates (2). The method described here utilizes a spectrophotometric va-
cancy technique that has been successfully applied to the quantitation of
inorganic ions by HPLC (3). Advantages of this method include short run
time, good resolution and sensitivity. We feel that this technique will
be very useful for the determination of pool sizes of phosphorylated
intermediates in whole leaves and the amount of radioactivity in those
pools after pulse labelling.

Materials and Methods
    The column (PRP-X100. Hamilton Co.) was 4.3 x 100 mm stainless steel
packed with a polymer based strong anion exchange resin. The mobile phase
for the separation of sugar monophosphates was 1 mM phthalic acid and
10 mM boric acid adjusted to pH 8.5 with LiOH. For the sugar bisphosphates the mobile phase was 1.5 mM trimesic acid and 10 mM boric acid
adjusted to pH 8.7 with LiOH. All runs were isocratic with a flow rate
of 1 ml/min. Samples were injected in a 50 µl sample loop using a
Rheodyne 7125 sample injector. Detection was accomplished using an LKB
2138 Uvicord absorbance detector set at 280 nm. All samples were solutions of standard sugar phosphates purchased from Sigma Chemical Co.

Results and Discussion
    The detection of the sugar phosphates as they elute from the column
is based on a decrease in the absorbance baseline of the eluent. All the
HPLC mobile phases described in Materials and Methods contain UV absorbing
anions which serve two roles. They compete with the anionic sugar phosphates for cationic sites on the column causing the sample ions to elute.
They also provide a baseline absorbance between 0.2 and 0.8 absorbance
units depending on the concentration and detector wavelength chosen. When
a negatively charged sugar phosphate elutes from the column there is a
decrease in the level of eluent so that an electrically neutral situation
is maintained. The peaks that are observed are actually negative absorbance peaks.
    A typical chromatogram of sugar monophosphates is shown in figure 1.
    An initial baseline disturbance occurs over the first 4 min. then
glucose-6-phosphate elutes followed by fructose-6-phosphate, ribose-5-
phosphate and 3-phosphoglyceric acid. Dihydroxyacetone phosphate and

Biggens, J. (ed.), Progress in Photosynthesis Research, Vol. III. ISBN 90 247 3452 5
© *1987 Martinus Nijhoff Publishers, Dordrecht. Printed in the Netherlands.*

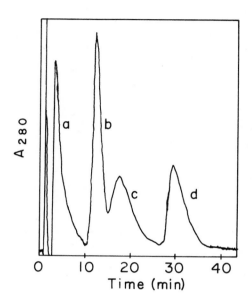

Figure 1.   Separation of sugar monophosphates, 50 µl of 1 mM each.
a) glucose-6-phosphate, b) fructose-6-phosphate, c) ribose-5-phosphate,
d) 3-phosphoglyceric acid.   Absorbance at 280 nm is in relative units.

inorganic phosphate coelute with glucose-6-phosphate. Shown in figure 2
is the resolution of the sugar bisphosphates. The linearity of the
detection is demonstrated in figure 3.   The system is linear between 10
and 50 nmol of fructose-6-phosphate.
     As is evident from the preceding discussion, it is important that the
monophosphates and bisphosphates be injected into the system under different chromatographic conditions. We have found that the monophosphates
can be separated as a class from the diphosphates by prefractionating them
on a short low capacity anion exchange resin using a step gradient of .05
to .15 N HCl to elute the mono and diphosphate respectively. This sample
preparation step also serves to concentrate the sample and to resolve the
sugar phosphates from other competing anions such as organic acids, NO_3^-
and Cl^-.

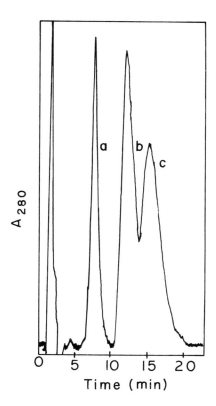

Figure 2. Separation of sugar bisphosphates, 50 μl of 1 mM each.
a) ribulose-1,5-bisphosphate, b) fructose-1, 6-bisphosphate, c) sedohep-
tulose-1,7-bisphosphate. Absorbance at 280 nm is in relative units.

The methodology described here should be useful for the detection of
any phosphorylated sugar that can be concentrated from leaf tissue down
to a level of 10 nmol per 50 μl of sample.

References

1 Giersch, C. (1979) J. Chromatog. 172, 153-161
2 Geiger, P.J., Shunwoo, A. and Bessman, S.P. (1980) in Methods in
Carbohydrate Chemistry (Whistler, R.L. and BeMiller, J.N., ed.), Vol. 8,
pp.21-32, Academic press, New York
3 Small, H. and Miller, T.E. (1982) Anal. Chem. 54, 462-469

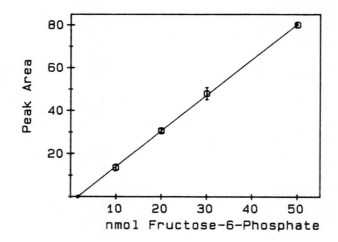

Figure 3. Standard curve for known amounts of fructose-6-phosphate. Solutions were injected in a volume of 50 μl. Results are expressed as mean +SD, n=3. Peak areas were calculated using a Varian 4270 integrator and areas are expressed in relative arbitrary units.

THE TIME COURSE FOR DEACTIVATION AND REACTIVATION OF RIBULOSE-1,5-BISPHOSPHATE CARBOXYLASE FOLLOWING CHANGES IN CO_2 AND O_2.

Rowan F. Sage, Jeffrey R. Seemann and Thomas D. Sharkey, Biological Sciences Center, Desert Research Inst. P.O. Box 60220 Reno, NV 89506, USA; College of Agriculture (JRS) and Biology Department (TDS) University of Nevada, Reno, NV 89507, USA

1. Introduction

In this paper we describe studies which address the role that RuBP carboxylase/oxygenase (rubisco) plays in regulating photosynthesis. Conflicting reports claim that rubisco activation limits photosynthesis even at low light or high CO_2 (1,2), and alternatively, that the RuBP regeneration capacity limits photosynthesis, even at low CO_2 levels (3). Following a decrease in light intensity, rubisco activation state declines in a parallel manner with the decline in photosynthesis (1,3). Recently, Sharkey et al. (4), observed that rubisco deactivates following a switch from 21% to 2% O_2 in plants which are photosynthetically O_2 insensitive. The degree of deactivation corresponded to that required to explain the lack of O_2 sensitivity, implying that rubisco deactivation is responsible for the O_2 insensitivity. RuBP levels, however, become rate limiting immediately after the decrease in O_2 pressure (5), indicating that rubisco activation does not limit photosynthesis, but is involved in the regulation of photosynthesis.

We have used step changes in CO_2, or both CO_2 and O_2 partial pressures to study the time course over which rubisco deactivates or reactivates and the corresponding changes in photosynthesis and RuBP levels. From this data we have identified conditions where rubisco activation limits photosynthesis and conditions where rubisco activation is regulated but does not limit photosynthesis.

2. Materials and Methods

2.1. Growth conditions and gas exchange measurements. Bean (Phaseolus vulgaris) plants were grown in a greenhouse and measured 4 weeks following germination. Photosynthesis was measured at 21°C at 1000 μmol $m^{-2} s^{-1}$ using an open type gas exchange system equipped with a modified leaf chamber which was incorporated into a high speed freeze clamp apparatus. This apparatus froze leaves within 0.25 s.

2.2. Deactivation experiments. The response of photosynthesis, rubisco activation and RuBP levels were followed from 3 to 25 min following a switch from an ambient CO_2 (C_a) of 400 μbars, at 180 mbars O_2 to a C_a of 1500 μbars at 180 mbars O_2. The initial condition was chosen to be that point where the sensitivity of photosynthesis to CO_2 dramatically declines; the high CO_2 point occurs in a region where photosynthesis is insensitive to changes in CO_2 and O_2 partial pressure. To see if a concomitant change in O_2 pressure would have an additional affect, the experiment was repeated, except that the switch was made from the same

Biggens, J. (ed.), Progress in Photosynthesis Research, Vol. III. ISBN 90 247 3452 5
© *1987 Martinus Nijhoff Publishers, Dordrecht. Printed in the Netherlands.*

initial condition to 1500 μbars CO_2 at 35 mbar O_2, and the response of photosynthesis, rubisco activation and RuBP levels were determined over the next 1 to 25 minutes. The gas exchange apparatus required 1 min to restabilize following a change in conditions.

2.3. Reactivation experiments. Leaves were initially equilibrated at 400 ubars C_a, 180 mbar O_2, and were then exposed to 1500 μbars C_a, 35 mbar O_2 for 25 minutes. The gas phase was then changed back to 400 μbars C_a, 180 mbars O_2. Photosynthesis, rubisco activation, and RuBP levels were followed until rubisco activation states had fully recovered.

2.4. Sample collection and analysis. Samples were collected at specified times (see figures) by freezing leaves with the freeze clamp apparatus. Rubisco activation states were measured by rapidly extracting and assaying rubisco and expressing this rate relative to the CO_2-Mg^{2+} activated rate. Rubisco catalytic sites were determined as the amount of CABP binding sites. RuBP was assayed in acid extracts by determining $^{14}CO_2$ incorporation.

3. Results

3.1. The time course for the change in photosynthesis and rubisco activation following a switch to high CO_2. Photosynthesis rapidly re-equilibrated at a slightly higher rate following a sudden increase in CO_2 pressure (Fig. 1). Photosynthesis in most leaves stabilized within 3 minutes. Those leaves requiring longer periods to re-equilibrate (3-6 minutes) exhibited pronounced oscillations in the rate of CO_2 assimilation. Rubisco activation state fell 50% over 25 minutes. The half time for deactivation was estimated to be 10-15 minutes.

Figure 1. The time course for the change in photosynthesis and rubisco activation following a switch from the initial low CO_2, high O_2 condition to high CO_2, high O_2. The arrow indicates the time of the switch. Error bars = S.E.M.

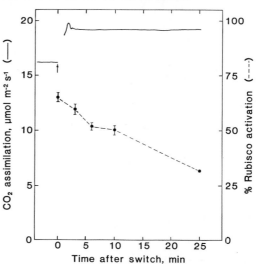

3.2. The time course for the change in photosynthesis and rubisco activation following a switch to high CO_2 and low O_2. As observed following the switch to high CO_2 alone, switching to high CO_2 and low O_2 produced a rapid re-equilibration of photosynthesis and a gradual 50%

decline in rubisco activation (Fig. 2A). Again, rubisco deactivation was not accompanied by a change in photosynthesis and the half time for deactivation was 10-15 minutes. O_2 did not have an effect independent of that caused by CO_2 alone. The differences in rubisco activation state between this and the above experiment were related to the fact that the two experiments were conducted on different sets of bean plants and not to experimental conditions.

3.3 The time course for the change in photosynthesis and rubisco activation following a return to low CO_2 and high O_2. Rubisco reactivated within 10 minutes following a return to the initial low CO_2, high O_2 condition from the high CO_2, low O_2 condition (Fig. 2B). Photosynthesis was severely depressed by the switch, but recovered at a rate which paralleled the rate of reactivation during the first 10 minutes after the switch. This indicates that rubisco limits photosynthesis during this initial 10 minute period. About ten minutes after the switch, the rate at which photosynthesis increased declined dramatically, and further increases in photosynthesis rate appear to result from an increase in stomatal aperture since intercellular CO_2 (C_i) values increased.

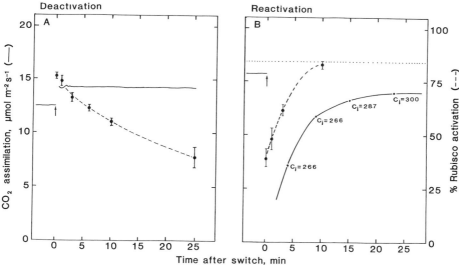

Figure 2. The time course for the change in photosynthesis and rubisco activation following A) a switch from low CO_2, high O_2 to high CO_2, low O_2, and B) a switch from high CO_2, low O_2 to low CO_2, high O_2. The arrow indicates the point of the switch. RuBP is expressed relative to CABP binding sites. Error bars = S.E.M.

3.4. RuBP pool size. RuBP data from both deactivation experiments (3.1, 3.2) were combined. RuBP pools fell to rate limiting levels (<1 mole RuBP per mole CABP binding sites) within 60 s after the switch (Fig. 3). RuBP levels recovered to the initial value during the next 24 minutes while rubisco deactivated. This pattern is similar to the change in RuBP observed by Mott et al. (6) in response to light reduction, and Sharkey et al. (5) following O_2 removal. During reactivation, no change in RuBP pool size was measured; it remained saturating during the entire reactivation period.

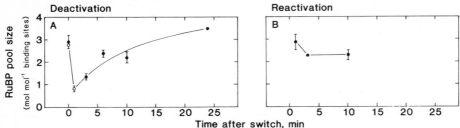

Figure 3. The change in RuBP levels following A) a switch from the initial low CO_2, high O_2 condition to high CO_2, high O_2 (●) or high CO_2 and low O_2 (○), and B) a return to low CO_2, high O_2 from the high CO_2, low O_2 condition. Error bars = S.E.M.

4. Discussion

In this study, we show that increasing CO_2, in addition to decreasing light (1,3) and O_2 pressure (4) induces rubisco deactivation. In all 3 cases, RuBP levels initially decline to rate limiting levels and completely recover as a result of rubisco reactivation. Rubisco deactivation under these conditions does not limit photosynthesis, but may participate in the regulation of photosynthesis. In contrast, during reactivation, RuBP levels remain saturating, and photosynthesis increases at the same rate as rubisco reactivation. We believe that rubisco limits photosynthesis at this point. The fact that RuBP levels do not climb during reactivation indicates that the limitation resulting from rubisco deactivation leads to a regulatory control on RuBP regeneration.

The deactivation of rubisco under these conditions indicates that rubisco may be involved in the regulation of metabolite pool sizes and inorganic phosphate levels. During the period when RuBP levels fall and recover, PGA and UDPG levels increased dramatically and then fell back toward the control levels as rubisco deactivated (Sage and Sharkey, unpublished). This regulation of metabolite pools could control the rate of starch and sucrose synthesis and ensure that the utilization of phosphorylated intermediates is balanced with their production.

References

1. Perchorowicz, J.T., Raynes, D.A. and Jensen, R.G. (1981) Proc. Natl. Acad. Sci. USA 78, 2985-2989
2. Dietz, K.J. and Heber, U. (1984) Biochim. Biophys. Acta 767, 432-443
3. Taylor, S.E and Terry, N. (1984) Plant Physiol. 75, 82-86
4. Sharkey, T.D., Seemann, J.R. and Berry, J.A. (1986) Plant Physiol. 81,788-791
5. Sharkey, T.D., Stitt, M., Heineke, D., Gerhardt, R., Raschke, K. and Heldt, H. (1986) Plant Physiol. in press
6. Mott, K.A., Jensen, R.G., O'leary, J.W. and Berry, J.A. (1984) Plant Physiol. 76,968-971

Supported by DOE grant DE-ECO8-84ER13234

A PARTICULATE CYTOPLASMIC FRACTION ENHANCING PHOTOSYNTHESIS OF ISOLATED CHLOROPLASTS BY ACTIVATION OF RUBPCase.
V. Kagan-Zur and S.H. Lips. Ben Gurion University. Beer Sheva. Israel.

Introduction
        The capacity of a microbody-rich fraction (MBL) to stimulate photosynthesis of isolated chloroplasts was previously reported (1). Leaf extracts from plants kept in darkness for 24 h prior to homogenization yielded, after density gradient centrifugation, two peaks of RuBPCase: one in the chlorophyll band and the second with catalase (2). Consequently we studied interactions between isolated chloroplasts and the (MBL) and the effect of the latter on the photosynthetic rates of isolated chloroplasts. MBL enhanced photosynthesis of chloroplasts isolated from pea seedlings, kept in darkness for 24 h before homogenization, by 300% (1,2). Chloroplasts isolated from pea seedlings kept in the light could be enhanced by MBL to a lesser extent (50%). Marked differences were observed in the staining properties of the outer membrane of chloroplasts in leaf sections from plants kept in the light or in darkness (3). These differences of chloroplast envelopes may reflect changes in membrane composition and permeability due to light conditions. Gigot et al (4), observed portions of the stroma envolved by a membrane leaving the chloroplasts during the process of protoplast isolation. Vaughn et al (5) observed association of vesicles containing nitrate reductase, with chloroplasts, suggesting a possible fusion with the chloroplast envelopes. Pederson et al (6) suggested that Calvin cycle enzymes may be attached to a membrane. Raghavendra et al (7) pointed out a specific attachment of RuBPCase to chloroplast membranes. We considered the possibility that particles recovered in the MBL fraction may have detached themselves from the chloroplasts during darkness and reattached to them in the light. Results of experiments to test this hypothesis are described here.

Materials and methods
        Pea (Pisum sativum L., var. Dan) seedlings were irrigated with half strength Hoagland solutions, and grown under 16 h light (230 μeinstein m^{-2} s^{-1}), and 8 h darkness at 25°; seedlings were used 10 days after sowing. L-plants were exposed to 550 μeinstein m^{-2} s^{-1} for 30 min (preillumination treatment) immediately prior to homogenization. D-plants were kept in darkness 24 h before homogenization. Chloroplast were isolated according to Bucke et al (8). The MBL fraction was obtained as described elsewhere (1). Chlorophyll was assayed according to Arnon (9). Chloroplast intactness was determined according to Heber and Santorious (10). CO_2 fixation was estimated as acid stable ^{14}C compounds (1). Photosynthetic products were fractionated on Dowex 50x8-400 cation and Dowex 1x8-400 anion exchange resins. Sample loading and elution procedures followed Atkins and Canvin (11). RuBPCase activity was assayed according to Slack and Hatch (12).

Results
        The MBL fraction contained mitochondria, peroxisomes, β-oxidation particles, disrupted chloroplast membranes and presumably other vesicles (5). Activities related to mitochondria, β-oxidation and glycolate or glyoxylate metabolism did not have any differential effect either on chloroplast photosynthesis or on the enhancement capacity of the MBL (13).

Biggens, J. (ed.), Progress in Photosynthesis Research, Vol. III. ISBN 90 247 3452 5
© *1987 Martinus Nijhoff Publishers, Dordrecht. Printed in the Netherlands.*

Addition of MBL to chloroplasts resulted in a marked increase of labelled amino acids from $^{14}CO_2$ (Table 1) although no difference in response was observed using L- or D-leaves. A lower level of sugar monophosphates and organic acids was also observed as well as an enhanced labelling of sugar diphosphates in D-chloroplasts than in L-chloroplasts. Addition of MBL to chloroplasts during photosynthesis elliminated these differences in label distribution. MBL may have had the capacity to return to the D-chloroplast a missing factor, allowing them to behave as L-chloroplasts.

Table 1. ^{14}C-distribution among products of $^{14}CO_2$-photosynthesis. Figures indicate % of total label recovered from columns.

Fractions	Chloroplasts		Chloroplasts+MBL	
	Light pre-treatment	Dark pre-treatment	Light pre-treatment	Dark pre-treatment
Cations (amino acids)	13	10	23	25
Weak acids (Organic acids, sugar phosphates)	30	20	29.5	29
Neutral (sugars)	11	9	9.5	10
Strong acids (sugar diphosphates, PGA, RUBP)	46	61	38	36

Photosynthesis took place in 10 mM CO_2 (3.5×10^5 Bq . ml^{-1}) and 1 mM ribose-5-phosphate for 15 min. The reaction was stopped by addition of ethanol (8 ml-tube^{-1}). The resulting solution was then dried at low temperature. The dry residue was dissolved in distilled water and passed first through the cation exchange resin, then through the anion exchange resin. Label not retained on either column was the neutral fraction.

Both subcellular fractions, chloroplasts and MBL, were combined for 15 min in the absence of CO_2 or other substrates; this was followed by a second separation of the combined chloroplasts and MBL and the photosynthetic rates of the chloroplasts were determined (Table 2). Another part of the isolated chloroplasts underwent a similar procedure without mixing with MBL, as a control. Preincubation in light or in darkness lasted 15 min. Illumination under non-photosynthetic conditions accelerates chloroplast deterioration. Similar observations have been described (14, 15). The joint preincubation of MBL with the chloroplasts enhanced photosynthetic activity of the latter. D-chloroplasts were enhanced to a larger extent by 15 min preincubation with MBL than L-chloroplasts. However, continuous presence of MBL throughout the photosynthesis period seemed essential to obtain the high levels of enhancement frequently observed (about 300%)(1).

The accumulation of sugar diphosphates by D-chloroplasts (Table 1) could be the result of a block in the flow of assimilated $^{14}CO_2$. We tried, therefore, to assess whether low RuBPCase activity was the cause of lower CO_2 fixation rates of the D-chloroplasts and whether addition of this enzyme as part of the MBL fraction could account for the photosynthetic enhancement observed. Jensen and Bahr (16) reported that RuBPCase was rapidly inactivated when isolated from chloroplasts. Later (17) they described assay conditions for the activation of the enzyme in vitro which were mostly followed in this work. Heldt et al (18) reported a 2-3 fold light activation of RuBPCase of isolated chloroplasts.

Table 2. Effects of preincubation period on the photosynthesis capacity of isolated chloroplasts (incubated by themselves or in the presence of MBL and reisolated after 15 min preincubation). Results are presented as % of CO_2 fixed by isolated chloroplasts kept in darkness during preincubation.

Fractions	Preincubation (15 min) conditions	Treatment of seedlings before extraction	
		24 h darkness	30 min light
Chloroplasts	Light	85	80
	Darkness	100*	100*
Chloroplasts reseparated after incubation with MBL	Light	106	77
	Darkness	140	116
* Photosynthesis rates: μmoles CO_2 mg chl^{-1} h^{-1}		31	73

Chloroplasts and MBL were obtained and treated as described for table 2. Each subcellular fraction was assayed for RuBPCase activity under two different conditions: high photon flux (950 μeinstein m^{-2} s^{-1}) and darkness (Table 3).

Table 3 Activation of RuBPcase in MBL. Expressed as % of activity of MBL incubated for 15 min in darkness.

Fraction	Preincubation (15 min) conditions	Conditions of seedlings before extract.			
		24 h darkness		30 min light	
		Reaction light	Reaction darkness	Reaction light	Reaction darkness
MBL	Light	230	92	200	89
	Darkness	261	100*	261	100**
MBL reseparated after incubation with CHL	Light	1515	677	333	155
	Darkness	1230	438	300	150

RuBPcase activity: (*) 64 μmoles CO_2 fixed mg chl^{-1} h^{-1} or 0.7 μmoles CO_2 fixed mg protein^{-1} h^{-1}. (**) 20 μmoles CO_2 fixed mg chl^{-1} h^{-1} or 0.24 μmoles CO_2 fixed mg protein^{-1} h^{-1}.

MBL preincubated with chloroplasts and subsequently reisolated, exhibited higher RuBPCase activity than MBL which were not preincubated with chloroplasts. D-MBL after preincubation with chloroplasts, increased their own RuBPCase activity by 4-6 fold. RuBPCase activity in all MBL fractions could be doubled by illumination during the assay. Overall activation of RuBPCase in D-MBL was more than 10 fold, while RuBPCase in L-MBL was enhanced only by a factor of 3. Light activation of chloroplast RuBPCase in the absence of MBL was negligeable.

Discussion

Although the MBL fraction contained mitochondria, broken chloroplasts and particles with β-oxidation activity, none of these subcellular components seemed to be directly related to the capacity of MBL to enhance photosynthesis of isolated chloroplasts (1,2,13). Peroxisomes which were also present in the MBL may have taken part in the enhancement phenomenon by stimulating synthesis of amino acids (Table 1). Diversion of ^{14}C flow to amino acids may constitute a significant component of the limited L-MBL stimulated photosynthesis by L-chloroplasts, but was presumably unrelated to the large stimulation of photosynthesis of D-chloroplasts by D-MBL. This conclussion is based on two observations: (a) the two chloroplast preparations (L and D) showed differences in label distribution in the anion but not in the cations fractions while exhibiting different rates of CO_2 fixation; (b) L-MBL could not enhance photosynthesis of D-chloroplasts to the same extent as D-MBL (1), while both MBL preparations had a similar effect on the extent of amino acid synthesis.

The main reason for photosynthesis enhancement by MBL seemed to be the activation of RuBPCase. This enzyme was activated more than 10 fold in D-MBl following their preincubation with chloroplasts for 15 min at 25°, followed by reisolation of the subcellular components. Activation of MBL-RuBPCase, therefore, required the presence of chloroplasts as well as light (Tables 2 and 3). This activation cannot be the result of adsorption of the enzyme from newly disrupted chloroplasts onto MBL particles for two reasons: (a) resuspended chloroplasts exhibited a higher capacity for CO_2 fixation after resuspension, which could hardly indicate loss of enzyme; (b) Chloroplasts from preilluminated leaves showed a higher RuBPCase activity, but less enhancement of this activity was observed in MBL jointly incubated with these chloroplasts.

References
1 Kagan-Zur V. and S.H. Lips, 1983. Isr. J. of Bot. 32, 203-212
2 Kagan-Zur V. and S.H. Lips, 1975. Eur. J. Biochem. 59, 17-23
3 Kagan-Zur V., M. Friedlander and S.H. Lips, 1980. Planta 149, 427-432.
4 Gigot C., M. Kopp, C. Schmitt and R.C. Milne, 1975. Protoplasma
   84, 31-41
5 Vaughn C., S.O. Duke and E.A. Funkhouser, 1983. Plant Physiol.(S)
   72,130
6 Pederson T.A., M. Kirk and J.A. Bassham, 1966. Biochim. Biophys. Acta
   112, 189-203
7 Raghavendra A.S., N. Carrillo and R.H. Vallejos, 1980. In Proc. 5th.
   Int. Photosynth. Cong. Ed.: G. Akoyunoglou. Balaban Int. Sci.
   Services, Philadelphia, Vol IV, 31-38
8 Bucke C., D.A. Walker and C.W. Baldry, 1966. Biochem. J. 101, 636-641
9 Arnon D.I., 1949. Plant Physiol. 24, 1-15
10 Heber U. and K.A. Santorius, 1970. Z. Naturforschung 296, 718-728
11 Atkins C.A. and D.T. Canvin, 1971. Can. J. Bot. 49, 1225-1234
12 Slack C.R. and M.D. Hatch, 1967. Biochem. J. 103, 660-665
13 Kagan-Zur V.,1979. Ph. D. Thesis. Ben Gurion University. Israel
14 Krause G.H., M. Kirk, U. Heber and C.B. Osmond, 1978. Planta
   142, 229-233
15 Whitehouse D.G., L.J. Ludwig and D.A. Walker, 1971. J. Exp. Bot. 22,
   772-791
16 Jensen R.G. and J.T. Bahr, 1974. In Proc. 3rd. Int. Cong. Photosynth.
   Ed.: M. Avron. Elsevier Sci. Publ. Amsterdam. Vol II, 1411-1420
17 Jensen R.G. and J.T. Bahr, 1977. Ann. Rev. Plant Physiol. 28, 379-400
18 Heldt H.W., W. Laing, G.H. Lorimer, M. Stitt and W. Wirtz, 1980. In
   Proc. 5th. Int. Photosynth. Cong. Ed.: G. Akoyunoglou. Balaban Int.
   Sci. Services, Philadelphia, Vol IV, 213-216

ASSIMILATORY FORCE IN RELATION TO PHOTOSYNTHETIC FLUXES

U. Heber, S. Neimanis, K.J. Dietz and J. Viil*)

Institute of Botany and Pharmaceutical Biology of the University of Würzburg, Germany, and Institute of Experimental Biology of the Estonian Academy of Sciences, Harku, Estonian SSR*)

1. INTRODUCTION

In 1958, Arnon et al. introduced the term assimilatory power for the ATP and NADPH generated in the light by the chloroplast electron transport chain, as both compounds cooperate to reduce the CO_2 fixed during carboxylation of ribulose bisphosphate (RuBP) to the sugar level (1). Relevant reactions may be summarized as

$$RuBP^{4-} + CO_2 + H_2O \rightarrow 2\ PGA^{3-} + 2\ H^+ \qquad [1]$$

and

$$PGA^{3-} + ATP^{4-} + NADPH + H^+ \rightarrow DHAP^{2-} + ADP^{3-} + NADP^+ + P_i^{2-} \qquad [2]$$

where PGA and DHAP are 3-phosphoglycerate and dihydroxyacetone phosphate respectively. The mass action ratio R of reaction [2]

$$\frac{(PGA^{3-})}{(DHAP^{2-})} \frac{(ATP^{4-})}{(ADP^{3-})(P_i^{2-})} \frac{(NADPH)}{(NADP^+)} (H^+) = R \qquad [3]$$

clearly shows that the product of phosphorylation potential $(ATP/ADP)(P_i)$ and redox ratio $(NADPH)(NADP)$ is the driving force of the reduction of PGA to the sugar level. In previous publications (2,3), we have adopted Arnon's term assimilatory power for this product which we now prefer to call assimilatory force

$$F_A = \frac{(ATP)}{(ADP)(P_i)} \frac{(NADPH)}{(NADP)} \quad **) \qquad [4]$$

for two reasons: (1) Laisk has used the term assimilatory power in a rather different context (4). He described it as the total pool of RUBP and other compounds responsible for the post-illumination CO_2 uptake by leaves. (2) Hill has recently drawn attention to the difference between force and power (5). The latter is energy per unit time. As in F_A time is not involved, we clearly deal with a force, not with power.

We were interested to determine the maximum force which the photosynthetic apparatus can generate on illumination and compare it with the

**) Actually, a thermodynamic driving force is a gradient in free energy, i.e. the logarithm of F_A. However, for the arguments developed in this work, F_A is more instructive than its logarithm into which it can easily be transformed.

Biggens, J. (ed.), Progress in Photosynthesis Research, Vol. III. ISBN 90 247 3452 5
© *1987 Martinus Nijhoff Publishers, Dordrecht. Printed in the Netherlands.*

force that is sufficient to drive photosynthesis in a leaf. Clearly, the two are different. The maximum force can only be developed when it is not used to do work. When photosynthesis proceeds, force is produced by light, and it decreases as work is done. The relationship between the magnitude of force as defined above and power production, that is work per unit time or rate of photosynthesis, was of interest.

2. PROCEDURE
2.1. Materials and methods
Spinach was grown in a green house (10 hours light and 15 hours dark). Chloroplasts were isolated aqueously by a standard procedure (6). After osmotic rupture, photophosphorylation and NADP reduction were measured as described before (3,7). Intact chloroplasts were illuminated with added PGA, and PGA and DHAP were determined in the chloroplast suspension after oxygen evolution had ceased (3). Leaves were gassed with different concentrations of CO_2 in air and illuminated as described in the legend to Fig. 1. During illumination, they were frozen in liquid nitrogen and then freeze-dried. Chloroplasts were isolated from the dry material in mixtures of petrolether and carbon tetrachloride and photosynthetic intermediates were measured enzymatically in the chloroplast fraction (2,3).

3. RESULTS AND DISCUSSION
3.1. Phosphorylation potentials, redox ratios of NADPH to NADP and assimilatory force FA generated by broken chloroplasts
Maximum phosphorylation potentials produced by broken chloroplasts under high intensity illumination in the presence of a cofactor of cyclic photophosphorylation or of an electron acceptor may vary with the permeability properties of the thylakoids liberated during chloroplast rupture. Kraayenhof (8) has reported maximum phosphorylation potentials close to 30 000 (M^{-1}) and Giersch et al. (7) have obtained values as high as 80 000 (M^{-1}). We have recently even observed values above 100 000 (M^{-1}). At a phosphate concentration of 2 mM, this corresponds to ATP/ADP ratios above 200.

When coupled broken chloroplasts are illuminated with NADP in the presence of ferredoxin, oxygen is evolved until most of the added NADP is reduced. Enzymatic reduction of the residual NADP revealed maximum NADPH/NADP ratios to be higher than 25 (3). Multiplication of NADPH/NADP ratios with maximum phosphorylation potentials then yields maximum F_A values above $2.5 \cdot 10^6$ (M^{-1}). It should be noted that such high force can be generated by thylakoids only, if cyclic electron transport and /or electron transport to oxygen remain effective after added NADP has been reduced, since maintenance of a high phosphorylation potential requires maintenance of electron transport.

3.2. Assimilatory force generated by intact chloroplasts
Maximum phosphorylation potentials in intact chloroplasts calculated from adenylate and phosphate data were much lower than those measured in suspensions of broken chloroplasts. Values were as low as 200 to 500 (M^{-1}) (7). ATP/ADP ratios were rarely higher than 4. Redox ratios of NADPH/NADP varied considerably and were usually 2 to 3, very rarely 9 (9,10). F_A calculated from these values was at the most 4500 (M^{-1}), i.e. less than 0.2 % of the maximum values calculated for broken chloroplasts. However, the significance of direct measurements of adenylates, phosphate and pyridine nucleotides in intact chloroplasts may be doubted. The protein concentration in the chloroplast stroma is

very high. A main protein of the stroma is RuBP carboxylase. The concentration of RuBP binding sites of this enzyme may be as high as 4 mM. When the enzyme is not saturated in respect to RuBP, it can bind other phosphate esters or even phosphate (11). Bound components of F_A, however, do not contribute to F_A. It should be noted that this complication does not apply to F_A measurements in suspensions of broken chloroplasts, where the components of F_A are added to the medium in excess so that binding becomes insignificant. It was therefore of interest to determine F_A in intact chloroplasts by a method which circumvents possible drawbacks of metabolite measurements in the chloroplast stroma. Advantage was taken of the fact that at or close to equilibrium of the reactions catalyzed by phosphoglycerate kinase, glyceraldehydephosphate dehydrogenase and triosephosphate isomerase, ratios of DHAP to PGA permit calculation of F_A, if pH is known

$$\frac{(DHAP) \ K}{(PGA)(H^+)} = \frac{(ATP)}{(ADP)(P_i)} \ \frac{(NADPH)}{(NADP)} = F_A \qquad [5]$$

The equilibrium constant K is $9.8 \cdot 10^{-6}$. Intact chloroplasts were illuminated with added PGA in the absence of CO_2 until oxygen evolution was replaced by oxygen uptake which indicated oxygenation of RuBP. PGA and DHAP were then measured in the chloroplast suspension as a function of time. From ratios of DHAP to PGA outside the chloroplasts F_A inside the chloroplasts was calculated on the basis of equation [5]. This avoids artifacts arising from binding of metabolites in the chloroplast stroma. The method is based on the following considerations: In the light, external PGA is exchanged as the divalent anion by the phosphate translocator of the chloroplast envelope against the $DHAP^{2-}$ produced during reduction of PGA inside the chloroplasts. When the PGA reducing reaction reaches equilibrium (see equation [3]) in the chloroplast stroma, flux of PGA and counterflux of DHAP must cease. At flux equilibrium, the external and internal $DHAP^{2-}/PGA^{2-}$ ratios must be identical. From

$$(PGA^{2-}) = K' \ (PGA^{3-}) \ (H^+) \qquad [6]$$

it follows that $(DHAP^{2-})/((PGA^{3-})(H^+))$ are also identical inside and outside the chloroplasts. These relations justify calculation of F_A inside the chloroplasts from metabolite and H^+ concentrations outside the chloroplasts provided reaction and flux equilibrium is established. Unfortunately, this condition is not easily met. Oxygenation of RuBP results in the production of PGA from previously formed DHAP. Reduction of this PGA inside the chloroplasts consumes ATP and NADPH thereby lowering maximum F_A values. Extrapolation of F_A values calculated from measured DHAP/PGA ratios yielded maximum F_A at air levels of oxygen of about 4000 (M^{-1}) (3). This value is in rather close agreeement with the maximum F_A calculated from direct measurements of phosphorylation potentials and NADPH/NADP ratios. However, as stated in the introduction, maximum F_A values can be expected only under conditions, when force is not used to do work. In the presence of DHAP, RuBP is formed and oxygenation of RuBP produces glycolate and PGA (12). Reduction of the latter reduces F_A. Oxygenation of RuBP can be decreased efficiently by a reduction in O_2 levels. However, anaerobiosis must be avoided, as under nitrogen over-reduction of the electron transport chain inhibits electron flow and photophosphorylation (13). Experiments were therefore performed at 25 μM O_2 (about 2 % oxygen in the gas phase in equilibrium with a chloroplast suspension). Maximum F_A values observed under these conditions were, as expected, higher than maximum F_A in the presence of air levels of oxygen. They were somewhat variable in different experiments depending on chloroplast ageing. The highest values observed did not exceed 25 000 (M^{-1}), i.e. 1 % of the maximum F_A calculated for broken chloroplasts (3). It must be concluded that even slow consumption of NADPH and ATP considerably decreases the magnitude of F_A below its maximum. It follows that production of F_A is very slow in intact chloroplasts when the stromal NADP system is reduced. Under these conditions, it is usually thought that cyclic electron transport is activated in the presence of oxygen which serves to prevent over-reduction of the electron transport chain. The fact that isolated intact chloroplasts were unable to increase F_A much above 1 % of the maximum calculated for broken chloroplasts, when electron transport and photophosphorylation were efficient, shows clearly that electron transport to oxygen and cyclic electron transport cannot be very effective in isolated chloroplasts. The observations make it necessary to postulate additional proton pumping during linear electron transport in addition to the four protons usually thought to be deposited in the thylakoid lumen when two electrons are transfered from water to NADP. Additional proton pumping, perhaps via a Q cycle mechanism (14, 15), appears necessary to satisfy the ATP requirements of photosynthesis at a H^+/ATP ratio of 3.

When during illumination of intact chloroplasts the light intensity was reduced, F_A decreased. The decrease was indicated by the oxidation of external DHAP which was imported into the chloroplasts in exchange for PGA (3). Reversal of the PGA reducing reaction during the transition from high to low intensity illumination explains why in leaves photosynthesis is transiently inhibited upon a sudden reduction in light intensity (16).

3.3. Assimilatory force of chloroplasts in photosynthesizing leaves

For the determination of F_A in leaves it is necessary to separate chloroplasts from the remainder of the leaf tissue so that their metabolite distribution remains as it was when metabolism was arrested. Leaves were permitted to photosynthesize under various conditions. While illumination was continued, they were rapidly frozen in liquid nitrogen. After freeze-drying, chloroplasts were isolated in organic solvent mixtures in which photosynthetic intermediates were insoluble. Metabolites were determined in the chloroplast fraction (17). As in isolated chloroplasts, ATP/ADP ratios were rarely above 3 in nonaqueous chloroplasts even when the leaves has been light-saturated. NADPH/NADP ratios decreased after a transient increase when photosynthesis approached steady state (18). The highest F_A values calculated from adenylate, phosphate and pyridine nucleotide measurements were observed during the lag phase of photosynthesis, when the Calvin cycle was not yet in operation, i.e. about 20 sec after illuminating a predarkened leaf. Even then they remained below 1000 (M^{-1}). Apparently, in leaves turnover of ATP and NADPH prevented F_A from reaching the maximum observed in isolated chloroplasts.

There was the question whether DHAP/PGA ratios in nonaqueous chloroplasts could be used to indicate F_A during photosynthesis of leaves. Analysis of the components of the PGA reducing system of the

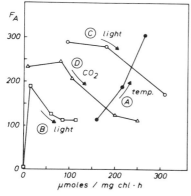

FIGURE 1. Assimilatory force F_A as a function of photosynthesis.
(A) Temperature was increased: 12, 20, and 30 °C, from left to right. (B) Light intensity was increased: dark, 8, 16, 50, 85 and 320 Wm^{-2}, from left to right; air with 330 ppm CO_2, temperature 26 °C. (C) Light intensity was increased: 35, 105, 400 Wm^{-2}, from left to right; air with 20 000 ppm CO_2; temperature 25 °C. (D) CO_2 was increased in 21 % oxygen/79 % nitrogen at 200 W m^{-2}: 50, 350, 550, 1100, 2750 ppm CO_2, left to right; temperature 25 °C.

chloroplasts revealed that reaction [3] was sufficiently close to equilibrium to warrant calculation of F_A from ratios of DHAP to PGA even when photosynthetic flux was large (17). Determination of FA according to equation [5] requires knowledge of pH. The stroma pH increases on illumination owing to the pumping of protons into the intrathylakoid space. Alkalization of the stroma is largely saturated at light intensities above 10 Wm^{-2} (19). The stroma pH may therefore be considered to be more or less constant in illuminated chloroplasts except when light intensities are very low. In two previous publications, we assumed the stroma pH to be 7.8 in the light (2,3). This is perhaps a slight underestimate. Recent evidence suggests that it is close to 8.1 (20). Owing to the pH dependence of F_A, values calculated for pH 7.8 must then be doubled to be valid for pH 8.1. Fig. 1 shows F_A as a function of rate of photosynthesis in 4 different experiments with spinach leaves. All F_A values except the value for zero photosynthesis (dark) were calculated for pH 8.1. For the dark value, pH was assumed to be 7.4. As expected, F_A increased with photosynthesis rate, when the rate was varied by varying the temperature (expt. A). Naturally, F_A also increased when a darkened leaf was illuminated (expt. B). However, in contrast to expectation, F_A did not increase but actually decreased as photosynthesis increased with increasing light intensity in experiment B, when leaves were illuminated in air. A similar decrease in F_A with increasing rate of photosynthesis was observed, when the light intensity was increased in the presence of saturating levels of CO_2 (expt. C). At constant high light, F_A again decreased with increasing rate of photosynthesis, as CO_2 levels were increased from 35 ppm to 2750 ppm (2).

Flux F in any system is defined by

$$F = \frac{driving\ force}{resistance} \qquad [7]$$

When photosynthetic flux increases at a decreasing driving force F_A, it is obvious that resistance must decrease more than the driving force to satisfy the requirements of equation [7]. This is not very difficult to understand for experiment D in Fig. 1. Carboxylation resistance decreases when the CO_2 concentration is increased.

Increased carboxylation increases PGA levels which help to shift reaction [2] towards reduction even when F_A is decreased as a consequence of increased consumption of ATP and NADPH.

A decrease in F_A with increasing light intensity is more difficult to understand. Clearly, resistance to flux must decrease more than F_A to permit increased carbon flux. We explain the decrease in resistance by enzyme activation. Enzymes of the Calvin cycle are known to be light-activated (21). The data of Fig. 1 show that they are carefully controlled by light. They further suggest that even at light and CO_2 saturation, photosynthesis is still limited by the activity of carbon cycle enzymes. Low F_A values facilitate high rates of electron transport and phosphorylation by flooding ATP synthetase and NADP reductase with substrate. High F_A values are not necessary for carbon reduction when enzyme activities are high, as the irreversible reactions of the Calvon cycle exert the pull necessary for carbon flow.

As long as photosynthesis is clearly limited by light, NADPH and ATP synthesized by the electron transport chain should be expected to be immediately consumed during carbon reduction. In contrast, F_A is increased considerably at very low light (Fig. 1 B). This increase is reflected by a large increase in DHAP/PGA ratios. Through the phosphate translocator of the chloroplast envelope, the shift in the DHAP/PGA ratio is communicated to the cytosol. It causes inhibition of glycolysis (22) and may explain the Kok effect which has been interpreted to indicate inhibition of respiration (23).

4. CONCLUSIONS

If it is permissible to compare the incomparable, then power production in photosynthesis (i.e. rate of photosynthesis) may be compared with the economy of power in a modern traffic system. Quite obviously, uphill transport of a load in a traffic system will be slow at reasonable energy input. Much more distance will be covered at the same expense of energy, if tunnels make climbing unnecessary. F_A couples carbon reduction to electron flow. The maximum capacity of the thylakoid system to generate F_A is very high. It is never used in photosynthesis. At high F_A, generation of ATP and NADPH is slow. Cyclic photophosphorylation is not efficient in vivo. ATP and NADPH can be made very fast only at low F_A. The Calvin cycle is capable of catalyzing rapid carbon flux at low F_A by activating regulated enzymes. This represents tunnelling. Limits to activation appear to contribute to the overall limitation of photosynthesis at CO_2 and light saturation.

Acknowledgement
We are grateful to Dr. Laisk, Tartu, and Professor Walker, Sheffield, for stimulating discussions.

REFERENCES
1 Arnon, D.J., Whatley, F.R. and Allen, M.B. (1958) Science 127
2 Dietz, K.J. and Heber, U. (1986) Biochim. Biophys. Acta 848, 392-401
3 Heber, U., Neimanis, S., Dietz, K.J. and Viil, J. (1986) Biochim. Biophys. Acta, in press
4 Laisk, A. (1985) in Kinetics of Photosynthetic Carbon Metablism in C_3 Plants (Viil, J., Grishina, G., Laisk, A., eds.) pp. 21-34, Valgus, Tallinn
5 Hill, R. (1985) Physiol. Veg. 23, 545-554
6 Jensen, R.G. and Bassham, J.A. (1966) Proc. Natl. Acad. Sci. U.S.A. 56, 1056-1101

7 Giersch, Ch., Heber, U., Kobayashi, Y., Inoue, Y., Shibata, K. and Heldt, H.W. (1980) Biochim. Biophys. Acta 590, 59-73

8 Kraayenhof, R. (1969) Biochim. Biophs. Acta 180, 213-215

9 Takahama, U., Shimizu-Takahama, M. and Heber, U. (1981) Biochim. Biophys. Acta 637, 530-539

10 Heber, U. and Santarius, K.A. (1965) Biochim. Biophys. Acta 109, 390-408

11 Ashton, A.R. (1982) FEBS-Letters 145, 1-7

12 Kirk, M.R. and Heber, U. (1976) Planta 132, 131-141

13 Ziem-Hanck, U. and Heber, U. (1980) Biochim. Biophys. Acta 591, 266-274

14 Bendall, D.S. (1982) Biochim. Biophys. Acta 683, 119-151

15 Hartung, A. and Trebst, A. (1985) Physiol. Veg. 23, 635-648

16 Prinsley, R.T., Dietz, K.J. and Leegood R.C. (1985) Annual Report of the Research Institute of Photosynthesis, pp. 42-45, University of Sheffield

17 Dietz, K.J. and Heber, U. (1984) Biochim. Biophys. Acta 767, 432-443

18 Heber, U. and Santarius, K.A. (1965) Biochim. Biophys. Acta 109, 390-409

19 Enser, U. and Heber, U. (1980) Biochim. Biophys. Acta 592, 577-591

20 Oja, V., Laisk, A. and Heber, U. (1986) Biochim. Biophys. Acta 849, 355-365

21 Buchanan, B.B. (1980) Annu. Rev. Plant Physiol. 31, 341-374

22 Kandler, O. and Haberer-Liesenkötter, I. (1957) Z. Naturforschg. 12 b, 271-279

23 Hoch, G. Owens, O.v.H. and Kok, B. (1963) Arch. Biochem. Biophys. 101, 171-180

MULTIPLE ROLES OF OXYGEN: THE USE OF NEW TECHNIQUES IN THE STUDY OF THE EFFECTS OF OXYGEN ON PHOTOSYNTHESIS.

MIRTA N. SIVAK, RESEARCH INSTITUTE FOR PHOTOSYNTHESIS, THE UNIVERSITY, SHEFFIELD S10 2TN, U.K.

1. INTRODUCTION

The increase in photosynthetic CO_2 uptake observed when O_2 concentration is decreased (e.g. from 21 % to 2 % or so) can be easily explained in terms of the supression of photorespiratory metabolism. Oxygen, however, plays other roles, and some effects brought about by changes in oxygen concentration cannot be identified so easily. Oxygen is involved in the Mehler reaction [1] (pseudocyclic electron transport, in which O_2 rather than NADP acts as the terminal electron acceptor) and some O_2 is required for the maintenance of an appropriate redox state of the electron transport chain [2]. Oxygen also seems to play a major role in a less well understood process, the xanthophyll cycle [3,4]. Also, it is still a matter of argument at what rate dark respiration proceeds in the light. It is likely that the relative importance of these many roles of O_2 varies with the plant material and environmental conditions.

2. THE RESPONSE TO DECREASED CO_2 CONCENTRATION

Electron tranport associated with the Mehler reaction and with the photorespiratory pathway has been implicated in the complex responses of chlorophyll fluorescence to changes in the gas phase [5,6] under unchanged illumination, the so called "gas-transients" (for a review see [7]). In the air to CO_2-free air transients, CO_2 is removed. Chlorophyll fluorescence emitted by the leaf displays characteristic kinetics which result from fast relaxation of q_Q quenching and a slower increase in q_E quenching (see [6] and refs. therein). Oxygen, at a concentration of 21 %, is likely to act in these circumstances as an alternative electron acceptor to CO_2 (in pseudo-cyclic electron transport), contributing to the conservation of some q_Q quenching [5,6]. Removal of CO_2 results in a situation which resembles, in a way, photosynthetic induction, where photosynthetic carbon assimilation is slow and pseudocyclic electron transport comes to play a major role, running at higher rates than when steady-state rates of photosynthesis are achieved [8]. Also, when CO_2 is removed from the gas-phase, some degree of photorespiration can still go on, although ribulose bisphosphate will, eventually, fall to very low levels.

When at the same time as CO_2 is removed, O_2 concentration is decreased to 2% (from 21%), the initial fluorescence excursion is always larger than when only CO_2 is removed (Figure 1, and see [5,6]). This larger rise has been attributed [5,6] to the fact that at these low [O_2] the alternative routes for electron transport provided by O_2 are denied. Decreased [O_2] should diminish q_Q quenching of fluorescence, thus contributing to the larger rise in fluorescence observed. Fig. 1 shows that, during this gas-transient, q_E quenching is also affected. The rise in q_E is preceded by a dip and the fall is preceded by a rise, indicating that relaxation of q_E contributes to the initial rise in fluorescence and that the proton gradient actually *decreases*. These data strenghten the following hypothesis, proposed previously on the basis on light-scattering data [6]. The decrease in energization (which occurs as CO_2 is removed and O_2 concentration is decreased) indicates that depletion of the proton gradient is faster than its replenishment. The principal mechanism for accumulation of H^+ in the thylakoids is linear electron transport from water to CO_2. When this can no longer occur, because NADPH reoxidation ceases in the absence of CO_2, the proton-gradient can be built-up by linear electron-transport to oxygen in the Mehler reaction (or other acceptors, e.g. nitrate, oxaloacetate), or by cyclic electron transport. The transient fall in q_E (and in

Biggens, J. (ed.), Progress in Photosynthesis Research, Vol. III. ISBN 90 247 3452 5
© *1987 Martinus Nijhoff Publishers, Dordrecht. Printed in the Netherlands.*

light-scattering [6]) suggests that cyclic electron transport is either slower and/or requires a relatively long time to adapt to the new circumstances, particularly when redox poising is inadequate [2]. This explanation is consistent with the effect of nitrate feeding [6] and oxaloacetate feeding (Sivak, unpublished), which gradually reduce the size of the initial rise in the transient comprising a change from air to CO_2-free, decreased $[O_2]$. Nitrate or oxaloacetate would act as alternative electron acceptors, partly taking over from CO_2 (absent) and O_2 (diminished to only 2 %). Light-scattering measurements corroborate this hypothesis [6].

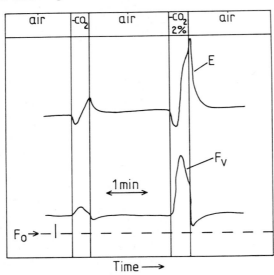

Figure 1. Effect of decreased $[O_2]$ on the response of a spinach leaf to the removal of CO_2 from the gas-phase during unchanged illumination. Kinetics of chlorophyll fluorescence and of non-photochemical quenching displayed during changes in the composition of the gas-phase as indicated show the significant contribution of q_E to the large rise in F_v. "E" denotes the difference between $(F_v)_s$, the fluorescence yield when the primary acceptor of PS II was completely reduced by ssaturating, repetitive, flashes (every 5 sec), and F_m (for details). Temperature, 20°C (for details see ref. [34].

3. RESPONSE TO DECREASED O_2 IN THE PRESENCE OF CO_2

The increase in photosynthetic CO_2 uptake observed when O_2 concentration is decreased (e.g. from 21 % to 2 % or so) can be easily explained in terms of the supression of photorespiratory metabolism. In certain circumstances a transition from air to 2% O_2 may result in a diminished response, i.e. lower stimulation than expected from the supression of photorespiratory metabolism [9-13]. Photorespiration proceeds more or less unchanged in conditions that favour this anomalous response, so called "oxygen insensitivity" (i.e. the post-illumination burst, the high compensation point and CO_2 efflux into CO_2-free air, which are all features of photorespiration, are retained [11]). When allowance is made for photorespiration, which has been supressed by low $[O_2]$, the loss of sensitivity amounts to inhibition and, if inhibition is strong, a net *decrease* in CO_2 uptake may be observed when $[O_2]$ is *decreased*. Inhibition by decreased $[O_2]$ has been attributed [9,10] to the role of O_2 in photophosphorylation (for a review see [14]). An alternative explanation [11-13] is that it may be related to P_i release in photorespiration. Formation and hydrolysis of 2-phosphoglycollate within the chloroplast stroma releases P_i and is favoured by high $[O_2]$. One manifestation of this role of photorespiration is that the P_i optimum for isolated chloroplasts varies with $[O_2]$ in the assay medium [15]. Several pieces of evidence support the involvement of photorespiration in its relation to P_i re-cycling in the phenomenon of oxygen insensitivity.

Inhibition of photosynthesis by decreased $[O_2]$ is usually associated with oscillatory behaviour [10,11] and it is enhanced by mannose feeding [11]. Enhanced oscillatory behaviour may be taken as an indicator of inadequate P_i supply [16,17]. Also, the response of several aspects of the photosynthetic process to increased $[O_2]$ resembles the response to increased P_i (by feeding), and P_i feeding can compensate for decreased

[O₂]. For example, oscillatory behaviour initiated by decreasing [O_2] from 20 % to 2 % can be supressed by a few minutes of P_i feeding (Fig 2 and see [6]). After P_i feeding, the full extent of the supression of photorespiration by decreased [O_2], masked previously by the inhibitory effect of decreased oxygen, becomes apparent.

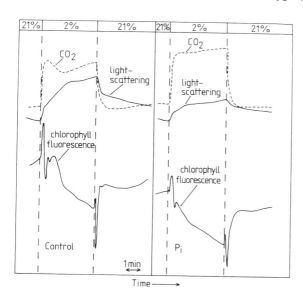

Figure 2. P_i feeding alleviates, and eventually eliminates, inhibition by low O_2 concentrations and this effect is associated to decreased oscillatory behaviour. Kinetics of CO_2 uptake, light-scattering and chlorophyll fluorescence displayed by a spinach leaf when [O_2] was decreased from 21 % to 2 % (illumination, temperature, 20^oC and [CO_2], 360 $\mu l/l$, were unchanged) before and after 5 min Pi feeding (5 mM). Increased P_i supply diminished oscillatory behaviour and the extent of energization initiated by decreased [O_2]. Also, the full extent of supression of photorespiration by diminished [O_2] becomes apparent.

Energization of the chloroplast as indicated by light-scattering or q_E quenching of chlorophyll fluorescence increases when [O_2] is diminished (Fig. 2). After P_i feeding, some energization still occurs when [O_2] is decreased, but the effect is much smaller (Fig.2). Krause *et al.* [18] observed similar responses of light-scattering to changes in [O_2] and interpreted them as a manifestation of phosphorylation energy in the chloroplast being dissipated by photorespiratory processes. It is believed, however, that the ratios of ATP/NADPH required in photosynthesis and in photorespiration are very similar [19]. On the basis of the P_i feeding experiments, the changes in light-scattering and q_E may reflect changes in the availability of P_i (assuming that q_E and light-scattering are good indicators of ΔpH also in these experimental conditions). Decreased O_2 concentration would then limit the availability of P_i, ΔpH would rise and photosynthesis would fall. This interpretation [17] assumes that [P_i] can limit photophosphorylation *in vivo*, a possibility that was considered unlikely but now has some experimental support ([16,17] and R.T. Furbank and C.H. Foyer, personal communication).

It is worth noting that if inhibition by low [O_2] were related to the role of oxygen in redox poising of the electron transport chain or as electron acceptor in the Mehler reaction, decreased oxygen would be expected to result in *decreased* energization. This can only be observed in some special circumstances (Sivak and Heber, unpublished) but increased energization when [O_2] is decreased seems to be the rule.

Inhibition by low O_2 can be observed at relatively low [CO_2] when plants grown in warm conditions are transfered to low temperatures [20,21,22]. Decreased temperature has similar effects to sequestration of cytosolic P_i in terms of oscillatory behaviour and effect of oxygen. P_i feeding or increased temperature have similar effects (Fig. 3 and see [22]). Isolated chloroplasts require higher [P_i] for maximum photosynthesis at low temperatures [23]. The lower thresholds in [CO_2] needed to initiate oscillatory behaviour at low temperatures suggest that this is also true for chloroplasts *in situ*. Alternatively (or possibly, in addition to) the overall rate of P_i recycling (by starch and sucrose synthesis, movement from the vacuole, see e.g. [24]) is slower at these temperatures.

In conditions in which inhibition by low $[O_2]$ is observed, the experimental evidence indicates that supression of photorespiration, instead of increasing the rate of photosynthesis as expected, is only decreasing the light and CO_2 thresholds at which P_i limits photosynthesis, replacing one limitation by another, as less P_i is made available by photorespiration and as increased carboxylation makes higher demands on the supply of P_i.

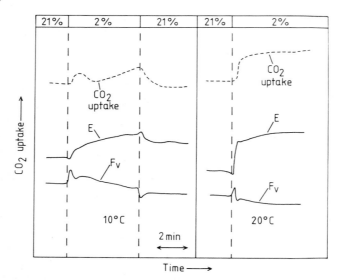

Figure 3. Increased temperature has similar effects to Pi feeding (c.f. Fig. 2) in terms of oscillatory behaviour, effect on photosynthetic CO_2 uptake, and on energization (here as indicated by non-photochemical quenching, E, see legend to Fig. 1). Temperature was varied as indicated.

4. THE RESPONSE TO O_2 IN THE PRESENCE OF HIGH CO_2

Extensive use of the leaf-disc electrode [27] has shown that high $[CO_2]$ (up to 5 % or so) are not deleterious ([17,25,26] and Demmig & Björkman, personal communication). On the contrary, quantum yields near the theoretical maximum, and maximum photosynthetic rates (near light saturation) are *higher* [26] that those measured using IRGA at $[CO_2]$ in the range around 1000 $\mu l/l$ (0.01 %) or below. This indicates that high $[CO_2]$ are not necessarily inhibitory (although long dark periods should be avoided) and also contradicts the widespread belief that leaf pieces should perform less well than the intact attached leaf. It seems that, in saturating $[CO_2]$ and saturating light intensity, the leaf can express its maximum photosynthetic capacity. It is in these conditions that one would expect P_i to become limiting, an expectation now confirmed [6,16].

Although the explanation (see above) of the inhibitory effect of oxygen in $[CO_2]$ near air levels is consistent with our understanding of photorespiratory metabolism and its role in the recycling of P_i, this interpretation cannot account for the fact that O_2 concentration seems to have rather similar effects at $[CO_2]$ which *do not* favour photorespiration [16,17,28]. Increased O_2 concentration from 2% to 21% *increased* the rate of photosynthesis displayed by barley leaves in saturating in 0.55 % CO_2 (Fig. 4 and see [17]). Because these effects were similar in magnitude to the increases observed after P_i feeding, it was proposed that the stimulation by 21% O_2 was consistent with an increase in P_i supply (Fig. 4 and see [17]). Strong similarities between the effects of low $[O_2]$ and restricted P_i supply were observed at these very high $[CO_2]$ (as at $[CO_2]$ in which photorespiration is known to occur, see above). Thus, low $[O_2]$ promotes oscillatory behaviour [17,25,28] and increases chloroplast energization, as indicated by light-scattering (Fig. 4 and see [28]) and by q_E (Fig. 5). P_i feeding, again, seems to compensate for decreased $[O_2]$.

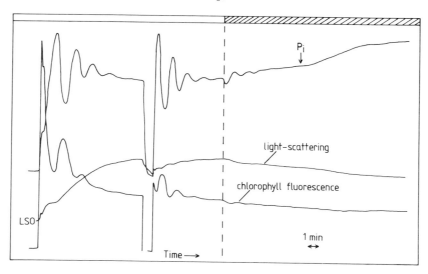

Figure 4. Stimulation of photosynthetic CO_2 uptake in high $[CO_2]$ (0.68 %) by increased O_2 concentration (from 2 %, clear, to 20 %, hatched) and by subsequent P_i feeding. Stimulation was correlated to decreased energization (as indicated by light-scattering). Photosynthetic rate after the increase in $[O_2]$ and in P_i status reached the high rates displayed during oscillations.

There are good grounds for believing that more P_i is made available by photorespiration at normal $[O_2]$ and in air concentrations (or so) of CO_2, but no such mechanism has been proposed to explain how changes in $[O_2]$ could affect P_i supply at high concentrations of CO_2. Concentrations of CO_2 of 0.5% or higher (as used in the experiments of Fig. 4 to 6, and see [16,17]) are believed to largely supress photorespiration by themselves, even at high $[O_2]$. Experimental evidence supports this belief. At high CO_2 concentrations, decreased $[O_2]$ *inhibits* CO_2 uptake (not illustrated), and increased $[O_2]$ increases CO_2 uptake, as shown in Fig. 4. If rates of photorespiration were still substantial in these high $[CO_2]$, this should become apparent after P_i feeding, once the hypothetical differences in P_i supply were compensated for. However, after P_i feeding CO_2 uptake rates are similar in $[O_2]$ in the range from 2% to 40% (not illustrated). In short, the cluster of effects of low $[O_2]$ has rather similar characteristics at low and high $[CO_2]$.

5.   THE EFFECT OF O_2 DURING ADAPTATION TO HIGH CO_2
    When a leaf is subjected to changes in the environment, it reaches steady-state photosynthesis in a number of ways (for a review see [7]) and, adding to the complexity of induction phenomena, there are often slow changes which can sometimes be observed after 20-30 min or so of illumination. Rabinowich [29] called this slow increase in photosynthetic rate "slow induction" to differentiate it from the large changes which occur during the first few minutes of re-illumination [7]. It has been proposed [16,17] that this slow induction could be the result of internal adjustment of P_i supply. The slow increase in photosynthetic CO_2 uptake is accompanied by other changes. For example, as slow induction proceeds, oscillatory behaviour initiated by re-illumination after short dark intervals (e.g. 1 min) becomes less and less pronounced:

amplitude and frequency of the oscillations diminishes, damping is faster. The response of the leaf to changes in $[O_2]$ (as described above) and oscillatory behaviour are also modified during this period of adaptation. It is worth noting that these changes are not associated with deterioration of the plant material, as photosynthetic rate *increases* during the course of the experiment. Leaves and leaf pieces can be illuminated in the presence of high $[CO_2]$ for hours (some experiments involved continuous illumination for as long as 7 hours) without showing signs of deterioration. Also, if the high $[CO_2]$ gas-phase is replaced by air, i.e. the leaf is illuminated for 10 min or so in air, adaptation to high $[CO_2]$ is reversed, i.e. oscillatory behaviour, rate, etc., are as before slow induction.

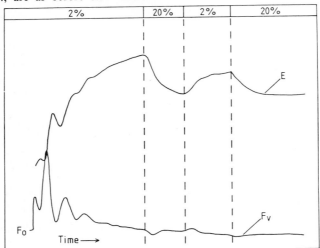

Figure 5. At high CO_2 concentrations (0.68 %) an increase in $[O_2]$ from 2 % to 20 % decreased energization in a spinach leaf as indicated by non-photochemical quenching of chlorophyll fluorescence and decreased $[O_2]$ increased energization.

Figure 6 illustrates the complexity of induction phenomena that can be observed when a leaf is re-illuminated in high $[CO_2]$ and near saturating light intensity. The "CO_2 gulp" (see also [7]) is over in a few seconds and gives way to dampening oscillations. After oscillations dampen out, photosynthetic rate increases very slowly for more than 1 hour until it eventually approaches the rate displayed during the first peak(s) of the oscillations. The rates measured during oscillations can be much higher than the *quasi* steady-state reached soon after oscillations dampen out (values 50 % higher than the *quasi* steady-state are common [7,16,17]). The importance of this point becomes apparent when the experiment is run long enough to establish a *true* steady-state, which may then approach (as in Fig. 6) the high rates reached during oscillations.

Oxygen concentration affects adaptation to high $[CO_2]$ in several ways:
1) the time required by the leaf to reach the steady-state depends on $[O_2]$: low concentrations extend the length of slow induction, i.e. the time to reach the *true* steady-state in 40 % O_2 may be as short as 30 min while in 2% O_2 may take up to 7 hours.
2) low $[O_2]$ extend the period of oscillatory behaviour (i.e. reduces dampening) that preceeds or is superimposed to the slow increase in rate.
3) during the early stages of slow induction a decrease in $[O_2]$ (e.g. from 20 to 2%) initiates oscillations superimposed on a decrease in photosynthetic rate (see above). Conversely, increased $[O_2]$ brings about an increase in rate. When slow induction is well advanced and as the maximum photosynthetic rate is approached changes in $[O_2]$ do not result in changes in CO_2 uptake.

There is a clear relationship between slow induction and P_i status. The effects of P_i feeding as shown above (and see also [16,17]) amount to shortening of the adaptation process from hours to a few minutes. This is not only in respect to photosynthetic rate but also as far as the response to $[O_2]$ and oscillatory behaviour are concerned. Feeding of 2-deoxyglucose slows down or stops slow induction (depending on the conditions such as concentration of the sugar and duration of feeding), i.e. the rate remains constant and the effect of $[O_2]$ on rate and oscillatory behaviour is retained (not illustrated). The effect of temperature is also consistent with this explanation and adaptation is slower at lower temperatures. At 8 ^{o}C, for example, slow induction was so slow that photosynthetic rate did not reach the maximum rate displayed during oscillations after 6 hours or so of illumination. This is consistent with the view that, in many circumstances, overall rates of P_i recycling are inadequate at low temperatures [17].

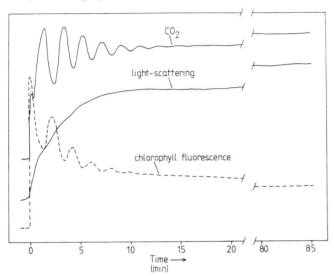

Figure 6. Induction phenomena in a spinach leaf, illuminated for several hours in high $[CO_2]$ (0.68%) and 20 % O_2 (temperature 20°C, red light, near saturation, 150 W. m^{-2}). After the CO_2 "gulp" was over and oscillations dampened out, photosynthetic rate increased very slowly to a value that approximated the high rates displayed during oscillations.

There seems to be is a tight association between $[O_2]$ and adaptation to high $[CO_2]$, on the one hand, and between P_i and adaptation on the other, and these relationships bear a strong similarity to the effects observed in the low CO_2 concentrations that favour photorespiration. There is no straightforward explanation for this similarity.

An appropriate P_i supply in the chloroplast stroma is the result of recycling of P_i by sucrose synthesis, the activity of the P_i translocator and the maintenance of a controlled flow to and from the vacuole (which contains much of the cellular P_i). The first two mechanisms are now fairly well understood, and there seems to be no space for a major role for O_2 in either of them. About the flow of P_i to and from the vacuole, little is known [31,32] except that some control mechanism seem to exist. However, from what is known about the general transport mechanisms in the tonoplast (for reviews see [33]), it seems unlikely that $[O_2]$ might have a direct effect on the transport of P_i to and from the vacuole. Effects of P_i on protein phosphorylation or on the ATPase itself cannot be ruled out.

The violaxanthin (an epoxy carotenoid) cycle is a photosynthetic O_2 uptake pathway, mediated by H^+ transport [3] and sudden changes in $[O_2]$ are likely to affect the rate of the epoxidation reaction to a larger extent than the rate of de-epoxidation. If this were so, changes in energization of the thylakoid could follow. The violaxanthin cycle also occurs at high $[CO_2]$ [4] and although a role for P_i in the pathway is not apparent, it seems worthy of further examination in this context. Whatever the answer, the similarity between the phenomena observed at low and high $[CO_2]$ suggests that if further light is shed on the latter, re-examination of the hypotheses proposed for the former should follow.

Acknowledgements
I wish to thank U. Schreiber for bringing to my attention the less well understood role of oxygen and I am grateful to him, U. Heber, G. Kaiser and D.A. Walker for many helpful discussions.

REFERENCES
1   Mehler A.H. (1951) Arch. Biochem. 33, 65-77
2   Heber, U., Egneus, H., Hanck, U., Jensen, M., and Köster, S. (1978) Planta 143, 41-49.
3   Yamamoto, H.Y. and Takeguchi, C.A. (1971) In: Proc. 2nd. Int. Congress on Photosynthesis, Stressa. 621-627.
4   Siefferman, D. (1971) In: Proc. 2nd. Int. Congress on photsynthesis, Stressa. pp. 629-635.
5   Sivak, M.N., Prinsley, R.T. and Walker, D.A. (1983) Proc. R. Soc. Lond. B. 217, 393-404.
6   Sivak, M.N. & Walker, D.A. (1985) Ann. Proc. Phytochem. Soc. Eur. 26, 29-44.
7   Sivak, M.N. & Walker, D.A. (1985) Plant Cell Environ. 8, 439-448.
8   Furbank, R.T., Badger, M.R. & Osmond, C.B. (1982) Plant Physiol, 70: 927-931.
9   Viil, J., Laisk, A., Oja, V. and Parnik, T. (1977) Photosynthetica (Prague) 11, 251-259.
10 Canvin, D.T. (1978) In: Photosynthetic Carbon Assimilation (H.W. Siegelman and G. Hind, eds.) pp. 61-76, Plenum Press, New York.
11 Harris, G.C., Cheesbrough, J.K. & Walker, D.A. (1983) Plant Physiol. 71, 108-111.
12 Badger, M.R., Sharkey, T.D. amd von Caemmerer, S. (1984) Planta 160, 305-313.
13 Sharkey, T. (1985) Plant Physiol. 78, 71-85.
14 Heber, U. (1985) In: Kinetics of photosynthetic carbon metabolism in C_3 plants, Part II, pp. 9-22. Valgus, Tallin.
15 Usuda, H. & Edwards, G.E. (1982) Plant Physiol. 69, 469-473.
16 Walker, D.A., & Sivak, M.N. (1985) Physiol. Veg. 23, 829-841.
17 Sivak, M.N. & Walker, D.A. (1986) New Phytol. 102, 499-512.
18 Krause, G.H., Lorimer, G.H., Heber, U. & Kirk, M. 1977. In: Proc. IVth Congress on Photosynthesis. pp 299-310.
19 Edwards, G.E. & Walker, D.A. (1983) C3, C4: Mechanisms and cellular and environmental regulation of photosynthesis. Oxford: Blackwell Scientific Press.
20 Arrabaca, M.C., Keys, A.J. and Whttingham, C.P. (1981) In: Proc. 5th Congress on Photosynthesis (Akoyunoglou, G., ed.) pp. 463-470. Balaban Int. Serv., Philadelphia.
21 Rabinowich, E.I. (1956) Photosynthesis and related processes. Vol 2 Part 2, Interscience, New York.
22 Leegood, R.C. & Furbank, R.T. (1986) Planta (in press).
23 Leegood, R.C. and Walker, D.A. (1983) Biochem. Soc. Trans. 11, 74-76.
24 Housley, T.L. and Pollock, C.J. (1985) New Phytol. 99, 67-80.
25 Walker, D.A., Sivak, M.N., Prinsley, R.T. & Cheesbrough, J.K. (1983) Plant Physiol. 73, 534-541.
26 Walker, D.A. and Osmond, C.B. (1986) Proc. R. Soc. Lond. B. 227, 267-280.
27 Delieu, T.J. & Walker, D.A. (1983) Plant Physiol. 73, 542-549.
28 Sivak, M.N., Dietz, K.J., Heber, U. & Walker, D.A. (1985) Arch. Biochem. Biophys. 237, 513-519.
29 Schnyder, H., Mächler, F. & Nösberger, J. (1984) J.Exp. Bot. 35, 147-156.
31 Rebeille, F., Bligny, R., Martin, J.B. and Douce, R. (1983) Arch. Biochem. Biophys. 255, 143-148.
32 Foyer, C.H., Walker, D.A., Spencer, C. and Mann, B. (1981) Biochem. J. 202, 429-434.
33 Marin, B.P. (1985) Biochemistry and function of vacuolar adenosine triphosphatase in fungi and plants. Springer Verlag, Berlin.
34 Schreiber, U., Bilger, W. and Schliwa, U. (1986) Photos. Res. in press.

THE INTERACTION OF RIBULOSE-1,5-BISPHOSPHATE WITH THYLAKOID AND STROMAL REACTIONS

CHRISTINE FOYER, ROBERT FURBANK AND DAVID WALKER, Research Institute for Photosynthesis, University of Sheffield, Sheffield S10 2TN, U.K.

The chloroplast stroma contains a high concentration of the enzyme ribulose-1,5-bisphosphate(RuBP)carboxylase [1]. The measured concentration of this protein is 0.4-0.5 mM which is equivalent to 3-4 mM active sites [2] although the theoretical maximum concentration is 1.2 mM carboxylase protein (or 9 mM active sites, [3,4]). The following experiments demonstrate that the high binding capacity of this protein has significant repercussions not only on the activity of the enzyme itself but also on other important reactions of photosynthesis.

Thylakoid reactions

Table 1A. The effect of added pure activated RuBP carboxylase ($2.5 mg.ml^{-1}$) on O_2 evolution in the reconstituted chloroplast system in the absence or presence of 10 mM NH_4Cl with either NADP, ferricyanide or glycerate 3-phosphate as electron acceptor. The reconstituted chloroplast system contained 10 mM $MgCl_2$, 10 mM $NaHCO_3$ 0.33 M sorbitol, 2 mM EDTA, 10 mM KCl, 50 mM Hepes/KOH buffer (pH 8.1), 4 mM ascorbate, 1 mM dithiothreitol and thylakoid and stromal protein equivalent to 50 $\mu g.ml^{-1}$ chlorophyll. For NADP reduction 2 mM NADP, 2 mM ADP, 2 mM Pi and 100 $\mu g. ml^{-1}$ ferredoxin were used. For glycerate 3-phosphate reduction, similar mixtures were used in which ADP and NADP were decreased to 0.2 mM, and glycerate 3-phosphate was 5 mM. For ferricyanide reduction the assay contained 1.5 mM ferricyanide, 2 mM ADP, 2 mM Pi and 10 mM D,L-glyceraldehyde (dithiothreitol was omitted).

Electron Acceptor	Rate of O_2 evolution ($\mu mol \cdot h^{-1}.mg^{-1}$ chlorophyll)			
	−RuBP carboxylase		+RuBP carboxylase	
	− NH_4Cl	+ NH_4Cl	− NH_4Cl	+ NH_4Cl
NADP	200	220	120	220
Ferricyanide	172	187	95	187
Glycerate 3-phosphate	116	−	12	−

Table 1B. The effect of adding pure activated RuBP carboxylase ($2.5 mg.ml^{-1}$) in the light after addition of Pi on glycerate 3-phosphate-dependent O_2 evolution in the reconstituted chloroplast system.

Additions	$\mu mol\ O_2$ evolved. $h^{-1}. mg^{-1}$ chlorophyll
10 mM Pi	142
10 mM Pi + 2.5 mg.ml^{-1} RuBP carboxylase	90
2 mM Pi + 2.5 mg.ml^{-1} RuBP carboxylase	50

Biggens, J. (ed.), Progress in Photosynthesis Research, Vol. III. ISBN 90 247 3452 5

Table 1A shows that the addition of RuBP carboxylase inhibited photophosphorylation but did not inhibit electron transport directly. Both fully activated carboxylase and inactivated carboxylase inhibited PGA reduction. This effect could also be demonstrated with stromal protein at limiting light intensities. When the amount of thylakoid membrane was varied with a constant level of added RuBP carboxylase the inhibition of the rate of coupled PGA-dependent O_2 evolution relative to controls decreased with the increasing thylakoid concentration. The inhibition of PGA reduction by RuBP carboxylase was highly dependent on $MgCl_2$ concentration in the reaction medium.

The inhibition of glycerate 3-phosphate reduction by ribulose-1,5-bisphosphate carboxylase could not be reversed by adding high concentrations of ADP, P_i, NADP or glycerate 3-phosphate during illumination. With glycerate 3-phosphate, a linear rate of O_2 evolution was obtained which declined within 1 to 2 min in the presence of ribulose-1,5-bisphosphate carboxylase, particularly at low Pi concentrations. The addition of saturating NADP or an ATP-generating system (creatine phosphate and creatine phosphokinase) seemingly restored O_2 evolution but only to the basal non-phosphorylating rate of NADP reduction seen in the absence of ADP and P_i. This is consistent with the view that ribulose-1,5-bisphosphate carboxylase affects the operation of thylakoid ATPase. When ribulose-1,5-bisphosphate carboxylase was added (during illumination) after the addition of P_i, partial protection of the rate of glycerate 3-phosphate-dependent O_2 evolution against inhibition was afforded (Table 1B). 10 mM P_i was sufficient to maintain 62% of the control rate of O_2 evolution and prevented the decline in rate observed after 1 to 2 min at lower Pi concentrations (Table 1B). These data implicate Pi as a protecting agent against ribulose-1,5-bisphosphate carboxylase inhibition of photophosphorylation. This possibility was explored by examining the Pi requirement for coupled NADP reduction with isolated thylakoids in the presence and absence of carboxylase or stromal protein. When the stromal protein concentration was increased from 0.4 to 4.0 mg the Pi optimum for photophosphorylation was also increased (the $K_{1/2}$ increased from 50 to 130 μM). In addition the rate of coupled NADP reduction increased 25% between 1 and 10 mM P_i in the treated sample while the control showed no response over this P_i range. A similar response was seen when purified ribulose-1,5-bisphosphate carboxylase was added.

Stromal reactions

Certain Calvin cycle metabolites are known to inhibit RuBP carboxylase activity (5). The effect of physiological concentrations of glycerate 3-phosphate on the activity of RuBP carboxylase was measured over a range of RuBP levels. In assays were low RuBP was required it was sucessfully maintained at a constant low level using an RuBP-generating system consisting of 5 mM R5P, 10 mM creatine phosphate and creatine phosphate kinase (0.1 unit per ml). Under conditions where RuBP was maintained at 200 μM in the presence of 20 mM glycerate 3-phosphate, severe inhibition of CO_2 fixation was observed. The inhibition of carboxylation showed hyperbolic kinetics with respect to glycerate 3-phosphate concentration. At the high concentrations (10 to 20 mM) of glycerate 3-phosphate (which may be encountered *in vivo*) inhibition was severe, with approximately 85% inhibition of carboxylation at 20 mM glycerate 3-phosphate. The apparent Km for RuBP increased by approximately 10 fold in the presence of 20 mM glycerate 3-phosphate. Physiological concentrations of triose phosphate and FBP also caused significant inhibition of RuBP carboxylase activity. In order to simulate the conditions which might be encountered *in vivo*, these experiments were repeated, varying the concentration of glycerate 3-phosphate from 2 to 20 mM in the presence of 0.5 mM FBP and 1.5 mM triose phosphate.

Changes in glycerate 3-phosphate concentration in this range occur during transitions in light intensity and gas phase composition in intact leaves [6,7]. Despite the presence of triose phosphate and FBP (both of which must compete with glycerate 3-phosphate for the RuBP binding site) a five fold increase in the apparent Km for RuBP was observed (Table 2). When a steady state RuBP concentration was maintained in stromal extracts at high protein concentration (2 mg.ml^{-1}) using triose phosphate as substrate, the addition of a "pulse" of glycerate 3-phosphate (10) reduced the rate of CO_2 uptake almost to zero, although the ATP level was relatively constant and the ATP/ADP ratio never fell below 4. This suggests that the inhibition observed did not result from reduced conversion of Ru5P to RuBP (due to a lower ATP/ADP ratio). Consequently, it appears that the rate of carboxylation of RuBP could be significantly affected, *in vivo*, by changes in stromal glycerate 3-phosphate concentration.

Table 2. The effect of pH and Benson-Calvin cycle metabolites on the apparent Km of RuBP carboxylase for RuBP.

Conditions	Apparent Km for RuBP
pH 7.0	0.25
pH 7.9	0.25
pH 7.0, 20 mM glycerate 3-phosphate	2.50
pH 7.9, 20 mM glycerate 3-phosphate	2.50
pH 7.9, 0.5 mM FBP	0.28
pH 7.9, 1.5 mM DHAP	0.22
pH 7.9, 2 mM glycerate 3-phosphate, 0.5 mM FBP, and 1.5 mM triose phosphate	0.60
pH 7.9, 20 mM glycerate 3-phosphate, 0.5 mM FBP, and 1.5 mM triose phosphate	3.00

Conclusions

The data shown here indicate that RuBP carboxylase acts as a classical energy transfer inhibitor of photophosphorylation. The dependence of the inhibition on Mg^{++} concentration suggests that inhibition of photophosphorylation is associated with binding of the RuBP carboxylase to the thylakoid membrane. This situation may be pronounced in situations where the volume of the chloroplast is decreased such as osmotic stress. These results may explain the fact that the chloroplast behaves as though it were Pi-deficient in concentrations of Pi considerably higher than those which would affect photophosphorylation in isolated thylakoids. It is therefore of considerable significance when considering the manner in which cytosolic Pi concentration influences the rate of photosynthesis.

Glycerate 3-phosphate will be an effective competitive inhibitor of RuBP carboxylation *in vivo* when the RuBP level is low. This effect may be particularly important when considering the effects of changes in input on photosynthesis and carboxylation rate such as transitions in light intensity (7). At such times when the activation state of the enzyme remains relatively constant the activity may be directly modulated by the action of Calvin cycle intermediates such as glycerate 3- phosphate. Similarly, there will be times when activites of other calvin cycle enxzymes such as phosphoglycerate kinase will be limited because of binding of their substrates to RuBP carboxylase. These interactions with RuBP Carboxylase (Figure 1) need to be considered if the activity of the Calvin cycle is to be accurately modelled since the effective concentrations of metabolites may be substantially lower than those measured (8).

Figure 1. A digrammatic representation of the interactions of RuBP carboxlyase in the chloroplast environment.

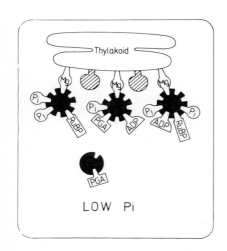 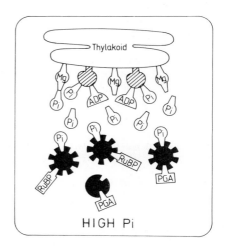

KEY : RuBP carboxylase -

Glycerate-3-phosphate kinase -

Thylakoid ATPase -

References
(1) Ellis R.J. (1979) *Trends Biochem. Sci. 4* 241-244.
(2) Jensen, R.G. and Bahr, J.T. (1977) *Ann. Rev. Pl. Phys. 28,* 379-400.
(3) Leegood, R. C., Sivak, M.N. and Walker, D.A. (1986) *Phil. Trans. Roy. Soc.* (London) In Press.
(4) Lilley, R. McC., Fitzgerald, M.P. Rienits, K.G. and Walker, D.A. (1975) *New Phytol. 75,* 1-10.
(5) Badger, M. R. and Lorimer, G.H. (1981) *Biochemistry, 20,* 2219-2225
(6) Furbank, R.T. and Foyer, C.H. (1986) *Arch. Biochem. Biophys. 246,* 240-244.
(7) Prinsley, R.T., Dietz, K-J. and Leegood, R.C. (1986) *Biochim. Biophys. Acta, 849,* 254-263.
(8) Ashton, A.R. (1982) *FEBS Lett. 145* 1-6.

CHLOROPLAST ENERGIZATION DURING PHOTOSYNTHETIC INDUCTION

MIRTA N. SIVAK and DAVID A. WALKER, RESEARCH INSTITUTE FOR PHOTOSYNTHESIS, UNIVERSITY OF SHEFFIELD, SHEFFIELD S10 2TN, U.K.

1. INTRODUCTION

Although *photosynthetic induction* [1] and *fluorescence induction* (Kautsky effect) are not synonymous, experimental evidence indicates that fluorescence kinetics are closely related to the complexities of induction and its termination. Early studies of chlorophyll fluorescence kinetics in a physiological context were mainly descriptive (for reviews see e.g. [2,3]), given the complexity and ambiguity of the fluorescence signal. More recently, a better understanding of photosynthesis in general and of the mechanisms involved in fluorescence quenching allowed further interpretation of slow fluorescence kinetics *in vivo*. First, an association between photosynthetic induction and a slow transient in chlorophyll fluorescence was demonstrated, and an explanation of the relevant events proposed [4]. Multiple simultaneous measurements (i.e. CO_2, light-scattering, etc. see e.g. [2,5]) corroborated this association and provided a sound basis for interpretation. A similar process led to an interpretation of dampening oscillations in chlorophyll fluorescence and of other features of the fluorescence induction curves [2,6,7]. Now, further innovations in techniques facilitate the testing of hypotheses previously proposed on the basis of "circumstancial" evidence.

The main quenching components of fluorescence (for reviews see e.g. [2,3,8]) are q_Q (related to photochemical energy conversion at PS II reaction centres) and q_E (non-photochemical or "energy" quenching, related to increased radiationless energy dissipation). In the following experiments, chlorophyll fluorescence was measured using a pulse amplitude modulation fluorometer recently developed by Schreiber *et al.* [9]. The pulse modulation technique is based on the rationale that at any moment during the measurement of chlorophyll fluorescence, a short pulse of sufficiently strong light will cause complete reduction of the PSII acceptor Q_A, transiently removing q_Q quenching. Any remaining quenching is assumed to be non-photochemical. If these short flashes are applied repetitively, this approach supplies continuous information about the two main quenching components, q_E and q_Q.

2. FLUORESCENCE DURING PHOTOSYNTHETIC INDUCTION

Chlorophyll fluorescence, q_E, and light scattering were followed during re-illumination after a dark interval (Fig. 1). The kinetics displayed by q_E was very similar to those of light-scattering. The "M-peak" as defined previously is clearly associated with the duration of the dark interval and is most readily observed at very low light-intensities [2,4,5]. For spinach leaves, in air, the slow transient associated with induction could only be observed at the lowest light intensities used. At higher light intensities, where energization was high, the clear peak became a "shoulder" [2,5]. Although q_E and light-scattering showed very similar trends, kinetics were not identical, and q_E and light-scattering could peak (Fig. 1) at different times after re-illumination.

3. INDUCTION AFTER A CHANGE FROM HIGH TO LOW LIGHT INTENSITY

It is widely accepted that photosynthetic induction derives from the slow build-up of the pool of intermediates and activation of enzymes of the Benson-Calvin cycle [1]. Induction caused by transition from high to low light (Steeman-Nielsen) is another matter, as metabolite pools had been primed and enzymes activated during the period of high light intensity. Nevertheless, an abrupt decrease in light intensity may be followed by an appreciable delay before photosynthesis reaches its new, slower rate. It

Biggens, J. (ed.), Progress in Photosynthesis Research, Vol. III. ISBN 90 247 3452 5

has recently been proposed that this type of induction is the result of changes in the activation status of RuBP carboxylase/oxygenase and the rate of RuBP regeneration, and of an overshoot in sucrose synthesis [10,11]. Previously it has been suggested, on the basis of light-scattering and chlorophyll fluorescence data [12], that the *primary* cause resides in the proton gradient which is established across the thylakoid membrane as a consequence of light-driven electron transport. The pulse-modulation technique data support this view. Fig. 2 shows a time course of photosynthetic oxygen evolution and chlorophyll fluorescence emission by an ivy leaf illuminated in moderate light and it can be seen that when light intensity was decreased, O_2 evolution decreased to zero before picking up to a new, lower, steady-state value after about two minutes. Following the decrease in actinic light the proton gradient, as indicated by q_E, is still high. As time passes, the saturated fluorescence yield at 100% Q_A^- becomes higher and higher, indicating slow relaxation of q_E. As q_E and light-scattering (indicators of chloroplast energization) fall, so the rate of electron transport and O_2 evolution rises.

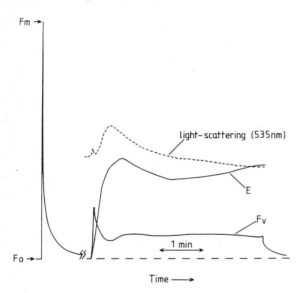

Figure 1. Kinetics of chlorophyll fluorescence and of two indicators of chloroplast energization, the non-photochemical quenching of chlorophyll fluorescence and light-scattering. The M peak in chlorophyll fluorescence is associated with relaxation of non-photochemical quenching. Spinach leaf re-illuminated (in air) at low light intensity. "E" represents the difference between F_m and $(F_v)_s$ (fluorescence emission when q_Q is supressed by saturating flashes). Chlorophyll fluorescence displays the transient leading to the "M peak", E and light-scattering increase after re-illumination and then relax.

4. FLUORESCENCE KINETICS DURING OSCILLATIONS

When it became possible to follow chlorophyll a fluorescence, CO_2 uptake, and O_2 evolution [2,6] during oscillations in photosynthetic carbon assimilation it became clear that although there was a broadly reciprocal or anti-parallel relationship between fluorescence on the one hand and CO_2 and O_2 on the other, there was also a phase-shift. This was so large in some circumstances (with fluorescence anticipating O_2 evolution and CO_2 uptake by as much as 15-25 seconds) that the possibility of artefacts of measurement seemed remote. Indeed, evaluation of response times showed that the phase shifts were an order of magnitude larger than delays in gas measurement. It was thought that a separation in time between the evolution of oxygen and the reduction of "Q" of the order observed was unlikely [2,6,7] and that thylakoid energization was the factor most likely to contribute to the phase shift between fluorescence and oxygen. Previously, a hypothesis was proposed which explained oscillatory behaviour in terms of an imbalance of the [ATP]/[ADP] ratio occasioned by some pertubations of the steady-state [2,6]. This proposal implied the existence of oscillations in the proton gradient. Measurement of light-scattering showed that light-scattering did indeed oscillate, and so did an indicator of the transthylakoid membrane potential, the electrochromic shift [7]. Fig. 3 shows that q_E oscillates in advance of fluorescence just as fluorescence oscillates in advance of CO_2 uptake. These data strengthens the view [2,6,7] that fluorescence anticipates O_2 because that part of the fluorescence signal which is quenched by the establishment of chloroplast energization (non-photochemical quenching) is affected, during oscillatory behaviour, *before* that part which is governed by the oxidation state of Q (q_Q quenching). The part of the signal determined by q_Q quenching would then be synchronous and anti-parallel with O_2 evolution and CO_2 uptake (synchronous within the degree of resolution, about 1 sec, afforded by our measuring system.

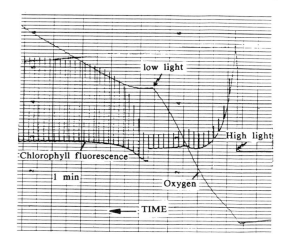

Figure 2. Time course of photosynthetic oxygen evolution and fluorescence saturation pulse kinetics displayed by an Hedera helix leaf illuminated in moderate (160 $\mu E.m^{-2}.s^{-1}$) light. At the time indicated, the photon flux was decreased to 36 $\mu E.m^{-2}.s^{-1}$. O_2 evolution decreased to zero before picking up to a new, lower, steady-state value after about two minutes and this photosynthetic induction was associated with a slow decrease in non-photochemical quenching.

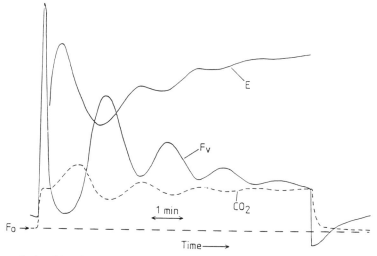

Figure 3. The relationship between chlorophyll fluorescence (F_v), non-photochemical fluorescence quenching ("E", see legend to Fig. 1) and the rate of CO_2 fixation during oscillations in photosynthesis. The broken horizontal marks the dark values of the CO_2 uptake rate and of F_0. Non photochemical quenching oscillates in advance of fluorescence and fluorescence oscillates in advance of CO_2 uptake. Spinach leaf photosynthesizing at 20^oC, in 2 % O_2 and 0.55 % CO_2 (balance N_2).

It should be noted, however, that NADP is not the only electron acceptor and q_Q and O_2 evolution do not necessarily have to be synchronous if the rate of electron transport to other acceptors (O_2 in the Mehler reaction, nitrate, oxaloacetate, etc.) were significant and *not* constant.

5. CONCLUSIONS

Previously [2,5], we have discussed the difficulties in establishing a widely accepted nomenclature for fluorescence induction kinetics. The new techniques eliminate much of the "guessing", i.e. the interpretation of the complex chlorophyll fluorescence signals on the basis of other aspects of the photosynthetic process. It is clear, however, that the present understanding of the regulatory processes and of the processes which affect quenching of chlorophyll fluorescence was in itself sufficient to enable prediction of the kinetics followed by the main quenching mechanisms during the complexities of photosynthetic induction [2,4,7]. It is also clear that light-scattering not only provides a good assessment of energization during the steady-state but is also a good indicator of it during photosynthetic induction.

Besides the two main quenching mechanisms, q_E and q_Q, there are other mechanisms known to affect the yield of chlorophyll *a* fluorescence, some of them known to operate *in vivo* ("state I-state II" transitions, photoinhibition and reversible quenching by high temperature, etc. [8]). In many experimental conditions these latter mechanisms can be safely disregarded and the results may then be interpreted in terms of the two main quenching mechanisms, q_E and q_Q, described above. Recent data, however, show that under conditions that favour energization (e.g. relatively high light intensity) F_0 may itself be quenched [13] and it is essential to check whether or not significant changes in F_0 have occurred during the course of the experiment (see e.g. Fig. 3). Another complication arises from the fact that the saturation pulse method *is not* a "non-intrusive" probe. Particularly at low light intensities, the saturation flashes applied every few seconds do reflect in higher photosynthetic rates (see e.g. Fig. 2). Surprisingly, although the overall kinetics *are* changed, the *basic* shape of the curve remains the same and meaningful analysis is still possible.

Acknowledgements
We wish to thank U. Heber, U. Schreiber and W. Bilger for helfpul discussions and for practical and theoretical instruction.

REFERENCES
1   Walker, D. A. (1973) New Phytol. 72, 209-235.
2   Sivak M.N. and Walker, D.A. (1985) Plant Cell Environ., 8, 439-448.
3   Renger, G. and Schreiber, U. (1986) In: Govindjee, Amesz, J. and Fork, D.C. (eds) Light emission by plants and bacteria, Academic Press, New York, in press.
4   Walker, D.A. (1981) Planta 153, 273-278.
5   Sivak, M.N., Heber, U. and Walker, D.A. (1985) Planta 163, 419-423.
6   Walker, D.A., Sivak, M.N., Prinsley, R.T. and Cheesbrough, J.K. (1983) Plant Physiol. 73, 542-549.
7   Sivak, M.N., Dietz, K.-J., Heber, U. and Walker, D.A. (1985) Arch. Biochem. Biophys. 237, 513-519.
8   Krause, G. and Weis, E. (1984) Photos. Res. 5, 139-157.
9   Schreiber, U., Bilger, W. and Schliwa, U. (1986) Photosynth. Res. (in press).
10   Mott, K.A., Jensen, R.G., O'Leary, J.W. & Berry, J.A. (1984) Plant Physiol. 76, 968-971.
11   Prinsley, R.T., Dietz, K.-J. & Leegood, R.C. (1986) Biochim. Biophys. Acta 849, 254-263.
12   Walker, D.A., Sivak, M.N. & Cerovic, Z. (1983) In: Advances in Photosynthesis Research (C. Sybesma, ed.), pp. 645-652, Martinus Nijhoff/ Dr. W. Junk, The Hague.
13   Bilger, W. and Schreiber, U. (1986) Photos. Res. (in press).

DIURNAL VARIATIONS IN THE CAPACITY FOR FORMATION OF TRANSIENT PEAKS IN THE
DELAYED LUMINESCENCE FROM SCENEDESMUS OBTUSIUSCULUS AT HIGH AND LOW CO_2
CONDITIONS.

STAFFAN MELLVIG, DEPT OF BOTANY, UNIV. OF STOCKHOLM, 106 91 STOCKHOLM
SWEDEN.

1. INTRODUCTION

Delayed luminescence results from back reactions in darkness between
positive and negative charges in the reaction centre of photosystem 2.
Positive charges (the higher S-states on the oxygen evolving enzyme)
form charge pairs with the negative charges (the reduced QA/QB) and
recombine (11).

The transient peaks in delayed luminescence from cells of the unicellular alga Scenedesmus have been shown to be induced by white light
saturating for CO_2 assimilation when the CO_2 availability is low but
also during certain stages of phosphorus starvation (1). Transient
peaks appear to form, in response to an altered availability of ATP.
As metabolic changes directly or indirectly affects the reduction of
QA/QB, it can affect delayed luminescence. Delayed luminescence
decay kinetics were therefore measured at periodic intervals throughout the 24 h cell cycle.

2. MATERIAL AND EXPERIMENTAL

Cells of the unicellular green alga Scenedesmus obtusiusculus were
synchronously cultured. They were flushed with air +2.5% CO_2 (high
CO_2 cells) at a diurnal regime of 15 h light (PF 130 μmol·m^{-2}·s^{-1},
400-700 nm) and 9 h darkness, at 30°C. Cells were sampled every
second hour starting at "light on" in the cell cycle. The chlorophyll
content of the samples were adjusted to about 8.5 μg chl·ml^{-1} with
fresh medium and HCO_3^- in excess were given to the cells before the
measurements.

The samples of cell suspension (20 ml) were excited for varying
periods of time with white light (PF 1000 μmol·m^{-2}·s^{-1}). The decay
kinetics of the delayed luminescence (DL) and the accumulated output
of photons from the cells for 4 min were recorded with a photon-
counting device as described by Mellvig and Tillberg (2). The samples
were thereafter allowed to equilibrate to normal air (low CO_2 cells)
for 2 h in light (PF 80-100 μmol ·m-2.s^{-1}) before new excitations
and measurements of DL again were made. The pH varied between 6.2-
6.4 (high CO_2 cells) and between 6.5-6.9 (low CO_2 cells). The
accumulated output pf photons from high CO_2 cells measured for 4 min
(DL_{240}) was divided into three periods depending on when the transient
peaks in the DL decay curve appeared. 1) The early output, DL_{10},
included the photons accumulated after 10 s in darkness following
excitation. 2) The late output of photons, DL_{10-60}, included the

Biggens, J. (ed.), Progress in Photosynthesis Research, Vol. III. ISBN 90 247 3452 5
© *1987 Martinus Nijhoff Publishers, Dordrecht. Printed in the Netherlands.*

photons accumulated between 10 and 60 s in darkness following excitation.   3) The accumulated output of photons between 60 and 240 s, DL_{60-240}.

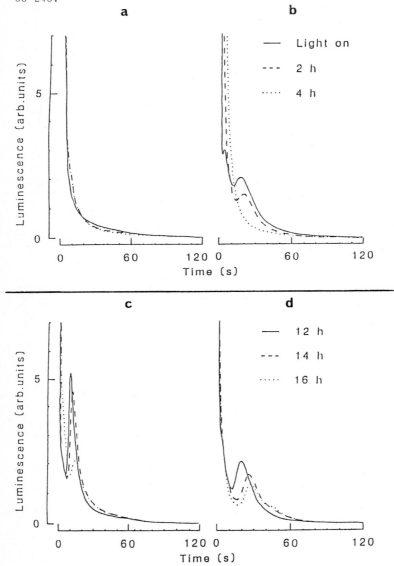

Fig. 1 a-d
Effects of white light excitation for 30 s on the delayed luminescence decay kinetics in high CO_2 cells (Figs 1a and c) and low CO_2 cells (Figs. 1b and d) at "light on" and after certain periods of time in the cell cycle as indicated.

3. RESULTS AND DISCUSSION

High CO_2 cultured Scenedesmus cells form a transient peak in the delayed luminescence decay kinetics after a 30 s excitation period at certain developmental stages (at the end of the 15 h light period of the cell cycle and at the beginning of the 9 h dark period as e.g. in Fig. 1). In cells sampled during the period when the photosynthetic capacity is high (3) (during the first 8 h of the cell cycle) no transient peak in the delayed luminescence could be induced. When the cells were exposed to only a 5 s excitation period, no transient peak in the DL decay kinetics was inducible, although a shoulder was observed in 14-16 h cells (not shown).

In low CO_2 cells white light excitation for 30 s induced a transient peak in the delayed luminescence during the whole cell cycle except in 4 h cells (Fig. 1).

The data presented shows that the capacity to form transient peaks in DL depends on cell age (photosynthetic capacity) and CO_2 availability. The amplitude of the transient peaks depends on the length of exposure to the saturating white light. There seems to be an inverse correlation in high CO_2 cells during the 24 h cell cycle between the amount of photons released early (0-10 s) and the amount released late (10-60 s) after a 30 s excitation period with saturating white light (Fig. 2) but not after a 5 s excitation period (not shown). Since DL_{240} does not change much during the cell cycle, this inverse correlation seems mainly to be a change from a high early to a high late output of photons.

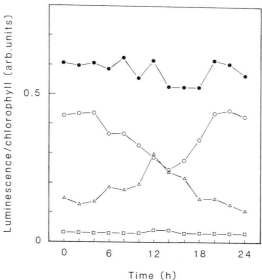

Fig. 2.
The output of photons$\cdot$chl^{-1} during different periods in the dark after the excitation with white light for 30 s.DL_{10} (O), DL_{10-60} ($\triangle$), DL_{60-240} ($\square$) and DL_{240} ($\bullet$).

During light excitation of low CO_2 cells the demand for ATP for CO_2 assimilation is high (4). A low ATP availability leads to increased levels of NADPH, which in turn increase the closing of PS2 traps and increase the rate of cyclic photophosphorylation (5). Cyclic photophosphorylation probably provides the extra ATP needed for the CO_2 assimilation (active bicarbonate transport (4)). In this way the cloroplasts can adjust the ratio of ATP/NADPH to the changing needs by the competitive interaction between non-cyclic and cyclic electron transport (6, 7).

During excitation changed turnover rates of one or several components in the Calvin cycle due to e.g. decreased CO_2 availability will induce changes in the amounts of ATP and NADPH. When turning off the excitation light the carboxylation reactions in the Calvin cycle cease. Reversal of some of the Calvin cycle reactions in darkness will give rise to different amounts of ATP depending on the conditions before and during excitation. According to Schreiber and Avron (8, 9) and Schreiber (10) ATP is hydrolyzed by the light-activated ATPase in darkness. Differencies in the pH which then develops over the thylakoid membrane, by the splitting of ATP, change the rate of reverse electron flow to the PQ-pool and to QA/QB (9). The recombination deactivations will then be enhanced and as a consequence of this, DL increases.

During the 24 h cell cycle the demand for CO_2, ATP and NADPH by the Calvin cycle changes due to the change in metabolic activity imposed by e.g. nuclear and cell division. Changes in the balance between formation and consumption of ATP and NADPH affect the capacity for reversed electron flow to QA/QB, thereby giving rise to different DL decay kinetics.

REFERENCES

1. Mellvig, S. and Tillberg, J. -E. Physiol. Plant. (in press).
2. Mellvig, S. and Tillberg, J. -E. (1984) Photochem. Photobiol.39:817-22.
3. Senger, H. (1970) Planta 90:243-266.
4. Spalding, M.H., Critchley, C., Govindjee and Ogren, W.L. (1984) Photosynthesis research 5:169-176.
5. Arnon, D.I. and Chain, R.K. (1977) In Photosynthetic Organelles,Special issue of Plant and Cell Physiology: 129-147.
6. Mills, J.D., Crowther, D., Slovacek, R.E., Hind, G. and McCarthy, R.E. (1979) Biochim. Biophys. Acta 547:127-137
7. Slovacek, R.E., Crowther, D. and Hind, G. (1980) Biochim. Biophys. Acta 592:495-505.
8. Schreiber, U. and Avron, M. (1977) FEBS Lett. 82:159-162.
9. Schreiber, U. and Avron, M. (1979) Biochim. Biphys. Acta 546:436-447.
10. Schreiber, U. (1980) FEBS Lett. 122: 121-124.
11. Rutherford, A.W., Crofts, A.R. and Inoue, Y. (1982) Biochim. Biophys. Acta 682:457-465.

ANALYSIS OF REGULATION AND LIMITATIONS OF PHOTOSYNTHETIC CARBON METABOLISM

JEFFREY R. SEEMANN AND THOMAS D. SHARKEY

BIOLOGICAL SCIENCES CENTER, DESERT RESEARCH INSTITUTE, P.O BOX 60220, RENO, NEVADA 89506 (JRS, TDS) and COLLEGE OF AGRICULTURE (JRS) and BIOLOGY DEPARTMENT (TDS), UNIVERSITY OF NEVADA, RENO, NEVADA 89507

INTRODUCTION

The pathways which serve to regenerate ribulose 1,5-bisphosphate (RuBP) (electron transport, photophosphorylation, photosynthetic carbon reduction cycle) and the pathways of starch and sucrose synthesis form an integrated system where a rate limitation in one pathway may limit the overall capacity for CO_2 assimilation. We are particularly interested in the limiting factors and regulatory responses of photosynthesis in leaves of plants exposed to various environmental stresses. Our approach has been to feed intact leaves with chemical compounds which depress the rate of CO_2 assimilation through known effects on metabolism. After feeding these compounds until photosynthetic capacity was decreased approximately 50%, leaves were freeze-clamped. Characteristics of RuBP carboxylase (RuBPCase) and RuBP pool size were measured and the biochemical basis for the reduction in photosynthetic capacity determined. These responses were compared to those which occur in leaves whose photosynthetic capacity has been reduced by stress (e.g. water deficits, salinity) or the feeding of endogenous compounds produced during stress, such as abscisic acid (ABA). We have found that mild water stress, salinity, ABA and vanadate feeding all reduce RuBPCase activity and photosynthetic capacity by some mechanism other than substrate availability, CO_2-Mg^{2+} activation state or inhibitor formation.

MATERIALS AND METHODS

Phaseolus vulgaris L. (var. Tendergreen) was grown as described by Sharkey et al (1). Leaflets were detached underwater and the petiole inserted in distilled water (DI). Compounds to be fed (typically dissolved in 95% ethanol) were added to DI (to a maximum ethanol concentration of 5%) and fed through the transpiration stream. Feeding times averaged 1hr. Prior to inhibitor feeding, the response of photosynthetic CO_2 assimilation (A) to intercellular partial pressure of CO_2 (C_i) was determined on the detached leaflet (measured as described by Sharkey et al [1]). The compound of interest (or 5% ethanol control) was then fed until photosynthetic capacity was reduced. The reduction is expressed as a percent of the rate at the same C_i prior to feeding. Thus all reductions in photosynthesis described below are 'nonstomatal' in nature. RuBPCase concentration and RuBP pool size were measured as described by Seemann and Sharkey (2). RuBPCase activation (initial activity/CO_2-Mg^{2+} activated activity) was measured as described by Seemann et al (3).

Biggens, J. (ed.), Progress in Photosynthesis Research, Vol. III. ISBN 90 247 3452 5

RESULTS

Feeding up to 5% ethanol for one hour did not reduce the rate of photosynthetic CO_2 assimilation (Table 1). RuBPCase activation state was $73\pm3\%$ for these control experiments. Results for each compound are referenced to control values determined at the same time, since activation state varied somewhat from day to day. The high $k_{cat}s$ indicate that under conditions used here, little or no inhibitor was present. Control plants had over 3-fold more RuBP than CABP binding sites.

Table 1. Effects of chemical feedings on photosynthetic characteristics of leaflets of Phaseolus vulgaris ($\pm$S.E.M.).

Treatment	Fed/Control % Ps	RuBPCase Act % of Control	kcat s^{-1}	RuBP mol/mol[1]
Control (5% EtOH), n = 10	99 ±1	100 ±4	20.0 ±1.3	3.21 ±0.27
Cyanazine (100uM), n = 6	45 ±2	91 ±4	19.4 ±1.8	1.03 ±0.14
Methyl viologen (2mM), n = 6	40 ±4	100 ±4	21.8 ±1.8	0.63 ±0.07
Sodium Azide (1mM), n = 3	47 ±3	91 ±2	22.3 ±2.8	0.42 ±0.07
Nigericin (120um), n = 5	56 ±4	76 ±6	23.3 ±0.5	1.57 ±0.26
Sodium Cyanide (5mM), n = 3	57 ±9	93 ±0	10.0 ±2.6	N.A.
Abscisic Acid (60uM), n = 6	69 ±5	93 ±1	21.6 ±1.6	3.12 ±0.50

[1] mol RuBP mol^{-1} CABP binding sites

The electron transport inhibitor cyanazine (a triazine like atrazine) and the artificial electron acceptor methyl viologen both reduced photosynthetic capacity (Table 1), presumably by limiting the reducing power available for RuBP regeneration. In both cases RuBPCase activation state and k_{cat} were unaffected and the pool size of RuBP was markedly reduced. Sodium azide had an effect similar to these inhibitors.

Nigericin is an ionophore which promotes K^+/H^+ exchange across membranes and uncouples photophosphorylation from electron transport. After feeding this compound RuBPCase was substantially deactivated and the RuBP pool size was reduced, though not to the same extent as with the electron transport inhibitors. Nigericin had no effect on the k_{cat} of

RuBPCase.

Cyanide may have a number of effects on metabolism. Once inside the chloroplast cyanide would be expected to combine with RuBP to form carboxyarabinatol (CABP) and carboxyribitol 1,5-bisphosphate, potent tight-binding inhibitors of RuBPCase. Cyanide feeding reduced both the photosynthetic capacity and total activatible RuBPCase activity of bean leaflets, with no effect on [14]CABP binding or activation state of RuBPCase. This is seen as a reduction in the k_{cat} of the enzyme (Table 1). This result indicates that the inhibition of RuBPCase activity was not due to the final product of a cyanide reaction being CABP, since [14]CABP binding would be reduced by the presence of unlabeled CABP. Rather, the data are entirely consistent with the final product being the in vivo inhibitor carboxyarabinitol 1-phosphate (3, JA Berry and GH Lorimer, personal communication), which can be displaced by [14]CABP. We hypothesize that the in vivo pathway for inhibitor synthesis may be by the reaction of endogenous cyanide with RuBP and subsequent removal of the 5-phosphate by a specific phosphatase.

The endogenous stress hormone ABA caused a substantial decline in photosynthesis but had very little effect on any of the biochemical parameters measured in this study.

DISCUSSION

A reduction in photosynthetic capacity must ultimately result from a decrease in the rate at which RuBPCase fixes CO_2 in the carboxylation reaction. This can result from either changes in substrate concentrations available to the enzyme or a reduction in the activity of the enzyme (the known mechanisms being reduced CO_2-Mg^{2+} activation and reduced k_{cat} by inhibitor synthesis [4]). Reduced RuBPCase activation state was measured in response to nigericin, reduced k_{cat} was seen after feeding cyanide and reduce RuBP pool size was seen in response to cyanazine, methyl viologen and nigericin. The pool size of RuBP after feeding cyanazine and methyl viologen is likely to limit the rate of photosynthesis (2). The deactivation of RuBPCase in response to the uncoupler nigericin is similar to the deactivation seen under conditions where either P_i supply or light limits ATP formation (1,4). Activation state regulation thus appears to be well correlated with changes in the extent to which ATP supply limits photosynthesis. It has been hypothesized that the reduced activation state of RuBPCase is not the ultimate rate limiting reaction in photosynthesis, but rather it is regulated when ATP formation limits photosynthesis (1,4)

Feeding ABA (Table 1), water stress (Sharkey and Seemann, unpublished), salt stress (2) and feeding sodium (ortho) vanadate (specific inhibitor of plasma membrane ATPases [5]) (Seemann and Sharkey, unpublished) all caused reductions in photosynthetic capacity which could not be explained by either a reduction in RuBPCase substrate availability or a measurable change in the activity of RuBPCase. However, in each instance the activity of RuBPCase must have been reduced, otherwise the rate of CO_2 assimilation would not have fallen. Therefore, we conclude that (1) environmental stress can reduce the activity of RuBPCase in vivo; (2) the mechanism is not reduced substrate availability, RuBPCase deactivation or inhibitor synthesis, and; (3) the mechanism involves a rapidly reversible phenomenon, since the change in activity cannot be detected following rapid extraction and assay.

ABA is known to affect K^+ exchange of cells. This is apparently the result of the binding of cis [+] ABA to specific proteins located on the plasmalemma (6), with a subsequent amplification and transmission of a signal to the chloroplast (7), since ABA has no direct effect on the photosynthetic capacity of isolated chloroplasts (8). This hypothesis is consistent with the large depression of photosynthesis at ABA concentrations well below that of specific target sites in the chloroplast, such as the catalytic site of RuBPCase.

ABA is also known to reduce the activity of leaf plasma membrane K^+/H^+ ATPases (9,10). The fact that vanadate caused similar effects on photosynthesis as ABA strongly suggests that a reduction in plasma membrane ATPase activity can lead to a reduction in RuBPCase activity and hence a reduction in photosynthetic capacity. This conclusion is consistent with those of Marre (11), who found that a wide variety of plant physiological and biochemical processes were sensitive to the rate of the plasmalemma K^+/H^+ ATPase. The mechanistic link between the activities of plasma membrane ATPases and RuBPCase is unknown.

We suggest that the common link between different types of stress which reduce photosynthetic capacity by this unknown mechanism may well be ABA. It is known that ABA is produced during water stress (7), salt stress (12) and flooding (13). Environmental stresses of these sorts are known to have a significant impact on the potential yield of agriculturally-important species (14) and this stress-induced reduction in RuBPCase activity and photosynthetic capacity that we have described may be an important contributor to such yield reductions.

Supported by USDA grant #85-CRCR-1-1656 (JRS), DOE grant #DE-FG08-84ER132234 (TDS) and by the USDI Geological Survey through the Nevada Water Resources Research Institute (JRS).

REFERENCES

1 Sharkey, T.D., Seemann, J.R. and Berry, J.A. (1986) Plant Physiol. 81, 788-791
2 Seemann, J.R. and T.D. Sharkey (1986) Plant Physiol. in press
3 Seemann, J.R., Berry, J.A., Freas, S.M. and Krump, M.A. (1985) Proc. Nat. Acad. Sci. (USA) 82, 8024-8028
4 Seemann, J.R. (1986) in Biological Control of Photosynthesis (Marcelle, R., ed.), pp. 71-82, Martinus Nijhoff, Dordrecht
5 Sze, H. (1985) Ann. Rev. Plant Physiol. 36, 175-208
6 Hornberg, C. and Weiler, E.W. (1984) Nature 310, 321-324
7 Raschke, K. (1982) in Plant Growth Substances (Wareing, P.F., ed.), pp. 581-590, Acedemic Press, New York
8 Kriedemann, P.E., Loveys, B.R. and Downton, W.J.S. (1975) Aust. J. Plant Physiol. 2, 553-567
9 Itai, C. and Roth-Bejerano (1986) Physiol. Plant. 66, 664-668
10 Lurie, S. and Hendrix, D.L. (1979) Plant Physiol. 63, 936-939
11 Marre, E. (1977) in Plant Growth Regulation (Pilet, P.E., ed.), pp. 54-66, Springer-Verlag, Berlin
12 Tal, M. (1977) Bot. Gaz. 138, 119-122
13 Wright, S.T.C. and Hiron R.W.P (1972) in Plant Growth Substances, 1970 (Carr, D.J., ed.), pp. 291-298, Springer-Verlag, Berlin
14 Boyer, J.S. (1982) Science 218, 443-448

THE ROLE OF ORTHOPHOSPHATE IN THE REGULATION OF PHOTOSYNTHESIS IN VIVO

I. MADHUSUDANA RAO, JAVIER ABADIA AND NORMAN TERRY
DEPT. PLANT & SOIL BIOLOGY, UNIVERSITY OF CALIFORNIA,
BERKELEY, CA 94720, U.S.A.

1. INTRODUCTION

The view that orthophosphate (Pi) is an important regulator of the rate of photosynthesis, and of the partitioning of triose phosphates between starch and sucrose biosyntheses, is to a large extent based on research carried out with in vitro systems (1,2). Recently, attempts to confirm the role of Pi in the regulation of photosynthesis in vivo have been made using whole plant systems in which leaf Pi was varied nutritionally (3-5). Using a nutritional approach to perturb leaf P_i concentration, we determined how changes in leaf phosphate status affect the rate of photosynthesis, levels of adenylates and Calvin cycle intermediates, and the activities of certain enzymes involved in photosynthetic carbon metabolism.

2. MATERIALS AND METHODS

Sugar beets (cv. F58-554H1) were cultured hydroponically in growth chambers at 25°C, 500 μmol PAR quanta m^{-2} s^{-1} photon flux density and a 16 h photoperiod (6). Low-P and control plants were grown as described in Rao et al. (5). In some experiments, Pi supply to low-P plants was increased by raising the P concentration in the culture medium (5). Measurements were made of net CO_2 uptake of individual attached leaves; activities of ribulose bisphosphate (RuBP) carboxylase (RuBPCase), ribulose-5-phosphate (Ru-5P) kinase, 3-phosphoglyceric acid (PGA) kinase, NADP-glyceraldehyde-3-phosphate dehydrogenase, fructose-1,6-bisphosphate (FBP) aldolase, FBP phosphatase, sedoheptulose-1,7-bisphosphate (SBP) phosphatase, triose-phosphate isomerase, and transketolase; and leaf metabolites including RuBP, PGA, and adenylates.

3. RESULTS AND DISCUSSION

When plants were supplied with 1/20th of the amount of Pi supplied to control plants (the "low-P" treatment), the acid soluble leaf P and total leaf P decreased by about 77% in a period of 3 weeks. This had a much greater effect on growth (especially total leaf area) than on photosynthesis. Total plant dry weight decreased by 77% with low-P treatment while the rate of light-saturated photosynthesis/area decreased by 32% (Table 1).

Biggens, J. (ed.), Progress in Photosynthesis Research, Vol. III. ISBN 90 247 3452 5
© *1987 Martinus Nijhoff Publishers, Dordrecht. Printed in the Netherlands.*

TABLE 1. Effect of low-P treatment on photosynthetic and growth characteristics of 4-5 week-old sugar beet plants.

Characteristics	Control	Low-P	% Control
Photosynthesis/area (μmol CO_2 m^{-2} s^{-1})	30.3	20.6	68
Chl/area (nmol cm^{-2})	47.5	58.8	124
Soluble protein (mg cm^{-2})	0.74	0.83	112
Acid-soluble P (% dry wt)	0.41	0.09	22
Total leaf area (cm^2)	2269	450	20
Total dry wt (g $plant^{-1}$)	23.5	5.3	23

Low P treatment increased the extractable activities of FBPase in the light (2-fold compared to the control values) and FBP aldolase (2.8-fold) and decreased the activities of transketolase (to 67% of control) (data not shown), NADP-G3P dehydrogenase (to 80%), and SBPase (to 65%) (Table 2). Low P treatment did not change (or had very slight effects on) the total activities of triose phosphate isomerase, RuBPCase, PGA kinase, and Ru-5P kinase.

TABLE 2. Effect of low-P treatment on the initial and total activities (nmol cm^{-2} min^{-1}) of Calvin cycle enzymes from leaves of 4-5 week-old plants. Plants were dark adapted for 8 h prior to irradiation for 1 h in the growth chamber.

Enzyme		Control	Low-P	% Control
RuBPCase	-initial	104	77	74
	-total	165	177	107
PGA kinase	-initial	174	99	57
	-total	371	348	94
NADP-G3P dehydrogenase	-initial	145	35	24
	-total	592	476	80
Ru-5P kinase	-initial	55	46	84
	-total	58	52	90
SBPase	-initial	26	10	38
	-total	26	17	65

Low-P treatment decreased RuBPCase activation by light, the initial activity being reduced by 26%. Low-P treatment also decreased the initial activities (without DTT in extraction and assay media) of NADP-G3P dehydrogenase, SBPase, PGA kinase, and Ru-5P kinase (Table 2).

RuBP and PGA in low-P leaves were only 9 and 33%, respectively, of the values observed in control leaves (Table 3). When Pi supply to low-P plants was increased, the rate of photosynthesis at an irradiance of 500 µmol quanta $m^{-2} s^{-1}$ (equivalent to growth chamber light intensity) increased to 98% of the control in 5 h (Table 3). Over the same period, the acid-soluble Pi of the leaf increased from 22 to 43% (of control), RuBP increased 5-fold, PGA 2-fold, and adenylates 4-fold. Thus, the rapid recovery of the rate of photosynthesis with the increase in Pi supply to low-P leaves was associated with large increases in RuBP, PGA and adenylate levels.

The 91% reduction in RuBP with low-P treatment suggests that RuBP regeneration, rather than RuBPCase kinetics, was limiting photosynthesis. Brooks (4) arrived at a similar conclusion. Despite the low level of RuBP, photosynthesis still functioned at 65% of the control value suggesting that RuBP concentration has to drop to remarkably low levels before photosynthetic rate is affected by RuBP regeneration. The regeneration of RuBP may be affected by several processes, including light harvesting, electron transport, and photophosphorylation. However, from our studies of the effect of low-P treatment on the structure and function of the thylakoid membrane (7) as well as those of Brooks (4), we conclude that the RuBP regeneration capacity of low-P leaves was not limited by photochemical capacity. Low P treatment may diminish RuBP levels by decreasing the activity of SBPase while increasing the activity of chloroplastic FBPase (Table 2), and ADP-glucose pyrophosphorylase (8). This would have the effect of increasing the amount of carbon stored as starch and decreasing the amount of carbon present in the sugar phosphate pools of the Calvin cycle (from SBPase to RuBPCase).

An alternative reason for the small amount of RuBP in low-P leaves may be that low-P treatment decreased the total amount of adenylates; ATP was decreased by 90% which may have contributed to the decrease in sugar phosphates in the Calvin cycle (Table 3). The lower adenylate levels observed in low P leaves may be due to the combined effect of lower rates of adenylate synthesis and higher rates of adenylate degradation (9). One surprising observation was that the total adenylate level of control leaves increased 6-fold after being illuminated for 6 h, i.e., from 20 (dark leaf value) to 120 nmol mg^{-1} Chl (in light). In low-P leaves, total adenylate levels also increased on illumination but to

TABLE 3. Effects of increasing Pi supply to low-P leaves. Data expressed as % control. Time points: 0 (i.e., low-P leaf), 2.5 and 5 h after Pi resupply to low-P leaf.

Parameter	0 h	2.5 h	5 h
Photosynthesis/area	77	87	98
Acid-soluble P	22	34	43
RuBP	9	26	44
PGA	33	37	68
ATP	14	18	53
Total adenylates	10	16	37

a lesser extent (1.7 x dark value). Thus, after 6 h of illumination low-P treatment decreased total adenylates by 90%, a value similar to the reduction in RuBP with low P treatment.

4. CONCLUSIONS

Low P treatment appears to reduce the rate of photosynthesis in vivo by diminishing the rate of regeneration of RuBP (the level of which was reduced by 91%); low-P had less effect on RuBPCase activity. Low P treatment may have decreased RuBP levels by decreasing the activity of SBPase, thereby diverting carbon to starch. Alternatively, low-P may have diminished RuBP by causing a 90% reduction in total leaf adenylates, especially ATP.

REFERENCES

1 Walker, D. A. (1980) In: Physiological Aspects of Crop Productivity, pp. 195-207, "Der Bund" AG, Bern, Switzerland
2 Herold, A. (1980) New Phytol. 86:131-144
3 Dietz, K-J. and Foyer, C. (1986) Planta 167:376-381
4 Brooks, A. (1986) Aust. J. Plant Physiol. 13:221-237
5 Rao, I. M., Abadia, J., and Terry, N. (1986) Plant Sci. 44:133-137
6 Terry, N. (1980) Plant Physiol. 65:114-120
7 Abadia, J., Rao, I. M., and Terry, N. (1986) Proc. VIIth International Congress on Photosynthesis, Martinus Nijhoff/Dr. W. Junk Publishers
8 Rao, I. M., Abadia, J., and Terry, N. (1986) Proc. VIIth International Congress on Photosynthesis, Martinus Nijhoff/Dr. W. Junk Publishers
9 Miginiac-Maslow, M., Nguyen, J. and Hoarau, A. (1986) J. Plant Physiol. 123:69-77

CONTROL FUNCTION OF HEXOSEMONOPHOSPHATE ISOMERASE AND PHOSPHOGLUCOMUTASE IN STARCH SYNTHESIS OF LEAVES

KARL-JOSEF DIETZ
Lehrstuhl Botanik I der Universität, Mittl. Dallenbergweg 64, 8700 Würzburg W.-Germany. Present Address: Biological Laboratories, Harvard University, 16 Divinity Avenue, Cambridge, MA 02138, USA.

1. INTRODUCTION
    Within biochemical reaction sequences the overall reaction rate is a function not only of the thermodynamic driving force of the chemical reaction, but also a function of the amount of catalysts and the degree of substrate saturation of the catalysts involved in the reactions. Thus, every single enzyme within the sequence may contribute to the rate limitation imposed on the reaction, i.e. may exert control function; the values which describe this control function vary between 0 (no control function) and 1 (solely rate controlling). The sum of the control values of all enzymes involved in the considered reaction sequence is 1 (1). It is rather difficult to describe the control function of individual photosynthetic reactions in vivo and in vitro by numerical values. Control functions vary with changes in environmental parameters of photosynthesis.
    Starch synthesis in chloroplasts is in a side path of the reactions regenerating ribulose-1,5-bisphosphate from triosephosphate and F6P in the Calvin cycle. F6P is isomerized by hexosemonophosphate isomerase to G6P which is transphosphorylated by phosphoglucomutase to G1P. The latter is converted to ADP glucose by ADP glucose pyrophosphorylase. This enzyme is known to be regulated by allosteric interactions with inorganic phosphate and 3-phosphoglycerate and is therefore considered to control carbon flux into starch synthesis (2, 3).
    The aim of this work was to determine whether hexosemonophosphate isomerase and phosphoglucomutase may assume colimiting functions in starch synthesis in vivo. First evidence for such a role of the isomerase has been described previously (4). In this report, measurements of chloroplast levels of G1P provide further insight into the extent of colimitation.

2. MATERIAL AND METHODS
2.1. Material: Spinach was grown in a green house. Biochemicals were obtained from Boehringer (Mannheim).
2.2. Methods: Cut leaves (Spinacia oleracea) were illuminated under conditions as indicated in the legends to the figures. When steady state of photosynthesis was established, metabolism was arrested by pouring liquid nitrogen into the cuvette and thereby freezing the leaves rapidly in the light. Chloroplasts were isolated by non-aqueous techniques (5) from freeze dried leaf material. Metabolites were measured in the chloroplast fractions as described previously (6). For accurate determinations of G1P (7) non-aqueous chloroplasts equivalent to upto 300 µg chlorophyll were used in one assay. CO_2 gas exchange and chlorophyll content of leaves were determined as described in (6).

ABBREVIATIONS: F6P: fructose-6-phosphate, G1P: glucose-1-phosphate, G6P: glucose-6-phosphate

3.   RESULTS AND DISCUSSION

3.1. Photosynthetic CO_2 fixation and levels of hexosemonophosphates
The rates of photosynthetic CO_2 fixation and of starch synthesis varies with environmental conditions. Photosynthesis increases with increasing CO_2 concentration (Fig. 1A) and increasing temperature (Fig. 2A). But whereas stromal levels of all three hexosemonophosphates involved in starch synthesis (fructose-6-phosphate, glucose-1-phosphate, glucose-6-phosphate) fall when temperature increases (Fig. 2B), levels of these metabolites reveal more complex changes with changes in CO_2 concentrations (Fig. 1B). The levels of the glucosemonophosphates first decreased and then increased with increasing CO_2 concentrations while F6P levels rose steadily at lower and remained constant at higher CO_2 concentrations (cf. also (4), where F6P first decreased at very low CO_2 levels). In parallel, the ratio of F6P to G6P passed through a maximum at about 330 $\mu l \cdot l^{-1}$ CO_2 (occasionally values

FIGURE 1:   FIGURE 2:

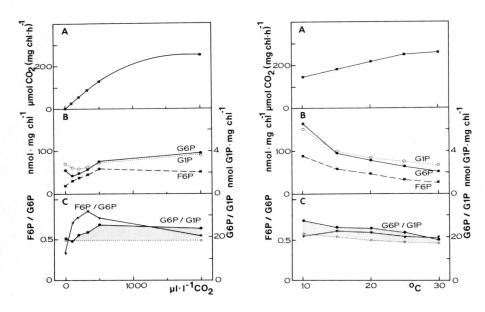

Figure 1: CO_2 fixation rate of spinach leaves in dependence of the CO_2 concentration (leaf temperature: $25^{\circ}C$; light intensity: 200 Wm^{-2}) (A). (B) shows levels of hexosemonophosphates in chloroplasts prepared non-aqueously from leaves treated as in (A) with the exception that the leaf temperature was $20^{\circ}C$. In (C) ratios of F6P to G6P ($\blacktriangledown$) and G6P to G1P ($\bullet$) were computed from the data in (B). The open circles show the G6P to G1P ratio in equilibrium, thus the dotted area indicates displacement of the phosphoglucomutase reaction from equilibrium.

Figure 2:   Temperature dependence of CO_2 fixation rate (A), chloroplast metabolite levels (B) and metabolite ratios (C). CO_2 concentration was 2000 $\mu l \cdot l^{-1}$, light intensity 200 Wm^{-2}. See also legend to Figure 1.

higher than 1 were observed) but fell to lower values when the CO_2 concentration was close to saturating photosynthesis. An optimum curve could also be seen in dependence of the temperature (Fig. 1C and Fig. 2C): at intermediate temperature the ratio of F6P to G6P was highest. This displacement of the hexosemonophosphate isomerase reaction from equilibrium indicates carbon drainage into starch synthesis.

As the maximal extractable activity of chloroplast hexosemonophosphate isomerase exceeds the actually observed rates of leaf starch synthesis this finding had been rather surprising. But it could be shown that 3-phosphoglycerate and other stromal metabolites, and ions in concentrations as present in the stroma caused a drastic inhibition of the isomerase in vitro. The inhibition had competitive properties (4) and explained the occurance of the displacement of the reaction catalyzed by the isomerase in vivo.

3.2. State of the reaction G6P ⟷ G1P

The phosphoglucomutase converts G6P into G1P. Depending on the temperature, the G6P concentration exceeds the G1P concentration by a factor ranging from 22 (at $10^{\circ}C$) to 17 (at $30^{\circ}C$) at equilibrium (see lower limit of the dotted area in Fig. 2C). Except of low CO_2 concentrations, the measured ratios of G6P to G1P were higher than equilibrium ratios indicating that the phosphoglucomutase reaction was also displaced from equilibrium by ongoing starch synthesis. However, the extent of displacement was much smaller than that of the isomerase reaction.

3.3. Supply of ADP glucose pyrophosphorylase with substrate

The K_M (G1P) of ADP glucose pyrophosphorylase is reported to be 40 $\mu mol \cdot l^{-1}$. Assuming a ratio of stromal volume to chlorophyll of 30 $\mu l \cdot mg^{-1}$ chlorophyll (8), stromal metabolite levels can be converted into concentrations. The lowest G1P levels were measured at high temperature and low CO_2 concentrations. Under these conditions the G1P concentration was close to 50 $\mu mol \cdot l^{-1}$ (data not shown). At $20^{\circ}C$ and intermediate CO_2 concentrations (330 $\mu l \cdot l^{-1}$) the G1P concentration exceeded hardly 80 $\mu mol \cdot l^{-1}$. Thus, the ADP glucose pyrophosphorylase was not saturated by substrate. A simple consideration reveals that the isomerase and the mutase contribute to the rate limitation in starch synthesis under these conditions: Assuming that both reactions were in equilibrium with F6P the G1P concentration would be 270 $\mu mol \cdot l^{-1}$ (instead of 80 $\mu mol \cdot l^{-1}$) and the ADP glucose pyrophosphorylase would be close to saturation by G1P. This illustrates the significance of rate limitations in the reactions F6P ⟷ G6P ⟷ G1P for the substrate saturation of the ADP glucose pyrophosphorylase. It should be mentioned that the control function of the isomerase and mutase would be even stronger if the K_M (G1P) of the phosphorylase is higher in vivo than in vitro as is likely in the ionic and metabolic environment of the chloroplast stroma.

Fig. 3 reveals the relationship between displacement of the isomerase reaction and stromal levels of either G6P or G1P. The displacement was highest when the stromal hexosemonophosphate levels were low. A possible explanation is the following: there are two main factors which govern the displacement of the isomerase reaction:
i.   the rate of starch synthesis, i.e. mainly the activation state of ADP glucose pyrophosphorylase
ii. the degree of inhibition of the isomerase by competitive interactions; 3-phosphoglycerate, glucose-6-phosphate and ions compete with fructose -6-phosphate for the active binding site of the enzyme (4).

FIGURE 3: Ratio of fructose-6-phosphate to glucose-6-phosphate as a function of glucose-6-phosphate (upper figure) and glucose-1-phosphate levels (lower figure). Each point is the mean value of 5 or 6 measurements; the data were obtained from Fig. 1 and Fig. 2 and from measurements in 330 $\mu l \cdot l^{-1}$ CO_2 at various temperatures. A F6P to G6P ratio of 0.25 to 0.33 corresponds to reaction equilibrium at the various temperatures.

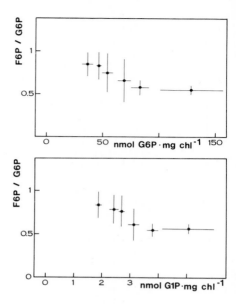

The rates of CO_2 fixation and starch synthesis were high under almost all conditions used to calculate the values in Fig. 3. Therefore a high degree of displacement of the reaction should be expected. But at high levels of hexosemonophosphates the displacement is smaller. This simply shows that high substrate concentrations overcome the competitive inhibition of the isomerase, for example by 3-phosphoglycerate.

4.   CONCLUSIONS
     Analysis of metabolites involved in plastid starch synthesis shows that the overall rate of the reaction sequence is a function of shared control by the enzymes involved. Therefore, the ADP glucose pyrophosphorylase is the major but not solely rate controlling factor in starch synthesis.

5.   ACKNOWLEDGEMENTS: This work was supported by the DEUTSCHE FORSCHUNGS-GEMEINSCHAFT. Prof. Heber's continuous support of this work is gratefully acknowledged.

6.   References
1 Kacser, H. and Burns, J.A. (1979) Biochem.Soc.Trans. 7, 1149-1160
2 Gosh, H.P. and Preiss, J. (1965) Biochem. 4, 1354-1361
3 Preiss, J. (1984) Trends in Biochem. Sci. 9, 24-27
4 Dietz, K.-J. (1985) Biochim.Biophys.Acta 839, 240-248
5 Heber, U. and Willenbrink, J. (1964) Biochim.Biophys.Acta 82, 313-324
6 Dietz, K.-J. and Heber, U. (1984) Biochim.Biophys.Acta 767, 432-443
7 Bergmeyer, H.-U. and Klotzsch, H. (1963) in: Methods of Enzymatic Analysis (Bergmeyer, H.-U. ed.) Verlag Chemie, Weinheim, 131-133
8 Heldt, H.F. and Sauer, F. (1971) Biochim.Biophys.Acta 234, 83-91

THE REGULATION OF PHOSPHOGLYCERATE KINASE IN THE CHLOROPLASTS AND
CYTOPLASM OF BARLEY LEAVES.

EILEEN M.McMORROW and J.WILLIAM BRADBEER,
DEPARTMENT OF BIOLOGY, KING'S COLLEGE LONDON, 68, HALF MOON LANE, LONDON
SE24 9JF.

INTRODUCTION
        The chloroplast and cytoplasmic Phosphoglycerate kinase (PGAK)
[EC 2.7.2.3.] isoenzymes have important roles in photosynthetic carbon
reduction and glycolysis respectively. Most of the previous work on PGAK
regulation has been performed on the yeast (1) or the mammalian enzymes
see e.g. (2). Although it was previously thought that the plant enzymes
were specifically regulated by energy charge levels (3), the barley
isoenzymes appear to have more complex regulatory mechanisms and to show
some similarity to the yeast enzyme as previously shown in preliminary
studies on the enzymes from spinach and wheat (4).

MATERIALS AND METHODS
        The experiments were performed on the separate barley PGAK
isoenzymes which were isolated as described in the accompanying paper in
this volume (5).
        The enzyme was assayed by a linked reaction with
glyceraldehyde-3-phosphate dehydrogenase (TD), [EC 1.2.1.12.] The
consumption of NADH was followed by measuring the decrease in absorbance
at 344nm on a CECIL CE 505 spectrophotometer. One unit of activity is
defined as the amount of enzyme which consumes one µmol of substrate per
minute at 30°C. The standard 1ml assay mixture contained 3µmol $MgATP^{2-}$,
1 or 10µmol free Mg^{2+}, 5µmol PGA, 50µmol Tris pH 7.8 and approximately 2.7
units of TD. Following discussions with Dr.M.Larsson-Raznikiewicz who has
been largely responsible for the work on the kinetics of the yeast enzyme,
the assay conditions were set up to comply with the following conditions:
1.The true substrates of the PGAK reaction in the 'backward' direction
(in the direction of photosynthetic carbon reduction), are PGA and $MgATP^{2-}$.
It is essential to allow for the dissociation constant of $MgATP^{2-}$ and add
the appropriate amounts·of Mg^{2+} and ATP to give a variable amount of $MgATP^{2-}$
with a fixed level of Mg^{2+} or ATP^{4-}. The dissociation constant used for
$MgATP^{2-}$ was 0.08 mM (1). 2.The reaction is inhibited by the presence of
heavy metal ions. All reagents were prepared in distilled and deionised
water. The buffers and substrates were extracted with dithizone in CCl_4 to
remove heavy metals. All glassware (pyrex) was washed in dilute nitric
acid followed by exhaustive rinsing in heavy metal-free water. 3.Sulphate
is known to activate the PGAK reaction at low concentrations and to
inhibit at high concentrations (6). The ammonium sulphate in the
commercially obtained TD was removed by centrifugation followed by 'rapid
dialysis' of the resuspended pellet on a small column of Sephadex G25.

RESULTS AND DISCUSSION
        The results of the kinetics experiments are shown in Figs. 1,2
and 3. Each point is the mean of four assays which were performed within

the same experiment. The data were plotted by a regression programme on a BBC B plus microcomputer, which allowed both linear and non-linear regression analysis. The 'best-fit' for all the data has been shown to be non-linear with evidence of substrate activation for both substrates. Earlier work on the yeast (1) and human erythrocyte enzyme (2) also gave non-linear double-reciprocal plots. The yeast enzyme only produced non-linear plots at high levels of free Mg^{2+}.

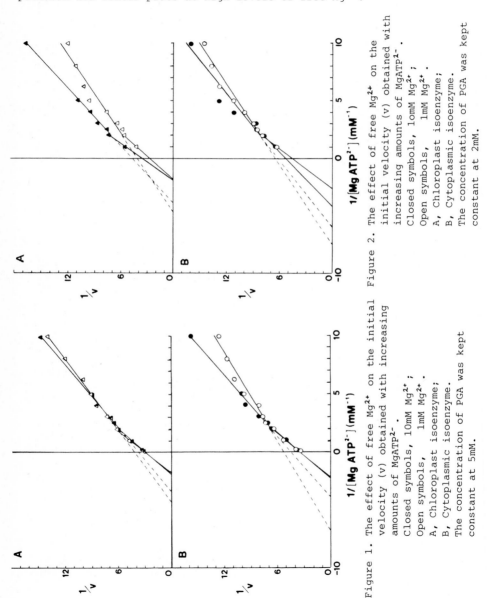

Figure 1. The effect of free Mg^{2+} on the initial velocity (v) obtained with increasing amounts of MgATP^{2-}.
Closed symbols, 10mM Mg^{2+};
Open symbols,   1mM Mg^{2+}.
A, Chloroplast isoenzyme;
B, Cytoplasmic isoenzyme.
The concentration of PGA was kept constant at 5mM.

Figure 2. The effect of free Mg^{2+} on the initial velocity (v) obtained with increasing amounts of MgATP^{2-}.
Closed symbols, 10mM Mg^{2+};
Open symbols,   1mM Mg^{2+}.
A, Chloroplast isoenzyme;
B, Cytoplasmic isoenzyme.
The concentration of PGA was kept constant at 2mM.

In this presentation of the kinetic data for the barley isoenzymes, the double-reciprocal plots have been divided by linear regression analysis of the two parts of the curves corresponding to the high (0.5-10mM), and low (0.1-0.5mM) substrate concentrations. This procedure allows a comparison of the apparent Km values for both substrates over the two substrate concentration ranges. The effect of free Mg^{2+} on the reaction velocities for both isoenzymes was also evaluated. Although the PGAK isoenzymes do not conform to Michaelis Menten kinetics, they have been analysed in a conventional fashion (7).

Figure 1A shows that free Mg^{2+} appears to be a mixed or partially competitive inhibitor of the chloroplast enzyme with respect to $MgATP^{2-}$ at low substrate concentration, (with 5mM PGA), whilst at higher concentrations of $MgATP^{2-}$ there appears to be a slight activation by Mg^{2+}. These double-reciprocal plots are quite close together at high substrate concentrations, the Km values being 0.53 and 0.50mM at 1 and 10mM Mg^{2+} respectively. The cytoplasmic enzyme appears to show a similar pattern of inhibition by Mg^{2+} on the $MgATP^{2-}$ kinetics at 5mM PGA, without the very slight activation just described for the chloroplast enzyme (see Fig. 1B). The apparent Km values for $MgATP^{2-}$ determined for the cytoplasmic enzyme are all lower than the corresponding values for the chloroplast isoenzyme.

As shown in Figure 2A, the effect of Mg^{2+} on the $MgATP^{2-}$ kinetics was altered when the PGA concentration was lowered to 2mM. The chloroplast enzyme showed evidence of mixed inhibition by Mg^{2+} with respect to $MgATP^{2-}$ at low substrate concentrations. At high substrate concentrations there appears to be mixed inhibition, although the double-reciprocal plots do not intersect above the 1/s axis but suggest a convergence point below this line. Comparing Fig.1A with Fig.2A (both for the chloroplast enzyme), there is a much greater difference in the reaction velocities obtained at low levels of $MgATP^{2-}$ with 1 or 10 mM Mg^{2+} when the PGA concentration is lowered to 2mM, compared with the rates

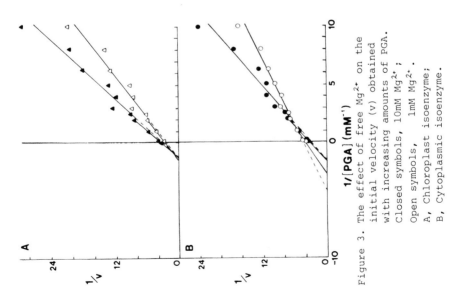

Figure 3. The effect of free Mg^{2+} on the initial velocity (v) obtained with increasing amounts of PGA. Closed symbols, 10mM Mg^{2+}; Open symbols, 1mM Mg^{2+}. A, Chloroplast isoenzyme; B, Cytoplasmic isoenzyme.

obtained at 5mM PGA. The cytoplasmic enzyme however maintains the same mixed or partially competitive inhibition pattern for Mg^{2+} with respect to $MgATP^{2-}$ at low and high substrate concentrations even at the lower level of PGA, (as seen by comparing Fig.1B with Fig.2B).

At low levels of substrate, Mg^{2+} appears to behave as a non-competitive inhibitor with respect to PGA for the chloroplast enzyme. At higher levels of substrate, the linear plots intersect just above the abscissa, which may still qualify for non-competitive inhibition, but the Km values diverge slightly, so this is probably mixed inhibition (Fig.3A).

The cytoplasmic enzyme shows a more complex inhibition pattern for Mg^{2+} with respect to the PGA kinetics, as shown in Fig.3B. At substrate concentrations <1mM there appears to be some kind of mixed or possibly partially competitive inhibition. The latter type of inhibition describes a situation in which an inhibitor affects the binding of a substrate to an enzyme without completely preventing it from occuring. Consequently the Km for that substrate is increased whilst the Vmax is not altered. The inhibition patterns obtained for Mg^{2+} for the cytoplasmic enzyme with respect to PGA do not fit exactly into any of those described by Dixon and Webb (7). At concentrations of PGA >1mM, Mg^{2+} appears to behave as an activator by decreasing the Km and also increasing Vmax. The double reciprocal plots therefore intersect to the right of the 1/v axis. The Km for PGA is much lower at 1mM Mg^{2+} than at 10 mM for both high and low substrate concentrations compared with the corresponding values obtained for the chloroplast enzyme.

All of the experiments indicate that free Mg^{2+} behaves predominantly as an inhibitor with respect to both $MgATP^{2-}$ and PGA with some slight activation effects.

The effect of ADP and AMP on the activity of the two isoenzymes was investigated at both (10mM) and (1mM) free Mg^{2+}, with a fixed level of 1mM $MgATP^{2-}$, and 2mM PGA (data not shown). The chloroplast enzyme was inhibited to a greater extent by both AMP and ADP than the cytoplasmic isoenzyme. The degree of inhibition by AMP was greater at 10mM Mg^{2+} for the chloroplast isoenzyme, whilst the reverse was true for the cytoplasmic isoenzyme except at 10mM AMP where there was no effect of free Mg^{2+}. At concentrations of less than 0.5mM ADP, free Mg^{2+} appeared to increase the inhibition of the chloroplast enzyme by ADP. At concentrations greater than 0.5mM ADP, the effect of Mg^{2+} on the ADP inhibition appeared to be reversed. The cytoplasmic isoenzyme was always inhibited to a greater extent at 10mM Mg^{2+} than at 1mm Mg^{2+} at all concentrations of ADP.

Similar studies on the yeast enzyme gave a greater degree of inhibition by both ADP and AMP at 10mM than at 1mM Mg^{2+}(8).

The differences in the regulation of the PGAK isoenzymes on either side of the chloroplast envelope and the effects of Mg^{2+} on both substrate binding and inhibition by ADP and AMP, may prove to be important in the control of carbon partitioning in green plants.

REFERENCES

1 Larsson-Raznikiewicz, M. (1964)   Biochim. Biophys. Acta. 85, 60-68.
2 Ali, M. and Brownstone, Y.S.(1976) Biochim, Biophys. Acta. 445, 89-103.
3 Pacold, I. and Anderson, L.E. (1975)Plant Physiol.55, 168-171.
4 Larsson-Raznikiewicz, M.(1983) Acta Chem.Scand. B37, 657-659.
5 McMorrow, E.M. and Bradbeer, J.W. (1986) this volume.
6 Scopes R.K. (1978) Eur.J. Biochem. 85, 503-516.
7 Dixon M. and Webb (1979) in Enzymes 3rd ed. pp.332-354.Longman, London
8.Larsson- Raznikiewicz, M. and Arvidsson, L. (1971) Eur.J. Biochem. 22, 506-512.

EFFECTS OF K+ ON THE MgATP STIMULATION OF CO_2 AND PGA-SUPPORTED O_2 EVOLUTION BY INTACT SPINACH CHLOROPLASTS

Yung-sing Li and Martin Gibbs, Institute for Photobiology, Brandeis University, Waltham, MA 02254, U.S.A.

INTRODUCTION
    Results of an investigation of the MgATP stimulation of photosynthesis in intact spinach chloroplasts in a tris-tricine buffer (pH 8.1) containing less than 10mM monovalent cation, are reported. Our results suggest that in tris-tricine buffer, stromal pH is suboptimal in the light, and that MgATP in the presence of KCl increases stromal pH.

    A lowering of stromal pH produces two secondary effects on the functioning of intact chloroplasts: (i) inhibition of the enzymes of Calvin cycle (1); (ii) formation of an outward directed H^+ gradient (medium pH 8.1 is higher than stromal pH in the light) which decreases the Vmax of PGA uptake and the Km of Pi uptake (2). The first effect results in an inhibition of the CO_2 supported O_2 evolution. The second effect results in an inhibition of the PGA supported O_2 evolution due to the limitation of PGA uptake, even though PGA supported O_2 evolution is stromal pH insensitve (in the case of reference 3, a lowering of stromal pH in the light by decreasing medium pH did not form a H^+ gradient across the envelope). The second effect also makes both the CO_2 and the PGA supported O_2 evolution sensitive to Pi inhibition.

    The postulate that MgATP with KCl increases stromal pH to release the two secondary effects explains the stimulatory effect of MgATP.

MATERIAL AND METHODS
    Spinach (<u>Spinacia oleracea</u>) was cultivated under controlled conditions (8h light, 20°C/16h dark, 15°C). Chloroplast isolation and O_2 evolution assay were performed according to (4). The buffer systems contained: 0.33M sorbitol, 2mM EDTA, 1mM $MgCl_2$, 1mM $MnCl_2$, 5mM $KHCO_3$, 500-1000 unit/ml catalase, Pi as indicated in each experiment and 50mM tricine-NaOH or tris (15mM)-tricine (50mM), both at pH 8.1. MgATP was prepared from either tris- or Na-ATP (Sigma Chem. Co.) The ratio of Mg^{2+}/ATP is 2:5.

RESULTS AND DISCUSSION
    In initial experiments, we found that in tricine-NaOH buffer, monovalent cations are as effective as ATP or MgATP in stimulating CO_2 supported O_2 evolution. For instance, the relative rates of O_2 evolution were 100, 193, 190 and 180 for samples with no addition, 5mM ATP, 5mM MgATP and 70mM NaCl respectively. It was possible that Na+ carried by the Na-salt of

Table 1 MgATP stimulation of CO_2 supported O_2 evolution as a function of KCl concentration. Tris-tricine buffer; 0.2mM Pi, tris-ATP.

Number in parentheses indicates the number of samples

KCl, mM	0	20	33	66	100	150
			umol/mg Chl·h			
No MgATP	26	20		20	21	26
MgATP, 2.5mM	18	15	12	22	40±4(3)	35

ATP might contribute to part of the ATP effect. To eliminate the Na+ effect we used tris-tricine buffer.

In tris-tricine buffer, MgATP (prepared from tris-ATP) alone does not but with KCl it stimulates CO_2 supported O_2 evolution (Table 1).

There exists two K^+/H^+ exchange systems in the chloroplast envelope, an ATPase mediated system functional at low KCl, and an ATP independent system at high KCl (5). It has been known that K^+ stimulates CO2 supported O_2 evolution by increasing stromal pH (5). The MgATP-KCl dependent stimulation of O_2 evolution in Table 1 suggests that MgATP-KCl stimulates O_2 evolution by increasing stromal pH. The ATP dependent K^+/H^+ exchange system may be involved here. However, the stimulatory effect of MgATP at high KCl is not sensitive to oligomycin (not shown). But the postulate that MgATP-KCl affects stromal pH is strengthened by the following reasoning and experiments.

If the MgATP-KCl effect in tris-tricine buffer is stromal pH related, tris-tricine must have prevented stromal pH from rising to an optimal value for photosynthesis. As a result, an outward directed H^+ gradient is formed across the envelope (with a medium pH 8.1 and a lower stromal pH). Such a gradient decreases the Km of Pi for the Pi translocator (2). Table 2 shows that the MgATP effect is more pronounced at 1mM than 0.5mM Pi. An outward directed H^+ gradient also decreases the Vmax of PGA uptake (2). A tris-tricine inhibited PGA supported O_2 evolution is then expected.

Table 2 Pi concentration dependency of the MgATP stimulation of CO_2-supported O_2 evolution. Tris-tricine buffer; 100mM KCl, tris-ATP.

	umol/mg Chl·h	
MgATP, mM	0.5mM Pi	1mM Pi
0	44± 0920	14.3± 1.1(2)
2.5	53.6± 6.9(3)	15.3
5	60.1± 1.6(2)	31.5
10		34.5

Table 3   Comparative study of the effects of two buffers on CO_2 and PGA
supported O_2 evolution.

Pi		umol/mg Chl·h		rate ratio
	Acceptor			Tris-tricine
mM		Na-tricine	tris-tricine	Na-tricine
0.25	CO_2	85	19	0.22
1.25	PGA	56	26	0.47
0.25	PGA	54	28	0.52
0.1	PGA	44	29	0.66

Table 3 shows that tris-tricine inhibits PGA supported O_2 evolution; the inhibition is less with less Pi, in agreement with the fact that Pi competes
with PGA (6). MgATP stimulates PGA supported O_2 evolution, which is KCl-
independent (Table 4). The use of Na-salt PGA may contribute to the KCl-
independency in this case (in some experiments, 2.5mM KCl promotes a MgATP
stimulation of CO_2 supported O_2 evolution). Tris-tricine inhibits methylviologen dependent uncoupled electron transport in intact chloroplasts, the
inhibition is relieved by KCl, independent of MgATP (not shown). The MgATP
effect on PGA supported O_2 evolution is independent of the PGA concentration (Fig. 1), suggesting that the effect is not due to an increase
in stromal ATP. An increase in stromal ATP increases the ratio of [PGA]
[ATP]/[ADP], which favors PGA reduction (7). In such a case, the MgATP
effect on O_2 evolution should be less at higher PGA concentration, for
then the ATP effect on the ratio is less.

These results suggest that the MgATP effect is related to stromal pH.
Because of the MgATP dependency of the effect, the involvement of envelope ATPase is anticipated, yet the effect is oligomycin insensitive. Perhaps, MgATP modifies the properties of the K^+/H^+ exchanger, instead of
providing substrate for the ATPase.

Table 4   Stimulation of PGA supported O_2 evolution by MgATP. Tris-tricine
buffer; 0.4mM pi; 1.25mM PGA (Na-salt, pH 8); tris-ATP.

	umol/mg Chl·h	
	Control	MgATP
KCl, mM		2.5mM
0	60.6± 3.8(3)	81.5
66	62.5	84.2
100	55.3	72.7

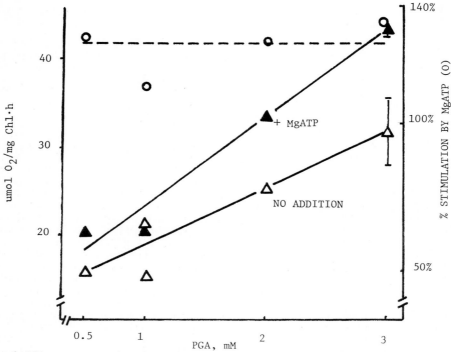

Fig. 1 PGA concentration dependency of MgATP effect on PGA supported O_2
evolution. 1mM Pi, 5mM MgATP (tris-salt).

REFERENCES
1. Kelly, G.J., Latzko, E. and Gibbs, M. (1976) Ann. Rev. Plant Physiol.
   27, 181-205
2. Flügge, U.-I., Gerber, J. and Heldt, H.W. (1983) Biochim. Biophys.
   Acta 725, 229-237
3. Werdan, K., Heldt, H.W. and Milovancev, M. (1975) Biochim. Biophys.
   Acta 396, 276-292
4. Berkowitz, G.A. and Gibbs, M. (1982) Plant Physiol. 70, 1143-1148
5. Maury, W.J., Huber, S.C. and Moreland, D.E. (1981) Plant Physiol. 68,
   1257-1263
6. Fliege, R., Flügge, U.-I., Werdan, K. and Heldt, H.W. (1978) Biochim.
   Biophys. Acta 502,232-247
7. Robinson, S.P. and Walker, D.A. (1979) Biochim. Biophys. Acta 545,
   528-536

ACKNOWLEDGMENT
   The authors thank NSF (PCM83-04147) for support.

GLYPHOSATE INHIBITS PHOTOSYNTHESIS AND ALLOCATION OF CARBON TO STARCH IN
SUGAR BEET LEAVES.

DONALD R. GEIGER, SHELLY W. KAPITAN, and MICHELLE A TUCCI, DEPARTMENT OF
BIOLOGY, UNIVERSITY OF DAYTON, DAYTON, OHIO 45469-0001

1. INTRODUCTION
    Glyphosate [(N-phosphonomethyl) glycine], a non-selective,
post-emergence herbicide, affects a number of major processes in plants.
At low dose rate, glyphosate affects growth and the partitioning of carbon,
without killing the plant. We believe that similar responses occur shortly
after application of lethal doses of glyphosate and that these early effects
may reveal mechanisms of glyphosate action. Several hours after its
application, glyphosate caused a decrease in accumulation of foliar starch,
altered partitioning of carbon among sinks but had no observable effects on
export (1). The present study was conducted to study details of the action
of glyphosate on net carbon exchange (NCE) and on allocation of newly-fixed
carbon in source leaves during the first 24 h after application.

2. PROCEDURE
2.1. Materials and methods
    2.1.1. Labeling and NCE Measurements: Experiments were conducted on 5-
        to 6-week-old sugar beet plants (Beta vulgaris L. Klein E
        multigerm). Growth of plants and experiments were carried out
        under a photon flux density of 450 μmol m^{-2} s^{-1} at leaf blade
        level. Steady-state labeling with $^{14}CO_2$ was carried out in a
        closed system as described by Geiger and Fondy (2). Export from
        a leaf was determined as the difference between the amount of
        carbon fixed and the amount accumulated in the source leaf during
        a given period. Stomatal conductance was measured with a steady
        state porometer (Licor Model LI-1600C).
    2.1.2. Glyphosate application: Solutions of analytical grade glyphosate,
        99% purity (Monsanto Agricultural Products Co., St. Louis), were
        adjusted to pH 5.5 with isopropylamine and diluted to a final
        concentration of 17-mM. Unless stated otherwise, Tween-20
        surfactant was added to give a final concentration of 0.01%
        (v/v). Sollutions were distributed evenly over the leaf surface
        with an atomizer.
    2.1.3. Leaf Sampling: To provide leaf material for replicate samples
        within leaves and among similar leaves, experiments were carried
        out on three or four adjacent source leaves, each successive
        leaf being approximately 2d younger. Sampling was done by
        removing sets of four leaf discs (total area 0.64 cm^2), one disc
        from each quadrant of the source leaf blade, at intervals during
        and after labeling. Immediately upon collection, samples to be
        fractionated were extracted at 65 °C in chloroform:methanol (1:4,
        v/v). Samples to be used for measuring total dry weight and
        labeled carbon were frozen on solid CO_2, dried under vacuum
        while frozen, weighed and prepared for scintillation counting.
    2.1.4. Analysis of Soluble Extract: The extract was separated by
        Sephadex ion-exchange columns into neutral, organic and amino
        acid fractions according to the method of Redgwell (3). Labeled

Biggens, J. (ed.), Progress in Photosynthesis Research, Vol. III. ISBN 90 247 3452 5
© *1987 Martinus Nijhoff Publishers, Dordrecht. Printed in the Netherlands.*

carbon present in the chloroform:methanol-extract and from each fraction was determined by liquid scintillation counting. Shikimate was separated and measured with a Waters HPLC (Milford, MA) equipped with a Model 450 variable wavelength UV detector. Samples were separated on a 5-μm C-18 Waters Radial Pak column, 250 x 4.6 mm, by elution with 50 mM KH_2PO_4 buffer adjusted to pH 2.4 with phosphoric acid or on a 5-μm silica-bonded NH_2 column, Merck Lichrosorb, 250 x 4.6 mm by elution with acetonitrile:water:phosphoric acid, 95:4:1 v/v.

2.1.5. <u>Analysis of Insoluble Residue</u>: Starch was analyzed by a method modified from Outlaw and Manchester (4) following hydrolysis with amyloglucosidase. Labeled carbon present in the residue remaining after alcohol extraction of a second set of samples was measured by scintillation counting. The label present in samples after extraction by alcohol and KOH was calculated by subtracting the label in the KOH extract, following centrifugation, from the label present in these alcohol-extracted samples. This value likely included the radioactivity present in some proteins, as well as in cell-wall and other structural materials.

3. RESULTS AND DISCUSSION

3.1. <u>Glyphosate lowered net carbon exchange and stomatal conductance</u>: From 4 to 5 h after glyphosate application, NCE rate began to decline and continued to do so for the rest of the light period. Approximately 4 h after treatment, stomatal conductance began to decrease in leaves treated with 17-mM glyphosate, either with or without surfactant. The marked decrease in conductance was not observed in leaves treated with surfactant solution or with water. The effect on stomatal conductance was reported previously (5,6).

3.2. <u>Glyphosate halted accumulation of newly-fixed carbon in starch</u>: Glyphosate inhibited allocation of newly-fixed carbon to starch within 4 to 6 h after application to sugar beet leaves, a result previously reported by Gougler and Geiger (1). By the end of the light period, the amount of starch stored in treated leaves was 30% less than in control leaves, a difference amounting to approximately 45 μg C cm^{-2}. Steady-state labeling with $^{14}CO_2$ throughout the light period confirmed the decrease in starch accumulation and also revealed that treated leaves contained less labeled carbon than control leaves (Table 1), a result consistent with the observed decrease in stomatal conductance and NCE. The difference in ^{14}C-starch, 150 nCi cm^{-2} or 60 μg C in [^{14}C]-starch cm^{-2}, was the same as the difference in the ^{14}C-content between treated and control leaves. In contrast, the ^{14}C-content of the chloroform:methanol-soluble fraction and of the residue remaining after starch extraction was similar in treated and control leaves, as was daytime export.

3.3. <u>Inhibition of starch not a result of direct competition with shikimate</u>: Shikimate accumulation increased dramatically in glyphosate-treated plants as reported in previous studies (7). In the 24 h following application, shikimate accumulated to 325 μg g^{-1} fresh wt, an 80-fold increase over the level found in control leaves. The carbon diverted to shikimate during 10 h after glyphosate application was approximately 2 μg C cm^{-2}, only 4% the size of the approximately 50 μg C cm^{-2} difference in foliar starch accumulation at that time. These data show that inhibition of starch accumulation was not the direct result of competition with shikimate for newly-fixed carbon.

TABLE 1.   Distribution of radioactivity in glyphosate-treated and
untreated leaves following steady-state labeling for 14 h.   At 4 h
after the beginning of the light period 17 mM glyphosate was applied to
leaves.   Values listed are the mean of data from 2 experiments followed
by the relative size of the difference between the 2 values [(100 x
absolute difference)/mean].

Fraction		^{14}C-content	
		Control	Glyphosate
		nCi cm^{-2}	
Insoluble	(total)	370 (4.2)	213 (5.9)
Starch		370 (3.2)	154 (8.7)
Non-starch		61 (9.1)	68 (8.2)
Soluble	(total)	238 (5.0)	230 (1.8)
Water-soluble		175 (4.0)	172 (3.2)
Neutral		127 (4.4)	130 (7.5)
Organic Acid		28 (13.0)	31 (17.6)
Amino Acid		23 (15.0)	24 (11.3)
Chloroform-soluble		62 (6.7)	63 (6.4)
Total Leaf		613 (4.7)	454 (6.2)

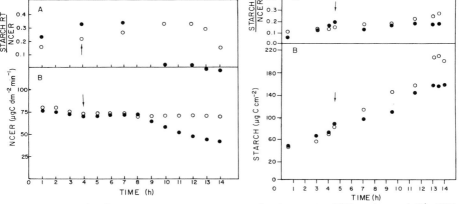

FIGURE 1.   (left) A) Ratio of starch accumulation rate:NCE rate and B) NCE
rate during light period for glyphosate-treated (●) and control leaves (o).
Arrow indicates time of treatment with 17 mM glyphosate.

FIGURE 2. A) Ratio of starch accumulation rate:NCE rate and B) starch
accumulation rate during light period for treated (●) and control (o)
leaves.   Arrow indicates time when NCE was lowered by approximately 20%
by lowering CO_2 concentration.

3.4. Glyphosate induced a change in allocation of carbon: Inhibition of NCE, by itself, is not able to account for the marked inhibition of starch accumulation following glyphosate application. A change in allocation of newly-fixed carbon to starch and to sucrose also occurred. This additional effect of glyphosate can be seen by comparing the ratio of rates of NCE and starch storage in glyphosate-treated (Fig. 1A.) and in control leaves under lowered CO_2 (Fig. 2A.). In leaves treated with glyphosate, not only was NCE inhibited (Fig. 1B.) but the proportion of newly-fixed carbon allocated to starch fell to zero as NCE decreased several hours after treatment (Fig. 1A.). When NCE was decreased in control leaves by lowering CO_2 concentration, accumulation of starch continued to follow the usual pattern and fell only to the extent that NCE decreased (Fig. 2B). As a consequence of glyphosate treatment, virtually no carbon accumulated in starch, while daytime export continued nearly unchanged (1). The small amount of carbon diverted to shikimate may have come at the expense of starch but most of the decrease in the latter resulted from a change in allocation of newly fixed carbon.

REFERENCES
1 Gougler, J.A. and Geiger, D.R. (1984) Weed Sci. 32, 546-551
2 Geiger, D.R. and Fondy, B.R. (1980) Plant Physiol. 64, 361-365
3 Redgwell, R.J. (1980) Anal. Biochem 107, 44-50
4 Outlaw, W.H. Jr. and Manchester, J. (1979) Plant Physiol. 64, 79-82
5 Brecke, B.J. and Duke, W.B. (1980) Plant Physiol. 66, 656-659
6 Shaner, D.L. and Lyon J.L. (1979) Plant Sci. Lett. 15, 83-87
7 Amrhein, N., Deus, B., Gehrke, P. and Steinrucken, H.C. (1980) Plant Physiol. 66, 830-834

CALCIUM METABOLISM IN CHLOROPLASTS AND PROTOPLASTS

G. KREIMER, B. SUREK, K. HEIMANN, M. BURCHERT, L. LUKOW,
J.A.M. HOLTUM, I.E. WOODROW*, M. MELKONIAN, E. LATZKO

Botanisches Institut der Westfälischen Wilhelms-Universität
D-4400 Münster, Schloßgarten 3, Federal Republic of Germany,
*CSIRO Division of Plant Industry, PO Box 1600, Canberra 2601,
Australia

1.   INTRODUCTION
     In eucaryotic cells, a large number of physiological
processes have been shown to be regulated by calcium, either
through a direct interaction of the calcium ion with its
target molecules or through, for example, a calcium/calmodulin
complex (c.f.1,2,3).
     High concentrations of free calcium, either in the
cytosol or within organelles, severely affect important
cellular functions (1). Since the plant protoplast is
surrounded by the calcium-rich cell wall, which leads to an
apoplastic calcium concentration in the millimolar range (4),
an effective regulatory system is required to maintain the
cytosolic free calcium concentration at 10^{-6} - 10^{-8} M (5, 6).
Mechanisms involved in the maintenance of low free calcium
levels in the cytosol include sequestration in certain
organelles, extrusion by the plasmamembrane Ca^{2+}-ATPase and
calcium-binding by, for example, phospholipids and proteins
(1,2,3,7).
     In chloroplasts, total calcium contents of 100 - 900 nmol
mg^{-1} chlorophyll and even higher have been determined,
corresponding to 4-40 mM (8). Since light induces uptake of
calcium into isolated chloroplasts (9,10), a gradual
accumulation of calcium in chloroplasts during leaf
development can be expected. Indeed, higher calcium contents
have been observed in chloroplasts isolated from ageing
tissue (8,11). In addition, high calcium contents are also
present in chloroplasts from younger leaves which were
isolated without the addition of calcium chelators and when
the isolation was not carried out in darkness. Since high
concentrations of free calcium drastically inhibit
photosynthetic CO_2 assimilation (12,13), most of the total
calcium has to be tightly bound. It has been shown that
thylakoid membranes possess a high calcium binding capacity
and could therefore serve as calcium sinks (14,15). Since
plants also contain soluble calcium binding proteins (1), we
have investigated the calcium binding properties of isolated
spinach stroma preparations. As a large proportion of the
stroma proteins exhibits acidic isoelectric points, a capacity
for calcium binding can be considered. In addition spinach
ferredoxin was chosen as a representative of this group; the
isolated protein was examined with regard to its presumptive
calcium binding capacity.

Biggens, J. (ed.), Progress in Photosynthesis Research, Vol. III. ISBN 90 247 3452 5
© *1987 Martinus Nijhoff Publishers, Dordrecht. Printed in the Netherlands.*

2.   MATERIAL & METHODS
     Spinach was either grown hydroponically (for protoplast isolation) according to (16) or was purchased at a local market. Peas were grown in vermiculite and moistened with a nutrient solution.
     Protoplasts were isolated from 5-10 weeks old spinach leaves, with slight modifications of the method described by (17). Isolation of intact chloroplasts from pea shoots was based on the procedure proposed by (18). Etioplasts were prepared as described by (19). Intact spinach chloroplasts and stroma extracts were purified according to (20), using sorbitol as osmoticum. Stroma preparations were concentrated by ultrafiltration (Amicon PM 10 membrane). Ferredoxin was isolated according to (21).
     For calcium binding studies, isolated stroma was transferred to 50 mM Hepes NaOH pH 7.8 and deionized by passage through a Sephadex G 25/Chelex 100 column. Calcium binding was measured according to the titration method of Ogawa & Tanokura (22). The assay medium was modified as follows: 25 mM KCl, 50 mM Hepes-NaOH pH 7.8 (previously Chelex-100 treated), 0.125 µM tetramethylmurexide. Differential absorbance was determined with a dual-wavelength spectrophotometer. Calcium binding data were analyzed according to (23).
     For measurement of calcium influx into protoplasts the metallochromic indicator arsenazo III was used. Calcium influx was determined as in (10), with a modification of the assay medium: 400 mM sorbitol, 50 mM Hepes-Tris pH 7.0, 20 mM KCl, 10 µM arsenazo III, 5 mM $NaHCO_3$, 10 µM ruthenium red, 20 µM $CaCl_2$, 4 µg chlorophyll.
     Total calcium contents of isolated chloroplasts were determined by atomic absorption spectroscopy. Data were calibrated with internal standards. Sample digestion was performed with hot nitric acid as in (8). Intactness of isolated chloroplasts was determined according to (24), and PGA-dependent oxygen evolution as in (25).
     Stroma proteins and ferredoxin were electrophoretically separated under non-denaturing (modified after 26) and denaturing (27) conditions. Gels were stained with the cationic dye "stains-all" according to (28).
     NAD kinase activity was measured as in (29), at pH 7.8. The test was supplemented with sorbitol when intact plastids were assayed.

3. RESULTS AND DISCUSSION
3.1. Light-induced calcium uptake into spinach protoplasts
     The chloroplast has seldom been considered as a possible regulator of cytosolic calcium concentration. In an attempt to relate the observed light-induced calcium influx of isolated chloroplasts (9,10) to cellular functions, uptake studies with isolated green spinach protoplasts were initiated.
     Isolated protoplasts take up calcium upon illumination. A significant calcium influx was induced by very low energy fluence rates (half maximal calcium influx is observed at~

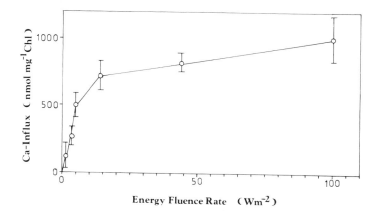

Fig. 1: Light induced calcium influx into spinach protoplasts
as a function of energy fluence rate. The experiments
were carried out under safe-green light. Different
energy fluence rates were obtained using neutral
density filters (Schott & Gen., Mainz, FRG). Results
are expressed as mean +/- S.E. of 4 different
protoplast preparations.

5 Wm^{-2}, fig. 1). The action spectrum is characterized by two
regions of enhanced calcium influx, one in the red, the other
in the blue wavelengths. Calcium uptake was abolished by DCMU
concentrations > 0.5 μM (fig. 2). Low concentrations of the
uncoupler FCCP (0.5 μM) stimulated light-induced calcium
uptake, whereas higher concentrations, known to inhibit
electron transport, suppressed this process.
    The action spectrum and the effects of DCMU and FCCP
indicate a connection to photosynthetic electron transport,
which has been postulated to be involved in light-induced
calcium uptake into isolated chloroplasts as well (10,30).
Since the light-induced calcium influx into protoplasts was
not inhibited by ruthenium red, it differs from the
chloroplast influx system which has the characteristics of an
electrogenic uniport (10).
    Light-induced calcium influx was neither inhibited by
0.5 μM FCCP nor by 6 μM vanadate, an inhibitor of
plasmamembrane ATPases. Therefore calcium influx into spinach
protoplasts is probably not driven by Δ pH.
    Low energy fluence rates, which are sufficient to
generate the proton motive force over the thylakoids and the
chloroplast envelope (31,32), induced calcium influx into
protoplasts. Light-induced signal transduction from the
thylakoids over the chloroplast envelope to the plasmamembrane
is probably mediated electrically (33,34). In summary, we
suggest that a transmembrane electric potential difference at
the plasmamembrane might serve as the driving force for light-
induced calcium influx into spinach protoplasts.

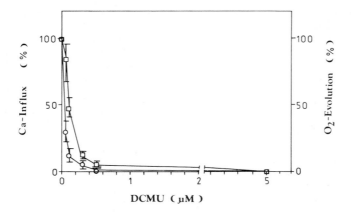

Fig. 2: Effect of DCMU on light-induced calcium influx into
spinach protoplasts. O = calcium influx, results are
expressed as mean of +/- S.E. of 5 different protoplast preparations; □ = CO_2-dependent oxygen
evolution, results are expressed as mean +/-
deviation of 3 different protoplast preparations.

The average rate of calcium influx into protoplasts is
17 μmol mg^{-1} chl h^{-1}. On a chlorophyll basis, the rate of
light-induced calcium uptake by isolated spinach chloroplasts
is $\sim$ 40 % of that observed with protoplasts. Chloroplasts may
therefore play an important role in maintaining a low level of
free calcium in the cytosol during the light-dependent calcium
influx across the plasmamembrane.

3.2. Total calcium content of isolated spinach chloroplasts
    The reported values on total calcium in chloroplasts vary
considerably (8,35,36) and range between $\sim$100 and $\sim$3000 nmol
mg^{-1} chl. Taking into consideration the effective calcium
uptake system in chloroplasts and the fact that chelating
agents can prevent the entry of calcium into chloroplasts
(11), we re-examined the total calcium content of spinach
chloroplasts.
    For calcium determinations only chloroplast preparations
with an average intactness greater than 90 % were used.
Photosynthetic activity of the isolated chloroplasts, measured
as PGA-dependent oxygen evolution, was $\sim$ 40 μmol O_2 mg^{-1} chl
h^{-1}. When chloroplasts were isolated without addition of
chelators to Chelex-100 pretreated media, to remove
contaminating calcium, calcium contents ranging from 150 - 350
nmol mg^{-1} chl were measured. These values correspond to a
concentration of 6.5 - 15 mM. Calcium contents could be even
further reduced by 30 - 50 % when 2 mM EDTA was present in the
homogenisation medium. The lowest measured amount of calcium
was 59 nmol mg^{-1} chl, corresponding to 2.6 mM, when isolated

chloroplasts were washed with medium containing 2 mM EGTA.
These results indicate that the calcium content of
chloroplasts in vivo is probably lower than the total calcium
which is actually measured within the isolated organelles.
These observations further illustrate the potential of
chloroplasts to sequester calcium.

3.3. Calcium binding by spinach stromal proteins
Information on calcium binding by soluble chloroplast
proteins is very limited. Thylakoids, however, have been
reported to exhibit a high capacity to bind calcium (14,37).
Furthermore, it has been described that photosystem II
possesses specific calcium binding sites (38) and a calcium
binding, calmodulin-like protein was isolated from PS II
particles (39). High contents of total calcium in isolated
chloroplasts indicate the presence of numerous calcium binding
sites. Since it is not known if, and to what extent, the
chloroplast stroma participates in buffering free calcium, we
investigated the calcium binding properties of isolated stroma
preparations.

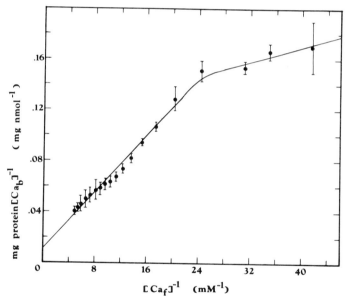

Fig. 3: Calcium binding by spinach stromal proteins. The
results are expressed as mean +/- S.E. of 4
determinations. The total number of binding sites was
90.9 nmol mg^{-1} protein with an "average" binding
constant of 2.0 mM^{-1}.

Our calcium binding studies indicate the existence of at
least two, probably independent classes of binding sites, with
significantly different affinities (fig. 3). In all stroma

preparations, two essentially linear regions could be resolved in double-reciprocal plots which related the concentrations of bound to free calcium. However, due to a large standard error in the range of calcium levels below 50 μM, only data points at higher calcium levels were used for the calculation of the total number of calcium binding sites and the "average" binding constants of these sites for calcium.

In stroma preparations assayed at pH 7.8 the total number of calcium binding sites varied between 90 and 155 nmol mg^{-1} protein, with "average" binding constants of 1.1 - 2.7 mM^{-1}. Binding experiments were also performed at pH 7.1 in order to detect and quantify possible changes in the calcium binding capacity of stromal proteins that might occur during the light/dark transition. In a typical experiment the total number of binding sites decreased from 90 nmol mg^{-1} protein to 59 nmol mg^{-1} protein, the "average" affinity, however, increased from 2.7 mM^{-1} to 4.5 mM^{-1} (fig. 4). These changes may be related to conformational changes of certain stromal proteins. Reversible pH-induced conformational transitions have been reported for, amongst other proteins, chloroplast FBPase (40,41). However, these changes in binding parameters could also reflect proton concentration induced changes in the ionziation state of charged protein molecules.

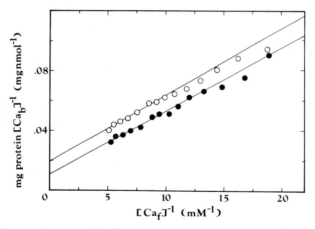

Fig. 4: pH effects on calcium binding by spinach stromal proteins. Stromal fractions were preincubated for 15 min. in the assay medium at the indicated pH values.

        ● = pH 7.8   O = pH 7.1

The specificity of calcium binding to stromal proteins was examined by adding increasing amounts of Mg^{2+} and La^{3+}, a probe for calcium binding sites, to the reaction mixture. Both Mg^{2+} and La^{3+} inhibit calcium binding competetively, with "average" inhibitor constants of 0.26 mM^{-1} and 39.4 mM^{-1}, respectively (fig. 5). The monovalent K^+, however, did not

affect either the "average" binding constant or the number of
binding sites for calcium at concentrations up to 50 mM.

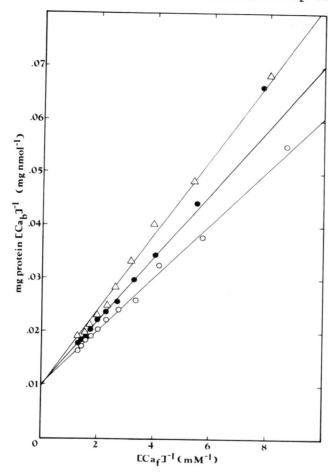

Fig. 5: Effect of $LaCl_3$ on Ca^{2+} binding by spinach
stromal proteins. $LaCl_3$ was added to the protein-
containing media 5 min. prior to the addition of the
indicator.

　　　　○ = 0 μM $LaCl_3$　　　● = 7.5 μM $LaCl_3$
　　　　△ = 20 μM $LaCl_3$
The total number of binding sites varied between
102.3 - 124.7 nmol mg^{-1} protein. The "average"
inhibitor constant is 39.4 mM^{-1}.

On the basis of these results, we suggest that calcium
binding of stromal proteins is not due to unspecific
interactions with negatively charged amino acids, but reflects

the existence of specific binding sites.

The existence of specific calcium binding sites on single proteins was further investigated by the electrophoretic separation of stromal proteins. The isolated proteins were separated by native and SDS polyacrylamide gel electrophoresis and stained with the cationic dye "stains-all", which forms characteristic blue reaction products with phosphoproteins (42) and a number of authentic calcium binding proteins like calmodulin (28), while other proteins stain red. Several stromal proteins, mainly in the low molecular mass range, stain blue or purple with this dye, even after phosphatase treatment, suggesting their ability to interact with calcium ions.

3.4. Calcium-binding chloroplast proteins

In addition to the chloroplast FBPase, which has been claimed to bind 16 moles calcium per mole enzyme protein (43), ferredoxin was considered to be a possible candidate for binding calcium, since it is a small, acidic protein. Therefore we have purified ferredoxin from spinach leaves and have investigated its reactivity with "stains-all". After electrophoresis under non-denaturing and denaturing conditions, in the absence and presence of alkaline phosphatase, purified ferredoxin stained blue with "stains-all". The colour intensity was more pronounced with the native protein. Reduced and oxidized ferredoxin showed different electrophoretic mobilities with native PAGE, the oxidized protein being the form that migrated faster towards the anode. The redox status of the protein had no influence on the reaction with the cationic dye, since both forms stained blue. It remains to be investigated, however, if quantitative differences exist.

Another potential calcium binding protein is chloroplastic NAD kinase. NAD kinase is the only enzyme known which catalyzes the phosphorylation of NAD with ATP as phosphoryldonor (44). Green leaves contain at least two different NAD kinase activities, one calcium/calmodulin (CaM)-dependent and one CaM-independent (45,46). However, there is some discrepancy as to the location of these enzyme activities within the cell and the CaM-regulation in the different compartments, especially in the chloroplast. In peas, a CaM-dependent activity has been found at the envelope and a CaM-independent activity in the stroma (45). Other authors, however, claim that stromal NAD kinase is regulated by chloroplastic CaM (47). In order to resolve these discrepancies, we re-examined the calcium/CaM dependence of NAD kinase activities in isolated chloroplasts of green pea plants and in etioplasts from dark grown pea shoots.

A calcium/CaM-dependent NAD kinase activity was detectable with chloroplasts when the assay was performed under isosmotic conditions, locating this form at the envelope membrane. Calcium/CaM stimulation of this NAD kinase activity was highest with younger plant material (see table 1). Lysis of chloroplasts yielded an additional CaM-independent

Table 1: NAD kinase activities in isolated chloroplasts from light-grown pea shoots

age of shoots	plastid state	specific activity (nmol mg^{-1} prot.)		CaM-dep./ CaM-indep.
		CaM-indep.	CaM-dep.	
7 days	intact	0.27	1.62	
	lysed	0.71	1.58	2.22
9 days	intact	0.14	1.15	
	lysed	0.52	1.08	2.08
14 days	intact	0.13	0.42	
	lysed	0.50	0.36	0.72
35 days	intact	0.11	0.20	
	lysed	0.44	0.25	0.57

CaM-independent activity = measured in the presence of 1 mM EGTA

CaM-dependent activity = total activity (measured in the presence of 1 mM CaCl$_2$ + 400 nM CaM) minus CaM-independent activity

activity, which was located in the chloroplast stroma (table 1). It is noteworthy that the specific activity of the CaM-dependent enzyme drastically decreased with the age of the plants, while the specific activity of the CaM-independent form remained more or less constant.

When etioplasts were assayed under isotonic conditions, a calcium/CaM-dependent NAD kinase activity could be measured. However, no additional activity appeared under hypotonic conditions (table 2). Since the enzyme activity inside the organelle was detectable only after greening of the plastids, we suggest that light could be involved in the development of this activity.

In order to examine the calcium/CaM-dependence of the stromal NAD kinase activity of pea chloroplasts, ion exchange chromatography of an isolated stroma preparation was performed to remove any calmodulin present from its target proteins. This experiment yielded an NAD kinase activity which could not be stimulated by CaM, but showed a 10 - 20 % increase of activity in the presence of 40 - 50 μM free calcium. This activation is reflected by changes in the sizes of NAD and NADP-pools which occur following chloroplast illumination (30,46,48). It seems possible that a transient increase of free calcium in chloroplasts originating from light-induced calcium influx (30,49) is responsible for the observed activation of stromal NAD kinase.

Table 2: NAD kinase activities in isolated etioplasts

age of shoots	plastid state	specific activity (nmol mg^{-1} prot.)		CaM-dep./ CaM-indep.
		CaM-indep.	CaM-dep.	
7 days	intact	0.44	8.76	
	lysed	0.47	8.69	18.52
9 days	intact	0.75	10.09	
	lysed	0.78	9.23	11.76

Intactness of etioplasts was estimated from NADP-dependent glyceraldehyde-3-phosphate dehydrogenase activity under isotonic and hypotonic conditions

4.   CONCLUSIONS
     The observation that in addition to chloroplasts, protoplasts from spinach leaves also exhibit a light-induced calcium influx supports the assumption that changes in the concentration of free calcium might act as a second messenger in plants. These changes convey information from endogenous and exogenous stimuli (e.g. hormones and light) to the final physiological response (1,2,3). The chloroplast could not only act as a sink for calcium, but it might also affect the cytosolic concentration of free calcium since both calcium influx into isolated spinach chloroplasts and spinach protoplasts is linked to photosynthetic electron transport.The significant calcium influx into protoplasts upon illumination could, at least transiently, increase the cytosolic free calcium concentration thus triggering calcium-dependent processes in the cytosol. This transient increase in the cytosolic free calcium level is most likely transmitted to the stroma of the chloroplasts.
     Within the chloroplast a number of physiological processes including photochemical and enzymic reactions, are affected by calcium (c.f. 11,43,50,51,52,53). Our study has provided the first evidence for the existence of specific calcium binding sites in the stroma. The binding of calcium by stromal proteins could affect chloroplast metabolism in two ways: (i) It may assist in adjusting calcium levels in the stroma during dark/light transitions and thus function as a calcium buffer. (ii) It may be involved in the regulation of stromal enzyme activities. Possible targets for the regulation of enzyme activities include enzymes of the photosynthetic carbon reduction cycle (FBPase, SBPase) (51,52,53). Furthermore, the specificity of the effector thioredoxin is modulated by calcium (54). Our observation that ferredoxin, a donor for multiple electron acceptors, might also be a calcium binding protein further underlines the importance of calcium metabolism.

4. ACKNOWLEDGEMENT

This work was supported by grants from the Deutsche Forschungsgemeinschaft.

REFERENCES

1 Hepler, P.K. and Wayne, R.O. (1985) Annu. Rev. Plant Physiol. 36, 397-439
2 Marmé, D. (1983) in Inorganic Plant Nutrition (Läuchli, A. and Bieleski, R.L., eds.), Encycl. Plant Physiol. NS, Vol. 15 B, pp. 599-625, Springer Verlag Berlin
3 Roux, S.J. and Slocum, R.D. (1982) in Calcium and Cell Function (Cheung, W.Y., ed.), Vol. 3, pp. 409-453, Academic Press, London
4 Zid, E. and Grignon, C. (1985) Phys. Vég. 23, 895-903
5 Williamson, R.E. and Ashley, C.C. (1982) Nature (London) 296, 647-651
6 Gilroy, S., Hughes, W.A. and Trewavas, A.J. (1986) FEBS Lett 199, 217-221
7 Moore, A.L. and Åkerman, K.E.O. (1984) Plant Cell Environ. 7, 423-429
8 Jones, H. and Halliwell, B. (1984) Photobiochem. Photobiophys. 7, 293-297
9 Muto, S., Izawa, S. and Miyachi, S. (1982) FEBS Lett. 139, 250-254
10 Kreimer, G., Melkonian, M. and Latzko, E. (1985) FEBS Lett. 180, 253-258
11 Bouthyette, P.-Y. and Jagendorf, A.T. (1981), in Photosynthesis, Chloroplast Development (Akoyunoglou, G., ed.), Vol. 5, pp. 599-609, Balaban International Sciences Services, Philadelphia
12 Portis, A.R. jr. and Heldt, H.W. (1976) Biochim. Biophys. Acta 449, 434-446
13 Demmig, B. and Gimmler, H. (1979) Z. Naturforsch. Teil C 34, 233-241
14 Gross, E.L. and Hess, S.C. (1974) Biochim. Biophys. Acta 339, 334-346
15 Prochaska, L.J. and Gross, E.L. (1975) Biochim. Biophys. Acta 376, 126-135
16 Lilley, R.McC. and Walker, D.A. (1974) Biochim. Biophys. Acta 368, 269-278
17 Edwards, G.E., Robinson, S.P., Tyler, N.J.C. and Walker, D.A. (1978) Plant Physiol. 62, 313-319
18 Cline, K., Andrews, J., Mersey, B., Newcomb, E.H. and Keegstra, K. (1981) Proc. Natl. Acad. Sci. USA 78, 3595-3599
19 Steup, M., Schächtele, C. and Melkonian, M. (1986) Physiol. Plant. 66, 234-244
20 Douce, R. and Joyard, J. (1982) in Methods in Chloroplast Molecular Biology (Edelman, M., Hallick, R.B. and Chua, N.-H., eds.), pp. 239-256, Elsevier Biomedical, Amsterdam, New York, Oxford
21 Yocum, C.F. (1982) in Methods in Chloroplast Molecular

Biology (Edelman, M., Hallick, R.B. and Chua, N.-H., eds.), pp. 973-981, Elsevier Biomedical, Amsterdam, New York, Oxford

22 Ogawa, Y. and Tanokura, M. (1984) J. Biochem. 95, 19-28
23 Klotz, I.M. and Hunston, D.L. (1971) Biochem. 10, 3065-3069
24 Lilley, R.McC., Fitzgerald, M.P., Rienits, K.G. and Walker, D.A. (1975) New Phytol. 75, 1-10
25 Enser, U. and Heber, U. (1980) Biochim. Biophys. Acta 592, 577-591
26 Maurer, H.R. (1971) in Disc Electrophoresis and Related Techniques of Polyacrylamide Gel Electrophoresis (Fischbeck, K., ed.), pp. 222, Walter de Gruyter, Berlin, New York
27 Laemmli, U.K. (1970) Nature (London) 227, 680-685
28 Campbell, K.P., MacLennan, D.H. and Jorgensen, A.O. (1983) J. Biol. Chem. 258, 11267-11273
29 Harmon, A.C., Jarrett, H.W. and Cormier, M.J. (1984) Anal. Biochem. 141, 168-178
30 Kreimer, G., Melkonian, M., Holtum, J.A.M. and Latzko, E. (1985) Planta 166, 515-523
31 Heldt, H.W., Werdan, K., Milovancev, M. and Geller, G. (1973) Biochim Biophys. Acta 314, 224-241
32 Robinson, S.P. (1985) Biochim. Biophys. Acta 806, 187-194
33 Bulychev, A.A., Andrianov, V.K., Kurella, G.A. and Litvin, F.F. (1972) Nature (London) 236, 175-177
34 Trebacz, K. and Zawadzki, T. (1985) Physiol. Plant. 64, 482-486
35 Nakatani, H.Y., Barber, J. and Minsky, M.J. (1979) Biochim. Biophys. Acta 545, 24-35
36 Yamagishi, A., Saton, K. and Katoh, S. (1981) Biochim. Biophys. Acta 637, 252-263
37 Davis, D.J. and Gross, E.L. (1975) Biochim. Biophys. Acta 387, 557-567
38 Barr, R., Troxel, K.S. and Crane, F.L. (1983) Plant Physiol. 73, 309-315
39 Sparrow, R.W. and England, R.R. (1984) FEBS Lett. 177, 95-98
40 Gontero, B., Meunier, J.-C., Buc, J. and Ricard, J. (1984) Eur. J. Biochem. 145, 485-488
41 Buc, J., Pradel, J., Meunier, J.-C., Soulie, J.-M. and Ricard, J. (1980) FEBS Lett. 113, 285-288
42 Green, M.R. and Pastewka, J.V. (1974) J. Histochem. Cytochem. 22, 774-781
43 Hertig, C.H. and Wolosiuk, R.A. (1983) J. Biol. Chem. 258, 984-989
44 McGuinness, E.T. and Butler, J.R. (1985) Int. J. Biochem. 17, 1-11
45 Simon, P., Bonzon, M., Greppin, H. and Marmé, D. (1984) FEBS Lett. 167, 332-338
46 Muto, S. and Miyachi, S. (1986) in Molecular and Cellular Aspects of Calcium in Plant Development (Trewavas, A.J., ed.), pp. 107-114, Plenum Publishing Corporation
47 Jarrett, H.W., Brown, C.J., Black, C.C. and Cormier, M.J.

(1982) J. Biol. Chem. 257, 13795-13804
48 Muto, S., Miyachi, S., Usuda, H., Edwards, G.E. and Bassham, J.A. (1981) Plant Physiol. 68, 324-328
49 Muto, S. (1982) FEBS Lett. 147, 161-164
50 Nakatani, Y. (1984) Biochem. Biophys. Res. Commun. 121, 626-633
51 Hertig, C.H. and Wolosiuk, R.A. (1980) Biochem. Biophys. Res. Commun. 97, 325-337
52 Charles, S.A. and Halliwell, B. (1981) Cell Calcium 2, 211-224
53 Wolosiuk, R.A., Hertig, C.H., Nishizawa, A.N. and Buchanan, B.B. (1982) FEBS Lett. 140, 31-35
54 Schürmann, P., Roux, J. and Salvi, L. (1985) Physiol. Veg. 23, 813-818

CALCIUM FLUXES ACROSS THE PLASMA MEMBRANE OF PEA LEAF PROTOPLASTS

MICHAEL O. PROUDLOVE AND ANTHONY L. MOORE, DEPARTMENT OF BIOCHEMISTRY, UNIVERSITY OF SUSSEX, FALMER, BRIGHTON BN1 9QG, U.K..

1. INTRODUCTION

Stimulation or excitation of numerous animal cells leads to a rapid increase in the cytosolic free calcium and the activation of certain enzyme systems, either directly or in a calmodulin-dependent fashion. Maintenance of low $[Ca^{2+}]_{cyt}$ is acheived by extrusion across the plasma membrane, by a calmodulin-dependent (Ca^{2+}/Mg^{2+})-ATPase and by internal buffering, by mitochondria, endoplasmic reticulum and others (1, 2). In plant cells there is now mounting evidence that Ca^{2+} may play a similar role (see 3 for a recent review). Whilst there is no direct evidence that cytosolic calcium levels change in plant cells, there do appear to be systems available which may be involved in buffering $[Ca^{2+}]_{cyt}$, again associated with organelles and the plasma membrane (4). In the monocotyledon, wheat, a preliminary study has shown that there is an inwardly directed Ca^{2+} gradient across the plasma membrane, sustained by a growth regulator/red/far red light-insensitive influx and an ATPase inhibitor-sensitive efflux (5). The susceptibility of chloroplasts of this plant to inhibition by calcium (6), however, prompted us to investigate the kinetics of mesophyll plasma membrane control of cytosolic Ca^{2+} in the dicotyledon, pea. Chloroplasts from this tissue can accumulate external calcium and the mitochondrial external NADH dehydrogenase is activated by changes in pCa.

2. PROCEDURE
2.1. Materials and methods
2.1.1. Preparation of protoplasts: Seeds were imbibed in aerated tap water for 18h at 20°C, planted in John Innes No. 3 and grown under a 16h light (20°C)/ 8h dark (10°C) cycle for 8 days (wheat) or 12-14 days (pea). Protoplasts were prepared by incubating leaf slices for 3h in digestion medium (0.5M sorbitol; 1mM $CaCl_2$; 5mM MES-KOH, pH 6.0; 0.2% (w/v) pectolyase; 2% (w/v) cellulase) followed by purification on sucrose/ sorbitol gradients, essentially as described in (7). For pea, the lowest layer was supplemented with 10% (v/v) Percoll and was overlayed with 2cm³ medium 3 (7). Protoplasts were collected from the gradients by Pasteur pipette, diluted fivefold with 0.4M sorbitol; 40mM KCl; 10mM HEPES-KOH, pH 7.6 (HKM), supplemented with 2mM EGTA, and centrifuged for 5min at 150g. Final pellets were resuspended in HKM to give a concentration of approximately 500µg chlorophyll.cm⁻³.

2.1.2. Measurement of $^{45}Ca^{2+}$ fluxes across the plasma membrane: Incubations were carried out at 20°C in the light or in the dark (foil wrapped tubes). Illumination was provided by a slide projector shining at a distance of 10cm through a water heat shield, the average intensity being 1.5-1.6mmole.m⁻².s⁻1.

Calcium steady-state accumulation and net uptake were followed as previously described (5).

3. RESULTS AND DISCUSSION

3.1. Protoplasts isolated from wheat and pea leaves both showed a time dependent uptake of $^{45}Ca^{2+}$, equilibrium being reached in approximately 30-40min (Figure 1). It is possible, however, that the influx of calcium into these cells may be due to a "wounding" response [8] following isolation. Results in Figure 1 suggest that this is not so because there is no change in the net accumulation of $^{45}Ca^{2+}$, by protoplasts from either tissue, over a period of 24h following isolation. Addition of the calcium ionophore, A23187, caused a rapid and substantial increase in the net accumulation of $^{45}Ca^{2+}$, indicating that there is an inwardly directed calcium gradient in mesophyll cells from the dicotyledenous plant, pea, as there is for wheat (Figure 1; [6]).

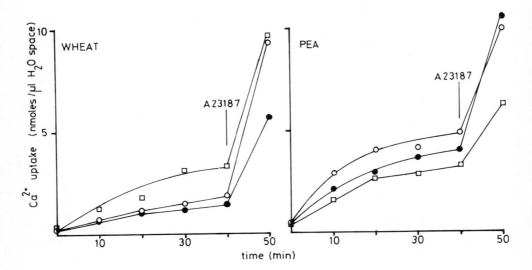

Figure 1. Calcium uptake by wheat and pea leaf protoplasts 1h (●——●), 4h (○——○) and 24h (□——□) after isolation. Experiments were performed in HKM, containing 1mM $CaCl_2$, $5\mu Ci.cm^{-3}$ $^{45}Ca^{2+}$ and $10\mu Ci.cm^{-3}$ ^3H_2O, as described in Procedure and $4\mu M$ A23187 was added after 40min, as indicated.

3.2. Low concentrations of diethylstibesterol (DES), a specific inhibitor of plasma membrane ATPases [9] and N,N'-dicyclohexylcarbodiimide (DCCD), an inhibitor of numerous membrane bound ATPases [10] caused an increase in the rate of calcium accumulation and in the final steady-state level (Figure 2). This suggests that calcium levels within the cell are controlled, at least at the plasma membrane, by an outwardly directed ATPase, influx being in response to the gradient of calcium across this membrane.

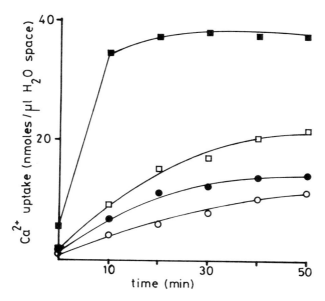

Figure 2. Calcium uptake by pea leaf protoplasts in the presence of A23187 (4µM; ■—■), DES (200 µM; □—□) or DCCD (10 µM; ●—●). Control incubations (O—O) contained an equal volume of dimethylsulphoxide and all experiments were performed as described in Figure 1.

3.3. When the net accumulation of calcium by pea protoplasts was investigated under illuminating and non-illuminating conditions, all previous experiments having been performed in the dark, it was found that steady-state levels were greater under non-illuminating conditions (Table 1). This suggests, albeit indirectly, that the $[Ca^{2+}]_{cyt}$ is lower in the light than in the dark. An increase in $[ATP]_{cyt}$ or in the activity of the outwardly directed ATPase could account for these results. The present hypothesis is that ATP/ADP

Table 1. Effect of various compounds on the accumulation of calcium by pea leaf protoplasts in the light and the dark. Experiments were performed as described in Figure 1 and results are expressed as percentage rates compared to the light control (100%).

Addition	light	dark
-	100	115
1mM $LaCl_3$	55	25
1mM $MgCl_2$	88	80
100µM fusicoccin	127	124
1µg cm^{-3} valinomycin	nd	86
1mM KCN	63	61

levels do not alter much between light and dark (11), particularly at high CO_2 (10mM $KHCO_3$ was present in all incubations),and this would support the idea that there is a change in the activity of the ATPase. The effects of various compounds on the net calcium accumulation are also presented in Table 1. These support the suggestion that calcium infux is in response to concentration and electrical potentials across the plasma membrane, movement being fairly calcium specific (stimulation by fusicoccin and inhibition by valinomycin, $LaCl_3$ and $MgCl_2$). The effect of 1mM KCN is difficult to interpret, however, as this may have been expected to stimulate net accumulation, by reducing the supply of ATP for Ca^{2+} export. One explanation could be that, by lowering $[ATP]_{cyt}$, it may have decreased the potential across the plasma membrane and so limited calcium uptake or prevented some form of ATP-dependent calcium sequestration within the cell. Results with calmodulin antagonists and a range of herbicides proved inconclusive (results not shown) as they inhibited CO_2-dependent oxygen evolution at concentrations which appeared to have little or no direct affect on calcium uptake (50-100µM).

4. CONCLUSIONS
Protoplasts isolated from mature pea leaves maintain a gradient across their plasma membrane. Calcium accumulation does not appear to be a "wounding" effect of isolation and results from light/ dark experiments suggest that cytosolic calcium levels may be lower in illuminated cells compared to non-illuminated cells.

AKNOWLEDGEMENTS
Financial support of this work was provided by an SERC grant to ALM.

REFERENCES
1. Akerman, K.E.O. (1982) Medical Biol. 60, 168-182
2. Borle, A.B. (1981) Rev. Physiol. Biochem. Pharmacol. 95, 150-201
3. Hepler, P.K. and Wayne, R.O. (1985) Ann. Rev. Plant Physiol. 36, 397-439
4. Moore, A.L. and Akerman, K.E.O. (1984) Plant Cell Environ. 7, 423-429
5. Akerman, K.E.O., Proudlove, M.O. and Moore, A.L. (1983) Biochem. Biophys. Res. Commun. 113, 171-177
6. Edwards, G.E., Robinson, S.P., Tyler, N.J.C. and Walker, D.A. (1978) Arch. Biochem. Biophys. 190, 421-433
7. Edwards, G.E., Robinson, S.P., Tyler, N.J.C. and Walker, D.A. (1978) Plant Physiol. 62, 313-319
8. Hanson, J.B., de Quintero, M.R., Ladror, U. and Rogers, S.A. (1986) in Molecular and Cellular Aspects of Calcium in Plant Development (Trewavas, A.J. ed.), in press
9. Balke, N.E. and Hodges, T.K. (1975) Plant Physiol. 56, S221
10. Linnet, P.E. and Beechey, R.B. (1979) Methods Enzymol. 55, 472-518
11. Gardestrom, P. (1986) in Plant Mitochondria: Functional, Structural and Physiological Aspects (Moore, A.L. and Beechey, R.B. eds.), in press

SOME RESULTS OF THE PHOTOSYNTHETIC SYSTEM OF MIXOTROPHIC CARROT CELLS (DAUCUS CAROTA).

A. Kumar*, L. Bender and K.H. Neumann
*Department of Botany, University of Rajasthan, Jaipur 302004, India.
Institut Fur Pflanzenernahrung, Justus Liebig Universitat, 6300 Giessen
F.R. Germany.

INTRODUCTION
Most cultured cells and tissue of higher plants turn green if cultured in the light and as was shown earlier, light fixation exceeds dark fixation 5 to 6 fold. (1) Such cultures contain chloroplasts with a well developed thylakoidal system (2-4). Also in mixotrophic conditions with sugar in the nutrient medium, photosynthetic carbon fixation contributes considerably to dry weight production (5). Out of such mixotrophic cultures autotrophic strains were isolated with the ability to grow for periods up to several years in autotrophic conditions (4, 6, 7, 8, 11).

Some of these cultures are able to grow at ambient CO_2 concentrations though with a low dry matter increment (6,7), however, photosynthetic activity and growth can be strongly increased coming to the range of intact leaves by an elevation of the atmospheric CO_2 concentration of up to 2% (8). In the present paper a summary will be given of experiments to characterize the photosynthetic system of cultured carrot root explants at some stages during the cultural cycle.

MATERIALS AND METHODS

Carrot root explants obtained from the secondary phloem of surface sterilized carrot roots were cultured in a nutrient medium supplied with 2% sucrose and IAA (2ppm), m inositol (50 ppm) and kinetin (0.5 ppm) at the beginning of the experiment as described previously (9), up to three weeks of culture. Methods to determine the number of cells per explant, for ultrastructural investigations, to determine ^{14}C fixation and its distribution were already described (3).

RESULTS AND DISCUSSIONS
In cells of freshly isolated carrot root explants besides some amyloplasts, the majority of the plastid population can be classified as chromoplasts containing carotinoid crystals. After about three days of culture in the light intermediate stages containing starch grains besides remains of carotin crystals, and amyloplasts could be observed. The latter were classified as stage 1 of a developmental sequence leading to chloroplasts which consists of 5 stages of chloroplasts development (3). At various periods during the growth cycle an accumulation of two (or 3) of these developmental stages could be observed, indicating some degree of synchronization. Measurements of low temperature absorption spectra and fluorescence induction profiles (Kautsky effect) were in agreement

Biggens, J. (ed.), Progress in Photosynthesis Research, Vol. III. ISBN 90 247 3452 5
© 1987 Martinus Nijhoff Publishers, Dordrecht. Printed in the Netherlands.

with the developmental stages of chloroplasts at these stages (3). The
number of plastids was basically determined by the sucrose in the medium
with a six to eight fold increase following a sucrose supplement (2%) as
compared to the sugar free medium, independent of a hormonal supply.
The development of amyloplasts into the various stages of chloroplast
development however required IAA, m inositol and/or kinetin in the
medium. Apparently the initiation of chloroplast development from
amyloplasts is under hormonal control whereas the number of plastids,
that is in later stages of the cultural cycle, the number of chloroplasts
seems to be basically determined by sucrose. During the culture period
of 10 days in a nutrient medium containing 2% sucrose, IAA, m inositol
and kinetin, a photosynthetic system develops sufficient to support
autotrophic growth (1). In the dark, plastid development is arrested at
the amyloplast stage, and etioplast could not be observed.

In table 1 influences of light on growth as measured in terms of
cells per explant at the end of a three week culture period are
summarized. The number of cells in the light is almost doubled,
indicating a strong contribution of illumination to grwoth. Still, here
also the phytochrome system and interactions of light with phytohormones
may have to be considered.

Table 1. The influence of light on cell number per explant ($.10^3$) and %
dry matter of cultured carrot explants in a liquid culture
medium originally supplied with 2% sucrose. (Harvest after
3 weeks of culture).

	DARK		LIGHT (Approx. 5000 Lux)	
	Cell Number per explant	% Dry Matter	Cell Number per explant	% Dry matter
d_0	13	11.5	13	11.5
d_{21}	444	13.9	752	14.3

As reported elsewhere (5) about 40% of carbon taken up by the
cultures from the 15th to the 21st day of the experiment was derived
from inorganic carbon and about 80% of this was due to light fixation.
Though the carrot is considered to be a C_3 plant, interestingly the
fixation of carbon by PEP carboxylation contributes heavily to total
fixation of inorganic carbon supplied as $H^{14}CO_3$. This can be derived
from the data in table 2 in which the total ^{14}C in malate (up to 40% of
total C^{14}), aspartate and citrate were summarized as derivatives of PEP
carboxylation. The difference to the total C^{14} fixed is assumed to come
from Calvin cycle fixation. About 80 percent of ^{14}C in malate could be
localised in the C_4 position (10). A high contribution of PEP
carboxylation to total fixation of inorganic carbon was also reported for
other cell culture systems (4, 7, 8, 11). In particular close ratios
of Rubisco/PEPC activity can be usually observed during the exponential
phase of the growth cycle when the cultures grow by a high cell division
activity (3). This was reported for mixo and autotrophic cell culture

Table 2.
The contribution of "C$_3$" and "C$_4$" fixation to total incorporation of inorganic carbon of cultured carrot explants (3 weeks of culture, 30 minute labelling).

	dpm/gFr. Wt. $.10^4$	%
Total Fixation	821	100
C$_3$ derivatives	316	38
C$_4$ derivatives total	505	62
in light	285	35
in dark	220	27

systems, however, in mixotrophic cultures this tendency is usually more accentuated. Highest PEPCase activity was reported for mixotrophic cultures during the exponential growth phase (8). During the stationary phase of the growth cycle, however, the PEPCase activity decreases and the Rubisco activity is strongly elevated. Employing an immuno assay NATO et al (4) could show for tobacco cultures that the concentration of these two enzymes follows closely the activity. As shown by SATO et al (11) in tobacco PEPCase is of the C$_3$ type and malate produced is not utilized to fuel the Calvin cycle as in C$_4$ or CAM plants. The main function of PEP carboxylation in these cultured cells seems to be to initiate an anaplerotic reaction to fuel the Krebs cycle when a high demand for its intermediates for synthetic purposes occurs as it is typically for cultures growing with high cell division activity during the exponential phase of the growth cycle (10).

The ratio of labelled C$_3$/C$_4$ derivatives is also influenced by a limited supply of phosphate to the nutrient medium (See Table 3). Here, at phosphorous deficiency the percentage of ^{14}C in C$_4$ derivatives almost compensates the reduction in C$_3$ derivatives, usually quite low at limited phosphorous supplies. A similar tendency can be observed following a high sucros supplement to the nutrient medium (10). It is hypothesized that P deficiency impairs the phosphate translocator followed by a reduced export of triosephosphates from the chloroplast to the cytoplasm. Then possibly a dicarboxylate shuttle (12) may substitute for this mechanism to export photosynthetic energy as malate derived from OAA as the product of PEP carboxylation catalized in the cytoplasm. The reduction of OAA after its uptake by the chloroplast to malate could be achieved using photosynthetic NADPH (6). This is in agreement with an increase in labeling of C$_4$ derivatives due to illumination.

Table 3
The influence of various phosphorus levels on growth, P concentration in the tissue and fixation and distribution of ^{14}C from $H^{14}CO_3$ of culture carrot cells in light (20 days of culture, 30 minutes of labelling)

mg P/l nutrient medium

	11.6	46.5
mg Dr. Wt./explant	19	30
mg P/g dr. wt.	2.8	7.1
dpm $.10^4$ g dr. wt.	612.5	782.0
% C_3 derivatives	29	52
% C_4 derivatives	71	48

REFERENCES
1. Neumann, K.H. and Raafat, A. (1973) Plant Physiol. 51: 685–690.
2. Israel, H.W. and Steward, F.C. (1967) Ann. Bot. 31: 1–18.
3. Kumar, A., Bender, L., Pauler, B., Neumann, K.H., Senger, H. and Jeske, C. (1983). Plant Cell Tissue Organ Culture 2:161–177.
4. NATO, A., Hoarau, J., Brangeon, Jr., Hirel, B., Suzuki, A. (1985) In Primary and Secondary Metabolism of plant cell cultures (K.H. Neumann, et al. ed.) pp. 43–57. Springer-Verlag Berlin, Heidelberg New York, Tokyo.
5. Neumann, K.H., Bender, L., Kumar, A., and Szegoe, M. (1982). Proc. 5th Intl. Cong. Plant tissue and cell culture, Tokyo, 251–252.
6. Bender, L., Kumar, A., Neumann, K.H. (1985). In: Primary and Secondary metabolism of Plant Cell Cultures. (Neumann, et al ed.), pp. 24–42. Springer-Verlag Berlin, Heidelberg, New York, Tokyo.
7. Rogers, S.M.D. and Widholm, J.M. (1986). Proc. VI Intl. Cong. Plant tissue and Cell Culture. Minneapolis, U.S.A. 1986, pp. 407.
8. Huesemann, W. Herzbeck, H. and Robenck, H. (1984) Physiol. Plant 62: 249–355.
9. Kumar, A., Bender, L., and Neumann, K.H. (1980) In: Plant tissue culture: Genetic manipulation and somatic hybridisation of plant cels (Rao, PS et al eds) pp. 169.
10. Neumann, K.H., and Bender, L. (1986). In: Proc. VI Intl. Cong. Plant Tissue and Cell Culture, Minneapolis, U.S.A. (in press).
11. Sato, F., Koizumi, N. Takeda, S. and Yamada, Y. (1986). VI Int. Conf. Plant Tissue and Cell Culture, Minneapolis (in press).
12. Heber U. and Heldt, H.W. (1981) Ann. Rev. Plant-Physiol. 32:139–168.

INHIBITION OF O_2 FIXATION IN SPINACH CHLOROPLASTS BY PHOSPHONATES AT
DIFFERENT TEMPERATURES

BRURIA HEUER[a] AND A. R. PORTIS, JR[b], INSTITUTE OF SOILS AND WATER, THE
VOLCANI CENTER, POB 6, BET DAGAN 50250, ISRAEL, PRESENTLY DEPARTMENT OF
AGRONOMY, UNIVERSITY OF ILLINOIS[a] AND U.S. DEPARTMENT OF AGRICULTURE,
AGRICULTURAL RESEARCH SERVICE[b], URBANA, ILLINOIS 61081

1. INTRODUCTION.
   Considerable information is available on the inhibition of the light
reactions of photosynthesis including electron transport and photophosphorylation (1,2). Compounds that specifically inhibit enzymes involved
in carbon metabolism are much less common.
   The phosphorous compounds involved in biochemical processes in the
plants are orthophosphates ($PO_4{}^{-3}$), chemical derivatives of phosphoric
acid. As an exception to these compounds, the phosphonates have a carbon
to phosphorus bond (C-P) which is known to be very stable. The structural
similarity of the phosphonates to the orthophosphates led to their
classification as effectors of the CO_2 fixation system (3). The possible
effects of phosphonic acids on the photosynthetic carbon metabolism of
isolated chloroplasts have not been extensively studied.
   The purpose of this study was to determine the effect of different
temperatures on the CO_2 fixation by intact chloroplasts and their interaction with the effect of three phosphonates on CO_2 fixation and PGA-
reduction.

2. PROCEDURES
2.1. Plant Material: Spinach (Spinacea oleracea L. cv. American Hybrid
   #424) was grown hydroponically under artificial light in full
   strength Hoagland solution. Intact chloroplasts were isolated from
   fully expanded leaves and their chlorophyll content was determined by
   measuring the absorbance at 652 nm.
2.2. CO_2 Fixation by chloroplasts: Isolated chloroplasts (30 μg
   chlorophyll) were added in the dark to the assay media which
   contained 0.33 M sorbitol and 50 mM Hepes buffer, pH 7.8, 5 mM
   $NaH^{14}CO_3$ (0.2 μCi/μmol), 0.3 mM PGA, 1 mM KH_2PO_4 and 172 units of
   catalase in a final volume of 1 ml. At time zero the lights were
   turned on and 0.2 ml aliquots were removed at time intervals.
2.3. O_2 evolution: PGA-dependent CO_2 evolution of chloroplasts was
   measured with a Clark-type CO_2 electrode (Hansatech, UK) at 15, 20,
   and 25° C. Light intensity of 1400 μE $m^{-2}s^{-1}$ was provided by a slide
   projector with a 500 W incandescent-type lamp.
2.4. Pi transport and exchange: Chloroplasts (1.5 mg chlorophyll) were
   incubated with 2 mM KH_2PO_4 and 20 μCi ^{32}Pi for 5 min at 0 °C, then
   transferred to sorbitol-hepes buffer and centrifuged 2000 RPM for 2
   min. The pellet obtained was resuspended and the chloroplasts were
   layered on 100 μl silicone oil AR 20/AR 200 (1:4), reacted with Pi or
   phosphonates for 10 sec and centrifuged into 20 μl of 10% $HClO_4$
   (w/r).

Biggens, J. (ed.), Progress in Photosynthesis Research, Vol. III. ISBN 90 247 3452 5

3. RESULTS AND DISCUSSION

Since the phosphonates are analogues of phosphate and phosphate esters, they have to enter the stroma to be effective. The access of Pi and other phosphorylated intermediates to the stroma is controlled at the chloroplast envelope by a specific transport protein, the phosphate translocator. At 25 °C, rapid exchange of phosphonoformic acid (PFA) and phosphonopropionic acid (PPA) with internal Pi occurred (Fig. 1). When added at a concentration of 2 mM, PFA and PPA were transported at rates of about 11 and 20 µmol/mg chl·h. The translocation rate was too low to be measured at 4 °C (data not shown). Although PPA entered the stroma more rapidly than PFA, both reached a similar level of exchange after 2 min.

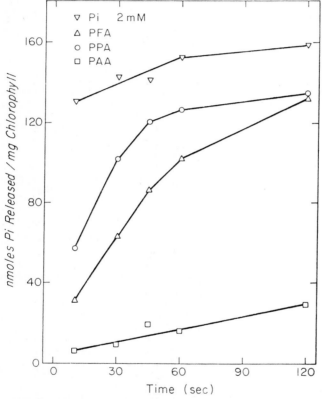

Figure 1. The time course of backexchange of the phosphonates with ^{32}Pi at 25 °C.

PFA was previously shown to strongly inhibit CO_2 fixation of intact chloroplasts at 25 °C (4). PPA was slightly less inhibitory at low concentrations, 2 and 4 mM. A positive correlation was found between the ability of the phosphonates to be translocated into the chloroplasts and the percentage of inhibition. The apparent reduced effect of temperatures on the translocation of these phosphonates indicated that they might be much less effective at inhibiting CO_2 fixation at lower temperatures. A drop of 5 °C, from 25 °C to 20 °C, yielded a decrease of 32% in CO_2 fixation in the absence of phosphonates. Exposing the chloroplasts to 15 °C further reduced the CO_2 fixation to only 42% of the control rates. The ability of PFA and PPA to inhibit CO_2 fixation was not greatly affected by decreasing the temperature when expressed as the percent of the inhibited rates (Fig. 2).

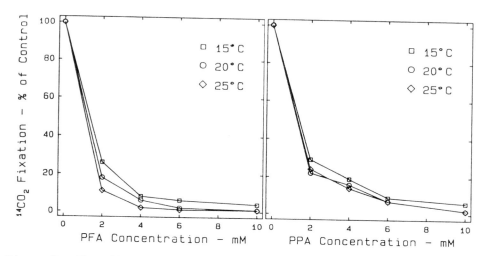

Figure 2. The effect of the phosphonates on CO_2 fixation by intact chloroplasts at three different temperatures. a. PFA; b. PPA.

The inhibition of CO_2 fixation by the phosphonates was further characterized by examining their effects on PGA reduction. At 25 °C, PGA reduction was inhibited by 41% with 2 mM PFA and 90% with 10 mM (Fig. 3a). Also, in this case, PPA was a little less effective than PFA (Fig. 3b).

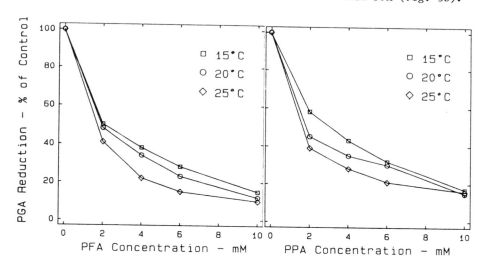

Figure 3. The effect of the phosphonates on the PGA reduction at three different temperatures. a. PFA; b. PPA.

The temperature effect on PGA reduction was similar to that on CO_2 fixation, although PGA reduction was somewhat more protected from inhibition by the phosphonates if the temperature was decreased.

4. CONCLUSIONS

It can be concluded that the inhibition of CO_2 fixation and PGA reduction by the phosphonates is relatively unaffected by the low temperatures.

REFERENCES
1. Izawa, S. 1977. Encyclopedia of Plant Physiology, Vol. 5, pp 266-282
2. Moreland, D.E. 1980. Annu. Rev. Plant Physiol. 31: 597-638
3. Hatch, A.L. 1980. Ph.D. Thesis, Univ. of Arizona, Tucson, AZ.
4. Heuer, B. and A.R. Portis, Jr. 1986. Suppl. Plant Physiol.

RUBISCO ACTIVASE: A NEW ENZYME IN THE REGULATION OF PHOTOSYNTHESIS

A.R. PORTIS, JR.[a,c], M.E. SALVUCCI[a,b], W.L. OGREN[a,c] AND J. WERNEKE[c], U.S. DEPARTMENT OF AGRICULTURE/AGRICULTURAL RESEARCH SERVICE[a] AND DEPARTMENTS OF AGRONOMY, UNIVERSITY OF KENTUCKY[b], LEXINGTON, KENTUCKY 40546 AND UNIVERSITY OF ILLINOIS[c], URBANA, ILLINOIS 61801

Rubisco has been one of the most studied enzymes in the photosynthetic process, since its discovery in 1956 (1). However our biochemical knowledge of the mechanism of the enzyme in vitro has been inadequate to account for its ability to function in vivo as measured by the photosynthetic rate of a leaf relative to its rubisco content. It was not until the middle 1970's that sufficient activity could be achieved in vitro to account for in vivo rates of photosynthesis (2,3). However, as soon as this problem was apparently resolved, another became more evident as research progressed. Namely, discrepancies exist between measurements of enzyme capacity in vivo and our knowledge of how the capacity of the enzyme for catalysis can be maintained in vitro (4-9). With the discovery and initial characterization of rubisco activase, many of these discrepancies have been resolved and a more detailed understanding of rubisco functioning should be possible in the near future.

Rubisco can exist in catalytically active and inactive forms and changes in the relative amount of these forms represent changes in specific activity (4). These changes in the specific activity (Vmax) occur without changes in affinities for the substrates, RuBP and CO_2. Detailed studies of the activation process with the isolated enzyme indicated that the degree of activation was determined by the pH, CO_2 and Mg concentrations (3). A simple model that accounted for the observed kinetics was:

$$E-NH_3^+ \xrightarrow[H^+]{} E-NH_2 + CO_2 \xrightarrow[H^+]{} E-NH-CO_2^- + Mg^{2+} \longrightarrow E-NH-CO_2^- \cdot Mg^{2+} \quad (1)$$

However, full activation in vitro required very high CO_2 and Mg concentrations at a pH near 8. When incubated with physiological substrate concentrations, 10 μM CO_2 and 5 mM Mg, the enzyme was only 25-30% activated. In contrast, estimates of the extent of activation of the enzyme in illuminated leaves are at least 2-3 fold higher under saturating light (6,8-11).

The catalytic binding site has a high affinity for RuBP but has a low specificity in that numerous phosphorylated compounds can also bind there (4,5). The binding of these compounds has been found to influence the activation of the enzyme at given CO_2 and Mg concentrations, but they also affect the activity. It has been difficult to define a role for the binding of such compounds in activation in vivo because of three factors: (a) the concentration of active sites (4 mM) is large relative to the stromal concentration of the most effective compounds (4); (b) increases

Biggens, J. (ed.), Progress in Photosynthesis Research, Vol. III. ISBN 90 247 3452 5
© *1987 Martinus Nijhoff Publishers, Dordrecht. Printed in the Netherlands.*

in the amount of activated enzyme which is inaccessible to RuBP is of no
consequence (5); (c) these compounds appear to be ineffective in the
presence of RuBP (12, Streusand and Portis, these Proceedings).

RuBP has another effect on the activity of the isolated enzyme which
presents difficulties (12-14). Inactive enzyme binds RuBP very tightly
(Kd = 0.02 M) (14) and cannot subsequently be significantly activated by
physiological CO_2 and Mg concentrations. Further, it is activated only
very slowly and to a much to lower extent than in the absence of RuBP by
even high CO_2 and Mg concentrations. Most importantly, enzyme which has
been activated by CO_2 and Mg in the absence of RuBP slowly loses activity
after it is added to initiate catalysis. In contrast, RuBP levels in vivo
can be considerably above binding site concentration and a constant
activation state can be observed (6,8).

Activation in vivo responds to light (6,9-11) and O_2 concentration
(7,8), but not to CO_2 levels above ambient (8,9). These observations are
inexplicable in terms of any of the in vitro experiments. Generally the
increased activation of the enzyme in the light vs. the dark was
attributed to light induced changes in Mg and pH (4). However it has been
shown that in several plants illumination with low intensity light results
in a decrease in the activation of the enzyme to levels considerably below
that measured even in the dark (6,8-15).

In 1982, Somerville et al. (16) reported the discovery of a mutant of
Arabidopsis whose phenotype was an inability to survive at atmospheric
levels of CO_2 because rubisco was not maintained in an activated form in
the light. The mutation did not appear to reside in the rubisco enzyme
itself and the conclusion reached was that "the characteristics of the rca
mutant are not explicable on the basis of our current understanding of the
properties of RuBP carboxylase and the chloroplast milieu". Recently the
mutant strain was characterized more extensively by Salvucci et al. (9)
and some of the discrepancies between the requirements for activation in
vivo and in vitro became even more defined.

During this characterization, a polypeptide was found to be absent
when SDS gel electrophoresis of chloroplast stromal proteins of the mutant
and wildtype were compared (16). Genetic analysis of an F2 population
containing the activation lesion demonstrated that the absence of this
polypeptide cosegregated with the physiological phenotype. Subsequent
analysis of the chloroplast stromal proteins by two-dimensional gel
electrophoresis indicated that two polypeptides between 40 and 50 kDa were
actually missing in the mutant. It was therefore clear that a soluble
chloroplast protein was essential for the activation of rubisco in vivo,
and the protein was called rubisco activase (17).

Biochemical evidence for the participation of a soluble protein in
rubisco activation was obtained from the development of a reconstituted
assay system in which a light dependent activation of rubisco could be
seen in vitro (17). The essential components of the system are isolated
chloroplast thylakoids, an electron acceptor, rubisco, RuBP, CO_2, Mg and a
protein fraction that contains rubisco activase. Using this system (17),
partially purified chloroplast extracts from spinach and wild-type
Arabidopsis stimulated rubisco activity in the light, but an extract of
the mutant did not. The ability to activate rubisco by illumination with

methyl viologen in vitro (17) and in leaves in which photosynthesis has
been partially inhibited by methyl viologen treatment (9), clearly
distinguished the mechanism from the thioredoxin-linked mechanism involved
the activation of other stromal enzymes.

With the development of an in vitro assay system, the further
purification of rubisco activase was pursued by using ion exchange and gel
filtration (Table 1) (Salvucci et al., these Proceedings). However the
activity of the protein is rather labile and evidence of proteolysis has
often been observed with gel electrophoresis. There has not been much
success with eliminating these problems and therefore preparations only
carried thru the ion-exchange step have been utilized in experiments
reported here.

TABLE 1. Purification of rubisco activase from spinach chloroplasts

Purification Step	Total Protein (mg)	Total Activity (μmol/mg)	Specific Activity (U/mg)	Recovery (%)	Purification (-fold)
Chloroplast lysate	159.4	45.8	0.3	100	--
FPLC-ion exchange	5.2	26.9	5.3	58	18
FPLC-Superose	1.0	17.7	17.9	38	62

After determining the optimal amounts of thylakoids (100 μg chl/ml),
rubisco (100 μg/ml), and rubisco activase (300 μg/ml) necessary to achieve
a high rubisco specific activity in the reconstituted assay system, the
concentration dependence of each of the other components was determined.

An unexpectedly high optimal RuBP concentration was required, with
maximum activation at about 3 mM RuBP (Fig 1). Without RuBP, the
spontaneous CO_2/Mg activation resulted in about 40% of the maximal
possible under the assay conditions used as compared to 80% with light and
3 mM RuBP. Preincubation with 0.1 mM RuBP completely inhibited rubisco
activation, and in the absence of activase or illumination, activity
remained low at all RuBP concentrations. From this experiment it is clear
that rubisco activase can function in the presence of the high RuBP
concentrations found to occur in illuminated leaves.

The maintenance of the activated state requires constant
illumination (Fig. 2). Upon illumination of the system, activation
occurred with a $t_{1/2}$ of about 2 min. Deactivation began immediately after
the light was turned off with a $t_{1/2}$ of about 4 min. Activation again
occurred during a second period of illumination. The deactivation process
in the dark appears to occur regardless of the presence of activase since
the kinetics were simulated by the addition of RuBP to rubisco that has
been previously allowed to activate by using CO_2 and Mg (not shown). The
relationship between activation/deactivation in the reconstituted system
and the extent of carbamate formation as defined in the CO_2 and Mg process
(5), has not yet been determined.

FIGURE 1. Dependence of rubisco
activation on RuBP concentration
in the light (o) or dark (●).
(Portis, Salvucci and Ogren,
submitted).

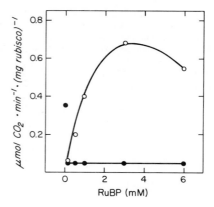

FIGURE 2. Time course of
rubisco activation level during
sequential light (o) or dark
(●) periods. (Portis, Salvucci,
and Ogren, submitted).

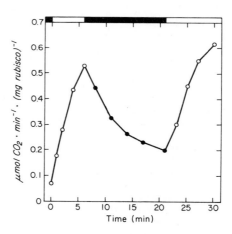

The effects of pH and Mg on the extent of activation in the
reconstituted system were of interest in view of the light-dependent
changes in pH and Mg that occur in the stroma. With 11.5 μM CO_2 and 6 mM
free Mg, a gradual increase in activation occurred as the pH increased
when RuBP was either present in the dark or absent (Fig 3). This increase
reflects the influence of pH on carbamate formation in the CO_2/Mg activation process (Eq. 1). In contrast, the pH optimum of the light and
rubisco activase dependent process is rather broad with an optimum of
about 8.2, which is near the optimal stromal pH for CO_2 fixation. The
system is rather insensitive to Mg (not shown) as decreases in activity
only occurred at concentrations below 3 mM.

FIGURE 3. Dependence of rubisco activation on pH. No RuBP (■), or 3mM RuBP in the light (o) or dark (●). (Portis, Salvucci, and Ogren, submitted).

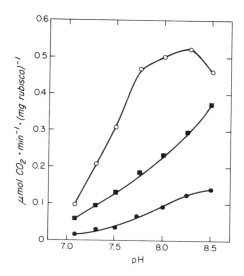

The CO_2 requirement for activation with rubisco activase in the reconstituted system is of special importance since near maximal activation of rubisco can occur <u>in vivo</u> in C_3 plants even at atmospheric CO_2 levels. In the presence of RuBP, the Kact (CO_2) was found to be about 4 µM, which is well below the estimated atmospheric concentration of about 10 µM (Fig 4). In the absence of RuBP, the Kact (CO_2) was about 23 µM which agrees well with the response of the CO_2/Mg process. In the dark with RuBP, little activity was found over the range of CO_2 concentrations examined.

FIGURE 4. Dependence of rubisco activation on CO_2 concentration. No RuBP (■) or 3 mM RuBP in the light (o) or dark (●). (Portis, Salvucci, and Ogren, submitted).

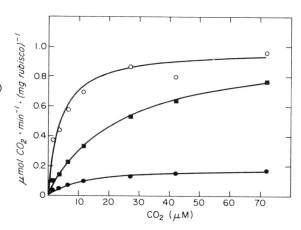

In vivo studies indicate that light intensity is the most significant
factor determining the extent of rubisco activation, and the response is
similar to that of photosynthesis itself (6,8-10). Light limits photosynthesis thru the rate of NADPH formation (rate of electron transport) and
ATP synthesis (magnitude of ΔpH) but ATP has been suggested to be more
important (18). The role of stromal pH changes is unclear but the
available evidence indicates that because of buffering, low light is
sufficient to establish the optimal stromal pH in vivo (19). With the
reconstituted system, the complications of changes in stromal pH are not
present. We therefore examined the effects of changes in light intensity
on activation and ΔpH established across the thylakoid membranes.
Significant activation only occurred with a ΔpH above about 2.5 units,
with maximal activation attained at 3.0 units or above (Table 2). These
results indicate that light is sensed by the activation system as the
establishment of a pH gradient across the thylakoid membranes. Similarly
uncouplers, which decrease ΔpH but increase electron transport, were also
found to inhibit activation (not shown). Variations in activation in vivo
may therefore reflect changes in the pH gradient across the thylakoids in
some cases. It may be very significant that the strong dependence of
rubisco activation on ΔpH somewhat resembles the response of
phosphorylation to ΔpH (20). Unfortunately, techniques to quantitatively
measure ΔpH in vivo need to be developed in order to explore this
hypothesis in detail.

TABLE 2. Dependence of activation on light intensity and thylakoid ΔpH.

Electron Acceptor	Light Intensity (μE m^{-2} s^{-1})	Activation (μmol mg^{-1} min^{-1})	Thylakoid ΔpH
Pyocyanine	500	0.53	3.2
	100	0.41	3.1
	40	0.18	2.7
	20	0.07	2.5
	Dark	0.06	-
Methyl Viologen	500	0.49	3.0
	40	0.27	2.8
	20	0.15	2.7

The existence of the Arabidopsis mutant and the biochemical characteristics of the reconstituted system established thus far are consistent
with an essential role for rubisco activase in the functioning of rubisco
and the regulation of rubisco activity in C$_3$ plants. Recently several
laboratories (21-23) have reported on the occurence of an inhibitor of
rubisco activity in the leaves of numerous plant species, including
soybean, tobacco and kidney bean. However, the inhibitor is apparently
not ubiquitous, since it has not been detected in many other species
(21,23). Therefore it was important to determine if rubisco activase was
present in other plant species, particularly those known to have the
inhibitor and those that have a CO$_2$ concentrating mechanism, species in
which CO$_2$ would not limit activation.

Antibodies to a rubisco activase preparation were raised and used to
probe for cross-reacting polypeptides in leaf extracts following

SDS-PAGE. Two polypeptides in the 40-50 kDa region were detected in spinach and wild-type Arabidopsis plants, but were completely missing in the activation mutant (see Salvucci et al., these Proceedings). The absence of cross-reacting material in this region confirmed that no rubisco activase protein was present in the mutant. Cross-reacting polypeptides were observed in this area in all C_3 and C_4 species examined including several known to have the inhibitor. A reaction was also obtained with Chlamydomonas. The presence of activase in all of the higher plants that we have examined is additional support for the concept that rubisco activase plays an essential role in the regulation of rubisco. The detection of activase in several C_4 species and in Chlamydomonas is particularly significant since the CO_2 concentration is elevated in the stroma in these cases (24,25). This indicates that in addition to an essential role for activation at limiting CO_2 levels in C_3 plants, activase may also be required to counteract the RuBP induced deactivation of rubisco which would occur regardless of the high CO_2 concentration in C_4 plants and algae.

The discovery of rubisco activase and initial characterization of the activation process have provided some answers to current discrepancies between knowledge of rubisco functioning in vivo and in vitro. The reconstituted activation system accounts for the activation of rubisco to high specific activities at atmospheric CO_2 concentrations and the ability to maintain activation in the presence of high RuBP concentrations. Activation is dependent on the establishment of a pH gradient across the thylakoid membranes. In this way the system links the utilization of RuBP by rubisco, and therefore the utilization of ATP and NADPH for the regeneration of RuBP by the other reductive pentose phosphate enzymes, rather directly to the capacity of the thylakoids to provide ATP and NADPH. Some features of the activation process in vivo, such as the effects of O_2 remain unexplained, but new ways to approach these questions are now evident. Moreover, most of the biochemical aspects of the process are essentially unknown at present, particularly the nature of the interactions between chloroplast thylakoid membranes, rubisco activase, and rubisco. More information about these processes may provide additional insights into the functioning of rubisco in relation to photosynthesis.

REFERENCES
1 Weissbach, A., Horecker, B.L. and Hurwitz, J. (1956) J. Biol. Chem. 218, 795-810
2 Laing, W.A., Ogren W.L. and Hageman, R.H. (1975) Biochemistry 14, 2269-2275
3 Lorimer, G.H., Badger, M.R. and Andrews, T.J. (1976) Biochemistry 15, 529-536
4 Jensen, R.G. and Bahr, J.T. (1977) Ann. Rev. Plant Physiol. 28, 379-400
5 Miziorko, H.M. and Lorimer, G.H. (1983) Ann. Rev. Biochem. 52, 607-535
6 Perchorowicz, J.T., Raynes, D.A. and Jensen, R.G. (1981) Proc. Nat. Acad. Sci. USA 78, 2985-2989
7 Schnyder, H., Mächler, F. and Nösberger, J. (1984) J. Exp. Bot. 35, 147-156
8 Perchorowicz, J.T. and Jensen, R.G. (1983) Plant Physiol. 71, 955-960
9 Salvucci, M.E., Portis, A.R. Jr. and Ogren, W.L. (1986) Plant Physiol. 80, 655-659
10 Taylor, S.E. and Terry N. (1984) Plant Physiol. 75, 82-86

11 Mächler, F. and Nösberger, J. (1980) J. Exp. Bot. 31, 1485–1491
12 Chu, D.K. and Bassham, J.A. (1975) Plant Physiol. 55, 720–726
13 Laing, W.A. and Christeller, J.T. (1976) Biochem. J. 159, 563–570
14 Jordan, D.B. and Chollet, R. (1983) J. Biol. Chem. 258, 13752–13758
15 Sharp, R.E. and Portis, A.R. Jr. (1984) Plant Physiol. 75S, 92
16 Somerville, C.R., Portis, A.R. Jr. and Ogren, W.L. (1982) Plant Physiol. 70, 381–387
17 Salvucci, M.E., Portis, A.R. Jr. and Ogren, W.L. (1985) Photosynthesis Res. 7, 193–201
18 Heber, U. (1973) Biochem. Biophys. Acta 305, 140–152
19 Leegood, R.C., Walker, D.A. and Foyer, C.H. (1985) in Photosynthetic Mechanisms and the Environment (Barber, J. and Baker N.R. eds.), pp. 189–258, Elsevier, Amsterdam
20 Portis, A.R. Jr. and R.E. McCarty (1976) J. Biol. Chem. 251, 1610–1617
21 Vu, C.V., Bowes, G. and Allen, L.H. (1984) Plant Physiol. 76, 843–845
22 Servaites, J.C. (1985) Plant Physiol. 78, 839–843
23 Seeman, J.R., Berry, J.A., Freas, S.M. and Krump, M.A. (1985) Proc. Natl. Acad. Sci. USA 82, 8024–8028
24 Hatch, M.D. (1971) Biochem. J. 125, 425–431
25 Badger, M.R., Kaplan, A. and Berry, J.A. (1980) Plant Physiol. 66, 407–413

RUBISCO ACTIVASE; PURIFICATION, SUBUNIT COMPOSITION AND SPECIES
DISTRIBUTION

MICHAEL E. SALVUCCI[a,b] JEFFREY M. WERNEKE[c], WILLIAM L. OGREN[a,c] AND ARCHIE
R. PORTIS, JR[a,c], U.S. DEPARTMENT OF AGRICULTURE/AGRICULTURAL RESEARCH
SERVICE[a] AND DEPARTMENTS OF AGRONOMY, UNIVERSITY OF KENTUCKY[b], LEXINGTON,
KENTUCKY 40546 AND UNIVERSITY OF ILLINOIS[c], URBANA, ILLINOIS 61801

1. INTRODUCTION
    The activity of RuBP carboxylase/oxygenase (rubisco) in leaves in-
creases upon illumination (1,2). Somerville et al. (3) provided evidence
that light activation of rubisco is a prerequisite for catalytic competence in vivo by isolating a rubisco light activation mutant of Arabidopsis, designed rca, in which rubisco appeared to be poorly activated in
vivo. Electrophoretic analysis of chloroplast proteins from mutant and
wild-type Arabidopsis demonstrated that two polypeptides were missing in
the mutant (4). A causal relationship between the missing polypeptides in
the mutant and the inability to activate rubisco in vivo was established
by genetic analysis (4). Using a reconstituted assay for rubisco activation, Salvucci et al (4) demonstrated that rubisco activation required a
protein present in chloroplast extracts of spinach and Arabidopsis wild
type but absent in extracts of the rca mutant. Based on these data it was
proposed that rubisco activation is catalyzed in vivo by a specific
stromal protein, rubisco activase (4,5). This paper reports the
purification of rubisco activase from spinach, its subunit composition,
and the identification of this protein in several plant species.

2. PROCEDURES
2.1. Assay of rubisco activase: Rubisco activase was assayed in a two
    stage assay by measuring the light-dependent increase in rubisco
    specific activity in a reconstituted chloroplast system (4). Activ-
    ity was dependent on the inclusion of thylakoid membranes and RuBP
    (Portis et al., this volume). One unit (U) = 1 μmol CO_2/mg (rubisco
    protein) min.
2.2. Purification of rubisco activase: Rubisco activase was purified from
    spinach chloroplasts by ion-exchange (6) and gel filtration FPLC.
2.3. Antibody preparation and immunoblotting: Polyclonal antibodies were
    obtained from tumor-induced mouse ascites fluid (7) after injection
    with 20-40 μg of activase. Plant material was extracted at 4 C in 20
    mM Tris-HCl, pH 8.0, 10 mM $MgCl_2$, 10 mM DTT and 20 μM leupeptin and,
    in some cases, 0.1% SDS. After centrifugation, the supernatant was
    separated by SDS-PAGE and the proteins were transferred electrophore-
    tically to nitrocellulose, probed with anti-activase antibodies, and
    visualized with an anti-mouse IgG alkaline phosphatase conjugate.

3. RESULTS AND DISCUSSION
3.1. Activase activity in crude and partially purified plant extracts: In
    preliminary experiments with lysed spinach chloroplasts, the specific
    activity of endogenous rubisco was found to increase upon illumina-
    tion if relatively high concentrations of RuBP were added to the
    lysate. This increase was ascribed to the activity of a soluble

protein present in the mixture, and led to the development of a reconstituted assay for rubisco activase (4). Rubisco activase activity was measured in lysed chloroplasts from spinach and Arabidopsis and in lysed tobacco protoplasts (Table 1). Activity was not detected in lysed chloroplasts from mutant Arabidopsis (Table 1). Rubisco activase activity could also be measured when spinach and Arabidopsis chloroplast extracts were freed of low molecular weight components and endogenous rubisco, and then added back to purified spinach rubisco and washed spinach thylakoids (Table 1.) In contrast, extracts prepared from chloroplasts of the Arabidopsis mutant were ineffective in stimulating light activation. These results indicated that rubisco can be activated in a light-dependent manner by a soluble protein component present in spinach, tobacco, and wild-type Arabidopsis, but absent in the Arabidopsis rca mutant.

TABLE 1. Rubisco activase activity in lysed chloroplasts and partially purified stromal extracts of spinach and Arabidopsis, and in lysed proto-plasts of tobacco.

Source of Extract	Rubisco Activity		Fold-Stimulation by Light
	Light	Dark	
	(nmol CO_2/min)		
Lysed Chloroplasts			
spinach	84	28	3.0
Arabidopsis (wild type)	33	13	2.5
Arabidopsis (rca mutant)	18	16	1.1
Lysed Protoplasts			
tobacco	24	8	3.0
	(nmol/mg (rubisco) protein·min)[1]		
Stromal Extracts			
spinach	523	40	13.1
Arabidopsis (wild type)	119	36	3.3
Arabidopsis (rca mutant)	37	34	1.2

[1]Measured in a reconstituted chloroplast system containing spinach thylakoid membranes (30 µg Chl), purified spinach rubisco (47 µg), 4 mM RuBP, 8 mM $MgCl_2$, 3000 units catalase, 40 µM pyocyanine, 100 mM Tricine, pH 8.0 and 3.1 mM NaH$^{14}CO_3$ in 300 µl total volume.

3.2. Purification of rubisco activase: Rubisco activase was purified from intact spinach chloroplasts. Following lysis, soluble chloroplast extracts were centrifuged, filtered and fractionated by ion-exchange FPLC (6). Activase activity eluted in a single peak with 0.17 M KCl (Fig. 1a). Peak fractions from this column were pooled, concentrated by ultrafiltration, and fractionated on a Superose-12 gel filtration FPLC column. Activase activity was measured in a single peak (Fig. 1b). The preparation purified through the two FPLC steps exhibited a 62-fold increase in specific activity to 18 U/mg protein with a 38% recovery. Analysis of the final preparation by SDS-PAGE identified

three major polypeptides at 44, 41 and 29 kDa (data not shown). The
44 and 41 kDa polypeptides, which co-migrated on native gels,
corresponded in molecular weight to the missing polypeptides in the
Arabidopsis rca mutant (4). The 29 kDa polypeptide was identified as
carbonic anhydrase on the basis of subunit and holoenzyme mass,
copurification with carbonic anhydrase activity, and localization
within the chloroplast.

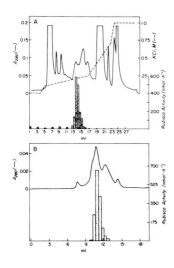

FIGURE 1. Isolation of rubisco
activase from spinach stromal
extracts. a) Elution profile of
rubisco activase activity from
a Mono Q (0.5x5.0 cm) anion
exchange FPLC column. b)
Elution profile of rubisco
activase after gel-filtration
on a Superose-12 (1.0x30 cm)
FPLC column.

3.3. Detection of activase on immunoblots: Polyclonal antibodies were
produced against the final activase preparation. Ascites fluid from
tumor-induced mice was collected early in the immune response in
order to lower the titer against the less abundant 29 kDa
polypeptide contaminant. The antibody-containing ascites fluid was
used to probe nitrocellulose replicas of SDS-PAGE gels of soluble
proteins from a number of plant species (Fig. 2). Two polypeptides
similar in molecular weight to those of the 44 and 41 kDa polypeptides of spinach were detected in leaf extracts of Arabidopsis wild
type, pea, soybean, kidney bean, celery, tobacco, maize, pigweed,
purslane, barley, oat and Chlamydomonas, indicating that activase was
present and that its structure was highly conserved. The
polypeptides were also found in extracts of tomato, dandelion,
sorghum, and crabgrass (not shown). For several of the species
examined, only a single band of about 41 kDa was visualized when
tissue was extracted without leupeptin, DTT, and/or SDS. This
observation suggests that the larger activase polypeptide is
particularly susceptible to proteolytic degradation in homogenized
leaf extracts and could explain why attempts to measure activase
activity in leaf extracts have been unsuccessful.

The activase polypeptides were not detected in leaf extracts of
the Arabidopsis rca mutant (Fig. 2). On two-dimensional gels (IEF x
SDS-PAGE), the polypeptides recognized in wild-type Arabidopsis

extracts by antibodies prepared against a spinach activase preparation migrated to the same location as the polypeptides identified as missing in the mutant (data not shown).

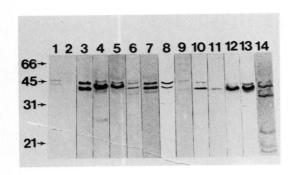

FIGURE 2. Immunological detection of activase polypeptides in leaf extracts separated by SDS-PAGE. Species examined were 1, Arabidopsis wild type; 2, Arabidopsis rca mutant; 3, spinach; 4, pea; 5, soybean; 6, kidney bean; 7, celery; 8, tobacco; 9, maize; 10, pigweed; 11, purslane; 12, barley; 13, oat; 14, Chlamydomonas.

3.4. CONCLUSIONS: Physiological studies with the Arabidopsis rca mutant (3,5) have revealed several interesting aspects of rubisco activation consistent with the concept of rubisco activase. Purification of rubisco activase provided biochemical confirmation of previous genetic studies linking the two polypeptides missing in the Arabidopsis rca mutant with rubisco activation (4). The altered physiology of the mutant and the widespread occurrence of the protein suggest a fundamental role for activase in the control of rubisco activity. By modulating the activity of rubisco, activase provides a mechanism to explain the coordinate regulation of photon flux rate through the electron transport chain with the fixation of CO_2 by the reductive pentose phosphate cycle.

REFERENCES
1. Mächler, F.M. and Nösberger, J. (1980) J. Expt. Bot. 31, 1485-1491
2. Perchorowicz, J.T., Raynes, D.A., and Jensen, R.G. (1981) Proc. Natl. Acad. Sci. U.S.A. 78, 2985-2989
3. Somerville, C.R., Portis, A.R. Jr., and Ogren, W.L. (1982) Plant Physiol. 70, 381-387
4. Salvucci, M.E., Portis, A.R. Jr., and Ogren, W.L. (1985) Photosynth. Res. 7, 191-203
5. Salvucci, M.E., Portis, A.R. Jr, and Ogren, W.L. (1986) Plant Physiol. 80, 655-659
6. Salvucci, M.E., Portis, A.R. Jr, and Ogren, W.L. (1986) Anal. Biochem. 153, 97-101
7. Lacy, M.J. and Voss, E.W. (1986) J. Immunol. Meth. 87, 169-177

EFFECTS OF 6-PHOSPHOGLUCONATE AND RUBP ON RUBISCO ACTIVATION STATE AND
ACTIVITY.

V. J. STREUSAND[a] AND A. R. PORTIS, JR[b]. DEPARTMENT OF AGRONOMY,
UNIVERSITY OF ILLINOIS[a] AND U.S. DEPARTMENT OF AGRICULTURE, AGRICULTURAL
RESEARCH SERVICE[b], 1102 S. GOODWIN AVE. URBANA, IL 61801.

1. INTRODUCTION
    Ribulose-1,5-bisphosphate carboxylase/oxygenase (Rubisco) catalyzes
the first step in photosynthetic carbon fixation and appears to have a
regulatory role in light activated photosynthesis (1). For purified
rubisco to be catalytically competent, it must form a complex with CO_2 and
magnesium (2). A number of chloroplast metabolites have been shown to
stimulate rubisco activity when preincubated with the enzyme before
assaying for activity (3,4,5,6,7). These positive effectors are thought to
stimulate rubisco activation by stabilizing the catalytically competent
enzyme-CO_2-Mg^{2+} complex, not by increasing binding of CO_2 and magnesium to
the enzyme (3,6,7). This stabilization apparently occurs when the
phosphorylated effectors bind to the active site of the enzyme so that the
effectors act as competitive inhibitors in the presence of substrate RuBP
(3,5,7). RuBP itself is known to inhibit activation of rubisco by binding
to the inactive form of the enzyme, preventing the addition of CO_2 and
magnesium (8,9).
    Since chloroplast RuBP levels are assumed to be close to binding site
concentration, these data caused difficulties in understanding the
activation of Rubisco in vivo until the discovery by Salvucci et al (10)
that Rubisco activation was catalyzed enzymatically in vivo by rubisco
activase. This enzyme stimulates activation of rubisco in the light in
the presence of RuBP. The mechanism of stimulation is not yet understood,
but it can be postulated that rubisco activase may work by producing a
positive effector of activation such as 6-phosphogluconate. The
experiments presented here with purified rubisco, 6-phosphogluconate and
RuBP suggest that is not the case.

2. MATERIALS AND METHODS
2.1 RuBP was synthesized from ribose-5-phosphate (Sigma) by M. E.
     Salvucci following the procedure of Jordan and Ogren (11).
     Carboxyarabinitol bisphosphate was synthesized from RuBP by the
     method of Pierce et al (12). Spinach (Spinacia oleracea L.) was grown
     hydroponically as described by Portis (13). Chloroplasts were
     isolated from spinach leaves as described by Salvucci et al (14).
2.2. Rubisco Purification. Rubisco was purified from the chloroplast
     stromal proteins by FPLC using the method of Salvucci et al (14).
     The enzyme was used directly from the ion-exchange column in 20 mM
     Tris, 2 mM DTT and approximately 0.3 M KCl. For short term storage
     the enzyme was stored at 4° C; for long term storage the enzyme
     solution was made 10 mM $NaHCO_3$, and 10 mM $MgCl_2$ and stored in
     aliquots in liquid nitrogen.
2.3. Rubisco Activation. Unless otherwise stated, rubisco 0.12 - 1.2
     mg/ml was activated at 25° C in 100 mM Tricine-NaOH, pH 8.0, 2-4 mM
     DTT, 1 mM $NaHCO_3$, and 20 mM $MgCl_2$. Other additions are as described

Biggens, J. (ed.), Progress in Photosynthesis Research, Vol. III. ISBN 90 247 3452 5

in the text.

Enzyme after short or long term storage was activated and deactivated before use to overcome cold deactivation effects. This enzyme was activated by the addition of 4 mM DTT, 10 mM $NaHCO_3$, and 10 mM $MgCl_2$ for 15 min. at 45° C, deactivated either on a desalting column (Ultrogel ACA44) equilibrated with 100 mM CO_2-free Tricine, 2 mM DTT, pH 8.0, or by sequential concentration and redilution in centricon 30's (Amicon).

2.4 Carboxylase Assays. RuBP carboxylase activity was measured by 0.5 min. assays at 25° C as RuBP dependent incorporation of $^{14}CO_2$ into acid stable products. The contents of the assay medium were 100 mM Tricine-NaOH, pH 8.0, 10 mM $MgCl_2$, 0.4 mM RuBP, 10 mM $NaH^{14}CO_3$, 1 µci/µmol. Assays were initiated by addition of enzyme activated as described above and were terminated by the addition of 4 N formic acid in 1 N HCl. Disintegrations per minute were determined by liquid scintillation spectroscopy. Background counts from zero-time assays were subtracted from all assay counts.

2.5 Activator CO_2 Binding. To measure activator CO_2 binding, enzyme activation was carried out as described above, but the activation mixture was made approximately 7 µci/µmol $NaH^{14}CO_3$. At the appropriate times, aliquots of activated enzyme were added to a quench containing 100 mM Tricine pH 8.0 and 300 µM CABP (approximately 200 fold excess of enzyme sites). The quenches remained at room temperature for one hour and then were either frozen in liquid nitrogen or used. Activator CO_2 binding was measured after gel filtration on an ultrogel Aca 44 (LKB) column to separate protein-- bound $^{14}CO_2$ from unbound CO_2. The columns were equilibrated with 100 mM Tricine, pH 8.0 and 100 mM $NaHCO_3$. Activator CO_2 binding was calculated from the NaOH stable, HCl labile counts associated with the high molecular weight fraction.

3. RESULTS AND DISCUSSION

6-phosphogluconate stimulates rubisco activation and activator CO_2 binding almost 2-fold over control values, but the rate of exchange of bound activator CO_2 is inhibited, although the final amount of bound CO_2 is the same in the presence and absence of 6-phosphogluconate (Fig. 1, A, B, C).

Because 6-phosphogluconate stimulates activation, but inhibits exchange of activator CO_2 in catalytically competent enzyme complexes, the hypothesis that 6-phosphogluconate and other positive effectors of rubisco activation act as stabilizers of the activated enzyme-CO_2-Mg^{2+} complex (3, 6, 7) is supported.

When deactivated rubisco is incubated in the presence of RuBP before being assayed, rubisco activation is strongly inhibited and 6-phosphogluconate does not increase activation (Figure 2 A, B). This indicates that RuBP binds rapidly to the enzyme, decreasing the rate of activation and, therefore, preventing the stimulatory effect of 6-phosphogluconate.

Figure 1. Time courses of Rubisco
Activation, Activator CO_2 Binding
and Activator CO_2 Exchange in the
presence and absence of 6-phospho-
gluconate. Rubisco was activated
in 1 mM $NaHCO_3$ and 20 mM $MgCl_2$ at
25° C. Values shown are percentages
of the control value at 20 min.
Squares - without 6-phosphogluconate
in the activation mix; circles - with
0.1 mM 6-phosphogluconate in the
activation mix. A, Time course of
activation. The control rate at 20
minutes was 0.6 µmol/min/mg. B,
Time course of activator CO_2 Binding.
The control value at 20 min. was 2.5
nmol/mg (1.4 mol/mol). C, Time
course of activator CO_2 exchange.
After 5 minutes of activation,
labelled bicarbonate was injected
into the activation mixes and
aliquots were removed and added to
CABP quenches at the times indicated.
The control value at 20 minutes was
3.6 nmol/mg (2.0 mol/mol).

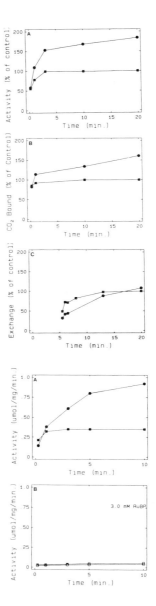

Figure 2. Time Course of
Rubisco Activation in the
presence and absence of RuBP
and 6-phosphogluconate.
Squares - without 6-phospho-
gluconate; circles - with
0.1 mM 6-phosphogluconate.
Rubisco was activated in the
absence (A) or presence (B)
of 3 mM RuBP and aliquots were
removed and assayed as described
in the methods section.

RuBP, even at 0.1 mM concentration, causes the activity of deactivated
Rubisco to remain low, and the addition of 6-phosphogluconate decreases
activity even further (Figure 3 A, B, C), presumably due to competitive
inhibition with RuBP.

Figure 3. Time course of
Rubisco activity in the
presence and absence of
6-phosphogluconate at 3 mM
RuBP concentrations. Rubisco
was incubated with 1 mM
$NaH^{14}CO_3$, 20 mM $MgCl_2$ with the
RuBP concentrations indicated
with (circles) or without
(squares) 0.1 mM 6-phospho-
gluconate. At the times shown,
aliquots were removed and added
to 4 N formic acid in 1 N HCl.

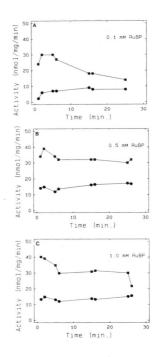

4. CONCLUSION
    The data presented here indicate that, under _in vivo_ conditions, the
presence of metabolites which stimulate activation of isolated rubisco are
unlikely to have a regulatory effect on rubisco activity.

REFERENCES
 1. Perchorowicz, J.T., D.A. Raynes, and R.G. Jensen. 1981. Proc. Nat.
    Acad. Sci. USA 78:2985-2989.
 2. Miziorko, H.M. and G.H. Lorimer. 1983. Annu. Rev. Biochem. 52:507-535.
 3. Badger, M.R. and G.H. Lorimer. 1981. Biochem. 20:2219-2225.
 4. Chu, D.K. and J.A. Bassham. 1974. Plant Physiol. 54:556-559.
 5. Chu, D.K. and J.A. Bassham. 1975. Plant Physiol. 55:720-726.
 6. Jordan, D.B., R. Chollet and W.L. Ogren. 1983. Biochem. 22:3410-3418.
 7. McCurry, S.D., J. Pierce, N.E. Tolbert, and W.H. Orme-Johnson. 1981.
    J. Biol. Chem. 256:6623-6628.
 8. Jordan, D.B. and R. Chollet. 1983. J. Biol. Chem. 258:13752-13758.
 9. Laing, W.A. and J.T. Christeller. 1976. Biochem. J. 159:563-570.
10. Salvucci, M.E., A.R. Portis, Jr. and W.L. Ogren. 1985. Photosynthesis
    Res. 7:193-201.
11. Jordan, D.B. and W.L. Ogren. 1981. Plant Physiol. 67:237-245.
12. Pierce, J., N.E. Tolbert and R. Barker. 1980. Biochem. 19:934-942.
13. Portis, A.R., Jr. 1981. Plant Phys. 67:985-989.
14. Salvucci, M.E., A.R. Portis, Jr. and W.L. Ogren. 1986. Anal. Biochem.
    153:97-101.

Isolation, Identification, and Synthesis of Carboxyarabinitol-1-Phosphate, A Diurnal Regulator of Ribulosebisphosphate Carboxylase Activiy†

Joseph A. Berry[1], George H. Lorimer[2], John Pierce[2]
James Meek[2], and Suzan Freas[1]
[1]Carnegie Institution of Washington, Department of Plant Biology, Stanford, CA
[2]E.I. Du Pont de Nemours & Co., Central R&D., Wilmington, DE

The activity of ribulose bisphosphate carboxylase of *Phaseolus vulgaris* (garden bean) and several other species (1, 2, 3, 4, 5) is influenced by mechanisms which control the concentration present *in vivo* of an inhibitor that binds tightly to the active site of the enzyme rendering it catalytically inactive. This mechanism appears to function together with the Mg^{2+} and CO_2 activation in regulating the activity of RuBP carboxylase *in vivo* (6). Previous work has demonstrated that the inhibitor: is a phosphorylated compound (1, 2); is a noncompetitive inhibitor (with respect to RuBP) when bound to the enzyme (2); is compartmented in the chloroplast (2); varies in concentration in nature with the diurnal pattern of light (2, 6). We report here the structural identity of the inhibitor present in leaves of *Phaseolus vulgaris*; its synthesis *in vitro*, and some kinetic parameters associated with its interaction with RuBP carboxylase.

Materials and Methods

RuBP carboxylase. Purified spinach RuBP carboxylase was used for all kinetic studies of the inhibitor. This was prepared from market spinach essentially as in (7) The concentration or RuBP carboxylase was determined by radiolabelling with [2- ^{14}C]carboxyarabinitol-1,5-bisphosphate (CABP), and catalytic activity as described (8).

Inhibitor. Crude preparations were obtained from leaves of *P. vulgaris* plants kept overnight in darkness. These were extracted (typically 1 Kg. per batch), and processed as described in (2).

Results

Assays of Inhibitor. The inhibitor was determined by titration of the activity of known concentrations of purified spinach RuBP carboxylase. During 30 min preincubation the inhibitor froms a stable enzyme-complex (EI). The amount of EI formed is indicated by the reduction of the rate of carboxylation of RuBP in a 1 min assay (v/v_o, Fig. 1) following the preincubation. The data was analyzed according to a quadratic equation relating EI to: E_{tot}, total enzyme concentration; I_{tot}, total inhibitor concentration, and K_d, the equilibrium constant. The unknowns, I_{tot} and K_d, of this equation were fitted to the data, yielding the solid line (Fig. 1) with $K_d = 1.1 \cdot 10^{-7}$M and a concentration of inhibitor in the preparation of 8.4 mM.

Purification of Inhibitor. Crude extracts were first fractionated on columns of Dowex 50 (H$^+$ form) and Dowex 1 (formate form). The inhibitor was retained on Dowex-1, and was eluted with a linear gradient (0-8 M) of formic acid. The fractions were assayed, combined and dried in a rotary evaporator under vacuum. In some cases, a further purification was achieved

---

†CIW-DPB Publ. No. 957

Biggens, J. (ed.), Progress in Photosynthesis Research, Vol. III. ISBN 90 247 3452 5
© *1987 Martinus Nijhoff Publishers, Dordrecht. Printed in the Netherlands.*

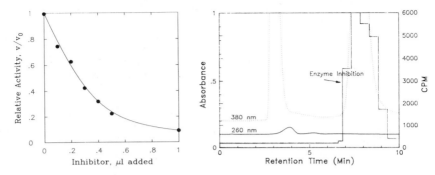

Figure 1. (left) Inhibition of the rate of catalysis of RuBP carboxylase as a result of preincubation with aliquots of an inhibitor preparation prepared from darkened leaves of *Phaseolus vulgaris*. The assays contained 0.66 μM purified spinach RuBP carboxylase in 500 μl assay mixture. The mixtures were preincubated for > 30 min, and the assay started by addition of RuBP.

Figure 2. (right) HPLC chromatography of an inhibitor preparation: UV absorbance of the column eluant (solid line) at 260 nm; post column development of the phosphomolybdate complex (dotted line), and inhibitory activity (histogram).

by binding inhibitor to purified spinach RuBP carboxylase and washing in an ultrafiltration cell (Amicon PM-30 membrane) with several changes of buffer. The enzyme was then denatured, and the released inhibitor was separated from salts, concentrated, and precipitated with BaCl$_2$ and ethanol at pH 8.5.

Analysis and final purification was achieved by HPLC chromatography on a column of Aminex A-15 developed with 10mM HEPES pH 7.4 and 0.08M Na$_2$SO$_4$. The column effluent flowed through a UV detector (260 nm), and could be directed to a fraction collector or to a system for on line detection of phosphorylated compounds (9). Two phosphate containing peaks were separated from the single peak of UV absorbing material (Fig. 2), and activity coincided with the second peak. This material was collected from repeated HPLC runs, separated from buffer and salts, and concentrated.

Characterization. The purified material had one organic phosphate per mol inhibitor. Titration (not shown) indicated the presence of an additional group with a pK$_a$ near 4.5. Preparations taken to dryness under acidic conditions *in vacuo* gave low activity when first taken up in water. High activity could be restored by saponifying the preparations at pH> 10, indicating the presence of a lactone.

Gas Chromatography-Mass Spectrometry. The highly purified inhibitor from the HPLC separation was treated with alkaline phosphatase to hydrolyse the phosphate ester, and silylated (10). Gas chromatography of the silylated mixture revealed 3 major peaks; one being TMS$_3$-PO$_4$, and the mass spectra of the others are essentially identical to spectra of authentic TMS-derivatives of 2-carboxyarabinitol in its free acid and its 1,4-lactone forms (Fig. 3).

NMR Analysis. A spectrum obtained of Tris salt of the lactone form of purified inhibitor (not shown) revealed resonances for each of the six carbon atoms expected for a 2-carboxy-pentitol derivative. The chemical shifts of these carbons were very similar to those of authentic carboxyarabinitol 1,5- bisphosphate lactone (spectra not shown)(10), excepting that [13]C- [31]P coupling constants indicated only one phosphate ester, at C-1. The chemical shift of C-3 is particularly diagnostic of the steriochemistry (10), and the spectra indicate that the compound is the lactone form of carboxyarabinitol 1-phosphate (i.e., [2-C-phosphohydroxymethyl]-ribonic

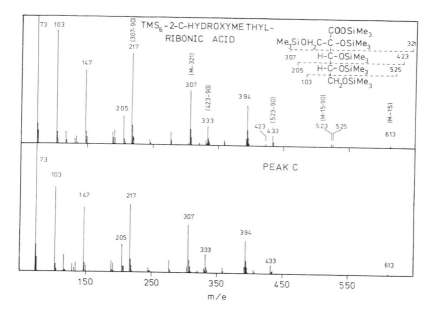

Figure 3. The mass spectra of a major peak from the GLC separation of TMS-derivatives of the inhibitor and of authentic persilyl-carboxyarabinitol (free acid).

acid). This expectation was confirmed by comparison of the ^1H spectrum of the dephosphory-lated inhibitor with the spectra of the four authentic 2-carboxy-pentitols. The spectra of the authentic compounds were easily distinguishable (one from the other), and the spectrum of the dephosphorylated inhibitor conformed with that of 2-carboxyarabinitol (not shown).

Synthesis of the inhibitor. Reaction of ribulose-5-phosphate with CN^- was use to prepare CA-5-P and its corresponding diastereoisomer. This material did not form a stable inhibitor-enzyme complex when preincubated with RuBP carboxylase (Fig. 5). A mixture of CA-1-P, CA-5-P was obtained by ion exchange chromatography of the products of partial phosphatase digestion of CABP. Since CA-5-P does not form a tight complex with RuBP carboxylase, the CA-1-P was purified from CA-5-P by ligation to RuBP carboxylase. This material had inhibitory properties very similar to that of the natural inhibitor (Fig. 5); its binding constant was identical to that of a preparation of natural inhibitor, and as shown (Fig. 6) the kinetics of release of inhibition by phosphatase treatment for the natural inhibitor and the synthetic CA-1-P are similar. The latter presumably indicates the rate of dissociation of the enzyme-inhibitor complex, releasing the free inhibitor which is attacked by the phosphatase. CABP (the starting material for the synthesis) is also a strong inhibitor of RuBP carboxylase, but shows much slower kinetics of inhibitor release (2).

DISCUSSION

We conclude that the naturally occurring compound that is associated with diurnal variations in RuBP carboxylase activity is the D form of carboxyarabinitol 1-phosphate. This compound has been concurrently identified by Gutteridge *et al.* (12) to be present in tobacco leaves at light-dependent levels. Diurnal regulation of RuBP carboxylase activity by a tight binding inhibitor is widespread among higher plants. This compound has very close structural similarity to CABP and to the six carbon intermediate formed in carboxylation. It is possi-

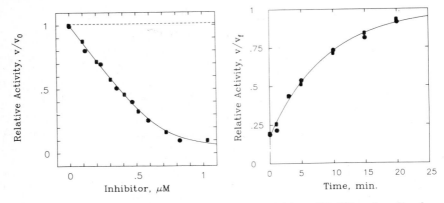

Figure 4. (left) Comparison of the extent of inhibition of RuBP carboxylase by a preparation of inhibitor obtained from *Phaseolus vulgaris* (•) and synthetic preparations of CA-5-P (...) and CA-1-P (x). The dashed line drawn for CA-5-P is a linear regression for 7 data points extending to 22 μM. The solid line is drawn using $K_d = 32$ nM.

Figure 5. (right) The time course of recovery of RuBP carboxylase activity upon treatment of the E-I complex with a saturating activity (10 units) of alkaline phosphatase. Enzyme-inhibitor complexes were prepared with inhibitor isolated from *Phaseolus vulgaris* (•) and with synthetic CA-1-P (x). The relative rate of catalysis was determined by taking aliquots from the reaction mixture and assaying for RuBP carboxylase activity at the indicated times. The $t_{1/2} = 6.2$ min.

ble that other compounds may play a similar role in other plant species, but the structural constraints for binding to the active site of RuBP carboxylase greatly restricts the possibilities.

The mechanisms regulating the concentration of this inhibitor *in vivo* are as yet unknown. The demonstration here, that the inhibitor may be synthesized, opens the way to prepare labeled material for studies of its metabolism by chloroplasts *in vitro*.

ACKNOWLEDGMENTS

We wish to thank Drs. F. Davidson and G.S. Reddy for assistance in obtaining the [13]C (F.D.) and [1]H (G.S.R.) NMR spectra

LITERATURE CITED

1. Servaities, J.C. (1985) *Plant Physiol* **78**, 839-843.
2. Seemann, J.R., Berry, J.A., Freas, S.M. & Krump, M.A. (1985) *PNAS, USA* **82**, 8024-8028.
3. Servaites, J.C., Torisky, R.S. & Chas, S.F. (1984) *Plant Sci. Lett.* **35**, 115-121.
4. Vu, C.V., Allen, L.H. & Bowes, G. (1984) *Plant Physiol.* **76**, 843-845.
5. Vu, C.V., Allen, L.H. & Bowes, G. (1983) *Plant Physiol.* **73**, 729-734.
6. Sharkey, T.D., Seemann, J.R. & Berry, J.A. (1986) *Plant Physiol.* (in press).
7. Hall, N.P. & Tolbert, N.E. (1978) *FEBS Lett.* **96**, 167-169.
8. Seemann, J.R., Badger, M.R. & Berry, J.A. (1984) *Plant Physiol.* **74**, 791-794.
9. Meek, J.L. & Nicoletti, F. (1986) *J. Chromatogr.* **351**, 303- 311.
10. Pierce, J., Tolbert, N.E. & Barker, R. (1980) *Biochemistry* **19**, 934-942.
11. Gutteridge *et al.* (1986) submitted to *Nature*.
12. Scloss, J.V. & Lorimer, G.H. (1982) *J. Biol. Chem.* **257**, 4691-4694.

PURIFICATION OF AN ENDOGENOUS INHIBITOR OF RIBULOSE-1,5-
BISPHOSPHATE CARBOXYLASE/OXYGENASE

JEROME C. SERVAITES, DEPARTMENT OF BIOLOGY,
UNIVERSITY OF DAYTON, DAYTON, OH 45469-0001 USA

1. INTRODUCTION
    In the leaves of some species, ribulose-1,5-bisphosphate
carboxylase/oxygenase (rubisco, EC 4.1.1.39) activity is
substantially lower when measured following a night period than
when measured after the leaves have been subjected to high
light conditions for a period of time (1,2,3,). This phenomenon
of predawn or dark inhibition of rubisco has been attributed to
the presence of a small mol wt phosphorylated molecule which
binds tightly at the enzyme's active site (2,4) and contains 1
mol of P/mol of inhibitor (4). Binding of the inhibitor is
sufficiently tight such that following purification of the
enzyme most of the inhibitor remains bound (4). This fact was
exploited to afford purification of the endogenous inhibitor
from tobacco. Using a similar procedure the inhibitor has been
purified from predawn leaves of potato and identified using NMR
spectroscopy (see Gutteridge et al., these proceedings).

2. PROCEDURE
2.1. Materials and methods
    2.1.1. Leaves of tobacco (Nicotiana tabacum cv. virginia
           gold) were collected 1 h before the light period
           from either field-grown or chamber-grown plants and
           stored in liquid nitrogen. Leaf material (400g) was
           extracted in 0.8 L of an activation buffer and
           rubisco was purified using PEG precipitation and
           Sepharose-6B chromatography as described previously
           (4). The purified rubisco was concentrated by PEG
           precipitation and dissolved in 40 ml of 50 mM
           NH_4HCO_3.
    2.1.2. Methanol (1 vol) was added and the precipitated
           protein removed by centrifugation. The methanol
           extract was reduced to dryness at 40°C under vacuum.
           The residue was dissolved in 1 ml of H_2O and
           phosphorylated compounds precipitated by the
           addition of barium acetate (20 μmol) followed by 5
           vol of ethanol (5).
    2.1.3. The barium phosphate salts were converted to the Na^+
           salts by addition of excess Dowex-50-Na^+ and the
           resulting solution was then immediately added at a
           flow rate of 0.3 ml/min to a 1.5 X 60 cm column of
           DEAE-Sephacel (formate form) equilibrated in 10%
           (v/v) isopropanol (IP). The column was washed with
           0.1 L of 10% IP and then eluted with a 0.5 L linear

gradient of 0 to 0.5 M triethylamine formate (pH
6.5) in 10% IP. Fractions (5 ml) were collected.
Aliquots were removed for determination of total P
(6). Ability of the eluted substances to inhibit
rubisco activity was determined following a 30 min
incubation with the enzyme.

3. RESULTS AND DISCUSSION
3.1. Two peaks containing phosphate were separated following
     DEAE-Sephacel chromatography of the barium precipitate
     (Fig. 1). Only the second and larger peak was able to
     inhibit in vitro rubisco activity and was therefore
     assumed to be the endogenous inhibitor. The first
     phosphate peak was later determined to be inorganic
     phosphate and was assumed to have arisen from the partial
     hydrolysis of the inhibitor during purification.

3.2. Increasing amounts of the inhibitor (peak 2, Fig 1.) were
     able to inhibit in vitro rubisco activity in a
     stoichiometric fashion (Fig. 2) indicating the
     tight-binding nature of the inhibitor for activated sites
     on the enzyme (2). Rubisco activity was inhibited about
     90% at a ratio of 10 moles of inhibitor/mole of rubisco
     active sites.

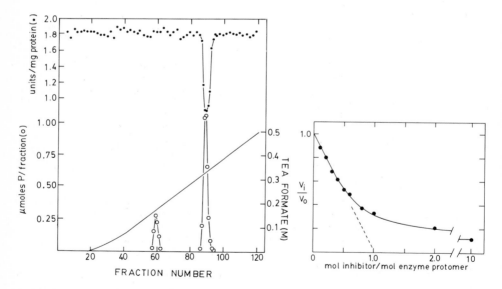

FIGURE 1. (left) Elution profile of phosphorylated compounds
           from a DEAE-Sephacel column.
FIGURE 2. (right) Variation of the steady-state rubisco
           velocity (v_i/v_o) as a function of the molar ratio of
           endogenous inhibitor to enzyme protomer.

3.3. Overall yield of inhibitor was 65% (Table 1) based upon an estimated amount of 4 mole of inhibitor/mole of rubisco in the initial extract (4).

TABLE 1. Typical recovery of the endogenous rubisco inhibitor from tobacco leaves.

Fraction	Total Protein[a]		Inhibitor[b]	
	mg	(%)	μmol	(%)
Extract	2750	(100)	6.8[c]	(100)
PEG precipitate	905	(33)	6.8[c]	(100)
Sepharose-6B	740	(27)	5.9	(87)
Methanol extraction	--		5.6	(82)
Barium precipitation	--		5.5	(81)
DEAE-Sephacel	--		4.4	(65)

[a]Determined using a dye binding assay (7).

[b]Based upon total phosphate present.

[c]Estimated amount of inhibitor present based on the assumption that 1/3 of total protein is rubisco and 4 mole of inhibitor were bound/mole rubisco (4).

REFERENCES
1 Vu, C. V., Bowes, G. and Allen, L. H. (1984) Plant Physiol. 76, 843-845
2 Seemann, J. R., Berry, J. A., Freas, S. A. and Krump, M. A. (1985) Proc. Natl. Acad. Sci. USA 82, 8024-8028
3 Servaites, J. C., Parry, M. A. J., Gutteridge, S. and Keys, A. J. (1986) Plant Physiol. in press
4 Servaites, J. C. (1985) Plant Physiol. 78, 839-843
5 Wood, T. (1968) J. Chrom. 35, 352-361
6 Penny, C. L. (1976) Anal. Biochem. 75, 201-210
7 Esen, A. (1978) Anal. Biochem. 89, 264-273

The structure of the naturally occurring inhibitor of rubisco that accumulates in the chloroplast in the dark is 2'-carboxyarabinitol-1-phosphate.

S. Gutteridge*, M. A. J. Parry, A.J. Keys, J. Servaites** and J. Feeney***. *Central Research and Development Dept., E. I. Du Pont Company, Delaware. Rothamsted Experimental Station, U.K.   **Dept. of Biology, University of Dayton, Ohio. ***N.I.M.R., Mill Hill, U.K.

1.   INTRODUCTION
    A whole range of plants exhibit, to a greater or lesser extent, diurnal regulation of the activity of Rubisco (1-4). This process in at least three plant species, potato, tobacco and Phaseolus is due to the synthesis of a potent inhibitor of rubisco that rapidly accumulates in the chloroplast stroma with the onset of darkness. The inhibitor is a mono-phosphate (2,3) that preferentially binds to the activated form of the enzyme, retaining it in this state throughout the dark period. Equally dramatic is the rate of restoration of full catalytic activity within some 60 minutes with the exposure of the leaf to light.

    The inhibitor binds to the enzyme tightly enough (Kd<0.1uM) that it can be copurified with the protein from dark treated leaves. This method provided enough of the inhibitor to determine its structure as a 2'-carboxy pentitol-1-phosphate. A comparison by spectroscopic methods of the natural compound with related monophosphates has also established the stereo-chemistry of the pentitol as the arabinitol isomer. Thus the compound is 2'-carboxyarabinitol-1-phosphate, a molecule that resembles the transition state intermediate of the carboxylation of ribulose-P_2 and the irreversible 2'-carboxyarabinitol-1,5-bisphosphate.

2.   PROCEDURE
2.1. Materials and Methods
    2.1.1. Time course of inhibition: The time course of the onset of
           inhibition of rubisco and the restoration of the catalytic
           activity in the plant as a result of diurnal changes, was
           followed by removal of leaf material at different times during a
           24 hr period and freezing immediately in liquid nitrogen. The
           frozen material was then ground at low temperature into a fine
           powder before mixing with a combined extraction and activation
           buffer (1). Rubisco activity was determined using a coupled
           enzyme assay (5).

    2.2.2. Inhibitor purification: Leaf material, usually about 1 kg
           collected in the dark, was extracted in the same buffer. The
           inhibitor, bound to the carboxylase in these conditions was
           precipitated along with the enzyme using polyethylene glycol.
           The enzyme was isolated from most of the other small molecular
           weight components by gel filtration. At this stage, the
           inhibitor was released from the enzyme by treatment with
           methanol (80%) and the soluble fraction evaporated to dryness.
           The residue was then further purified by ion-exchange
           chromatography in a volatile buffer(6).

Biggens, J. (ed.), Progress in Photosynthesis Research, Vol. III. ISBN 90 247 3452 5
© *1987 Martinus Nijhoff Publishers, Dordrecht. Printed in the Netherlands.*

3. RESULTS AND DISCUSSION

3.1. Figure 1 shows the dramatic changes to the activity of rubisco with the onset of darkness.

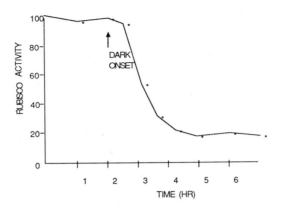

The rate of inhibition in e.g, <u>Phaseolus vulgaris</u>, reaches a maximum after about 30 min with at least 80% inhibition of the activity after 1 hr. Potato and tobacco only achieve about 50% inhibition over the same period. Exposure of the plant to light produces as rapid a return to maximum activity as the onset of inhibition. These diurnal changes occurred in fully expanded leaves, young material as well as plants grown in a controlled environment or in the field(1).

3.2. The low molecular weight compound copurified with rubisco and then released by solvent precipitation of the protein, is grossly contaminated with nucleic acid and peptides. Sequential ion-exchange chromatography after gel filtration removes many of the contaminants leaving a monophosphate containing compound as the major component and associated with the inhibitory function. Figure 2 shows a proton NMR spectrum of the purified inhibitor that provides significant details about the structure of the molecule.

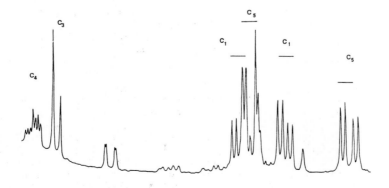

There are six protons evident in the spectrum attached to four carbon centres. Four of these protons reside on primary carbons and are non-equivalent, whereas the other two are on secondary carbons. The frequencies of the resonances are consistent with an organic alcohol, most likely a pentitol containing a quaternary carbon that is devoid of protons and thus not detected by the spectrometer. Clearly, one of the carbons must accommodate a phosphate group. Selective irradiation established that the protons of the primary carbon adjacent to the quaternary carbon are also split by phosphorus (C_1 of figure 3).

FIGURE 3

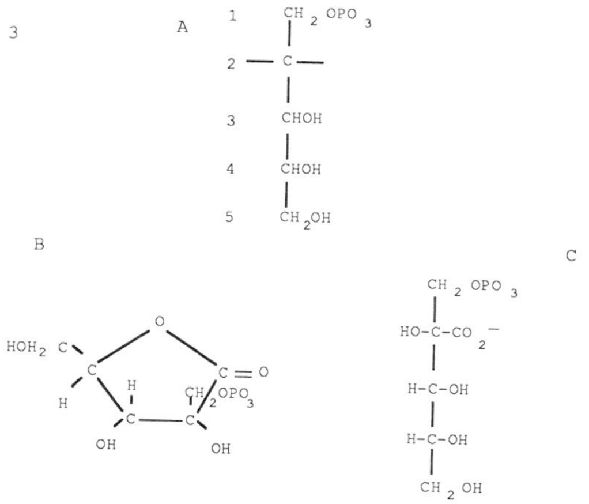

A

$$
\begin{array}{ll}
1 & CH_2 \, OPO_3 \\
2 & -C- \\
3 & CHOH \\
4 & CHOH \\
5 & CH_2OH
\end{array}
$$

B

C

$$
\begin{array}{l}
CH_2 \, OPO_3 \\
HO-C-CO_2{}^- \\
H-C-OH \\
H-C-OH \\
CH_2 \, OH
\end{array}
$$

3.3. The structure that can be proposed based on this spectrum is shown in figure 3a. Further NMR studies also established the structure at the quaternary carbon; paradoxically that centre without any protons. For example, the spectrum of figure 2 is at acid pH and the protons of the C_3 and C_4 centres have frequencies similar to those of lactone structures. Figure 3b satisfies these assignments. At higher pH the ring should hydrate to adopt the conformation of figure 3c and indeed the change in the frequencies of C_3 and C_4 protons suggests an open chain molecule. This is strong evidence that a group at the quaternary carbon is carboxyl. Finally, the stereoisomerism around C_2 and C_3 was established by comparison of the proton spectra of synthesized pentitol monophosphate isomers and their dephosphorylated derivatives with the natural inhibitor, completing the identification of the molecule as 2'-carboxyarabinitol-1-phosphate (2CA1P).

3.4. For some time a function proposed for phosphate esters in the chloroplast stroma that resemble ribulose-P_2 was that they favoured the activated state of rubisco. i.e. the enzyme metallo-carbamate ternary complex. However, this is the first time that one has been identified in this capacity and furthermore one with a unique structure. For a number of reasons 6 phosphogluconate has often been proposed as an effector/inhibitor that might function as a regulator of rubisco activity. It sequesters itinerant Mg ions at the active site of rubisco and hence favours activation and does accumulate in the dark. However, binding is not particularly tight and thus the concentrations of the molecule to inhibit the enzyme effectively need to exceed the active site concentration. The superior Kd of 2CA1P and its design ensures that it is not required in large excess over the number of enzyme active sites.

REFERENCES
1. Servaites, J. C., Parry, M. A., Gutteridge, S., Keys, A. J. (1986) Plant Physiol., in press.

2. Seeman, J. R., Berry, J. A., Freas, S. M., Krump, M. A., (1985) Proc. Natl. Acad. Sci. 82 8024-8028.

3. Servaites, J. C. (1985) Plant Physiol. 78 839-843.

4. Vu, C. V., Allen, L. H., Bowes, G. (1984) Plant Physiol. 76 834-845.

5. Pierce, J., Gutteridge, S. (1985) App. Env. Microbiol. 1094-1100.

EFFECT OF LIGHT INTENSITY DURING SOYBEAN GROWTH ON THE ACTIVATION STATE OF RuBP CARBOXYLASE FROM LEAVES IN THE LIGHT AND DARK.

GABRIEL P. HOLBROOK, WILLIAM J. CAMPBELL, and GEORGE BOWES.
Department of Botany (G.P.H., G.B.) and Department of Agronomy (W.J.C.), University of Florida, Gainesville, FL. 32611, U.S.A.

1. INTRODUCTION

The activity of RuBP carboxylase (RuBPC'ase) in leaves is a major factor controlling photosynthetic rates. It is well established that the activation state of this enzyme in vivo is a function of light intensity [1,2]. In Nicotiana tabacum and Phaseolus vulgaris leaves, regulation of carboxylation rates may be complicated by the presence of a phosphorylated inhibitor, which binds to RuBPC'ase during dark periods and reduces the maximal activity attainable in vitro after full activation by Mg^{++} and CO_2 [3,4]. Effects indicative of the interaction of a similar inhibitor with RuBPC'ase had previously been observed in a number of other species including soybean [5,6,7]. The mode of synthesis and degradation of the inhibitor is not known, and neither is its physiological relevance to the regulation of CO_2 fixation. We report here on the influence of light/dark transitions on 'initial' and 'total' activities [2] of RuBPC'ase in extracts of Soybean leaves from plants growing at high and low irradiance levels. The implications of the results are discussed in terms of metabolism of the inhibitor.

2. MATERIALS AND METHODS

2.1 Plant Material. Soybeans (Glycine max L. Merr. cv. Bragg) were grown in controlled environment chambers at a range of constant irradiance levels between 15 and 690 μmol photons/m^2.s PAR during a 16-h light/8-h dark period ($25^{o}C/20^{o}C$). At appropriate sampling times, the incident irradiance was measured at the surface of fully expanded trifoliate leaves, which were then excised from the plant and immediately immersed in liquid N_2 (LN_2). Frozen leaf samples were ground to a powder and stored in LN_2 [5].

2.2 Extraction and assay of RuBP carboxylase. Weighed quantities of LN_2-frozen leaf powder were homogenized inside an N_2-flushed glove bag at $0^{o}C$ with CO_2-free medium containing 100 mM Tris (pH 8.0), 5 mM $MgCl_2$, 5 mM dithiothreitol, 10 mM isoascorbic acid, and 1.5% w/v PVP-40 (Buffer A). Relative tissue to buffer (w/v) ratios were adjusted to give similar levels of soluble protein (~2.5 mg/ml) for leaf samples from plants grown at each light intensity. The homogenate was centrifuged for 1 min at 13,000 x g in a microcentrifuge and

10 µl aliquots of the supernatant liquid (25-35 µg protein) were immediately assayed for initial (EC and ECM forms) and total (E, EC, and ECM forms) RuBPC'ase activity essentially as described previously [5]. A 0.5 ml sample of the supernatant fluid was fractionated between 35% and 60% $(NH_4)_2SO_4$ and desalted by centrifugation through a 6-ml column of Sephadex G-25 equilibrated with Buffer A minus PVP-40.

3. RESULTS AND DISCUSSION

Previous studies [5,7] with soybeans grown under natural light have established that initial and total activities of RuBPC'ase in leaf extracts change in a manner closely related to the diurnal increase and decrease in solar irradiance levels. Less is known about changes in the enzyme's activity in darkened soybean leaves, or during rapid light/dark or dark/light transitions. The data in Fig.1 shows the effect of such transitions on RuBPC'ase activity in plants growing at relatively high (500-700 µmol photons/m^2.s) and low (30-40 µmol photons/m^2.s) irradiances. With high light grown plants (Fig. 1A), the total activity of RuBPC'ase declined slowly when leaves entered the dark period. Maximal reduction of total activity to 52% of values in the light did not occur until 6 hours had elapsed. In contrast, within 10 min, reillumination restored total activity to levels equivalent to those at the end of the previous light cycle (cf. Fig. 2 in Ref.5). The maximum Mg^{++}/CO_2 activated activity of the same samples treated with 35%-60% $(NH_4)_2SO_4$ remained relatively constant, showing no evidence of dark/light modulation. Incubation of RuBPC'ase from predawn samples of Tobacco leaves with $(NH_4)_2SO_4$ is known to remove the phosphorylated inhibitor from the enzyme [3]. Thus, changes in the total activity shown in Fig.1A are likely to represent a time course for the synthesis and binding of the inhibitor to Soybean RuBPC'ase during the dark period, and its subsequent removal in the light. Changes in the initial activity, reflecting the proportion of uninhibited enzyme in the EC and ECM form in vivo, paralleled changes in the total activities.

With P. vulgaris grown under natural irradiance, exposure to relatively high light intensities (1000 µmol photons/m^2.s) is necessary to completely degrade inhibitor bound to RuBPC'ase during an overnight dark period; lower light levels facilitate only a partial degradation of the inhibitor, and total activities remain below the maximal levels attainable at higher irradiance [4]. RuBPC'ase in plants grown under constant low light intensities might therefore exhibit activities permanently reduced by the presence of inhibitor. The results in Fig. 1B suggest that this is not the case with Soybean. Changes in total RuBPC'ase activity during dark/light transitions were very similar to those observed in Fig. 1A for plants growing at higher light intensity. In addition, total activity of the enzyme from leaf samples taken midway through the 16-h light period was not increased by treatment with $(NH_4)_2SO_4$, after correction was made for protein removed

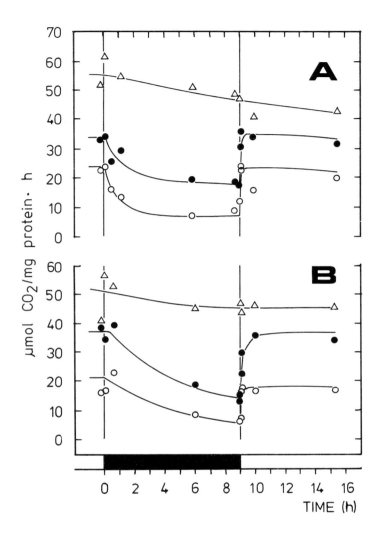

FIGURE 1. Effect of light/dark transitions on initial activity
(O—O) and total activity (●—●) of RuBPC'ase in centrifuged
crude extracts of Soybean leaves from plants growing at high
(A) and low (B) irradiance levels. The Mg^{++}/CO_2 activated
carboxylation rates were also measured for extracts treated
with 35-60% $(NH_4)_2SO_4$ (△—△). Mean incident photon flux
densities during collection of light-period leaf samples was
550 ± 28 (S.E.) μmol photons/m^2.s for (A), and 39 ± 1.1 μmol
photons/m^2.s for (B). Data are means of triplicate assays of
samples comprising one (A) or two (B) trifoliate leaves from
plants 30 days (A) and 102 days (B) after sowing.

by 35%-60% fractionation with the salt, and subsequent desalting. This indicates that the inhibitor is absent from the enzyme during the light period, even when plants are grown continuously at unnaturally low irradiance levels.

We confirmed the recent results of Vu et al. [8], showing that the Km(RuBP) is the same for RuBPC'ase in extracts from light and dark sampled leaves. Values measured were 53 ± 4 (S.E.) µM RuBP in light samples of leaves growing at 690 µmol photons/m^2.s, and 52 ± 2 µM RuBP from dark samples. No change in these values occurred upon treatment of extracts with $(NH_4)_2SO_4$. The Vmax of RuBPC'ase in dark samples was 58% of that for enzyme in light samples. This is consistent with the inhibitor acting noncompetitively with respect to RuBP on RuBPC'ase from darkened leaves [4].

4. REFERENCES

1 Mächler, F., and Nösberger, J. (1980) J. Exp. Bot. 31, 1485-1491.
2 Perchorowicz, J.T. et al. (1981) Proc. Natl. Acad. Sci. USA 78, 2985-2989.
3 Servaites, J.C. (1985) Plant Physiol. 78, 839-843.
4 Seemann, J.R. et al. (1985) Proc. Natl. Acad. Sci. USA 82, 8024-8028.
5 Vu, C.V. et al. (1983) Plant Physiol. 73, 729-734.
6 Vu, C.V. et al. (1984) Plant Physiol. 76, 843-845.
7 Servaites, J.C. et al. (1984) Plant Sci. Lett. 35, 115-121.
8 Vu, C.V. et al. (1986) Plant Sci. (in press).

5. ACKNOWLEDGEMENT

Supported in part by NSF grant no. DMB 8504856

RIBULOSE BISPHOSPHATE CARBOXYLASE/OXYGENASE: PARTITIONING OF AN ALTERNATE SUBSTRATE.

Henry M. Miziorko, Medical College of Wisconsin, Milwaukee, WI 53226 USA

SUMMARY

Ribulose 1-phosphorothioate, 5-phosphate (RuBP(S)) can substitute for ribulose bisphosphate (RuBP) as a substrate for RuBP carboxylase. Upon reaction of RuBP(S) with CO_2, glycerate 3-phosphate and glycerate 3-phosphorothioate are formed. The alternate substrate exhibits a V/K that is about 18% of that measured using RuBP. RuBP(S) also supports the oxygenase reaction, which results in formation of the phosphorothioate of glycolate as well as glycerate 3-phosphate. Partition of substrate between carboxylase and oxygenase reactions has been measured with RuBP and RuBP(S), using both spinach and R.rubrum enzymes. In both cases, the phosphorothioate analog exhibits enhanced partitioning into the oxygenase reaction in comparison with RuBP. Steric factors afford one possible explanation for these observations. This possibility prompted examination of the stability of binding of transition state analogs that contain a carboxyl group and either a phosphoryl group on C1 (i.e. carboxyarabinitol bisphosphate, CABP) or a phosphorothioate on C1 (i.e. carboxyarabinitol 1-phosphorothioate, 5- phosphate, CABP(S)). When R.rubrum enzyme is used to form a quaternary enzyme-activator CO_2-cation-transition state analog complex, the CABP(S) containing complex is observed to decay much more rapidly ($t_{1/2}$ = 2.5 hr) than the CABP containing complex ($t_{1/2}$ = 75 hr). Steric hindrance, i.e. unfavorable interaction between the carboxyl moiety and the bulky phosphorothioate group, may explain this effect; such an explanation is compatible with the enhanced partitioning of RuBP(S) into the oxygenase pathway.

INTRODUCTION

Since the balance between photosynthesis and photorespiration is determined by the reactions catalyzed by ribulose bisphosphate carboxylase/oxygenase, a great deal of attention has been focused on factors that affect the balance between these two reactions. Ogren and his colleagues (1,2) have described the differences in partitioning of sugar phosphate substrate observed with the enzyme from a variety of plant, algal, and bacterial species. In addition, the influence of the activating cation on partitioning has been reported (2,3). Until now, it has not been possible to evaluate whether structural features of the carbohydrate substrate influence partitioning, since there have been no reports of suitable alternate substrates which could be compared to RuBP. Recently, in determining the stereochemical course of the

Biggens, J. (ed.), Progress in Photosynthesis Research, Vol. III. ISBN 90 247 3452 5
© *1987 Martinus Nijhoff Publishers, Dordrecht. Printed in the Netherlands.*

ribulose 5-phosphate kinase reaction (4), we demonstrated that enzymatic synthesis of ribulose 1-phosphorothioate, 5-phosphate (RuBP(S)) can be accomplished. Furthermore, this compound was shown to be metabolized by RuBP carboxylase. In order to evaluate whether the chemical nature of the sugar phosphate substrate affects partitioning, RuBP(S) has now been isolated and its function as an alternate substrate in the carboxylase and oxygenase reactions is being investigated. The formation of a carboxylated transition state analog from RuBP(S) has been accomplished and evaluation of its properties is also in progress. This report describes the results of some initial experiments with these novel compounds and discusses their potential utility in exploring the molecular steps involved in photosynthesis and photorespiration.

EXPERIMENTAL PROCEDURES

Materials. Homogeneous RuBP carboxylase was prepared from spinach leaf by the method of paulsen and Lane (5); the R.rubrum enzyme, prepared as described by Pierce and Reddy (6), was a generous gift of J. Pierce and G. Lorimer. Ribulose 5-phosphate kinase was prepared in a homogeneous form by the method of Krieger and Miziorko (7). ATP-γ-[^{35}S] was supplied by New England Nuclear; unlabeled material was purchased from Boehringer Mannheim. Beckman Instruments supplied the Ultrasphere-IP HPLC column (4.6 x 25 cm; 5 micron) used in separation of the RuBP carboxylase/oxygenase reaction products.

Methods. [-^{14}C] RuBP is prepared and isolated by the method of Wishnick and Lane (8). [^{35}S] RuBP(S) is prepared as described by Miziorko and Eckstein (4) except that RuBP carboxylase is omitted from the reaction mixture. Purification is performed on a Dowex 1-X8 column (2.0 x 25 cm) using a triethylammonium bicarbonate gradient (10-500 mM). [3-^{14}C] glycerate 3-phosphate and [^{35}S]-glycerate 3-phosphorothioate were enzymatically prepared from [1-^{14}C] RuBP and [^{35}S] RuBP(S) respectively. [^{3}H] glycolate phosphate and [^{3}H] glycolate phosphorothioate were prepared by reaction of ATP or ATP-γ-S with [^{3}H] glycolate in the presence of pyruvate kinase, as described by Kayne (9). Carboxypentitol 1-phosphorothioate, 5-phosphates were prepared from RuBP(S) by the method of Pierce et al. (10).

Separation of the reaction products formed by RuBP carboxylase/oxygenase from [1-^{14}C] RuBP or [^{35}S] RuBP(S) was performed by isocratic elution (50 mM tetrabutylammonium phosphate (pH 2.6)/methanol; 80:20) of reaction mixtures loaded on an Ultrasphere-IP C-18 HPLC column. Quantitation of isolated products was accomplished using a Radiomatic Instruments flow detector (0.5 ml liquid detection cell). Paired reaction mixtures (air saturated) contained Tris-Cl (100 mM, pH 8.0), 10 mM $KHCO_3$, and 2 mM $MnCl_2$, to support easily measurable levels of oxygenase product. Reactions were performed using identical saturating levels of RuBP or RuBP(S) and were taken to completion prior to quenching with excess EDTA and storage at -20°C to prevent degradation of glycolate phosphorothioate.

Formation of quaternary complexes of enzyme-activator CO_2-Mg-transition state analog (either CABP or CABP(S)) as well as measurement of exchange of analog from the isolated complex was performed as previously described (11).

RESULTS AND DISCUSSION

By analogy with the enzyme catalyzed reactions of RuBP with CO_2 and O_2, it was anticipated that C1 and C2 of RuBP(S) should give rise to glycerate 3-phosphorothioate due to the carboxylase reaction and glycolate phosphorothioate as a result of the oxygenase reaction (Scheme I). Previous work (4) verified the identity of the carboxylation product. On the basis of chromatographic behavior, ^{31}P NMR spectrum and predicted chemical lability, the assignment of the oxygenase reaction product as glycolate phosphorothioate also seems to be straightforward.

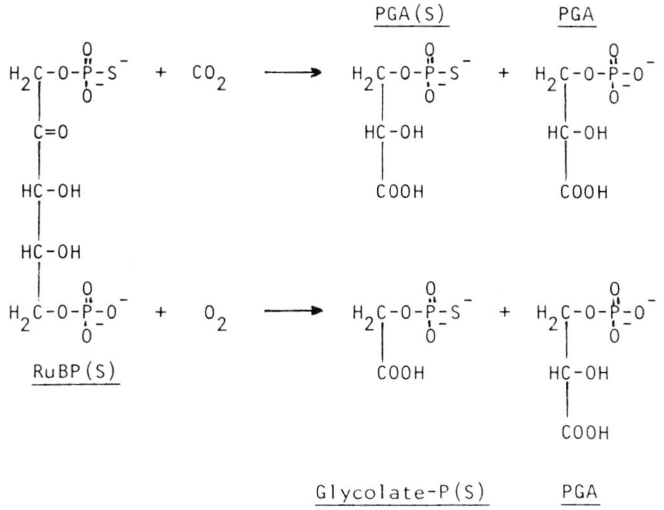

Scheme I. Partitioning of ribulose 1-phosphorothioate, 5-phosphate

In order to conveniently monitor formation of these phosphoro-thioate containing reaction products, which are not metabolized as smoothly as the corresponding phosphorylated compounds by the enzymes used in the typical spectrophotometric assay, an HPLC-based procedure has been developed. This method is also suitable for measuring RuBP-derived reaction products and represents a substantial improvement over earlier approaches that had been developed to simultaneously measure progress of carboxylase and oxygenase reactions. Jordan and Ogren (12) utilized a double-label method that required phosphatase treatment of phosphoglycolate prior to its isolation and measurement of radioactivity. Kent and Hemming (13) developed an HPLC procedure which involved gradient elution and did not detect unreacted substrate under conditions that were necessary to adequately resolve phosphoglycolate from phosphoglycerate. The methodology reported here does not require any post-reaction processing of products. Moreover, a simple isocratic elution produces baseline separation of the peaks due to the reaction products and any unreacted substrate (Fig. 1). For separation of

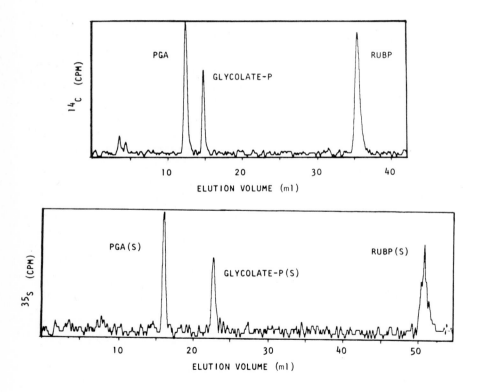

Fig. 1. HPLC separation of RuBP carboxylase/oxygenase reaction products.

glycerate-P, glycolate-P, and RuBP, less than 40 minutes are required (Fig. 1, top). The separation of RuBP(S) from the reaction products glycerate-P(S) and glycolate-P(S) was performed in less than an hour (Fig. 1, Bottom) under comparable conditions. Clearly, with such a large separation of product phosphorothioate peaks, running time could be substantially shortened by increasing the percentage of organic phase without sacrificing the ability to cleanly measure oxygenase/carboxylase ratios. Due to the isocratic elution, no time is required to re-equilibrate the column. Thus, several aliquots of reaction mixtures can be analyzed without a major time investment. Accurate estimates of oxygenase/carboxylase partitioning can be obtained; error level is typically in the 4-7% range.

Partitioning of RuBP was compared with RuBP(S) using paired samples. Air saturated reaction mixtures were used; manganese was used as the activating cation in order to generate easily detectable levels of oxygenase product. The spinach enzyme supports a substantially higher (33% increase) oxygenase/carboxylase ratio for turnover of RuBP(S) in comparison with RuBP. In similar experiments with the enzyme from R.rubrum, the oxygenase/carboxylase ratio measured for RuBP(S) turnover is 45% higher than measured with RuBP. A definitive explanation for these observations is not possible on the basis of these preliminary data, but one possible explanation involves steric arguments. A bulky phosphorothioate group could add to the crowding predicted at the reaction center. While not large in an absolute sense, the difference in molecular dimensions of CO_2 vs. O_2 is appreciable. Thus, when RuBP(S) is bound at the active site, occupancy of the substrate CO_2 site may be impaired.

If the preceding speculation is correct, it seemed likely that there should also be differences in the affinity with which a phosphorothioate containing analog of CABP would bind to the enzyme's active site. Preparation of carboxyarabinitol 1-phosphorothioate, 5-phosphate (CABP(S)) is possible by substitution of RuBP(S) for RuBP in the procedure of Pierce et al. (10). Kinetic experiments with the spinach enzyme verified that this compound is also a slow, tight binding inhibitor. As observed earlier with CABP (11), the phosphorothioate-containing transition state analog binds in an exchange-inert fasion. Complexes formed using the spinach enzyme do not dissociate at an appreciable rate, therefore formation of RuBP(S)-containing quaternary complexes with R.rubrum enzyme were investigated. When CABP[^{35}S] was used to form the enzyme-activator CO_2-cation-transition state analog complex, the G-75 isolated complex was found to contain stoichiometrically bound inhibitor. In parallel experiments, quaternary complexes were formed with [^{14}C]-CABP or CABP[^{35}S], isolated free of unbound ligands, and incubated with an excess of unlabeled CABP. Aliquots of these mixtures were subjected to gel filtration at various times and the depletion of initially bound transition state analog was measured. The $t_{1/2}$ for exchange of CABP(S) was 2.5 hr; this estimate is 30 fold faster than the $t_{1/2}$ measured for CABP under identical conditions. Since the association rate of CABP(S) with enzyme is not expected to be drastically different from the rate for CABP, a substantially more rapid dissociation rate must characterize the CABP(S) binding phenomenon. Since effective occupancy of both carboxyl and phosphoryl binding pockets is required for tight binding of transition state analogs, any increase in dissociation rate may reflect unfavorable steric interactions between carboxyl and phosphorothioate functionalities on CABP(S). Such an explanation is not only quite plausible, but is compatible with the enhanced partitioning of RuBP(S) into the oxygenase pathway that has been discussed above.

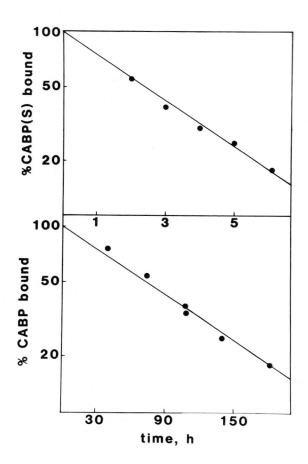

Fig. 2. Exchange of CABP(S) (top) and CABP (bottom) from isolated enzyme-activation CO_2-Mg^{2+}-transition state analog complexes.

Model building studies in which structures of the unstable intermediates (scheme II) formed by reaction of RuBP or RuBP(S) with CO_2 are compared with those intermediates formed upon reaction with O_2 are also worth considering in discussion of the steric hindrance hypothesis. Without imposing any constraints on the intermediates, it is certainly

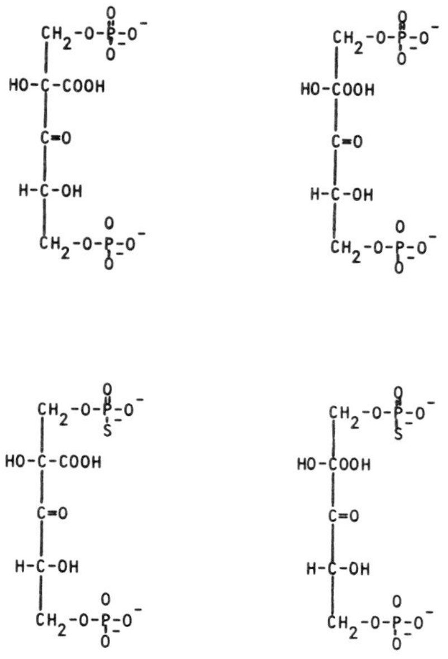

Scheme II. Intermediates formed upon reaction of CO_2 (left) or O_2 (right) with RuBP and RuBP(S).

possible to assemble a model in which a phosphate or phosphorothioate moiety on C1 lies relatively far in space from the carboxyl group that is derived from substrate CO_2. If the intermediate were really bound to the active site in such a fashion, the rationale outlined above to account for our observations with RuBP(S) and CABP(S) would not seem very attractive. However, substantial constraints on the orientation of these portions of the bound intermediate have been imposed on the basis of both ESR (14) and NMR (6) studies. The substrate CO_2 derived carboxyl is directly bound to cation and a phosphoryl group is only slightly further away. These data suggest a structure for the bound intermediate in which these groups are in close proximity. When a model for 3-keto-2-carboxyarabinitol bisphosphate is constructed with these functionalities in closest proximity, they approach each other but are not in contact. In contrast, when the model is constructed using a phosphorothioate at C1, contact between the carboxylate and the phosphorothioate can be demonstrated. This contact does not occur in a model of the corresponding oxygenase reaction intermediate, which contains a phosphorothioate at C1 and a hydroperoxide at C2. Thus, while the phosphorothioate may narrow by slightly less than one Angstrom the passage through which CO_2 and O_2 must approach the carbohydrate substrate, this may be a sufficient perturbation in the crowded catalytic domain to affect partitioning. Additional studies on how an alternate substrate such as RuBP(S) affects Michaelis constants for CO_2 and O_2 as well as solution and crystallographic studies on appropriate model complexes will clearly be required in order to more firmly establish the details of active site structure that have been predicted on the basis of this preliminary work.

ACKNOWLEDGEMENTS

I am grateful to Drs. John Pierce and George Lorimer for the gift of purified R.rubrum RuBP carboxylase. This work was supported in part by the USDA Competitive Research Grants Office.

REFERENCES

1. Jordan, D.B., and Ogren, W.L. (1981) Nature 291, 513-515.

2. Jordan, D.B., and Ogren, W.L. (1983) Arch. Biochem. Biophys. 227, 425-433.

3. Christeller, J.T. (1981) Biochem. J. 193, 839-844.

4. Miziorko, H.M., and Eckstein, F. (1984) J. Biol. Chem. 259, 13037-13040.

5. Paulsen, J.M., and Lane, M.D. (1966) Biochem. 5, 2350-2357.

6. Pierce, J., and Reddy, G.S. (1986) Arch. Biochem. Biophys. 245, 483-493.

7. Krieger, T.J., and Miziorko, H.M. (1986) Biochem. 25, 3496-3501.

8. Wishnick, M., and Lane, M.D. (1969) J. Biol. Chem. 244, 55-59.

9. Kayne, F.J. (1974) Biochem. Biophys. Res. Comm. 59, 8-13.

10. Pierce, J., Tolbert, N.E., and Barker, R. (1980) Biochem. 19, 934-942.

11. Miziorko, H.M. (1979) J. Biol. Chem. 254, 270-272.

12. Jordan, D.B., and Ogren, W.L. (1981) Pl. Physiol. 67, 237-245.

13. Kent, S.S., and Hemming, B.C. (1979) J. Chromatog. 177, 372-375.

14. Miziorko, H.M., and Sealy, R.C. (1984) Biochem. 23, 479-485.

EXPRESSION OF GENES FOR PLANT-TYPE RUBISCO IN *CHROMATIUM* AND *ES-CHERICHIA COLI*

HIROKAZU KOBAYASHI, ESTELA VALLE[*+], ALEJANDRO M. VIALE[*+], AND TAKASHI AKAZAWA[*]

RADIOISOTOPE CENTER, AND [*]RESEARCH INSTITUTE FOR BIOCHEMICAL REGULATION, SCHOOL OF AGRICULTURE, NAGOYA UNIVERSITY, CHIKUSA, NAGOYA 464, JAPAN

[+] PERMANENT ADDRESS: CENTRO DE ESTUDIOS FOTOSINTETICOS Y BIOQUIMICOS (CEFOBI), CONICET, F. M. LILLO, UNIVERSIDAD NACIONAL DE ROSARIO, SUIPACHA 531, 2000 ROSARIO, ARGENTINA

1. INTRODUCTION

Molecular mechanisms governing the biosynthesis of two constituent subunits (A and B) of RuBisCO and their functional roles in the enzyme reaction have received a great deal of attention from many investigators (1). RuBisCO is well known as one of light-inducible enzymes in higher plants (2). Detailed regulatory mechanisms of expression of RuBisCO genes have not been clarified, although upstream DNA sequences of the genes have been reported to be responsible for light induction (3). Since genes for the subunits are encoded by chloroplast and nuclear genomes in higher plants (2), it is frequently argued whether or not a synchronous mechanism operates in the biosynthesis of two subunits to make up the enzymically active holoenzyme in chloroplasts. In order to solve these problems, *Chromatium vinosum*, a photosynthetic purple sulfur bacterium, is thought to be a feasible material from the following reasons. (i) *Chromatium* is one of the most primitive organisms, yet containing a plant-type RuBisCO molecule (A_8B_8) (1,4). (ii) Inducible formation of RuBisCO molecules occurs under autotrophic growing condition after the initial heterotrophic culture.

It has previously been demonstrated that *Chromatium* synthesizes a large amount of RuBisCO under the autotrophic condition (5), the biosynthesis of both subunits proceeds in a tightly coordinated manner (6), and the genes for the subunits are present in the

---

Abbreviations: RuBisCO, ribulose-1,5-bisphosphate carboxylase/oxygenase; IPTG, isopropyl-β-D-thiogalactoside; HPLC, high performance liquid chromatography; PMSF, phenylmethylsulfonyl fluoride; ELISA, enzyme-linked immunosorbent assay; RuBP, ribulose-1,5-bisphosphate; SDS, sodium dodecyl sulfate; kb, kilobases; kbp, kilobase pairs; kd, kilodaltons

Biggens, J. (ed.), Progress in Photosynthesis Research, Vol. III. ISBN 90 247 3452 5
© *1987 Martinus Nijhoff Publishers, Dordrecht. Printed in the Netherlands.*

chromosomal DNA (7). The work presented in this article shows the experimental evidences that the induction of RuBisCO is transcriptionally regulated in *Chromatium* and that genes for the subunits of RuBisCO are expressed using expression vectors in *E. coli* and the products are processed and assembled in *E. coli* to make up the A_8B_8 type molecule of RuBisCO.

2. PROCEDURE

2.1. Materials and Methods

      Chromatium cells collected at the early logarithmic phase (2 hr) under different growth conditions were used for isolating RNA. One gram of the cells were disrupted by mortar in the presence of 1 g of autoclaved fine sea sand and 1 mM aurintricarboxylic acid as an RNase inhibitor. Homogenized cells were mixed with 1% SDS and 0.2 M Na borate (pH 9.0) and extracted 3 times with phenol-chloroform (1:1, v/v) and then 3 times with ether. After ethanol precipitation, nucleic acids were treated 5 times with 2.0 M LiCl, and RNA was recovered in the pellet.

      For preparing RuBisCO from *E. coli* cells which produce the enzyme molecules by expression vectors, the cells were grown in LB liquid medium containing ampicillin until absorbance at 550 nm reached 0.5. IPTG (final 1 mM) was added and the incubation continued for an additional 2 hr at 37℃. In both *Chromatium* and *E. coli*, the cells were collected by centrifugation, washed with 10 mM Tris-Cl (pH 7.0), 0.1 M NaCl, and 1 mM EDTA, and disrupted by Kubota 200M Insonator at maximal power for 10 min in the same medium containing 10% glycerol and 1 mM PMSF. Cell debris was removed by centrifugation. In order to purify *Chromatium* RuBisCO, the crude extracts were subjected to the next step as reported (8). For purifying RuBisCO from *E. coli* cells, the crude extracts were centrifuged for 2 hr at 220,000 g at 4℃. The supernatant fraction containing RuBisCO activity was treated with polyethylene glycol 4000 (final 10%). The mixture was stirred in an ice bath for 30 min and centrifuged at 40,000 g for 20 min. The pellet was resuspended in the extraction medium. The sample solution was then loaded on a column of gel filtration HPLC system (Superose 6, Pharmacia), and eluted by the running buffer, 50 mM Na phosphate (pH 6.9), 1 mM EDTA, 0.3 M NaCl, and 0.5 mM PMSF at a flow rate of 0.3 ml/min. After concentration by Millipore filter SJGC, the enzyme was dissociated with 1% SDS at 90℃ for 8 min. Both subunits were separated by gel filtration HPLC (Superose 12, Pharmacia) with a running buffer consisting of 50 mM Na phosphate, 0.1% SDS, and 0.15 M NaCl. RuBisCO content (1 - 100 ng) was measured by ELISA (9). The Edman degradation of peptides was performed according to the methods reported by Wada (10).

3. RESULTS AND DISCUSSION

3.1. Expression of RuBisCO genes in *Chromatium*

The biosynthesis of RuBisCO in *Chromatium* was found to be regulated by carbon sources, *e.g.*, CO_2 and pyruvate, as well as by the reduced sulfur compounds (5). In order to study the regulatory mechanism(s) operating in the overall processes at the molecular level, we have isolated RNA from the *Chromatium* cells grown under different conditions.

Employing the Northern blot analysis, *Chromatium* RNA was able to hybridize with a DNA fragment containing genes for both subunits of *Chromatium* RuBisCO. We have found that the transcription of RuBisCO genes is inhibited by CO_2 and pyruvate but stimulated by the sulfur compounds (Fig. 1). By comparing the level of these mRNAs, it is likely that the induction of RuBisCO (5) is regulated at the level of RNA, presumably at the transcriptional level. Among RNA species (Fig. 1), 2.0-kb RNA was hybridized with DNA probes for both subunits A and B,

FIGURE 1. Northern blot analysis of transcripts for RuBisCO genes in *Chromatium* cells grown under different conditions. Nick-translated RuBisCO genes (*Bam*HI-*Bam*HI fragment, see Fig. 2) was used as the DNA probe. RNA (10 μg) purified from *Chromatium* cells grown under different conditions were denatured by glyoxal and electrophoresed (30 V, 14 hr) on 1.2% agarose gel in 10 mM Na phosphate (pH 7.0) (11). The RNA was then transferred to GeneScreen (New England Nuclear) and subjected to hybridization. Lane 1, fluorescence of RNA from autotrophically grown cells after ethidium bromide staining of agarose gel. rRNA (23S and 16S) are indicated on the left margin. Lane 2, Northern blotting of RNA

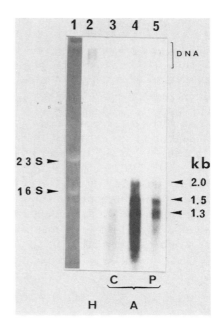

from heterotrophically grown cells (H); lane 3, RNA from autotrophically grown cells (A) with CO_2 (C); lane 4, RNA from autotrophically grown cells (A) minus CO_2; and lane 5, RNA from autotrophically grown cells (A) minus CO_2 plus pyruvate (P).

whereas 1.5- and 1.3-kb RNAs were shown to be hybridized only with a probe for subunit A. In the DNA sequence of *Chromatium* RuBisCO genes, we have found possible sequences of one promoter of the genes and 3 different termination signals of transcription. The sizes of transcripts estimated from the sequence correspond to 2.0, 1.5, and 1.3 kb. It is thus conceivable that these transcripts are the primary products and 2.0-kb species contains information of structures of two subunits.

3.2. Expression of *Chromatium* RuBisCO genes in *E. coli*

It has been found that cloned *Chromatium* RuBisCO genes in plasmid pCUB1 (see Fig. 2) (12) are not expressed in the cells of *E. coli* as demonstrable by either enzyme assay or ELISA using the antibodies against *Chromatium* RuBisCO.

For the expression of *Chromatium* RuBisCO genes in *E. coli*, expression vectors were employed. As shown in Fig. 2, the insertion containing genes for RuBisCO was treated with Bal31 exonuclease and ligated with expression vectors possessing a fused promoter of *trp-lacUV5* (*tac*) and an insertion site just after their ribosome-binding region (Shine-Dalgarno sequence), *i.e.*, pKK223-3 (Pharmacia) or ptac12 (13). The bacterial colonies

FIGURE 2. Diagrammatic representation of a strategy for introducing *Chromatium* RuBisCO genes into an *E. coli* expression vector. DNA fragments derived from *Chromatium* are represented by thick lines in the plasmids. After transforming *E. coli* JM105 with the produced plasmids, expressers of RuBisCO were screened (12). Two of these, pCKS1 and pCKS3, were analyzed by several restriction enzymes for making construction maps.

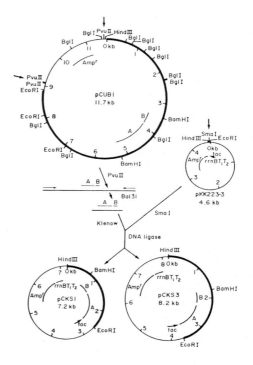

producing RuBisCO molecules were screened by immunoassay on LB agar employing an affinity column-purified antibody (IgG) raised against *Chromatium* RuBisCO after induction with IPTG (12). Among 3,700 colonies tested which were resistant to ampicillin, only 10 were found to produce high signals to noise level. The specific enzyme activities of RuBisCO in the positive colonies were determined by assaying the enzyme activity as well as its content by ELISA. Among 10 positive transformants, the specific activity was nearly the same as that of the original RuBisCO in *Chromatium*. The highest amount of RuBisCO produced in the cells of *E. coli* was approximately 15% of the total soluble protein after induction by IPTG for 8 hr.

In order to further analyze the mechanism involved in the transcriptional regulation of RuBisCO genes, we performed the *in vitro* transcription of plasmids containing RuBisCO genes: pCUB1 (no *E. coli* promoter, see Fig. 2), pCKS1 (*tac* promoter, see Fig. 2), and pCKS101 (*tac* promoter 0.8 kbp upstream of RuBisCO genes). The transcription of these plasmids was carried out by *E. coli* RNA polymerase (New England Biolabs) with $[\alpha^{-32}P]$UTP and the transcripts were subjected to the hybridization experiments (14). *E. coli* RNA polymerase initiated the transcription of RuBisCO genes from *tac* promoter. However, *E. coli* RNA polymerase appeared to be inactive for *Chromatium* RuBisCO promoter.

3.3. Processing and assemblage of two subunits of *Chromatium* RuBisCO in *E. coli*

The immuno-blotting analysis showed the presence of the two constituent subunits (A and B) in *E. coli* cells transformed by pCKS1 as well as in other *E. coli* expressers. It was observed that the ratio of subunit A to subunit B was nearly equal in the cellular extracts from *E. coli* and *Chromatium*, indicating that an equimolar amount of the two constituent subunits is present in the cells of *E. coli*.

For the purpose of conclusively defining the quaternary structure of *E. coli* RuBisCO, an aliquot of the crude extracts from *E. coli* cells transformed by pCKS1 was subjected to gel filtration HPLC. As presented in Fig. 3 a and b, extracts from both *E. coli* and *Chromatium* cells showed identical elution profiles with respect to the maximal RuBisCO activity as well as the enzyme content determined by ELISA. The estimated molecular weight of the peak fraction employing several molecular weight standards was 520 kd, strongly indicating that the molecular form of RuBisCO is A_8B_8 in both cases.

FIGURE 3. Analysis of RuBisCO produced by *E. coli* cells (pCKS1) by gel filtration HPLC. The 26 μg of purified *Chromatium* RuBisCO (a) and approximately 300 μg of fresh crude extract prepared from *E. coli* (b) were subjected to gel filtration HPLC using Superose 6 column equilibrated with 50 mM Na phosphate (pH 6.9), 1 mM EDTA, and 0.3 M NaCl. Elution was performed at a flow rate of 0.3 ml/min, and fractionated at 0.9 ml. The photographs superimposed in the figure represent the immuno-blotting analysis of the main active fractions separated by the HPLC.

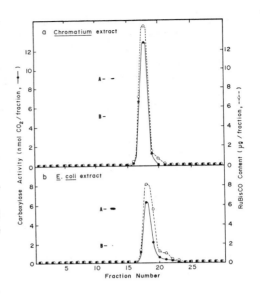

FIGURE 4. Electrophoretic pattern of RuBisCOs purified from *Chromatium* and *E. coli* (pCKS1). Purified RuBisCO from *E. coli* (lane 2) and purified *Chromatium* RuBisCO (lane 3) were subjected to SDS-polyacrylamide gel electrophoresis with 10-20% linear gradient (SDS-PAGE Plate 10/20, Daiichi Pure Chemicals) and stained with Coomassie brilliant blue R-250. Lane 1 is marker proteins (Electrophoresis Calibration Kit, Pharmacia). Subunits A (A) and B (B) are indicated.

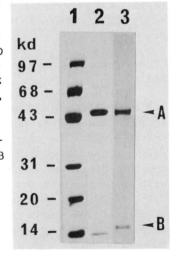

RuBisCO produced in *E. coli* JM109 transformed by pCKS1 was purified by using polyethylene glycol and gel filtration HPLC. The molecular weight of subunit B from *E. coli* was 1.3 kd lower than subunit B from *Chromatium* (Fig. 4). We determined the N termini of subunits A of RuBisCOs from *Chromatium* and *E. coli*. Three amino acids of subunit A including the first residue, methionine (initiation codon), were shown to be processed in

Chromatium RuBisCO. Some amino acids were also removed from the N terminus of subunit A in *E. coli*. These results indicate that *E. coli* contains proteases which can process subunits of RuBisCO to produce the enzymically active holoenzyme A_8B_8. In the extra sequence of subunit B molecule, a characteristic sequence for the transit peptide in higher plants was found, in spite of the fact that subunit B of *Chromatium* RuBisCO does not need to pass through membranes.

3.4. Concluding remarks

The induction of RuBisCO in *Chromatium* is regulated at the transcriptional level. A specific RNA polymerase distinguishable from that in *E. coli* or activating factor(s) is necessary for the transcription of *Chromatium* RuBisCO genes. Both subunits A and B are processed in *E. coli* at different positions in the RuBisCO molecule from those in *Chromatium*, eventually leading to the assembly of two subunits to produce an enzymically active A_8B_8 structure.

We have succeeded in expressing RuBisCO genes as unfused peptides free from β-galactosidase. An obvious advantage of our experimental system is the production of RuBisCO as high as 15% of the soluble protein in *E. coli* cells, being approximately 10 times higher value than other investigations (15-18). The presently employed system may provide us a chance of further clarifying the molecular mechanisms underlying the interaction of the two subunits in RuBisCO molecules and the regulation operating therein.

ACKNOWLEDGMENTS

We are much indebted to T. Takabe, M. Sugiura, K. Shinozaki, and H. Fukuda for their work in the cloning of genes for *Chromatium* RuBisCO. We also thank K. Wada for help in the Edman degradation experiment.

REFERENCES

1 Akazawa, T. (1979) in Encyclopedia of Plant Physiology: Photosynthesis II (Gibbs, M., and Latzko, E. eds.) pp. 208-229, Springer-Verlag, Berlin
2 Ellis, R.J. (1981) Ann. Rev. Plant Physiol. 32, 111-137
3 Morelli, G., Nagy, F., Fraley, R.T., Rogers, S.G. and Chua N.-H. (1985) Nature 315, 200-204
4 Takabe, T. and Akazawa, T.(1975) Biochemistry 14, 46-50
5 Kobayashi, H. and Akazawa, T. (1982) Arch. Biochem. Biophys. 214, 531-539
6 Kobayashi, H. and Akazawa, T. (1982) Arch. Biochem. Biophys. 214, 540-549

7 Kobayashi, H. and Akazawa, T. (1983) Arch. Biochem. Biophys. 224, 152-160

8 Brown, H. M., Bowman, L. H. and Chollet, R. (1981) FEMS Microbiol. Lett. 12, 105-109

9 Engvall, E. (1980) in Methods in Enzymology 70 (Van Vunakis, H. and Langone, J. L. eds.) pp. 419-439, Academic Press, New York

10 Oh-oka, H., Tanaka, S., Wada, K., Kuwabara, T. and Murata, N. (1986) FEBS Lett. 197, 63-66

11 Maniatis, T., Fritsch, E.F. and Sambrook, J. (1982) in Molecular Cloning, A Laboratory Manual, Cold Spring Harbor laboratory, New York

12 Viale, A.M., Kobayashi, H., Takabe, T. and Akazawa, T. (1985) FEBS Lett 192, 283-288

13 Amann, E., Brosius, J. and Ptashne, M. (1983) Gene 25, 167-178

14 Macherel, D., Kobayashi, H., Valle, E. and Akazawa, T. (1986) FEBS Lett. 201, 315-320

15 Gatenby, A.A., van der Vies, S.M. and Bradley, D. (1985) Nature 314, 617-620

16 Christeller, J.T., Terzaghi, B.E., Hill, D.F. and Laing, W.A. (1985) Plant Mol. Biol. 5, 257-263

17 Tabita, F.R. and Small, C.L. (1985) Proc. Natl. Acad. Sci. U.S.A. 82, 6100-6103

18 Gurevitz, M., Somerville, C.R. and McIntosh, L. (1985) Proc. Natl. Acad. Sci. U.S.A. 82, 6546-6550

ASSOCIATION OF RuBisCO SMALL SUBUNITS WITH A MEMBRANE FRACTION FROM CHROMATIUM VINOSUM

J. A. TORRES-RUIZ AND B. A. McFADDEN, BIOCHEMISTRY AND BIOPHYSICS PROGRAM, WASHINGTON STATE UNIVERSITY, PULLMAN, WASHINGTON 99164-4660

INTRODUCTION

Ribulose 1,5-bisphosphate Carboxylase/Oxygenase (RuBisCO) is the key enzyme of the Calvin-Benson cycle. RuBisCO in many prokaryotes and in all eukaryotes is present as a hexadecamer composed of eight copies of two types of polypeptides: a large subunit (L) of molecular weight 50,000 - 55,000 daltons which harbors the catalytic activity, and a small subunit (S) of molecular weight 12,000 - 16,000 daltons whose function is not well understood (1). In addition to the predominant L_8S_8 form of RuBisCO, the characterization of other structural forms of the enzyme, some lacking and some containing S subunits, has been previously described (2, 3). Since the function of S subunits remains unclear, the isolation of catalytically active RuBisCO preparations that are free of these subunits has stimulated considerable interest. In this context, more than a decade ago, RuBisCO was isolated from the purple non-sulfur bacterium Rhodospirillum rubrum as an L_2 enzyme (4) with turnover number comparable to that for the enzyme from higher plants.

In a previous paper we reported the isolation of active L_8 and L_8S_8 forms of RuBisCO from single cultures of the purple sulfur bacterium Chromatium vinosum (5). In light of those results, we postulated that RuBisCO must be associated with membrane-derived vesicles via the S subunits. Presumably, under our experimental conditions, membrane-associated S subunits are quantitatively removed by ultracentrifugation at 175,000 g in a dense medium containing 10% polyethylene glycol (PEG-6000). As a consequence, the catalytic L_8 core is detected in the supernatant. In the present report we provide evidence for an interaction between the S subunits and a membrane fraction from C. vinosum.

MATERIALS AND METHODS

Enzyme purification

Isolation of the L_8 and L_8S_8 enzymatic forms of RuBisCO was achieved from single cultures of C. vinosum strain D as described previously (5).

Membrane Isolation

A highly pigmented membrane fraction was isolated by sucrose flotation gradients (6). The 175,000-g pellet after polyethylene glycol treatment of the crude extract obtained from C. vinosum cells was dispersed in 12.9 ml of MEMMB buffer (50 mM MOPS, 1 mM $MgCl_2$, 0.1 mM EDTA, 1 mM β-mercaptoethanol, 1 mM PMSF, pH 7.3) containing 50% (W/V) sucrose. This suspension was placed in a 40-ml "quick seal" centrifuge tube from Beckman and was successively overlaid with 6.9 ml 40% (W/V) sucrose, 14.9 ml 30% (W/V) sucrose, and 4.6 ml 20% (W/V) sucrose. Flotation gradients were subjected to centrifugation at 45,000 rpm (175,000 g) at 4°C for 3 hr in a VTi 50 vertical rotor. After centrifugation, 1.2-ml fractions were collected from the bottom of the gradients. The OD_{280} as well as enzyme activity were estimated for every fraction.

Biggens, J. (ed.), Progress in Photosynthesis Research, Vol. III. ISBN 90 247 3452 5
© *1987 Martinus Nijhoff Publishers, Dordrecht. Printed in the Netherlands.*

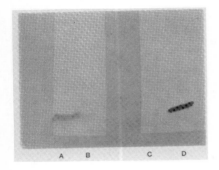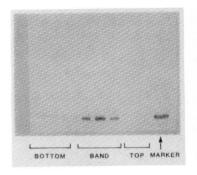

Fig. 1 (left): Immunoblot analysis, using anti-S subunits, or the 175,000-g pellet (A) and Supernatant (C), and the 35,000-g pellet (B) and supernatant (D). Fig. 2 (right): Immunoblot analysis after fractionation of the 175,000-g pellet on the sucrose flotation gradient.

SDS-polyacrylamide gel Electrophoresis and Immunoblotting
Protein samples (50-100 µg) were subjected to polyacrylamide gel electrophoresis (PAGE) in the presence of 0.1% $NaDodSO_4$ for 6 hr and 20 mA until the tracking dye (0.001% bromophenol blue) reached the bottom of the gel (5). The unstained gel was placed adjacent to a sheet of nitrocellulose paper (Bio-Rad) and electroblotted for 15h at 30 V(7). Polyclonal antibodies developed in rabbits against the S subunits of RuBisCO from C. vinosum (1:500 dilution) and biotinylated anti-rabbit IgG raised in goats (1:750 dilution) were used as primary and secondary antibodies respectively. The specific binding of antibodies was detected using the avidin/-biotinylated-horse radish peroxidase (HRP) System (Vector laboratories) followed by an HRP substrate, 4-chloro-1-naphthol (0.1 mg/ml) in Tris-saline buffer and 0.05% (v/v) H_2O_2.

Analysis of the release of S subunits from Membranes
The release of small subunits from the membrane fraction of C. vinosum, isolated from flotation gradients, was analyzed by salt treatment (KCl) or by pH variation. In the case of salt treatment, membranes were incubated with MEMMB buffer, pH 7.3, containing KCl at 50 mM, 100 mM, 200 mM, 400 mM or 600 mM. After incubation for 20 min at the specified temperature, the membranes were collected at 4°C by centrifugation for 5 min at 12,000 rpm in a Beckman microfuge. Proteins in the supernatant fraction were precipitated by adding trichloroacetic acid to 5% (W/V), collected by centrifugation (12,000 rpm for 5 min) and dried by lyophilization. The membrane fractions were washed thrice with 80% ice-cold acetone, collected by centrifugation as before, and dried under a nitrogen stream. For pH extraction, the membranes were incubated with a series of buffers: 25 mM glycine, pH 2.5; 25 mM acetate, pH 5; 25 mM Tris, pH 7 and 8; 25 mM $NaHCO_3/Na_2CO_3$, pH 10; 25 mM sodium phosphate, pH 12. All buffers contained 100 mM KCl and 1 mM PMSF. After incubation for 20 min at 4°C, the membrane and soluble proteins were collected and treated as described before. Soluble and membrane proteins were resuspended in sample buffer (containing 0.1% and 1% SDS respectively) and analyzed by SDS-PAGE and Western immunoblotting.

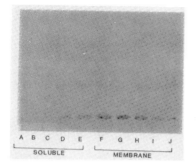
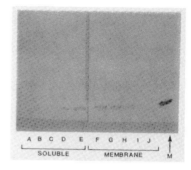

Fig. 3 (left): Immunoblot analysis of the S-subunits released after incubation of the membrane fraction band (Fig. 2) at 4°C or 25°C (Fig. 4, right) with KCl: lanes A and F, 50 mM; B and G, 100 mM; C and H, 200 mM; D and I, 400 mM; and E and J, 600 mM.

RESULTS AND DISCUSSION

The isolation of two structural forms of RuBisCO, L_8 and L_8S_8, from a single organism immediately raised the question of whether these L forms evolved as products of two different genes as in the case of R. sphaeroides (8). However, the apparent identity of the large subunits of the two forms of RuBisCO in C. vinosum has been confirmed by partial amino acid sequence analysis. Moreover, immunodiffusion of the L_8 and L_8S_8 form of RuBisCO from C. vinosum has revealed immunological identity of the two species (J. Torres-Ruiz and B. A. McFadden, unpublished results). These results have strengthened our conclusion that the L_8 form is derived from the L_8S_8 form by centrifugal fractionation.

In the present work, the possibility of an association of S subunits with membrane derived-vesicles has been tested by Western immunoblotting. Since the L_8 form of RuBisCO may be liberated when the speed of centrifugation after PEG-6000 treatment is increased from 35,000 g to 175,000 g, we have analyzed the pellets as well as the supernatants resulting from both centrifugations using anti-IgG against the S subunits of RuBisCO from C. vinosum as a probe. The results show (Figure 1) an enrichment of S subunits in the 175,000-g pellet whereas almost undetectable amounts were found in that supernatant. In contrast, when centrifugation at 35,000 g was attempted most of the S subunits were found in the supernatant. Nevertheless, a positive signal was observed in the 35,000-g pellet. Approximately 1% of the total carboxylase activity was found in both 35,000-g and 175,000-g pellets suggesting that in the case of the 35,000-g pellet the observed positive signal may reflect the presence of L_8S_8 molecules. Figure 2 shows the abundance of S subunits in representative fractions derived from the 175,000-g pellet banded in the sucrose flotation gradient. Most of the S subunits were present in the membrane-like fraction. While low but detectable carboxylase activity (<0.05% of total) can be measured at the bottom of the gradient, the membrane fraction contains none. S subunits could be detected in fractions at the bottom of

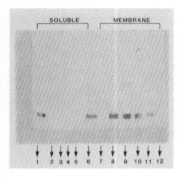

Fig. 5: Immunoblot analysis of the S-subunits released from the membrane fraction (band, Fig. 2) as a function of pH. Lanes 1 and 7, pH 2.5; 2 and 8, pH 5; 3 and 9, pH 7; 4 and 10, pH 8; 5 and 11, pH 10; 6 and 12, pH 12.

the gradient. We attribute the presence of those subunits to the presence of a small amount of L_8S_8 molecules in the 175,000-g pellet.

The nature of the association of S subunits with a membrane fraction was analyzed by salt extraction of the membrane fraction (isolated from the flotation gradient) or at diferent pH's. When the membrane fractions were incubated with buffer containing 600 mM KCl at 4°C or 400 mM KCl at 25°C the S subunits are quantitatively released from the membranes and subsequently recovered in the soluble fractions (Figure 3 and 4). A pH in the range of 5-10 did not affect significantly the association of S subunits with the membrane fraction (Figure 5). The release of S subunits from membranes at a pH of 2.5 or at pH 12.0 may be due to protein denaturation under those conditions.

To recapitulate, the association of S subunits to membranes in C. vinosum may account for the isolation of two catalytically active forms of RuBisCO. At this point we cannot deduce the nature of the molecular interactions between the S subunits and the membranes, i.e., hydrophobic or electrostatic. However, the detachment of those subunits at relatively low KCl (400 mM at 25°C or 600 mM at 4°C) suggests that the S subunit is a peripheral protein, and therefore presumably bound through electrostatic forces. Further experimental work is in progress to determine whether this interaction is specific and therefore of biological significance.

Acknowledgement. We acknowledge the partial support of this research by NIH grant 19,972.

REFERENCES

1. McFadden, B.A. (1978) Ornston L.N. and Sokatch J.R. (eds.), The Bacteria a Treatise on Structure and Function, Vol. VI. Bacterial Diversity. pp. 219-304, Academic Press. New York
2. McFadden, B.A. (1980) Acc. Chem. Res. 13, 394-399.
3. McFadden, B.A., Torres-Ruiz, J.A., Daniell, H. and Sarojini, G. (1986) Phil. Trans. Royal Soc., B. (in press).
4. Tabita, F.R., McFadden, B.A. (1974) J. Biol. Chem. 249, 3459-3464.
5. Torres-Ruiz, J.A. and McFadden, B.A. (1985) Arch. Microbiol. 142, 55-60.
6. Conder, M.J. and Lord, J.M. (1983) Plant Physiol. 72, 547-552.
7. Towbin, H., Staehelin, T. and Gordon, J. (1979) Proc. Natl. Acad. Sci. USA 76, 4350-4354.
8. Gibson, J.L. and Tabita, F.R. (1977) J. Biol. Chem. 252, 943-959.

INHIBITION OF RIBULOSE BISPHOSPHATE CARBOXYLASE ASSEMBLY BY
ANTIBODY TO A BINDING PROTEIN

SUSAN CANNON, PAULINE WANG, AND HARRY ROY

Ribulose-1,5-bisphosphate carboxylase/ oxygenase
(RuBisCO) (E.C. 4.1.1.39) catalyses CO_2 fixation in the
Calvin cycle of photosynthesis. It also catalyses the
competing oxygenation of ribulose bisphosphate to initiate
photorespiration (Lorimer, 1981). At saturating light
intensity and ambient CO_2, the activity of this enzyme is
rate limiting for photosynthesis. In most photosynthetic
organisms the enzyme consists of eight "large" subunits
(Mr=55,000) and eight "small" subunits (Mr=14,000). In
higher plants and green algae, the large subunits are
encoded in chloroplast DNA and synthesized in chloroplasts,
while the small subunits are synthesized as precursor
polypeptides in the cytoplasm, whence they are taken up by
the chloroplasts, processed to their final molecular
weight, and assembled into intact enzyme molecules. This
process is of special interest because it appears to
involve a protein which does not comprise part of the final
structure of the enzyme.

About 75% of large subunits synthesized in organello
(i.e., in isolated pea chloroplasts) occur in a soluble 29S
(Mr ca 720,000) particle in chloroplast extracts (Cannon et
al., 1986). The 29S particle consists of one 55 kD RuBisCO
large subunit and probably six 60 and six 61 kD "binding
protein" subunits each (Hemmingsen and Ellis, 1986). The
remainder of the newly synthesized large subunits sediment
at 7S (still faster than expected for a monomer). The
association of large subunits with the 29S complex is not
an artifact of in organello protein synthesis, because the
same result occurs when the subunits are synthesized in
vivo before chloroplast isolation. The complex is specific:
other proteins in the chloroplast extract do not bind to
the 29S complex, and the newly synthesized large subunits
are not associated with any other proteins (Hemmingsen and
Ellis, 1986). The incorporation of large subunits derived
from this complex into intact RuBisCO was demonstrated with
an in vitro reaction developed in this laboratory (reviewed
in Cannon et al., 1986). The reaction mechanism involves
release of large subunits from the 29S complex, followed by
their incorporation into RuBisCO, presumably dependent on
the endogenous small subunit pools. The dissociation of the
29S particle is accomplished in vitro by adding ATP and

Biggens, J. (ed.), Progress in Photosynthesis Research, Vol. III. ISBN 90 247 3452 5
© *1987 Martinus Nijhoff Publishers, Dordrecht. Printed in the Netherlands.*

MgCl₂, and is believed to be reversible in isolated chloroplasts.

The large subunit binding protein is immunologically detectable in tobacco, wheat, and barley leaves, in extracts of plastids from castor bean endosperm, in a large number of photosynthetic bacteria which contain RuBisCO (Hemmingsen and Ellis, 1986), and in spinach leaves (S.E. Hemmingsen, personal communication). We have detected the protein in maize.

If assembly of RuBisCO requires the binding protein, an antibody to the binding protein should inhibit assembly. Antibody to the binding protein subunits was made available by S.E. Hemmingsen and R.J. Ellis (Hemmingsen and Ellis, 1986). This antibody specifically inhibited the assembly of large subunits into RuBisCO.

RESULTS AND DISCUSSION

The large subunit of RuBisCO represents over 90% of the radioactive protein recovered in the soluble fraction of pea chloroplasts after in organello protein synthesis in the presence of radioactive amino acid. This permits routine analysis of incorporation of large subunits into RuBisCO by one-dimensional PAGE and fluorography. In this assay, the large subunits in the 29S complex migrate as a sharp, intensely radioactive band and can be resolved easily from intact RuBisCO. A disadvantage of the assay (compared with sucrose gradient analysis or gel filtration) is that the 7S large subunits migrate in a polydisperse fashion or precipitate at the top of stacking gel lanes. The gel lanes therefore have background radioactivity which is due to the 7S large subunits. This behavior is not alleviated by altered buffers or the addition of nonionic detergents to the electrophoresis media (unpublished). Despite this disadvantage, the one dimensional assay is used because it is the only assay which permits analysis of the effects of multiple treatments on the incorporation of radioactive large subunits into RuBisCO. The unassembled 7S and 29S large subunits have been identified previously by immunoprecipitation and two dimensional electrophoresis (references in Cannon et al., 1986, and Hemmingsen and Ellis, 1986). Assembly proceeds efficiently after preincubation of chloroplast extracts with 5 mM ATP, or after depletion of endogenous ATP by addition of AMP in the presence of adenylate kinase (Cannon et al., 1986).

We used a batch of antiserum which had been raised against the pure binding protein subunits. This reacts strongly with the binding protein but not with large subunits (Hemmingsen and Ellis, 1986). It cross reacts weakly (less than 1%) with small subunits, as judged by a sensitive immunoblotting assay employing ^{125}I-conjugated protein A. This extent of cross reactivity should be negligible when the serum is employed at equivalence with the binding protein, as in the experiments described here.

As shown in Figure 1, incorporation of radioactive large subunits into RuBisCO (lane A, preimmune serum control) was strongly inhibited by preincubation of the chloroplast extract with antibody to the binding protein (lane B). Inhibition was at least 50% in this experiment, as judged by densitometric analysis of the film. In other experiments, inhibitions as high as 70% were observed. The variation in the extent of inhibition observed is believed to reflect differences in concentration of the binding protein from one chloroplast extract to the next. Pretreatment of the antibody with excess concentrations of small subunits (lanes C and D) did not affect the ability of the antibody to inhibit the assembly reaction. However, preincubation of antibody with purified binding protein (lane E) strongly inhibited the ability of the antibody to inhibit assembly (lane A). This experiment was conducted in the presence of 20 mM AMP, so that the large subunits participating in assembly were derived from the 7S subunits present in the original extract (Cannon et al., 1986). Antibody to the binding protein also inhibited assembly when the assembly reaction was carried out after incubation in the presence of 5 mM ATP, which permits 7S large subunits derived from the 29S complex to participate in the assembly reaction (Cannon et al., 1986).

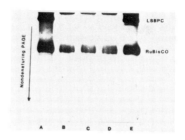

Figure 1

The most reasonable interpretation of the data is that the great majority of assembly-competent large subunits are associated with binding protein. It is clear therefore that no model for assembly of RuBisCO in pea chloroplasts can exclude the binding protein. Our current working hypothesis is illustrated in Figure 2. In this model, large subunits are bound to monomers of the binding protein in 7S complexes, ($LS_1 LSBP_1$), or in dodecamers of binding protein in 29S complexes ($LS_1 LSBP_{12}$). ATP in the mM range (large letters) can mediate the release of 7S complexes and binding protein monomers ($LSBP_1$). Since antibodies to LSBP inhibit assembly, and radioactive large subunits sediment at 7S, and since

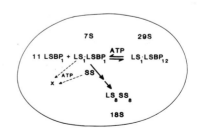

Figure 2

under no circumstances have we observed direct transfer of radioactivity from the 29S complex into 18S RuBisCO (LS_8SS_8), assembly of RuBisCO is depicted as a small subunit (SS) reaction with LS_1LSBP_1, followed by more than one unknown step. To account for the stimulatory effects of ATP removal on assembly (Cannon et al., 1986) we propose an ATP-dependent step (arrows pointing to X, designating unknown products) which inhibits the participation of small subunits or LS_1LSBP_1 in the assembly process. Since this can occur at very low ATP concentrations, the ATP is printed in smaller capitals in the drawing. We think that at low ATP, the restriction mechanism dominates and little assembly is observed. At zero ATP the restriction mechanism is not operative and assembly can be seen. At high ATP the large subunit pool is so much enlarged that assembly outpaces the restriction mechanism. Although the mechanism of this restrictive activity is not known, one possibility is that it is due to an ATP-dependent protease.

What is the function of the binding protein? It could be concerned primarily with regulating the supply of large subunits available for assembly in the light and the dark. It could be required for assembly of RuBisCO in higher plant chloroplasts; for example, the binding protein might confer assembly competence on large subunits by maintaining their solubility until they are capable of interacting with small subunits. Since the 7S large subunits sediment more rapidly than the monomeric binding protein subunits (Hemmingsen and Ellis, 1986), and have an estimated molecular weight of 117,000, and since their ability to assemble is inhibited by antibody to the binding protein, it appears that they are heterodimers containing one large subunit and one binding protein subunit. If this is the case, perhaps the next step in the assembly process involves association of small subunits with the heterodimer, or release of large subunit from the heterodimer.

Acknowledgment

Research supported by Grant # 5-RO1 GM33469 from the National Institutes of Health.

REFERENCES
1 Cannon, S.V., Wang P., and Roy, H. (1986) J Cell Biol. 103, In press.
2 Hemmingsen, S.E., and Ellis, RJ (1986) Plant Physiol. 80, 269-276.
3 Lorimer, G.H. (1981) Annu. Rev. Plant. Physiol. 32, 349-383.

RIBULOSE-1,5-BISPHOSPHATE CARBOXYLASE PROTEIN IN HIGHER PLANTS BY A NEW
RAPID HPLC METHOD

MASSACCI, A., * CORRADINI,D., DI MARCO,G. AND GIARDI,M.T.
INSTITUTE OF PLANT BIOCHEMISTRY AND ECOPHYSIOLOGY AND (*) INSTITUTE OF
CHROMATOGRAPHY, NATIONAL RESEARCH COUNCIL, VIA SALARIA KM 29.300,
P.O.BOX 10, MONTEROTONDO SCALO, ROME, ITALY

1.   INTRODUCTION
In the context of a study on the relationships between biochemical and
biophysical parameters of photosynthesis of the leaf we needed a rapid and
sensitive method for quantification of RuBPCase protein. The methods
hitherto described all appear time consuming and most are multistep
processes. Here we suggest a new fast method of determination of RuBPCase
by Hydrophobic Interaction Chromatography (HI-HPLC) which allows the
quantitative determination of this enzyme without apparently affecting its
specific activity.

2.   PROCEDURE
2.1. Materials and methods
2.1.1. Plant material: Seeds of pea and spinach were sown in pots
containing soil, sand and peat. The plants were grown in
controlled environment at 300 uE m^{-2} s^{-1}, 24 C and 60%
relative humidity, using a Hewitt nutrient solution with
various nitrogen contents.
2.1.2. Chemicals and reagents: The chemicals were obtained from their
respective commercial sources. RuBPCase, as partially purified
powder from spinach, D-Ribulose-1, 5-bisphosphate acid and
Bicine were from Sigma (St. Louis, MO, USA). Labeled sodium
bicarbonate from Amersham (England). Water HPLC grade reagent
and all other chemicals not mentioned here were from Carlo Erba
(Milan, Italy).
2.1.3. Enzyme extraction: Fresh leaf blades were homogenized in a
chilled mortar with quartz sand in 50 mM Bicine at a leaf to
buffer ratio of 0.25 (w/v). The homogenate was centrifuged at
1700 x g at 4 C for 10 min. The supernatant was in part
directly injected into the chromatographic column for the
quantitative determination of RuBPCase and in part assayed
for the enzymatic activity and soluble proteins content.
2.1.4. Enzymatic assay: RuBPCase activity was measured in triplicate
at 25 C by incorporation of $^{14}CO_2$ in the presence of RuBP.
The final reaction mixture contained 0.1 M Bicine, 200 mM MgCl,
0.5 mM RuBP and 20 mM NaH $^{14}CO_3$ in a total volume of 0.6 ml.
The reaction was started by addition of 50 ml of 6.0 mM RuBP
after preincubation of the extract (50 ul) in the other
constituents for 5 min in the scintillation vials. The reaction
was stopped after 1 min by adding 200 ul 2N HCl. After drying,
the vials were counted in a LKB scintillation counter. Control
samples assayed in absence of RuBP were also counted. Soluble
protein content was measured according to Lowry et al.(1).

Biggens, J. (ed.), Progress in Photosynthesis Research, Vol. III. ISBN 90 247 3452 5
© *1987 Martinus Nijhoff Publishers, Dordrecht. Printed in the Netherlands.*

2.1.5. <u>HPLC analysis</u>: Decreasing salt gradient linear in salt
concentration was used in net aqueous solutions. The
chromatographic runs consisted of three segments: a, isocratic
elution with 1.2 M $MgSO_4$ 50 mM Bicine pH 6.2 as starting eluent
for 2 min; b, linear gradient from 0 to 100% of 50 mM Bicine
pH 6.2 as gradient former in 20 min; c, isocratic elution with
the gradient former proper. The flow rate was 1 ml/min. All
solutions were filtered through a millipore type HA 0.45 u
membrane filter and degased by spraying with Helium prior to
use as eluents. The column effluent was analyzed at 280 nm in
UV or at 340 nm in fluorescence excited at 295 nm. The equipment
used was a Perkin-Elmer Model Series 4 Liquid Chromatograph with
a Model LC 75 autocontrol variable wavelength spectrophotometer
detector, a Model 203 fluorescent detector and a Sigma 15
Chromatography Data Station. A TSK Pheynl 5 PW (75 x 7.5 m.m.
I.D.) column from Bio Rad was used.

2.1.6. <u>Electrophoresis</u>: Electrophoresis was carried out in presence
of sodium dodecyl sulphate (SDS) 0.1% on Polyacrylamide gels
(PAGE) gradient from 5 to 25% in Tris 25 mM- Glycine 192 mM
buffer, pH 8.3, at 6 mA for 24 hours. Gels were fixed and
stained overnight in a solution containing 0.09% Coomassie
blue, 48% (v/v) methanol and 4.3% (v/v) acetic acid. Gels were
destained in 7% (v/v) acetic acid. Electrophoresis was also
performed at pH 8.2 in absence of SDS.

3.   RESULTS AND DISCUSSION

3.1 FIGURE 1. Elution profile of leaf extracts of Spinach and Pea (B)
from HI-HPLC column. The arrows indicate RuBPCase peak. The continuous
line represents the absorbance at 280 nm, the dotted line the
fluorescence at 340 nm at very low sensitivity.

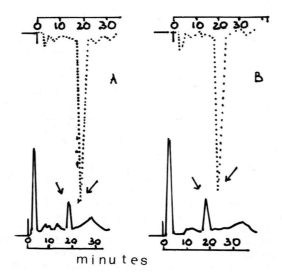

minutes

Figure 1 represents a typical separation by Hydrophobic Interaction
Chromatography with weakly hydrophobic microparticulate stationary
phase and decreasing salt gradient of RuBPCase in spinach and pea leaf
extract. The analyses were carried out under conditions of linear
gradient elution, preceded by isocratic elution in the same
chromatographic run. Such an elution strategy is frequently used in
practice in order to separate sample components of widely different
retention behavior in a single chromatographic experiment comprising
both isocratic and gradient elution steps. Elution of RuBPCase occurs
during gradient development following the initial isocratic elution
step with a retention time of 18.4 min. As seen in fig 1, this was the
only fluorescent peak detected at 340 nm. In a separate study it has
been found that this protein is particularly suitable for this
separation technique since it shows a good response to variations in
salt concentration and in column temperature (2). The identity of the
RuBPCase peak in spinach leaf extract was established by its enzymatic
activity and further confirmed by several criteria. The optical density
ratio at 280/260 nm was normally greater than 1.80 which is consistent
with that of trice cristallized tobacco RuBPCase (3) and of the
spinach enzyme purified by the classical techniques (4). PAGE analysis
of collected material from the peak gave a single slow moving narrow
band in 5-25% gradient gels. SDS-PAGE of the same material performed
in 5-25% gradient gels gave a major band with a molecular mass of
56 KDa and a minor band with a molecular mass of 14 KDa. The mass
recovery of the RuBPCase from the chromatographic column was greater
than 95% as established by rechromatographying a known amount of
RuBPCase purified from spinach leaves by the proposed method and
quantified by the Lowry method. Besides, the enzymatic activity was
also almost completely recovered. This was established with spinach
leaf extract by completely collecting the chromatographic peak and
comparing the activity of this solution with that of a solution
prepared by diluting in a buffer of the same volume and composition as
the collected peak the same amount of extract used for the injection.
A linear dependence of the area of the well resolved RuBPCase
chromatographic peak on protein concentration in spinach leaf extracts
up to 400 ug of column loaded enzyme was observed.
The homogenization and assay of replicate samples of spinach showed
an extremely reproducible result. The standard deviation of 7
determinations was as low as 2% of the mean value. A second
homogenization of 30 g of pellets showed that more than 95% of the
extractable RuBPCase is obtained in the first supernatant.

3.2   TABLE 1 shows the results obtained by applying this method to the
determination of RuBPCase protein in pea and spinach grown with
different nitrogen supply in order to obtain different concentrations
of RuBPCase. The amount of RuBPCase protein reported in the table was
derived as follows. A given amount of commercial RuBPCase was purified
by ammonium sulphate precipitation and resolubilized. The protein
content of this solution, which gave a single chromatographic peak and
a single electrophoretic band, was determined by Lowry method. This
solution was used as a standard. The enzymatic activity was measured
on the extract before injection. Assuming for pea the same mass
recovery of spinach, i.e. greater than 95%, the specific activity can
be calculated. The average value of six experiments for spinach and
five for pea, all characterized by a different soluble protein content,

is respectively 2.6 and 2.1 umol/mg RuBPCase min. The relative content of RuBPCase respect to total soluble protein differs slightly with the different protein content laying around 50% both in spinach and in pea. Preliminary attempts to use this method with other plant sources have been effected. With some modifications, it appears promising with tomato, sunflower, alfalfa and tobacco (data not shown).

TABLE 1.   Total Soluble Proteins, RuBPCase content in leaf extracts of spinach and pea and RuBPCase activity of leaf extracts of spinach and pea grown with different nitrogen supply.

Nitrate ml	Proteins mg/g.f.w.		RuBPCase activity umol/g.f.w. min		RuBPCase mg/g.f.w.	
	Spinach	Pea	Spinach	Pea	Spinach	Pea
0.6	8.1	10.6	7.8	12.8	3.2	5.0
1	11.6	17.0	10.6	19.5	4.2	9.8
6	13.4	21.0	17.3	21.3	6.6	10.9
12	16.0	29.0	21.9	30.2	8.2	13.6
18	20.1	31.3	28.0	34.7	10.8	17.5
24	23.4		31.0		11.9	

REFERENCES

1 Lowry, O.H., Rosebrough, N.J., Farr, A.L., and Randall, R.J. (1951) J. Biol. Chem. 193, 265
2 Corradini, D., Giardi, M.T., and Massacci, A. Journal Chromatography (in press)
3 Kung, S.O., Chollet, R., and Marsho, T.W. (1980) in Methods in Enzymology (San Pietro, A., ed.), Vol. 69, pp. 326–336, Academic Press, New York
4 Paulsen, C. (1982) in Methods in Chloroplasts Molecular Biology (Edelman, M., Lellick, R.B., and Chua, N.H., eds.), pp. 767–781, Elsevier, New York

Research work supported by C.N.R., Special Grant I.P.R.A. Sub-project 1 No. 947.

CYCLIC AMP SYSTEM AND PHYTOCHROME IN THE REGULATION OF THE RIBULOSE BISPHOSPHATE CARBOXYLASE ACTIVITY

N.G.Doman, E.P.Fedenko, K.K.Kassumov and V.K.Jaworskaya

The Bach Institute of Biochemistry, Moscow, Leninsky pr.33, USSR

The Institute of Plant Physiology, Kiev, Wassilkowskaya 31/17, USSR

The functioning of Calvin cycle and in the first of all its key enzyme – ribulose bisphosphate carboxylase (RBPC) – mainly determines the photosynthetic efficiency of plants. The synthesis of this enzyme is controlled by the photomorphogenic receptor of higher plants – by phytochrome, wich reacts specifically on red-far red light. Up to now the mode of phytochrome action was unclear. Our data indicate what a messenger in the phytochrome action is may be cAMP – an universal regulator of biochemical processes in every living cell. In the first of all it was raised a qi question, if phytochrome can activate the adenylate cyclase – an enzyme, which catalyses the synthesis of cAMP from ATP (ATP-pyrophosphate lyase (cyclising) EC 4.6.1.1).

MATERIALS AND METHODS

The work was performed using 5-7 days old ethiolated mayz seedlings. The seedlings were illuminated 3-5 min with red or far red light after red light by means of a monochromator or interferential filters 661 nm and 730 nm. The seedlings were homogenized in a medium containing 0.4 M sorbitol, 5 mM $MgCl_2$, 20 mM mercaptoethanol, 3 mM Na-EDTA and tris bis pH 7,6. The homogenate was filtered through 8 cloth layers, centrifuged at 500 g 5 min and the sediment was away. All procedures were performed in a cold room at weak green light. The supernatant was once more centrifuged at 30000 g 10 min. The supernatant pellet was resuspended and this membrane fraction was assayed on the adenylate cyclase activity. The adenylate cyclase activity was assayed radiometrically in the medium containing 0.04 M tris-HCl (pH 8.7), 0.06 mM cAMP, 5 mM $MgCl_2$, 50 mM NaF, 4 mM creatinephosphate, 0.1 mg/ml creatine kinase, 1 mM ATP and 0.1 mM ATP, -14 -CATP 40 mkl (26 mCi/mmole), membrane fraction 0.2mg. The nucleotides were separated on silufol plates ("Silufol UV-254","Kavalier" CSSR) in a solution system n-butanol:acetone:25%ammiak (Д.Р. Бериташвили и Кафиани,1975 once at ratio 8:2:5, then at ratio 8:2:1 (Н.М.Гулиев и др., 1978). The nucleotide spots were cuted out and the radioactivity was counted on a liquid scintillation counter ("Intertechnik", France).

The cAMP phosphodiesterase activity was assayed in supernatant 70000 g also radiometrically (M.L.Elks et al., 1983).

Biggens, J. (ed.), Progress in Photosynthesis Research, Vol. III. ISBN 90 247 3452 5
© *1987 Martinus Nijhoff Publishers, Dordrecht. Printed in the Netherlands.*

RESULTS AND DISCUSSION

As can see from fig.I, the illumination of cut ethiolated mayz seedlings with red light increases the adenylate
cyclase activity as compared with the dark experiment. In
seedlings illuminated with far red light after red light the
adenylate cyclase activity was lower than in the dark variant.
The values of the activities were as follows dark:red:far red
=2I.5:30.I:I2.9 pmole cAMP per mg protein min (February-May,
3 min illumination) and 2I.5:II0.0:23.3 (August,4 min illumination).

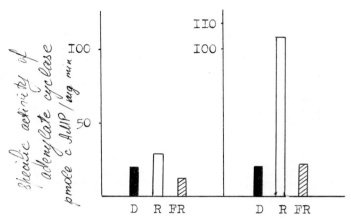

Figure I. The adenylate specific activity by monochromatic
illumination of ethiolated cut mayz seedlings. D - dark, R -
red light 66I nm, FR - far red light 730 nm after red light
a - February, 3 min illumination, b - August, 4 min illumination

The activation of adenylate cyclase by red light and the
release of the effect by far red light indicate that adenylate cyclase is activated by the phtochrome absorbed light.
The increase of phytochrome "pelletability" by red illumination of mayz seedlings - i.e. increase of phytochrome amounts
connected with membrane fraction (R.Ju,1978) - confirms too
the activation of adenylate cyclase - as a membrane bound
enzyme - by phytochrome. By far red illumination the most of
phytochrome will be again soluble and the adenylate cyclase
activation will be away.

These data show for the first time the connection of two
regulatory systems of eucaryotic organisms - of cAMP systems
and of phytochrome systems.

So far as the cAMP level in the cell is regulated not
only by adenylate cyclase but also by cAMP phosphodiesterase
(PDE), which catalyses the hydrolysis of cAMP to adenosine
monophosphate, it was raised a question how the phytochrome
absorbed light acts on PDE activity.

The work was performed with the I fraction of PDE
received after 50% $(NH_4)_2SO_4$ saturation of 70000 g supernatant.

PDE I displays high activity at pH 5.6 and 7.6.
The red illumination of ethiolated mayz seedlings inhibits
the PDE I activity roughly half as a much again (fig.2).

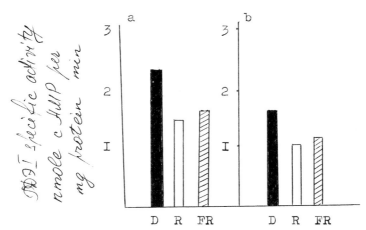

Figure 2. The PDE I specific activity by monochromatic illumination of ethiolated cut mayz seedlings. D,R,FR - see fig.I.
a - PDE I activity at pH 5.6, b - PDE I activity at pH 7.6

The far red illumination of ethiolated seedlings after
red illumination almost takes not away the inhibitory action
of red light on PDE I. This observed at pH 5.6 and at pH 7.6
(fig. 2 a and b).
Thus, the action of red light on PDE I is opposite to
its action on adenylate cyclase. Such combination of PDE I
inhibition with adenylate cyclase activation must lead to an
essential increase of cAMP levels by red illumination and then
in accordance with our data (В.К.Яворская и др., 1986   ) -
to an increase of isolated plant nuclei RNA-polymerase activity, i.e. to alteration of genome expression and to alteration
of corresponding physiological functions.
As showed our prolonged investigations, the described
inhibition of PDE I by red light is observed only in spring,
in autumn it is replaced by by the activation, but already
in January one can see the inhibition of PDE I by red illumi-
nation.These data can explane to a certain extent the different behaviour of plants at the same light conditions in
spring and in autumn. See table.
The action of inhibitors and activators of adenylate
cyclase and PDE I on RBPC activity by red illumination is
discussed.
The mode of action of cAMP itself is connected with
functioning of specific cAMP binding proteins. We also isolated such proteins from pea seedlings. We also isolated
and purified to homogeneity by the use of ion-exchange and

affinity chromatography a cAMP and cytokinine binding protein. The existence of certain connection between cAMP system, phytohormones and phytochrome in RBPC activity regulation is discussed.

TABLE The PDE_2I specific activity by red illumination (661 nm, 100 erg/sm^2sec, 4 min illumination) of cut ethiolated mayz seedlings in different months of year

888

	PDE I specific activity nmole/mg protein min							
Month	January		February		March		November	
Illumination	D	R	D	R	D	R	D	R
pH 5.6	0.16	0.12	2.31	1.46	5.02	4.41	0.10	0.16
pH 7.6	0.05	0.05	1.63	1.05	3.77	2.75	–	–

	The degree of activation-inhibition of PDE I by red illuminatin of seedlings			
pH 5.6	-1.34	-1.58	-1.36	+1.30
pH 7.6	1.00	-1.55	-1.13	–

REFERENCES

1 Бериташвили, Д.Р. и Кафивни, К.А. (1975) Вопросы мед. химии 31, 322-324
2 Гулиев,Н.М., Феденко, Е.П., Комарова, Т.И. и Доман, Н.Г. (1978) Биохимия 43, 928-934
3 Ju, R. (1978) Intern. J. Plant Physiol. 89, 95-103
4 Elks, M.L., Manganiello, V.C. and Vaughan, M. (1983) J. Biol. Chem. 258, 8582-8587
5 Яворская, В.К., Калинин, Ф.Л., Драговоз, И.В., Выплова, М.Б., Феденко,Е.П. и Доман, Н.Г. (1986) Физиология растений (в печати)

THE INTERACTION OF FERREDOXIN WITH CHLOROPLAST FERREDOXIN-LINKED ENZYMES

M. HIRASAWA[a], M. BOYER[a], K. GRAY[a], D. DAVIS[b] AND D. KNAFF[a], [a]DEPT. OF CHEMISTRY AND BIOCHEMISTRY, TEXAS TECH UNIV., LUBBOCK, TX 79409 AND [b]DEPT. OF CHEMISTRY, UNIV. OF ARKANSAS, FAYETTEVILLE, AR 72701, U.S.A.

1. INTRODUCTION

Ferredoxin, reduced by Photosystem I during oxygenic photosynthesis, then serves as the electron donor for the reduction of $NADP^+$ [1], nitrite [2], and the reductive conversion of 2-oxoglutarate plus glutamine to glutamate [3]. Evidence exists that two of these ferredoxin-dependent reductions, that of $NADP^+$ (catalyzed by ferredoxin:$NADP^+$ oxidoreductase, hereafter abbreviated FNR) and of nitrite to ammonia (catalyzed by ferredoxin:nitrite oxidoreductase, hereafter referred to as nitrite reductase) involve electrostatically-stabilized complexes between ferredoxin and the enzymes [4,5]. Glutamate synthase, the enzyme responsible for the ferredoxin-dependent synthesis of glutamate, like FNR [6] and nitrite reductase [7], can be purified by ferredoxin affinity chromatography [8]. This suggested that glutamate synthase can form an electrostatic complex with ferredoxin. Evidence for such a complex [9] is presented below.

^{13}C-NMR investigations [10] and cross-linking studies [11] have implicated carboxyl groups on ferredoxin in complex formation with FNR. Chemical modification [12,13] of ferredoxin carboxyl groups by treatment with glycine ethyl ester (GEE) in the presence of the water-soluble carbodiimide, 1-ethyl-3-(3-dimethylaminopropyl) carbodiimide (EDC), gave almost complete inhibition of the protein's ability to form a complex with FNR without affecting ferredoxin reduction by Photosystem I [12]. We have investigated the effect of carboxyl group modification on ferredoxin's ability to form complexes with nitrite reductase and glutamate synthase and demonstrated considerable inhibition of ferredoxin binding to both enzymes, suggesting that a set of carboxyl groups on ferredoxin may be part of a common binding site for ferredoxin-dependent enzymes.

2. PROCEDURE

Ferredoxin [5], FNR [6], ferredoxin-linked nitrite reductase [5] and glutamate synthase [8] were prepared from field grown spinach. Difference spectra were measured at 4°C in a Perkin-Elmer Lambda 5 spectrophotometer using a split cell technique [14].

3. RESULTS

Fig. 1 shows evidence for electrostatic complex formation between ferredoxin (Fd) and three chloroplast ferredoxin-dependent enzymes. Nitrite reductase, glutamate synthase and FNR [15] each co-migrate with ferredoxin during gel filtration at low ionic strength. No co-migration was observed at high ionic strength.

Biggens, J. (ed.), Progress in Photosynthesis Research, Vol. III. ISBN 90 247 3452 5
© *1987 Martinus Nijhoff Publishers, Dordrecht. Printed in the Netherlands.*

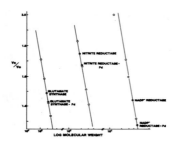

Fig. 1. Gel filtration chromatography of enzyme•ferredoxin complexes
at low ionic strength. Open circles are molecular weight
standards. From Ref. 9.

Additional evidence for complex formation between ferredoxin and
glutamate synthase or nitrite reductase comes from the observation of
spectral perturbations produced when ferredoxin was mixed with either
enzyme. Fig. 2A and 3A show the difference spectra resulting from
complex formation between ferredoxin and glutamate synthase or nitrite
reductase, respectively.

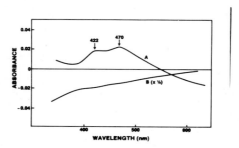

Fig. 2. Effect of glutamate synthase-ferredoxin interaction on the
visible spectra of the proteins. (A) Native Ferredoxin. The
sample contained 25 μM glutamate synthase and 50 μM native
ferredoxin. (B) Modified Ferredoxin. Native ferredoxin was
replaced by (GEE + EDC)-modified ferredoxin. From Ref. 9.

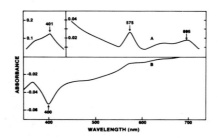

Fig. 3. The effect of nitrite reductase-ferredoxin interaction on the visible spectra of the proteins. (A) Native Feredoxin. As in Fig. 2 except that 30 µM nitrite reductase and 30 µM native ferredoxin were present. (B) Modified Ferredoxin. From Ref. 9.

Plots of the absorbance changes of Fig. 2A and 3A vs. ferredoxin concentration fit single hyperbolic binding isotherms with K_d values of 14.5 µM and 0.63 µM for glutamate synthase and nitrite reductase, respectively (Table 1). No absorbance changes were observed when the experiments of Fig. 2A and 3A were repeated in the presence of 200 mM NaCl, providing additional evidence for the electrostatic nature of the ferredoxin-enzyme interactions.

The role of ferredoxin carboxyl groups in complex formation with glutamate synthase and nitrite reductase was investigated using ferredoxin reacted with GEE plus EDC to modify 3-4 carboxyl side chain groups [12]. The modified ferredoxin neither co-migrated with FNR during gel chromatography nor produced spectral perturbations on mixing with FNR. As shown in Fig. 2B and 3B, the spectral perturbations produced by mixing the modified ferredoxin with either enzyme differed considerably from those obtained with native ferredoxin, pointing to altered protein:protein interactions. Carboxyl group modification also decreased the affinity of ferredoxin binding to the two enzymes: Formation constants decreased 7-fold for glutamate synthase and 160-fold for nitrite reductase (Table 1).

TABLE 1. THE EFFECT OF FERREDOXIN MODIFICATION ON KINETIC AND BINDING PARAMETERS FOR NITRITE REDUCTASE AND GLUTAMATE SYNTHASE

Enzyme	Electron donor	Relative V_{max}	K_m^{app} (µM)	K_d (µM)
Glutamate synthase	native ferredoxin	100	2.0	14.5
	modified ferredoxin	20	84	100
Nitrite reductase	native ferredoxin	100	20	0.63
	modified ferredoxin	20	20	100

Table 1 also shows that carboxyl group modification of ferredoxin impaired its ability to serve as an electron donor to both nitrite reductase and glutamate synthase.

4. DISCUSSION

The above results provide evidence that nitrite reductase and glutamate synthase, like FNR, form electrostatic complexes with ferredoxin. Similar complexes may be involved in the reactions of all plant enzymes that use ferredoxin as an electron donor. The data implicate ferredoxin carboxyl groups in the binding to all three chloroplast enzymes, raising the possibility of a common binding site on ferredoxin for all ferredoxin-dependent enzymes. Peptide mapping experiments [13] have identified the chemically modified amino acid residues in (GEE + EDC)-modified ferredoxin in the three regions of the protein shown in Fig. 4.

Fig 4. Location of possible sites of modified carboxyl groups on ferredoxin. The backbone of ferredoxin is based on the structure of <u>Spirulina platensis</u> ferredoxin but altered to bring it into agreement with the spinach ferredoxin sequence. The carboxyl groups most likely to have been modified by GEE + EDC are shown in black. Other carboxyls are cross-hatched. From Ref. 13.

It may be that one or more of these regions of ferredoxin define, at least in part, the binding site for ferredoxin-linked enzymes.

5. ACKNOWLEDGEMENTS
Supported by a grant (to D.K.) from the U.S. Dept. of Agric. (85-CRCR-1-1574)

REFERENCES
1 Shin, M. and Arnon, D. I. (1965) J. Biol. Chem. 240, 1405-1411
2 Hewitt, E. J. (1975) Ann. Rev. Plant Physiol. 26, 73-100
3 Lea, P. J. and Miflin, B. J. (1974) Nature 251, 614-616
4 Foust, G. P., Mayhew, S. G. and Massey, V. (1969) J. Biol. Chem. 244, 1932-1936
5 Hirasawa, M. and Knaff, D. B. (1985) Biochim. Biophys. Acta 830, 173-180
6 Shin, M. and Oshino, M. (1978) J. Biochem. 83, 357-361
7 Ida, S., Kobayakawa, K. and Morita, Y. (1976) FEBS Lett. 65, 305-38
8 Hirasawa, M. and Tamura, G. (1984) J. Biochem. 95, 983-994
9 Hirasawa, M. Boyer, J. M., Gray, K. A., Davis, D. J. and Knaff, D. B. (1986) Biochim. Biophys. Acta, in press
10 Chan, T.-M., Ulrich, E. L. and Markley, J. L. (1983) Biochem 22, 6002-6007
11 Zanetti, G., Aliverti, A. and Curti, B. (1984) J. Biol. Chem. 259, 6153-6157
12 Vieira, B. and Davis, D. (1985) Biochem. Biophys. Res. Commun. 129, 467-471
13 Vieira, B. and Davis, D. (1986) Biochim. Biophys. Acta, in press
14 Knaff, D. B., Smith, J. M. and Malkin, R. (1978) FEBS Lett. 90, 195-197
15 Shin, M. (1973) Biochim. Biophys. Acta 292, 13-19

THE FERREDOXIN-NADP REDUCTASE/BINDING PROTEIN COMPLEX : IMMUNOLOGICAL AND OTHER PROPERTIES

VALLEJOS,R.H., CHAN,R.L., CECCARELLI,E.A., SERRANO,A., SONCINI,F.C.

CENTRO DE ESTUDIOS FOTOSINTETICOS Y BIOQUIMICOS,CEFOBI(CONICET F.M. LILLO, U.N.ROSARIO). SUIPACHA 531, 2000 ROSARIO. ARGENTINA.

1. INTRODUCTION

Ferredoxin-NADP reductase is the final enzyme of the photosynthetic electron transport chain. It is associated with a trimer of a 17.5 kDa polypeptide in spinach and lettuce chloroplasts (1,2). The complex was purified and can be reversibly dissociated. It was suggested that the reductase is bound to thylakoids through the 17.5 kDa polypeptide acting as a binding protein. The reductase in the complex conserves allotopic properties of the membrane-bound enzyme.

In this work we show another allotopic property of the reductase, the temperature dependence of its diaphorase activity. We also report properties of antibodies against the reductase binding protein and evidence indicating that the binding protein is protruding from the thylakoid membrane.

2. PROCEDURE

2.1. Materials and methods

Spinach thylakoids (3), ferredoxin-NADP-reductase (4), ferredoxin (5), the reductase-binding protein complex and the binding protein (1) were prepared as described.

Diaphorase activity was measured essentially as described (6). NADP photoreduction was measured at 28ºC in a medium containing 5 mM $MgCl_2$ 250 mM sucrose, 0.5 mM NADP, 20 µM ferredoxin, 50 mM Tricine-NaOH (pH 8) and chloroplasts corresponding to 10 µg chlorophyll. The reaction was followed spectrophotometrically by the increase in absorbance at 359 nm.

Fab fragments of IgG anti binding protein were obtained according to (7). For the agglutination tests, thylakoids (5 µg of chlorophyll) were incubated with 5 µl of IgG (titer 1:16) for 5 min at 25ºC in a humidity chamber and observed in a microscope.

All reagents were of analytical grade.

3. RESULTS AND DISCUSSION

3.1. Diaphorase activity of the spinach ferredoxin-NADP reductase was measured as a function of temperature (Fig.1). Soluble reductase shows a linear Arrhenius plot (Fig.1A) while the purified complex reductase binding protein (Fig.1B) and the thylakoid membranes (Fig.1C) present a break in the Arrhenius plot between 25-30ºC. ΔH* values for thylakoids and the complex are lower than the value for the soluble enzyme, particularly above the transition temperature. These results suggest that the conformation of the membrane-bound enzyme and the complex is more favourable to the reaction. Almost the same difference is observed when comparing the ΔH* values below and above the transition temperature for the thylakoids and the complex meaning that the transition is similar in both cases and does not require the membrane structure, that is the break in the Arrhenius plot seems to be a consequence of the interaction between the reductase and its binding protein.

Biggens, J. (ed.), Progress in Photosynthesis Research, Vol. III. ISBN 90 247 3452 5
© *1987 Martinus Nijhoff Publishers, Dordrecht. Printed in the Netherlands.*

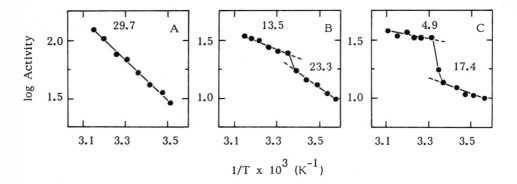

Figure 1 : Arrhenius plots of diaphorase activity of soluble ferredoxin-NADP reductase (A), purified reductase-binding protein complex (B), and thylakoid-bound reductase (C). Activity measurements were done with 1 µg FNR, 3 µg complex, or thylakoids corresponding to 3.5 µg chlorophyll respectively. The values adjacent to the plotted lines are the calculated ΔH* in kJ/mol.

3.2. Spinach thylakoids were agglutinated by the anti reductase serum forming dense agglutination stacks (Table I). These data are consistent with the localization of the flavoprotein on the outer surface of the thylakoids as shown by Bohme (8). When anti binding protein serum was mixed with thylakoids very little microagglutination was observed. However, when thylakoids were washed with 1 mM EDTA and thus depleted of reductase, agglutination with formation of dense stacks was observed (Fig.2). These results indicate that the absence of reductase increases the exposition of the binding protein, providing additional evidence that the 17.5 kDa polypeptide is the reductase binding protein.

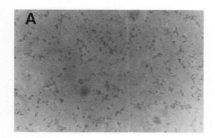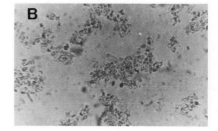

Figure 2 : Microphotography of EDTA particles treated with (A) anti FNR, and (B) anti reductase binding protein (1 x 400).

TABLE I : Agglutination tests of thylakoids with antibodies against reductase and its binding protein.

Experiment	Addition	Agglutination reaction
1	non immune serum	–
	anti FNR	+ + +
	anti binding protein	+/–
2	non immune serum	–
	anti FNR	–
	anti binding protein	+ + +

Experiment 1 : thylakoids were washed with 50 mM Tris-HCl (pH 8), 5 mM magnessium chloride. Experiment 2 : thylakoids were washed with 1 mM EDTA. Thylakoids were incubated with antibodies as described in Materials and methods.

3.3. Fab fragments of IgG anti binding protein inhibited NADP photoreduction by thylakoids in a time and concentration dependent way (Table II). This inhibition was affected by the concentration of ferredoxin present. The final extent of inactivation was the same for all the ferredoxin concentrations tested, but the rate increased when ferredoxin concentrations decreased.

Dissipation of the proton gradient by the addition of NH_4Cl slowed down the rate of inactivation of NADP photoreduction.

When thylakoids were preincubated with Fab in the dark the rate of inactivation was slower than in the light (not shown).

TABLE II : Inhibition of NADP photoreduction by Fab of IgG anti binding protein.

Fab added (µg)	Inhibition (%) after Fab addition	
	25 s	200 s
47	13	26
100	20	55
105	22	48
352	43	75
100, NH_4Cl (4 mM)	8	35
105, Fd (0.7 µM)	50	50

NADP photoreduction by spinach thylakoids was measured as described in the text for 1 min (controls were above 80 µmol NADPH/h.µg of chlorophyll). Then, Fab was added and the inhibition observed after 25 and 200 s is reported. When stated, Fd concentration was lower and NH_4Cl was added.

We can conclude that the reductase binding protein is protruding into the stroma, as shown by the agglutination experiments. Furthermore, even in the presence of reductase the binding protein is exposed to Fab fragments as revealed by inactivation of NADP photoreduction. Since the rate but not the extent of inactivation of NADP photoreduction by Fab depends on the concentration of ferredoxin, it is tempting to speculate that the reaction of Fab with the binding protein interferes with the formation of the ternary complex involving photosystem I-ferredoxin and the reductase-binding protein complex. Such a complex was proposed by Wagner et al. (9). Due to the effect of light it is reasonable to assume that energization of the membrane increased the accessibility of the binding protein to Fab, since this effect was abolished by the presence of ammonium chloride. These effects could be due to conformational changes of the binding protein caused by a light induced proton gradient.

4. ADDENDUM

4.1. Acknowledgements

This work was supported by research grants from the Consejo Nacional de Investigaciones Científicas y Técnicas, CONICET, Argentina. RHV is a member of the Career Investigator and RLC, EAC and AS are Fellows of the same Institution.

REFERENCES

1 Vallejos, R.H., Ceccarelli, E.A. and Chan, R.L. (1984) J. Biol. Chem. 259, 8048-8051
2 Ceccarelli, E.A., Chan, R.L. and Vallejos, R.H. (1985) FEBS Lett. 190, 165-168
3 Vallejos, R.H. (1973) Biochim. Biophys. Acta 292, 193-196
4 Carrillo, N. and Vallejos, R.H. (1983) Biochim. Biophys. Acta 742, 285-294
5 Buchanan, B.B. and Arnon, D.I. (1971) In Methods Enzymol. (Collowick, S.P. and Kaplan, O.K. eds.), Vol. 23, pp 413-440 Academic Press, New York
6 Zanetti, G. and Curti, B. (1980) In Methods Enzymol. (San Pietro, A. ed.) Vol. 69, pp 250-254, Academic Press, New York
7 Mage, M.G. (1980) In Methods Enzymol. (Collowick, S.P. and Kaplan, O.K. eds) Vol. 70, pp 142-150, Academic Press, New York
8 Bohme, H. (1978) Eur. J. Biochem. 84, 87-93
9 Wagner, R., Carrillo, N., Junge, W. and Vallejos, R.H. (1982) Biochim. Biophys. Acta 680, 317-330

The Ferredoxin/Thioredoxin System from the Green Alga, Chlamydomonas reinhardii

H.C. Huppe and B.B. Buchanan, Division of Molecular Plant Biology, University of California, Berkeley, CA   94720
F. Christophe de Lamotte-Guéry and J.-P. Jacquot, Physiologie Végétale Moléculaire, Universite de Paris-Sud, Orsay, France   91405

1.   INTRODUCTION
    Efficient utilization of light energy by plants involves the coordination of overall metabolism with the light environment. One of the ways in which plants couple light to metabolism is via the ferredoxin-thioredoxin system (FTS) (1). In chloroplasts and cyanobacteria, photoreduced ferredoxin reduces thioredoxins via ferredoxin-thioredoxin reductase (FTR). Reduced thioredoxins regulate a number of chloroplastic enzymes including members of the reductive pentose phosphate cycle. Two types of ferredoxin linked thioredoxins, m and f, have been designated according to their ability to activate chloroplastic NADP-malate dehydrogenase (NADP-MDH) and chloroplastic fructose-1,6-bisphosphatase (FBPase), respectively. At night, green plants utilize carbon stored in chloroplasts during daylight. A tight coordination of metabolism to light environment via FTS is critical to prevent futile cycling of carbon resources. An interesting question arises concerning green algae which can grow both phototrophically and heterotrophically. Previous work on green algae has shown the presence of an NADP-thioredoxin system in which thioredoxin is reduced via NADP-thioredoxin reductase (NTR). However, the FTS has not been investigated in these organisms (2,3). We have purified the thioredoxins and FTR from Chlamydomonas reinhardii and compared these proteins to those of higher plants and cyanobacteria. We have, also, investigated the properties of the target enzyme for thioredoxin m, Chlamydomonas NADP-MDH.

2.   METHODS
    Phototrophically grown Chlamydomonas cells were cultured in 20 liter carboys for one week in continuous light. The heterotrophic cultures were maintained for 14 days in the dark on media supplemented with acetate. Cell-free extracts were prepared by sonication in a solution containing 100 mM Tris-HCl (pH=7.9) or 20 mM potassium phosphate (pH=7.0), 3 mM MgCl2, and 1 mM PMSF. Purification was accomplished by ion exchange, affinity and molecular sieving columns using gravity flow and Pharmacia Fast-Phase Liquid Chromatography (FPLC). Thioredoxin f and m activities were monitored by spinach chloroplast FBPase and corn chloroplast NADP-MDH assays, respectively. All thioredoxins, enzymes and antibodies used were prepared in our laboratories by published methods, except for E. coli thioredoxin which was a gift from Dr. D. LeMaster (4). Reagents used in electrophoresis, Western blot, and the enzyme-linked immunosorption assay (ELISA) were purchased from BioRad.

3.   RESULTS AND DISCUSSION
    Separation of Chlamydomonas cell-free extracts on DE52 cellulose at pH 7.9 revealed two m thioredoxins and one f thioredoxin (Fig.1). The thioredoxin profile did not change significantly when cells were grown under heterotrophic conditions, but some quantitative differences may exist.

Biggens, J. (ed.), Progress in Photosynthesis Research, Vol. III. ISBN 90 247 3452 5
© 1987 Martinus Nijhoff Publishers, Dordrecht. Printed in the Netherlands.

The thioredoxin <u>m</u> which passed through DE52 (pH= 7.9) was designated <u>m1</u> and the adherent activity, <u>m2</u>. After molecular sieving and FPLC chromatography, both thioredoxins <u>m1</u> and <u>m2</u> appeared as single bands at 10.5 kDa on SDS PAGE.

FIGURE 1.
Elution Profile of <u>Chlamydomonas</u> Thioredoxins

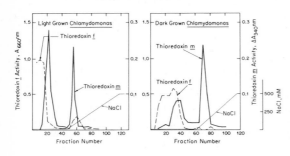

The <u>f</u> activity divided into two peaks when chromatographed on the cation exchange column, FPLC Mono S (pH= 6.0). Because of instability we have not completely purified the thioredoxin <u>f</u>. On SDS PAGE, the <u>f</u> thioredoxins from Mono S had a major component at 11 kDa. The behavior of the <u>Chlamydomonas</u> <u>f</u> thioredoxins on the molecular sieving column Sephadex G-75 supports an estimation in this size range. The size of the <u>Chlamydomonas</u> <u>f</u> is typical of thioredoxins, but indicates that it is different from the larger <u>f</u> thioredoxins reported for the green alga, <u>Scenedesmus</u>, and the cyanobacterium, <u>Anabaena</u> (28,000 and 26,000, respectively) (5,6).

Immunological studies of the thioredoxins are summarized in Table 1. The two <u>m</u>-type <u>Chlamydomonas</u> thioredoxins were similar to their counterparts in higher plants and cyanobacteria, cross reacting with antibodies to higher plant thioredoxin <u>m</u> and to <u>E. coli</u> thioredoxin in both ELISA and Western blot.. Despite their similarity in size with the thioredoxins of higher plants, the <u>Chlamydomonas</u> <u>f</u> thioredoxins did not react with antibodies to either thioredoxin <u>f</u> or <u>m</u> from higher plants. It reacted slightly with anti-<u>E. coli</u> thioredoxin antibody in ELISA. The cross reaction of the anti-<u>E. coli</u> antibody with <u>Chlamydomonas</u> <u>f</u> was not sufficient to be visualized by Western blot.

TABLE 1: Immunological Cross-Reactivity of Thioredoxins from <u>Chlamydomonas</u>, Higher Plants and <u>Nostoc</u>

	THIOREDOXINS					
ANTIBODIES	Higher Plant		Chlamydomonas			Nostoc
	<u>f</u>	<u>m</u>	<u>m1</u>	<u>m2</u>	<u>f</u>	<u>m</u>
anti-Thioredoxin <u>m</u>	−	+	+	+	−	+
anti-Thioredoxin <u>f</u>	+	−	−	−	+	−
anti-<u>E. coli</u> thioredoxin	−	+	+	+	+/−	+

The FTR from <u>Chlamydomonas</u> was purified to homogeneity by ion exchange and affinity chromatography. The absorbance spectrum peaks at 278 nm and 410 nm indicate that the <u>Chlamydomonas</u> FTR has an Fe-S center. As in chloroplasts and cyanobacteria, the <u>Chlamydomonas</u> FTR showed two bands on SDS-PAGE (7). The larger subunit of the algal protein had a molecular size similar to that of the large subunit of the <u>Nostoc</u> FTR and the small subunit of the spinach FTR (Fig. 2). Western blots with antibodies to corn and <u>Nostoc</u> FTR showed antigenic homology between these subunits.

SDS-PAGE of FTR

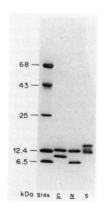

kDa Stds <u>C</u> <u>N</u> S

FIGURE 2. The Laemmli gradient gel (10-17%) was loaded with 4 ug each of purified FTR from <u>Chlamydomonas</u> (C), <u>Nostoc</u> (N), and spinach (S). The molecular weight standards are bovine serum albumin (68 kDa), ovalbumin (43 kDa), chymotrypsinogen (25 kDa), cytochrome c (12.4 kDa) and bovine trypsin inhibitor (6.4 kDa). Gel was stained with Coomasie blue.

We have established that <u>Chlamydomonas</u> thioredoxins and FTR are active in the light activation assays; thioredoxin $\underline{m}$ activated the <u>Chlamydomonas</u> NADP-MDH and thioredoxin $\underline{f}$ activated the spinach chloroplastic FBPase (Table2). The NADP-MDH, presumably of chloroplastic origin in <u>Chlamydomonas,</u> differs from its counterpart in higher plant in that it was partially activated by the monothiols 2-mercapthoethanol and reduced glutathione (8). The autologous FBPase system has not yet been studied.

Table 2: Ferredoxin-Thioredoxin Light Activation System of <u>Chlamydomonas</u>

	NADP-MDH Activity umoles NADPH ox.·min^{-1}·mg chl^{-1}	FBPase Activity umoles FBP·min.$^{-1}$·mg chl^{-1}
Light		
Complete	12.50	1.55
minus Ferredoxin	0.36	0.31
" Thioredoxin	0.59	0.30
" FTR	0.98	0.06
" enzyme	0.39	n.d.
Dark		
Complete	0.20	0.26

4. CONCLUSION

The <u>Chlamydomonas</u> FTS, like that of other oxygenic photosynthetic organisms, consists of an FTR, ferredoxin and both thioredoxins $\underline{m}$ and $\underline{f}$. Unlike previous results with higher plant tissue culture, the <u>Chlamydomonas</u> thioredoxins did not qualitatively change when cells were grown heterotrophically (9).

The proteins of the FTS in <u>Chlamydomonas</u> resembled their counterparts in other oxygenic photosynthetic organisms. The size and pattern of immunological reactivity with antibodies to thioredoxins was similar for the m thioredoxins from Chlamydomonas, cyanobacteria and plants. The Chlamydomonas f thioredoxins differed from f thioredoxins which have been reported for higher plants, cyanobacteria and other algae. The f thioredoxins of Chlamydomonas did not react with antibodies to spinach f and the size estimate of 11,000 indicates a smaller size than the f thioredoxins of Scenedesmus and Anabaena (5,6). Sequence and immunological analysis of the bacterial and m thioredoxins supports a close relation among these proteins. The f thioredoxins apparently vary substantially from the m-type proteins and from each other (10). More complete comparisons of the f thioredoxins from various organisms will be necessary to understand the basis for the similar function of these structurally diverse group of proteins.

FTR from Chlamydomonas resembles previously descibed FTR proteins (7). It has an Fe-S center and consists of two subunits, one of which reacts with the antisera to corn FTR. The subunit sizes of the Chlamydomonas FTR more closely resembles the pattern of Nostoc than of the higher plant reductase.

An interesting variation from most thioredoxin regulated systems is the activation of the target enzyme chloroplast NADP-MDH from Chlamydomonas by monothiols. Activation of the NADP-MDH from higher plants requires dithiols. Whether this property extends to other target enzymes in Chlamydomonas is under investigation.

We have found that Chlamydomonas has an FTS similar to that of other oxygenic photosynthetic systems. The importance of the FTS to Chlamydomonas is indicated by the fact that the alga maintains all components of the system even when grown heterotrophically. Because of its ability togrow both heterotrophically and phototrophically, Chlamydomonas can survive in environments where light availabity is variable. Retention of the FTS would enable the alga to respond quickly to light should it become available, allowing efficient utilization of energy resources.

References
1. Buchanan, BB. (1984) Biosc. 34, 378- 393.
2. Tsang, M.L.-S. (1981) Plant Physiol. 68, 1098- 1104.
3. Munavalli, S., Parker, D.V. and Hamilton F.D. (1975) Proc. Nat. Acad. Sci. USA. 72, 4233- 4237.
4. Crawford N.A., Yee B.C., Hutchinson S.W., Wolosuik R.A. and Buchanan B.B. (1986) Arch. Bioch. Biop. 244, 1- 15.
5. Langlotz P., Wagner W., and Follman H. (1986) Z. Naturforsch. 41c, 275- 283.
6. Whittaker M.W. and Gleason F.K. (1984) J. Biol. Chem. 259, 14088-.
7. Droux, M., Jacquot, J.-P., Miginiac-Maslow, M.,Gadal, P., Crawford, N.A. Yee,B.C. and Buchanan, B.B. (1986) Seventh International Congress on Photosynthesis, Providence, RI, USA, Abstract.
8. Jaquot, J.-P. (1983) Thèse Doct. Etat. Université Paris-Sud, Orsay.
9. Cao, R.Q.,Johnson, T.C. and Buchanan, B.B. (1983) Sixth International Congress on Photosynthesis, Brussels, Belgium, Abstract 415-9.
10. Tsugita, A., Maeda K., and Schürmann, P. (1983) Biochem. Biophys. Res. Commun. 115, 1- 7.

(Supported by a grant from NASA)

THERMODYNAMICS AND KINETICS OF THE INTERACTIONS OF THIOREDOXIN F_b WITH
FRUCTOSE BISPHOSPHATASE FROM SPINACH CHLOROPLAST.

Jean BUC, Jean-Michel SOULIE, Mireille RIVIERE, Brigitte GONTERO and
Jacques RICARD. Centre de Biochimie et de Biologie Moléculaire, CNRS,
31, Chemin Joseph Aiguier, 13402 Marseille Cedex 9 (France).

One of the most important process involved in the photoregulation of
chloroplast enzymes is their oxidoreductive modification by protein mediators known as thioredoxins (1). Different molecular species of thioredoxins
from Spinach chloroplast have been purified and characterized (2). The reductive activation of fructose bisphosphatase is due to the reduction of two
disulfide bridges followed by a conformation change of the protein (3). Very
little was known, till recently about molecular interaction occuring between
the enzyme and its protein effector. We discuss here the binding properties
and the kinetics of inactivation of fructose bisphosphatase (FBPase) by
thioredoxin f_b.

RESULTS AND DISCUSSION
125 I-Thioredoxin f_b was prepared without loss of activity according to
Greenwood and al. (4) and binding experiments were performed with the Hirose
and Kano method (5). A typical binding isotherm is shown in figure 1.

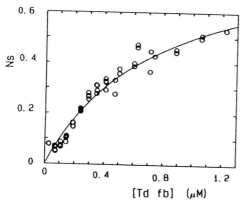

Figure 1 : 125 I-Thioredoxin f_b
isotherm to chloroplastic FBPase.
Experimental points are obtained
using phosphate buffer 5 mM, EDTA
0.5 mM, pH 7.5 and 422 nM FBPase.
The continuous line is the theoretical curve corresponding to the
best fit of the experimental data
to the hyperbola equation.

The experimental results show that
at saturating ligand concentrations
one molecule of dimeric thioredoxin f_b is bound per tetrameric
FBPase. This stoichiometry is constant whatever the medium conditions
(Table 1). Kinetically saturating
concentrations of fructose bisphosphate (FBP) do not change significantly
the affinity of thioredoxin for the enzyme. The dissociation constant of the
thioredoxin FBPase complex decreases more than 6 times, upon increasing
the pH from 7 to 8 which are the pHs of chloroplast stroma in dark and light
conditions. Dissociation constants of the protein-protein complex increase
upon raising the ionic strength. However the variation of the dissociation
constant is linear with respect to neither the ionic strength nor to its
square root. This suggests that the interaction between thioredoxin and
FBPase involves both ionic and hydrophobic contribution. The existence of
this non ionic contribution may be confirmed by studying the effect of to-
luydinyl naphtalene sulfonate (TNS) on enzyme activation. This hydrophobic

Biggens, J. (ed.), Progress in Photosynthesis Research, Vol. III. ISBN 90 247 3452 5
© *1987 Martinus Nijhoff Publishers, Dordrecht. Printed in the Netherlands.*

probe does not affect FBPase activity but significantly slowers its activation by thioredoxin (Figure 2). If reduced thioredoxin may activate oxidized FBPase, oxidized thioredoxin may deactivate its target enzyme. Tian and Tsou (6) have recently presented a treatment for the determination of the rate constant of enzyme irreversible chemical modification by measuring the substrate reaction in the presence of the modifier. By applying this technique to thioredoxin f_b and FBPase we reduced a simple model of deactivation of FBPase by thioredoxin f_b. A typical progress curve of product appearance, during enzyme reaction in the absence and in the presence of oxidized thioredoxin f_b is shown in figure 3.

Conditions	pH	Ionic strength (M)	NS max	Kd (nM)
1452 nM FBPase	7.5	0.0085	1.031 ± 0.122	713 ± 129
422 nM FBPase	7.5	0.0085	0.877 ± 0.068	769 ± 100
3.8 mM Mg	7.5	0.0091	1.108 ± 0.102	1108 ± 282
2 mM FBP	7.5	0.0117	0.727 ± 0.108	872 ± 185
0.13 mM TNS	7.5	0.0085	0.916 ± 0.100	1365 ± 320
	7	0.0085	0.979 ± 0.134	1636 ± 390
	8	0.0085	1.104 ± 0.107	244 ± 49
	7.5	0.0160	0.953 ± 0.123	1098 ± 257
	7.5	0.0504	1.020 ± 0.322	1830 ± 1050
	7.5	0.0807	1.447 ± 0.492	2399 ± 1117
	7.5	0.1784	1.178 ± 0.422	2776 ± 1378

Table 1 : Binding of thioredoxin f_b to FBPase.

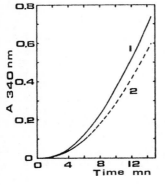

Figure 2 : Effect of TNS on the reductive activation of FBPase by thioredoxin f_b. Thioredoxin prereduced was brought to a concentration of 414 nM in a reaction mixture containing in 100 mM Tris/Cl buffer pH 7.5 at 30°C, 0.5 mM EDTA and 2.5 mM FBP, 10 mM $MgCl_2$, 0.7 U/ml phosphoglucoisomerase, 0.35 U/ml glucose-6-phosphate dehydrogenase, 0.13 mM NADP and 0.5 mM DTT (curve 1) plus 0.13 mM TNS (curve 2). FBPase was then added to a concentration of 1.47 nM and the reaction monitored at 340 nm. Controls under similar conditions, but at pH 8.8 in the absence of thioredoxin and DTT, show no lag and no difference between the assays with or without TNS.

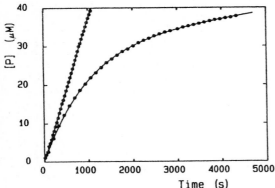

Figure 3 : Time course of product formation by reduced FBPase. The reaction mixture contained 0.34 nM FBPase, $MgCl_2$ 10 mM and FBP 0.375 mM at pH 7.5 and 30° C. Open circles : in the absence of thioredoxin the experimental data fit a slope of 37.2 nM s^{-1}. Full circles : in presence of 2.27 µM oxidized thioredoxin. The parameters fitted to $P = at + b (1 - e^{-\lambda t})$ are a = 1.75 nM s^{-1}, b = 30.6 µM, $\lambda = 1.01$ 10^{-3} s^{-1}.

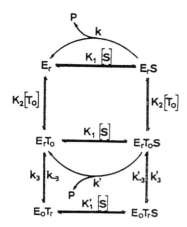

Figure 4 : Minimum kinetic model for deactivation of reduced FBPase by oxidized thioredoxin f_b. E_O and E_r stand for oxidized and reduced enzyme, T_O and T_r for oxidized and reduced thioredoxin. S and P are the substrate and product.

Figure 5 : Initial velocity of FBPase versus oxidized thioredoxin concentration. Initial velocity are calculated from parameters fitted to equation of figure 3 for inactivation progress curves in presence of 1.875 mM and varying oxidized thioredoxin f_b concentration. The simulated curve is the best fit of experimental data to
$$V = (1 + a \, [T_o])/(b + c \, [T_o])$$

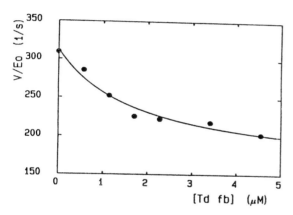

In the time scale investigated in the absence of thioredoxin, the progress curve is always a straight line. If oxidized thioredoxin is present in the reaction mixture, the progress curve can be fitted neither to a straight line nor to a exponential but by a combination of these two functions. The simplest model that may generate this kind of behaviour is shown in figure 4. This model embodies the view that the enzyme and the enzyme substrate complex bind thioredoxin at a thioredoxin binding site with the same affinity constant.

By estimating the initial velocity of FBPase in the presence of oxidized thioredoxin, we determined the effect of the protein effector on the enzyme under quasi equilibrium conditions. When FBPase follows Michaelis-Menten kinetics thioredoxin behaves like a noncompetitive inhibitor of the reaction. This implies that at fixed substrate concentration, the reaction rate, under quasi equilibrium conditions, hyperbolically declines as a function of oxidized thioredoxin concentration (Figure 5). Integration of the model (Figure 4) allows to derive the equation of product appearance :

$$\frac{[P]}{[E]_o} = \frac{K_1 [S] (k + k' K_2 [T_o])}{(1 + K_2 [T_o])(1 + K_1 [S])} \frac{\lambda_{-1}}{\lambda} t + \frac{\lambda_1}{\lambda^2} (1 - e^{-\lambda t})$$

$$\lambda = \lambda_1 + \lambda_{-1}$$

$$\lambda_1 = \frac{K_2 [T_o] (k_3 + k'_3 \, K_1 \, [S])}{(1 + K_2 \, [T_o])(1 + K_1 \, [S])} \qquad \lambda_{-1} = \frac{k_{-3} + k'_{-3} \, K'_1 \, [S]}{1 + K'_1 \, [S]}$$

$k = 498 \text{ s}^{-1}$	$k' = 256 \text{ s}^{-1}$	$k_3 = 2.7 \times 10^{-3} \text{ s}^{-1}$
$K_1 = 844 \text{ M}^{-1}$	$K_2 = 495 \times 10^3 \text{ M}^{-1}$	$k'_3 = 0.8 \times 10^{-3} \text{ s}^{-1}$
	$k_{-3}, k'_{-3} \ll k_3, k'_3$	

Table 2 : Kinetic and equilibrium parameters for inactivation of reduced FBPase by oxidized thioredoxin. These data are calculated from the presented equations by using the parameters fitted to experimental data.

The plot of time constant for a given substrate concentration as a function of thioredoxin (Figure 6) shows that for $[T_o] = 0$ the λ value of is close to 0 which implie that $\lambda \gg \lambda_{-1}$. This result is consistent with our previous proposal that deactivation of FBPase by thioredoxin is thermodynamically favoured with respect to the reverse process.

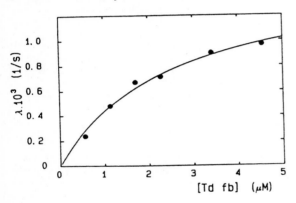

Figure 6 : Variation of the time constant of inactivation exponential as a function of thioredoxin concentration. The simulated curve is the best fit of the experimental data to an hyperbola.

It also supports the view that the deactivation of enzyme is preceded by a fast binding of thioredoxin to the enzyme. All the other predictions of the presented model have been verified experimentally and its kinetic and equilibrium parameters have been determined (Table 2). The present results show that thioredoxin binds to a specific site distinct from that of FBP and may be viewed as a regulatory subunit of FBPase.

REFERENCES

1 Soulié, J-M. Buc, J., Meunier, J-C., Pradel, J. and Ricard, J. (1981) Eur. J. Biochem., 119, 497-502.
2 Buc, J., Rivière, M., Gontero, B., Sauve, P., Meunier, J-C. and Ricard, J. (1984) Eur. J. Biochem., 140, 199-202.
3 Pradel, J., Soulié, J-M., Buc, J., Meunier, J-C. and Ricard, J. (1981) Eur. J. Biochem., 113, 513-520.
4 Greenwood, F.C., Hunter, W.M. and Clover, J.S. (1963) Biochem. J., 89, 114-123
5 Hirose, M. and Kano, Y. (1971) Biochim. Biophys. Acta, 251, 376-379.
6 Tian, W.X. and Tsou, C.L. (1982) Biochemistry, 21, 1028-1032.

THE THIOREDOXIN BINDING SITE OF PHOSPHORIBULOKINASE OVERLAPS THE
CATALYTIC SITE

Michael A. Porter and Fred C. Hartman, Biology Division, Oak Ridge
National Laboratory and The University of Tennessee-Oak Ridge Graduate
School of Biomedical Sciences, Oak Ridge, Tennessee 37831 U.S.A.

1. INTRODUCTION

Phosphoribulokinase (PRK[1], E.C. 2.7.1.19) is one of a class of
enzymes whose activity is linked to the light/dark cycle by thioredoxin,
ferredoxin-thioredoxin reductase, and ferredoxin (1). Although enzyme
activation is known to require disulfide reduction, the location of the
disulfide at the primary structural level has not been established for PRK
or for any other plant enzyme that is regulated in this fashion. Previous
studies with BrAcNHEtOP, a potential active-site probe, identified an
essential cysteinyl residue within the ATP-binding domain of PRK. Two
observations suggested a connection between the active-site cysteine and a
thioredoxin-sensitive cysteine. Oxidized PRK is not alkylated by
BrAcNHEtOP (2) and ATP retards the oxidative deactivation of PRK (3,4).
More recent data, summarized here, confirm that the active-site sulfhydryl
also participates in the regulation of kinase activity.

2. PROCEDURES

2.1. Materials: PRK was purified from spinach leaves using successive gel
   filtration, ion exchange chromatography, and affinity chromatography
   on reactive red agarose (4). BrAcNHEtOP was synthesized as described
   earlier (5). [^{14}C]IAA from ICN was diluted with unlabeled IAA
   (Aldrich) as needed.

2.2. Methods: PRK activity was determined by a coupled spectrophotometric
   assay (4). PRK was alkylated at 4°C in 45 mM bicine-KOH, 4.5 mM
   MgCl$_2$, 0.9 mM EDTA, 10% v/v glycerol with the adjustments described.
   Oxidations were conducted in the same basic buffer but at room
   temperature. Prior to oxidation experiments, the kinase was desalted
   by gel filtration in the desired buffer to remove the exogenous DTT
   present in the storage buffer. Tryptic digestions, HPLC chromatography, amino acid compositions and automated sequence analyses were
   accomplished by standard methods (4). The carboxymethylation
   procedure (6) was altered to use IAA in only a 10 mM excess over
   exogenous thiol, and guanidine was not included in the reaction
   mixture when carboxymethylation of the native enzyme was desired.

---

[1]Abbreviations: PRK, phosphoribulokinase; Ru5P, ribulose 5-phosphate;
IAA, iodoacetic acid; DTT, dithiothreitol; BrAcNHEtOP, bromoacetyl-
ethanolamine phosphate; DHA, dehydroascorbate.

Biggens, J. (ed.), Progress in Photosynthesis Research, Vol. III. ISBN 90 247 3452 5
© *1987 Martinus Nijhoff Publishers, Dordrecht. Printed in the Netherlands.*

3. RESULTS AND DISCUSSION

BrAcNHEtOP completely inactivates PRK concomitant with the incorporation of only one molar equivalent of reagent, corresponding to the alkylation of a cysteinyl residue. This implied high degree of specificity is
confirmed by HPLC of a tryptic digest and by gel electrophoresis under
non-denaturing conditions of the partially modified dimeric protein.
Because ATP provides protection against alkylation and because the
concentration dependency of the protection indicates a K_D (29 µM) similar
to its K_m (50 µM), the reactive cysteine appears to be located within the
ATP-binding domain of the active site. The tryptic peptide containing the
reactive cysteine has been purified by ion exchange chromatography
followed by reverse phase HPLC. Edman degradation on a gas phase
sequencer (Applied Biosystems) establishes the following primary structure
(the asterisk denotes the modified residue),

```
                5                    10                    15
    Ser-Gln-Gln-Gln-Thr-Ile-Val-Ile-Gly-Leu-Ala-Ala-Asp-Ser-Gly-

    Cys*-Gly-Lys
```

which corresponds exactly to the amino terminal region of the protein (4),
in which cysteine was positively identified at position 16. Prior
oxidation of the kinase prevents incorporation of BrAcNHEtOP, suggesting
that the reactive cysteine is a possible thioredoxin-regulatory target
residue. This supposition is supported by the observation that ATP
retards the oxidation of the kinase by oxidized glutathione, molecular
oxygen (3,4), and oxidized thioredoxin. We investigated this effect in
detail by determining the oxidation rate of the kinase by DHA in the
presence of various concentrations of ATP. The calculated K_D for ATP is
23 µM, a value similar to that revealed by the kinetics of ATP protection
against the alkylation of PRK. A single ATP binding site is thus invoked
to account for the protective effects observed in both processes. Because
both alkylation and oxidation of sulfhydryls require the ionized form,
these reactions will show pH dependencies which reflect the pK_a of the
target sulfhydryl. Both alkylation and oxidation rates of PRK were
determined from pH 6.8-8.3 at fixed ionic strength. Both processes reveal
an inflection centered at about pH 7.8, reinforcing the above postulate
that the same sulfhydryl group is the target of both oxidation and
alkylation. The parallels between the alkylation and oxidation of PRK are
summarized in Table 1.

TABLE 1. Parallels between alkylation and oxidation

Measurement	Reaction examined	
	Alkylation	Oxidation
K_D (ATP) of protection	29 ± 4 µM	23 ± 5 µM
Apparent pK_a of reactive group	7.9 ± 0.2	7.7 ± 0.15
ADP protection	complete	complete
Ru5P protection	partial	partial

The proposal that the alkylating reagent and oxidant are reacting at a single sulfhydryl in PRK was tested directly by observing the carboxy-methylation pattern after various treatments. Carboxymethylation of the enzyme under denaturing conditions with [^{14}C]IAA reveals four peaks of radioactivity upon HPLC of a tryptic digest (Fig. 1B). Prior alkylation with BrAcNHEtOP prevents the appearance of peak II (Fig. 1F), establishing that this peak represents the active-site cysteinyl residue. As expected from the protection pattern, inclusion of ATP during carboxymethylation prevents labeling of the cysteinyl residue represented by peak II (Fig. 1D). Prior oxidation of the kinase by DHA also diminishes the level of peak II (Fig. 1C,E), confirming a role of one sulfhydryl group in both catalytic and regulatory functions. The level of peak III in tryptic digests is also less in the DHA-oxidized kinase (Fig. 1E), thereby identifying the second cysteinyl residue involved in disulfide bond formation. This finding excludes the possibility that the disulfide bond is comprised of Cys-16 from two different subunits. A final interesting observation is that one of the regulatory sulfhydryls (Cys-16, represented by peak II) is exposed to solvent (_i.e._ accessible to IAA, Fig. 1A) while the other (represented by peak III) is buried (_i.e._ not accessible to IAA, Fig. 1A).

FIGURE 1. Radioactivity profiles of [^{14}C]carboxymethyl-PRK after tryptic digestion. PRK was alkylated with [^{14}C]IAA, digested with trypsin, and subjected to reverse phase HPLC (4,6). Treatments were: A, no guanidine, 10 mM DTT; B, guanidine, 10 mM DTT; C, no guanidine, 10 mM DTT (carboxymethylated in the presence of ATP); E, guanidine, prior oxidation; F, guanidine, 10 mM DTT, (prior alkylation with BrAcNHEtOP).

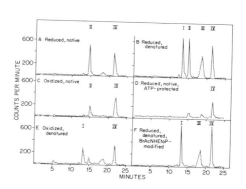

The finding that two cysteines per subunit undergo oxidation during deactivation of PRK (Fig. 1E) is consistent with earlier evidence (4) that the disulfide bond formed is intrasubunit, rather than intersubunit or intermolecular. Further support excluding intermolecular oxidation is found in the failure of gel filtration in the presence of 6 M urea to resolve oxidized and reduced kinase.

4. CONCLUSIONS

4.1. BrAcNHEtOP alkylates a single essential sulfhydryl (Cys-16) in PRK.

4.2. The reactive sulfhydryl is within the ATP-binding domain of the enzyme and is exposed to solvent.

4.3. The reactive sulfhydryl is involved in redox regulation of kinase activity.

4.4. The catalytic/regulatory site is comprised, in part, of the amino-terminal region of the protein.

4.5. The regulatory disulfide is intrasubunit.

5. REFERENCES
1 Buchanan, B.B. (1980) Annu. Rev. Plant Physiol. 31, 341-374
2 Omnaas, J., Porter, M.A. and Hartman, F.C. (1985) Arch. Biochem.
  Biophys. 236, 646-653
3 Ashton, A.R. (1983) in Thioredoxins, Structure and Functions. Proceedings of the conference held June 21-24, 1981, pp. 245-250, Berkeley
4 Porter, M.A., Milanez, S., Stringer, C.D. and Hartman, F.C. (1986) Arch.
  Biochem. Biophys. 245, 14-23
5 Hartman, F.C., Suh, B., Welch, M.H. and Barker, R. (1973) J. Biol. Chem.
  248, 8233-8239
6 Schloss, J.V., Stringer, C.D. and Hartman, F.C. (1978) J. Biol. Chem.
  253, 5707-5711

6. ACKNOWLEDGMENT
   Research sponsored jointly by the Science and Education Administration
of the U.S. Department of Agriculture under Grant No. 84-CRCR-1-1520 from
the Competitive Research Grants Office and by the Office of Health and
Environmental Research, U.S. Department of Energy under contract DE-AC05-
84OR21400 with the Martin Marietta Energy Systems, Inc. M.A.P. is a
Postdoctoral Investigator supported by the U.S. Department of Agriculture
grant through Subcontract No. 3322 from the Biology Division of the
Oak Ridge National Laboratory to The University of Tennessee.

CHARACTERIZATION OF THE THIOREDOXIN SYSTEM IN RHODOBACTER SPHAEROIDES Y.

J.D. CLEMENT-METRAL

Thioredoxin, a small ubiquitous protein having two redox-active half cystine residues in an exposed active center, with the sequence -Cys-Gly-Pro-Cys-, participates in redox reactions through the reversible oxidation of its active center dithiol to a disulfide, and catalyses dithiol-disulfide exchange reactions. (1).

The first evidence of the existence of the thioredoxin system (NADPH, thioredoxin, thioredoxin reductase) in a photosynthetic organism came from Rhodobacter sph. Y (2). The presence of the thioredoxin system in a photosynthetic bacterium is of great interest, as the regulation of pigment synthesis of Rhodobacter sph. has been discussed for a number of years in terms of the existence of a regulatory substance sensitive to the oxido-reduction state of the cell (3). In fact, the activity of 5-aminolevulinic acid synthetase, the key enzyme in tetrapyrrole synthesis has recently been shown to be possibly regulated by dithiol-disulfide interchange mediated by the thioredoxin system (4).

MATERIAL AND METHODS

Cells were grown and enzyme activities determined as described in (5, 6). The complete purification procedure of the two proteins : thioredoxin and thioredoxin reductase, is given in (6).

Fluorescence spectroscopy

Fluorescence of E. coli and Rhodobacter sph. thioredoxins 1.67 µM in 50 mM Tris-HCl, pH 7.0, 1mM EDTA was analyzed at 25°C in a modified Zeiss spectrofluorometer by excitation at 280 nm. Reduction of the oxidized proteins was achieved by addition of 3 mM DTT preincubated for 3 minutes. Solutions of denatured thioredoxin were equilibrated in 5 M guanidine-HCl , 50 mM Tris, pH 7.0, 1 mM EDTA, 25°C.

Peptides analysis

Rhodobacter sph. thioredoxin was reduced in 6 M guanidine-HCl and carboxymethylated with iodo (^{14}C) acetate, excess reagents removed by Sephadex G-25 chromatography in 0.5 % ammonium bicarbonate. Identification of the active site peptide containing the disulfide group was carried out by chymotryptic digestion of the carboxymethylated protein for 4 hours at 37°C in 0.1M ammonium bicarbonate at an enzyme-to-substrate ratio of 1 : 50. Peptides were separated by reverse-phase HPLC on a Waters µBondapack C 18 using acetonitrile gradient in 0.1 % trifluoroacetic acid for elution. Amino acid sequence of the peptide was determined by the dimethylazobenzene isothiocyanate method.

Biggens, J. (ed.), Progress in Photosynthesis Research, Vol. III. ISBN 90 247 3452 5
© *1987 Martinus Nijhoff Publishers, Dordrecht. Printed in the Netherlands.*

RESULTS AND DISCUSSION

The assay of Rhodobacter sph. thioredoxin during the purification used its cross-reactivity with E. coli NADPH thioredoxin reductase. Similarly, pure E. coli thioredoxin was used as the substrate in the assay for Rhodobacter sph. thioredoxin reductase.

An identical purification procedure was used for thioredoxin and thioredoxin reductase up to the affinity chromatography on 2'5'ADP Sepharose (disruption of the cells in French pressure cell; 2 centrifugations 40,000 g for 30 minutes, then 257,000 g for 3 hours; 80 % saturation ammonium sulfate precipitation; dialysis; adsorption on ADP Sepharose).(6).

Purification and characterization of thioredoxin

The effluent from the ADP Sepharose column was subjected to 3 successive chromatographies : DEAE cellulose, G 50 and Mono Q HR 5/5 anion exchange column. Pure fractions of Mono Q run were homogeneous as judged by native polyacrylamide gel electrophoresis.

The molecular mass calculated from the amino acid analysis based on 2 half cystine residues is 10,800.

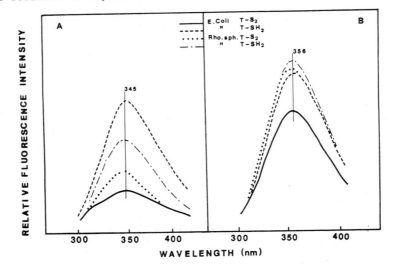

Fig.1: Fluorescence emission spectra of E. coli and Rhodobacter sph. thioredoxins (T) in oxidized (-S$_2$) and reduced (-SH$_2$) forms.
(A) native proteins in Tris-HCl, 1mM EDTA, pH 7.0
(B) same solvent containing 5 M guanidine-HCl.

The emission spectrum of the native oxidized Rhodobacter sph. protein (fig 5 A) has a maximum at 345 nm, in the range anticipated for tryptophan residues. The fluorescence intensity of these residues is quenched in E. coli thioredoxin; on reduction it increases about 3 fold. In Rhodobacter sph. thioredoxin a much smaller increase (1.5) is observed.However the presence of 5 M guanidine-HCl results, as in E. coli,in the complete exposure of the two tryptophan. These data indicate that the second tryptophan residue is not in the same environment in the two proteins and that the conformational change in the vicinity of the S-S bridge on reduction does not lead to the exposure of the second tryptophan to the solvent in Rhodobacter sph. native thioredoxin.

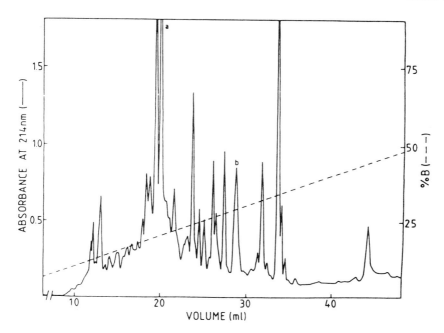

Fig.2: HPLC of the peptides from a chymotryptic digest of ^{14}C-carboxymethy-
lated thioredoxin from <u>Rhodobacter</u> <u>sph</u>. Flow rate 1ml.min[-1]
Peak a contains the tripeptide Ala-Glu-Trp
Peak b the active center peptide.

The amino acid sequence around the active center was determined after
reverse-phase HPLC separation of the chymotryptic peptides (fig.2). The 2
chymotryptic peptides are aligned after the amino acid sequence of E. <u>coli</u>
thioredoxin (fig.3). The corresponding sequence of <u>Anabaena</u> <u>sp</u>. thioredoxin
has been added for comparison.

<pre>
                                   S ─────────S
E. coli   (Trp)-Ala-Glu-Trp-Cys-Gly-Pro-Cys-Lys-Met-Ile-

                                   S ─── ─────S
Rhodo. sph.   (Trp)-Ala-Glu-Trp-Cys-Gly-Pro-Cys-Arg-Gln-Ile-

                                   S───────── S
Anabaena sp.   (Trp)-Ala-Pro-Trp-Cys-Gly-Pro-Cys-Arg-Met-Val-
</pre>

Fig.3: <u>Comparison of active site sequences of thioredoxins.</u>

The almost completed sequence (99 amino acid our of 104) of <u>Rhodobac-
ter</u> <u>sph</u>. thioredoxin shows a very high degree of homology with E. <u>coli</u>
(46 %), <u>Anabaena</u> <u>sp</u>. (48 %) and <u>Corynebacterium</u> <u>nephridii</u> (52 %).

Purification and characterization of thioredoxin reductase

Thioredoxin reductase activity was adsorbed on 2'5' ADP Sepharose column and eluted with 1 M NaCl. After dialysis and concentration two successive chromatographies (MonoQ column) led to pure thioredoxin reductase. The
gradients were : first linear Tris-HCl, pH 7.5, 1 mM EDTA, 100 to 750 mM,
    second non linear Tris-HCl, pH 7.5, 1 mM EDTA, 200 to 750 mM. Fractions
of the second anion exchange chromatography were homogeneous as judged by
SDS polyacrylamide gel electrophoresis.
    Rhodobacter sph. thioredoxin reductase has structural and fonctional
similarities to E. coli thioredoxin reductase : it has a molecular mass of
68 kDa, consists of two probably identical subunits. Each subunit has one
bound FAD molecule. The enzyme is highly specific for NADPH; it is also highly specific for Rhodobacter sph. thioredoxin with a Km value of 3.3 $\pm$ 0.6
μM. It is well known that the E. coli thioredoxin reductase has a very
pronounced specificity for the disulfide of E. coli thioredoxin (Km = 2.4 $\pm$
0.3 μM), when compared with other disulfide containing substances. It is interesting to note that the affinity of each reductase for the thioredoxin of
he other system is of the same order of magnitude; the two systems have therefore a high degree of cross-reactivity.

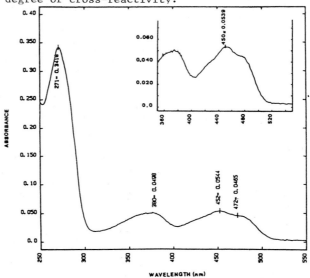

Fig.4: Absorption spectrum of Rhodobacter sph. thioredoxin reductase in 50
    mM Tris-HCl, 1mM EDTA, pH 7.5. A_{271}/A_{452} = 6.3

REFERENCES
1 Holmgren, A. (1985) Ann. Rev. Biochemistry 54, 237-271
2 Clement-Metral, J.D. (1979) FEBS Letters 101, 116-120
3 Cohen-Bazire, G., Sistrom, W.R. and Stanier, R.Y. (1957) J. Cell. Comp.
  Physiol. 49, 25-68
4 Clement-Metral, J.D. and Holmgren, A. (1983) in "Thioredoxins, structure
  and functions" (Gadal,P. ed) p. 59-68, CNRS, Paris
5 Clement-Metral, J.D. (1985) in "Thioredoxin and Glutaredoxin systems:
  structure and function"(Holmgren, A. et al. ed.) Raven Press, in press
6 Clement-Metral, J.D., HOOG, J.O. and Holmgren, A. (1986) Eur. J. Biochem.
  in press

ACTIVATION OF CHLOROPLAST NADP-GLYCERALDEHYDE-3-P DEHYDROGENASE BY CHAOTROPIC ANIONS AND DETERGENTS

Ricardo A. Wolosiuk, Mariana Stein and Liliana Busconi
Instituto de Investigaciones Bioquimicas, Antonio Machado 151, (1405)
Buenos Aires, Argentina

1. INTRODUCTION

Studies in vitro have shown that the NADP-glyceraldehyde-3-P dehydrogenase activity is enhanced by components indigenous to chloroplasts: low molecular weight effectors and reduced thioredoxin (1-4). We recently showed that, as had been found for chloroplast fructose-1,6-bisphosphatase, the modification of the milieu constitutes another way for changing the kinetic properties of the enzyme (5-7); organic solvents miscible in water (cosolvents) are functional in increasing the NADP-linked activity. Moreover, the increase of enzyme specific activity correlates with the hydrophobic character of the respective organic solvent; high concentrations of cosolvent are required for maximal stimulation when low is the octanol/water partition coefficient. These results suggest that slight modifications of the enzyme hydrophobic interactions have substantial effects in its activity. To explore this possibility the enzyme was (i) subjected to treatments aimed at modifying its intramolecular interactions, and (ii) analyzed for kinetic changes. In this context we studied on chloroplast glyceraldehyde-3-P dehydrogenase the influence of salts and detergents that are often used for perturbing the balance of interactions in proteins (8,9).

2. EXPERIMENTAL PROCEDURES

Chloroplast glyceraldehyde-3-P dehydrogenase was purified as described in (3). Although in our preparations (a) the ratio of A_{280} to A_{260} was 1.36 which indicated a 1% of nucleic acids (10), and (b) we did not detect RNA when the enzyme was electrophoresed in agarose gels (11), we devised further purification steps to circumvent any contaminating RNA (12). Therefore, we subjected chloroplast glyceraldehyde-3-P dehydrogenase to the following sequential treatments:
(i) Fractions of molecular weight 145.000, originated in a chromatography in a column of Sephacryl S-300 eluted with buffer A (50 mM Tris-HCl (pH 7.9), 10 mM EDTA, 1.4 mM 2-mercaptoethanol) containing 1 mM NAD, were pooled, concentrated by ultrafiltration, and
(ii) treated with RNAase,
(iii) dialyzed against buffer A containing 1 mM NADP, and
(iv) chromatographed in a column of Sephacryl S-300 (elution solution: buffer A containing 1 mM NADP).
(v) Fractions of molecular weight 600.000 were pooled, concentrated by ultrafiltration, dialyzed against buffer A, and assayed for chaotrope- and cetyltrimethylammonium bromide- mediated activation.

Biggens, J. (ed.), Progress in Photosynthesis Research, Vol. III. ISBN 90 247 3452 5

Chloroplast glyceraldehyde-3-P dehydrogenase was assayed at 23°C in a spectrophotometric cuvette (1-cm light path) which contained 3 units of 3-phosphoglycerate phosphokinase and the following (in micromoles): Tris-HCl buffer (pH 7.9), 50; MgCl$_2$, 10; 3-phosphoglycerate, 1, ATP, 1; and NAD(P)H, 0.1. Following the injection of chloroplast glyceraldehyde-3-P dehydrogenase changes in absorbance at 340 nm were followed with a Gilford 2000 spectrophotometer. When other metabolites were carried with the enzyme in the injection, the assay solution was arranged according to (13).

3. RESULTS AND DISCUSSION

3.1. Effect of chaotropic anions on chloroplast glyceraldehyde-3-P dehydrogenase

Changes in water organization by high concentration of ions of low density charge favor the transfer of apolar regions of macromolecules to the solvent (14). Thus, chaotropes, (as Hatefi and Hanstein named these ions (15,16)) constitute one way for modifying intramolecular non-covalent interactions in a protein. In testing the effect of chaotropic anions on chloroplast glyceraldehyde-3-P dehydrogenase we first preincubated the enzyme with different Na-salts and subsequently assayed the 3-phosphoglycerate- and ATP-dependent oxidation of NADPH. Experiments reported in Table I show that the effectiveness of anions in stimulating NADP-glyceraldehyde-3-P dehydrogenase followed an order that resembled the lyotropic (Hofmeister) series. Similar ordering was observed in diverse phenomena with different molecules (14,16) and it was attributed to anion-mediated changes in the structure of the solvent.

TABLE I. Effect of neutral salts on the activation of chloroplast glyceraldehyde-3-P dehydrogenase

Salt (0.1 M)	NADPH oxidized . min^{-1}
NaSCN	26
Na-OOC-CCl$_3$	24
NaI	15
NaBr	7
NaCl	5
none	5

Chloroplast glyceraldehyde-3-P dehydrogenase (5 ug) was preincubated at 23°C in 0.1 ml of 0.1 M Tris-HCl buffer (pH 7.9) containing 10 umoles of the indicated salt. Following 5 min. of preincubation an aliquot was withdrawn and injected into the solution for assaying NADP-glyceraldehyde-3-P dehydrogenase.

3.2. Effect of cetyltrimethylammonium bromide on chloroplast glyceraldehyde-3-P dehydrogenase

As shown in Table II the NADP-linked activity of chloroplast glyceraldehyde-3-P dehydrogenase was enhanced by incorporating a cationic detergent, cetyltrimethylammonium bromide, to the assay solution. On the contrary, other positive bearing compounds with short aliphatic chains were ineffective in stimulating the specific activity of this enzyme (not shown). On kinetic basis, present stimulation differs from other forms of activation observed previously. In comparison to effectors (3), reduced thioredoxin (4), organic solvents (6,7) and chaotropes (this report), cetyltrimethylammonium bromide elicited in chloroplast glyceraldehyde-3-P dehydrogenase both faster responses and much higher values of specific activity. Moreover, since the concentration of cetyltrimethylammonium bromide used in the assay was lower than its critical micelle concentration (17,18) we concluded that the enzyme reacted with the monomer form of the detergent.

TABLE II. Activation of NADP-glyceraldehyde-3-P dehydrogenase by cetyltrimethylammonium bromide

Assay conditions	NADPH oxidized . min^{-1}
Complete	150
minus 3-phosphoglycerate	1
minus ATP	1
minus 3-phosphoglycerate phosphokinase	1
minus enzyme	0
minus cetyltrimethylammonium bromide	7

The NADP-dependent activity of chloroplast glyceraldehyde-3-P dehydrogenase was assayed with 5 ug of enzyme according to the procedure outlined under Experimental Procedures. The concentration of cetyltrimethylammonium bromide was 0.17 mM.

Present results constitute two new alternatives for stimulating the specific activity of NADP-glyceraldehyde-3-P dehydrogenase (cf. Table I in ref. 19). Although unphysiological in its nature, the activation reported herein and in previous publications (3,4,6,7) are congruent with the view that the modification of hydrophobic intramolecular interactions are crucial for determining the NADP-dependent activity.

4. REFERENCES

1.- Muller, B., Ziegler, I., and Ziegler, H. (1969) Eur. J. Biochem. 9, 101-106.
2.- Pupillo, P., and Giuliani-Piccari, G. (1975) Eur. J. Biochem. 51, 475-482.
3.- Wolosiuk, R.A., and Buchanan, B.B. (1976) J. Biol. Chem. 251, 6456-6461.
4.- Wolosiuk, R.A., and Buchanan, B.B. (1978) Plant Physiol. 61, 669-671.
5.- Corley, E., and Wolosiuk, R.A. (1985) J. Biol. Chem. 260, 3978-3983.
6.- Wolosiuk, R.A., Corley, E., Crawford, N.A., and Buchanan, B.B. (1985) FEBS Lett. 189, 212-216.
7.- Wolosiuk, R.A., Hertig, C.M., and Busconi, L. (1986) Arch. Biochem. Biophys. 246, 1-8.
8.- Nemethy, G., Peer, W.J., and Scheraga, H.A. (1981) Ann. Rev. Biophys. Bioeng. 10, 459-497.
9.- Nozaki, Y., Reynolds, J.A., and Tanford, C. (1974) J. Biol. Chem. 249, 4452-4459.
10.- Layne, E. (1957) in: Methods in Enzymology (Colowick, S.P. and Kaplan, N.O., eds.), Vol. III, pp. 447-454, Academic Press, New York.
11.- Maniatis, T., Fritsch, E.F., Sambrook, J. (1982) in: Molecular Cloning, pp. 150-172, Cold Spring Harbor, New York.
12.- Levy, L.M., and Betts, G.F. (1985) Biochim. Biophys. Acta 832, 186-191.
13.- Wolosiuk, R.A., and Hertig, C.M. (1983) in: Thioredoxins, Structure and Functions (Gadal, P., ed.) pp..167-173, Editions du CNRS, Paris.
14.- von Hippel, P.H., and Wong, K.Y. (1964) Science 145, 577-580.
15.- Hamaguchi, K., and Geiduschek, E.P. (1962) J. Am. Chem. Soc. 84, 1329-1338.
16.- Hatefi, Y., and Hanstein, W.G. (1969) Proc. Nat. Acad. Sci. U.S. 62, 1129-1136.
17.- Fendler, J.H., and Fendler, E.J. (1975) in Catalysis in Micellar and Macromolecular Systems, Academic Press, New York.
18.- Cline Love, L.J., Habarta, J.G., and Dorsey, J.G. (1984) Anal. Chem. 1132A-1148A.
19.- Wara-Aswapati, O., Kemble, R.J., and Bradbeer, J.W. (1980) Plant Physiol. 66, 34-39

OBLIGATE CO-ACTIVATION OF NADPH-DEPENDENT GLYCERALDEHYDE-3-PHOSPHATE DEHYDROGENASE AND PHOSPHORIBULOKINASE ON DISSOCIATION OF A MULTIMERIC ENZYME COMPLEX.

ROY POWLS, SYLVIA NICHOLSON and JOHN S. EASTERBY

Department of Biochemistry, University of Liverpool, P.O. Box 147, Liverpool, L69 3BX, Great Britain.

1. INTRODUCTION

Several enzymes of the reductive pentose phosphate cycle have been shown to be activated by light [see 1]. There is evidence that for NADPH-dependent glyceraldehyde-3-phosphate dehydrogenase (NADPH-G3PDH) and phosphoribulokinase (PRK) [2-4] activation occurs as a result of depolymerization of a high-molecular weight proenzyme with latent [4] or very low activity [5]. These proenzyme forms have been isolated from the green alga, Scenedesmus obliquus [4,5]. Depolymerization and activation of the proenzymes was promoted by incubation with dithiothreitol and NADPH. It is now apparent that a single multimeric protein serves as a common proenzyme for both NADPH-G3PDH and PRK.

2. METHODS

Scenedesmus obliquus (Cambridge 276/6a) was grown in medium supplemented with glucose and yeast extract [6]. The multimeric enzyme was isolated as the latent PRK [4], the active form of PRK [4] and NADPH-G3PDH [3] as described previously. For the time-course of activation, the multimeric enzyme was incubated with NADPH (lmM) and Tris HCl pH 7.5 (100mM) in a total volume of 400µl, in the presence of various concentrations of dithiothreitol and algal thioredoxin (if required). At specified times, 10µl aliquots were removed and their PRK and G3PDH activities determined. Activation of the multimeric enzyme for sedimentation velocity analysis has been described [3].

3. RESULTS AND DISCUSSION

Both latent PRK and G3PDH activity, characteristic of the regulatable G3PDH (linked to both pyridine nucleotides but predominantly to NADH) were found to be present in a single homogeneous multimeric protein. This had previously been isolated as the latent form of PRK [4]. The two activities could not be separated by DEAE-cellulose, gel-filtration and hydroxylapatite chromatography.

Incubation of the multimeric enzyme with 20mM dithiothreitol and lmM NADPH promoted the co-activation of PRK and NADPH-G3PDH with an associated partial deactivation of NADH-G3PDH (Figure 1).

The multimeric protein had a molecular weight of 560000 by sedimentation equilibrium analysis. During a sedimentation velocity study, activation promoted by 18mM dithiothreitol and 1 mM NADPH caused the replacement of the boundary sedimenting at 14.2S (characteristic of the multimeric protein) by two sedimenting boundaries of 7.4S and 4.4S (Figure 2). These have been previously shown to be due to NADPH-G3PDH and PRK respectively [4,5]. Activation with dithiothreitol alone promotes PRK activity in a similar way, but activation of NADPH-G3PDH is

Biggens, J. (ed.), Progress in Photosynthesis Research, Vol. III. ISBN 90 247 3452 5
© *1987 Martinus Nijhoff Publishers, Dordrecht. Printed in the Netherlands.*

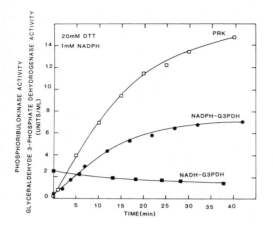

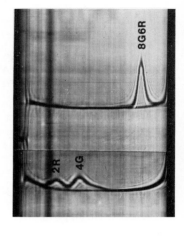

Figure 1

Co-activation of NADPH-G3PDH and PRK on incubation of the multimeric enzyme with dithiothreitol and NADPH.

Figure 2

Sedimentation velocity analysis
Top. Native multimeric enzyme.
Bottom. Multimeric enzyme activated with dithiothreitol and NADPH.

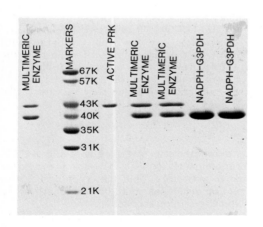

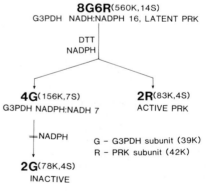

Figure 3

SDS-PAGE.

Figure 4

Molecular changes on activation of the multimeric enzyme.

transient and only a single sedimenting boundary of 4.6S is observed [3]. This is now considered to be due to over-reduction of the enzyme to produce an inactive dimer of G3PDH. Over-reduction and the associated deactivation is prevented by nucleotide. The multimeric protein was shown by SDS-PAGE (Figure 3) to be composed of two subunits of Mr 39K(G) and 42K(R). That of 39K co-migrated with the subunit of NADPH-G3PDH, whereas that of 42K was identical to the subunit of the active PRK. In view of the measured molecular weight the subunit composition of the multimeric enzyme is 8G6R. Scanning of the Coomassie blue stained peptides on SDS-PAGE gave a G:R ratio of 8:5. This is reasonably consistent with a 8G6R composition, as the proportion of the R subunit has been found to decrease on storage, a phenomenon previously reported for the analogous subunit of plant chloroplast G3PDH [7]. On activation, dissociation of the multimeric enzyme (8G6R) occurs to produce the active form of PRK (2R) and NADPH-G3PDH (4G) (Figure 4).

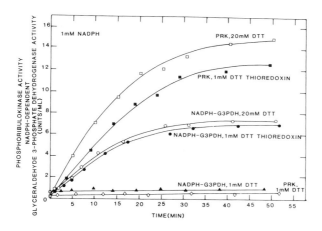

Figure 5. Thioredoxin stimulation of activation.

Algal thioredoxin was obligatory for the co-activation of PRK and NADPH-G3PDH when the multimeric enzyme was incubated with 1mM dithiothreitol and 1mM NADPH (Figure 5). We suggest that in vivo light activation occurs by a similar depolymerization promoted by photoreduced thioredoxin. This is supported by the finding of high concentrations of both the active form of PRK and NADPH-G3PDH in photoheterotrophically-grown algae, whereas in dark-grown algae the multimeric enzyme predominates. The role of the multimeric enzyme is in the oxidation of the triose phosphate imported into the chloroplast. This would provide the ATP necessary for starch synthesis. Obviously this ATP requirement could not be met by photophosphorylation when the alga is grown heterotrophically. Illumination promotes rapid formation of NADPH and ATP by photosynthetic electron transfer and hence of 1,3-bisphosphoglycerate. NADPH and 1,3-bisphosphoglycerate [2,3] are the most effective metabolites in promoting activation in a reducing environment (dithiothreitol or photoreduced thioredoxin in vivo). Hence illumination would dissociate the multimeric enzyme to give the active form of PRK and NADPH-G3PDH, both essential for the operation of the Calvin cycle.

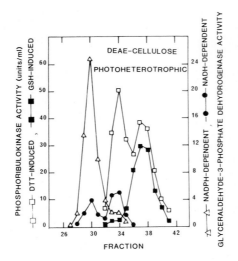

Figure 6.   DEAE-cellulose chromatography of extracts of photoheterotophically-grown S. obliquus.

Substantial evidence favours a similar association of G3PDH and PRK in plants. High molecular weight forms of G3PDH [7-9] and PRK [10] have been isolated and are eluted together from DEAE-cellulose and on gel-filtration [10]. Plant G3PDHs are characterised by NADP-promoted depolymerization [7,9] and possess two subunits of similar size to those of the algal enzyme [7,11]. NAD-promoted repolymerization of the purified low molecular weight G3PDH is dependent upon a binding fraction with a subunit of Mr 42K [12], we would now suggest that the binding fraction is the active form of PRK.

REFERENCES

1.   Buchanan, B.B. (1980) Annu.Rev. Plant Physiol. 31, 341-374.
2.   O'Brien, M.J., Easterby, J.S. & Powls, R. (1976) Biochim. Biophys. Acta 449, 209-223.
3.   O'Brien, M.J., Easterby, J.S. and Powls, R. (1977) Biochim. Biophys. Acta 481, 349-358.
4.   Lazaro, J.J., Sutton, C.W., Nicholson, S. and Powls, R. (1986) Eur. J. Biochem. 156, 423-429.
5.   O'Brien, M.J. and Powls, R. (1976) Eur. J. Biochem. 63, 155-161.
6.   Kessler, E., Arthur, W. and Brugger, J.E. (1957) Arch. Biochem. Biophys. 71, 326-335.
7.   Pupillo, P. and Faggiani, R. (1979) Arch. Biochem. Biophys. 194, 581-592.
8.   Wolusiak, R.A. and Buchanan, B.B. (1976) J. Biol. Chem. 251, 6456-6461.
9.   Cerff, R. (1978) Eur. J. Biochem. 82, 45-53.
10.   Wolusiak, R.A. and Buchanan, B.B. (1978) Arch. Biochem. Biophys. 189, 97-101.
11.   Cerff, R. (1979) Eur. J. Biochem. 94, 243-247.
12.   Cerff, R. (1978) Plant Physiol. 61, 369-372.

S-ADENOSYL-L-METHIONINE-DEPENDENT METHYLATION OF CHLOROPLAST PROTEINS

MICHAEL T. BLACK, WILLIAM R. WIDGER, DANIEL MEYER, and WILLIAM A. CRAMER, Dept. of Biological Sciences, Purdue University, West Lafayette, Indiana 47907, U.S.A.

Introduction
    It is well-known that the functional activity of proteins can be altered by post-translational modification, perhaps most notably by reversible phosphorylation (1,2). One such modification is the incorporation of methyl groups into proteins from S-adenosyl-L-methionine (SAM) resulting in carboxyl methylester formation (for reviews, see 3-6). Methyltransferase activity appears to be an almost ubiquitous phenomenon in both procaryotes and eucaryotes. At least two classes of methyltransferase exist which catalyse the methylation of distinct amino acids and which have different physiological roles. Class I enzymes catalyse the methylation of carboxylic acid residues of normal proteins and have thus far only been detected in bacteria where they catalyse the formation of L-glutamyl γ-methylesters on specific membrane chemoreceptors involved in chemotaxis. In these cases the differential action of methylated and demethylated chemoreceptors controls the processing of sensory input. Demethylation is catalysed by L-glutamyl γ-methylesterases; such a protein has been isolated in pure form from Salmonella typhimurium (7) and from Bacillus subtilis (8). Class II enzymes, which are active in both procaryotes and eucaryotes, recognise structurally altered carboxylic acids such as D-aspartate and L-isoaspartate and it has been proposed that methylation of these residues is involved in metabolism of altered protein or, more likely, in a possible repair mechanism (5). Class II enzymes are also characterised by the fact that they display low substrate specificity and low stoichiometry of methylation in the range of 10^{-6} to 10^{-2} methyl groups per polypeptide (6). Class II methyltransferases have been purified from brain tissue (9) and from human erythrocytes (10). To date, no enzymatic esterase activity, at least in erythrocytes, appears to be associated with Class II activity.
    Although many tissue types from a variety of organisms have been screened for the presence of protein carboxyl group methylation, there have been no reports of this type of activity in the photosynthetic tissues of higher plants, although protein methylation has been ascribed a role in the chemosensory and photosensory responses of the archaebacterium Halobacterium halobium (11,12). The present communication represents the first report of SAM-dependent methylation of chloroplast proteins. The enzymatic activity displays some Class II characteristics. A novel feature is that the activity appears to be under light or redox control.

Materials and Methods
    Intact chloroplasts were isolated from spinach leaves according to the method of Lilley & Walker (13), and resuspended in a reaction medium

Biggens, J. (ed.), Progress in Photosynthesis Research, Vol. III. ISBN 90 247 3452 5
© *1987 Martinus Nijhoff Publishers, Dordrecht. Printed in the Netherlands.*

(RM) consisting of 50 mM HEPES, pH 7.0, 5 mM $MgCl_2$ and 0.33 M sorbitol; chloroplasts were generally 40-50% intact. Chloroplasts were osmotically shocked in water, and an equal volume of double-concentration RM added, setting the final chloroplast concentration to an equivalent of 1 mg chl/ml. S-Adenosyl-L-[methyl-^{3}H] methionine (20 μM) was added at a specific activity of 2-4 μCi/nmol (or 20 μCi/nmol for PAGE) and suspensions were either illuminated or maintained in darkness at room temperature. Aliquots (generally 250 μl) were taken at the various times indicated, resuspended in 0.8 ml of 0.1 M pyrophosphate, pH 6.5 (PP), in Eppendorf microcentrifuge tubes, and the thylakoids sedimented by centrifugation. Stromal proteins were carefully removed from above the thylakoids and precipitated by addition of perchloric acid (PCA). Thylakoids were washed a second time with PP, the supernatant removed, and any stromal proteins again precipitated with PCA. Precipitated proteins were resuspended in PP and again precipitated and washed twice in acetone. The corresponding thylakoid fractions were washed with PCA (2%) and twice in acetone. All fractions were dissolved in 10% SDS and Budget-Solve scintillation cocktail and the incorporation of ^{3}H measured on a Beckman LS7000 Scintillation Counter. Base-labile incorporation was estimated essentially according to the procedure of Freitag & Clarke (14). Samples were suspended in 1.5 M NaOH for 1 min and subsequently precipitated with 15% PCA before solubilisation in 10% SDS. PAGE was carried out according to Laemmli (15).

Results and Discussion
    Dark incubation of freshly broken whole chloroplasts in the presence of 20 μM SAM results in the methyl esterification of both thylakoid and stromal proteins. The extent of incorporation into both stromal and thylakoid proteins was irreversibly increased by illumination, although the incorporation was usually below 100 pmol CH_3 per mg protein (Figs. 1,2). PAGE revealed that the methylation was apparently non-specific,

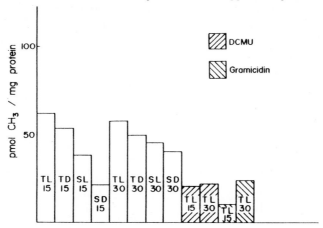

Figure 1. Histogram showing incorporation of (-CH_3) from SAM into thylakoid and stromal proteins after 15 or 30 min in illuminated and unilluminated chloroplasts. DCMU (20 μM) or Gramicidin (1 μM) were added as indicated. See text for further details. (T = thylakoid protein, S = stromal protein, L = illuminated, D = unilluminated.)

as autoradiograms indicated that highly radioactive regions of
the gel correlated with regions of high protein concentration (e.g.,
RuBisCO large and small subunits) (data not shown). The time course of
incorporation has been found to be variable in different chloroplast
suspensions prepared at different times; methylation appears to approach
completion at various times from 15 to > 60 min but is invariably at a
low level relative to protein. Assuming an average relative molecular
mass for chloroplast proteins of 30,000, 1 mg protein would be equivalent
to 33 nmol; typical incorporation would then be equivalent to < 1 to ~5
methyl groups per 1000 polypeptides.

In addition to the apparent irreversibility, low stoichiometry of
methylation, and non-specificity of substrate protein, thylakoid enzymes
demonstrate another Class II characteristic, that of catalysing the
methylation of exogenous protein. The illumination of thylakoids (1 mg
Chl) in the presence of 10 mg ovalbumin and 20 μM SAM resulted in the
incorporation of ~18 pmol ($-CH_3$)/mg ovalbumin and also ~18 pmol ($-CH_3$)/mg
thylakoid protein.

It is known that ($-CH_3$) incorporation is frequently base-labile (6).
As much as ~90% of the incorporated ($-CH_3$) can be removed by treating
methylated proteins with 1.0 M NaOH, and this has been of some concern
due to the alkaline pH (~8.6) used for PAGE. The treatment of
chloroplast proteins with 1.5 M NaOH according to the procedure of
Freitag and Clarke resulted in a loss of 60-70% of ($-CH_3$) from all
protein fractions. Preliminary data suggests that this loss does not
change the apparent non-specificity of chloroplast protein methylation.
In addition, mild base treatment equivalent to that employed during PAGE
decreases the incorporation by only ~20%. It is of interest to note that
experiments conducted at low concentrations (~0.3 μM) of SAM resulted in
specific incorporation into the small subunit of RuBisCO, and no
detectable methylation of the large subunit was apparent (16). The data
is currently insufficient to determine whether this results from the
presence of more than one enzyme with different substrate affinities or
whether the observations are accounted for by one multi-substrate
methyltransferase.

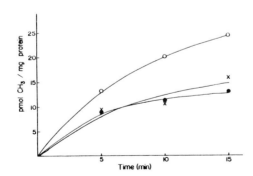

Figure 2. Time course of incorporation of ($-CH_3$) into illuminated (O) or
unilluminated (●) thylakoids. (X) represents incorporation into
thylakoids illuminated in the presence of 20 μM DCMU.

Methylation of thylakoid proteins is sensitive to the electron transport inhibitor, DCMU, as shown in Figure 2. DCMU reduces the level of incorporation into proteins of illuminated thylakoids to that of unilluminated samples. Addition of the uncoupler gramicidin results in a similar inhibition of methyl incorporation (Fig. 1); a reduction in methylation level in response to dissipation of $\Delta\tilde{\mu}_H+$ previously has been reported for Halobacteria (12).

The functional significance of these reactions is unclear at present. All data currently available indicate that the methyltransferase activity in chloroplasts is characteristic of class II enzymes. If, as has been proposed (5), this activity is involved in the repair of proteins containing racemised amino acids, then it remains unclear why this activity should benefit from light/redox control or control through $\Delta\tilde{\mu}_H+$. SAM has no effect on whole chain electron transport (Hill reaction), or on O_2 evolution as measured in the reconstituted chloroplast system of Robinson and Walker (17).

Acknowledgment
This research has been supported by NSF grant PCM-84 3308. We thank Ed Bjes for technical assistance and Ms. Janet Hollister for her work on the manuscript.

References
1 Recently Discovered Systems of Enzyme Regulation by Reversible Phosphorylation (1980) Cohen, P. (ed.), Elsevier Biomedical Press, Amsterdam
2 Enzyme Regulation by Reversible Phosphorylation - Further Advances (1984) Cohen, P. (ed.), Elsevier Biomedical Press, Amsterdam
3 Paik, W.K. and Kim, S. (1980) in Protein Methylation, pp. 202-231, New York, Wiley
4 O'Dea, R.F., Viveros, O.H., and Diliberto, E.J., Jr. (1981) Biochem. Pharmacol. 30, 1163-1168
5 Clarke, S. and O'Connor, C.M. (1983) Trends Biochem. Sci. 8, 391-394
6 Clarke, S. (1985) Ann. Rev. Biochem. 54, 479-506
7 Snyder, M.A., Stock, J.B., and Koshland, D.E., Jr. (1984) Methods Enzymol. 106, 321-330
8 Goldman, D.F., Nettleton, D.O., and Ordal G.W. (1984) Biochemistry 23, 675-680
9 Aswad, D.W. and Deight, E.A. (1983) J. Neurochem. 40, 1718-1726
10 Kim, S., Choi, J., Jun, G.-F. (1983) Biochem. Biophys. Methods 8, 9-14
11 Schinz, A. (1981) FEBS Lett. 125, 205-207
12 Bibikov, S.I., Baryshev, V.A., and Glagolev, A.N. (1982) FEBS Lett. 146, 255-258
13 Lilley, R.Mc. and Walker, D.A. (1974) Biochim. Biophys. Acta 368, 260-278
14 Freitag, C. and Clarke, S. (1981) J. Biol. Chem. 256, 6102-6108
15 Laemmli, U.K. (1970) Nature (London) 227, 680-685
16 Black, M.T., Widger, W.R., Meyer, D., and Cramer, W.A. (1986) manuscript in preparation
17 Robinson, S.P. and Walker, D.A. (1979) Biochim. Biophys. Acta 545, 528-536

PARTIAL AMINO ACID SEQUENCE OF SPINACH CHLOROPLAST FRUCTOSE-1,6-BISPHOSPHATASE.

FRANK MARCUS AND PETER B. HARRSCH, DEPARTMENT OF BIOLOGICAL CHEMISTRY AND STRUCTURE, UNIVERSITY OF HEALTH SCIENCES/THE CHICAGO MEDICAL SCHOOL, NORTH CHICAGO, ILLINOIS 60064, USA.

Chloroplast fructose-1,6-bisphosphatase (FbPase) is an essential enzyme in the photosynthetic pathway of carbon dioxide fixation into sugars. The properties of the chloroplast enzyme are clearly distinct from cytosolic gluconeogenic FbPases. Light-dependent activation via a ferredoxin/thioredoxin system and insensitivity to inhibition by AMP are unique characteristics of the chloroplast enzyme (for refs. see 1-3). However, preliminary amino acid sequence data has demonstrated that a significant degree of amino acid sequence similarity exists between spinach chloroplast and mammalian gluconeogenic fructose-1,6-bisphosphatase. This conclusion (4) was based on the sequence analysis of one cyanogen bromide peptide and three tryptic peptides obtained from S-carboxymethylated spinach chloroplast FbPase and the data provided sequence information for a total of 78 amino acid residues.

We have now continued our primary structural analysis of spinach chloroplast FbPase by determining the amino acid sequence of: (a) most of the peptides (T-) obtained after reversed phase high performance liquid chromatography of a tryptic digest of S-carboxymethylated enzyme (see Fig. 1B, ref. 4); (b) three of the peptides obtained by cleavage of S-carboxymethylated spinach chloroplast FbPase with Staphylococcus aureus protease (SAP-); (c) a large tryptic peptide (T-i) obtained by cleavage of S-pyridylethylated spinach chloroplast FbPase; (d) an additional cyanogen bromide (CNBr-) fragment; and (e) a minor product (TS-11A) of limited digestion of spinach chloroplast FbPase with subtilisin. The results of these structural studies are summarized in Table 1 (next page) which gives the amino acid sequence of the above peptides. Table 1 also includes the tentative location of each of these peptides in the total structure using residue numbers according to their corresponding position in the established sequence of pig kidney FbPase (5).

The data given in Table 1 has extended our amino acid sequence information on spinach chloroplast FbPase to a total of 285 residues. All of the sequences, except the peptide designated as T-3, can be aligned with confidence using as a frame of reference the established amino acid sequence of pig kidney FbPase as shown below in Fig. 1. The comparison of the sequence data so far available for spinach chloroplast FbPase with that of the kidney enzyme reveals 46% amino acid identity. This high degree of homology strengthens our previous suggestion of a common evolutionary origin for a variety of gluconeogenic FbPases and the chloroplast FbPase (4,6), enzymes catalyzing the same reaction but having different functions and modes of regulation. The sequence information has disclosed the position of 3 out of 5 or 6 cysteine residues per

Biggens, J. (ed.), Progress in Photosynthesis Research, Vol. III. ISBN 90 247 3452 5
© *1987 Martinus Nijhoff Publishers, Dordrecht. Printed in the Netherlands.*

TABLE 1. Amino acid sequences of peptides from spinach chloroplast fructose-1,6-bisphosphatase.[a]

Peptide	Amino Acid Sequence	Tentative location (Residue numbers)
T-2	Y-L-A	330-332
T-3	A-A-V-G-E-A-A-T-E-T-K	160-170
T-4	M-W-D-D-K	218-222
T-7	(T-4)-L-K	218-224
T-8	N-I-A-S-L-V-G-R	43- 50
T-9	K-Y-M-D-D-L-K-E-P-G-E-S-Q-K-P-Y-S-S-R	225-243
T-10	(T-4)-L-K-K	218-225
+11	+ Y-M-D-D-L-K-E-P-G-E-S-Q-K-P-Y-S-S-R	226-243
T-12	I-Y-S-F-N-E-G-N-Y-K	208-217
T-13	I-L-D-I-Q-P-T-E-I-H-Q-R	302-313
T-14	A-G-I-S-N-L-T-G-I-Q-G-A-V-N-I-Q-G-E-D-Q-K	51- 71
T-15	Y-I-G-S-L-V-G-D-F-H-R	244-254
T-16	K-L-D-V-V-S-N-E-V-F-S-S-C-L-R	72- 86
T-17	T-L-L-Y-G-G-I-Y-G-Y-P-R	255-266
T-19	V-P-L-Y-I-G-S-V-E-E-V-K-L-E-K	314-329
T-20	(T-21)-G-S-D-G-H-Q-R	277-301
T-21	L-L-Y-E-C-A-P-M-S-F-I-V-E-Q-A-G-G-K	277-294
T-22	S-K-Y-E-I-E-T-L-T-G-W-L-L-K	6- 19
T-23	G-V-Y-A-F-T-L-D-P-M-Y-G-E-F-V-L-T-S-E-K	180-199
T-24	(T-22)-Q-P-M-A-G-V-I-D-A-P-L-T-I-V-L-	6- 34
SAP-12	V-F-S-S-C-L-R-S-S-G-R-T-G-I-I-A-S-E	80-107
SAP-19	S-Y-S-G-N-Y-I-V-V-F-D-P-L-D-G-S-S-N-I-D-	
	A-A-V-S-T-G-S-I-F-G-I-Y-	108-139
SAP-6	C-I-V-D-S-D-H-D-D-E	147-156
T-i	T-G-I-I-A-S-E-E-E-D-V-P-V-A-V-E-E-S-Y-S-	
	G-N-Y-I-V-V-F-D-P-L-	91-120
TS-11A	S-P-N-D-E-C-I-V-D-S-D-H-D-D-E	140-156
CNBr-4	Y-G-E-F-V-L-T-S-E-K-I-Q-I-P-K-A-G-K-(T-12)	190-217

[a] For a description of "Materials and Methods" see ref. 4. For peptide designations, see text. Residue numbers indicate their tentative location as aligned with the amino acid sequence of pig kidney FbPase (5).

```
                                     10                          20
-     -     -     -     -SER-LYS-TYR-GLU-ILE-GLU-THR-LEU-THR-GLY-TRP-LEU-LEU-LYS-GLN
                          30                          40
PRO-MET-ALA-GLY-VAL-ILE-ASP-ALA-PRO-LEU-THR-ILE-VAL-LEU-    -    -    -    -    -
                        50                          60
-    -GLN-ILE-ALA-SER-LEU-VAL-GLN-ARG-ALA-GLY-ILE-SER-ASN-LEU-THR-GLY-ILE-GLN
                           70                          80
GLY-ALA-VAL-ASN-ILE-GLN-GLY-GLU-ASP-GLN-LYS-LYS-LEU-ASP-VAL-VAL-SER-ASN-GLU-VAL
                         90                         100
PHE-SER-SER-CYS-LEU-ARG-SER-SER-GLY-ARG-THR-GLY-ILE-ILE-ALA-SER-GLU-GLU-GLU-ASP
                       110                         120
VAL-PRO-VAL-ALA-VAL-GLU-GLU-SER-TYR-SER-GLY-ASN-TYR-ILE-VAL-VAL-PHE-ASP-PRO-LEU
                     130                         140
ASP-GLY-SER-SER-ASN-ILE-ASP-ALA-ALA-VAL-SER-THR-GLY-SER-ILE-PHE-GLY-ILE-TYR-SER
                  150                         160
PRO-ASN--------ASP-GLU-CYS-ILE-VAL-ASP-SER-ASP-HIS-ASP-ASP-GLU-    -    -    -
                          170                         180
-    -    -    -    -    -    -    -    -    -    -    -    -    -    -    -GLY
                          190                         200
VAL-TYR-ALA-PHE-THR-LEU-ASP-PRO-MET-TYR-GLY-GLU-PHE-VAL-LEU-THR-SER-GLU-LYS-ILE
                      210                         220
GLN-ILE-PRO-LYS-ALA-GLY-LYS-ILE-TYR-SER-PHE-ASN-GLU-GLY-ASN-TYR-LYS-MET-TRP-ASP
                      230                         240
ASP-LYS-LEU-LYS-LYS-TYR-MET-ASP-ASP-LEU-LYS-GLU-PRO-GLY-GLU-SER-GLN-LYS-PRO-TYR
                      250                         260
SER-SER-ARG-TYR-ILE-GLY-SER-LEU-VAL-GLY-ASP-PHE-HIS-ARG-THR-LEU-LEU-TYR-GLY-GLY
                      270                         280
ILE-TYR-GLY-TYR-PRO-ARG-    -    -    -    -    -    -    -    -    -    -LEU-LEU-TYR-GLU
                      290                         300
CYS-ALA-PRO-MET-SER-PHE-ILE-VAL-GLU-GLN-ALA-GLY-GLY-LYS-GLY-SER-ASP-GLY-HIS-GLN
                     310                         320
ARG-ILE-LEU-ASP-ILE-GLN-PRO-THR-GLU-ILE-HIS-GLN-ARG-VAL-PRO-LEU-TYR-ILE-GLY-SER
                     330
VAL-GLU-GLU-VAL-GLU-LYS-LEU-GLU-LYS-TYR-LEU-ALA-
```

FIGURE 1.   Partial amino acid sequence of spinach chloroplast fructose-1,6-bisphosphatase. All peptide sequences given in Table 1 are included here, except that of peptide T-3. Numbers above residues indicate their corresponding position in the sequence of pig kidney FbPase (5). Amino acids are indicated by the three-letter code. Deletions are indicated by a line of dashes (----) and yet unknown regions are indicated by blanks between dashes, i.e., (- - -). Cys residues are shown shaded.

subunit, and these are located at positions corresponding to numbers 84, 147, and 281 (shown shaded in Fig. 1). The above cysteine residues, however, are probably not the cysteine residues involved in the light regulated activation of the enzyme. It would appear more likely that these cysteine residues will be part of a not yet found peptide (i.e., between residues 157 and 179, Fig. 1) containing two close-by cysteines, like in thioredoxins, glutaredoxin, protein disulfide isomerase, thioredoxin reductase, and disulfide oxido-reductases (7-10).

Acknowledgements - This work was supported by grants from the U.S. Department of Agriculture (83-CRCR-1-1299), and the National Institutes of Health (AM 26564).

REFERENCES
1 Buchanan, B.B. (1980) Ann Rev. Plant Physiol. 31, 341-374
2 Halliwell, B. (1981) in Chloroplast Metabolism, pp. 66-88, Oxford University Press, New York
3 Tejwani, G.A. (1982) Adv. Enzymol. 54, 121-194
4 Harrsch, P.B., Kim, Y., Fox, J.L. and Marcus, F. (1985) Biochem. Biophys. Res. Commun. 133, 520-526
5 Marcus, F., Edelstein, I., Reardon, I. and Heinrikson, R.L. (1982) Proc. Natl. Acad. Sci. USA 79, 7161-7165
6 Marcus, F., Gontero, B., Harrsch, P.B. and Rittenhouse, J. (1986) Biochem. Biophys. Res. Commun. 135, 374-381
7 Hoog, J., Jornvall, H., Holmgren, A., Carlquist, M. and Persson, M. (1983) Eur. J. Biochem. 136, 223-232
8 Edman, J.C., Ellis, L, Blacher, R.W., Roth, R.A. and Rutter, W.J. (1985) Nature 317, 267-270
9 Ronchi, S. and Williams, C.H. (1972) J. Biol. Chem. 247, 2083-2086
10 Fox, B.S. and Walsh, C.T. (1983) Biochemistry 22, 4082-4088

CHELATES OF FRUCTOSE 1,6-BISPHOSPHATE-IONS FUNCTION AS SUBSTRATES AND
FREE FRUCTOSE 1,6-BISPHOSPHATE-IONS AS INHIBITORS OF FRUCTOSE 1,6-BISPHOS-
PHATASE FORM B FROM SYNECHOCOCCUS LEOPOLIENSIS

K.-P. GERBLING, M. STEUP and E. LATZKO
Botanisches Institut der Westf. Wilhelms-Universität, Schloßgarten 3
D - 4400 Münster, Federal Republic of Germany

1. INTRODUCTION

The catalytic activity of fructose bisphosphatase (FbPase) form B from
Synechococcus leopoliensis (Anacystis nidulans) requires both, fructose
1,6-bisphosphate (FbP) and Mg^{++}. In principle, the free FbP or, alternatively, a FbP-Mg-complex may represent the actual substrate of the enzyme.
The two alternatives imply fundamental differences in the reaction mechanism of the enzyme. They also have important implications for the evaluation of hexose phosphate pool sizes determined under in vivo conditions.

In the present study, initial rate kinetics of purified FbPase form B
from Synechococcus leopoliensis were used to distinguish between these
two alternatives and to identify the actual substrate of the enzyme.

2. MATERIALS AND METHODS

Algal Material. Synechococcus leopoliensis strain No. 1402-1 was obtained
from the algal collection of the Institute of Plant Physiology, University
of Göttingen (FRG). Axenic cultures were grown as previously described (1).
Purification of FbPase Form B. The enzyme purification was performed as
described elsewhere (2).
Enzyme Activity Measurements. Fructose bisphosphatase activity was measured as previously described (1).

Biggens, J. (ed.), Progress in Photosynthesis Research, Vol. III. ISBN 90 247 3452 5
© 1987 Martinus Nijhoff Publishers, Dordrecht. Printed in the Netherlands.

3. RESULTS AND DISCUSSION

The initial velocity (v) of fructose 6-phosphate formation by FbPase
form B from Synechococcus leopoliensis was measured in a coupled enzyme
assay at varying FbP and Mg^{++} levels. When the initial velocity (v) was
plotted against the total FbP concentration added to the reaction mixture
a series of hyperbolic curves was obtained (Fig. 1a). Double reciprocal
plots (1/v versus 1/ FbP) resulted in straight lines with a common intercept but different slopes (Fig. 1b).

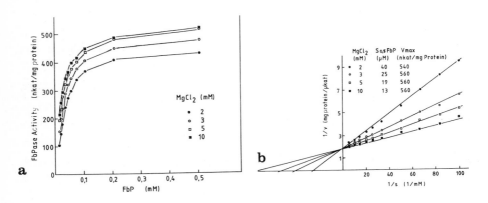

Fig. 1. Initial velocity (v) of fructose 6-phosphate formation by FbPase
form B measured at varying FbP and Mg^{++} levels. The velocity v was plotted
against the total FbP concentration added to the reaction mixture.
a: direct plot; b: double reciprocal plot

The actual concentrations of the different FbP-species existing under
assay conditions were then calculated. The FbP-species were:

$$(FbP)^{2-}, \ (FbP)^{3-}, \ \text{and} \ (FbP)^{4-} \ \text{as non-complexed forms}$$

and

$$(FbP-Mg), \ (FbP-Mg)^{-}, \ \text{and} \ (FbP-Mg)^{2-} \ \text{as complexed species.}$$

The calculation, performed according to McGilvery (3) and Gardemann and
Heldt (4), is based upon the assumption that the six FbP-species are in
equilibrium under assay conditions.

Under these conditions $(FbP)^{2-}$ and (FbP-Mg) represent less than 0.3%

of the total amount of FbP. Therefore, these two FbP-species were not considered as potential substrates of the enzyme.

If one assumes that only one of the residual four FbP-species functions as the actual substrate (and the other species do not exert any effect), this species can be identified by replotting the data shown in Fig. 1 against the calculated concentration of each of the four FbP-species. In case of the true substrate of the enzyme, all values of v, obtained at different FbP and Mg^{++} levels, should fit into a single saturation curve or, in the double reciprocal plot, on a single straight line. Replots of v against the concentration of each of the four FbP-species are shown in Fig. 2.

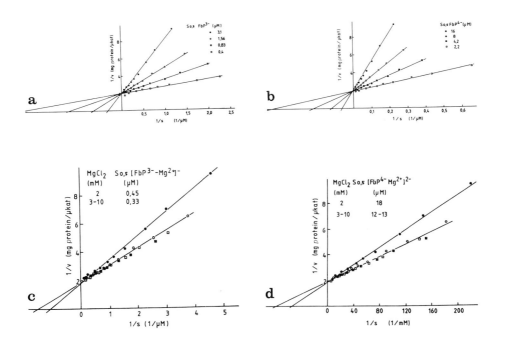

Fig. 2. Double reciprocal plot of v (Fig. 1) against the calculated concentration of each of the four FbP-species.

a: $(FbP)^{3-}$ b: $(FbP)^{4-}$ c: $(FbP-Mg)^{-}$ d: $(FbP-Mg)^{2-}$

The replots of v against the actual $(FbP)^{2-}$ or $(FbP)^{3-}$ concentrations do not fulfil the above criteria (Fig. 2a and b). However, a noticable (but incomplete) approximation of a single line was obtained when v was plotted against either of the two FbP-Mg-complexes (Fig. 2c and d). Therefore, only the two FbP-Mg-complexes are expected to function as substrate.

In contrast, the non-complexed free FbP-species function as inhibitors. This inhibitory effect is most noticeable when v is replotted against the calculated concentrations of the free FbP-species $\left[(FbP)^{3-} + (FbP)^{2-}\right]$ at fixed FbP-Mg levels (Fig. 3). It is reasonable to assume that due to this inhibitory effect the approximation to a single curve, as shown in Fig. 2c and d, was incomplete.

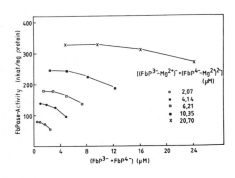

Fig. 3. The initial velocity v (Fig. 1) was plotted against the concentration of the free FbP-species $\left[(FbP)^{3-} + (FbP)^{4-}\right]$ at fixed levels of the complexed FbP-species $\left[(FbP-Mg)^{-} + (FbP-Mg)^{2-}\right]$

Thus, FbP exerts a dual effect on the FbPase reaction because distinct FbP-species function either as substrate or as inhibitor. Factors which affect the ratios between the FbP-species therefore affect the rate of fructose 6-phosphate formation by chainging substrate/inhibitor ratios.

REFERENCES

1 Gerbling, K.-P., Steup, M., Latzko, E. (1984) Arch. Microbiol. 137, 109–114

2 Gerbling, K.-P., Steup, M., Latzko, E. (1986) Plant Physiol. 80, 716–720

3 McGilvery, R.W. (1965) Biochemistry 4, 1924–1930

4 Gardemann, A. and Heldt, H.W. (1983) Hoppe-Seyler Z.Physiol. Chem. 364, 1127

COMPARTMENT-SPECIFIC PHOSPHORYLASE FORMS FROM HIGHER PLANTS

MARTIN STEUP, JUTTA CONRADS and JOACHIM van BERKEL
Botanisches Institut der Westf. Wilhelms-Universität,
Schloßgarten 3, D - 4400 Münster, Federal Republic of Germany

1. INTRODUCTION

In the highly compartmented plant cell many reactions of carbohydrate and nitrogen metabolism occur in more than one compartment. These reactions are usually catalyzed by compartment-specific enzyme forms which, in many cases, differ in some physical properties and are separable by electrophoretic or chromatographic techniques. From the point of view of comparative enzymology and cell biology, the existence of compartment-specific enzyme forms raises the question of how closely related these forms are in their kinetic, regulatory and structural properties. By definition, the various forms of an enzyme catalyze the same reaction and, therefore, one may expect at least partial structural similarity. In addition to these common features, however, the compartment-specific enzyme forms may possess a variety of distinct properties which enable them to fulfil a number of specific functions: to operate in different intracellular environments, to exert a selective compartment-specific control of metabolic pathways, and to ensure a precize intracellular topogenesis. These distinct properties may be directly related to structural differences between the compartment-specific enzyme forms. The proposal that in higher plant cells several enzymes of glycolysis and oxidative pentosephosphate cycle have a dual location in the cytoplasm and in the plastids is generally accepted (1). In contrast, evidence recently accumulated shows that even glucan metabolizing enzymes occur both inside and outside the chloroplast (2-4). These observations imply that a restriction of glucan

Biggens, J. (ed.), Progress in Photosynthesis Research, Vol. III. ISBN 90 247 3452 5
© *1987 Martinus Nijhoff Publishers, Dordrecht. Printed in the Netherlands.*

metabolism to the plastid is questionable.

In the present study, various compartment-specific forms of α-1,4-glucan phosphorylase (EC 2.4.1.1) have been compared with respect to their kinetic and structural properties.

2. MATERIALS AND METHODS

Plant Materials. Pea (Pisum sativum L. var. 'Kleine Rheinländerin') and spinach (Spinacia oleracea L. var. 'Wiremona') plants were grown as described elsewhere (2).
Enzyme Preparations. Potato tuber and spinach leaf phosphorylases were purified as previously described (5). For purification of the pea phosphorylases the procedures described by Conrads et al. (6) were applied.
Phosphorylase Assay. Phosphorolytic activity was measured in a coupled photometric assay (6). Kinetic measurements were performed as previously described (6).
Peptide Mapping and Polyacrylamide Gel Electrophoresis. For mapping, phosphorylases were cleaved with N-chlorosuccinimide/ urea as previously described (6).   For electrophoretic separation of the peptides a modified version (6) of the system of Laemmli was used.

3. RESULTS AND DISCUSSION

In this study cytoplasm- and chloroplast-specific phosphorylase forms from spinach and pea leaves were investigated. The intracellular location of these enzyme forms has been defined by indirect immunofluorescence (cf. 4, 6). The physical and kinetic properties of the compartment-specific enzyme forms are compiled in Table 1. The apparent molecular weights of the two cytosolic phosphorylase forms were lower than that of the chloroplastic counterparts; the values for the two pea enzymes were slightly higher than those of the respective spinach leaf enzyme forms. Phosphorylase forms residing in the same compartment exhibited the same glucan discrimination. Presumably, this reflects different characteristics of chloroplast- and cytosol-specific oligo/polysaccharide pools.

Table 1. Physical properties and glucan specificities of compartment-specific phosphorylase forms from pea and spinach leaves. Kinetic measurements were performed at saturating orthophosphate levels.

Enzyme Form	Monomer Size	Half-saturating Glucan Concentration (mmol glucose equivalents l^{-1})		
		Starch	Glycogen	Maltoheptaose
Cytosolic Form from Pea	88 700	0.017	0.055	1.4
Cytosolic Form from Spinach	84 000	0.017	0.077	2.67
Chloropl. Form from Pea (Form III)	109 000	0.435	52.6	0.224
Chloropl. Form from Spinach	105 000	0.9	30.0	0.4

The structural properties of the four compartment-specific enzyme forms were studied by peptide mapping. Proteins were cleaved with N-chlorosuccinimide in the presence of urea. For the purpose of comparison, the peptide patterns of rabbit muscle phosphorylase and of the dominant potato tuber enzyme form are included. The latter phosphorylase form has been localized previously in the amyloplasts.   The peptide patterns of the six phosphorylase forms are shown in Fig. 1. The cytoplasm- and chloroplast-specific phosphorylase forms from pea leaves were cleaved into different patterns (Fig. 1 lanes a and e). Likewise, the patterns of the two compartment-specific spinach leaf phosphorylases were different (Fig. 1 lanes b and f). This strongly suggests that the enzyme forms residing in dif-

ferent compartments within the same cell possess different primary structures and do not arise from modifications of a single
amino acid sequence.

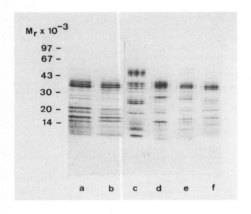

Fig. 1. Peptide patterns of
purified phosphorylase forms.
Peptides equivalent to 10 µg
uncleaved protein were applied
per lane. a and b: cytosolic
enzyme forms from pea and
spinach, respectively; c: rabbit muscle phosphorylase; d:
plastidic potato tuber phosphorylase; e and f: chloroplastic phosphorylase from pea
and spinach, respectively. The plastidic pea phosphorylase represents
form III (cf. 2).

The peptide pattern of the cytosolic pea phosphorylase form
was very similar to that of the cytoplasm-specific spinach enzyme (Fig. 1 lane a and b), thus suggesting a high degree of
homology. In contrast, the plastidic counterparts (Fig. 1 lanes
d-f) exhibited a higher inter-species diversity. Therefore, it
appears that the primary structure of the plastid-specific
enzyme forms is conserved to a lesser extent.

REFERENCES

1 Latzko, E. and Kelly, G.J. (1979) in Photosynthesis II, Encycl. Plant
   Physiol. Vol. 6 pp. 239-250
2 Steup, M. and Latzko, E. (1979) Planta 145, 69-75
3 Okita, T.W., Greenberg, E., Kuhn, D.N.and Preiss, J. (1979) Plant
   Physiol. 64, 187-192
4 Schächtele, C. and Steup, M. (1986) Planta 167, 444-451
5 Steup, M. and Schächtele, C. (1986) Planta 168, 222-231
6 Conrads, J., van Berkel, J., Schächtele, C. and Steup, M. (1986) Biochim. Biophys. Acta 882, 452-463

THE ISOLATION AND THE IMMUNOLOGICAL PROPERTIES OF CHLOROPLAST AND
CYTOPLASMIC PHOSPHOGLYCERATE KINASE FROM BARLEY.

EILEEN M.McMORROW and J.WILLIAM BRADBEER,
DEPARTMENT OF BIOLOGY, KING'S COLLEGE LONDON, 68, HALF MOON LANE, LONDON,
S.E.24. 9JF.

INTRODUCTION

In green plants phosphoglycerate kinase (PGAK) [EC 2.7.2.3] is
involved in the photosynthetic carbon cycle, glycolysis and
gluconeogenesis and is located in both chloroplasts and cytoplasm. Pacold
and Anderson (1) reported that pea leaves contained distinct chloroplast
and cytoplasmic isoenzymes of PGAK, but that the isoenzymes possessed
closely similar catalytic properties. The plant PGAK isoenzymes were
subsequently neglected until the procedure of Kuntz et al. (2) offered the
possibility of a several hundred-fold purification by affinity
chromatography. The technique was applied to barley in this laboratory and
subsequent separation of the PGAK isoenzymes was achieved by both
non-dissociating polyacrylamide gel electrophoresis and gel
isoelectrofocusing (3). The preparative use of these isoenzyme separation
techniques produced considerable difficulties and eventually the two
isoenzymes were separated from each other on a DEAE Sephacel column and
each of them was then separately purified (700-900 fold) by affinity
chromatography on ATP Sepharose.

Milligram quantities of the isoenzymes were obtained by these
procedures and were used in a comparative investigation of the kinetic
properties of the isoenzymes (4) and the comparative investigation of the
immunological properties of the isoenzymes described below.

MATERIALS AND METHODS

In the preparation of the isoenzymes, it was clear that the DEAE
Sephacel step should precede the use of the affinity column in order to
obtain a concentrated product from the final step. When not in use the
affinity gel was placed in 30% glycerol, 50mM Tris, pH 8.0, 5mM $MgSO_4$, and
stored at -20°C. This procedure allowed the gel to be re-used several
times.

In a typical preparation, 640g of leaves from 7-day old
light-grown Hordeum vulgare c.v. Golden Promise seedlings were homogenised
in 1 l of 0.3M Tris-HCl, pH 8.0. The homogenate was squeezed through 3
layers of butter muslin prior to 1 h centrifugation at 30,000 g at 5°C and
overnight dialysis of the supernatant against 5 l of a solution of 50mM
Tris, pH 8.0 at 5°C. After further centrifugation and addition of
5mM $MgCl_2$, the supernatant was applied to a 500ml column of DEAE Sephacel
equilibrated with 40mM Tris, pH 8.0, and 5mM $MgCl_2$ buffer. The cytoplasmic
PGAK was found to comprise the 10% of PGAK activity which did not bind to
the column. The chloroplast PGAK was eluted from the DEAE Sephacel by
means of a gradient from 0 to 0.4M NaCl, produced from a lower reservoir
containing 500ml of Tris/$MgCl_2$ buffer and an upper reservoir containing
500ml of Tris/$MgCl_2$ buffer with 0.4M NaCl. The chloroplast PGAK eluted as
a very sharp peak. Both isoenzymes were either frozen until required or
subjected to affinity chromatography immediately.

Biggens, J. (ed.), Progress in Photosynthesis Research, Vol. III. ISBN 90 247 3452 5
© *1987 Martinus Nijhoff Publishers, Dordrecht. Printed in the Netherlands.*

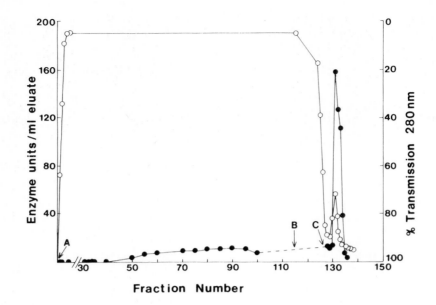

FIGURE 1. Affinity chromatography of a crude extract of barley leaves
on a 60ml ATP Sepharose column, fraction volume 10ml.
A, extract applied;
B, 50mM Tris, pH 8.0, and 5mM $MgSO_4$, buffer added;
C, 50mM Tris, pH 8.0, and 0.7M NaCl, added.
Closed circles, PGAK activity;
Open circles, % transmission at 280nm.

Each isoenzyme was separately purified by affinity
chromatography on ATP Sepharose (2). The fractions containing the
isoenzymes were dialysed overnight against 50mM Tris, pH 8.0, and made up
to 5mM $MgSO_4$ before being passed through the affinity column equilibrated
with the same buffer. The particular PGAK isoenzyme was eluted by a
solution comprising 50mM Tris, 40mM $MgSO_4$, and 40mM ATP, pH 8.0.
Fig. 1 illustrates the use of the affinity column in the
purification of the PGAK in a crude extract of barley leaves, in which
case PGAK was eluted with 0.7M NaCl in 50mM Tris, pH 8.0.
The purified PGAK isoenzymes were stored at -20°C. Prior to
further investigation, the ATP and $MgSO_4$ were removed by dialysis. When
necessary the concentration of each isoenzyme was increased with a
Centricon YM10 concentrator. PGAK activity was assayed as described
elsewhere in this volume (4).
Antibodies were raised in New Zealand white rabbits by
subcutaneous injection of an emulsion of 0.5ml enzyme (approximately 200µg
protein) and 0.5ml Freund's complete adjuvant every 14 days. The reactions
between the antisera and the isoenzymes were tested by Ouchterlony double
diffusion, crossed immunoelectrophoresis and immunoprecipitation.

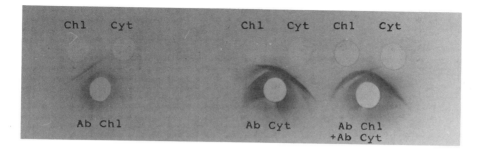

FIGURE 2. Ouchterlony double diffusion of chloroplast
and cytoplasmic isoenzymes (Chl and Cyt) against antisera to the
chloroplast and cytoplasmic isoenzymes (Ab Chl and Ab Cyt).
10µl was used in each well except in the case of the mixed
antisera where 5µl of each antiserum was used.
The gel was stained with Coomassie blue R250.

RESULTS AND DISCUSSION

The chloroplast isoenzyme amounts to approximately 90% of the
PGAK activity in barley leaves, the remainder being accounted for by the
cytoplasmic isoenzyme. This is similar to the situation reported for
fructose-1,6 bisphosphate aldolase in leaves of four species (5).

No evidence of the presence of contaminating polypeptides was
found in the course of gel electrophoresis of the purified PGAK
isoenzymes.

In Fig.2 the Ouchterlony double diffusion shows that the
chloroplast isoenzyme raised a low titre antiserum which gave a fairly
weak precipitin line with the chloroplast isoenzyme and a scarcely
discernible line against the cytoplasmic isoenzyme. In contrast the
cytoplasmic isoenzyme yielded a high titre antiserum which reacted most
strongly with the cytoplasmic isoenzyme. This antibody cross-reacted with
the chloroplast isoenzyme. The spur in Fig.2 provides evidence of a single
dissimilarity cross-reaction. This indicates that the antiserum recognises
common epitopes between the two isoenzymes and also that the cytoplasmic
isoenzyme has at least one additional epitope which is absent in the
chloroplast isoenzyme.

The mixed antisera gave evidence of a spur which indicates that
the chloroplast isoenzyme also contains an additional epitope which is not
present in the cytoplasmic isoenzyme.

The higher titre of the cytoplasmic antiserum shows up clearly
in Fig.3 in the immunoprecipitation titrations for each isoenzyme against
each antiserum. The greater efficiency of each antiserum in precipitating
its own isoenzyme is consistent with the isoenzymes being very similar
whilst possessing some different antigenic properties.

In crossed immunoelectrophoresis the isoenzymes migrated at
different rates in the first dimension so that clearly distinct but
related precipitin lines were produced. The largest peak representing the
chloroplast isoenzyme was seen to migrate further than the cytoplasmic
isoenzyme.

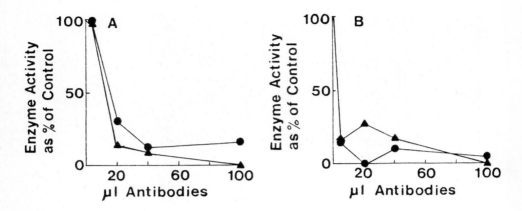

FIGURE 3. Immunoprecipitation titrations of two
antisera against the chloroplast and cytoplasmic isoenzymes.
Procedure a modification of that of Kruger <u>et al.</u> (6)
with protein A omitted.
A, chloroplast antiserum v chloroplast (▲) and cytoplasmic (●)
isoenzymes.
B, cytoplasmic antiserum v chloroplast (▲) and cytoplasmic (●)
isoenzymes.

Although the chloroplast and cytoplasmic isoenzymes possess some
similar kinetic and antigenic properties, they also must be presumed to
have differences in amino acid sequence which result in some distinct
differences in their kinetic (4), antigenic and physical properties. The
determination of their amino acid sequences is intended to be part of this
continued investigation.

ACKNOWLEDGEMENTS
Grateful acknowledgement is made to Professor F.E.G. Cox and Dr.
E.A. Ebringer for assistance in the production of antisera.

REFERENCES
1. Pacold, I. and Anderson, L.E. (1975) Plant Physiol. 55, 168-171.
2. Kuntz, G.W.K., Eber, S., Kessler, W., Krietsch, H. and Krietsch, W.K.G.
   (1978) Eur. J. Biochem. 85, 493-501.
3. Allsop, D., Atkinson, Y.E. and Bradbeer, J.W. (1981) in
   Photosynthesis. V. Chloroplast Development (Akoyunoglou, G.ed.)
   pp 787-795, Balaban, Philadelphia.
4. McMorrow, E.M. and Bradbeer, J.W. (1986) this volume.
5. Schnarrenberger, C. and Kruger, I. (1986) Plant Physiol. 80, 301-304.
6. Kruger I. and Schnarrenberger C. (1983) Eur. J. Biochem. 136, 101-106.
7. Porro, M., Viti, S., Antoni, G. and Saletti, M. (1982) Analytical
   Biochem. 127, 316-321.

A NOVEL TYPE OF PHOSPHOFRUCTOKINASE FROM PLANTS

Joshua H. Wong, Boihon C. Yee, and Bob B. Buchanan,
Division of Molecular Plant Biology, Hilgard Hall,
University of California, Berkeley, CA 94720

1. INTRODUCTION

Phosphofructokinase (PFK), one of two enzymes that phosphorylates Fru-6-P to Fru-1,6-P$_2$, catalyzes a key regulatory reaction in glycolysis. It is established that plants contain cytosolic and plastidic types of PFK and that these types differ from one another in certain respects, e.g., response to Pi, and are similar in others, i.e., inhibition by ATP, PEP, citrate and NADPH and deinhinhibition by Fru-6-P (1-6). The plastidic type of PFK has been purified to homogeneity from developing castor bean endosperm (1) and Chlorella (6), but the cytosolic type has not. Significantly, the regulation of plant PFK, especially the cytosolic type, is not clear, and the subunit molecular weight is not established.

We now report the purification of PFK to homogeneity from carrot roots. Purified PFK, presumably the cytosolic type, could exist in an array of forms that differed in regulatory properties and molecular weight (240,000 to greater than 5 million). Furthermore, the different forms of the cytosolic PFK could be reversibly converted from one to another by appropriate metabolites. A similar preparation was partially purified from spinach leaves. The results suggest that plant cytosolic PFK is novel in its subunit molecular weight, immunological properties, and effect of regulatory metabolites on activity, but resembles certain mammalian counterparts in its ability to undergo metabolite-dependent aggregation and disaggregation.

2. MATERIALS AND METHODS

PFK was purified at 4°C from 1-2 kg carrot roots that had been homogenized in 2-4 l of 50 mM Tris-HCl, pH 8.0, 10% glycerol, 2 mM EDTA, 14 mM 2-mercaptoethanol, 20 mM Na diethyldithiocarbamate, 5 mM MgCl$_2$, 1.5% w/v insoluble PVP, 50 mM KF, 0.1 mM ATP, 0.5 mM PMSF, 2 mM ε-amino-n-caproic acid, and 2 mM benzamidine-HCl. The 5-15% polyethylene glycol fraction, containing the bulk of the PFK activity, was dissolved in Buffer A (50 mM Tris-HCl, pH 8.0, 10% glycerol, 1 mM EDTA, 14 mM 2-mercaptoethanol, 80 mM KCl and 0.1 mM ATP), and was further purified by sequential chromatography on columns of DE-52 (2.5x47 cm, eluted with a linear 0.08 to 0.5 M KCl gradient in Buffer A), Sepharose 4B (3.5x60 cm, equilibrated and eluted with modified Buffer A - 5 mM Fru-6-P replacing ATP), and Blue Sepharose (eluted with 10 mM MgCl$_2$ and 5 mM ATP). For FPLC (Pharmacia) Superose 6 chromatography experiments, an aliquot of the dialyzed and concentrated PFK from Sepharose 4B was applied to a HR10/30 column that was equilibrated and eluted with the indicated metabolite in Buffer A without ATP. Native molecular weights were determined at 25°C by gel filtration with the FPLC Superose 6 column. Subunit M_r was determined in a 7.5-15% PAGE in the presence of SDS. PFK assay and protein determination were as described previously (7).

3. RESULTS AND DISCUSSION
The five step procedure resulted in an 800-fold purification and a yield of about 1%. The final specific activity of 79 units/mg protein compares favorably to corresponding values of 50-90 for rabbit liver PFK (8), 29 for algal plastid PFK (6) and 22 for endosperm plastid PFK (1). However, homogeneous carrot PFK was not stable and more than 50% of its activity was lost in 3 to 5 days when stored at − 20°C. Superose 6 gel filtration in the presence of 5 mM Fru-6-P, revealed a large form of PFK, with an average M_r greater than 5 milliom, and a smaller form, with an average M_r of about 300,000 (Fig. 1 upper panel). When 5 mM ATP replaced Fru-6-P, PFK was recovered in a form with an average M_r of about 600,000 (Fig. 1 lower panel).

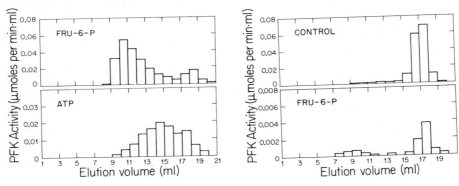

Fig. 1. Superose profiles showing dissociation of PFK (L).
Fig. 2. Superose profiles showing reasociation of PFK (R).

Significantly, the different forms of PFK were found to be interconvertible. The small forms generated with ATP-treatment and isolated by FPLC Superose chromatography (Fig. 2 upper panel) were converted in part (about 25%) to their larger counterparts when treated with 5 mM Fru-6-P (Fig. 2 lower panel). No such conversion was observed in parallel experiments conducted without Fru-6-P. Similar changes were observed with the cytosolic PFK from spinach leaves.
The three forms of carrot PFK all showed hyperbolic kinetics at pH 8.0, but differed in their response to ATP, citrate, Pi and pH. Thus, while the small and intermediate forms were inhibited by ATP and citrate, the large form was unaffected by both metabolites. In the case of Pi, activation was

observed at pH 6.5 but not at pH 8.0, and then only with the large and intermediate forms. Fructose-2,6-P_2 (1 μM) has only slight effect on the three forms of PFK. The three PFK forms also showed specific difference with respect to optimal pH and Mg^{+2} and ATP concentration, but resembled one another in being inhibited both by PEP and NADPH.

The M_r of native carrot PFK is heterogeneous as mentioned above. However, after electrophoresis in a 7.5-15% gradient polyacrylamide gel in the presence of SDS, the purified PFK showed a single silver-staining component (M_r = 60,000) that reacted in Western blots with anti carrot root PFK but not with control serum (Fig. 3).

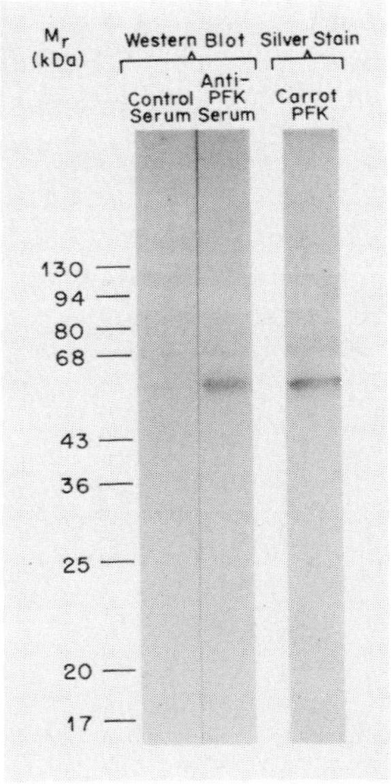

Fig. 3. Silver stain and Western blot analysis of carrot root PFK by SDS-PAGE.

In other experiments, the activity of this same PFK preparation was inhibited by the anti carrot PFK serum. We concluded that the 60 kDa component corresponds to the subunit of carrot root cytosolic PFK. The 68 kDa component, seen as a weak band in Fig. 3, was observed in all gel lanes irrespective of the protein added and in most cases even between the lanes. We concluded that this band is an artifact of the electrophoresis system.

Immunologically, the carrot root and rabbit muscle PFK enzymes were largely distinct. Each PFK reacted selectively with the corresponding antiserum as determined by both ELISA analysis (data not shown) and activity measurements. In the latter case, 25 μl of the anti carrot serum inhibited carrot root PFK activity by 80% and muscle PFK activity by only 14%. Similar, but opposite, results were obtained in parallel studies with anti rabbit muscle PFK serum from guinea pig. In both cases, control sera had no effect on PFK activity.

The subunit M_r of the carrot root PFK warrants further comment. While appreciably less than muscle PFK (60,000 vs. 80,000), the M_r of the carrot PFK subunit is similar if not identical to the smaller of the two subunits of the major form of PFP and to the only subunit of a reported second form of PFP (9). To our knowledge, such a M_r for PFK has not been reported previously.

It should be noted that purified carrot PFK had neglegible PFP activity. The earlier reported enhancement of PFP activity in PFK preparations by UDP-glucose (7,10) was found in the current study to be due to UTP that was generated in the reaction mixture in the presence of PPi by UDP-glucose pyrophosphorylase contaminating the α-glycerophosphate dehydrogenase coupling enzyme used for assay. Thus, as recently suggested by others (11), UDP-glucose served to generate UTP substrate for PFK and not to convert PFK to PFP as we had earlier concluded (7,10). In short, while the current findings suggest that PFK and PFP contain a subunit of similar M_r, further work is needed to determine if there are other fundamental similarities in these proteins.

4. CONCLUSION

Plants have a novel type of PFK, presumably in the cytosol. The homogeneous enzyme from carrot roots was found to differ from its mammalian counterpart in (a) subunit molecular weight, (b) immunological properties, and (c) response to fructose-2,6-bisphosphate. But similar to mammalian PFK, the plant enzyme also existed in a variety of aggregation forms. These forms were interconvertible: Fru-6-P promoted aggregation and ATP favored disaggregation. Significantly, each of the major forms showed a particular set of regulatory characteristics. The results suggest that metabolite-dependent aggregation-disaggregation is a mechanism whereby plants regulate the activity of cytosolic PFK and the accompanying rate of glycolytic carbon flux.

REFERENCES

1. Garland, W.J. and Dennis, D.T. (1980) Arch. Biochem. Biophys. 204, 302-309
2. Turner, J.F. and Turner, D.H. (1980) in The Biochemistry of Plants, (Davies, D.D., ed.), Vol. 2, pp. 279-316, Academic Press
3. Cséke, C. et al. (1982) Plant Physiol. 70, 658-661
4. Isaac, J.E. and Rhodes, M.J.C. (1986) Phytochem. 25, 339-343
5. Kombrink, E. and Wöber, G. (1982) Arch. Biochem. Biophys. 213, 602-619
6. Kelly, G.J. et al. (1985) Plant Cell Physiol. 26, 301-307
7. Balogh, Á., et al. (1984) FEBS Lett. 169, 287-292
8. Kemp, R.G. (1971) J. Biol. Chem. 246, 245-252
9. Yan, T.F.J. and Tao, M. (1984) J. Biol. Chem. 259, 5087-5092
10. Wong, J.H., et al. (1984) Biochem. Biophys. Res. Comm. 121, 842-847
11. Kruger, N.J. and Dennis, D.T. (1985) Biochem. Biophys. Res. Comm. 126, 320-326

[Supported by a grant from the National Science Foundation.]

STUDIES ON A MULTIENZYME COMPLEX CONTAINING RuBP CARBOXYLASE, PHOSPHORIBO-
ISOMERASE AND PHOSPHORIBULOKINASE

JAYASHREE K. SAINIS and GARY C. HARRIS
DEPARTMENT OF BIOLOGICAL SCIENCES, WELLESLEY COLLEGE, WELLESLEY, MA 02181

1. INTRODUCTION
In some prokaryotes it has been reported that the enzymes of the photo-synthetic carbon reduction cycle are associated in polyhedral bodies called carboxysomes (1). In eukaryotes where the high protein concentration in the stroma (2) makes protein-protein interactions inevitable, aggregates of photosynthetic enzymes have not been frequently reported. Recently we have observed that RuBPCase in pea chloroplasts was associated with phosphoribu-lokinase and phosphoriboisomerase (3) and had a molecular weight of 800,000 to 850,000 daltons. Here we report some of the kinetic properties of the RuBPCase complex and discuss the potential significance of the influence of the aggregated state of the enzymes on their properties.

2. PROCEDURE
Materials and methods:
Chloroplasts were isolated from 12-20 day old shoots of Pisum sativum (Little Marvel) (4). The stromal fraction was processed on 5-27% sucrose gradients in a vertical rotor (TV-865) in a Sorvall SA60 as detailed in (3). The enzyme activities were determined in the faster sedimenting protein peak. The RuBPCase complex was assayed for R-5-P, Ru-5-P and RuBP depend-ent CO_2 fixation after preincubation for 30 min with 20 mM bicarbonate, 10 mM $MgCl_2$ and 10 mM DTT in 50 mM Bicine buffer pH8. The reaction was initiated with RuBP (2 mM), Ru-5-P (2 mM) or R-5-P (10 mM) and ATP (2 mM). The reaction was terminated by 6 N acetic acid and ^{14}C labelled product was measured according to Lorimer (5).

3. RESULTS AND DISCUSSION
Previously we had observed that RuBPCase purified by sedimentation on sucrose gradients showed R-5-P, Ru-5-P and RuBP dependent CO_2 fixation activities (3). However, the R-5-P and Ru-5-P dependent activities re-presented only 4-5% of total RuBP dependent activity. In these experiments GSH was used as sulfhydryl agent. There was also a lag of 5-7 min with R-5-P and Ru-5-P as substrates and maximum velocities were low. In the experiments reported here we preincubated the enzyme with DTT before assay. Fig. 1 shows the R-5-P and Ru-5-P dependent activities compared to RuBP dependent CO_2 fixation activity. It was observed that Ru-5-P dependent CO_2 fixation rates were equivalent to RuBP dependent rates after a 30 sec lag (Fig. 1). R-5-P showed lesser activity. ATP, Ru-5-P and R-5-P were found to inhibit RuBP dependent CO_2 fixation activity (Table 1). This assay has, therefore, these inherent limitations. The kinetic constants were measured using this assay mixture and the values are reported in Table 2. The Km value for R-5-P was 0.83 mM compared to 2.3 mM for the purified enzyme (6). The Km values for ATP and Ru-5-P were 0.083 mM and

Biggens, J. (ed.), Progress in Photosynthesis Research, Vol. III. ISBN 90 247 3452 5
© *1987 Martinus Nijhoff Publishers, Dordrecht. Printed in the Netherlands.*

0.5 mM respectively whereas purified kinase was reported to have Km values of 0.069 mM for ATP and 0.17 mM for Ru-5-P (7). Ru-5-P dependent activity showed Vmax of 1.33-1.66 umoles min^{-1} mg^{-1} protein. If a correction is applied for the inhibitory effects of ATP and Ru-5-P, the theoretical Vmax would be much higher. R-5-P dependent rates were consistently lower than Ru-5-P dependent rates. This may be due to either less isomerase associated with the complex or may be due to incorrect assay conditions for the complexed isomerase.

TABLE 1. Inhibition of RuBP dependent activity in the complex by R-5-P, ATP and Ru-5-P.

	K_I	Type of Inhibition
R-5-P	15.0 mM	Comp.
ATP	6.4 mM	Comp.
Ru-5-P	5.2 mM	Comp.

TABLE 2. Kinetic constants of R-5-P and Ru-5-P dependent CO_2 fixation activity of the complex.

	K_M	Vmax
R-5-P	0.833 mM	0.142 umoles min^{-1} mg^{-1} protein
ATP	0.083 mM	1.330 " " " "
Ru-5-P	0.500 mM	1.660 " " " "

FIGURE 1. Time course of R-5-P (●---●), Ru-5-P (o---o) and RuBP (x---x) dependent CO_2 fixation by the complex after preincubation with 10 mM DTT in activation buffer. One unit is equal to 1 μmole of CO_2 fixed min^{-1} mg^{-1} protein.

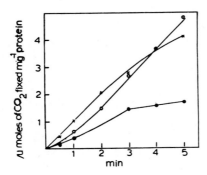

The complex showed a linear rate of Ru-5-P dependent CO_2 fixation for 15 min whereas RuBP dependent activity was linear only for 1-2 min (Fig. 1 and 2). RuBPCase has been shown to undergo rapid inactivation in the assay mixture (5). RuBP has also been demonstrated to inhibit activation of the enzyme (8). The RuBPCase complex may provide some protection from these inhibitory effects of RuBP and thus may be important in vivo. These en- zymes probably exist in the complex in a fixed stiochiometry such that RuBP is provided to RuBPCase at the proper rate to prevent inactivation. In our experiments it was observed that externally added kinase results in inhibition of Ru-5-P dependent CO_2 fixation when measured for 10-20 min periods (data not shown). When the chloroplast stroma or crude extracts of spinach were fractionated on sucrose gradients, RuBPCase showed R-5-P and Ru-5-P dependent CO_2 fixation activities (Table 3).

TABLE 3. Specific activities of R-5-P, Ru-5-P and RuBP dependent CO_2 fixation in RuBPCase recovered from sucrose gradient centrifugation of extracts from spinach leaves.

| | nmoles of CO_2 fixation min^{-1} mg^{-1} protein | | |
	R-5-P dependent	Ru-5-P dependent	RuBP dependent
Crude extracts	195.07	333.16	834.98
Stromal extracts	397.49	549.09	846.09

Crude extracts and stroma were loaded on sucrose gradients and pro- cessed as given in Methods. The activities were measured after pre- incubation with DTT.

FIGURE 2. Same as Fig. 1 except that reaction rates were measured for 1 hr

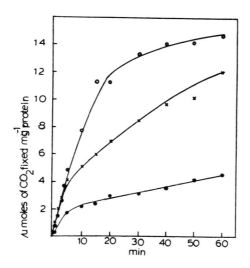

4. CONCLUSIONS

These observations and our earlier observations (3) have demonstrated that RuBPCase readily associates with phosphoribulokinase and phospho-riboisomerase. This association would have an advantage over the freely soluble enzyme systems in protecting the RuBPCase during catalysis _in vivo_. This association also imparts sulfhydryl sensitivity to RuBPCase because the Ru-5-P dependent CO_2 fixation is minimal in the absence of sulfhydryl agents. Thus, RUBPCase would become light sensitive by virtue of its association with the kinase.

Abbreviations: R-5-P: Ribose-5-phosphate; RuBP: Ribulose-1,5-bisphosphate; Ru-5-P: Ribulose-5-phosphate.

REFERENCES
1 Beudeker, R.F. and Keunen, J.G. (1981) FEBS Lett. 131, 269-274.
2 Ellis, R.J. (1979) Trends Biochem. Sci. 4, 241-244.
3 Sainis, J.K. and Harris, G.C.(1986) communicated to BBRC.
4 Cerovic, Z.G., Sivak, M.N. and Walker, D.A. (1984) Proc. R. Soc. Lond. 220, 327-338.
5 Lorimer, G.H., Badger, M.R. and Andrews, T.J. (1977) Anal. Biochem. 78, 66-75.
6 Anderson, L.E. (1971) Biochim. Biophys Acta 235, 245-249.
7 Anderson, L.E. (1973) Biochem. Biophys Acta 321, 484-488.
8 Jordan, D.B. and Chollet, R. (1983) J. Biol. Chem. 258, 13752-13758.

This work was supported by a National Science Foundation Grant to GCH DMB-8410.

SEPARATION OF CALVIN CYCLE ENZYMES, FROM SPINACH CHLORO-
PLASTS, BY AFFINITY PARTITIONING WITH TRIAZINE DYE LIGANDS

Lars-Olof PERSSON and Göte JOHANSSON, Dept. of biochemistry,
Chemical Center, POB 124, S-221 00 LUND, Sweden

1. INTRODUCTION
Partition between two aqueous phases is a technique for
protein purification which is both rapid and easy to
scale up. The partition of proteins between the phases
can be significantly and selectively influenced by covalently binding various chemical groups to one of the
polymers used to obtain the two phases (Johansson 1985).
Most widespread is the application of specific ligands,
so-called affinity partitioning. Especially the triazine
dyes which are cheap and easily bound to the phase-
forming polymers, dextran and polyethylene glycol, have
been used. We have explored the possibility of purification of a few Calvin cycle enzymes from spinach chloroplasts by using affinity partitioning with triazine dye
ligands together with batchwise DEAE-cellulose treatment.

2. MATERIALS AND METHODS
All manipulations were carried out at 4 ± 1 °C unless
otherwise indicated.
2.1 <u>Materials</u>: Dextran, M_r=500 000 was obtained from
Pharmacia, polyethylene glycol (PEG) from Union Carbide,
DEAE-cellulose from Whatman (DE-52) and biochemicals from
Sigma. Triazine dyes were bound to dextran and PEG
(Johansson et al 1985) using Procion dyes from ICI. All
other chemicals were of analytical grade.
2.2 <u>Preparation of chloroplast extracts</u>: Chloroplasts from
spinach (**Spinacia oleracia**) were prepared according to
Jensen and Bassham (1966) with some modifications, and
stored on ice as pellets. Chloroplast extracts were prepared immediately before use by suspending the pellets in
25 mM sodium phosphate buffer, pH 7.5, containing 0.1 mM
$MgCl_2$-Na_2EDTA and 1 mM 2-mercaptoethanol and treating the
suspension in a Potter-Elvehjem glass-teflon homogenizer
on ice (20 strokes by hand) followed by centrifugation
(Sorvall SE-12) at 20 000 rpm (41 500xg at r_{max}) for 10
min. The resulting clear supernatant showed a slight
green colour and was used without further manipulations.

Biggens, J. (ed.), Progress in Photosynthesis Research, Vol. III. ISBN 90 247 3452 5
© *1987 Martinus Nijhoff Publishers, Dordrecht. Printed in the Netherlands.*

2.3 <u>Counter-current distribution (CCD)</u>: Two-phase systems containing 8%(w/w) dextran, 6%(w/w) PEG, 25 mmole/kg sodium phosphate buffer, pH 7.5, 0.1 mmole/kg $MgCl_2$-Na_2EDTA, 1 mmole/kg 2-mercaptoethanol and, occasionally (Fig. 1) 0.06 %(w/w) Procion Turquoise MX-G PEG (1% of total PEG) were made up from stock solutions (Johansson 1984). 2.05 ml of the mixed system were loaded into each of the 60 chambers of a centrifugal CCD apparatus (Åkerlund 1984). Lower phases and small portions of the upper phases were stationary. In Fig. 2 the lower phases also contained a concentration gradient of Procion Yellow HE-3G dextran (0-4% of total dextran). Samples were included in chambers no 0-2 (Fig. 1, chloroplast extract) or no 0-4 (Fig. 2, material extracted from chloroplast extract with Procion Turquoise PEG, see section 2.4) After the desired number of transfers (shaking for 5 min, transfer and centrifugation at 80xg for 5 min) 2.5 ml of buffer were added to each chamber giving one-phase solutions to be analyzed.

2.4 <u>Batch procedures</u>: The chloroplast extract was included in 10 g of a two-phase system containing Procion Turquoise PEG. The phases were mixed (60 s) and separation was enhanced by low speed centrifugation. The upper phase, containing material extracted by the ligand-PEG was washed with pure lower phase and either used in CCD (Fig. 2) or diluted 5 times with buffer(as before except 5mM phosphate) and treated with DEAE-cellulose (0.5 ml packed gel/μmole min^{-1} aldolase activity), directly or after partitioning with lower phase containing Procion Yellow HE-3G dextran (4% of total dextran). In the latter case the resulting lower phase was also treated with DEAE-cellulose. The DEAE-cellulose was then washed with the same volume buffer and treated on a glass filter funnel with buffer containing various concentrations of KCl.

2.5 <u>Assays</u>: Enzyme activities were measured at 340 nm and 25°C; phosphoglycerate kinase according to Scopes(1975), except that albumin was omitted and the initial velocity was determined, and aldolase according to Horecker (1975). Protein was determined according to Bradford (1976) at 595 nm using bovine serum albumin as standard and KCl concentration by conductivity measurements.

3. RESULTS AND DISCUSSION
The two Calvin cycle enzymes phosphoglycerate kinase and aldolase exhibited a higher partitioning to the upper, PEG-enriched phase than the main material (Fig. 1) when CCD was performed on spinach chloroplast extracts using a

two-phase system with PEG-bound Procion Turquoise MX-G ligand (1 % of total PEG). This shows the possibility of separating the two enzymes from the bulk protein by a rapid partitioning followed by washing with pure lower phase. Two strategies have been explored to separate the two enzymes. The first involves competition by a second dextran-bound dye ligand enriched in the lower phase. This is illustrated in Fig. 2 where CCD was performed on the material extracted to the upper phase by a single partitioning in a system with Procion Turquoise PEG (1 % of total PEG). The CCD was carried out with two-phase systems where the lower phases contained a concentration gradient of dextran-bound Procion Yellow HE-3G. The two enzyme activities were separated and the protein gave a broad peak with maximum coinciding with aldolase. The second strategy is ion-exchange chromatography which is illustrated in Fig. 3. Here the upper phase from the single partitioning was diluted with buffer and treated with DEAE-cellulose as described in section 2.4 above. As can be seen in Fig. 3 the two enzymes are eluted in different positions when a KCl-gradient is applied. The same procedure was performed on upper and lower phases after batch partitioning with Procion Turquoise PEG followed by extraction with a new lower phase containing Procion Yellow dextran as described in section 2.4 (data not shown). From these results preliminary batch procedures, including single partitioning steps and batchwise DEAE-cellulose treatment have been developed. The preparation time is only 60-90 min and the procedure can easily be scaled up.

ACKNOWLEDGMENTS

This work has been supported by grants from the National Swedish Board for Technical Development.

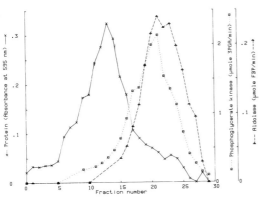

Fig. 1: CCD of chloroplast extract in a two-phase system containing Procion Turquoise PEG (1% of total PEG).
27 transfers.

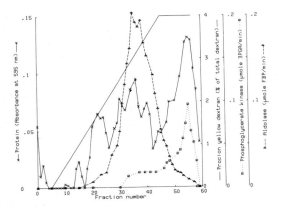

Fig. 2: CCD of material extracted from chloroplast extract with Procion Turquoise PEG in two-phase systems containing a concentration gradient of Procion Yellow dextran (0-4 % of total dextran). 55 transfers.

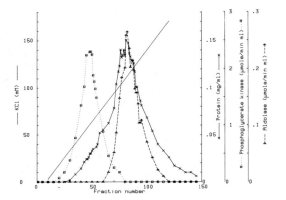

Fig. 3: Ion-exchange chromatography on DEAE-cellulose of material extracted from chloroplast extract with Procion Turquoise PEG.

REFERENCES

1. Johansson, G. (1985) J. Biotechnol. 3, 11-18
2. Johansson, G., Joelsson, M. and Åkerlund, H.-E. (1985) J. Biotechnol. 2, 225-237
3. Jensen, R.G. and Bassham, J.A. (1966) Proc. Nat. Acad. Sci. U.S.A. 56, 1095-1101
4. Johansson, G. (1984) Methods Enzymol. 104, 356-364
5. Åkerlund, H.-E. (1984) J. Biochem. Biophys. Methods 9, 133-141
6. Scopes, R.K. (1975) Methods Enzymol. XLII, 127-134
7. Horecker, B.L. (1975) Methods Enzymol. XLII, 234-239
8. Bradford, M.M. (1976) Anal. Biochem. 72, 248-254

ISOLATION OF INTACT CHLOROPLASTS FROM CHLAMYDOMONAS REINHARDTII WITH BECKMAN CENTRIFUGAL ELUTRIATION SYSTEM:

Robert K. Togasaki[1], Karen Brunke[2], Masahiko Kitayama[1], and O. Mitchel Griffith[3]. [1]Department of Biology, Indiana U., Bloomington, IN 47405, [2]Zoecon Co., 975 California Ave., Palo Alto, CA 94304, [3]Beckman Instrument Inc., Palo Alto, CA 94304.

INTRODUCTION:

The unicellular green alga Chlamydomonas reinhardtii has been widely used in genetic, molecular, and biochemical studies of photosynthesis (1,2). Isolation of intact chloroplasts from this alga has been difficult, due to the large size of the single chloroplast present within each cell. Klein et al. (3) have reported a method of isolating functional chloroplasts from wild type cells of Chlamydomonas, utilizing autolysine (4) to remove the cell wall and detergents to differentially solubilize the plasma and organellar membranes. Belknap (5) used autolysine for cell wall removal and pressure shock with a Yeda press (6) for plasma membrane disruption to isolate functional intact chloroplasts from both the wild type and a mutant, F60, lacking phosphoriulokinase. L. Mendiola-Morgenthaler et al. (7) combined the Yeda press method and cell wall deficient mutant, CW-15-2, to isolate intact chloroplasts with high rates of both photosynthesis and light dependent organellar protein synthesis. In all of the above published procedures, Percoll step gradients were used to separate intact chloroplasts from unbroken cells and other subcellular components. Recently, Lin et al. (8) applied centrifugal elutriation to the isolation of intact chloroplasts from wheat. In this paper, we describe the application of the same Beckman centrifugal elutriation system to the isolation of intact chloroplasts from CW-15 cells.

MATERIALS AND METHODS:

Culture Condition: A cell wall deficient mutant strain of Chlamydomonas reinhardtii, CW15 (CC1616), was obtained from the Chlamydomonas National Collection at Duke University. Cells were cultured in Tris/acetate/phosphate (TAP) medium as described in Gorman and Levine (9), and cultures with cell densities between 1-2 x 10^6 cells/ml were used.

Preparation of pressate: Cell pressatee were prepared following the procedure of Belknap (5) with minor modifications. Cells were harvested with a Sorvall GSA rotor at 4000 RPM x 3 min. After two washes with HEPES buffer (20 mM, pH 7.0, 25°C), 1 x 10^9 cells were resuspended in 10 ml of Breaking Media (300 mM sorbitol, 50 mM Mes-Tris buffer (pH 7.5), 2 mM EDTA, 1mM $MgCl_2$, and 1% (w/v) BSA (Sigma A6003), 3 mM KH2PO4, 25°C, final pH 7.2). This suspension was rapidly cooled to 5°C in an ice bucket containing an ice/water slurry. The chilled cell suspension was placed in the ice-cold Yeda Press, and equibrated under N_2 gas pressure (100 psi) for 3 minutes. The needle valve of the Yeda press was turned rapidly to a preset position, and the cell suspension was released into a chilled polypropylene tube. This preparation, or pressate (10) was further fractionated by cetrifugal elutriation.

Biggens, J. (ed.), Progress in Photosynthesis Research, Vol. III. ISBN 90 247 3452 5
© *1987 Martinus Nijhoff Publishers, Dordrecht. Printed in the Netherlands.*

Elutriation: The centrifugal elutriation system consists of the following components as shown in the Figure 1: A beaker containing Washing Buffer (Breaking media with 0.1% BSA), a 25 ml plastic syringe attached to a 3-way injection valve located between the buffer reservoir and a peristaltic pump (Cole Parmer Master-Flex Pump with digital readout, model R-7523-10, with 7016 standard head and size 16 tubing), Beckman Elutriation Rotor Model JE-6B, with a standard chamber in Beckman Model J21C centrifuge and a sample receptacle.

1. 50 mls of ice cold Washing Buffer were passed through the elutriation system, with the elutriation rotor in a Beckman J21C centrifuge, running at 4000 RPM, with the 3-way valve set as seen in Figure 2B. The initial flow rate of the peristaltic pump was set at 13 ml/min.

2. 10 mls of pressate (or 1×10^9 cells) was poured into a chilled 25 ml plastic syringe attached to the 3-way injection valve with the reservoir and the pump still connected (Figure 2B).

3. The sample was pulled into the flow system when the 3-way valve connecting the syringe and pump was opened (Figure 2A). When the pressate syringe was nearly empty, the valve was turned again to connect the Washing Buffer reservoir to the pump (Figure 2B).

4. 100 ml of Washing Buffer was allowed to flow through while 50 ml fractions were collected. If the initial Yeda Press breakage was gentle enough, we have seen very little chlorophyll (and hence thylakoid fragments) here.

5. The buffer flow rate was increased to 23 ml/min. and another 100 ml of washing buffer was allowed to flow through. First 50 ml fraction contained most of chlorophyll coming off at this flow rate setting and it contained intact chloroplasts.

6. At higher flow rate, unbroken cells began to emerge.

Analysis of isolated chloroplasts: Each fraction coming out of elutriation system was examined by phase contrast microscopy, and those fractions showing cup shaped chloroplasts were concentrated by centrifugation (Sorvall SS34 rotor, 3000 G x 30 sec), and resuspended in Reaction Medium (5) (300 mM sorbitol, 50 mM Mes-Tris buffer (pH 7.5), 2 mM EDTA, 1 mM $MgCl_2$, 1 mM $MnCl_2$, 8 mM $NaHCO_3$, 3 mM KCl, 0.15% (w/v) BSA, and 1 mM KH_2PO_4) 3PGA was added where indicated with final concentration of 1 mM. PEP carboxylase activities were determined as described (5).

RESULTS AND DISCUSSION:

The Beckman centrifugal elutriation system separates suspended particles under constant G force by the flow rate gradient of washing buffer, generated by the elutriation chamber. Since washing buffer enters the narrow end of the conical elutriation chamber at a constant flow rate and then moves with decreasing flow rate as the cross section of the chamber widens, a flow rate gradient is generated, and suspended particles will be distributed according to their shape, size and density. The smallest and lightest particles will be eluted immediately while the heaviest and the largest particles will remain in suspension at the narrow end of the chamber, due to the fast flow rate of the incoming washing buffer. Thus this system fractionates particles in the centrifugal field

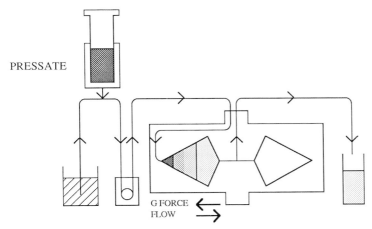

PRESSATE

G FORCE
FLOW

WASH   PUMP ELUTRIATION ROTOR   SAMPLE
BUFFER

FIGURE 1.   CHLAMYDOMONAS CHLOROPLASTS
ISOLATION BY CENTRIFUGAL ELUTRIATION

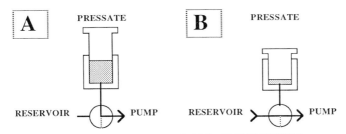

A   PRESSATE   **B**   PRESSATE

RESERVOIR ──→ PUMP   RESERVOIR ──→ PUMP

Figure 2. **FLOW PATTERNS FOR ELUTRIATION**

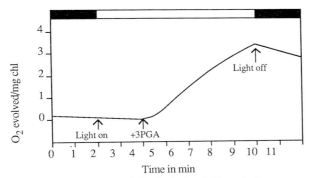

Figure 3. Oxygen photoevolution by isolated chloroplasts:
30 umole of 3PGA is injected in 3.0 ml of reaction mixture.
at the time indicated.

without ever pelleting the particles, a distinct advantage when dealing with fragile but sticky particles such as isolated intact chloroplasts from many sources. Since suspended particles are constantly washed by the buffer flowing through the chamber, we can isolate chloroplasts in buffers similar to the final reaction mixture for photosynthesis assay, eliminating washing procedure to remove Percol. We can start the photosynthetic assay with isolated chloroplasts within one hour after the initial Yeda press disruption of CW-15 cells.

Intact chloroplasts isolated by the current procedure carry out 3PGA dependent O_2 photoevolution (Figure 3) at the rate of 40 umole/mg chl/hr. The similar preparation showed very low rate of K-ferricyanide dependent O_2 photoevolution prior to the osmotic disruption and the greatly increased rate after the disruption (data not shown). They show the uniformly cup shaped appearance of typical Chlamydomonas chloroplasts under phase contrast microscope. Measurement of PEP carboxylase, a cytoplasmic enzyme

Table 1. Distribution of phosphoenolpyruvate carboxylase

	CPM/sample	Chl/sample in ug	CPM/mg chl.
Chloroplasts	62	23	2,700
Boiled chloroplasts	59	"	2,600
Chloroplasts (-PEP)	61	"	2,650
Pressate	1090	33	33,000

Samples from pressate chloroplast were sonicated and assayed for PEP carboxylase according to (5). 3PGA dependent O_2 photoevolution rate for this chloroplasts preparation was 46 umol/mg chl/hr.

marker, shows the preparation to be essentially free of cytoplasmic enzymes (Table 1). Thus, the current procedure appears to produce functional intact chloroplasts with very little intact cell contamination with ease and rapidity useful for many applications.

REFERENCES

1. Gillham, N.W. (1978). Organelle Heredity (Raven, New York).
2. Sommerville, C.R. (1986). Ann. Rev. Plant Physiol. 37:467-5-7.
3. Klein, U., C. Chen, M. Gibbs, K. Platt-Alola (1983). Plant Physiol., 72:481-487.
4. Schlosser, U.G., H. Sachs, D.G. Robinson (1976) Protoplasma 88:51-64.
5. Belknap, W.R. (1983). Plant Physiol. 72:1130-1132.
6. Schneyour, A., M. Avron. (1970). FEBS Lett. 8:164-166.
7. Mendiola-Morgenthaler, L., S. Leu, A. Boschetti. (1985). Plant Sci. 38:33-39.
8. Lin, J-T., O.M. Griffith, J. Corse (1984) Plant Physiol. 75:S-837.
9. Gorman, D., R.P. Levine (1965) Proc. Natl. Acad. Sci. USA 54:1665-1669.
10. Belknap, W.R., R.K. Togasaki. (1981). Proc. Natl. Acad. Sci. USA 78:2310-2314.

This research was supported by N.S.F. grant PCM 83-18174 to R.K.T. and by Nippon Zeon Company.

IN VITRO PHOSPHORYLATION OF MAIZE LEAF PHOSPHOENOLPYRUVATE CARBOXYLASE

RAYMOND J.A. BUDDE AND RAYMOND CHOLLET
DEPARTMENT OF AGRICULTURAL BIOCHEMISTRY, UNIVERSITY OF NEBRASKA, LINCOLN, NE 68583 U.S.A.

INTRODUCTION

Numerous polypeptides in a leaf have been shown to undergo covalent modification by phosphorylation, but only a few have been identified to date. The results described herein demonstrate an ATP-dependent seryl phosphorylation of maize leaf PEPCase in comparison to the previously described ADP-dependent threonyl phosphorylation of maize leaf PPDK (PEPCase and PPDK have similar native and subunit relative masses). Phosphorylation of PEPCase resulted in a partial inactivation of the enzyme, whereas effector modulation by L-malate and glucose-6-phosphate was relatively unaffected. Both of these C_4 mesophyll-cell enzymes are considered to catalyze key reactions during C_4 photosynthesis.

PEPCase from illuminated plants dissociated from its native tetrameric state when diluted, while enzyme from darkened plants remained associated as judged by immunoblots from native electrophoresis gels. One might be drawn to speculate that the phosphorylation of C_4 PEPCase may be related to the photoactivation reported for C_4 PEPCase [1,2], and changes in native aggregation state reported herein and recently for CAM PEPCase by Wu and Wedding [3].

MATERIALS AND METHODS

Sample Preparation: Illuminated laminar tissue was homogenized in cold 0.1 M Tris-HCl (pH 8.0) buffer containing 10 mM $MgCl_2$, 1% PVP, 2 mM Na_2HPO_4, 2 mM Na ascorbate and 7 mM β-mercaptoethanol. The homogenate was filtered through cheesecloth and centrifuged (5 min, 48,000g, 4°C). Soluble proteins were concentrated with 60% $(NH_4)_2SO_4$ and the pellet dissolved in 50 mM Tris-HCl (pH 8.0) containing 5 mM $MgCl_2$, 1 mM Na_2HPO_4 and 7 mM β-mercaptoethanol prior to desalting.

Phosphorylation Assays: The desalted protein sample was incubated (30°C) with 0.25 mM AP5A (an adenylate kinase inhibitor), 0.75 mM [γ-^{32}P]ATP, and a phosphocreatine (4 mM)/creatine phosphokinase (10 units) ADP-scavenging system for studies of ATP-dependent phosphorylation, or with 0.25 mM AP5A, 1 mM [β-^{32}P]ADP, and 0.2 mM ATP for ADP-dependent phosphorylation. After 15 min the reaction was quenched with either an acidic denaturing medium and neutralized prior to one-dimensional SDS-PAGE [4], or quenched in O'Farrell's lysis buffer for two-dimensional electrophoresis.

Aggregation State of Native PEPCase: Plants were placed for at least 4 h in the dark or light prior to homogenization of laminar tissue in cold 0.1 M Tris-HCl (pH 8.0) buffer containing 10 mM $MgCl_2$, 2% PVP, 25% glycerol and 7 mM β-mercaptoethanol. The homogenate was filtered through cheesecloth and centrifuged (10 min, 48,000g, 4°C). Samples were immediately electrophoresed (4°C) on a non-denaturing 5-10% polyacrylamide gradient gel. Electrode buffers contained 5 mM $MgCl_2$ and 10 mM Na

Biggens, J. (ed.), Progress in Photosynthesis Research, Vol. III. ISBN 90 247 3452 5

thioglycolate. The antibodies against maize leaf PPDK and PEPCase were evaluated for any cross-reaction with purified PEPCase and PPDK, respectively, by immunoblotting following one- and two-dimensional electrophoresis and were found to be specific for their respective antigens.

RESULTS AND DISCUSSION

When the soluble maize leaf proteins are analyzed by one-dimensional SDS-PAGE, an intense band is visualized at 94-100 kD (Fig. 1, lane B) which is actually a doublet (Fig. 1, inset to lane B) composed of the subunits of PPDK and PEPCase. When the soluble leaf protein sample was incubated with either $[\beta-^{32}P]$ADP plus ATP or $[\gamma-^{32}P]$ATP alone, <u>acid-stable</u> labeling of the 94-100 kD band resulted (Fig. 1, lanes C and D). Phosphorylation of PPDK by ADP has been well established and is absolutely dependent upon its prior phosphorylation by ATP to form the <u>acid-labile</u> E-HIS-P catalytic intermediate (5,6); thus, the acid-stable ATP-dependent phosphorylation of the 94-100 kD band (Fig. 1, lane D) cannot be assigned to the E-HIS-P form of PPDK. Two-dimensional gel electrophoresis and immunoblotting established PEPCase as the polypeptide phosphorylated by ATP (data not shown).

Two-dimensional thin-layer electrophoresis (4) indicated that the ^{32}P-label from ATP-phosphorylated PEPCase was located exclusively in P-serine (Fig. 2A). As expected (4,7), the ^{32}P-label in the ADP-phosphorylated PPDK was located specifically in P-threonine (Fig. 2B).

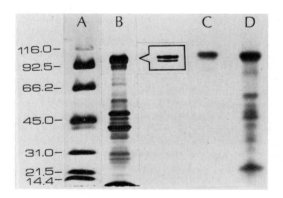

FIGURE 1: ADP- versus ATP-dependent phosphorylation of maize leaf soluble proteins. Lanes A and B are from gels stained with Coomassie blue R-250. Lanes C and D are autoradiographs. Lane A, molecular weight markers; Lane B, soluble leaf proteins (inset shows the doublet at 94-100 kD at one-fourth the specified protein load); Lane C, protein sample phosphorylated with $[\beta-^{32}P]$ADP (1×10^{8} dpm/μmol); Lane D, protein sample phosphorylated with $[\gamma-^{32}P]$ATP (1×10^{8} dpm/μmol). Sample load = 40 μg protein.

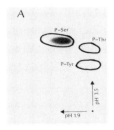 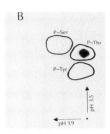

FIGURE 2: Two-dimensional thin-layer electrophoretic separation of phosphoamino acids. Encircled areas represent the location of the phosphoamino acid standards as detected by ninhydrin. A) 94-100 kD band from soluble leaf protein sample phosphorylated with [γ-^{32}P]ATP; B) 94-100 kD band from protein sample phosphorylated with [β-^{32}P]ADP.

Incubation with ATP consistently resulted in a partial inactivation (~12%) of PEPCase when measured at pH 8.0 and saturating PEP concentrations; this inhibitory effect of phosphorylation was doubled when assayed at pH 7.0 and subsaturating levels of PEP. A similar effect of pH and PEP concentration has been reported (1,2) in relation to the photoactivation of PEPCase in C_4 plants.

To determine if there are changes in the aggregation state of C_4 leaf PEPCase similar to those reported for CAM PEPCase by Wu and Wedding (3), samples were electrophoresed on nondenaturing gels, transferred to nitrocellulose and probed immunologically for PEPCase. Samples from darkened plants gave a single high molecular weight band irregardless of protein load (Fig. 4, lanes A and B). In contrast, samples from illuminated plants yielded a single high molecular weight band <u>only</u> when at least 4 µg of protein were loaded onto the gel (Fig. 4, lanes C and D). While the dilution of protein necessary to effect dissociation under these conditions questions if such a phenomenon would occur <u>in vivo</u>, they demonstrate a physical difference between PEPCase from illuminated versus darkened maize plants.

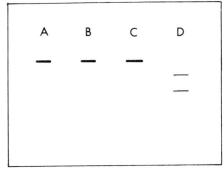

FIGURE 3: Dissociation of native PEPCase from illuminated maize leaves. PEPCase was localized on native gels by immunoblotting. Lanes A and B are samples from dark-adapted plants, Lanes C and D are samples from light-adapted plants. Sample loads: Lane A, 3 µg; Lane B, 1.5 µg; Lane C, 4 µg; and Lane D, 2 µg of protein.

Future research will hopefully provide insight into the following questions: Does phosphorylation of C_4 leaf PEPCase occur in vivo and if so, what is its physiological role?;[4] Is phosphorylation of the protein related to the photoactivation of C_4 PEPCase observed by Karabourniotis et al. (1), light-dark changes in effector sensitivity recently reported by Huber and Sugiyama (2), and/or changes in the native aggregation state?

REFERENCES

1. Karabourniotis, G., Y. Manetas and N.A. Gavalas (1985) Plant Physiol. 77, 300-302.
2. Huber, S.C. and T. Sugiyama (1986) Plant Physiol. 81, 674-677.
3. Wu, M.X. and R.T. Wedding (1985) Arch. Biochem. Biophys. 240, 655-662.
4. Budde, R.J.A., G.P. Holbrook and R. Chollet (1985) Arch. Biochem. Biophys. 242, 283-290.
5. Burnell, J.N. and M.D. Hatch (1984) Biochem. Biophys. Res. Commun. 118, 65-72.
6. Budde, R.J.A., S.M. Ernst and R. Chollet (1986) Biochem. J. 236, 579-584.
7. Ashton, A.R. and M.D. Hatch (1983) Biochem. Biophys. Res. Commun. 115, 53-60.

ABBREVIATIONS

PEPCase, phosphoenolpyruvate carboxylase; PPDK, pyruvate Pi dikinase; AP5A, P^1P^5-di(adenosine-5')pentaphosphate.

PHOTOSYNTHETIC CARBON METABOLISM IN C4 MAIZE LEAVES

H. Usuda, Laboratory of Chemistry, Teikyo University, Hachioji, Japan.

1. INTRODUCTION

Recently, we analyzed induction phenomena in maize leaves and indicated
that i) stomatal conductance does not limit photosynthesis during induction
period [(1), also Furbank & Walker (2)], ii) during an initial phase of
induction, light activation of pyruvate,Pi dikinase, NADP-malate dehydrogenase, and FBPase plays an important role in increasing the rate of CO_2
assimilation (3), and iii) build-up of metabolites in the C4 cycle and RPP
pathway is required to reach a steady-state photosynthesis (4).
During the course of these studies we found that there was a transient
peak in the level of RuBP during the initial phase of induction of photosynthesis in maize leaves (4).   The purpose of this study are to assess
the possible mechanism behind the transient peak of RuBP which occurs during the induction of photosynthesis and to measure metabolite levels in a
steady-state photosynthesis under various conditions in maize leaves.

2. MATERIAL and METHODS

The largest fully expanded leaves of Zea mays L. (variety Chuseishu B)
about 4-6 weeks old were used.   The procedures used in this study were
essentially the same as those described elsewhere (1,3-7).

3. RESULTS and DISCUSSIONS

3.1. Non-autocatalytic build-up of RuBP during the induction

There was a transient peak in the level of RuBP during the initial phase of
induction in maize leaves (Fig 1).   This transient peak is unique to maize
compared to C3 plants where the level of RuBP increased gradually and then
reached a steady-state level during induction (8).   One possible explanation for this transient peak of RuBP level is that RuBP carboxylase in
maize leaves is substantially deactivated in the dark.   Recently, in
assessing this possibility we found that 67 to 84% of the enzyme in darkened maize leaves was in the active form and denied this possibility (9).
In order to assess possible mechanism behind this transient peak of RuBP,
CO_2 was directly provided to bundle sheath chloroplasts by vacuum-infiltration of $NaHCO_3$ solution through vascular tissues according to the method

Abbreviations: Ci intercellular CO_2 concentration; DHAP, dihydroxyacetonephosphate; FBP, fructose 1,6-bisphosphate; F6P, fructose 6-phosphate;
G6P, glucose 6-phosphate; PEP, phosphoenolpyruvate; PGA, 3-phosphoglycerate;
RPP, reductive pentose phosphate; R5P, ribose 5-phosphate; Ru5P, ribulose-
5-phosphate; RuBP, ribulose 1,5-bisphosphate; Xu5P, Xylulose 5-phosphate.

Biggens, J. (ed.), Progress in Photosynthesis Research, Vol. III. ISBN 90 247 3452 5
© *1987 Martinus Nijhoff Publishers, Dordrecht. Printed in the Netherlands.*

of (10). In a control experiment with vacuum-infiltration of CO_2-free
buffer, there was an obvious transient peak of RuBP during initial phase of
induction (Fig 2A). In the experiment with vacuum-infiltration of $NaHCO_3$
solution, the levels of RuBP and PGA showed strikingly different changes
(Fig 2). The level of RuBP gradually increased up to 5 min of illumination and decreased slightly thereafter (Fig 2C). The rate of initial
increase of RuBP level in this experiment was obviously lower than that in
the control experiment and there was no transient peak of RuBP level during
the induction period (Fig 2A,C). The level of PGA rapidly increased
during the first 1 min of illumination and then continued to gradually
increase up to 10 min (Fig 2D). Throughout the induction period of 10 min
the level of PGA in this experiment was higher than that in the control
experiment (Fig 2B, D). These results suggest that a shortage of CO_2
supply to the bundle sheath cells is responsible for the transient peak of
RuBP during the initial phase of induction.

In previous studies the net increase of carbon among intermediates of
the C4 cycle and RPP pathway was far above the level of carbon input from
CO_2 fixation, and the increase in intermediates of the RPP pathway could
not be accounted for by decarboxylation of C4 acids during the initial
phase of induction in maize leaves (4,11). A direct supply of CO_2 to
bundle sheath chloroplasts lowered the initial increase in RuBP level (Fig
2A, C). The level of RuBP increased very rapidly during initial phase of
induction even under N_2 (Fig 3). The rate of initial increase of RuBP
level under N_2 was higher than that in air (Fig 3). These indicate that
the level of RuBP increased faster under conditions of limited CO_2 in the
bundle sheath chloroplasts than when there was an abundant supply of CO_2
to RuBP carboxylase. The most likely explanation of these findings is
that RuBP is partially formed through a non-autocatalytic pathway during
the initial phase of induction in maize leaves. Conversely, in C3 plant
intermediates of the RPP pathway are considered to be autocatalytically
built up during the photosynthetic induction period (12). A metabolite
which is accumulated in the dark could be a candidate of the precursor for
this non-autocatalytic build-up of RuBP. The total level of Ru5P, R5P,
and Xu5P in the dark was 20 to 30 nmol/mg Chl and this level increased
slightly during the initial phase of induction under N_2 and air (Fig 3).
The levels of PGA, DHAP, FBP, F6P, and G6P in the dark were lower than
those after 2 min of illumination during the induction in maize leaves in
air (4). These results indicated that none of these compounds is the
precursor of non-autocatalytic build-up of RuBP. PGA phosphatase is
active in maize mesophyll (13). Glycerate kinase mainly localized in
mesophyll chloroplasts in maize leaf (14) is known to be activated by
light (15). Therefore, if glycerate is accumulated in the dark and converted to PGA by ligh-activated glycerate kinase, glycerate could be the
precursor (Fig 4). However, the level of glycerate in the dark was 20
nmol/mg Chl and this is far below that required to be the precursor.
Thus, glycerate is not the precursor. 6-Posphogluconate was accumulated
in the dark and decreased during initial phase of induction in Chlorella
pyrenoidosa (16). However, 6-phosphogluconate is converted to Ru5P and
CO_2 (Fig 4). The transient peak of RuBP disappeared with the supply of
CO_2 to bundle sheath chloroplasts (Fig 2C). Thus, 6-phosphogluconate
could not be the likely precursor. Carbohydrate could be converted to F6P
and then to RuBP (Fig 4). This non-autocatalytic build-up of RuBP must be
a stand-by mechanism for operating the C4 cycle. Because, $NDAP^+$ is needed
for decarboxylation of malate and for continuous supply of $NADP^+$, RuBP
should be stood by to produce PGA using CO_2 from decarboxylation of malate.
Further studies are needed to determine the precursor.

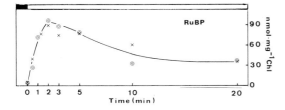

Fig 1   Changes in the level of
RuBP in an attached maize leaf
during induction.   Solid bar
indicates darkness and open bar
indicates light (1100 μE/m^2·s).

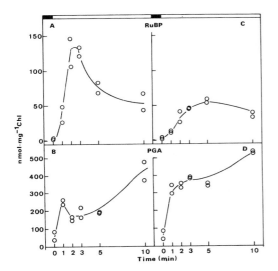

Fig 2   Changes in the levels
of RuBP and PGA in a maize
leaf strip during induction
with or without vacuum-
infiltration of NaHCO$_3$
solution.
CO$_2$-free, 60 mM HEPES-KOH (pH
8.5) (A,B) or 60 mM HEPES-KOH
and 0.4 M NaHCO$_3$ (final pH was
8.5) (C,D) were vacuum-
infiltrated through vascular
tissues for 15 s immediately
after onset of illumination.
Solid bar indicates darkness
and open bar indicates light
(1100 μE/m^2·s).

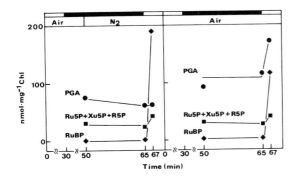

Fig 3   Changes in the
levels of RuBP, Ru5P, Xu5P,
R5P, and PGA in an attached
maize leaf during induction
under N$_2$ or air.
Solid bar indicates darkness
and open bar indicates light
(1300 μE/m^2·s)

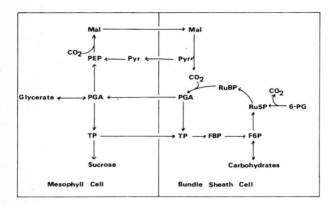

Fig 4 Carbon flow in C4 maize leaf illust-rating possible carbon sources for the non-autocatalytic build-up of RuBP during induction. For the discussions of regula-tion of partitioning of PGA, see text.

3.2. Metabolite levels in steady-state photosynthesis under various conditions

Levels of intermediates of the C4 cycle and RPP pathway under various con-ditions of steady-state photosynthesis are shown in Fig 5, 6, and 7. The levels of RuBP decreased substantially with increasing Ci (Fig 5). Its levels stayed almost constant with increasing light intensity (from 50 to 1900 $\mu E/m^2 \cdot s$) under 350 $\mu l/l$ CO_2 (Fig 6). Under this concentration of CO_2, the rate of CO_2 supply to RuBP carboxylase through C4 cycle (highly related to the rate of CO_2 assimilation) could increase with increasing light intensity (Fig 6). The rate of RuBP regeneration is believed to increase with increasing light intensity. Thus, both the rate of RuBP consumption and that of its regeneration should increase with increasing light intensity and therefore, the level of RuBP could stay almost constant throughout light intensities used in this experiment. In contrast to this, in C3 plant the level of RuBP increased with increasing light intensity (17). In C3 plant, Ci decreased with increasing light intensity (18). Thus, these differences could be due to the different modes of dependence of CO_2 concentration around RuBP carboxylase on light intensity in C3 and C4 plant. The level of RuBP under 133 $\mu l/l$ CO_2 stayed almost constant over the range of light intensity of 50 to 300 $\mu E/m^2 \cdot s$ where both the rate of CO_2 supply to RuBP carboxylase and that of RuBP regeneration are believed to increase with increasing light intensity (Fig 7). In contrast to this, the level of RuBP under 133 $\mu l/l$ CO_2 increased substantially with increasing light intensity from 300 to 1300 $\mu E/m^2 \cdot s$ where only the rate of RuBP regeneration but not the rate of CO_2 supply to RuBP carboxylase is believed to increase (Fig 7).

There was a distinct change in the FBP level: it decreased from more than 200 nmol/mg Chl to 15nmol/mg Chl with increasing Ci (Fig 5). This change in the level of FBP in maize leaves was totally different from that in C3 bean (17) and wheat (19) leaves under various Ci. The level of FBP in maize leaves under low Ci was significantly higher than those reported in C3 plants (17, 19). The basis for these differences is cur-rently unknown, but it is worth noting in this regard that cytoplasmic FBPase in maize leaves has a lower substrate affinity than that found for the enzyme from C3 species, especially in the presence of inhibitor like fructose 2,6-bisphosphate (20).

3.3. Tentative estimation of CO_2 concentration in bundle sheath cells

The CO_2 concentration in maize bundle sheath chloroplasts is believed to be
several times higher than that in solution equilibrated with atmospheric
CO_2 (21) but little study has been made on this subject (22).   Tentative
estimation of CO_2 concentration in the bundle sheath cells were made by
considering the RuBP content in C3 plants relative to Ci and the RuBP content in the C4 plant maize.   It was assumed that the relation between the
decrease in [RuBP] and the increase in [CO_2] observed in C3 plants is
applicable to the C4 plant maize.   For this comparison we assumed that 40%
of the Chl in maize leaves is in the bundle sheath chloroplasts (23) and
RuBP is retained in bundle sheath chloroplasts, and calculated nmols RuBP
per mg bundle sheath Chl.   The comparisons are shown in Fig 8, using the
RuBP levels reported (17).   In maize leaves under atmospheric CO_2 concentration where Ci was 175 µl/l, the CO_2 concentration in the bundle sheath
compartment is estimated to be 600 µbar by comparing RuBP levels in C3 and
C4 maize leaves (Fig 8).   This concentration of 600 µbar is equivalent to

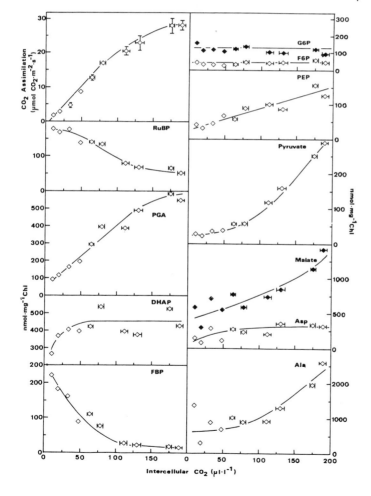

Fig 5   Changes in the rate of CO_2 assimilation and levels of metabolites of the C4 cycle and RPP pathway in an attached maize leaf as a function of Ci.   Three separate measurements were made on the rate of CO_2 assimilation at each CO_2 concentration utilized. Three samples were taken at each CO_2 concentration and combined for metabolite assays. Mean values ±SE are shown for the rates of CO_2 assimilation and Ci.

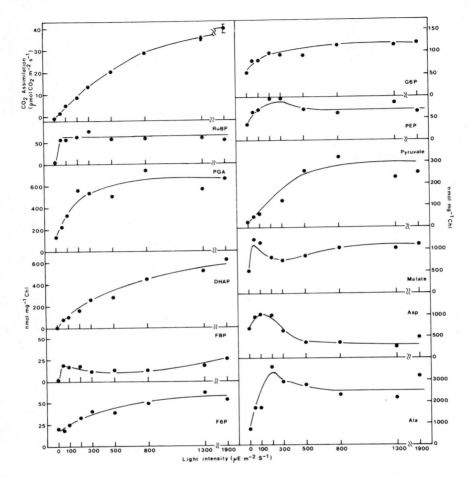

Fig 6   Changes in the rate of CO_2 assimilation and levels of metabolites of the C4 cycle and RPP pathway in an attached maize leaf as a function of light intensity under 350 µl/l CO_2.   Three separate samples were taken as described in the legend of Fig 5.

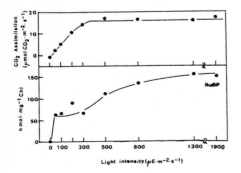

Fig 7 Changes in the rate of CO_2 assimilation and level of RuBP in an attached maize leaf as a function of light intensity under 133 µl/l CO_2.
Three separate samples were taken as described in the legend of Fig 5.

about 0.8 mM of inorganic carbon (HCO_3^- + CO_2) at 30°C, at 1 atm, and at pH
8. Previously, Hatch and Osmond using another approach estimated a concentration of inorganic carbon in the cytoplasm and chloroplasts of maize
bundle sheath cells to be 0.6 mM (21). Thus, results from the present
analysis are in good agreement with their interpretations. From these
comparisons, we suggest that the CO_2 atmosphere in the bundle sheath compartment in maize leaves can be concentrated roughly 2 to 10 times higher
than the C_i over the range of C_i investigated, and that the CO_2 atmosphere
in the bundle sheath compartment can be roughly 1.5 to 9 times higher than
the ambient CO_2 concentration used in this study (21 to 388 µl/l CO_2).

3.4. Supplement of carbon into the C4 cycle from the RPP pathway to
enhance photosynthesis rate

Previously, we suggested that supplement of carbon into the C4 cycle from
the RPP pathway is required to increase the photosynthesis rate in maize
leaves during induction period (4). The total amount of intermediates in
the C4 cycle (malate, aspartate, pyruvate, PEP, and alanine) increased with
increasing C_i (Fig 5). However, the total amount of intermediates in the
C4 cycle did not increase with increasing light intensity (Fig 6). The
concentration of malate in maize leaves differed significantly from leaf to
leaf under a given condition (4). A large fraction of the total malate
pool was photosynthetically inactive (24). Thus, photosynthetically inactive pools of malate, aspartate, and alanine could mask real changes in
levels of the photosynthetically active pools of these compounds and made
it difficult to estimate the real changes in the total amount of the intermediates of the C4 cycle. The level of pyruvate may reflect the total

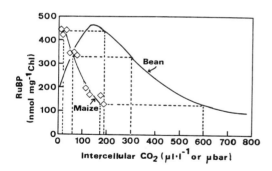

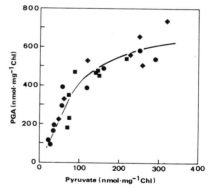

Fig 8 Comparison of RuBP levels in
the chloroplasts where RuBP caboxy-
lase is functioning in C3 bean and
C4 maize leaves. The results of
RuBP level in bean leaves were taken
from (17) and expressed on the basis
of µbar. For a calculation of the
values of the RuBP on a bundle
sheath Chl in maize leaves, see text.
Its values were expressed on the
basis of C_i of µl/l.

Fig 9 Relatioship between the
levels of PGA and pyruvate in an
attached maize leaf under a
steady-state photosynthesis under
various conditions. Data were
taken from Fig 5 (●) and Fig 6 (◆)
and from the experiment under
various light intensities with
constant CO_2 of 133 µl/l (■).

amount of photosynthetically active intermediates of the C4 cycle.
Pyruvate levels increased with increasing Ci (Fig 5) or increasing light
intensity (Fig 6).   Photosynthesis rates also increased with increasing
Ci or increasing light intensity (Fig 5, 6).   The relationship between the
levels of PGA and pyruvate in a steady-state photosynthesis under various
conditions were shown in Fig 9.   There is a distinct positive relation
between the levels of PGA and pyruvate.   These results suggest that there
is a supplement of carbon into the C4 cycle from the RPP pathway to enhance
photosynthesis rate.   One possible way in which this may occur is by conversion of PGA of the RPP pathway to PEP of the C4 cycle (Fig 4).   An
elevated concentration of PGA could stimulate the conversion of PGA to PEP
in the C4 cycle and consequently, the level of total intermediates of the
C4 cycle could be raised.   The mechanism of the regulation of the partitioning of PGA in mesophyll into PEP and into DHAP (which is converted into
sucrose in mesophyll and also into RuBP and starch in bundle sheath chloroplasts) to maintain an appropriate levels of intermediates of both the C4
cycle and RPP pathway is open to further studies (Fig 4).

I would like to thank G.E. Edwards and E. Hase for their stimulating discussions, R.C. Leegood for his advice on measuring metabolites, M.Samejima
for his guidance on a vacuum-infiltration experiment, and also T.D.
Sharkey for constructing the high speed freeze clamp equipment.

References

1 Usuda H, Edwards GE (1984) Plant Sci Lett 37, 41-45
2 Furbank RT, Walker DA (1985) Planta 163, 75-83
3 Usuda H, Ku MSB, Edwards GE (1984) Plant Physiol 76, 238-243
4 Usuda H (1985) Plant Physiol 78, 859-864
5 Usuda H (1984) Plant Cell Physiol 25, 1297-1301
6 Usuda H (1986) Plant Cell Physiol 27, 745-749
7 Usuda H , submitted to Plant Physiol
8 Prinsley RT, Leegood RC (1986) Biochim Biophys Acta 849, 244-253
9 Usuda H (1985) Plant Cell Physiol 26, 1455-1463
10 Samejima M, Miyachi S (1978) Plant Cell Physiol 19, 907-916
11 Leegood RC, Furbank RT (1984) Planta 162, 450-456
12 Edwards GE, Walker DA (1983) C3, C4:Mechanisms and cellular and environmental regulation of photosynthesis. Blackwell Sci.
13 Randoll DD, Tolbert NE, Gremel D (1971) Plant Physiol 48, 480-487
14 Usuda H, Edwards GE (1980) Plant Physiol 65, 1017-1022
15 Kleczkowski IA, Randoll DD (1986) Plant Physiol 81, 656-662
16 Bassham JA, Kirk M (1968) in Comparative Biochemistry and Biophysics of
Photosynthesis (Shibata K et al, ed) Univ. of Tokyo Press, pp 365-378
17 Badger MA, Sharkey TD, Von Caemmerer S. (1984) Planta 160, 305-313
18 Sharkey TD, Imai K, Farquhar GD, Cowan IR (1982) Plant Physiol 69, 657-
659
19 Woodrow IE, Furbank RT, Brooks A, Murphy DJ (1985) Biochim Biophys Acta
807, 263-271
20 Stitt M, Heldt HW (1985) Planta 164, 179-188
21 Hatch MD, Osmond CB (1976) in Encyclopedia of Plant Physiol (Stocking
CR, Heber U, ed) Vol 3, Springer-Verlag pp, 144-184
22 Hatch MD (1971) Biochem J 125, 425-432
23 Kanai R, Edwards GE (1973) Plant Physiol 51, 1133-1137
24 Hatch MD (1979) Arch Biochem Biophys 194, 117-127

ENZYME REGULATION IN C_4 PHOTOSYNTHESIS: PURIFICATION, PROPERTIES AND REGULATION OF CHLOROPLAST INORGANIC PYROPHOSPHATASE FROM SORGHUM VULGARE.

V. Ananda Krishnan and A. Gnanam

1. INTRODUCTION

Inorganic pyrophosphatase (PPase) (E.C.3.6.1.1) is known to be present in yeast, mammalian tissues, plant leaves and bacteria (1). PPases from plants (including maize) were studied to a less extent in comparison to their mammalian and bacterial counterparts (2). Up to now no report has been published about the possible regulatory role of this mesophyll chloroplast PPase in C_4 photosynthesis. Hence the enzyme has been purified to homogeneity and an evidence for its role in the C_4 pathway of photosynthesis is presented.

2. MATERIALS AND METHODS

2.1 Purification of chloroplast PPase. The enzyme from 2 weeks old sorghum plants was purified according to (2). The 40-55% $(NH_4)_2SO_4$ precipitate was first fractionated through Sephadex G-75 column and then was separated by passing through DEAE-cellulose column using a linear gradient of 20-120 mM $MgCl_2$.

2.2 SDS-PAGE was carried out as described by Laemmli (3).

2.3 Inorganic pyrophosphatase was assayed in a reaction mixture containing 50 or 100 mM Tricine-KOH (pH 8.0), 2 mM $Na_4P_2O_7$, 10 mM $MgCl_2$ and 2 mM mercaptoethanol and requisite amount of the enzyme to a total volume of 1 ml at 23 ± 2 C. After incubation for 1 min the reaction was stopped by adding 0.1 ml of 50% trichloroacetic acid. The liberated Pi was estimated according to the method of Fiske and Subbarow (4). One unit of enzyme activity is defined as that amount which catalyses the hydrolysis of 1 μmole of PPi per min. Specific activity is defined as units per mg protein.

2.4 Protein determinations were carried out according to (5), using crystalline bovine serum albumin as standard.

3. RESULTS
3.1 Purification.

Inorganic pyrophosphatase was purfieid by 75 fold (Table 1) to an apparent homogeneity as judged by SDS-PAGE (Fig. 1). About 25% of the original activity was recovered in the fractions from DEAE cellulose.

TABLE 1. Purification of sorghum plastid inorganic pyrophosphatase.

Step	Total protein (mg)	Activity Total (units)	Specific units mg^{-1} protein	-fold	% Yield
Crude	264*	460.9	1.743	1.0	100
40-55% $(NH_4)_2SO_4$	47	207.4	3.665	2.1	45
G-75	6.7	174.8	10.792	6.1	38
DE-52	0.6	115.24	129.8	74.4	25

*From approximately 50g of leaf tissue.

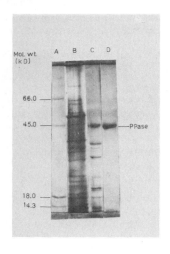

FIGURE 1. SDS-PAGE (10%) of PPase during various stages of purification. Lane A - molecular weight standards; Lane B - 150 μg of proteins from 40-55% ammonium sulfate fractionation; Lane C - 60 μg of proteins from the eluant of G-75 column; Lane D - 20 μg of PPase from DE-52 eluant.

3.2 Properties

3.2.1 Various organic and inorganic phosphoesters were tested as substrates (Table 2). Inorganic pyrophosphate was by far the only active substrate indicating a high specificity.

TABLE 2. Substrate specificity of plastid PPase at pH 8.0 in the presence of Mg^{2+}

Substrate	$\mu mol. \ PPi \ hydrolyzed \ min^{-1}$
Sodium pyrophosphate	0.2524
Tripolyphosphate	0.0011
Tetrapolyphosphate	0.0009
Adenosine 5' triphosphate	Nil
Glucose-6-phosphate	Nil

The assay mixture contained 5 mM of the respective substrates. Approximately 0.8 µg of purified enzyme was used.

3.2.2 Several metabolites involved in various phases of C_4 metabolism related to the operation of PPase were tested for possible effects on the enzyme. No effect was observed with PEP, 3-PGA or OAA up to 5 mM. In addition to the inhibition of the enzyme Pi, malate too inhibited the enzyme. From Fig. 2 it is evident that malate inhibits the enzyme in an allosteric type . A Hill plot of malate (Fig. 3) inhibition gave a slope (n_H) of 1.31. Thus it is concluded that two molecules of malate bind to each molecule of the enzyme.

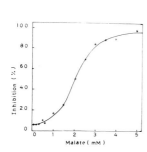
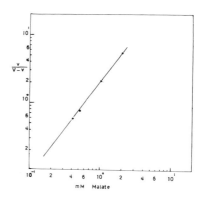

FIGURE 2 (left). Inhibition of purified PPase by L-malate. Control activity is expressed as 100% (1.2115 umol PPi hydrolyzed min^{-1}).

FIGURE 3 (right). Hill plot of malate inhibition.

4. DISCUSSION

Although phosphoryl-transfer enzymes are widespread and of central importance to cellular metabolism, they remain incompletely understood with respect to their regulatory role in the C_4 pathway photosynthesis. Our reporrt that this enzyme exhibits[4] a high specificity for $Na_4P_2O_4$ as substrate confirms its primary role of mediating rapid hydrolysis of PPi formed in the reaction catalyzed by pyruvate, Pi dikinase. The substrate specificity of the enzyme resembles the enzyme from other sources (1,2,6,7).

3-PGA was reported to inhibit PPase from leaves of Pennisetum americanum (7) while it did not inhibit sorghum enzyme. Although the potential PPase activity is more than 5 times the maximum photosynthetic rate, this enzyme may have a regulatory role in C_4 pathway of CO_2 assimilation unless its activity in vivo is reduced by allosteric inhibition of malate alone or along with some other, yet, unidentified feedback regulatory mechanisms.

Acknowledgements. Financial assistance from the USDA - PL480 grant no. FG-IN-557 (IN-SEA-172) is gratefully acknowledged.

Abbreviations used. PPase - inorganic pyrophosphatase; PPi - inorganic pyrophosphate; 3-PGA - 3-phosphoglyceric acid; OAA - oxalacetic acid.

REFERENCES.

1. Butler, L.G. (1971) In: The Enzymes (Boyer, P.D., ed.) Vol. 4, pp. 529-541. Academic Press, New York.
2. Mukherjee, J.J. and Pal, P.R. (1983) Agric. Biol. Chem. 47, 1341.
3. Laemmli, U.K. (1970) Nature 227, 680.
4. Fiske, C.M. and Subba Row, Y. (1925) J. Biol. Chem. 66,375.
5. Bradford, M. (1976) Anal. Biochem. 72, 248.
6. Bucke, C. (1970) Phytochemistry 9, 1303.
7. Lavergne, D. and Haorau, A. (1984) In: Advances in Photosynthesis Research (Sybesma, C. ed.) Vol. III, PP.497-500, Junk Publishers, The Hague, Boston, Lancaster

Authors' address : Department of Plant Sciences
                   School of Biological Sciences
                   Madurai Kamaraj University
                   Madurai 625021, INDIA

IS THERE A RELATION BETWEEN CRASSULACEAN ACID METABOLISM AND ENDOPOLYPLOIDY IN BRYOPHYLLUM CRENATUM ?

ANABELA BERNARDES DA SILVA AND MARIA CELESTE ARRABAÇA, DEPARTMENT OF PLANT BIOLOGY, FACULTY OF SCIENCE, UNIVERSITY OF LISBON, PORTUGAL

INTRODUCTION

Crassulacean acid metabolism (CAM) is characterized by a daily rythm of CO_2 fixation (1), associated with changes on the activity of phosphoenol-pyruvate carboxylase (PEPC) and on the content of malate (2), in leaves. In some plants, like Kalanchöe, Sedum, Bryophyllum, etc., the establishment of CAM is induced by the change from long day (LD) to short day (SD) conditions (3, 4). Namely in Kalanchöe blossfeldiana , this change results in an exponential increase in PEPC activity, due to "de novo" synthesis of an isoenzyme (5). Concomitant with biochemical and physiological changes, induced by SD, it can be shown that in some species, like Bryophyllum crenatum, morphological changes may occur (6), such as shortening of internodes and leaf petioles, and decrease in leaf blade area. However, the leaf blades are thicker due to larger cells and nuclei, probably as a consequence of increased endomitotic activity (endopolyploidy). It has been shown that, from the 12th day (young leaf) to the 24th day (mature leaf) of development, leaves of B. crenatum, grown under short day conditions, suffer an increase of about five times on the level of endopolyploidy. This increased level is maintained until the 72nd day (7). There is also indication that the mechanisms of CO_2 fixation by Kalanchöe pinnata, changes during development, from the C_3 to the CAM pathway (8).

MATERIAL AND METHODS
1. Plant material
    Bulbils of Bryophyllum crenatum, individually potted, were kept under short day conditions (eight hours light) for about three months. The temperature was 17 $^{\circ}C$ during the night and about 30 $^{\circ}C$ under natural light conditions, with a relative humidity of about 85%. The leaf pairs were numbered from the apex, considering the first pair, the one with zero days of development. The plastochrome of B. crenatum under these conditions was six days.
2. Enzyme extracts and assay
    PEPC was assayed in extracts of the 2nd, 3rd and 5th leaf pairs, as described in (9).
3. Malate determination
    The content of malate of the leaf pairs was measured in extracts, according to (10).

RESULTS AND DISCUSSION
We tried to establish whether the increase in endopolyploidy found in (7), correlates with the change from C_3 metabolism to CAM, described in (8). Results shown in Fig. 1 indicate a decrease in malate content throughout the day and an increase during darkness. The activity of PEPC is at its highest at the beggining of the dark period and decreases continuously during the night. Maximal activity of PEPC is concomitant with minimum ma-

Biggens, J. (ed.), Progress in Photosynthesis Research, Vol. III. ISBN 90 247 3452 5
© *1987 Martinus Nijhoff Publishers, Dordrecht. Printed in the Netherlands.*

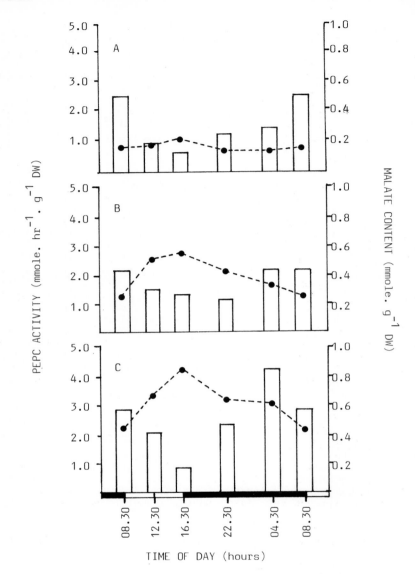

TIME OF DAY (hours)

FIG. 1: THE EFECT OF LEAF AGEING ON THE ACTIVITY OF PEPC AND ON THE
CONTENT OF MALATE ALONG THE DIURNAL CYCLE

        PEPC activity   -----
        malate content   bar
        A - 2nd leaf pair (6 days old)
        B - 3rd leaf pair (12 days old)
        C - 5th leaf pair (24 days old)

late content. Leaf ageing induces an increase, both in malate content and on PEPC activity, as well as in the amplitude of the circadian rythm characteristic of CAM. We may conclude that the appearance of CAM in B. crenatum as shown in Fig. 1, is not age-dependent, in contrast with the results obtained with K. pinnata (9). No relation is evident between the increase of endopolyploidy observed from the 12th to the 24th day of leaf development and the induction of Crassulacean acid metabolism, as the present results indicate that even the 2nd leaf pair (6 days of development) already shows CAM. We should stress, however that CAM characteristics became more accentuated during leaf development, in parallel with the increase in endopolyploidy.

REFERENCES
(1) Brulfert, J., Guerrier, D. and Queiroz, O. (1982)
    Planta, 154, 326-331
(2) Queiroz, O. (1976)
    Physiol.Vég. 14, 629-639
(3) Klüge, M. and Osmond, C. B., (1972)
    Pflanzenphysiol., 66, 97-105
(4) Queiroz, O. and Brulfert, J. (1982)
    in "Crassulacean Acid Metabolism", I. P. Ting and M. Gibbs, eds.,
    Waverly Press, Baltimore, 208-230
(5) Brulfert, J., Arrabaça, M. C., Guerrier, D. and Queiroz, O. (1979)
    Planta 146, 129-133
(6) Resende, F. and Catarino, F. (1963)
    Proc. XI Int. Cong. Gen. I, The Hague, 112-124
(7) Warden, J. and Catarino, F. (1977)
    Port. Acta Biol., serie A, XV, 39-58
(8) Thompson, A., Vogel, J. and Edward Lee, R. (1977)
    J. Exp. Bot. 28, 1037-1041
(9) Gavalas, N. A., Caravatas, S. and Manetas, Y. (1982)
    Photosynthetica 16, 49-58
(10)Hohorst, H. J. (1970)
    in "Methods in Enzymatic Analysis", H. U. Bergmeyer ed., Verlag Chemie,
    Weinheim, 2, 1544-1548

ACKNOWLEDGMENTS

The present work was partially supported by the Instituto Nacional de Investigação Científica, Portugal (CEB/LA 6).

PHOTOSYNTHETIC PROPERTIES OF THE THREE MAJOR TISSUE LAYERS OF THE CAM
PLANT, PEPEROMIA CAMPTOTRICHA

JOHN N. NISHIO, ANDREW N. WEBBER, LONNIE J. GURALNICK, ROBERT L. HEATH &
IRWIN P. TING. DEPARTMENT OF BOTANY AND PLANT SCIENCES, UNIVERSITY OF
CALIFORNIA, RIVERSIDE, CA   92521

INTRODUCTION

   C_4 plants are generally characterized by a spatial separation of the
C_3 and C_4 modes of carbon fixation into bundle sheath cells and mesophyll
cells, respectively. In contrast CAM plants typically exhibit a temporal
separation of the C_3 and C_4 modes with initial carbon fixation by the
C_4 pathway at night and decarboxylation of the C_4 products and refixation
of the released CO_2 via the reductive pentose phosphate pathway during the
day.
   Various modes of photosynthesis operate in Peperomia camptotricha
depending on growth conditions and leaf age (1). Young leaves exhibit CAM
cycling (diurnal acid flux and C_3-type gas exchange). Mature leaves are
more CAM-like, with a diurnal acid flux and gas exchange occurring at
night, but they also exhibit gas exchange during the light period.
   The leaves of P. camptotricha contain three distinct upper tissue
layers and a one cell thick lower epidermis. The upper layer is a water
storing multiple epidermis (ME), which may also serve as a "window" (2).
The median tissue is composed of a densely packed, palisade mesophyll (PM)
rich in Chl, and the bottom layer consists of a spongy mesophyll (SM) with
anatomical features typical of CAM plants (3,4).
   The distinct delineation of tissue types in P. camptotricha leaves
suggests the possibility that a spatial separation of photosynthetic modes
also exists in this CAM plant. To test this, acid fluctuation, dark CO_2
fixation and the activities of RuBP carboxylase and several C_4 enzymes in
the different tissue layers were investigated. In addition we summarize
here some of the information about the photosynthetic apparatus in the PM
and SM, which combined contain 97% of the total leaf Chl.

METHODS

   Plants and Tissue. P. camptotricha plants were grown in pots in a
glasshouse at the University of California, Riverside. Tissue layers were
separated by hand using a single edged razor blade—aided by 70X
magnification.
   Enzyme Activity. RuBP carboxylase activity was measured by
incorporation of $^{14}CO_2$ in tissue extracts. PEP carboxylase and PEP
carboxykinase activities were assayed spectrophotometrically by following
the oxidation of NADH at 340 nm. NADP-malic enzyme was assayed spectrophotometrically by following NADP reduction at 340 nm.
   Dark CO_2 Fixation. Dark CO_2 fixation was measured by
incorporation of $^{14}CO_2$ into intact layers of tissue.
   Titratable Acidity. Samples ground in water were titrated to an
endpoint of pH 7.0 with 0.01 N KOH.

Biggens, J. (ed.), Progress in Photosynthesis Research, Vol. III. ISBN 90 247 3452 5
© 1987 Martinus Nijhoff Publishers, Dordrecht. Printed in the Netherlands.

Electron Transport. PS1: Ascorbate-DCPIP —> MV; PS2: H_2O —> phenyl-pBQ; PS1 + PS2: H_2O —> MV.

Fluorescence. Fluorescence emission at 680 nm was determined in intact tissue infiltrated with DCMU.

Chl-Proteins. Chl-proteins were separated on SDS gels (5). SDS:Chl=10.

P700. P700 was determined from the redox difference spectrum at 698 nm using an extinction coefficient of 64 mM^{-1} cm^{-1}(6).

Chlorophyll. Chl was determined in 80% acetone (7).

Light Transmission. Transmission of photosynthetically active radiation through the tissue layers was measured by placing tissue on a plastic wrap covered sensor of a Li-Cor model LI-185A photometer.

RESULTS

On an areal basis the majority of the activity of the C_4 enzymes, PEP carboxylase, PEP carboxykinase (the major decarboxylating enzyme in this plant, accounting for 93% of the measured decarboxylation activity), NADP-malic enzyme, and dark CO_2 fixation occurred in the SM, whereas the least activity was found in the PM (Table I). On a Chl basis the activity of these enzymes was greatest in the ME (due to the low amount of Chl), and lowest in the PM. In contrast the activity of the carboxylating enzyme of the reductive pentose phosphate pathway, RuBP carboxylase, was highest in the PM on an areal basis, and on a Chl basis was highest in the ME and lowest in the SM.

TABLE I. Enzyme Activity in Different Tissue Layers

ENZYME*	ME	PM	SM
C_3 FIXATION			
RUBPC/CHL	75.3 ± 32.9	24.7 ± 7.9	14.0 ± 2.0
RUBPC/AREA	9.6 ± 2.9	71.7 ± 3.4	18.8 ± 2.8
C_4 FIXATION			
PEPC/CHL	287.7 ± 138.3	8.1 ± 3.5	56.3 ± 5.0
PEPC/AREA	30.1 ± 6.0	19.4 ± 4.7	50.5 ± 2.1
DARK CO_2 FIXATION/CHL	184.1 ± 10.8	3.2 ± 1.5	26.8 ± 8.1
DARK CO_2 FIXATION/AREA	35.2 ± 2.1	13.7 ± 6.4	51.2 ± 15.7
DECARBOXYLATION			
PEPCK/CHL	288.9 ± 36.1	10.8 ± 1.0	73.2 ± 5.1
PEPCK/AREA	22.9 ± 3.1	19.1 ± 1.4	58.0 ± 3.4
NADP-MALIC ENZYME/CHL	20.0 ± 2.3	n.d.	6.2 ± 1.6
NADP-MALIC ENZYME/AREA	24.9 ± 4.7	n.d.	75.1 ± 4.7

*Per CHL is expressed as mmol mol Chl^{-1} s^{-1}, and per AREA is expressed as % of total activity per area. n.d.--not detected.

TABLE II. Titratable Acidity in the Different Tissue Layers

TISSUE	TITRATABLE ACIDITY (μEq cm^{-2})					
	MORNING		AFTERNOON		$\Delta\mu$Eq cm^{-2}	% CHANGE	% TOTAL FLUX/AREA
ME	7.48	2.12	2.50	0.34	4.98	299	37.2
PM	3.22	0.29	1.83	0.31	1.39	176	9.8
SM	9.90	1.09	2.79	0.26	7.11	355	53.1

The titratable acidity data corresponded to the C_4 enzyme activity and dark CO_2 fixation data. The greatest flux of titratable acidity per area occurred in the SM and the least in the PM (Table II). The percent change in acidity was also highest in the SM and smallest in PM.

The lowest concentration of Chl was found in the upper ME, and the highest in the median PM (Table III). The chl a/b ratio was highest in the ME and lowest in the lower SM, suggesting that there is a greater amount of LHC in the lower, shaded portion of the leaf.

TABLE III.   Chlorophyll Content and Chl a/b ratio in Different Tissue

TISSUE	CHLOROPHYLL		
	Total a+b ($nmols\ cm^{-2}$)	% Total	a/b
ME	2.4 ± 0.6	3.1 ± 0.5	2.85 ± 0.40
PM	51.3 ± 13.4	67.2 ± 2.9	2.66 ± 0.05
SM	22.5 ± 5.7	29.9 ± 2.5	2.37 ± 0.09

The value of Fv/Fm, which represents an estimate of the quantum efficiency of PS2 primary photochemistry, was the same in the PM and SM (Table IV). Since Fv/Fm values were equal, the lower half-rise time in the SM suggests a larger light harvesting apparatus is associated with the PS2 complexes in the SM. The value of βmax has been used to estimate the extent of connectivity between LHC2 and PS2 within the photosynthetic apparatus (8). The lower value of βmax found in the SM suggests more effective coupling between the LHC2 and PS2 in the lower tissue layer.

TABLE IV.   Fluorescence Parameters in the Different Tissue Layers

TISSUE	FLUORESCENCE		
	Fv/Fm	$t_{\frac{1}{2}}$ (ms)	β max
PM	0.62	36	0.955
SM	0.60	20	0.825

The rates of electron transport per Chl were always lower in the SM. The P700/Chl ratio was also lower in the SM (1.546×10^{-3} in the SM and 1.981×10^{-3} in the PM--these numbers are equivalent to Chl/P700 ratios of 646 and 505, respectively), and preliminary Chl-protein analysis indicated a reduced amount of CPa on a Chl basis in the SM.

TABLE V.   Electron Transport in the PM and SM

TISSUE	ELECTRON TRANSPORT		
	($mmol\ O_2\ mol\ Chl^{-1}\ s^{-1}$)		
	PS1	PS2	PS1+PS2
PM	35.1 ± 9.6	20.1 ± 2.6	3.8 ± 1.0
SM	18.1 ± 1.7	7.2 ± 2.8	2.6 ± 0.5

DISCUSSION

The results of the enzyme studies suggest that in addition to a temporal separation, there is also some spatial separation of the C_3 and C_4 modes of carbon reduction in P. camptotricha. The majority (over 70%) of the pentose phosphate reduction occurred in the middle layer, where most of the Chl (67%) is concentrated. The lower, shaded portion of the leaf contained the majority of the C_4 pathway enzymes, while the median PM accounted for less than 20% of the total activity of these enzymes. This organization of the C_3 and C_4 pathways in these tissue layers would result in a CO_2 concentration gradient, with CO_2 from the atmosphere, respiratory CO_2, and CO_2 released after decarboxylation of malate moving to the middle layer during the day.

The amount of acid flux in each layer appeared to be related to vacuolar size. Vacuoles on a percentage of cell volume basis are 2.4 times bigger in the ME and SM than in the PM (9), which correlated to the percent change in titratable acidity, which was 2.0 and 1.7 times larger in the SM and ME, respectively, than in the PM. Acid flux into vacuoles is limited thermodynamically by the ability of the vacuoles to maintain a high proton gradient (10). Thus a large vacuole with tonoplast properties similar to a small vacuole would be able to contain more malate than the small one. The smaller vacuolar size and lower C_4 metabolic activity in the PM suggests the possibility that large vacuoles are requisite for CAM activity (see 11).

The ME transmitted only 70% of the incident PAR to the PM, and the ME and PM combined transmitted only 10% of the incident PAR to the SM. The decrease in RuBP carboxylase activity per Chl, the Chl a/b ratio (the amount of LHC2 increased), and the P700/Chl ratio from the top to bottom leaf sections suggests that in addition to the spatial separation of the different modes of carbon fixation in P. camptotricha, the photosynthetic apparatus in the lower leaf sections has undergone adaptions similar to shade adapted plants (12,13). Studies on the photosynthetic apparatus in spongy mesophyll and palisade mesophyll cells of spinach and in different planes (top to bottom) of the spinach leaf (14,15) are in agreement with the findings of this study.

It has been suggested that an upper, water storage tissue does not increase the leaf surface area, yet provides a "window" for light to reach the underlying photosynthetic tissue (2). It has also been shown that the transmittance of the water storing layer decreases during water stress ( ). Since many plants with "windows" (including P. camptotricha) grow in high light environments and often experience drought conditions, our light transmission results indicate that the "window" may serve to protect the photosynthetic tissue from light damage.

REFERENCES

1 Sipes, D.L. and Ting, I.P. (1986) Plant Physiol. 77: 59-63
2 Krulik, G.A. (1980) Can. J. Bot. 58: 1591-1600
3 Kluge, M. and Ting, I.P. (1978) Crassulacean Acid Metabolism, Springer-Verlag, Berlin
4 Gibson, A.C. (1982) In Crassulacean Acid Metabolism. Proceedings of the Fifth Annual Symposium in Botany (Ting, I.P. and Gibbs, M., ed.), pp. 1-17, Waverly Press, Baltimore
5 Anderson, J.M., Waldron, J.C., and Thorne, S.W. (1978) FEBS Lett. 92: 227-233
6 Hiyama, T. and Ke, B. (1972) BBA 267: 160-171
7 Arnon, D.I. (1949) Plant Physiol. 24: 1-15
8 Percival, M.P., Webber, A.N., and Baker, N.R. (1984) BBA 767: 582-589
9 Gibeaut, D.M. (1986) Ph.D. Thesis, University of California, Riverside
10 Luttge, U., Smith, A.C., and Marigo, G. (1982) In Crassulacean Acid Metabolism. Proceedings of the Fifth Annual Symposium in Botany (Ting, I.P. and Gibbs, M., ed.) pp. 69-91
11 Black, C.C., Carnal, N.W., and Kenyon, W.H. Ibid., pp. 51-68
12 Anderson, J.M. (1986) Ann. Rev. Plant Physiol 37: 93-136
13 Barber J. (1985) in Photosynthetic Mechanisms and the Environment (Barber, J. and Baker, N.R., eds.), Elsevier, Amsterdam
14 Terashima, I. and Inoue, Y. (1985) Plant Cell Physiol. 26(1): 63-75
15 Terashima, I. and Inoue, Y. (1985) Plant Cell Physiol. 26(4): 781-785

CO_2, NOT HCO_3^-, IS THE INORGANIC CARBON SUBSTRATE OF NADP MALIC ENZYME FROM MAIZE AND WHEAT GERM

RAINER E. HÄUSLER, JOSEPH A.M. HOLTUM, ERWIN LATZKO

BOTANISCHES INSTITUT DER WESTFÄLISCHEN WILHELMS-UNIVERSITÄT
SCHLOSSGARTEN 3, 4400 MÜNSTER, GERMANY (BRD)

1. INTRODUCTION

NADP malic enzyme catalyses the reversible oxidative decarboxylation of malate in the presence of Mg^{2+} or Mn^{2+}. The reaction yields pyruvate, NADPH and inorganic carbon. Although the equilibrium favours oxidative decarboxylation, $K_{eq} = 19.6$ l $mole^{-1}$ at pH 7.0 and 22-25 ^{o}C (1), reductive carboxylation can be measured under the appropriate conditions.

NADP malic enzyme is widely distributed amongst organisms and can be divided into two groups; a cytosolic form is present in animal tissues, C-3 and CAM-plants and a chloroplastic form has a central role in C-4 metabolism (2). The NADP malic enzyme-catalysed decarboxylation of malate supplies the photosynthetic carbon reduction cycle with inorganic carbon in the bundle-sheath chloroplasts of C-4 plants such as <u>Zea mays</u>.

The commonly held view that CO_2 is the inorganic carbon species which is released from malate is based upon a single report for the cytosolic NADP malic enzyme from wheat germ (3). A putative catalytic mechanism for the NADP malic enzyme from pigeon liver also assumes that CO_2 is released from malate during oxidatative decarboxylation (4).

In contrast, Asami et al. (5) purified and characterized the chloroplastic NADP malic enzyme from maize leaves and observed that HCO_3^- is the inorganic carbon substrate for this enzyme. If true, the reaction mechanisms of both enzymes are different.

To clarify the above uncertainties we partially purified the chloroplastic NADP malic enzyme from maize leaves and cytosolic NADP malic enzyme from wheat germ and have established the inorganic carbon substrate for both.

2. PROCEDURES

2.1. <u>Materials</u> <u>and</u> <u>methods</u>

2.1.1. Theory: The active inorganic carbon species were determined

Biggens, J. (ed.), Progress in Photosynthesis Research, Vol. III. ISBN 90 247 3452 5
© *1987 Martinus Nijhoff Publishers, Dordrecht. Printed in the Netherlands.*

according to (3, 6). The method is based upon the observation that the attainment of equilibrium between CO_2 and HCO_3 in solution is a slow process which depends upon temperature, pH and ionic strength (Fig. 1). The rate of enzymic carboxylation should reflect the concentration changes of the active inorganic carbon substrate when the reaction is initiated with either CO_2 or HCO_3. When equilibrium between CO_2 and HCO_3 is approached one should observe steady state rates. In the presence of carbonic anhydrase the attainment of equilibrium between CO_2 and HCO_3, and so the attainment of steady state rates, should be rapid. The patterns of progress curves for carboxylases which use CO_2 and for carboxylases which use HCO_3 are different (Fig.2).

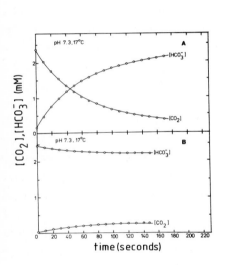

Fig. 1: Calculated changes in concentrations of CO_2 and HCO_3 during the
reversible hydration of CO_2 following
initiation with CO_2(A) or (HCO_3).
Total inorganic carbon concentration
was 2.5 mM, pH was 7.3 and temperat
ure was 17 °C.

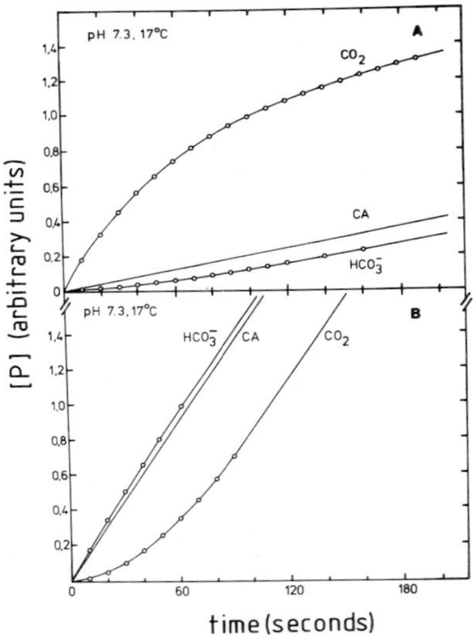

Fig. 2: Progress curves for carboxylases which use CO_2 (A) or
HCO_3 (B) when the reactions are
initiated with either CO_2 or HCO_3
in the presence or absence of
carbonic anhydrase (CA). Conditions were pH 7.3 and 17 °C.

2.1.2. Enzyme assay: The reactions were followed photometrically at 334 or 365 nm. Reductive carboxylation of pyruvate was assayed after (3,5). Assays

contained, in a final volume of 1 ml, 100 mM tricine pH 8.4 or HEPES-NaOH
pH 7.0 or triethanolamine-hydrochloride-NaOH pH 7.3; 30 mM pyruvate; 7.5 mM
$MgCl_2$; 0.15 mM NADPH and the appropriate inorganic carbon concentration. If
necessary, 20 Wilbur Anderson units (7) of carbonic anhydrase were added.
All operations were carried out under nitrogen in order to avoid
contamination by exogenous inorganic carbon. The reactions were initiated
with 100 ul of either 25 mM CO_2 or HCO_3^- solutions.

3. RESULTS AND DISCUSSION

The final preparations of both enzymes were free from ribulose-1,5-bisphos-
phate carboxylase, PEP carboxylase, NAD malic enzyme, pyruvate decarboxylase and carbonic anhydrase.
The patterns of progress curves we observed for both the NADP malic enzyme
from maize and from wheat germ are similar to those predicted for a CO_2 dependent carboxylase (Fig. 3). The difference between initial velocity and
steady state velocity is more pronounced at higher pH-values (pH 8.4) indicating that the bulk of inorganic carbon is in HCO_3^- .

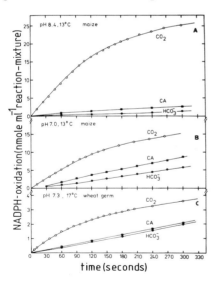

Fig. 3: Progress curves observed
for NADP malic enzyme from maize
and wheat germ. The reactions
were initiated under different
conditions with either CO_2 or
HCO_3^- . CA indicates the presence
of CO_2 or HCO_3^- plus carbonic
anhydrase.

The steady state rates observed in the presence of carbonic anhydrase are
similar to those observed when equilibrium between CO_2 and HCO_3^- is
approached in the absence of carbonic anhydrase.
If the progress curves do indeed reflect the reversible hydration of CO_2 ,
the slopes of the linearized forms should equal the summed derivative first

order rate constants for the reversible hydration of CO_2 (Fig. 4). For both enzymes, at pH 7.3 and 17 ^{O}C, the summed first order rate constants of 0.017 sec^{-1} are identical to those reported by (3). The half time of reaction under the above conditions is 41 sec. As one would expect the attainment of equilibrium between CO_2 and HCO_3^- is slower at a lower temperature. The summed first order rate constants are 0.011 sec^{-1} ($t_{0.5}$= 63 sec) and 0.009 sec^{-1} ($t_{0.5}$= 77 sec) at 13 ^{O}C for pHs 7.0 and 8.4, respectively.

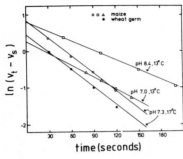

Fig. 4: Linearized progress curves from fig. 3. (v_t = momentary velocity; v_s = steady state velocity)

We conclude that CO_2, not HCO_3^-, is the product of the reaction catalysed by both the chloroplastic and the cytosolic NADP malic enzyme. In the case of Zea mays this is of special importance since there are indications that carbonic anhydrase activity in bundle-sheath chloroplasts is absent or very low (8). If HCO_3^- is released, then at stromal pH during the light the bulk of inorganic carbon will be present as HCO_3^-. When the rate of HCO_3^- - dehydration is low compared to that of CO_2 uptake, the efficiency of assimilation by ribulose-1,5-bisphosphate carboxylase will be retarded. The direct production of CO_2 within the bundle-sheath chloroplasts reduces this likelihood.

4. REFERENCES

1. Harary, I., Korey, S.R. Ochoa, S.(1953) J. Biol. Chem. 203, 595-603
2. Nishikido, T., Wada, T.(1974) Biochem. Biophys. Res. Com. 61, 243-249
3. Dalziel, K., Londesborough, J.C.(1968) Biochem. J. 110, 223-230
4. Schimerlik, M.I., Cleland, W.W. (1977) Biochemistry 16, 576-582
5. Asami, S., Inoue, K., Matsumoto, K., Murachi, A., Akazawa, T.(1979) Arch. Biochem. Biophys. 194, 503-510
6. Cooper, T.G., Tschen, T.T., Wood, H.G., Benedict, C.R.(1968) J. Biol. Chem. 243, 3857-3863
7. Wilbur, K.M., Anderson, N.G.(1948) J. Biol. Chem. 76, 147-154
8. Poincelot, R.P.(1972) Plant Physiol. 50, 336-340

THE IONIC STRENGTH CHANGES THE QUATERNARY STRUCTURE OF PHOSPHOENOLPYRUVATE CARBOXYLASE FROM MAIZE LEAVES.

WAGNER, R., GONZALEZ, D.H., PODESTA, F.E., ANDREO, C.S.

CENTRO DE ESTUDIOS FOTOSINTETICOS Y BIOQUIMICOS,CEFOBI (CONICET, F.M. LILLO, U.N.ROSARIO). SUIPACHA 531, 2000 ROSARIO, ARGENTINA.

1. INTRODUCTION

Phosphoenolpyruvate (PEP) carboxylase (E.C. 4.1.1.31) is the enzyme responsible for the primary CO_2 fixation in maize and other C_4 plants (1).

It is usually described as a homotetramer of MW 400,000 (1,2). Several metabolites, such as glucose-6-phosphate (3) and malate (4,5), modulate its activity, thus suggesting the allosteric nature of the carboxylase. Changes in the oligomeric structure of the enzyme of a Crassulacean plant have been reported which depend on the day-night cycle and in consequence also on the levels of metabolites (6).

In this work we present evidences which suggest that the enzyme from maize leaves may undergo dissociation into dimers and monomers under conditions of high ionic strength.

2. PROCEDURE

2.1. Materials and methods.

PEP carboxylase was purified from maize leaves by a procedure similar to that of Uedan and Sugiyama (2), described in (7).

Activity was measured by the coupled assay method, in a medium containing 50 mM Tris-HCl (pH 7 or 8), 5 mM $MgCl_2$, 10 mM $NaHCO_3$,0.16 mM NADH, malic dehydrogenase (2 I.U. per ml), PEP carboxylase (4 to 8 µg per ml) and the indicated concentrations of PEP.

Gel filtration was performed by High Performance Liquid Chromatography (HPLC) using a calibrated TSK 3000 column equilibrated with the corresponding buffer. Equilibration and elution were performed at 25ºC and at a flow rate of 1 ml per minute.

3. RESULTS AND DISCUSSION

When a sample of PEP carboxylase was subjected to gel filtration in the presence of a buffer which contained 200 mM NaCl and 50 mM Tris-HCl at pH 7 or 8, dissociation of the enzyme could be observed (Figure 1).

At pH 8, peaks corresponding to 400, 200 and 100 kDa appeared after treatment, while at pH 7 only the peaks corresponding to tetramer and dimer were present.

In both cases, the extent of dissociation increased with time and NaCl concentration, and decreased as the enzyme concentration raised.

The velocity of this process was much higher at pH 7 than at pH 8. Raising the pH over 8 caused almost complete conversion into monomers.

This process proved to be reversible as indicated by the appearance of the 400 and 100 kDa peaks when only the 200 kDa protein peak was rechromatographed.

Figure 2 shows that when PEP carboxylase was assayed in a reaction medium containing NaCl its activity decayed in a time- and NaCl-concentration-dependent manner. This decay was faster at pH 7. Similar results were

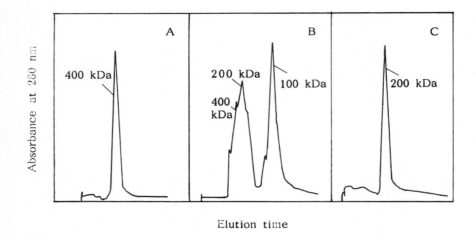

Figure 1 : Elution patterns of salt treated PEP carboxylase. Enzyme (1 mg) was chromatographed as described in the text in the absence of NaCl at pH 7 (A), or in the presence of 200 mM NaCl after 1 hour of incubation with the same concentration of salt at pH 8 (B) or 7 (C). Elution times were 13, 16 and 27 min for the 400, 200 and 100 kDa peaks, respectively.

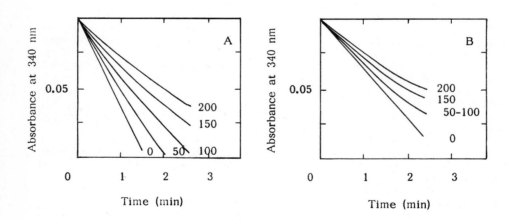

Figure 2 : Influence of NaCl on PEP carboxylase activity. Reaction was started by adding the enzyme to the assay medium containing NaCl in the mM concentrations indicated by the numerals. A : pH 7, B : pH 8. PEP concentration was 0.5 mM.

obtained with $(NH_4)_2SO_4$, Na_2SO_4 and KCl. The effect of salt treatment was diminished by increasing concentrations of PEP.

When the enzyme was incubated with 200 mM NaCl at pH 7 during 60 minutes, and then assayed in a medium without salt at pH 7 or 8 and different PEP concentrations, changes in Vmax and Km were observed (Figure 3). At pH 8 only Km varied, while at pH 7 the main effect was on Vmax.

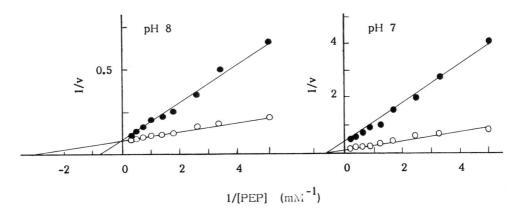

Figure 3 : Double reciprocal plot of enzyme activity versus PEP concentration. PEP carboxylase was incubated 1 hour with 200 mM NaCl at pH 7 and 30ºC (●) and then activity was measured in a medium without salt at the indicated PEP concentrations. (O) Control incubated in the absence of salt.

Since the gel filtration experiments indicate that under these conditions the carboxylase would be present essentially as a dimer, we attribute these kinetic parameters to this oligomeric form.

In a similar way, we can relate the time dependent changes in the MW to the decay in the activity at pH 7 or 8, in view of the similar responses to enzyme, NaCl and H^+ concentration.

Recently, Wu and Wedding (6) reported that in <u>Crassula</u> <u>argentea</u> PEP carboxylase exists in different oligomeric forms, in an equilibrium dependent on the diurnal cycle and metabolites concentration.

The results reported in this work suggest that in maize leaves this equilibrium would exist only in the presence of high ionic strength. Whether this equilibrium is affected or not by metabolites remains to be established, as also is its physiological significance.

4. ADDENDUM
4.1. Acknowledgements

This work was supported by a research grant from the Consejo Nacional de Investigaciones Científicas y Técnicas, CONICET, Argentina and by the Alexander von Humboldt Foundation, F.R. Germany. CSA is a member of the Career Investigator of the CONICET; DHG and FEP are Fellows of the same Institution. RW was a recipient of a Theodor Lynen Fellowship from the Alexander von Humboldt Foundation, F.R. Germany.

III.7.**534**

REFERENCES
1 O'Leary, M.H. (1982) Annu. Rev. Plant Physiol. 33, 297-315
2 Uedan, K. and Sugiyama, T. (1976) Plant Physiol. 57, 906-910
3. Coombs, J., Baldry, C.W. and Bucke, C. (1973) Planta 110, 95-107
4 Huber, S.C. and Edwards, G.E. (1975) Can. J. Bot. 53, 1925-1933
5 González, D.H., Iglesias, A.A. and Andreo, C.S. (1984) J. Plant Physiol. 116, 524-434.
6 Wu, M-X. and Wedding, R.T. (1985) Arch. Biochem. Biophys. 240, 655-662
7 Iglesias, A.A., González, D.H. and Andreo, C.S. (1986) Planta, in press.

INTEGRATION OF CARBON AND NITROGEN METABOLISM

C.A. ATKINS and J.S. PATE
BOTANY DEPARTMENT, UNIVERSITY OF WESTERN AUSTRALIA, NEDLANDS WA 6009,
AUSTRALIA

1. INTRODUCTION

    C and N are the two most significant elements in the nutrition of
the green plant. Both are obtained autotrophically and metabolic
pathways for their assimilation interact at a multitude of sites in
the many sequences where organic compounds of N are formed, transformed, or utilised to generate the fabric of the plant at all levels
of its organization. The C nutrition of the plant, through photosynthetic CO_2 fixation and the synthesis of "translocatable" photosynthates, is closely dependent on N supply for provision of the
enzymic catalysts and structural components which support the
process. Mechanisms of N assimilation are directly dependent on the
products of photosynthesis to supply energy for assimilatory reactions
themselves or to supply C skeletons for the synthesis of organic N
solutes. These solutes constitute the translocated forms of N which
are distributed from sites of their formation to the rest of the plant
through the agency of long-distance transport in xylem and phloem and
complementary shorter distance exchanges between xylem and phloem.
Based upon analyses of solutes in transport channels empirical models
of whole plant partitioning of C and N have been constructed,
providing a basis for outlining nutritional interactions among organs
for these elements and allowing the source/sink relationships of the
plant for N to be expressed in terms of the compounds involved,
assigning specific nutritional roles for particular translocated
solutes. Detailed C and N economies of individual organs has allowed
the production or consumption of organic solutes of N to be related to
the metabolic properties of the organs themselves as well as to their
efficiency of functioning in terms of C usage. The studies described
are largely restricted to two grain legumes, cowpea (Vigna unguiculata
L. Walp) and white lupin (Lupinus albus). As a nodulated plant the
grain legume provides an ideal experimental system for studying
mechanisms involved in the partitioning of C and N. Not only does the
single site for N assimilation simplify the nature of N inputs to the
plant, but species which form quite different translocated solutes of
N may be compared.

2. INTEGRATION AT SITES OF PRIMARY N ASSIMILATION

    Metabolic sequences for the primary assimilation of NO_3 or N_2 are
closely dependent on reducing power and ATP. While the extent to
which NO_3 assimilation in leaves is supported directly by reductant
and energy generated by photosynthesis is not yet clear,
heterotrophic assimilation of NO_3 in roots and leaves in darkness is
closely linked with sugar oxidation, both in the cytosol and in
plastids via the pentose-P pathway (1,2). There is a close
relationship between endogenous carbohydrate level and the capacity of

both leaf and root systems to reduce NO_3. This is also the case for N_2 fixation in which oxidizable C substrate(s) are transferred continuously (3) from the host plant cell to support the substantial energy demands of nitrogenase activity in bacteroids.

In addition to supplying the energy needs for N assimilation, carbohydrate also provides the C skeleton for the formation of organic solutes of N. These are many and varied but in terms of primary assimilation the need is for a supply of -ketoglutarate to form glutamate and glutamine (GLN). In most plant tissues malate is formed, to a greater or lesser extent, by the activity of PEP carboxylase providing C for anapleurotic functioning of the TCA cycle, thus indirectly generating the C skeleton needed for amino acid synthesis. There is a close relationship between malate metabolism and NO_3 assimilation in plants and, while this is in part due to a contribution of the C_4 acid to maintenance of ionic and pH balance (4), malate - C is readily utilised in amino acid formation both in leaves and roots (1). In legume nodules C for synthesis of amino acids is likewise derived from products of nodular PEP carboxylase activity (5) as well as from C of translocated photosynthate (6).

Plant species differ markedly in the extent to which NO_3 is reduced and assimilated by roots (7). Those which translocate a large amount of NO_3 to the shoot in xylem show high levels of NO_3 reductase in leaves and a low level in roots. The reverse is true for those which translocate principally organic solutes of N from the roots (8,9). These differences in site of NO_3 reduction are reflected in the relative C economies of the root systems of the two types of plant (9,10), but whether they relate specifically to a conservation of C by those with a shoot-based reduction system is not known. Plant roots readily absorb NH_4^+ but, unlike NO_3, NH_4^+ is not translocated in xylem being completely assimilated in the roots with N translocated to the shoot in an organic form.

3. INTEGRATION IN THE SYNTHESIS OF TRANSLOCATED N-SOLUTES

Long distance translocation of N is achieved from roots and nodules in xylem and from leaves in phloem. As indicated above xylem may carry currently absorbed inorganic N as well as products of N assimilation in the root, or in the case of the nodulated plant, products of N_2 fixation. Other N-solutes in xylem may simply be "cycling" in the translocation channels of the plant or be derived from protein degradation as a result of turnover or senescence of below-ground organs. In a number of cases in young seedlings xylem has been found to carry unusual N-solutes (4-methyleneglutamine, 1-canavanine or homoserine) which are derived from the seed and, though prominent in the translocation channel, are relatively inert, apparently "cycling" from xylem to phloem and contributing little to the N nutrition of the plant (11).

GLN, as the initial product of NH_3 assimilation, may also be a major constituent of both xylem and phloem. This is the case in a wide range of species, including many grasses (12), but more often GLN-N is further metabolised to form specific translocated solutes such as asparagine (ASN), citrulline, allantoin or allantoic acid.

Among legumes there appears to be two major groups in this respect (7,15); temperate species form ASN as a result of NO_3 or N_2 assimilation, while those of tropical origin form ASN from NO_3 reduction, but principally the ureides, allantoin and allantoic acid, as a result of N_2 fixation. Symbiotic soybean is somewhat intermediate, producing both ASN and ureides. ASN is formed both in roots and nodules by the activity of ASN synthetase. Although the cellular location of the enzyme has not been clearly established for each of these sites (13) in both cases the amide group is derived from that of GLN rather than from NH_3. Ureide synthesis on the other hand is rather more complex, involving de novo purine synthesis to form IMP with subsequent oxidation via IMP and xanthine dehydrogenases and urate oxidase to form allantoin (14,15). In cowpea and soybean nodules the pathway is highly compartmentalised, involving plastids, cytosol and microbodies of both infected and uninfected nodule cells (16), and incorporates N both as amino (aspartate, glycine) and amide (GLN) groups.

4. INTEGRATION IN THE REASSIMILATION OF N

There are two major areas where N already assimilated is released from metabolic sequences as NH_3 necessitating its prompt and efficient reassimilation. The first is photorespiration in leaves where, as a result of transamination of glyoxylate to form glycine and subsequent condensation to form serine, both NH_3 and CO_2 are released (17). The second involves the utilisation of translocated organic N at all sites in the plant where N is required for protein synthesis and growth. As mentioned above one or two solutes of N predominate, both in xylem and phloem, necessitating extensive N metabolism to provide the amino acids required to form protein. While to some extent this may be accomplished by aminotransferases both amide- and ureide-N appear to be released largely as NH_3.

The quantitative significance of NH_3 assimilation due to photorespiration and to release from amides utilised in the lupin leaf throughout development is shown in Table 1 (18) together with the level of enzymes likely to achieve the reassimilation of the NH_3 as amino-N (17). Although release in photorespiration is quantitatively the more significant in the development of the leaf, when protein synthesis is most rapid, deamidation of amides comprises a substantial component of the activity. Mechanisms involve both asparaginase and ASN aminotransferase (18) though in lupin the former is of greatest significance. Interestingly the enzyme responsible for ASN transamination is identical with peroxisomal serine:glyoxylate aminotransferase (19) and ^{15}N labeling studies in pea leaf indicate that substantial amino-N from ASN enters the photorespiratory pathway (20).

The need for breakdown of translocated N-solutes and reassimilation of NH_3 is especially evident in relation to fruit and seed development. In lupin incoming ASN both in xylem and phloem is largely deamidated by the activity of asparaginase (21) with, during early embryo development, free NH_3 accumulating to a concentration of 50-60 mM before its utilisation in embryo protein synthesis. A similar situation occurs in cowpea when ureides, again delivered to the fruit in both xylem and phloem, are metabolised in the pod wall and seed coat with the released N transferred to the embryo as amides and amino acids (22,23).

TABLE 1. Estimates of NH_3 release in photorespiration and in deamidation of ASN and GLN together with activities of GLN synthetase and glutamate synthase in the developing leaf of white lupin (18).

Days after initiation	0-11	12-20 (n mol/min. g FW)	1-38	39-66
1. Net photoperiod CO_2 fixation	1024	4096	3036	1596
2. NH_3 release in photorespiration	256	1024	759	399
3. NH_3 release from deamidation of amides	81	44	1	4
4. GS activity (in vitro)	955	1385	1283	1333
5. GOGAT activity (in vitro)	480	1141	1230	1527

5. PARTITIONING OF C AND N IN THE NODULATED GRAIN LEGUME

5.1 Whole plant studies

As the starting point for discussion a model of coupled C and N flow (24) for a one-week period of growth at the early flowering stage of white lupin may be considered. This model for C flow (Fig. 1A) incorporates information on the net fixation of C by four strata of leaves on the shoot, the C donated through phloem by these groups of leaves to other organs, and that cycling through the root system to form export products of xylem. The profile of C flow is dominated by downward transloction to the nodulated root, which at this stage attracts over half of the net photosynthate generated in the plant. The value for nodulated cowpea at a similar growth stage was close to 25% (9).

The model for coupled N flow (Fig. 1B) indicates that the uppermost leaf stratum of still expanding leaflets consumes almost 40% of the currently fixed N in xylem, and that these upper leaves receive through xylem an "extra" component of N due to progressive "xylem to xylem" transfer in the stem. This mechanism (marked by stars in Fig. 1B) consists of lateral withdrawal by stem nodal tissue of N initially directed towards leaves of lower strata (especially of C and B) but rerouted to the upward moving xylem stream. In this way xylem becomes more concentrated in N as it moves up the stem, so that apical organs receive significantly more and lower leaves less N than might be expected from their respective transpirational water losses. As a result of this abstraction from xylem, together with exchange of C and N with the phloem tissues of stem and leaves, adjacent petioles are able to continuously adjust their C and N balance, initially gaining these two elements from long distance transport channels moving up or down the stem and later from mobilization to younger parts of the plant. Lateral apices and the terminal inflorescence have access to three distinct sources of N, (1) that derived from xylem in accordance with transpiration, (2) that carried in phloem from the nearest stratum of leaves, and (3) that obtained through xylem to phloem transfer in the upper stem. The contribution of these three sources are estimated to provide 41%, 40% and 19% respectively of the total requirement of the apical tissues for N. As well as the short distance exchanges between translocation channels in stem and petioles, which together serve to progressively enrich xylem with N and phloem with N relative to C in the upper parts of the shoot, the flow pattern for N is dominated by N transfer from xylem to phloem in leaves (Fig. 1B). The form of xylem to phloem exchange will be considered below in relation to the detailed C:N economy of the lupin leaf.

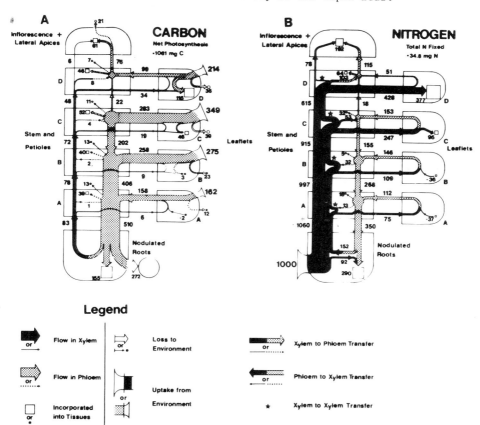

Fig. 1 Budgets for C and N flow in nodulated lupin (24)

These models of whole plant functioning, whether in terms of C:N balance or in relation to the transport of a single N-solute, depict events over only a short period of growth and do not therefore indicate how the plant reacts ontogenetically to changes in either absolute or relative inputs of C and N. Distribution profiles for N during the rapid vegetative growth of legumes show continuous high rates of investment of N in bacterial and plant protein of the nodules, and at the same time in chloroplastic protein of leaves, thereby allowing both photosynthesis and N assimilation to increase exponentially in a tightly coupled fashion. This tight relationship is however strained considerably at either end of the plant's life cycle, during early seedling growth when both nodule development and the establishment of the photosynthetic surfaces of the plant compete for N, and during reproductive growth when seed protein synthesis is accompanied by declining rates of N assimilation and, in some cases, leaf senescence (25).

5.2 Case studies of individual organs

The first organ to be considered is the leaf. Daily measurements of C and N contents of dry matter, continuously recorded exchanges of C as CO_2 (both day and night), together with analyses of xylem (tracheal) sap and phloem bleeding sap collected at the petiole, provide the base data from which the C:N economy of the upper main stem leaf of lupin has been constructed (18,26). A summary of the net turnover of C and N which the leaf makes during its 66 d life is shown in Fig. 2B, and the time course of rates of exchange made with the rest of the plant via xylem and phloem in Fig. 2A. After an initial period of import in both xylem and phloem the leaf rapidly developed the capacity for phloem export of both C and N. This occurred well before full leaf expansion as there was a net retention of imported N by the leaf up to day 20. From then, however, the leaf's activity was characterised by continuous cycling of N. As indicated in Fig. 2B 18 mg N was imported through xylem during the leaf's life, with 11.4 mg, or 63%, of this cycling back to the plant more or less immediately by xylem to phloem exchange; the rate of import of N in xylem after 20 d being roughly equal to that of export in phloem (Fig. 2A), except towards the end of the leaf's life as protein breakdown contributed N to phloem export (3.2 mg N in Fig. 2B). Maximum rates of C export in phloem were maintained for a surprisingly short time. After a sharp peak at 20 d phloem C level fell quite rapidly resulting in a phloem stream with a progressively narrowing C:N ratio (Fig. 2A).

Short distance transfer of N in leaves may be achieved by direct exchange of xylem-borne N-solutes to phloem in the minor vein network of the leaf or following their metabolism in the mesophyll with subsequent release of N back to phloem as N-solutes incorporating photosynthetic C. Apparently both processes contribute to the exchange in lupin leaves but the proportion of each varies markedly for different xylem solutes. ASN is the principal nitrogenous component of xylem entering the leaf throughout its life, accounting for around 60% of the N input (see Fig. 2B,C). As much as 90% of the ASN in phloem leaving a fully expanded lupin leaf is derived from direct xylem transfer, representing around 35% of phloem-N over the life of the leaf (Fig. 2C). Other amino acids such as GLN, aspartate and glutamate, which also enter the leaf in xylem, engage to a much smaller extent in direct exchange providing their N to phloem

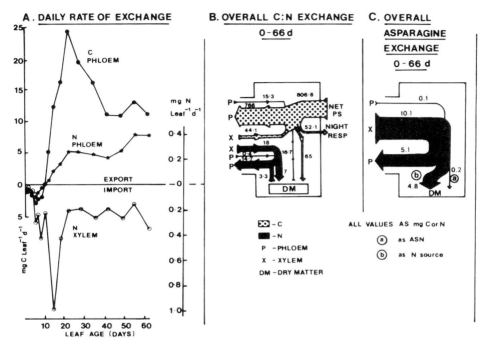

Fig. 2   C, N and ASN-N exchange of the developing lupin leaf.

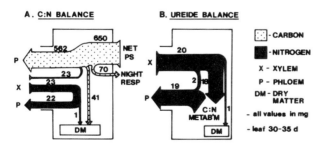

Fig. 3   C, N and ureide-N exchange of the cowpea leaf.

mainly via leaf metabolism (27,28).   In cowpea the major xylem-borne form of N reaching the leaf is ureide, and using "cryopuncture" sampling of phloem at peduncles and fruits subtended in leaf axils (29) it is possible to assess the role of these solutes in leaf functioning.   Fig. 3 indicates the C:N balance of a fully expanded cowpea leaf for a 5 d period 30-35 d after its initiation and the relative contribution of ureide import to the N balance.   In this case allantoin and allantoic acid comprised more than 80% of the xylem-borne N entering the

leaf but, unlike ASN in lupin, only 10% was transferred directly
to phloem with most of the ureide-N being utilised in the leaf and
re-exported in phloem as other solutes (principally ASN and GLN).
In cowpea shoots ^{14}C-ASN supplied in xylem is, as in lupin,
transferred largely unmetabolised to phloem (29), suggesting that
direct xylem:phloem transfer is not necessarily a feature of the
predominant N-solute but rather of ASN. NO_3 entering the leaf in
xylem is also largely metabolised, with a negligible amount, even
at greatly elevated levels supplied to the plant, being loaded
onto phloem. In both cowpea and lupin NO_3-N supplied to the leaf
is recovered in phloem principally as GLN and ASN (9,29).

The second case study to be considered is the root nodule. Nodule
development is wholly dependent on assimilates translocated in
phloem from the photosynthesizing shoot; sucrose providing C for
nodule functioning as well as for its structure. The entering
phloem stream also brings in N-solutes which, though providing N
essential for both bacteroid and plant protein synthesis during
initial establishment of the symbiosis (30), are insufficient to
maintain full nodule development. Thus nodules might be expected
to meet a large proportion of their requirement for N by directly
incorporating into their own structure NH_3 formed from N_2
fixation. This has been confirmed for nodulated cowpea seedlings
grown from seed which was uniformly labeled with ^{15}N (P. Sanford,
J.S. Pate and C.A. Atkins, unpublished). While cotyledon N was
recovered both in bacteroids and nodule plant cells this
dependence was restricted to the initial development of the nodule
with fixed N rapidly contributing the majority of N needed for the
organ's nutrition.

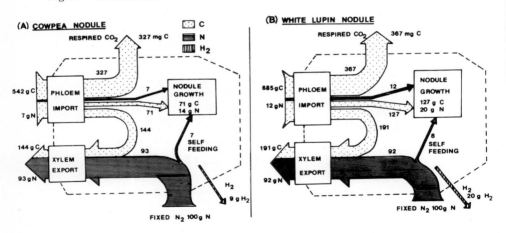

Fig. 4 C and N economy of nodules of cowpea and lupin.

Despite the considerably more complex pathway of ureide biosynthesis compared to that for ASN (7) considerations of the likely
energy costs of each as mechanisms for NH_3 assimilation (31)
suggest little difference. Furthermore the C and N economies of

nodules of the two species show that those of cowpea are more economical in terms of C usage than those of lupin (5.4 versus 6.9 gC/gN fixed in Fig. 4). While, on paper at least, this may appear to be the result of a lower C requirement for the skeleton of ureide (1C:1N) compared to that for ASN (2C:1N) there is no evidence that such a difference represents a real economy in terms of nodule or plant functioning. Nodules of the two types differ considerably in structure, growth habit, water economy, rates of H_2 evolution and of CO_2 fixation (and probably in other characters also) and it seems likely that these could contribute to their differing rates of C usage. Detailed analysis of the C:N enconomy of cowpea nodules indicates that the efficiency of C usage in nodule functioning is not constant but varies both diurnally (3) and in relation to nodule and plant development (32).

The third class of organ whose C:N economy may be considered is the developing fruit. Budgets, constructed using data on CO_2 exchange, H_2O exchange, increments of C, N and H_2O throughout development as well as analysis of the C:N contents of xylem and phloem streams serving the fruit, have been provided for both lupin (33) and cowpea (34). The model of functioning for lupin shows a near perfect match for the fruit's respective budgets for C, N and H_2O; a xylem:phloem mixture of 60:40 (by vol) satisfying precisely the C and N consumption of the fruit and providing an amount of H_2O almost exactly meeting transpiration and tissue water needs. Phloem supplies 89% of the N and 98% of the C emphasising the importance of phloem nutrition in this instance. The corresponding budget for cowpea is quite different in that the N requirement for development is met to a greater extent by xylem and, as a consequence, H_2O intake (in both xylem and phloem) greatly exceeds the fruit's needs. Anatomical and tracer studies indicate that the surplus H_2O returns to the parent plant (35), principally during the photoperiod when transpirational loss from the fruit is low compared to the mass inflow of H_2O in phloem. It is premature to tell whether diurnally-reversing xylem flow of H_2O is a common phenomenon amongst grain legumes, but preliminary evidence suggests that it may be especially prominent in fast growing, large-seeded fruits in which high rates of dry matter gain are supported by inordinately high rates of import of phloem-borne H_2O. Likewise it is premature, but tempting, to suggest that this "back-flow" carries from the site of water excess, the developing embryo, factors which affect the functioning of other organs on the plant.

REFERENCES
1 Naik, M.S. and Nicholas, D.J.D. (1986) Phytochemistry 25,571-576
2 Oaks,A. and Hirel, B. (1985) Ann. Rev. Plant Physiol. 36,345-365
3 Rainbird, R.M., Atkins, C.A. and Pate, J.S. (1983) Plant Physiol. 72,308-312
4 Smith, F.A. and Raven, J.A. (1979) Ann. Rev. Plant Physiol. 30,289-311
5 Vance, C.P., Boylan, K.L.M., Maxwell, C.A., Heichel, G.H and Hardman, L.L. (1985) Plant Physiol. 78,774-778
6 Herridge, D.F., Atkins, C.A., Pate, J.S. and Rainbird, R.M. (1978) Plant Physiol. 62,495-498

7 Pate, J.S. and Atkins, C.A. (1983) in Ecology and Nitrogen Fixation (Broughton, W.J., ed.) Vol. 3, pp245-298, Clarendon Press

8 Atkins, C.A., Pate, J.S. and Layzell, D.B. (1979) Plant Physiol. 64,1078-1082

9 Atkins, C.A., Pate, J.S., Griffiths, G.J and White, S.T. (1980) Plant Physiol. 66,978-983

10 Pate, J.S., Layzell, D.B. and Atkins, C.A. (1979) Plant Physiol. 64,1083-1088

11 Peoples, M.B., Pate, J.S., Atkins, C.A. and Bergersen, F.J. (1986) Plant Physiol. (in press)

12 Pate, J.S. (1971) in Nitrogen-15 in Soil-Plant Studies (Krippner, M., ed.) pp.165-188, International Atomic Energy Agency, Vienna, Austria

13 Shelp, B.J. and Atkins, C.A. (1984) Plant Sci. Lett. 36,225-230.

14 Atkins, C.A., Ritchie, A., Rowe, P.B., McCairns, E. and Sauer, D. (1982) Plant Physiol. 70,55-60

15 Atkins, C.A. (1982) in Advances in Agricultural Microbiology (Rao, N.S.S., ed.) pp.53-88, Oxford and IBH, New Delhi

16 Shelp, B.J., Atkins, C.A., Storer, P.J. and Canvin, D.T. (1983) Arch. Biochem. Biophys. 224,429-441

17 Ogren, W.L. (1984) Ann. Rev. Plant Physiol. 35,415-442

18 Atkins, C.A., Pate, J.S., Peoples, M.B. and Joy, K.W. (1983) Plant Physiol. 71,841-848

19 Ireland, R.J. and Joy, K.W. (1983) Plant Physiol. 72,1127-1129

20 Ta, T.C., Joy, K.W. and Ireland, R.J. (1985) Plant Physiol. 78,334-337

21 Atkins, C.A., Pate, J.S. and Sharkey, P.J. (1975) Plant Physiol. 56,807-812

22 Peoples, M.B., Pate, J.S. and Atkins, C.A. (1985) J. exp. Bot. 36,567-582

23 Peoples, M.B., Atkins, C.A., Pate, J.S. and Murray, D.R. (1985) Plant Physiol. 77,382-388

24 Layzell, D.B., Pate, J.S., Atkins, C.A. and Canvin, D.T. (1981) Plant Physiol. 67,30-36

25 Atkins, C.A. (1986) Outlook on Agric. (in press)

26 Pate, J.S. and Atkins, C.A. (1983) Plant Physiol. 71,835-840

27 Atkins, C.A., Pate, J.S. and McNeil, D.L. (1980) J. exp. Bot. 31,1509-1520

28 McNeil, D.L., Atkins, C.A. and Pate, J.S. (1979) Plant Physiol. 63,1076-1081

29 Pate, J.S., Peoples, M.B. and Atkins, C.A. (1984) Plant Physiol. 74,499-505

30 Atkins, C.A., Shelp,B.J., J. Kuo, Peoples, M.B. and Pate, J,S. (1984) Planta 162,316-326

31 Pate, J.S., Atkins, C.A. and Rainbird, R.M. (1981) in Current Perspectives in Nitrogen Fixation (Gibson and Newton, eds.) pp.105-116, Aust. Acad. Sci., Canberra

32 Atkins, C.A. (1984) Plant and Soil 82,273-284

33 Pate, J.S., Sharkey, P.J. and Atkins, C.A. (1977) Plant Physiol 59,506-510

34 Peoples, M.B., Pate, J.S., Atkins, C.A. and Murray, D.R. (1984) Plant Physiol. 77,142-147

35 Pate, J.S., Peoples, M.B., van Bel, A., Kuo, J. and Atkins, C.A. (1984) Plant Physiol. 77,148-156

CO_2 PHOTOFIXATION AND NO_2^- PHOTOREDUCTION IN LEAF MESOPHYLL CELL
ISOLATES FROM N_2-FIXING SOYBEAN PLANTS HELD IN THE ABSENCE OF NO_3^-

J. MICHAEL ROBINSON[1] and PETER van BERKUM[2], USDA-ARS, Plant
Photobiology Lab[1] and N_2 Fixation and Soybean Genetics Lab[2],
Beltsville Agricultural Research Center-West, Beltsville, Maryland, 20705

1. INTRODUCTION

In soybean species growing in soils low in inorganic nitrogen (e.g.,
NO_3^-), and where symbiotic N_2 fixation by Bradyrhizobium japonicum
supplies those plants with almost all of the organic nitrogen (1), there
is a dependence by the symbiont bacteriods upon photosynthate (sucrose
and organic acids), as well as upon high photosynthetic rates, both for
maximal success for the symbioses as well as for maximal and sustained
N_2 fixation rates (2,3,4). The fact that a soybean plant can be
manipulated to depend for its nitrogen nutrition either upon NO_3^-, or
upon the symbiosis with Rhizobium, provided a tool to ask the following
questions:
1) Does a N_2-fixing soybean plant, which has placed a high demand upon
   source leaves for photosynthate to support growth, as well as to
   support symbiosis (2,3,4), have a higher net CO_2 photofixation rate
   in its mesophyll leaf cells, when compared with the rate of net
   photosynthesis in leaf cells of a nonsymbiont, counterpart soybean
   plant which is absorbing and assimilating NO_3^-?
2) Is the nitrite photoassimilatory system in the chloroplasts of the
   mesophyll leaf cells of N_2-fixing soybean plants repressed relative
   to that process in the cells of the nitrate fed plants? Nitrate can be
   successfully withheld from a symbiont soybean plant, because it has an
   alternate source of organic N (1). However, the absence of NO_3^- in
   soybean leaves could result in the repression and degradation of both
   leaf cytoplasmic nitrate reductase and leaf chloroplast nitrite
   reductase as happens when NO_3^- is withheld from pea plants (5). If
   NO_2^- photoassimilation is repressed or absent in an N_2-fixing
   soybean plant (where NO_3^- has been withheld) then it would support
   the growing view (6,7,8) that it is the chloroplast nitrite
   photoassimilatory process in NO_3^- assimilating plants that must
   supply as much as 100% of the reduced N while N_2-fixing
   soybean-symbiont depends upon ureides synthesized in the root system(1).
We report and discuss: 1) that net CO_2 photoassimilation in
isolates of soybean leaf mesophyll cells prepared from N_2-fixing plants
(maintained in the absence of NO_3^-) display higher rates than those
isolates from counterpart NO_3^- fed plants, and 2) that leaf mesophyll
cell isolates from the same N_2 fixing symbionts displayed a repressed
NO_2^- photoassimilation when compared with isolates from NO_3^- fed
control plants.

2. MATERIALS AND METHODS

2.1. Propagation of Soybean and Bradyrhizobium. Glycine max cv
'Clark'-nodulating plants were propagated in the greenhouse (March and

Biggens, J. (ed.), Progress in Photosynthesis Research, Vol. III. ISBN 90 247 3452 5
© 1987 Martinus Nijhoff Publishers, Dordrecht. Printed in the Netherlands.

April) in sterilized vermiculate pots. The seeds were initially surface sterilized, and, at seeding, one-half of the test set seed pots were inoculated with 10ml of 10^9 cells/ml of Bradyrhizobium japonicum strain USDA 122 (Beltsville collection). Initially, inoculated pots were fed a nutrient solution in which NO_3 was withheld from the inoculated pots through the entire growth period. It required approximately 15-20 days for the symbiosis to become well established. (See reference 14 in van Berkum et al.(1) for composition of the plant growth nutrient solution, and ref. 21 of (1) for Bradyrhizobium culture procedures.)

Concurrent with the inoculated plant's growth, the other half of the test plants were propagated in the same environment, but were subirrigated through the growth period with the nutrient solution (ref. 14 in (1)) containing 6 mM NO_3. After approximately 40 days of growth, mature (source) leaves of both treatments were measured for net CO_2 assimilation; some plants were taken for mesophyll leaf cell isolates. Subsequent to gas exchange measurements, the plants were depotted and washed, roots and root-nodules were excised, and acetylene reduction was measured. All measurements were carried out in the same experimental period.

2.2. Acetylene Reduction. N_2 fixation capacity of nodulated roots was measured by acetylene reduction as previously described (ref. 10 in van Berkum et al. (1)).

2.3. Photosynthesis and NO_3 Photoreduction. All procedures employed for measurement of soybean trifoliolate (3 leaflets) net CO_2 photoassimilation have been described previously (9). Net CO_2 exchange was estimated in bright sunlight (1000 uE/m^2·s), but measurement leaf chamber conditions were held at 25°C, 65% relative humidity, and 350 ppm CO_2. Trifoliolates (4 and 5) were numbered acropetally from the first unifoliolate.

All procedures for the isolation of soybean leaf mesophyll cells, for the quantitation of cell isolate net $^{14}CO_2$ photofixation, as well as for quantitation of cell isolate NO_3 photoreduction to a -amino nitrogen have been described in previous reports(7,8). Cell isolate assay medium(7,8) additionally contained, in 3.5 ml, 5 mM ^{14}C-(CO_2 + HCO_3) (20 uCi); 0.5 mM Pi; where noted 0.5 mM $NaNO_2$; and 100% intact mesophyll cells equivalent to 150 ug Chl (isolates from NO_3 fed plants) or 131 ug Chl (N_2 fixing plants).

3. RESULTS AND DISCUSSION

3.1. N_2 Fixation and Plant Growth. After 40 days of growth, plants whose nitrogen nutrition was supported by N_2 fixation were slightly (10%) smaller in size to those plants fed with NO_3 (not shown). Also it was very clear that N_2-fixing plants were dependent, successfully, for N nutrition upon the bacteriod symbiosis (1). For example, the root-nodule mass of N_2-fixing plants displayed C_2H_2 reduction of average rates of 20.9 $\pm$ 4.5 umol/h·plant, while NO_3 fed plants displayed no detectable noduation and no C_2H_2 reduction (not shown). Additionally, although mature source leaves (trifoliolates 4 & 5) of N_2-fixing plants were approximately 25% smaller in total area/leaf than counterpart foliage (TFs 4 & 5) of NO_3 fed plants, comparable leaves of N_2-fixing plants often displayed the same or slightly higher specific leaf weights, as well as, approximately the same, or 25% larger concentrations of Chl/dm^2 relative to leaves of NO_3 fed plants (not shown).

3.2. Net CO_2 Photofixation In Leaves. With incident light at approximately 1000 uE/m^2·s, and CO_2 levels at 350 ppm, net CO_2 fixation was often observed to display slightly higher rates in the leaves of N_2-fixing compared with NO_3^- fed plants, especially when expressed on a leaf area basis. For example, average rates of net CO_2 fixation for the leaves (TFs 4 & 5) of N_2-fixing plants were 957.6 $\pm$ 55.4 umol/h·dm^2, and for comparable NO_3^- fed plants, rates were 883.6 $\pm$ 54.8 umol/h·dm^2 (not shown). In subsequent experiments, we also observed that foliar net photosynthesis, both on an area, as well as on a Chl basis, was higher in N_2-fixing plants. For example, in another study, for N_2-fixing plants, average net CO_2 photofixation rates in source leaves were, 1027.0 $\pm$ 65.7 umol/h·dm^2 and 158.2 $\pm$ 17.7 umol/h·mg Chl, while for NO_3^- fed plants they were, for area and Chl basis, respectively, 864.8 $\pm$ 49.7 and 131.7 $\pm$ 8.5 (not shown). We should also note, that in one study, the rates of photosynthesis were the same in both N nutrition regimes (not shown), although leaf cell preparation of the N_2-fixing plants did display higher CO_2 fixation rates than the isolates of NO_3^- fed plants (see below).

FIGURE 1. Net $^{14}CO_2$ photoassimilation (A) and NO_2^- photoreduction (B) in isolates of mesophyll leaf cells from NO_3^- fed compared with N_2-fixing soybean (cv. 'Clark'-nodulating) plants. Numerical values on traces represents the most maximal rates in umol substrate converted/h·mg Chl. Legends on the graphics identify cell source.

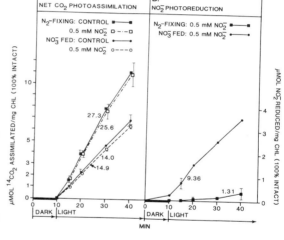

3.3. Net CO_2 Fixation in Mesophyll Cell Isolates. Interestingly, mesophyll leaf cells, isolated from N_2-fixing plants, always and routinely displayed a significantly higher rate of $^{14}CO_2$ photofixation, relative to cell isolates from NO_3^- fed plants (Fig. 1A). The rates of cell CO_2 fixation were always linear and sustained for the 30 min light period (Fig. 1A).

3.4. NO_2^- Photoreduction in Cell Isolates. We often observed that NO_2^- photoreduction in leaf cell isolates from NO_3^- fed plants displayed much higher rates than did that process in leaf cells isolates from comparable leaves of N_2-fixing plants. In some cases, it appeared that NO_2^- reduction was almost totally repressed in isolates of N_2-fixing plant foliage (Fig. 1B), but in other studies light dependent NO_2^- disappearance was only repressed by approximately 50% (not shown).

3.5. Conclusions. We surmised 1)That because enhanced sink organ strength often leads to stimulated photosynthetic metabolism (9), the higher rate of net CO_2 photofixation in the N_2-fixing plant foliage was a reflection of an enhanced sink strength conferred by the demand for photosynthate by the active N_2-fixing metabolism in the nodules of the symbiotic plants (1-4) (Fig. 1A).

Further, we concluded that 2a)N_2-fixing plants were able to subsist on lowered plastid NO_2^- photoreduction rate (Fig. 1B) because, N_2 fixation was actively ongoing and able to supply reduced N (1). Also, we concluded 2b)that it was the long duration of lack of NO_3^- which was responsible for the low rate of plastid NO_2^- photoreduction, since withholding NO_3^- from plants often results in repression and degradation of nitrate as well as nitrite reductase. Indeed, Gupta and Beevers (5) recently reported that withholding NO_3^- from pea plants for as much as 48 hrs. resulted in total repression and degradation of both cytoplasmic NO_3^- reductase and chloroplastic NO_2^- reductase. In our experiments we withheld NO_3^- from N_2-fixing plants over 30 days, certainly providing the possibility for repression of both NO_3^- and NO_2^- reductases.

Finally, we believe that the results of our experiments (Fig. 1B) strongly imply that if a plant does not have an alternate mechanism for obtaining reduced N (e.g., symbiotic N_2 fixation), then it must retain the chloroplast nitrogen photoassimilatory pathway (6-9).

4. REFERENCES

1 van Berkum, P., Sloger, C, Weber, D.F., Cregan, P.B., and Keyser, H.H. (1985) Plant Physiol. 77, 53-58.
2 Quebedeaux, B. (1979) in Photosynthesis II, Encyclopedia of Plant Physiology, New Series (Gibbs, M. and Latzko, E., eds.), Vol. 6, 472-480, Springer-Verlag, Berlin.
3 Schubert, K.R. (1982) in The Energetics of Biological Nitrogen Fixation (Schuber, K.R., ed.), pp. 3-23, American Society of Plant Physiologists, Rockville, Maryland, USA.
4 Sloger, C. (1985) in Nitrogen Fixation Research Progress (Evans, H.J., Bottomley, P.J., and Newton, W.E., eds.), p. 343, Martinus Nijhoff Publ., Dordrecht, The Netherlands.
5 Gupta, S.C. and Beevers, L. (1985) in Exploitation of Physiological and Genetic Variability to Enhance Crop Productivity (Harper, J.E., Schrader, L.E., and Howells, R.W., eds.), pp. 1-11, American Society of Plant Physiologists, Rockville, Md. USA.
6 Reed, J.A., Canvin, D.T., Sherrand, J.H., and Hageman, R.H. (1983) Plant Physiol. 71, 291-294.
7 Robinson, J.M. (1986) Plant Physiol. 80, 676-684.
8 Robinson, J.M. and Baysdorfer, C. (1985) in Regulation of Carbon Partitioning in Photosythetic Tissues (Heath, R.L. and Preiss, J. eds.) pp. 333-357, American Society of Plant Physiologists, Rockville, Maryland USA.
9 Robinson, J.M. (1984) Plant Physiol., 75, 397-409.

5. ACKNOWLEDGEMENTS

The authors thank Mr. Walter F. Stracke, Mr. Dave Lee, and Mr. Michael B. McMahon for many hours of dedicated technical assistance, and Ms. Lisa A. Motel for excellent typing of the manuscript.

STEREOSPECIFICALLY-TRITIATED GLYCERATE AS A PROBE OF PHOTORESPIRATORY METABOLISM

KENNETH R. HANSON AND RICHARD B. PETERSON

DEPARTMENT OF BIOCHEMISTRY AND GENETICS, THE CONNECTICUT AGRICULTURAL EXPERIMENT STATION, P.O. BOX 1106, NEW HAVEN, CT 06504, U.S.A.

1. INTRODUCTION

The metabolic pathways of photosynthesis and photorespiration have been defined as a result of extensive biochemical studies (1, 2). The selection of mutants in which enzymes of the photorespiratory pathway are inactive has confirmed the essential correctness of the proposed pathway (3). There is, however, a large gap between knowing the pathways and understanding how they work together in vivo. In this paper attention will be focused on the in vivo steady-state partitioning of key metabolic intermediates. Preceding papers in this symposium discussed the role of RuBP carboxylase-oxygenase in partitioning RuBP between carboxylation and oxygenation. Here emphasis will be placed on a second partitioning: that of glycolate.

In the photorespiratory pathway glycine, derives from glycolate, which derives from C-1 and C-2 of RuBP. Two molecules of glycine give one of CO_2 and one of serine. If this reaction is the sole source of photorespired CO_2, 25% of the glycolate carbon is photorespired. This can be treated as a conceptual partitioning of glycolate between complete oxidation to CO_2 (fraction $\underline{S}$ = 25%)[1] and complete conversion to PGA (fraction $1 - \underline{S}$ = 75%). It is convenient to refer to $\underline{S}$ as the stoichiometry of photorespiration. $\underline{S}$ need not be fixed at 25%. Although the peroxisomes have an abundance of catalase we cannot be certain that all of the H_2O_2 generated from the action of glycolate oxidase is destroyed before it has a chance to attack hydroxypyruvate, or other keto acids, or glyoxylate. Hydroxypyruvate derives from two glycolate molecules with one CO_2 released. Attack by H_2O_2 yields another CO_2 with the regeneration of glycolate. In effect, one glycolate has been completely oxidized to CO_2, i.e., $\underline{S}$ = 100%. This example suffices to show that the known biochemistry allows $\underline{S}$ to be between 25 and 100%. Factors which influence the amount and distribution of catalase and the pool concentrations of glycolate and hydroxypyruvate should influence $\underline{S}$.

[1]/ Symbols used: $\underline{S}$, stoichiometry of photorespiration (fraction of glycolate carbon photorespired); $\underline{g}$, fraction of ribulose 1,5-bisphosphate (RuBP) oxidized; $\underline{f}$, fraction of triose phosphates (TP) leaving the chloroplasts to form sugars; $\underline{t}$, fraction of ^3H in the TP committed to RuBP regeneration that is retained in the TP produced from RuBP; $\underline{h}$, ^3H leak (fraction of ^3H apparently lost from C-1 of ribose-5-P prior to RuBP formation); $\underline{r}$, ^3H retention (fraction of ^3H from the supplied and metabolized labeled glyceric acid that leaves the chloroplasts as TP); $\underline{p}_n$, ratio of photorespiration to net photosynthesis (PR/NPS).

Biggens, J. (ed.), Progress in Photosynthesis Research, Vol. III. ISBN 90 247 3452 5
© *1987 Martinus Nijhoff Publishers, Dordrecht. Printed in the Netherlands.*

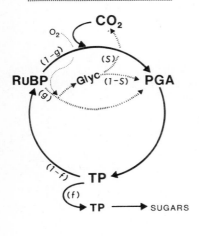

PARTITIONING MODEL

FIG. 1

STOICHIOMETRY				
	RuBP	+ CO$_2$↓ →	TP	+ CO$_2$↑
CARBOXYLATION	(1–g)	(1–g)	2(1–g)	
OXYGENATION				
Glycolate → CO$_2$	Sg		Sg	2Sg
Glycolate → PGA	(1–S)g		$\frac{5}{3}$(1–S)g	
SUM	1	(1–g)	$2 - \frac{g}{3} - \frac{2}{3}Sg$	2Sg

FIG. 2

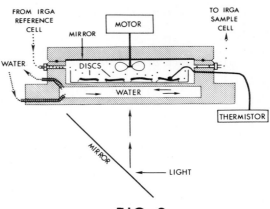

FIG. 3

2. THEORY

2.1. Partitioning model. In Fig. 1 the complex metabolic processes of photosynthesis and photorespiration are reduced to three key partitionings. The heavy circle represents the reductive photosynthetic cycle and the broken lines linking RuBP to PGA represent the reactions subsequent to the oxidation of RuBP (the "loop"). RuBP is partitioned between oxidation (fraction g) and carboxylation (fraction $1 - g$). The partitioning of glycolate has been explained above. The fraction of carbon leaving the system of cycle plus loop, other than as CO_2, can be defined by a third partitioning term, f. It is convenient to define f in terms of triose phosphates as the diversion of glucose-6-P to starch formation and PGA to malate formation can be treated as equivalent processes. Fraction f leaves the chloroplasts to form, ultimately, sucrose and fraction $1 - f$ is committed to the regeneration of RuBP. If the system is in a steady state carbon balance two partitioning terms suffice to define the system. This can be argued in detail by considering the stoichiometry of converting 1 mole of RuBP to triose phosphates. In Fig. 2, NPS = GPS - PR = CO_2↓ - CO_2↑, hence f = NPS/3TP = $[1 - (1 + 2S)g]/[6 - (1 + 2S)g]$. Fig. 2 also relates the partitioning terms g and S to the observable ratio PR/NPS:

$$p_n = PR/NPS = 2Sg/[1 - (1 + 2S)g] \qquad [1]$$

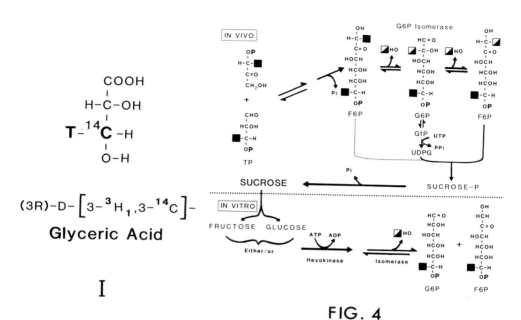

COOH
|
H-C-OH
|
T-^{14}C-H
|
O-H

(3R)-D-$\left[3-^3H_1,3-^{14}C\right]$-

Glyceric Acid

I

FIG. 4

One can only deduce g from p_n if S is assumed. To determine both g and S an independent source of information is needed.

2.2. Physical probe. In our standard experiment 10 inverted tobacco leaf discs (16 mm diam.) are placed in an illuminated chamber at constant temperature (Fig. 3) and the difference between the CO_2 inflow and outflow is monitored with an infrared gas analyzer. This allows the steady-state net photosynthetic rate to be calculated. Photorespiration is measured by following the postillumination burst. The CO_2 transient is calculated from the analyzer response by assuming a rate function and iteratively simulating the response until a best fit is obtained (4, 5; R.B. Peterson, Symposium paper, these volumes). It is assumed that the initial steady-state photorespiration rate approximates to the maximum observed rate of CO_2 evolution.

2.3. Stereochemical probe. The substrate for the second probe of the photosynthetic-photorespiratory system is (3R)-D-[3-$^3H_1$,3-^{14}C]glyceric acid (I) which can be prepared with high specific radioactivity. The leaf discs in the chamber rest on drops of water. Once a steady rate of net photosynthesis has been reached, the chamber is briefly opened and the substrate introduced (^3H:^{14}C = 20; ^{14}C = 500,000 dpm). A steady-state is soon reestablished. There is a linear uptake of substrate and after a lag period a linear release of $^{14}CO_2$. The $^{14}CO_2$ is trapped. After about 1 hour two measurements of photorespiration are made. The discs are then rinsed and killed. Sucrose is isolated by HPLC and then split into glucose and fructose by an HPLC procedure. The isolated glucose and fructose are further investigated.

The triose phosphates that leave the chloroplasts are labeled with ^3H at C-3 in the same configuration as in I and generally labeled with ^{14}C.

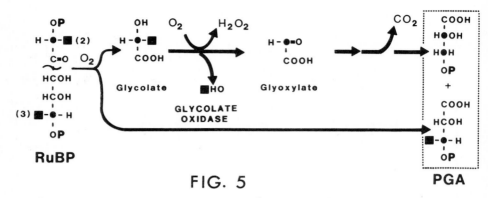

FIG. 5

RuBP

PGA

In the course of sucrose formation some ^3H is lost in the cytoplasm through the action of glucose-6-P isomerase (Fig. 4). To deduce the $^3H:^{14}C$ ratio of the triose phosphates it is necessary to complete this process in vitro by treating the glucose and fructose samples with ATP, hexokinase, and glucose-6-P isomerase. The hexose phosphates produced are ^3H labeled only at C-6, hence the $^3H:^{14}C$ ratio of the exported triose phosphates is twice this ratio. The glucose- and fructose-based estimates always agree closely and a further isomerase treatment of the hexose phosphate mixtures produces no change in the ratio. The final estimate is thus based on four $^3H:^{14}C$ measurements. The ^3H retention, r, is defined as the fraction of ^3H entering the system of cycle plus loop that leaves as triose phosphates. It is calculated as the $^3H:^{14}C$ ratio of the triose phosphates divided by the $^3H:^{14}C$ ratio of the supplied glyceric acid, times a correction for the fraction of metabolized ^{14}C lost as $^{14}CO_2$. The ^3H retention is not significantly influenced by the duration of substrate uptake. The 1 hour period adopted allows us to determine r under four different conditions in one day.

Once glyceric acid enters the chloroplasts it is phosphorylated and becomes part of the photosynthetic and photorespiratory metabolism. In the direct operation of the reductive cycle the methylene group remains intact. If there are no side losses of ^3H and no oxidation of RuBP, all of the ^3H entering the system leaves as triose phosphates and the ^3H retention, r, is 100%. The ^3H labeling of RuBP formed in the reductive cycle is shown in Fig. 5. On oxidation the glycolate formed is labeled with ^3H at C-2 in the same configuration as I, i.e., (R). Glycolate oxidase is pro-2R specific (6, 7), hence no ^3H can pass this point; the only ^3H reaching PGA is that from C-5 of RuBP. It follows that the more important the oxidation of RuBP the more ^3H is lost by this route and the lower r.

Fig. 6 summarizes the disposition of ^3H in the system of cycle plus loop. In deriving the relationship between r and the partitioning terms it is convenient to first ignore the small loss (leak) of ^3H that occurs through side reactions in the regeneration of RuBP from triose phosphates, i.e., h = 0. Of the ^3H that enters the system as PGA and is converted to triose phosphates, fraction f leaves and fraction 1-f is recycled. The fraction of ^3H retained on returning to triose phosphates, t, depends on the fraction of RuBP oxidized, g, and on the ^3H distribution in RuBP (2 at C-2 for every 3 at C-5): t = 1 - (2/5)g. After the first complete turn, fraction ft(1 - f) of the original ^3H leaves the system, after the second,

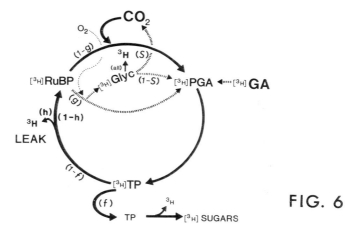

FIG. 6

$ft^2(1 - f)^2$, etc. On summing all the ^3H leaving the ^3H retention is obtained: $r = \bar{f}/[1 - t(1 - \bar{f})]$. As f is defined by g and S (see above), substitution for t and $\bar{f}$ gives

$$r = [1 - (1 + 2S)g]/[1 + (1 - 2S)g]. \tag{2}$$

Hence:
$$g = (1 - r)/[1 + (1 + 2p_n)r] \tag{3}$$

$$S = p_n \, r/(1 - r) \tag{4}$$

The same argument can be followed in order to allow for a ^3H leak in the regeneration of RuBP from triose phosphates. Any ^3H loss from C-1 of chloroplast fructose-6-P through the side action of glucose-6-P isomerase leads to less ^3H at C-1 of ribulose-5-P. The ^3H at C-1 of ribulose-5-P derives only from xylulose-5-P whereas that at C-5 derives from both ribose-5-P and xylulose-5-P. The side action of ribose-5-P isomerase leads to a loss of ^3H from ribulose-5-P so that RBP is depleted in ^3H at C-1 but not at C-5. If fraction h of the ^3H is lost from C-1 along the way, there is $2(1 - h)$ ^3H atoms at C-1 for every 3 at C-5 and $t = 1 - (2/5)g - (2/5)(1 - g)h$. The new expression for r obtained by substituting for t and f contains an additional $2(1 - g)h$ in the denominator. In combination with Eq. [1] this leads to

$$g = [1 - r - Ah]/[1 + (1 + 2p_n)r - Ah] \tag{5}$$

$$S = p_n \, r/[1 - r - Ah] \tag{6}$$

where $A = 2(1 + p_n)r$. On making a correction for a ^3H leak in RuBP regeneration g decreases and S increases. The correction corresponds to a movement along a p_n contour on a plot of S against g. The leak may be estimated under zero photorespiration conditions by measuring r. If $g = 0$, $p_n = 0$, hence, from Equation [5],

$$h = (1/r - 1)/2. \tag{7}$$

2.4. Development of theory. The above theory has evolved in four stages. It was first shown that p_n and g could be deduced from the fate of ^3H and ^{14}C derived from the labeled glyceric acid I provided $S = 25\%$ (8). When it was found that p_n so determined did not vary with changes in gas composition in the same way that the ratio determined with the infrared gas analyzer varied, it was concluded that $\underline{S}$ is not fixed (9). The theory was then reformulated to give equal weight to the two probes (10) and expanded to allow for isomerase-catalyzed losses of ^3H in the regeneration of RuBP (K.R. Hanson and R.B. Peterson, submitted for publication).

3. RESULTS
3.1. <u>Sample variation</u>. In most of our experiments with inverted tobacco-leaf discs, normal air, 32°C, and ca. 800 µE m^{-2}s^{-1} have been used as stan-dard reference conditions. Twenty measurements under these conditions show that there is considerable sample variation. Some of this may arise from seasonal changes in the greenhouse environment. The net photosynthetic rates ranged from 100 to 400 µmol gFW^{-1} h^{-1}. The mean ratios were $\underline{p}_n$ = PR/NPS = 35($\pm$8)% and $\underline{r}$ = 46($\pm$5)%. The partitioning terms, calculated from these ratios assuming no ^3H leak in RuBP regeneration, were $\underline{g}$ = 30($\pm$4)% and $\underline{S}$ = 30($\pm$8)%. Because of this variation most experiments have been carried out 3 times and the results have been examined by regression analysis.

3.2. <u>Irradiance</u>. For discs in normal air at 32°C increasing the irradiance from 190 to 1570 µE m^{-2} s^{-1} increased net photosynthesis and photorespira-tion three fold. The values for 1020 and 1570 µE m^{-2} s^{-1} were essentially the same. The increase in $\underline{p}_n$ was small and there was not detectable trend in $\underline{r}$. The only trend in the partition terms was a slight increase in $\underline{S}$ with irradiance.

3.3. <u>Gas composition</u>. In one set of experiments at 32°C and 1000 µE m^{-2} s^{-1} O_2 was varied from 16 to 40% at 340 µl CO_2/l and in a second set CO_2 was varied from 320 to 1,000 µl/l at 21% O_2. In Figs. 7, 8, and 10 the results are plotted against the external $[O_2]/[CO_2]$ ratios. There were significant differences between the normal air values in the two sets of experiments. Fig. 7 shows that increasing $[O_2]$ had more effect on photorespiration than decreasing $[CO_2]$, but the trends for net photosynthesis were similar. This difference is seen in the $\underline{p}_n$ curves in Fig. 8. The zero $\underline{p}_n$ value at zero $[O_2]/[CO_2]$ is assumed. The trends in $\underline{r}$ were similar in the two sets of experiments. The 80% values for $\underline{r}$ at zero $[O_2]/[CO_2]$ are taken from recent experiments. It appears that $\underline{r}$ does not approach 100% as zero photorespi-ration conditions are approached (Fig. 9). By Eq. [7] this result implies that there is a leak, $\underline{h}$, of about 13% in the regeneration of RuBP. Prior to these experiments we calculated the partition terms $\underline{g}$ and $\underline{S}$ as in the dots and error bars of Fig. 10. The fraction of RuBP, oxidized, $\underline{g}$, increased with increasing $[O_2]/[CO_2]$ as expected. The exact relationship cannot be predicted because the external gas concentrations differ from those at the active site of RuBP carboxylase-oxygenase. The stoichiometry of photorespiration, $\underline{S}$, also increased, however, many of the $\underline{S}$ values were below the theoretical lower limit of 25% defined by the sequence (2 glycine $\rightarrow CO_2$ + serine). The absence of any influence of irradiance on the calcu-lated ^3H leak, $\underline{h}$, under zero photorespiration conditions and other consid-erations suggest that at constant temperature a uniform value for $\underline{h}$ can be assumed and that $\underline{g}$ and $\underline{S}$ may be recalculated using Eqs. [5] and [6]. The corrected curves (dashed) derived from the corrected data points suggest that $\underline{S}$ approaches 25% as oxidation of RuBP approaches zero. The revision does not change the conclusion that at high O_2 concentrations the stoichio-metry of photorespiration substantially exceeds 25%. (For further discus-sion see I. Zelitch, these volumes).

For various reasons our methods appear to be inappropriate for conditions

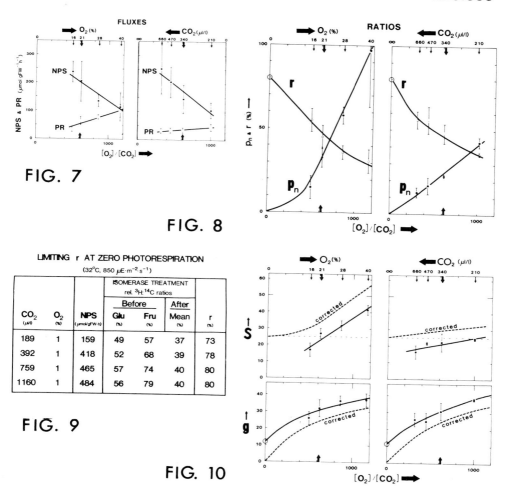

FIG. 7

FIG. 8

FIG. 9

FIG. 10

LIMITING r AT ZERO PHOTORESPIRATION

(32°C, 850 μE·m⁻²·s⁻¹)

CO₂ (μl/l)	O₂ (%)	NPS (μmol/gFW·h)	ISOMERASE TREATMENT rel. ³H:¹⁴C ratios			r (%)
			Before		After	
			Glu (%)	Fru (%)	Mean (%)	
189	1	159	49	57	37	73
392	1	418	52	68	39	78
759	1	465	57	74	40	80
1160	1	484	56	79	40	80

approaching the CO_2 compensation point. Recent studies by Brooks and Farquhar (11), extending earlier work by Jordan and Ogren (12), indicate that over a range of temperature S is close to 25% at the compensation point.

3.4. Temperature. For discs in normal air and 1000 μE m⁻² s⁻¹ net photosynthesis changed very little on going from 22 to 39°C but there was a doubling of photorespiration. There was no trend in r but the ³H leak increased. The partitioning terms were calculated with and without correction for the ³H leak. In either case g decreased slightly with increasing temperature and S increased substantially: S at 40°C appeared to be about double that at 30°C. If there were no such increase, net photosynthesis should increase with temperature instead of remaining constant.

3.5. C₄ photorespiration. When ³H retentions were determined for maize leaf discs in 21 and 1% O_2 (340 μl CO_2/l) the means were 82 and 84% respectively (P<0.001). If there is essentially zero photorespiration in 1% O_2

the ^3H leak is about 9%. Hence in normal air p_n is about 1.3% if $\underline{S}$ is about 25% (Eq. 6).

4. DISCUSSION

At the present time the methods described here are the only ones available for investigating the stoichiometry of photorespiration under conditions removed from the CO_2 compensation point. Some of our earlier estimates have been revised as we have come to understand better the importance of ^3H losses associated with isomerase action. The revisions have not changed our qualitative conclusions. Our ultimate concern is the stoichiometry of photorespiration of field-grown plants. The issue is of great practical importance. If under normal physiological conditions the stoichiometry is fixed at 25%, the only route to reduce photorespiration appears to be to modify the partitioning of RuBP by modifying the properties of RuBP carboxylase-oxygenase. If the stoichiometry can vary and if it can substantially exceed 25% under normal physiological conditions, even if only at higher temperatures, a second option is available. It should be possible to devise selection methods that will result in mutant plants in which the stoichiometry is reduced to near the 25% limit. Such a change should increase net photosynthesis. The gains of the two approaches to improving crop yields should be additive.

ACKNOWLEDGMENTS. We wish to thank C. Barbesino and E. Hennessey for technical assistance and I. Zelitch for helpful discussions.

REFERENCES

1 Tolbert, N.E. (1979) in Encyclopedia of Plant Physiology (Gibbs, M. and Latzko, E., eds.), Vol. 6, pp. 338-352, Springer-Verlag, Berlin/New York
2 Ogren, W.L. (1984) Annu. Rev. Plant Physiol. 35, 415-442
3 Somerville, C.W. (1986) Annu. Rev. Plant Physiol. 37, 467-507
4 Peterson, R.B. (1983) Plant Physiol. 73, 978-982
5 Peterson, R.B. and Ferrandino, F.J. (1984) Plant Physiol. 76, 976-978
6 Rose, I.A. (1958) J. Amer. Chem. Soc. 80, 5835-5836
7 Johnson, C.K., Gabe, E.J., Taylor, M.R., and Rose, I.A. (1965) J. Amer. Chem. Soc. 87, 1802-1804
8 Hanson, K.R. (1984) Arch. Biochem. Biophys. 232, 58-75
9 Hanson, K.R. and Peterson, R.B. (1985) Arch. Biochem. Biophys. 237, 300-313
10 Hanson, K.R. and Peterson, R.B. (1986) Arch. Biochem. Biophys. 246, 332-346
11 Brooks, A. and Farquhar, G.D. (1985) Planta 165, 397-406
12 Jordan, D.B. and Ogren, W.L. (1984) Planta 161, 308-313

INHIBITION OF GLYCOLATE AND D-LACTATE METABOLISM IN A MUTANT OF
CHLAMYDOMONAS REINHARDTII DEFICIENT IN MITOCHONDRIAL RESPIRATION

Diane W. Husic and N.E. Tolbert, Department of Biochemistry,
Michigan State University, East Lansing, Michigan 48824 U.S.A.

1. INTRODUCTION

Glycolate dehydrogenase, an enzyme of the oxidative photosynthetic
carbon cycle (C_2 cycle), catalyzes the oxidation of glycolate to
glyoxylate (1). This enzyme also reportedly catalyzes the oxidation of
D-lactate (1). Unlike glycolate oxidase of higher plants, the
dehydrogenase does not use molecular O_2 as the direct electron acceptor;
however, the natural acceptor is unknown. Both glycolate and D-lactate
dehydrogenase activities have been cytochemically localized in the
mitochondrial membrane (2) and glycolate-dependent reduction of
cytochrome c by Chlamydomonas mitochondrial fractions has been reported
(3). The present study describes the use of a mutant of Chlamydomonas
reinhardtii deficient in cytochrome oxidase activity (4) to provide
evidence that the metabolism of both glycolate and D-lactate in these
algae is linked to mitochondrial electron transport.

2. MATERIALS AND METHODS

Chlamydomonas reinhardtii UTEX 90 was from the R.C. Starr
collection, University of Texas, Austin, and the 137 strain was from
R.K. Togasaki, Indiana University. The mutant of Chlamydomonas deficient
in mitochondrial respiration (dk97) was provided by E. Harris at the
Chlamydomonas Culture Center, Duke University. [^{14}C] NaHCO$_3$ was from
New England Nuclear.

Algae were grown as described in (5) using minimal medium (6).
Preparation of algal suspensions and crude homogenates, and the
^{14}C-labeling studies were performed as described previously (5). For
chase experiments, after 30 s of photosynthesis, cells were centrifuged
into a tightly packed pellet in a microfuge allowing accumulation of
[^{14}C]lactate (5). The supernatant was removed and cells were resuspended
in fresh buffer, containing 1 mM H^{12}CO$_3^-$, and returned to the light.
At various times, 200 μl aliquots were removed and added to 0.5 ml of
methanol. Glycolate (D-lactate) dehydrogenase was assayed anaerobically
following the reduction of 2,6-dichlorophenolindophenol at 600 nm (7).
Cytochrome oxidase assays were performed by measuring the oxidation of
ferrocytochrome c at 550 nm (8). Photosynthesis was measured with a
suspension of intact algal cells in 25 mM Hepes, pH 7.5, either as the
rate of [^{14}C]NaHCO$_3^-$ (1 mM) incorporation or as the rate of O_2
evolution in the presence of 10 mM NaHCO$_3^-$. Dark respiration rates
were measured as the rate of O_2 uptake in the dark in an O_2 electrode
chamber. Glycolate was measured by the Calkins method (9).

Biggens, J. (ed.), Progress in Photosynthesis Research, Vol. III. ISBN 90 247 3452 5
© 1987 Martinus Nijhoff Publishers, Dordrecht. Printed in the Netherlands.

3.   RESULTS AND DISCUSSION

Characterization of the dk97 mutant of Chamydomonas

The dk97 mutant, isolated by Wiseman et al. (4) is unable to survive heterotrophically on acetate in the dark, but grows photoautotrophically with CO_2. The levels of cytochrome oxidase activity measured in extracts prepared from the mutant cells was 11% of that observed with wild type cells. However, the mutant had dark respiration rates similar to the wild type cells (20 to 65 μmoles O_2 uptake·hr^{-1}·mg Chl^{-1}). In the dk97 cells, this respiration rate was inhibited 70% by 5 mM salicylhydroxamic acid (SHAM), an inhibitor of the alternative (cyanide-insensitive) pathway of respiration. Dark respiration rates in dk97 cells were inhibited only 4% by 1 mM KCN. SHAM alone had no significant effect on the dark respiration rate in wild type cells, but the combination of SHAM and KCN inhibited respiration by 80 to 90%.

Inhibition of glycolate metabolism in SHAM-treated dk97 cells

Inhibition of glycolate metabolism in unicellular algae results in increased glycolate excretion from the cells into the surrounding medium (1). Such studies have been done with Chlamydomonas using the C_2 cycle inhibitors aminooxyacetate (AOA) and aminoacetonitrile (10-12). Our hypothesis was that, if glycolate dehydrogenase was linked to mitochondrial electron transport, then, in the mutant deficient in cyanide-sensitive respiration, SHAM should cause an increase in the rate of glycolate excretion. However, in SHAM-treated wild type cells, electrons from glycolate could still flow through the cytochrome oxidase pathway of respiration and little or no increase in glycolate excretion would be expected.

The results of such experiments are presented in Table I. In the absence of inhibitors, both wild type and dk97 cells exhibited low rates of glycolate excretion into the medium during photosynthesis. The rate of glycolate excretion by wild type Chlamydomonas cells was not significantly increased in the presence of SHAM (p > .4). In contrast, glycolate excretion was increased 7.4-fold over the controls in SHAM-treated dk97 cells. The rate of glycolate excretion observed with the mutant cells in the presence of SHAM was similar to the estimated maximal rates of glycolate flux through the C_2 cycle in Chlamydomonas (12), indicating that glycolate oxidation was completely blocked. SHAM had no effect on the glycolate (D-lactate) dehydrogenase activity in detergent-solubilized membrane fractions. Results of the excretion experiments with SHAM-treated cells can be compared with similar experiments with cells treated with the C_2 cycle inhibitor AOA which blocks the transamination of glyoxylate to glycine and also results in increased glycolate excretion (Table I and ref. 10-12).

Table I. Effect of SHAM and aminooxyacetate on the rate of glycolate excretion by air-grown <u>Chlamydomonas reinhardtii</u> wild type and dk97 cells. Values are the average ± the standard deviation for the number of experiments shown in parenthesis.

Rate of Glycolate Excretion

	Wild type (137)	Mutant (dk97)
	μmoles glycolate excreted$\cdot$hr$^{-1}\cdot$mg Chl^{-1}	
Control	0.03 ± 0.07 (5)	0.46 ± 0.46 (15)
+ 5 mM SHAM	0.12 ± 0.27 (5)	3.6 ± 1.8 (8)
+ 2 mM AOA	2.1 ± 0.0 (2)	1.7 ± 0.6 (7)

Inhibition of D-lactate metabolism in SHAM-treated dk97 cells

It has been previously reported that D-lactate accumulates in Chlamydomonas during brief periods of anaerobic conditions (5). This lactate is ^{14}C-labeled from newly fixed $^{14}CO_2$ via the photosynthetic sugar-phosphate pool. We observed that in air-grown wild type Chlamydomonas, the ^{14}C-label in lactate was rapidly turned over in $H^{12}CO_3$ chase experiments (Figure 1) only during aerobic conditions in the light. The ^{14}C-label in lactate also turned over in chase experiments performed with dk97 cells or with wild type cells treated with SHAM. However, D-lactate turnover was blocked in SHAM-treated dk97 mutant cells (Figure 1). Thus, in the absence of mitochondrial respiration, D-lactate oxidation does not occur.

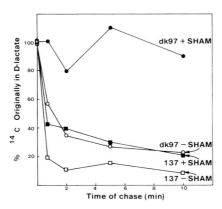

Figure 1. Inhibition of D-[^{14}C]lactate turnover in SHAM treated <u>Chlamydomonas</u> dk97 cells deficient in cytochrome oxidase activity.

4. CONCLUSIONS

1. In the absence of mitochondrial respiration, the metabolism of both glycolate and D-lactate is inhibited in the unicellular green alga, Chlamydomonas reinhardtii. This suggests that the mitochondrial membrane-bound glycolate (D-lactate) dehydrogenase is linked to the mitochondrial electron transport chain.

2. In plants, it has been proposed that the oxidative photosynthetic carbon cycle can dissipate excess photosynthetic energy, which under certain stress conditions, could help prevent against photoinhibition (1). We propose that in algae, electrons from glycolate could feed into the alternative, non-energy producing pathway of respiration to utilize excess reducing equivalents during photorespiration. Similarly, NADH produced during glycine oxidation in both algae and leaves could feed electrons into the alternative respiratory pathway for the same purpose.

3. D-Lactate which is produced anaerobically in Chlamydomonas (5) can only be oxidized under aerobic conditions.

REFERENCES

1. Husic, D.W., Husic, H.D. and Tolbert, N.E. (1986) CRC Critical Reviews in Plant Sciences (in press).
2. Beezley, B.B., Gruber, P.J. and Frederick, S.E. (1976) Plant Physiol. 58, 315-319.
3. Paul, J.S. and Volcani, B.E. (1976) Planta 129, 59-61.
4. Wiseman, A., Gillham, N.W. and Boynton, J.E. (1977) J. Cell Biol. 73, 56-77.
5. Husic, D.W. and Tolbert, N.E. (1975) Plant Physiol. 78, 277-284.
6. Sueoka, N. (1960) Proc. Natl. Acad. Sci. U.S.A. 46, 83-91.
7. Nelson, E.B. and Tolbert, N.E. (1970) Arch. Biochem. Biophys. 141, 102-110.
8. Wharton, D.C. and Tzagoloff, A. (1967) Meth. Enzymol. 10, 245-250.
9. Calkins, V.P. (1943) Anal. Chem. 15, 762-763.
10. Tolbert, N.E., Husic, H.D., Husic, D.W., Moroney, J.V. and Wilson, B.J. (1985) in Inorganic Carbon Uptake by Aquatic Photosynthetic Organisms (Lucas, W.J. and Berry, J.A., eds.), pp. 221-227, Amer. Soc. of Plant Physiologists.
11. Tolbert, N.E., Harrison, M. and Selph, N. (1983) Plant Physiol. 72, 1075-1082.
12. Moroney, J.V., Wilson, B.J. and Tolbert, N.E. (1986) Plant Physiol. (in press).

GLYCOLATE : DICHLOROPHENOL INDOPHENOL REDUCTASE ACTIVITY OF GLYCOLATE OXIDASE
IN THE PRESENCE OF OXYGEN: EXPERIMENTS WITH CRUDE ENZYME, PURIFIED ENZYME,
AND PEROXISOMES FROM PEA LEAVES

THOMAS BETSCHE, BOTANISCHES INSTITUT DER UNIVERSITÄT; D 4400 MÜNSTER,
F. R. GERMANY

INTRODUCTION
Glycolate is oxidized in photorespiring organisms by two different enzymes.
Higher plants and certain algae have glycolate oxidase. This enzyme is located in the peroxisomes and transfers electrons directly to O_2 (1). Many algae
possess glycolate dehydrogenase which is a mitochondrial enzyme, membrane-
associated, and transfers electrons to the respiratory chain (2). The two
enzymes have a remarkable immunological similarity (3). Nevertheless, the
enzymic properties of isolated glycolate oxidase and glycolate dehydrogenase
are different: Glycolate oxidase can oxidize L-lactate but not D-lactate, and
is not inhibited by KCN. Glycolate dehydrogenase can oxidize D-lactate but
not L-lactate, and is inhibited by KCN (4). Glycolate dehydrogenase readily
reduces dichlorophenol indophenol (DCPIP) in the presence of O_2, whereas the
DCPIP reductase activity of glycolate oxidase is supressed by O_2(4). This
observation supports the widely accepted opinion that O_2 is the only natural
electron acceptor for glycolate oxidase, and that no energy is conserved during glycolate oxidation. It is noteworthy in this respect that the inhibitory
effect of O_2 on the DCPIP reductase activity of glycolate oxidase from different organisms varies considerably (4). This suggests heterogeneity of the
enzyme, or that endogeneous compounds, or other factors, influence the electron acceptor specificity of glycolate oxidase. It has been shown that natural phenolic compounds can stimulate the activity of isolated glycolate oxidase (5). This paper reports of the capability of glycolate oxidase to reduce DCPIP under different environmental conditions and in the presence of
oxygen. Furthermore, results of experiments on the influence of inhibitors of
the O_2-metabolism in plants on glycolate-dependent DCPIP-reduction are presented.

MATERIAL AND METHODS
Plants. Pea (Pisum sativum L., cult."Kleine Rheinländerin") was grown at 23°C
in continuous light. Fully expanded leaves were used in the experiments.
Crude extract preparation. Leaves were ground with sand in 50 mM Tris/HCl
buffer, pH 8.0 (0 C). The homogenate was filtered through cheesecloth and centrifuged at 20 000 g for 15 min. The supernatant is referred to as crude extract.
Glycolate oxidase purification. After differential extraction of the leaves
at pH 6 and pH 8, anion and cation chromatographies resulted in apparent electrophoretic homogeneity of the enzyme (T. Betsche, R. Eising, M. Ruschke, unpublished).
Peroxisome purification. Peroxisomes were purified in a continuous Percoll
gradient (15-50 % , 10 min centrifugation at 20 000 g, fixed angle rotor). The
peroxisomes were free of contamination by mitochondria or chloroplasts (T.
Betsche, R. Eising, G. Müller unpublished).
Assays. Glycolate oxidase was determined by the phenylhydrazine method (6) in
50 mM Tris/HCl buffer, pH 8.0 . Glycolate:DCPIP reductase was determined in the
same buffer. The DCPIP-concentration was 0.12 mM, and the calculations were
done with the molecular absorption coeffizient of DCPIP at 600 nm of 21.9 x10^6
cm^2/ mol. In "isoosmotic" assays, the buffer contained 500 mM sucrose.

Biggens, J. (ed.), Progress in Photosynthesis Research, Vol. III. ISBN 90 247 3452 5
© *1987 Martinus Nijhoff Publishers, Dordrecht. Printed in the Netherlands.*

RESULTS AND DISCUSSION

Fig.1 shows spectrophotometric records from experiments on glycolate-dependent DCPIP-reduction with crude extract. Addition of crude extract to buffer
containing oxidized DCPIP leads to a continuing increase of the light absorption at 600 nm. Triton X-100 (0.1 %) diminishes the increase in light absorption, whereas Percoll (2.5 %)does not affect this blank reaction.(data not
shown). Both compounds were utilized in experiments with peroxisomes. Addition of glycolate (1 mM) to the reaction mixture gradually diminishes the increase in light absorption, and after about 20 min the light absorption
starts decreasing slightly (data not schown). Decreasing light absorption at
600 nm suggests that DCPIP is reduced. Glycolate oxidase activity (determined
in the absence of DCPIP) was readily detectable in the crude extract utilized.

The apparent lack of appreciable glycolate:DCPIP reductase activity in crude
extract may be caused either by the inability of glycolate oxidase to reduce
DCPIP in the presence of oxygen, or the reduced DCPIP is reoxidized. Peroxidase or phenoloxidase could oxidize DCPIP, and both enzymes are present in
crude pea leaf extract (T. Betsche, D. Schaller, unpublished). To scavenge
H_2O_2 from the glycolate oxidase reaction, catalase (200 ug/ml) was added to
the reaction mixture. There is still a long phase of increasing light absorption after the addition of glycolate, but then the light absorption decreases steadily (Fig 1 A, left). If, however, catalase is added, which was inactivated with 20 mM aminotriazole (15 min preincubation), the phase of increasing light absorption after the addition of glycolate is relatively short
and followed by a rapid
decrease of light absorp-
tion (Fig. 1 A, middle).
The influence of Na-azide,
an inhibitor of cata-
lase but not of peroxi-
dase, is shown in Fig.1A,
right. More work is required to understand the
events occurring in the
glycolate:DCPIP reductase
assay when catalase, ami-
notriazole, or azide are
added. The data available
so far does not support
the suggestion that per-
oxidase would play a
major role in the reoxi-
dation of reduced DCPIP
in crude pea-leaf-extract.

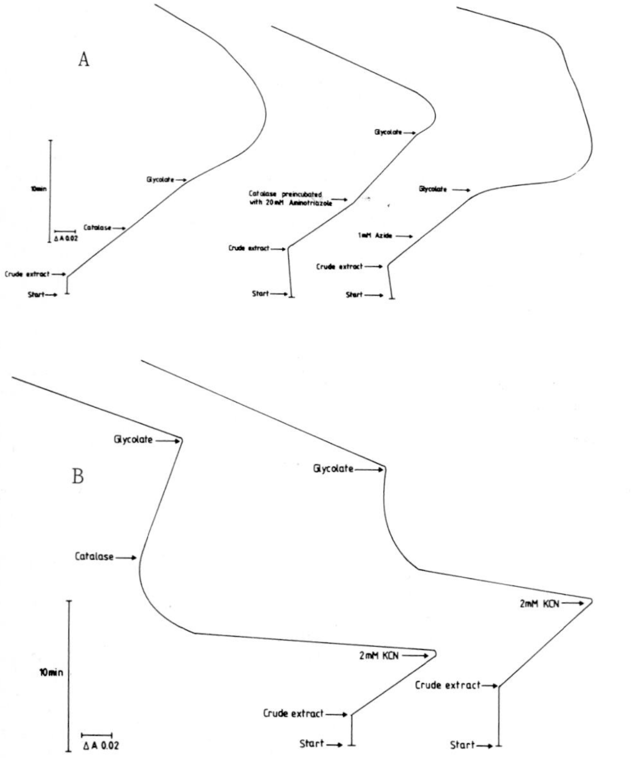

Fig. 1. Glycolate-depen-
dent DCPIP-reduction in
crude pea-leaf-extract
in the presence of O_2
(air) and various effec-
tors.

The figure depicts spec-
trophotometric records
obtained at 600 nm.

KCN inhibits catalase and phenoloxidase (8). Fig. 2 B shows the effect of KCN (2 mM) on the glycolate-dependent DCPIP-reduction in crude extract. KCN causes a strong but transient decrease of light absorption at 600 nm which is followed by a low and steady blank rate. On addition of glycolate, the light absorption decreases rapidly and steadily without a lag phase. This shows that in the presence of glycolate and KCN reduced DCPIP accumulates in crude extract under aerobic conditions. These and other results suggest that, without KCN, reduced DCPIP is reoxidized in crude extract, presumably by phenoloxidase. It is, however, worth noting that enriched and highly purified glycolate oxidase from pea leaves has no DCPIP-reductase activity under aerobic conditions unless KCN is present (Table 1). This indicates that KCN not only inhibits reoxidation of DCPIP in crude extract, but also increases the capability of glycolate oxidase to reduce DCPIP in the presence of O_2. (T. Betsche, D. Schaller, unpublished).

Based on the observations that the DCPIP reductase activity of purified glycolate oxidase can be induced by KCN and catalase (T. Betsche, D. Schaller, unpublished), one can speculate that the electron acceptor specificity of the isolated enzyme and the enzyme in its natural environment (peroxisome) is

Table 1   Glycolate oxidase and glycolate:DCPIP reductase activities in various fractions from pea leaves, and the effects of KCN (2 mM) and Triton X-100 (0.1 %) on glycolate:DCPIP reductase-

		Glyc.-DCPIP reductase (nkat/ml)	KCN-effect on Glyc.-DCPIP reduct.act. (%)	Triton-effect on Glyc.-DCPIP reduct.act. (%)	GO (glyox.form.) (nkat/ml)
Enriched GO		0.0	(1.28 nkat/ml)		5.9
Purified GO		0.0	(0.16 nkat/ml)	–	0.78
Fractions from Percoll gradient / 'soluble'	1	0.0	0	(0.36 nkat/ml)	1.22
	2	0.68	+ 123	+ 17	3.74
	3	1.32	+ 108	– 29	7.10
	4	0.56	+ 11	– 37	2.44
Peroxisomes	27	0.37	+ 19	– 39	1.07
	28	0.51	+ 39	– 17	1.15
	29	1.62	+ 79	– 30	3.23
	30	0.20	+ 130	– 41	0.62

different. To study this, peroxisomes were purified in Percoll, and the glycolate dependent reduction of DCPIP was determined in an isoosmotic medium. Table 1 shows that all but the uppermost fraction from the gradient have glycolate:DCPIP reductase activity in the presence of O_2. There is more glycolate:DCPIP reductase activity relative to glycolate oxidase activity in the peroxisome fractions than in the upper ("soluble") fractions containing broken peroxisomes and other contaminants. The ratio between these activities even varies from one fraction to the next (Table 1). Addition of KCN (2 mM) stimulates glycolate:DCPIP reductase activity in fractions except the uppermost one. The stimulation of the glycolate:DCPIP reductase activities by KCN is stronger in the "soluble" fractions than in the peroxisome fractions (Table 1). If the integrity of the organelles is destroyed by Triton X-100 (0.1 %), glycolate:DCPIP reductase (determined in the presence of KCN) is inhibited in all fractions except the uppermost "soluble" fraction (Table 1). There, the detergent stimulates the glycolate:DCPIP reductase activity. The glycolate oxidase activity (determined without KCN) of the peroxisome fractions is not affected by Triton X-100, whereas in some "soluble" fractions from the Percoll gradient, the glycolate oxidase activity is significantly increased after the addition of the detergent (data not shown).

The results presented above demonstrate that pea-leaf-glycolate oxidase is capable of reducing DCPIP at a high rate (up to 50 % of the glycolate oxidase activity) in the presence of oxygen, if the enzyme is located in its natural environment, i.e., the peroxisome. It is therefore possible to propose that glycolate oxidase has the capability of transferring electrons to acceptors other than oxygen in the living cell.

REFERENCES

1 Tolbert, N.E., Oeser, A., Kisaki, T. Hageman, R.H. and Yamazaki, R.K. (1968) J. Biol. Chem. 243, 5179-5184
2 Paul, S.C. and Volcani, B.E. (1974) Planta 129, 59-61
3 Codd, G.A. and Schmid, G.H. (1972) Plant Physiol. 50, 769-773
4 Frederick, S.E., Gruber, P.J. and Tolbert, N.E. (1973) Plant Physiol. 52, 318-323
5 Codd, G.A. and Schmid, G.H. (1971) Planta 99, 230-239
6 Baker, A.L. and Tolbert, N.E. (1966) in Methods in Enzymology (Wood, W.A. ed.), Vol. IX, pp.338-342, Academic Press, New York
7 Armstrong, J.D. (1964) Biochim. Biophys. Acta 86, 194-197
8 Butt, V.S. (1980) in The Biochemistry of Plants, (Davies,D.D. ed.), Vol. 2, pp.81-123, Academic Press, New York

IDENTIFICATION AND SOME CHARACTERISTICS OF TWO NADPH-DEPENDENT REDUCTASES INVOLVED IN GLYOXYLATE AND HYDROXYPYRUVATE METABOLISM IN LEAVES

LESZEK A. KLECZKOWSKI, DOUGLAS D. RANDALL AND DALE G. BLEVINS

1. INTRODUCTION

The peroxisomal NADH-hydroxypyruvate reductase (HPR-I) is one of a series of reactions comprising the photorespiratory pathway (1). The enzyme prefers NADH and hydroxypyruvate as substrates, but uses NADPH and glyoxylate (K_m = 20 mM) rather inefficiently (2). Besides HPR-I, leaves contain a specific glyoxylate reductase (GR-I) which utilizes only NADPH (3) and is localized in chloroplasts (4). The hydroxypyruvate- or glyoxylate-dependent rates with NADH and/or NADPH reported for crude leaf extracts have usually been attributed to either NADH-HPR-I or NADPH-GR-I or both of these enzymes. We report here the identification and partial characterization of two novel reductases utilizing hydroxypyruvate and/or glyoxylate as substrates and prefer NADPH as a cofactor. This is the first report describing leaf NADPH(NADH)HPR-II, while the purification and some characteristics of the second new reductase, NADPH(NADH)-GR-II, have already been published (5).

2. MATERIALS AND METHODS

Plant Material. Spinach (Spinacia oleracea) leaves were purchased in a local market.

Enzyme Assays. The 1.0 ml GR assay contained 100 mM Mops (pH 7.1), 0.2 mM NADPH/NADH, 2.5 mM glyoxylate and enzyme. The 1.0 ml HPR assay contained 100 mM Mes (pH 6.5), 0.2 mM NADPH/NADH, 1 mM hydroxypyruvate/glyoxylate and enzyme. NADPH/NADH oxidation was monitored spectrophotometrically at 340 nm. Assays were initiated with glyoxylate or hydroxypyruvate. A unit of activity was the amount of enzyme that oxidize 1 μmol NADPH/NADH per min.

Enzyme Purification. Spinach leaf NADPH(NADH)-GR-II was purified as previously described (5) using $(NH_4)_2SO_4$ precipitation (50-60% sat) and chromatography on DEAE-cellulose, Sephadex G-75SF, and two affinity columns. Spinach leaf NADPH(NADH)-HPR-II was purified by $(NH_4)_2SO_4$ precipiation (45-60% sat) followed by chromatography DEAE-cellulose, Sephadex G-75SF, Green A dye-ligand (Amicon) and affinity chromatography columns (L.A. Kleczkowski, unpublished). Spinach leaf NADH-HPR-I (Sigma) was purified to homogeneity by Sephadex G-75SF chromatography.

Other Methods. Immune blotting was according to Towbin et al. (7), as modified by Kleczkowski (8); protein and Chl determined by (9,10).

3. RESULTS

Identification of Leaf GR-II and HPR-II. NADPH/glyoxylate and NADPH/hydroxypyruvate-dependent activities of crude extracts were 60-100 and 100-200 nmol/min/mg protein, respectively. It has usually been assumed that these rates belong to GR-I (3), and to HPR-I (2). The two reductases can be easily separated one from another by $(NH_4)_2SO_4$ fractionation (HPR-I, 0-40% sat; GR-I, 50-60% sat) (3) with an almost complete recovery of enzymatic activity. We have repeated these experiments and monitored NADPH/glyoxylate-, NADH/hydroxypyruvate-,

NADH/glyoxylate- and NADPH/hydroxypyruvate-dependent rates. These data (Table 1) suggested that leaves contain at least one activity which is distinct from NADPH-specific GR-I and HPR-I. Further fractionation on DEAE-cellulose and affinity columns (where all the activities co-purified), followed by chromatography on Sephadex G-75SF (Fig. 1) strongly suggested the presence of two enzymes: A GR-II which utilized both NADPH and NADH, but not hydroxypyruvate, and HPR-II which used NADPH/NADH and hydroxypyruvate/glyoxylate as substrates. In both cases, the activity with NADPH was preferred over that with NADH. Further purification to homogeneity of the pooled fractions of either NADPH(NADH)-GR-II or NADPH(NADH)-HPR-II did not remove or change the relative activities associated with either protein.

TABLE 1. Separation of Spinach Leaf GRs and HPRs

Activity	Crude Extract		$(NH_4)_2SO_4$ ppt.*	
	Total U	Yield (%)	Total U	Yield (%)
NADPH/glyoxylate	69	100	73.5	106
NADPH/OH-pyruvate	157	100	31.0	20
NADH/glyoxylate	47	100	42.0	89
NADH/OH-pyruvate	906	100	31.5	3

*Proteins from crude leaf extract were precipitated with 45 to 60% sat'd. $(NH_4)_2SO_4$. Assays contained 0.2 mM NADPH/NADH and 2.5 mM glyoxylate or 1 mM hydroxypyruvate.

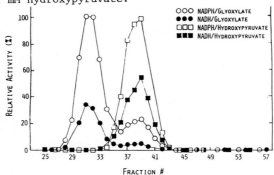
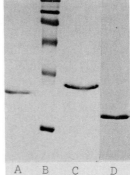

FIGURE 1. Separation of GR-II and HPR-II by Sephadex G-75SF chromatography.

FIGURE 2. SDS-PAGE of purified spinach leaf HPR-II (lane A), HPR-I (lane C) and GR-II (lane D). Lanes B and E, standard proteins. Gels were stained with Coomassie Blue.

Physical and Kinetic Properties of GR-II and HPR-II. The properties of the two new reductases are compared in Table 2 with the classical peroxisomal HPR-I. The purified enzymes gave single bands on SDS-PAGE (Fig. 2) with different M_rs. Purified GR-II was highly specific for glyoxylate, while HPR-II utilized glyoxylate at 40% of hydroxypyruvate rate. Neither enzyme used oxalacetate or pyruvate, and both preferred NADPH to NADH.

TABLE 2. Properties of Spinach Leaf GR-II and HPR-I.

	GR-II	HPR-II	HPR-I
Native M_r	12,500	70,000	90,000
Subunit M_r	33,000	38,000	41,000
pH Optimum	6.4 - 7.4	5.8 - 6.9	About 6.0
Glyoxylate-Dependent	+	+	+
OH-Pyruvate-Dependent	-[a]	+	+
K_m/Glyoxylate (mM)	0.085 (NADPH)	ND[b]	20
	1.10 (NADH)		
K_m/OH-pyruvate (mM)	-	0.75	0.1
Activity in Leaves (μmol/min/mg protein)	0.090 (NADPH)	0.040 (NADPH)	1.5 (NADH)
Activity of Purified enzyme (μmol/min/mg)	210[c] (NADPH)	265[d] (NADPH/OH-pyr.)	ND
	210[c] (NADH)	120[d] (NADH/OH-pyr.)	
		55[d] (NADPH/glyox.)	
		13[d] (NADH/glyox.)	
Abundance in leaves (% total proteins)	0.03	0.02	0.2
Cross-reactivity with Anti-NADPH(NADH)-GR-II	++++	-	+
Cross-reactivity with Anti-NADPH(NADH)-HPR-II	-	++++	-
Cross-reactivity with Anti-NADH(NADPH)-HPR-I	-	-	++++

[a] Only neglibile rates with hydroxypyruvate (5); [b] Not determined; [c] V_{max};

[d] Activity determined with 1 mM hydroxypyruvate/glyoxylate and 0.2 mM NADPH.

Immunological Comparison of Leaf GRs and HPRs. To compare the
reductases immunologically, we chose Western blot analysis because of
clear differences in subunit size of the three enzymes (Fig. 2) and
because of high sensitivity of the method. The high sensitivity was
essential to test cross-reactivity of antibodies with proteins from crude
spinach leaf extracts since the two novel reductases which constituted
only about 0.02% of total leaf proteins (Table 2). Immunoblots using
antibodes prepared against spinach GR-II recognized both GR-II and HPR-I,
although the cross-reactivity with the latter was much weaker. In
contrast, antibodies prepared against peroxisomal HPR-I did not cross-
react with GR-II. Antibodies prepared against HPR-II recognized only its
own antigen, suggesting that both HPRs do not share common determinants.
For all three reductases, the same results were obtained following immunoblotting of either purified enzymes and crude leaf proteins.

4. DISCUSSION
    Data presented in this paper indicate that leaves contain consider-
able activities of two apparently novel isozymes of GR and HPR, respectively, which prefer NADPH over NADH as a cofactor and show activity with
hydroxypyruvate and glyoxylate (HPR-II) or glyoxylate only (GR-II). Both
reductases are clearly distinct from the well-known NADH-HPR-I (Table 2)
and from the NADPH-specific GR-I described by Zelitch and Gotto (3).
Attempts to fractionate the $(NH_4)_2SO_4$-precipitated crude spinach leaf
proteins on several affinity matrixes failed to resolve a supposed NADPH-
specific GR from NADPH(NADH)-GR-II (5), perhaps because of instability of
the former enzyme. It has become clear, however, that GR-II is by far the
most active GR in leaves, with rates of 30 to 80 μmol/h/mg Chl for the

enzyme from leaves of spnach, pea, tomato, wheat and soybean. Rates of crude leaf HPR-II were more difficult to establish due to interference of the non-specific NADPH-dependent activity of HPR-I. For spinach about one fifth of the NADPH/hydroxypyruvate-dependent activity determined in leaf extracts belonged to the novel HPR-II (Table 1) and corresponded to the rate of about 20 μmol/h/mg Chl. Despite reactivity with the same substrates, the two HPRs do not share common immunogenic determinants (Table 2), indicating substantial differences in the molecular structure.

Because of their relatively high activity in leaf extracts, both GR-II and HPR-II could be regarded as important components of carbon metabolism in leaves, possibly involved in the flow of photorespiratory carbon GR-II could serve as a side reaction of photorespiration, shunting both glyoxylate and reducing power into the peroxisome. In nitrogen-fixing tropical legumes (e.g., soybeans), this reductase may also be involved in ureide breakdown, utilizing the glyoxylate produced during allantoate metabolism (11). The HPR-II may possibly compete with "classical" peroxisomal HPR-I for hydroxypyruvate. Detailed localization studies are in progress to properly assess the role of both reductases in leaf metabolism. An obvious difficulty in such studies will be lack of a reliable assay system for HPR-II due to high non-specific activity of HPR-I. On the other hand, the lack of common antigenic determinants between the two HPRs (Table 2) should allow rapid separation of each of these proteins by immunoaffinity chromatography and thus help to unequivocally establish subcellular localization of HPR-II.

REFERENCES
1. Tolbert, N.E., Yamazaki, R.K. and Oeser, A. (1970) J. Biol. Chem. 245, 5129-5136.
2. Kohn, L.D. and Warren, W.A. (1970) J. Biol. Chem. 245, 3831-3839.
3. Zelitch, I and Gotto, A.M. (1962) Biochem. J. 84, 541-546.
4. Thompson, C.M. and Whittingham (1967) Biochim. Biophys. Acta 143, 642-644.
5. Kleczkowski, L.A., Randall, D.D. and Blevins, D.G. (1986) Biochem. J., in press.
6. Kleczkowski, L.A. and Randall, D.D. (1985) Plant Physiol. 79, 274-277.
7. Towbin, H., Staehelin, T. and Gordon, J. (1979) Proc. Natl. Acad. Sci. USA 76, 4350-4354.
8. Kleczkowski, L.A. (1985) Leaf Glycerate Kinase, Ph.D. Thesis, University of Missouri, Columbia.
9. Bradford, M.M. (1976) Anal. Biochem. 72, 248-254.
10. Arnon, D.I. (1949) Plant Physiol. 24, 1-15.
11. Winkler, R.G., Polacco, J.C., Blevins, D.G. and Randall, D.D. (1985) Plant Physiol. 79, 787-793.

ACKNOWLEDGEMENTS
Supported by National Science Foundation and USDA-CRGO.

ADDRESS OF AUTHORS
Interdisciplinary Plant Research Group, University of Missouri, Columbia, MO 65211, USA. L.A.K. was on leave from the Department of Plant Physiology, Warsaw Agricultural University, 02-528 Warsaw, Poland.

GLYOXYLATE AND PYRIDOXAL 5-PHOSPHATE INHIBITION OF ISOLATED GLYCINE
DECARBOXYLASE

G. Sarojini and David J. Oliver, Department of Bacteriology and
Biochemistry, University of Idaho, Moscow, Idaho 83843 USA

Glyoxylate is an intermediate in the photosynthetic carbon oxidation
(glycolate) cycle that has been shown to inhibit a range of enzymes.
Exposure of glycine decarboxylase to millimolar concentrations of glyoxyate
resulted in the inhibition of the enzyme. A broad range of other organic
acids tested had no effect. Kinetic analysis showed mixed function
inhibition that exhibited both competitive and noncompetitive
characteristics with respect to glycine. In the presence of 10 mM glycine,
50% inhibition of the glycine decarboxyase reaction required about 7 to 9
mM glyoxylate. The enzyme was also inhibited by excess pyridoxal
phosphate. Both pyridoxal phosphate and glyoxylate may bind to an
essential lysine residue (possibly at the active site of the P protein) and
inhibit the enzyme activity. Due to the localization of glycine
decarboxylase within the mitochondria, the high concentrations of
glyoxylate needed for inhibition, and the peroxisomal site of glyoxylate
metabolism, it is unlikely that this inhibition is of physiological
importance.

INTRODUCTION

The glycine decarboxylase complex from pea leaves is composed of four
subunits, the 97.5 kD P-protein (binds pyridoxyl 5-phosphate (PLP)), the 15
kD H-protein (contains lipoamide), 45 kD T-protein (binds
tetrahydrofolate), and the 59 kD L-protein (lipoamide dehydrogenase) (8).
The P, L, and T subunits do not appear to interact with each other, but
each binds to the 15 kDa H-protein. This small protein acts as a shuttle
and carries intermediates between the three larger proteins of the reaction
complex.

Glyoxylate is formed in the photosynthetic carbon oxidation
(glycolate) cycle from the oxidation of glycolate. The reaction is
catalyzed by the enzyme glycolate oxidase and occurs within the peroxisome.
The glyoxylate is normally transaminated to glycine within this same
organelle. The restriction of glyoxylate metabolism to the peroxisome
prevents the distribution of this highly reactive organic acid throughout
the cell. Glyoxylate inhibits a number of reactions including glycine
oxidation by mitochondria (6) and RuBP carboxylase/oxygenase (1). The
inhibition of RuBP carboxylase has been studied in considerable detail and
results from the reaction of glyoxylate with an essential lysine residue in
the protein.

MATERIALS AND METHODS

A crude preparation of glycine decarboxylase was solublized from
isolated pea leaf mitochondria following acetone-extraction of the
mitochondrial lipids. The crude enzyme was separated into P, H, and T
protein-enriched fractions by ammonium sulfate precipitation. The

individual component enzymes were then purified (8). Following the
fractionation, the active enzyme was reconstituted from its component
enzymes by mixing P, H, and T proteins in a units of enzyme activity ratio
of 1:5:1, respectively. Lipoamide dehydrogenase activity was provided by
adding 50 ug of purified yeast enzyme (Sigma) per assay.

RESULTS AND DISCUSSION

The mitochondrial oxidation of glycine results in the reduction of NAD
to NADH. This NADH can be reoxidized by either the mitochondrial electron
transport chain or by using a substrate shuttle that is capable of removing
the electron from the mitochondrial matrix (4,5). In both cases the
glycine decarboxylase complex competes with the citric acid cycle enzymes
for available NAD. Several investigators have observed a preferential
oxidation of glycine (2,3). To date, no satisfactory explanation exists
for this observation. We were interested in determining if any of the
organic acids of the citric acid and photorespiratory cycles were capable
of inhibiting isolated glycine decarboxylase. The compounds tested
included oxaloacetate, pyruvate, isocitrate, aconitate, citrate,
ketoglutarate, succinate, fumarate, malate, glycolate, and glyoxylate.
Only glyoxylate inhibited glycine decarboxylation.

The inhibition of glycine decarboxylation by glyoxylate was
concentration-dependent but not time-dependent (Fig. 1). At low
concentrations of glyoxylate (approximately 0.2 mM to 1.0 mM), the rate of
carbon dioxide release was stimulated about 10%. This stimulation,
although reproducible, is not readily explained. At higher concentrations
of glyoxylate, glycine decarboxylation was inhibited. In the presence of
10 mM glycine, 50% inhibition required about 7 mM to 9 mM glyoxylate.
Increasing the preincubation time of the enzyme with the glyoxylate in the
absence of glycine from 0 min to 60 min had no effect on the extent of
inhibition. This suggests that either the binding of the glyoxylate to the
enzyme was readily reversible or that the formation of the covalent linkage
was very rapid. Enzymatic activity was restored by dialysis or desalting
by gel filtration thus confirming the reversibility of the binding (data
not presented).

Kinetic analysis of the inhibition of glycine decarboxylation at
increasing glycine concentrations showed a mixed type of inhibition with a
strong competitive component (Fig. 2). The untreated enzyme had a Km of
6.1 mM for glycine and a Vmax of 36.6 nmol released/min. In the presence
of 10 mM glyoxylate, the Km for glycine increased to 17.2 mM and the Vmax
decreased to 21.6 nmol/min.

Glyoxylate also inhibited the glycine bicarbonate exchange reaction
catalyzed by the isolated P and H proteins (Fig. 3). This inhibition also
showed mixed competitive and noncompetitive components. The control enzyme
had a Km for glycine of 2.2 mM and a Vmax of 25.0 nmol/min. In the
presence of 10 mM glyoxylate, the Km was 4.3 mM glycine and the Vmax was
4.6 nmol/min.

Glycine decarboxylase requires PLP for activity. This cofactor binds
to the P protein and forms a Schiff base with the amino group of glycine as
the first step of the reaction (8). While PLP is essential for the
reaction, superoptimal concentrations of the cofactor inhibit enzyme
activity (7). This inhibition likely results from formation of a Schiff
base with an essential lysine residue in the protein. The inhibition

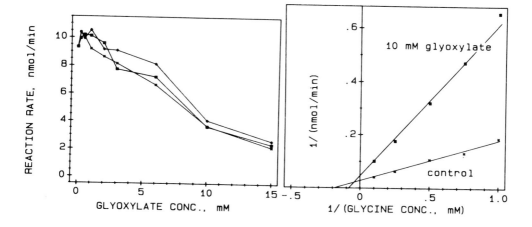

Fig. 1. Effect of Incubating Glycine Decarboxylase with Glyoxylate for (■) 0, (✗)15, or (◆)60 min.

Fig. 2. Kinetic Analysis of the Inhibition of Glycine Decarboxylation by Glyoxylate.

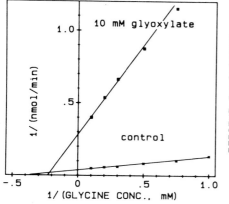

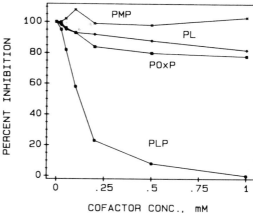

Fig. 3. Kinetic Analysis of the Inhibition of Glycine Bicarbonate Exchange Reaction by Glyoxylate.

Fig. 4. Inhibition of Glycine Decarboxylase by Pyridoxamine 5-P (PMP), Pyridoxal (PL), Pyridoxine 5-P (POxP), and Pyridoxal 5-P (PLP).

requires an available carbonyl group on the coenzyme. The other B-6 cofactors, pyridoxine phosphate (POxP) and pyridoxamine phosphate (PMP), were much less inhibitory (Fig. 4). This presumptive lysine residue may occur at the active site of the P protein as suggested by the requirement for a 5-phosphate group on the pyridoxal to cause full inhibition (Fig. 4). The combination of requiring both an active aldehyde group to effect the modification and a 5-phosphate group to direct the carbonyl to the correct location resulted in much stronger inhibition by PLP than with any of the other analogs tested.

CONCLUSIONS

Glyoxylate and PLP are both potent inhibitors of isolated glycine decarboxylase. Glyoxylate, under some circumstances, can react with the epsilon-amino groups of lysine residues within the protein and inhibit enzyme activity (1). A lysyl residue should exist within the active site of the P protein to form a Schiff base with PLP. The Schiff base formation between this lysine and glyoxylate may have resulted in inhibition that obscured the competitive interactions between glycine and glyoxylate. Earlier evidence for an essential lysine residue came from the observation that the enzyme was very sensitive to the PLP concentration (7). While this cofactor is essential for enzyme activity, superoptimal concentrations inhibit the enzyme. The inhibition was dependent on both the carbonyl group for reactivity and the phosphate group, presumably for inhibitor placement. At this point, there is no conclusive evidence that the same amino acid residue is involved in both inhibitions, although this may in fact be the case.

Although two photosynthetic enzymes, RuBP carboxylase/oxygenase (1) and the glycine decarboxylase complex, have been shown to be inhibited with glyoxylate, it is unlikely that these inhibition patterns are of physiological importance. Glyoxylate is normally sequestered within the plant peroxisome, the site of its synthesis and metabolism. This is necessitated by the very reactive nature of this keto acid. While a small amount of glyoxylate may escape the peroxisomes, it could not build up to the high levels needed to inhibit either enzyme.

REFERENCES

1. Cook, C.M., Spellman, M., Tolbert, N.E., Stringer, C.D., and Hartman, F.C. (1985) Arch. Biochem. Biophys. 240: 402-412
2. Day, D.A., Neuburger, M., and Douce, R. (1985) Aust. J. Plant Physiol. 12: 119-130
3. Dry, I.B. and Wiskich, J.T. (1985) Aust. J. Plant Physiol. 12: 329-339
4. Ebbighausen, C.J. and Heldt, H.W. (1985) Biochim. Biophys. Acta 810: 184-199
5. Oliver, D.J. and Walker, G.H. (1984) Plant Physiology 76: 409-413
6. Peterson, R.B. (1982) Plant Physiology 70: 61-66
7. Sarojini, G. and Oliver, D.J. (1983) Plant Physiology 72: 194-199
8. Walker, J.L. and Oliver, D.J. (1986) J. Biol. Chem. 261: 2214-2221

REGULATION OF GLYCINE DECARBOXYLASE BY SERINE

David J. Oliver and G. Sarojini. Department of Bacteriology and Biochemistry, University of Idaho, Moscow, Idaho 83843 USA

The glycine decarboxylase multienzyme complex was isolated from pea leaf mitochondria, the P, H, T, and L proteins were purified separately, and the active enzyme was reconstituted in vitro. A range of amino acids were tested for their effects on the enzymic decarboxylation of glycine. Only serine, the ultimate product of glycine metabolism within the mitochondria, inhibited the enzyme. With the isolated enzyme, serine was competitive with glycine. The Ki for serine was 4 mM, and the Km for glycine was 6 mM.

Serine also inhibited glycine decarboxylation by isolated mitochondria although not as effectively as for the isolated enzyme. Where 10 mM serine inhibited the oxidation of 10 mM glycine >95% with the isolated enzyme, it only inhibited glycine oxidation by intact mitochondria < 20%. Mitochondria became more sensitive to serine as the pH gradient across the membrane was decreased by the addition of ADP or uncouplers. Because the glycine decarboxylase complex converts glycine to serine, the matrix level of glycine would be expected to be lower than the concentration of serine. Potent inhibition of glycine decarboxylase is prevented in situ by the functioning of a glycine-serine transport system that removes newly formed serine in exchange for glycine.

INTRODUCTION

The photorespiratory conversion of glycine to serine within the mitochondria of plant leaves involves the sequencial reaction of two enzymes, the glycine decarboxylase multienzyme complex and serine hydroxymethyl transferase. The glycine decarboxylase complex from pea leaves is composed of four subunits, the 97.5 kD P-protein (binds pyridoxal 5-phosphate), the 15 kD H-protein (contains lipoamide), 45 kD T-protein (binds tetrahydrofolate), and the 59 kD L-protein (lipoamide dehydrogenase) (Walker and Oliver 1986a). The P, L, and T subunits do not appear to interact with each other, but each binds to the 15 kDa H-protein. Thus, this small protein acts as a shuttle and carries intermediates between the three larger proteins of the reaction complex. Although it is possible that the enzymes may exist as a physical complex at the very high concentrations of the component enzymes found within the mitochondrial matrix, following dissolution of the mitochondrial membrane, the component enzymes are dissociated and can be readily separated.

The four component proteins of the glycine decarboxylase complex are all nuclear encoded and translated on 80S cytosolic ribosomes. The individual proteins are present at low levels in dark-grown shoots and other nongreen tissues and increase 4 to 12 fold following exposure to light (Walker and Oliver 1986b).

During the course of our experiments on this enzyme, we became interested in determining the potential for metabolic control of this enzyme complex. Physiological studies did not suggest a control mechanism for the flux of carbon through the photosynthetic carbon oxidation (glycolate) pathway. The purpose of the pathway is to return as much carbon as possible back to the Calvin cycle following the undesirable but obligatory oxygenation of RuBP to phosphoglycolate. Even mechanisms that suggest a bifurcation of the pathway at glyoxylate (Oliver 1981, Hansen and Peterson 1986) do not involve regulatory mechanisms per se as much as changes in the concentrations of reactive intermediates, hydrogen peroxide and glyoxylate.

Following the purification of the component enzymes of the glycine decarboxylase complex, we have reconstituted the active enzyme complex and used this preparation to investigate potential metabolic regulators of this activity. Two photorespiratory intermediates, glyoxylate and serine, were shown to be potent inhibitors of glycine metabolism by the isolated enzyme complex. Of these two only serine appears to be a physiologically important regulator of the activity.

MATERIALS AND METHODS

A crude preparation of glycine decarboxylase was solublized from isolated pea leaf mitochondria following acetone-extraction of the mitochondrial lipids. The crude enzyme was separated into P, H, and T protein-enriched fractions by ammonium sulfate precipitation. The individual component enzymes were then purified from each of these enriched reactions (Walker and Oliver 1986a). Following the fractionation the active enzyme was reconstituted from its component enzymes by mixing P, H, and T proteins in a units of enzyme activity ratio of 1:5:1, respectively. Lipoamide dehydrogenase activity was provided by adding 50 ug of purified yeast enzyme (Sigma) per assay. The decarboxylation of glycine requires all four enzyme components in a reaction mixture composed of 20 mM MOPS-KOH (pH 7.2), 0.03 mM pyridoxal 5-phosphate, 0.5 mM tetrahydrofolate, 1 mM NAD, and 2 mM DTT (Sarojini and Oliver 1983).

RESULTS AND DISCUSSION

Glycine decarboxylase is a multienzyme complex that catalyzes glycine oxidation during photorespiration and glycine and serine interconversion at other times. Serine and several other amino acids were tested to determine if they had any effect on the rate of glycine decarboxylation by isolated mitochondria and by the reconstituted glycine decarboxylase complex. Alanine, aminoacetonitrile, valine, leucine, isoleucine, 2-aminobutyrate, proline, hydroxyproline, and threonine did not inhibit the rate of glycine oxidation by the isolated enzyme in the presence of equimolar glycine. Serine, however, was a potent inhibitor of glycine decarboxylation.

INHIBITION OF GLYCINE DECARBOXYLASE ENZYME BY SERINE: The inhibition of glycine decarboxylation was strongly dependent on the concentration of glycine. A kinetic analysis of the inhibition (Fig 1) shows that serine was a competitive inhibitor with respect to glycine. In the absence of serine, the Km for glycine was 6.3 mM and the Vmax was 12.8 nmol released/min. The addition of 2 mM serine increased the apparent Km to 8.7 mM and 10 mM serine increased the Km for glycine to 14.3 mM while the Vmax at both serine concentrations was virtually unchanged at 12.2 nmol/min. The Ki for serine was calculated to be about 4 mM, a value less than the apparent Km for glycine. The inhibition involved the glycine decarboxylase directly and was not mediated through common intermediates with serine hydroxymethyl transferase. This could be concluded because of the lack of detectable serine hydroxymethyl transferase in the enzyme preparation and the independence from the preincubation time with serine and the tetrahydrofolate concentration (data not presented).

INHIBITION·OF MITOCHONDRIAL GLYCINE DECARBOXYLATION BY SERINE: While serine was a potent competitive inhibitior of glycine decarboxylation by the isolated reconstituted glycine decarboxylase complex, it was a much weaker inhibitor of glycine oxidation by isolated. mitochondria. In the presence of 10 mM glycine, 10 mM serine inhibited the isolated enzyme >95%. Under the same conditions, the reaction catalyzed by intact mitochondria was inhibited <20% (Fig. 2). This shows that the inner mitochondrial membrane can provide an effective barrier to the penetration of exogenous serine. In addition, the high rates of glycine oxidation observed in leaf mitochondria suggests that a favorable glycine to serine ratio must be maintained within the matrix. As glycine is converted to serine, the matrix glycine level will drop while the serine level increases. Unless some mechanism is available for concentrating glycine within the matrix and/or removing the endogenously formed serine, the unfavorable glycine to serine ratio that must result would inhibit the enzyme.

The amount of inhibition of mitochondrial glycine oxidation caused by serine was dependent on the energy state of the mitochondrial membrane. The addition of ADP, 2,4-dinitrophenol, or oxaloacetate (all three of which partially deenergize the membrane by different mechanisms) resulted in an increased sensitivity of mitochondrial glycine decarboxylation to exogenous serine (Fig. 2). When the membrane was deenergized, either the ability to exclude added serine or the ability to accumulate glycine was diminished.

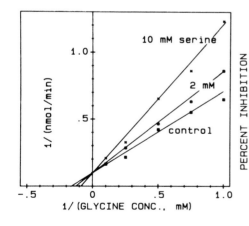

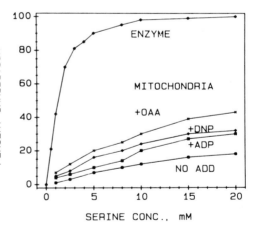

Fig.1. Kinetic Analysis of the Inhibition of Glycine Decarboxyla-tion by 2 mM and 10 mM Serine.

Fig.2. The Effect of Increasing Serine Concentrations on Glycine Decarboxyla-tion by Pea Enzyme and Mitochondria.

CONCLUSIONS

Serine was a potent inhibitor of glycine decarboxylation by isolated and reconstituted glycine decarboxylase multienzyme complex from pea leaves. Because serine is a major product of glycine oxidation within the mitochondria, the inhibition by serine is likely to be physiologically important. The inhibition was competitive with glycine. The Km for glycine was about 6 mM while the Ki for serine about 4 mM. During the course of the enzyme catalyzed reaction, glycine forms a Schiff base with the pyridoxal 5-phosphate cofactor of the P protein. There are no other known sites of interaction beteen glycine and the component enzymes of the complex. Glycine and serine are, therefore, both binding at the pyridoxal 5-phosphate site on the P protein. While the binding of serine to the catalytic site of the P protein might be strictly fortuitous, other amino acids and analogs that are structurally more similar to glycine (aminoacetonitrile and alanine) are not inhibitors of the reaction. Thus the binding of serine and the resultant inhibition of glycine metabolism may be of regulatory importance. This inhibition might prevent excess glycine metabolism under nonphotorespiratory conditions (in darkness or in storage tissues) and would decrease the consumption of the limited glycine pools which are needed for protein synthesis.

While the isolated enzyme was very sensitive to inhibition by added serine, mitochondria were not. The decreased sensitivity of mitochondria resulted from either an inability of serine to permeate the membrane rapidly or a higher than predicted internal glycine concentration. Whatever combination of these two events was responsible for the protection from exogenous glycine, it was dependent on the energization state of the membrane. Partial depolarization of the membrane caused increased sensitivity to serine. Glycine transport into leaf mitochondria has been reported (Walker et al. 1982, Yu et al. 1983) and the rate of glycine uptake was decreased by the addition of serine. This suggests that these two amino acids are using a common transporter. At low glycine concentrations, net accumulation of glycine within the mitochondria was also noted. This transporter is essential for rapid rates of glycine influx and serine efflux because these zwitterions would normally permeate the mitochondrial membrane very slowly. The inhibition of glycine decarboxylation by serine also suggest that the transporter would be needed to maintain a favorably high glycine to serine ratio within the mitochondrial matrix in order to allow rapid rates of glycine oxidation. Further, the effects of the membrane energization state on the sensitivity to serine suggest that the carrier is coupled to the mitochondrial pmf in such a way as to result in an active transport mechanism.

Hansen, K.R. and Peterson, R.B. (1986) Arch. Biochem. Biophys. 246: 332-346

Oliver, D.J. (1981) Plant Physiology 68: 1031-1034

Sarojini, G. and Oliver, D.J. (1983) Plant Physiology 72: 194-199

Walker, J.L. and Oliver, D.J. (1986a) J. Biol. Chem. 261: 2214-2221

Walker, J.L. and Oliver, D.J. (1986b) Arch. Biochem. Biophys. in press

Walker, G.H., Sarojini, G., and Oliver D.J. (1982) Biochem. Biophys. Res. Commun. 107:856-861

Yu, C., Claybrook, D.L., and Huang A.H.C. (1983) Arch. Biochem. Biophys. 227: 180-187

MODEL FOR THE RELATIONSHIP BETWEEN PHOTOSYNTHETIC CO_2 ASSIMILATION
AND GLYCOLATE SYNTHESIS IN Chlamydomonas reinhardtii[*]

Akiho YOKOTA, Kozi ASAMA, Shozaburo KITAOKA, Mitsunori TOMINAGA,
Kazuo MIURA, and Akira WADANO/Department of Agricultural Chemistry,
University of Osaka Prefecture, Sakai, Osaka 591, Japan

1. INTRODUCTION
    Ribulose 1,5-bisphosphate carboxylase/oxygenase (RuBisCO) acts at
the branching point of photosynthetic carbon reduction and oxidation cycles
(1). The properties of the enzyme and the CO_2 and O_2 concentrations around
the enzyme decide the metabolic fate of photosynthetically fixed carbon,
namely whether the carbon is directed to sugar synthesis or to the glycolate pathway to be released as CO_2 (2).
    Low-CO_2-grown microalgae concentrate extracellular dissolved inorganic
carbon (DIC) in the cells by a CO_2-concentrating mechanism (3,4). The
idea is now prevailing that a cooperation of the concentrating mechanism
with carbonic anhydrase enables the organism to concentrate low concentrations of extracellular DIC within the cells and consequently to suppress
glycolate synthesis and photorespiration (5).
    Aminooxyacetate (AOA) is a strong inhibitor of the algal glycolate
pathway, and glycolate formed during photosynthesis is excreted from algal
cells in its presence (6-8). We have found, using AOA, that low-CO_2-grown
green algae synthesize and metabolize glycolate at considerable rates even
during operation of the CO_2-concentrating mechanism (9). The rates of
glycolate synthesis by green algae are close to the rate in Euglena
gracilis in which the mechanism is absent (9). A possibility has been
proposed, based on these results, that photosynthetic CO_2 fixation enhanced
by the CO_2-concentrating mechanism causes the accumulation of O_2 within
chloroplasts, which unavoidably stimulates glycolate synthesis by RuBisCO.
However, the idea requires information on whether the rates of the glycolate synthesis by green algae is reasonable considering from the enzymic
properties of RuBisCO and the properties of the CO_2-concentrating mechanism.
    In the present study, we constructed a model for the relationship
of photosynthetic CO_2 fixation and glycolate synthesis in Chlamydomonas
photosynthesis by modifying the model of Spalding and Portis (10) for the
CO_2-concentrating mechanism and CO_2 assimilation of the same organism.
We analyzed the effect of the CO_2-concentrating mechanism on intrachloroplast O_2 concentration and glycolate synthesis in Chlamydomonas photosynthesis.

2. MATERIALS AND METHODS
    Chlamydomonas reinhardtii Dangeard was grown on air as reported pre-
viously (9). The rate of glycolate synthesis was calculated from the time-
dependent increase in the amount of glycolate excreted during photosynthesis at 310 μE s^{-1} m^{-2} in the presence of 1 mM AOA, various concentrations of DIC, and 240 μM O_2. DIC was quantified with a calibrated CO_2
analyzer.

---

[*] This is the 20th in a series on glycolate metabolism in Euglena gracilis.

Biggens, J. (ed.), Progress in Photosynthesis Research, Vol. III. ISBN 90 247 3452 5
© 1987 Martinus Nijhoff Publishers, Dordrecht. Printed in the Netherlands.

Modeling
A model for the photosynthesis and photorespiration of Chlamydomonas was constructed by modifying the model of Spalding and Portis (10). A Chlamydomonas cell was treated as a spherical, homogeneous chloroplast on the outer membrane of which a HCO_3^--transporter was present (Fig. 1).

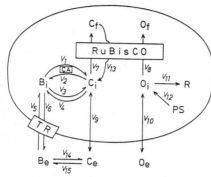

Fig. 1. Schematic diagram of Chlamydomonas photosynthesis model. TR, HCO_3^--transporter; CA, carbonic anhydrase.

Although the model of Spalding and Portis dealt only with the CO_2-concentrating mechanism and CO_2 fixation, ours included, in addition, intracchloroplast O_2 concentration (Oi) equilibrated with extracellular O_2 concentration (Oe) and the oxygenase reaction (Of) of RuBisCO. The rate of photosynthetic O_2 evolution from photochemical reactions (PS) followed the calculation of Farquhar and von Caemmerer (11);

$$v_{12} = (9 \times v_7 + 10.5 \times v_8)/8.$$

The equations for the RuBisCO carboxylase (Cf) and oxygenase reactions included a term for activation of the enzyme by CO_2 (1);

$$Cf = (Ci/(Ka + Ci)) \; Vc \; Ci/(Kc \; (1 + Oi/Ko) + Ci)$$

$$Of = (Ci/(Ka + Ci)) \; Vo \; Oi/(Ko \; (1 + Ci/Kc) + Oi),$$

where Ka, Kc, Ko, Vc, and Vo are Km for the activation for CO_2 (3.5 μM) (12,13), Km in the carboxylation reaction for CO_2, Km in the oxygenation reaction for O_2, Vmax of the carboxylation, and Vmax of the oxygenation, respectively. These equations are based on the assumption that the RuBisCO reactions are saturated with ribulose bisphosphate (RuBP) at any extracellular DIC concentration (14). We defined the rate of O_2 consumption by day respiration (R) as $0.5 \times v_8$ assuming that during photosynthesis O_2 is consumed during glycolate metabolism in mitochondria. v_{13} is the rate of CO_2 evolution during glycolate metabolism ($0.5 \times v_8$), and it lowers the CO_2 fixation rate in photosynthesis. Other parts of the model are the same as in the model of Spalding and Portis (10).

3. RESULTS AND DISCUSSION
Figure 2 shows calculated rates of photosynthetic CO_2 fixation of low-CO_2-grown Chlamydomonas and the actually measured rates of the organism from several different cultures in the presence of various concentrations of exogenous DIC. In the calculation, the initial exogenous O_2 concentration was set at 0 μM as in actual measurements of algal photosynthesis.

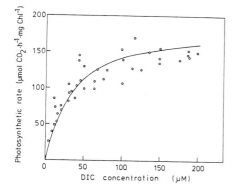

Fig. 2. Comparison of actual photosynthetic CO_2 fixation rates of
Chlamydomonas cells from different
cultures (open circles) with those
predicted by the model (solid line).

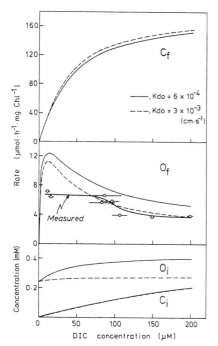

Fig. 3. Calculated rates of CO_2
fixation and glycolate synthesis
and calculated concentrations
of O_2 and CO_2 at two Kdo's and
actually measured rates of glycolate synthesis (open circles
with bold solid line) in Chlamydomonas photosynthesis. Horizontal bars represent fluctuations
of DIC in the incubation mixture
determined during glycolate
excretion experiments.

With Km for the HCO_3^--transporter
for $\overline{HCO}_3^-$ as determined by Spalding
and Ogren (15) and the intrachloro-
plast pH of 8.0, the calculated
rates of CO_2 fixation were in close
agreement with those actually measured at given DIC concentrations.
This indicates that the model and
the parameters used in our calculations were suitable for analysis
of Chlamydomonas photosynthesis.
   In Fig. 3 is shown the calculation of the changes in the rates
of photosynthetic CO_2 fixation and
glycolate synthesis, and the changes
of intrachloroplast CO_2 and O_2 concentrations at increasing concentrations of DIC in low-CO_2 -grown
Chlamydomonas. In this case, the
exogenous O_2 concentration was fixed
at 240 μM, since the actual measurements of glycolate synthesis were
made in the presence of exogenous
O_2 at this concentration. The rates
for glycolate synthesis are drawn
together with those actually measured in the presence of 1 mM AOA
(middle panel, bold line with open
circles). The CO_2 fixation rates
calculated with the reported diffusion constants of CO_2 (Kd = 5
x 10^{-4} cm s^{-1}) and O_2 (Kdo = 6 x
10^{-4} cm s^{-1}) in the cell membrane
were consistent with the actually
measured rates, but the calculated
rates of glycolate synthesis were
considerably higher than the measured ones. An increase of Kdo to
3 x 10^{-3} cm s^{-1} caused only a slight
stimulation of $CO2$ fixation and

decreased the rates of glycolate synthesis to the actually measured level. The intrachloroplast concentration of O_2 was calculated as 0.4 mM with Kdo of 6×10^{-4} cm s^{-1} and 0.27 mM with 3×10^{-3} cm s^{-1} in the presence of 0.2 mM exogenous DIC. In other words, the actually measured rates of glycolate synthesis ($3-4$ μmol h^{-1} mg chlorophyll^{-1}) require 0.27 mM intra-chloroplast O_2 in the presence of 0.2 mM exogenous DIC during the operation of the CO_2-concentrating mechanism.

A plateau was seen in the glycolate synthesis measured at low DIC concentrations, unlike the simulated results based on the assumption that the RuBisCO reactions were saturated with RuBP. This discrepancy may be partly because the RuBP level in chloroplasts is rate-limiting under these conditions, since the presence of AOA in the incubation medium stops salvaging glycolate carbon by the Chlamydomonas cells by glycolate pathway.

We are now reaching the following idea: The CO_2-concentrating mechanism facilitates the incorporation of extracellular DIC and enhances photosynthetic CO_2 fixation, which is accompanied with O_2 evolution. The O_2 produced photosynthetically stimulates the synthesis of glycolate by RuBisCO. Accordingly, it comes as no surprise that much glycolate is synthesized during the operation of the CO_2-concentrating mechanism in the presence of $100-200$ μM DIC. The fact that Euglena which does not have the concentrating mechanism (9,16) has similar rates of glycolate synthesis also indicates that the concentrating mechanism of green algae including Chlamydomonas stimulates photosynthetic CO_2 fixation but does not suppress the glycolate formation by RuBisCO. This is mainly due to elevated O_2 concentrations in chloroplasts during the operation of the concentrating mechanism.

4. REFERENCES
1 Lorimer, G.H. and Andrews, T.J. (1981) in The Biochemistry of Plants (Stumpf, P.K. and Conn, E.E., eds.), Vol. 8, pp. 329-374, Academic Press, New York
2 Tolbert, N.E. (1980) in The Biochemistry of Plants (Stumpf, P.K. and Conn, E.E., eds.), Vol. 2, pp. 488-525, Academic Press, New York
3 Badger, M.R., Kaplan, A. and Berry, J.A. (1980) Plant Physiol. **66**, 407-413
4 Kaplan, A., Badger, M.R. and Berry, J.A. (1980) Planta **149**, 219-226
5 Ogren, W.L. (1984) Ann. Rev. Plant Physiol. **35**, 412-442
6 Tolbert, N.E., Harrison, N.D. and Selph, N. (1983) Plant Physiol. **72**, 1075-1083
7 Krampitz, L.O. and Yarris, C.E. (1983) Plant Physiol. **72**, 1084-1087
8 Yokota, A. and Kitaoka, S. (1986) Submitted
9 Yokota, A. and Kitaoka, S. (1986) Submitted
10 Spalding, M.H. and Portis, A.R., Jr. (1985) Planta **164**, 308-320
11 Farquhar, G.D. and von Caemmerer, S. (1982) in Encyclopedia of Plant Physiology (Pirson, A. and Zimmerman, M.H., eds.), Vol. 12B, pp. 549-587, Springer-Verlag, Berlin
12 Lorimer, G.H., Badger, M.R. and Andrews, T.J. (1976) Biochemistry **15**, 529-536
13 Lorimer, G.H., Badger, M.R. and Andrews, T.J. (1977) Anal. Biochem. **78**, 66-75
14 Yokota, A. and Canvin, D.T. (1986) Plant Physiol. **80**, 341-345
15 Spalding, M.H. and Ogren, W.L. (1983) FEBS Lett. **154**, 335-338
16 Merrett, M.J. and Armitage, T.L. (1982) Planta **155**, 95-99

SEPARATION AND CHARACTERIZATION OF SEVERAL FORMS OF CATALASE IN LEAF
EXTRACTS OF NICOTIANA SYLVESTRIS.

EVELYN A. HAVIR AND NEIL A. McHALE, DEPT. OF BIOCHEMISTRY AND GENETICS,
THE CONNECTICUT AGRICULTURAL EXPERIMENT STATION, P.O. BOX 1106, NEW
HAVEN, CT 06504, U.S.A.

INTRODUCTION

Catalase is an abundant enzyme in the peroxisomes of leaves of C_3
plants (1), where it plays a vital role in decomposition of H_2O_2 arising
from the oxidation of photorespiratory glycolate. Although the leaves of
some C_3 plants appear to possess only one form of catalase (2, 3), multiple
forms of leaf catalase have been reported in various C_3 (4, 5, 6) and C_4
(7) plants. Two of the catalase forms in maize leaves exhibit cell-type
specificity, one residing in the cells of the bundle sheath, and the other
in the mesophyll (8). Since photorespiration in maize would be confined
to the bundle sheath, it appears that the mesophyll-specific catalase iso-
zyme must perform some other function. The physiological significance of
multiple forms of catalase in C_3 leaves remains unclear. We have undertaken
a characterization of leaf catalase in several C_3 plants, with particular
emphasis on Nicotiana, where multiple forms of the enzyme undergo marked
changes during seedling development and major shifts in relative distribution in response to manipulation of photorespiratory activity.

MATERIALS AND METHODS

Seedlings were grown under continuous illumination (80 μE m^{-2}s^{-1}) at
27°C in open plastic boxes on vermiculite saturated with culture medium.
Between 3 and 21 days post-germination, seedlings were extracted by grinding 2.0 g fwt in 10 ml K-phosphate buffer (0.05M, pH 7.0) containing 7.5 mg
DTT. Catalase forms were separated by chromatofocusing on columns prepared
from Polybuffer Exchanger 94 (Pharmacia) after equilibration to the starting pH. Fractions were assayed for catalase activity by measuring the
change in absorbancy at 240 nm of 0.0125M H_2O_2 in 0.05 K-phosphate buffer
(pH 7.0) at 30°C (one unit = one μmole H_2O_2 decomposed/min). Effects of
CO_2-enrichment on levels and distribution of catalase forms were examined
in seedlings grown in illuminated plexiglas chambers flushed with normal
air vs. high CO_2 (1% CO_2/21% O_2).

RESULTS AND DISCUSSION

The time courses for appearance of catalase activity in seedlings of
N. tabacum and N. sylvestris were similar, showing a peak about five days,
followed by a decline to a level that remained constant through 21 days and
which was similar to levels observed in mature leaves (Fig. 1). The elution
profiles of catalase activities separated by chromatofocusing in seedlings
of N. tabacum and N. sylvestris (Fig. 2) indicate that the enzyme exists in
several forms that undergo marked changes in distribution in the period
following germination. A form of catalase eluting at 10-20 ml (peak 1) was

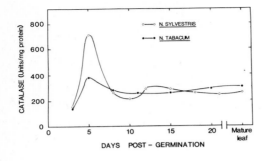

FIGURE 1

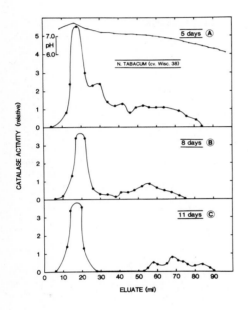

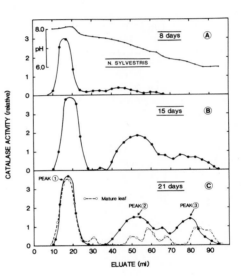

FIGURE 2

present at all stages of development in both species. Although this form was predominant in the early stages of seedling growth, other more acidic forms appeared at later stages. Losses of activity (50% or more) precluded accurate determination of the catalase profile in mature leaves of N. sylvestris, but there was clear evidence for peak 1 as well as the more acidic forms (Fig. 2, o-o-o).

The elution profile of catalase activity in mature leaves of N. tabacum is shown in Fig. 3. Losses of activity were negligible here permitting clear observation of peak 1, the predominant seedling form, and a substantial increase in the acidic forms eluting between 50 and 90 ml (peaks 2 and 3). These acidic forms represented 62% of the total activity in the mature leaf compared to 29% in 11 day seedlings (Fig. 2). Fractions representing each of the three peaks of activity were collected separately

and reapplied to individual columns to determine if they were in equilibrium (Fig. 3B, 3C). Peaks 1, 2 and 3 showed no evidence of interconversion, and the positions of peaks 2 and 3 were maintained in relation to
the marker proteins serine:glyoxylate aminotransferase and glutamate:
glyoxylate aminotransferase, respectively. These forms of catalase were
also shown to differ in regard to thermal stability at 55°C, with peaks 1,
2 and 3 exhibiting $t_{1/2}$ inactivation times of 41.5, 29.5 and 3.2 min.,
respectively. Values of $t_{1/2}$ inactivation for peaks 1, 2 and 3 from
N. sylvestris (Fig. 2) were 31.5, 11.5, and 3.0 min., respectively. Thus
each peak of catalase from N. tabacum is more thermostable than the corresponding peak from N. sylvestris.

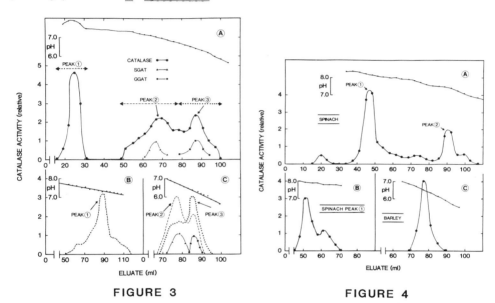

FIGURE 3   FIGURE 4

Catalase activities from leaves of two other C_3 plants (spinach,
barley) were separated by chromatofocusing, and the profiles are presented
in Fig. 4. The elution profile for spinach extract (Fig. 4A) and rechro-
matofocusing of peak 1 (Fig. 4B) gave evidence of at least four forms, but
the barley catalase (Fig. 4C) eluted as a single peak. The multiple forms
of catalase in tobacco (Fig. 3A; peaks 1, 2 and 3) and spinach (Fig. 4A;
peaks 1 and 2) also displayed differential mobility after electrophoresis,
while barley catalase migrated as a single band. Thus, results from two
independent methods of protein separation support the view that some C_3
plants contain multiple forms of catalase to decompose H_2O_2 in the leaves.

We had observed in N. sylvestris seedlings that atmospheric conditions
suppressing photorespiration (1% CO_2/21% O_2) produce a large reduction in
total catalase activity (96 units mg^{-1} protein) relative to seedlings grown
in normal air (159 units mg^{-1} protein). In order to determine whether
specific forms of catalase were affected, extracts were prepared from
seedlings grown in 1% CO_2 vs. normal air and applied to chromatofocusing
columns (Fig. 5). Catalase activity in air-grown seedlings (10 days
old) eluted primarily as peak 1. In contrast, the elution profile for
seedlings grown in 1% CO_2 revealed a sharp decline in peak 1 catalase,

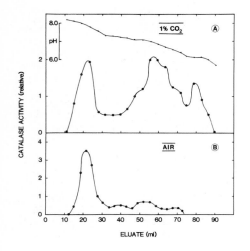

FIGURE 5

accompanied by the emergence of substantial activity of the more acidic forms (peaks 2 and 3). The decline in peak 1 activity in 1% CO_2 suggests a photorespiratory function for this form, in accordance with the observation that this was the only form present at all developmental stages examined. We are conducting organelle isolations to obtain direct identification of the peroxisomal form(s).

REFERENCES

1 Tolbert, N.E. (1971) Ann. Rev. Plant Physiol. 22, 45-74
2 Braber, J.M. (1980) Z. Pflanzenphysiol. 97, 135-144
3 Schiefer, S., Teifel, W., and Kindl, H. (1976) Z. Physiol. Chem. 357, 163-175
4 Eising, R. and Gerhardt, B. (1986) Phytochem. 25, 27-31
5 Galston, A.W., Bonnichsen, R.K. and Arnon, D.I. (1951) Acta Chem. Scand. 5, 781-790
6 Drumm, H. and Schopfer, P. (1974) Planta 120, 13-30
7 Scandalios, J.G. (1984) Develop. Genet. 4, 281-293
8 Tsaftaris, A.S., Bosabalidis, A.M. and Scandalios, J.G. (1983) Proc. Natl. Acad. Sci., USA 80, 4455-4459

PHOTORESPIRATORY DEVELOPMENT IN WHEAT. I AMMONIA CYCLING.

ALYSON K. TOBIN, [1]NAZIRA SUMAR, [1]MINA PATEL, MICHAEL O. PROUDLOVE,
[1]GEORGE R. STEWART AND ANTHONY L. MOORE, DEPARTMENT OF BIOCHEMISTRY,
UNIVERSITY OF SUSSEX, FALMER, BRIGHTON BN1 9QG, U.K. AND [1]DEPARTMENT OF
BOTANY AND MICROBIOLOGY, UNIVERSITY COLLEGE, LONDON WC1E 6BT, U.K..

1. INTRODUCTION

All the cells of a wheat leaf are derived from meristematic cell
division at the leaf base. Consequently, cell maturity increases
towards the leaf tip as does the size, number and photosynthetic
activity of the chloroplasts (1-3). Rapid changes in plastid morphology
and biochemistry within light-grown wheat leaves have received much
attention whereas the extent to which other leaf cell organelles
collaborate, during maturation of the mesophyll cell, has been largely
overlooked.

The photorespiratory ammonia cycle (4) requires interaction between
chloroplastic, cytosolic, peroxisomal and mitochondrial enzymes so
development of this pathway may involve co-ordinated organelle
biogenesis. We report here a study of the changes in photorespiratory
ammonia production and associated enzyme activities within leaf sections
at different stages of maturity.

2. PROCEDURE

2.1. Materials and methods

2.1.1. Measurement of photorespiratory NH_3 production: Wheat seeds
(Triticum aestivum cv Maris Huntsman) were imbibed for 18h at
$20^{\circ}C$, planted in John Innes No.3 compost and grown in a 16h
light ($300\mu mol.m^{-2}.s^{-1}$ PAR; $20^{\circ}C$)/ 8h dark ($10^{\circ}C$) cycle for
8 days. The primary leaf was cut into 0.5cm transverse
sections at 1cm intervals from the meristem (mean leaf length
= 10cm). For each treatment fifty, 0.5cm leaf sections were
floated on 2mM L-methionine sulphoximine (MSO; glutamine
synthetase (GS) inhibitor), water, 10mM pyrid-2-yl hydroxy-
methane sulphonate (HPMS; glycollate oxidase inhibitor) or
10mM $KHCO_3$ (all solutions adjusted to pH 7.0). After 30min
dark preincubation at $20^{\circ}C$, all the sections were illuminated
($200\mu mol.m^{-2}.s^{-1}$ PAR) for 30min then frozen in liquid N_2.
Leaf sections were extracted in chloroform: methanol: water
(5:12:13 (v/v)), fractionated through Dowex 50W, Na^+-form (5)
and assayed for ammonia according to McCullough (6).

2.1.2. Determination of enzyme activities: Freshly harvested transverse sections of wheat primary leaves were ground in liquid
N_2 with a pestle and mortar, extracted in appropriate buffers
(see references for details) and centrifuged for 20min at
10000 x g. The supernatants were assayed for glutamine
synthetase (3), nitrate reductase (7), glycollate oxidase
(8), NADH-glutamate dehydrogenase (9) and cytochrome oxidase
(10). GS isoenzymes were fractionated by FPLC (3). In vivo
nitrate reductase (5) was determined using fresh, unfrozen
leaf sections.

2.1.3. Determination of mesophyll cell number: Leaf sections (0.5 cm transverse) were incubated in 5% chromium trioxide for 7 days at 4°C (3) and mesophyll cells counted in a 0.2mm deep haemocytometer.

3. RESULTS AND DISCUSSION

3.1. Figure 1a shows that ammonia accumulates when wheat leaf sections are treated with the GS inhibitor, MSO. Extraction and assay of treated leaf sections showed that the 30min dark preincubation with MSO resulted in complete inhibition of GS in all sections (data not shown). The rate of ammonia accumulation increases ca. 15-fold from the base to the tip of 8 day old wheat primary leaves. This accumulation is light-dependent (data not shown) and is inhibited by elevated CO_2 concentrations (Figure 1b) and by treatment with 10mM HPMS[2] (glycollate oxidase inhibitor) and therefore appears to be of photorespiratory origin. Addition of 10mM glycine only stimulated NH_3 production when HPMS was also present, suggesting that glycine supply was not limiting its deamination under these conditions. This method of measuring photorespiration is known to underestimate rates for various reasons (see, for example, 11). Indeed, our maximum rate of ammonia accumulation (23mole $x 10^{-15}$.min^{-1}.$cell^{-1}$ in mature sections; equivalent to 12μmol.h^{-1}.g fwt^{-1}) is rather lower than other estimates in wheat leaves (80μmol.h^{-1}.g fwt^{-1}; 12), using gas exchange methods. It is, however, comparable to other MSO-dependent rates (13). Nevertheless, the data show that there is an increase in photorespiratory ammonia production during chloroplast and cell development and that maximum rates of photorespiration are attained when chloroplast division is virtually complete.

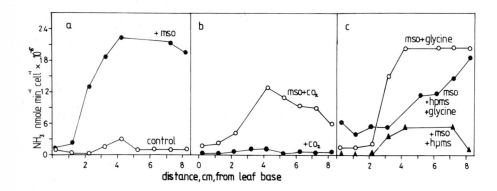

Figure.1. Ammonia accumulation in serial transverse sections of 8 day old wheat primary leaves. Note, the x-axis refers to the distance from the leaf base, i.e. sections increase in age towards the leaf tip.

3.2. There was no significant change in cellular nitrate concentration during incubation of the leaf sections (data not shown), indicating that nitrate reduction did not contribute significantly to ammonia accumulation under these conditions. This is supported by the relatively low activity of nitrate reductase (NR) in these sections (Figure 2). It is interesting to note that, whereas NR activity in extracts (in vitro) increases with leaf cell maturity, the in vivo NR activity remains constant beyond 3cm from the meristem, suggesting a physiological limitation on nitrate reduction in mature leaf cells. Furthermore, the pattern of development of NR activity (in vivo) closely follows that of cytosolic GS (GS1; Figure 2) and may indicate a role for the latter enzyme in primary ammonia assimilation.

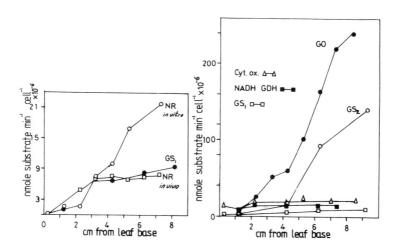

Figure.2. Changes in activities of NR (in vivo and in vitro) and cytosolic GS (GS1) during leaf development.

Figure.3. Changes in activities of chloroplast GS (GS2), glycollate oxidase, NADH glutamate dehydrogenase and cytochrome oxidase during leaf development.

3.3. Chloroplast GS activity (GS2; Figure 3) develops in parallel with peroxisomal glycollate oxidase, indicating a close control of synthesis of these photorespiratory enzymes at all stages of cell maturity. The pattern of development of these photorespiratory enzymes, similar to that of RubisCO (2), differs markedly from that of non-photorespiratory enzymes. For example, the activities of mitochondrial glutamate dehydrogenase and cytochrome oxidase remain relatively constant during leaf development (Figure 3).

This suggests that there is a controlled, co-ordinated synthesis of specific enzymes involved in photosynthesis and photo-respiration, rather than a general increase in all metabolic pathways, during wheat leaf and chloroplast development. The parallel development of GS2 and glycollate oxidase and the low activity of GS1, relative to photorespiratory activity, lend further support to the chloroplast being the principal site for the reassimilation of photorespiratory ammonia.

ACKNOWLEDGEMENTS
Financial support was provided by the Royal Society (I.C.I. Alfred Spinks Fellowship to AKT), AFRC (GRS) and SERC (ALM).

REFERENCES
1. Boffey, S.A., Ellis, R.J., Sellden, G. and Leech, R.M. (1979) Plant Physiol. 64, 502-505
2. Dean, C. and Leech, R.M. (1982) Plant Physiol. 69, 904-910
3. Tobin, A.K., Ridley, S.M. and Stewart, G.R. (1985) Planta 163, 544-548
4. Keys, A.J., Bird, I.F.,Cornelius, M.F., Lea, P.J., Wallsgrove, R.M. and Miflin, B.J. (1978) Nature 275, 741-742
5. Stewart, G.R., Lea, J.A. and Orebamjo, T.O. (1973) New Phytol. 72, 539-541
6. McCullough, H. (1967) Clinica Chimica Acta 17, 297-304
7. Emes, M.J. and Fowler, M.W. (1979) Planta 194, 249-253
8. Feierabend, J. and Beevers, H. (1972) Plant Physiol. 49, 28-32
9. Lees, E.M. and Dennis, D.T. (1981) Plant Physiol. 68,827-830
10. Tolbert, N.E., Oeser, A., Kisaki, T., Hageman, R.H. and Yamazaki, R.K. (1968) J. Biol. Chem. 243, 5179-5184
11. Walker, K.A. and Givan, C.V. (1984) J. Exp. Bot. 28, 1247-1257
12. Kumarasinghe, K.S., Keys, A.J., and Whittingham, C.P. (1977) J. Exp. Bot. 28, 1247-1257
13. Martin, F., Winspear, M., Macfarlane, J.D. and Oaks, A. (1983) Plant Physiol. 71, 177-181

PHOTORESPIRATORY DEVELOPMENT IN WHEAT.   II CHLOROPLAST DICARBOXYLATE TRANSPORT.

MICHAEL O. PROUDLOVE, ANTHONY L. MOORE AND ALYSON K. TOBIN, DEPARTMENT OF BIOCHEMISTRY, UNIVERSITY OF SUSSEX, FALMER, BRIGHTON BN1 9QG, U.K..

1. INTRODUCTION
    The increasing maturity of cells, from base to tip in the primary leaf of wheat, has been used as a model system to study chloroplast biogenesis. Chloroplast area, chlorophyll concentration and protein content and activity, notably RubisCO, all increase dramatically during the early stages of plastid development (1-3). Recent results also show that the activity of glutamine synthetase (GS), a central enzyme in amino acid biosynthesis and photorespiratory nitrogen cycling (4), increases acropetally and that this is due to an increase in the chloroplast isoenzyme (3).
    Such data suggest that the immature plastid is an heterotrophic organelle, requiring an import of numerous precursors from the rest of the cell. As it matures these requirements will change, reflecting the changing role of the chloroplast in cell metabolism.
    In particular, the movement of the amino acids, glutamate and glutamine, across the inner membrane of the chloroplast envelope, may vary according to their involvement in stromal metabolism during the development of photorespiratory activity. The kinetics of transport of these compounds have therefore been investigated in plastids isolated from serial, transverse sections of wheat primary leaves and compared to the activities of other enzymes involved in their metabolism within these cells.

2. PROCEDURE
2.1. Materials and methods
    2.1.1. Preparation of chloroplasts: Wheat seeds (Triticum aestivum cv Maris Huntsman) were imbibed for 18h at $20^{o}C$, planted in John Innes No. 3 and grown under a 16h light $(20^{o}C)$/ 8h dark $(10^{o}C)$ cycle for 8 days. Regions of primary leaves, at the specified distance from the basal meristem, were razor cut into 1-2mm sections and incubated for 3h in digestion medium (0.5M sorbitol; 1mM $MgCl_2$; 5mM MES-KOH, pH 6.0; 2% cellulase; 0.2% pectolyase) to prepare protoplasts. Following their purification on a sucrose/ sorbitol gradient (5) chloroplasts were isolated by passing cells once through 20 μm mesh, centrifuging at 500 x g for 2min and resuspending the pellet in 0.4M sorbitol; 10mM EDTA; 25mM TRICINE-KOH, pH 8.4. Chloroplasts were routinely prepared which were 90-95% intact and capable of CO_2-dependent oxygen evolution at rates of 50-60μmoles/mg chlorophyll/h (the same as seen for protoplasts from which they were prepared).
    2.1.2. Dicarboxylate uptake: Movement of [^{14}C]-labelled dicarboxylates into the stromal volume was followed using silicone oil centrifugal filtration, essentially as described by Proudlove and Thurman (6). Reactions, at $20^{o}C$ in the dark, were

Biggens, J. (ed.), Progress in Photosynthesis Research, Vol. III. ISBN 90 247 3452 5
© 1987 Martinus Nijhoff Publishers, Dordrecht. Printed in the Netherlands.

initiated by the addition of chloroplasts and terminated by centrifuging, for 6s in a Beckman microfuge B, through 70 μl of silicone oil (AR200:AR20; 5:1 (v/v)) into 20 μl of 10% (v/v) $HClO_4$. Radioactivity in supernatant and pellet was assessed by liquid scintillation spectrometry and protein (7) and chlorophyll (8) recoveries and measurement of stromal volume were performed for each experiment.

2.1.3. Determination of plastid number per cell and the chlorophyll content of plastids: The number of plastids per mesophyll cell was determined by counting the plastids in separated cells using a Zeiss photomicroscope fitted with Nomarski differential interference optics, as described in (1). Chlorophyll content of mesophyll cells was calculated by counting the number of protoplasts in an aliquot of each preparation, using a 0.2mm deep haemocytometer cell and light microscope, and estimating the chlorophyll concentration in each aliquot (8). The chlorophyll concentration per plastid was then calculated from the values determined for the plastids per cell and chlorophyll per protoplast.

3. RESULTS AND DISCUSSION

3.1. Wheat leaf chloroplasts are clearly able to transport the amino acids, glutamate and glutamine, through the inner membrane of the envelope and into the stroma (Table 1). The rates of uptake of both compounds, when expressed on the basis of chlorophyll content of each section, decrease with leaf development, that is with distance from the meristem. Also shown in Table 1, however, is

Table 1. Kinetics of glutamate and glutamine transport by wheat leaf chloroplasts. Values for Km and Vmax were estimated from Eadie-Hoftsee plots (better weighting to low concentration points), all other numbers were obtained as described in Procedure. Abbreviations used are: GS2- activity of chloroplast GS isoenzyme; chl- chlorophyll; pl- plastid; *- calculated from graph in (3).

Section cm from base	amino acid	Km (mM)	Vmax μmol/mg chl/h	Vmax pmol/cell/h	Vmax fmol/pl/h	GS2 fmol/pl/h	chl pg/pl	pl/cell
2	glu	0.44	39.89	1.89	24.3	4.6*	0.61	78
	gln	0.11	8.76	0.41	5.3			
4	glu	1.76	20.32	4.14	36.7	9.6	1.81	113
	gln	-	-	-	-			
6	glu	2.96	26.92	8.31	62.1	41.6	2.31	134
	gln	0.32	6.87	2.12	15.8			
8	glu	2.47	25.82	12.44	81.3	51.0*	3.15	153
	gln	0.41	4.02	1.94	12.7			

the finding that there is an increase in the chlorophyll content of plastids with maturity. Such an increase would therefore tend to overestimate transport rates in the younger sections and underestimate them in more mature sections. This is supported by expressing the data as the rate of uptake per plastid or per mesophyll cell. In both cases there is an increase in the rate of transport of glutamate and glutamine with plastid and cell maturity. Clearly, the parameters used when calculating the results can lead to differences in data interpretation, particularly when comparing different developmental stages. We also found that the Km values for both substrates increased significantly with cell age and further work is in progress to determine whether this is a reflection of a change in nature of the transporters.

3.2. Photorespiratory ammonia, produced during glycine deamination in the mitochondrion, must be reassimilated if the loss of cellular nitrogen is to be prevented (4). In the leaves of higher plants this process is catalysed primarily by the glutamine synthetase/ glutamate synthase (GS/GOGAT) enzyme pair (9), either in the cytosol, via the isoenzyme GS1, or in the chloroplast, via the isoenzyme GS2. If GS1 operates as the main means of reassimilating photorespiratory ammonia then glutamine must be transported into the chloroplast, to support glutamate production by GOGAT, and glutamate must efflux from the chloroplast, to maintain GS1 activity and to facilitate transamination in the peroxisome. If, however, GS2 acts as the major isoenzyme then the chloroplast no longer requires the import of cytosolic glutamine but glutamate must still be exported, to support peroxisomal reactions. Our preliminary data presented here indicate that glutamine transport is too slow to allow the measured rates of photorespiration to occur and, with other results (Tobin et al., this meeting), support the hypothesis that GS2 is the main isoenzyme involved in the reassimilation of ammonia, released during photorespiratory glycine deamination.

ACKNOWLEDGEMENTS
Financial support was provided by the Royal Society (I.C.I. Alfred Spinks Fellowship and EPA Cephalosporin fund to AKT) and by SERC (ALM).

REFERENCES
1. Boffey, S.A., Ellis, R.J., Sellden, G. and Leech, R.M. (1979) Plant Physiol. 64, 502-505
2. Dean, C. and Leech, R.M. (1982) Plant Physiol. 69, 904-910
3. Tobin, A.K., Ridley, S.M. and Stewart, G.R. (1985) Planta 163, 544-548
4. Keys, A.J., Bird, I.F.,Cornelius, M.F., Lea, P.J., Wallsgrove, R.M. and Miflin, B.J. (1978) Nature 275, 741-742
5. Edwards, G.E., Robinson, S.P., Tyler, N.J.C. and Walker, D.A. (1978) Plant Physiol. 62, 313-319
6. Proudlove, M.O. & Thurman, D.A. (1981) New Phytol. 88, 255-264
7. Lowry, O.H., Rosebrough, N.J., Farr, A.L. & Randall, R.J. (1951) J. Biol. Chem. 193, 265-275.
8. Arnon, D.I. (1949) Plant Physiol. 24, 1-15
9. Miflin, B.J. and Lea, P.J. (1977) Ann. Rev. Plant Physiol. 28, 299-329

TENTATIVE IDENTIFICATION OF THE PEA CHLOROPLAST ENVELOPE
GLYCOLATE TRANSPORTER

KONRAD T. HOWITZ & RICHARD E. McCARTY, SECTION OF BIOCHEMISTRY,
MOLECULAR AND CELL BIOLOGY, WING HALL, CORNELL UNIVERSITY,
ITHACA, NEW YORK 14853 USA

1. INTRODUCTION

The photorespiratory metabolites, glycolate and D-glycerate, cross the chloroplast envelope on the same carrier (1,2,3,4). The overall stoichiometry of this transport, during photorespiration, is proposed to be two glycolates plus one proton leaving the chloroplast and one D-glycerate entering it (3).

The sulfhydryl alkylating reagent, N-ethylmaleimide (NEM) inhibits the transporter, an inhibition prevented by the presence of glycolate or D-glycerate during the treatment (1,5). Employing this substrate protection, we have identified, by differential labeling with ^{14}C-NEM and ^{3}H-NEM, a probable glycolate carrier polypeptide of about 35 kD in inner envelope preparations from pea chloroplasts.

The substrates of the glycolate transporter appear to be limited to two or three carbon 2-hydroxymonocarboxylates. The D isomers of the chiral three carbon compounds, lactate and glycerate, are preferred (2). However, the L isomer of the substrate analogue mandelate (phenylglycolate) is a strong inhibitor of glycolate transport when present on the opposite side of the membrane. L-Mandelate can also serve as the protective agent for NEM differential labeling of the 35 kD band.

Both NEM and L-mandelate partially inhibit glycolate efflux from proteoliposomes prepared with cholate solubilized inner envelope proteins.

2. PROCEDURES

2.1 <u>General.</u> Intact pea chloroplasts were isolated as described (6). Inner envelopes were prepared by the method of Cline <u>et al</u> (7).

2.2 <u>NEM Labeling.</u> Envelopes were suspended in 20 mM HEPES/NaOH, pH 8.0, 0.1 M NaCl, 5 mM $MgCl_2$ at 1 mg protein /mL. The protective agents (glycolate, L-mandelate) were added to experimental samples and equal concentrations of non-protective salts were added to controls. The first labeling (^{14}C-NEM) was terminated after 15 min. at 4°C by addition of a 50% molar excess of N-acetylcysteine. Membranes were pelleted in a Beckman Airfuge (7 min., 100,000xg) and resuspended with a syringe. This was repeated until less than 0.01% of the initial ^{14}C remained. The second labeling (^{3}H-NEM) and subsequent washes were identical to the first round but without the protective compounds. Membranes were dissolved in a 5% SDS sample buffer and electrophoresed on 10, 11, or 12% polyacrylamide gels. Gels were stained with Coomassie blue, sliced and the slices

Biggens, J. (ed.), Progress in Photosynthesis Research, Vol. III. ISBN 90 247 3452 5
© *1987 Martinus Nijhoff Publishers, Dordrecht. Printed in the Netherlands.*

incubated with 90% NCS Tissue Solubilizer (Amersham), 50° C, 5 hours, before scintillation counting. Results are presented as the $^3H:^{14}C$ ratio in an experimental sample divided by the ratio for the comparable slice from a control ("LABEL RATIO"). Labeled NEM's were from New England Nuclear.

2.3 <u>Chloroplast Transport.</u> Intact chloroplasts (0.15 mg Chl/mL) were preilluminated at 4° C, as described (1), with 1 mM of the compounds listed in Table I. After 3 min., glycolate uptake was initiated by centrifugation of the chloroplasts through a 10% Percoll layer containing 1 mM ^{14}C-glycolate and ^3H-sorbitol in silicone oil microfuge tubes (2s incubation),(1).

2.4 <u>Reconstitution.</u> Proteoliposomes were prepared by mixing inner envelope protein (1 mg/mL) solubilized by either 0.75% cholate (NEM experiment) or 1.0% cholate (mandelate experiment) with soy azolectin (10 mg/mL), followed by removal of the cholate with Sephadex G-50 centrifuge columns (8). After freezing and thawing the liposomes with ^{14}C-glycolate and ^3H-sorbitol, they were assayed for retention of internal glycolate after passage through a second centrifuge column. ^3H-sorbitol was used as a marker for internal space. All buffers contained 10 mM Tricine/NaOH pH 8.0, 2 mM EDTA, 2 mM dithiothreitol.

3. RESULTS AND DISCUSSION

3.1 <u>Differential Labeling with NEM.</u> The presence of glycolate during a first round of NEM labeling (^{14}C) leads to an enhanced 2nd label:1st label ($^3H:^{14}C$) ratio in a 35 kD band on gels of inner envelope proteins (Fig. 1). The "LABEL RATIO" peak corresponds to the second major band above the phosphate translocator (Fig. 2). The peak is found in the same position when D-glycerate is used for protection (not shown). Absence of differential labeling effects other than in the 35 kD band has led us to adopt an abbreviated protocol consisting of slicing and counting the phosphate translocator band (P), the five bands immediately above it (1-5) and the Rubisco large subunit (R). The use of the substrate analogue/inhibitor L-mandelate as protective agent, and the D isomer as the control addition, leads to enhanced differential labeling in the same band as

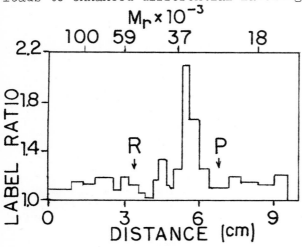

Figure 1. Label Ratio Histogram of Gel of Inner Envelope Protein 15 mM glycolate was used for protection. Both ^{14}C and ^3H NEM were 0.6 mM. See section 2.2 for labeling protocol and definition of "LABEL RATIO".

does glycolate protection (Fig. 2b). All results obtained to date on the differential labeling of the 35 kD band by NEM run parallel to and are predictable from the behavior of the transporter with respect to NEM inhibition of its function (1,2,5). It is therefore highly likely that the 35 kD polypeptide is, at least, a subunit of the glycolate transporter.

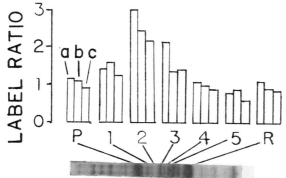

Figure 2. Comparison of Glycolate and L-Mandelate Protection in NEM Labeling: a) Protection by 100 mM glycolate; 5 mM NEM b) Protection by 25 mM L-mandelate, control addition 25 mM D-mandelate; 5 mM NEM c) Protection by 100 mM glycolate; 0.9 mM NEM.

3.2 Transmembrane Inhibition of Chloroplast Glycolate Uptake.

The effects on glycolate uptake, of preloading the stroma with a variety of 2-hydroxymonocarboxylates were tested (Table I). As reported previously (2,3), unlabeled glycolate, glyoxylate, D-glycerate and D-lactate displayed the trans stimulation of uptake rate characteristic of transported substrates of a so-called "mobile" carrier (1). L-lactate was slightly inhibitory and inhibition by the D,L mixtures of 2-hydroxybutyrate, 2-hydroxyvalerate and 2-hydroxycaproate increased as a function of chain length. Both mandelate isomers were weakly inhibitory when in direct competition with glycolate (20-25% inhibition, 5 mM external mandelate, 1 mM ^{14}C-glycolate, 2s uptakes), but L-mandelate completely eliminated glycolate uptake when present internally. D-mandelate was without effect in this assay.

Despite the lack of the pure L isomers of the 4-6 carbon aliphatic 2-hydroxymonocarboxylates, it appears probable that

TABLE I: Effect of Stromal 2-Hydroxymonocarboxylates on Chloroplast Glycolate Uptake Rates

Preloaded Compound	Rate (μmol/mg Chl h)	% of Control
None	15.6	100
Glycolate	55.6	356
D-Glycerate	43.7	280
L-Glycerate	19.1	122
D-Lactate	22.4	144
L-Lactate	13.2	85
D,L-2-Hydroxybutyrate	13.6	87
D,L-2-Hydroxyvalerate	8.5	54
D,L-2-Hydroxycaproate	1.7	11
D-Mandelate	16.4	105
L-Mandelate	0	0
1-Hydroxy-1-cyclo-propane carboxylate	9.4	60

a hydrophobic moiety in the L configuration is important for
trans inhibition. Possible explanations for the nature of the
L-mandelate inhibition include slow on and off rates and/or
interference with reorientation of the carrier. This new
class of inhibitor should provide interesting insights into
the structure of the substrate binding site and the carrier
mechanism. It also allows much more scope for the design of
affinity reagents than was possible when the only molecules
known to bind the carrier were 2 or 3 carbons long.

3.3 Reconstitution of Glycolate Efflux. Proteoliposomes, prepared as described in 2.4, were assayed for their retention of
internal glycolate after passage through Sephadex G-50 centrifuge columns (Table II). Although most of the internal glycolate had left the liposomes after a centrifuge column treatment, NEM pretreatment or external L-mandelate did retard the
efflux. While these results are quite preliminary, they do
indicate that cholate can solubilize a functional glycolate
carrier. It should be noted that the concentrations of cholate used do solubilize the differentially labeled 35 kD polypeptide (Howitz & McCarty, unpublished). Work is in progress
to purify the carrier and to optimize the reconstitution and
transport assay procedures.

TABLE II: Efflux of Glycolate from Proteoliposomes

Envelope Pretreatment	Liposome Formation[a]	Additions to Assay Column	Internal Glycolate (mM)	
			Initial	Final
None	FT	None	5.0	1.1
10 mM NEM	FT	None	5.0	1.7
None	FT	5 mM D-mandelate	2.4	0.63
None	FT	5 mM L-mandelate	2.4	0.92
None	FT+S	5 mM D-mandelate	2.4	0.37
None	FT+S	5 mM L-mandelate	2.4	0.62

[a] FT=Freeze/Thaw; S=10 s sonication.

REFERENCES
1 Howitz, K.T. and McCarty, R.E. (1985) Biochemistry 24, 2645-2652
2 Howitz, K.T. and McCarty, R.E. (1985) Biochemistry 24, 3645-3650
3 Howitz, K.T. and McCarty, R.E. (1985) Plant Physiol. 80, 390-395
4 Robinson, S.P. (1982) Plant Physiol. 70, 1032-1038
5 Howitz, K.T. and McCarty, R.E. (1983) FEBS LETT 154, 339-342
6 Mills, W.R. and Joy, K.W. (1980) Planta 148, 75-80
7 Cline, K., Andrews, J., Mersey, B., Newcomb, E.H., and Keegstra, K. (1981) Proc. Natl. Acad. Sci. USA 78, 3595-3599
8 Penefsky, H.S. (1977) J. Biol. Chem. 252, 2891-2899

ISOTOPE FRACTIONATION DURING OXYGEN PRODUCTION AND CONSUMPTION BY PLANTS.

ROBERT D. GUY[1], MARILYN F. FOGEL[2], JOSEPH A. BERRY[1] and THOMAS C. HOERING[2].
Carnegie Institution of Washington. (CIW-DPB Publ. No. 954)

1. Department of Plant Biology, 290 Panama St., Stanford, CA 94305.
2. Geophysical Laboratory, 2801 Upton St., Washington, DC 20008.

1. INTRODUCTION

Knowledge of isotope discrimination factors related to oxygen production and consumption by plants is important to an improved understanding of the global oxygen cycle. This information is also needed for the application of stable oxygen isotope methods, at natural abundance levels, to plant physiology and ecology.

The $^{18}O/^{16}O$ of atmospheric O_2 is 1.0235 times that of seawater (*i.e.* the $\delta^{18}O$ of air is +23.5 ‰ relative to standard mean ocean water [SMOW]). Known as the Dole effect (1), this difference is thought to result from isotope discrimination in photosynthesis and respiration. A small amount, perhaps 4 ‰, results from transpirational enrichment of leaf water (2). It is often assumed that there is no fractionation during the water-splitting reaction, but measurements of the $\delta^{18}O$ of photosynthetic O_2 range from slightly less than the source water (3) to 10 ‰ heavier or more (4). We addressed this question by examining O_2 evolution under conditions which eliminate O_2 uptake reactions. Normally, a large portion of photosynthetic O_2 is reabsorbed by photorespiration. The possibility that this process might fractionate isotopes during oxygenation of ribulose-1,5-bisphosphate (RuBP) by RuPB carboxylase/oxygenase (Rubisco; EC 4.1.1.39) or at glycolate oxidase (EC 1.1.3.1), is also dealt with here. Furthermore, although respiratory discrimination seems to vary among organisms (5), little is known for plants and no attention has been given cyanide-resistant respiration.

2. METHODS AND MATERIALS

2.1. General Procedure: Isotopic fractionation of molecular oxygen was studied by following the $\delta^{18}O$ of dissolved O_2 as it was produced or consumed by various preparations. Most experiments were within a leak-tight transparent Plexiglas reaction vessel which had a collapsible mid-section to ease sampling and provide an adjustable volume (100-400 ml). A magnetic impeller pumped aqueous media past an O_2 electrode and kept the volume well-mixed. Temperature was always 25°C. For experiments requiring light, irradiance was about 250 μE m^{-2} s^{-1}. Reagents were added through a septum. Dissolved O_2 was withdrawn in 10-150 ml aliquots through a separate port (also used for pre-gassing). After mixing with 5 ml H_3PO_4, with or without 1 g sulfosalicylic acid (to reduce foaming), O_2 was extracted by sparging with zero-grade He under a partial vacuum. Water vapour was removed cryogenically. Carbon dioxide was trapped on nine loops passing in and out of two liquid N_2 baths and could be saved for carbon isotope analysis. Oxygen was trapped on molecular sieve 5A at -196°C and purified of N_2 and Ar by gas chromatography on molecular sieve 5A at -60°C. The O_2 was reacted with graphite at 900°C to produce CO_2 for mass spectrometry. Yields were determined with calibrated pressure transducers (Omega Engineering, series PX236) and analyses were performed on Nuclide 6-60 isotope ratio mass spectrometers. Precision was better than ±0.1 ‰.

2.2. Oxygen Production: In one approach used to study O_2 production, isolated spinach thylakoids (6) were injected into degassed 50 mM phosphate buffer (pH 7.5) with 20 mM methylamine and 4 mM $K_3Fe(CN)_6$. Samples were removed following illumination for 3-12 min, a portion being kept for analysis of the source water. In a second approach, illuminated suspensions of the cyanobacterium *Anacystis nidulans* Richter were continuously sparged with He to remove O_2 as it was produced (20 mM HEPES, pH 8, with added nutrient salts and 20 mM $NaHCO_3$). Part of the sparging gas then entered the preparation line. Water samples were taken before and after each

Biggens, J. (ed.), Progress in Photosynthesis Research, Vol. III. ISBN 90 247 3452 5

experiment.

All buffers were prepared from a relatively heavy water with a $\delta^{18}O$ of +17.85 ‰. This helped minimize errors from potential leaks or carry-over of atmospheric O_2. Some of the thylakoid preparations, however, were made up in a water that was almost 30 ‰ lighter than the reaction buffer. This did not influence the results. Water oxygen isotopes were determined after conversion to CO_2 by the guanidine hydrochloride method (7).

2.3. Oxygen Consumption: Oxygen uptake by spinach glycolate oxidase (Sigma; 10-15 μg ml^{-1}) was in O_2-saturated 50 mM Tris (pH 8.3) that was 70 μM FMN and 2 mM glycolate. Glycolate was increased to 4 mM when catalase was included (10 μg ml^{-1}).

Spinach Rubisco was prepared according to Hall and Tolbert (8). Carbon and oxygen isotopes were studied together in some experiments and separately in others. Enzyme (to 250 μg ml^{-1}) was injected into O_2-saturated 50 mM Bicine buffer that was 20 mM MgCl$_2$ and 2-5 mM NaHCO$_3$, with 35 μg ml^{-1} carbonic anhydrase. Oxygen was excluded from carboxylation-only experiments. The pH was usually 8.5 but carboxylation experiments were also done at pH 7.6. Reactions were initiated with additions of degassed 240 mM RuBP synthesized according to Horecker et al. (9).

In one set of experiments, illuminated *Asparagus sprengeri* Regel mesophyll cells, isolated intact (10), were given enough NaHCO$_3$ to allow O_2 production to near air-saturation levels. Oxygen concentrations then remained fairly stable for up to 8 h. This created a "microcosm", favouring photorespiration, where O_2 was both produced and consumed at an equal rate. Samples were taken about once every 2 h. At the end of each experiment, to verify photosynthetic competence and confirm that dark respiration rate was low, additional NaHCO$_3$ was injected, followed by darkness a few minutes later.

Respiration experiments utilized O_2-saturated suspensions of baker's yeast (*Saccharomyces cerevisiae* Meyer), whole market purchased alfalfa sprouts (*Medicago sativa* L.), or *Asparagus* mesophyll cells. The yeast was supplied with sucrose (100 mM phosphate buffer, pH 7), but alfalfa sprouts (in water) and mesophyll cells (50 mM HEPES, pH 7.2) were simply allowed to metabolize their own reserves. Mesophyll cell studies were extended by including either 50 μM KCN, 20 mM salicylhydroxamic acid (SHAM) or 200 μM disulfiram.

2.4. Calculations: For all uptake reactions studied by depletion, per mil discrimination factors (*D*) were calculated from the "Rayleigh" equation,

$$D = - \frac{\ln R/R_o}{\ln f} \times 1000 \, ,$$

where R is the isotope ratio of the substrate at the time of sampling, R_o is the initial isotope ratio, and f is the fraction of substrate left. The R_o was determined from a reference sample taken at or near the beginning of each experiment, and f was calculated according to the gas yield. A plot of R/R_o against f indicates how the isotope ratio of the substrate changes as it is consumed (*eg.* Fig. 2). With the equation rearranged into its linear form, D becomes the slope of a regression through the origin (11). Standard error (SE) is then readily calculated and slopes compared by t testing.

3. RESULTS AND DISCUSSION

Illuminated thylakoids produced oxygen that was not different from the source water (Δ = +0.17 ‰ ±0.20 [SE], N=6). The inclusion of NaHCO$_3$ and carbonic anhydrase in some earlier experiments (not presented) was without influence. Continuously sparged *Anacystis nidulans* cultures also showed no isotope effect (Δ = +0.13 ‰ ±0.27 [SE], N=8). These results clearly show that there is no fractionation of oxygen isotopes in the photolysis of water by photosystem II. Stevens et al. (3) noted that this observation does not support theories invoking a role for CO_2 in this reaction. If such were the case, the substantial equilibrium isotope effect in the H_2O/CO_2 system (12) should have been expressed.

Other discrimination factors determined are summarized in Table 1. Respiration by baker's yeast discriminated against ^{18}O by 17.4 ‰. This is comparable to values reported for the bacterium *Escherichia coli* and another yeast, *Torulopsis utilis* (13). Whole alfalfa sprouts had a *D* of 21.2 ‰. Although the SE was large, it seems unlikely that the relatively long diffusion path in this tissue may have affected the results. Rates of uptake were fairly constant to almost zero oxygen. *Asparagus* mesophyll cells dark-respired O_2 with a D of 20.2 ‰, a value very similar to that obtained with the alfalfa sprouts.

TABLE 1. Summary of discrimination factors (D) determined for various uptake reactions, giving standard error (SE) and sample size (N).

Reaction or enzyme	D (‰)	SE	N
Respiration by baker's yeast	17.4	1.2	10
Respiration by alfalfa sprouts	21.2	1.9	6
Respiration by *Asparagus* mesophyll cells	20.2	1.3	12
with KCN	25.4	1.5	6
with disulfiram or SHAM	19.4	0.7	9
Glycolate oxidase	22.2	0.7	10
Rubisco: oxygenation	20.7	0.6	10
carboxylation (pH 8.5)	29.4	0.6	22
carboxylation (pH 7.6)	28.2	0.3	4

The D for *Asparagus* mesophyll cell respiration may reflect contributions from CN-sensitive (cytochrome oxidase) and CN-resistant (alternative) pathways. Respiratory discrimination against ^{18}O was modestly reduced to 19.4 ‰ in the presence of alternative path inhibitors (disulfiram or SHAM) but increased to 25.4 ‰ with KCN (significantly different at P<.005). This finding may be important for several reasons. It could account for some of the variability in respiratory discrimination factors apparent in the literature, perhaps resulting from differences in the balance between the two pathways. It also suggests that research applications may be found as a probe for this balance. Finally, it may provide clues as to the identity of the "alternate oxidase", an unsolved problem in plant biochemistry.

Data from the *Asparagus* mesophyll cell microcosm experiments are presented in Fig. 1. As the CO_2 compensation point was approached, promoting photorespiration, the $\delta^{18}O$ of dissolved O_2 began to diverge rapidly from the source water. It continued to rise well after O_2 concentrations had stabilized but, eventually, a steady difference of about +21.5 ‰ was achieved. This value should reflect a weighted mean D for the uptake reactions in these cells which, in large part, should represent photorespiration. Both Rubisco and glycolate oxidase, from spinach, were characterized by a D remarkably close to this estimate; 22.2 ‰ for glycolate oxidase (with or without added catalase) and 20.7 ‰ for Rubisco (Fig. 2). Although a carbon isotope effect on CO_2 fixation by Rubisco is well recognized, this is to our knowledge the first report of an oxygen isotope effect associated with this enzyme. *In vivo*, catalase promotes the decomposition of H_2O_2, produced by glycolate oxidase, to H_2O and O_2. Consequently, net O_2 uptake by Rubisco is twice that of glycolate oxidase. The expected overall D for photorespiration is therefore 21.2 ‰.

Carbon isotope fractionation by Rubisco (Fig. 2) had a D of 29.4 ‰ at pH 8.5, in exact agreement with Roeske and O'Leary (14). This was measured in the presence or absence of oxygenation. The moderately lower value at pH 7.6 (Table 1) was not statistically different. The experimental design also allowed an assessment of the effects on D of different CO_2 concentrations, covering an effective range of 2-20 μM. In agreement with Christeller et al. (15), no relationship was found.

4. CONCLUSION

Experiments are planned to measure oxygen isotope fractionation by Rubisco under different conditions and from different organisms, as well as to scale observations up to whole leaves. The difference in isotope discrimination associated with CN-resistant respiration deserves further attention. Another important O_2 consuming reaction in plants, the Mehler reaction, remains to be explored. We can conclude, however, that photorespiration must contribute to the isotopic balance of the atmosphere. Further studies will address its quantitative significance.

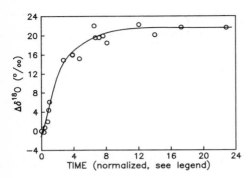

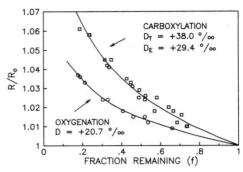

FIGURE 1. Divergence in $\delta^{18}O$ of dissolved O_2 from that of the source water in *Asparagus* mesophyll cell microcosm experiments. Data are from six experiments and are normalized according to the initial rate of O_2 production and final O_2 concentration (one unit of time being 20-60 min).

FIGURE 2. Substrate enrichment during oxygenation and carboxylation of RuBP by Rubisco at pH 8.5. The values D_T and D_E are total and enzyme-only discrimination factors for carboxylation before and after correcting for the CO_2/HCO_3^- equilibrium isotope effect (16).

5. REFERENCES

1 Dole, M. (1935) J. Amer. Chem. Soc. 57, 2731
2 Förstel, H. (1978) in Environmental Biogeochemistry and Geomicrobiology (Krumbein, W.E., ed.), Vol. 3, pp. 811-824, Ann Arbor Science, Ann Arbor
3 Stevens, C.L.R., Schultz, D., Van Baalen, C. and Parker, P.L. (1975) Plant Physiol. 56, 126-129
4 Metzner, H., Fischer, K. and Bazlen, O. (1979) Biochim. Biophys. Acta 548, 287-295
5 Lane, G.A. and Dole, M. (1956) Science 123, 574-576
6 Nolan, W.G. and Smillie, R.M. (1976) Biochim. Biophys. Acta 440, 461-475
7 Dugan, J.P., Borthwick, J., Harmon, R.S., Gagnier, M.A., Glahn, J.E., Kinsel, E.P., MacLeod, S. and Viglino, A. (1985) Anal. Chem. 57, 1734-1736
8 Hall, N. and Tolbert, N.E. (1978) FEBS Lett. 96, 167-169
9 Horecker, B.L., Hurwitz, J. and Stumpf, P.K. (1957) Methods Enzymol. 3, 193-195
10 Colman, B., Mawson, B.T. and Espie, G.S. (1979) Can. J. Bot. 57, 1505-1510
11 Mariotti, A., Germon, J.C., Hubert, P., Kaiser, P., Letolle, R., Tardieux, A. and Tardieux, P. (1981) Plant Soil 62, 413-430
12 O'Neil, J.R., Adami, L.H. and Epstein, S. (1975) J. Res. U.S. Geol. Survey 3, 623-624
13 Schleser, G.H. (1979) Radiat. Environ. Biophys. 17, 85-93
14 Roeske, C.A. and O'Leary, M.H. (1984) Biochemistry 23, 6275-6284
15 Christeller, J.T., Laing, W.A. and Troughton, J.H. (1976) Plant Physiol. 57, 580-582.
16 Winkler, F.J., Kexel, H., Kranz, C. and Schmidt, H.-L. (1982) in Stable Isotopes (Schmidt, H.-L., Förstel, H. and Heinzinger, K., eds.), pp. 83-89, Elsevier, Amsterdam

6. ACKNOWLEDGEMENTS

We thank Laura Green for supplying algal cultures and Dr. David DesMarais, NASA-Ames, for equipment access. This work was supported by U.S. Department of Energy grants to MFF and JAB. RDG is an NSERC (Canada) postdoctoral fellow.

CO-EVOLUTION OF RUBISCO AND CO_2 CONCENTRATING MECHANISMS.

M.R. Badger and T.J. Andrews

Dept. of Environmental Biology, Research School of Biological Sciences, Australian National University, PO Box 475, Canberra City, ACT, 2601, Australia.

1. INTRODUCTION.
The deleterious effects of oxygen on photosynthesis in C_3 higher plants can be attributed to the oxygenase properties of the primary CO_2-fixing enzyme ribulose bisphosphate carboxylase/oxygenase (Rubisco). Rubiscos isolated from organisms of all levels of complexity possess oxygenase activity and consequently all photosynthetic organisms have the potential to be adversely affected by O_2. It is apparent, however, that the potential for the expression of oxygenase activity has been greatly modified by two major factors; the evolutionary adaptation of the enzyme to its gaseous environment and its frequent location in natural or cellular environments, where the CO_2/O_2 ratio is higher than in the atmosphere.

2. A SINGLE PHYLOGENY FOR RUBISCO.
Rubisco is a unique catalyst upon which all life depends and it is clear that Rubisco evolved uniquely, all present day large subunits being descended from a common ancestral protein. Early suspicions that the single-subunit Rubisco in the purple bacteria might have an independent phylogeny (16,19) have been allayed following the determination of the complete sequence of the Rhodospirillum rubrum enzyme (17). Although this sequence differs markedly from that of the large subunits of the two-subunit Rubiscos, several regions of homology may be identified and these contain all of the amino acid residues of the R. rubrum sequence that have been identified by affinity labeling as being close to the active site.
    The origins of the small subunit are less clear because its sequence is more divergent and its function more mysterious. Nevertheless, several regions of its sequence are highly conserved (3), as is its overall size, and it might also have a monophyletic origin. We do not know whether the ancestral Rubisco had small subunits as well as large subunits.

3. EVOLUTIONARY PRESSURES ON RUBISCO.
    3.1 Rubisco Arose in Non-Photorespiratory Environments. It seems safe to assume that Rubisco must have arisen only once in the evolution of life, presumably appearing with the chemolithotrophs, which arose from primitive fermentative bacteria more than 3.5×10^9 years ago, before the genesis of even the oldest rocks on this planet (9). In these environments, the CO_2 level was high and O_2 was very low. Rubisco may have operated at close to CO_2 saturation in these organisms and there would have been no oxygenase activity. Obviously, photorespiration and the deleterious effects of O_2 were not a problem.

Biggens, J. (ed.), Progress in Photosynthesis Research, Vol. III. ISBN 90 247 3452 5
© *1987 Martinus Nijhoff Publishers, Dordrecht. Printed in the Netherlands.*

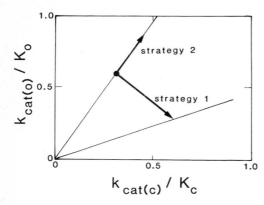

Figure 1. Two strategies for adapting Rubisco to a decreasing ratio of CO_2/O_2. Strategy 1 increases the s_{rel} of the enzyme ($[k_{cat(c)}/K_c]/[K_0/k_{cat(o)}]$), and may increase or decrease the $k_{cat(c)}/K_c$. Strategy 2 increases the K_{cat}/K parameters for each reaction proportionally, leaving s_{rel} the same.

3.2 The Declining CO_2/O_2 Ratio. Eventually, however, the operation of CO_2 consuming processes, particularly Rubisco carboxylation itself, led to a decline in the level of CO_2. Furthermore, with the advent of cyanobacteria, this was coupled to oxygenic photosynthesis and the CO_2/O_2 ratio would have become progressively less favourable for carboxylation. Eventually the positive carbon balance of photosynthesis would have been threatened if primitive Rubiscos had kinetic properties similar to those of present-day, anaerobic photochemotrophs such as R. rubrum. These Rubiscos have low affinities for CO_2 and highly effective oxygenase activities.

4. ADAPTIVE RESPONSES TO THE DECLINING CO_2/O_2 RATIO.
Photosynthetic organisms have adapted to the worsening atmospheric conditions in two ways. First, Rubisco itself shows some limited ability to adapt to lower CO_2/O_2 ratios. Second, organisms have developed mechanisms to insulate Rubisco from the external environment in compartments where the CO_2/O_2 ratio is more favourable.
   4.1 Changes in the Kinetic Properties of Rubisco. To improve the performance of Rubisco at low CO_2/O_2 ratios, the kinetic properties of the enzyme can change in two ways. First (strategy 1), the V_c/K_c term for the carboxylase reaction can increase relative to the V_0/K_0 term for the oxygenase reaction, increasing the relative specificity for CO_2 as opposed to O_2 ($s_{rel} = [V_c/K_c]/[V_0/K_0]$). This increases the rate of carboxylation relative to oxygenation. Alternatively (strategy 2), the V/K terms for both reactions could increase in proportion, leaving s_{rel} unaltered. This would allow the carboxylation rate per unit enzyme to be maintained at falling CO_2 concentrations. The latter alternative would falter, however, if the products of the oxygenase reaction were lost to the cell through such processes as photorespiration and glycolate excretion, since photosynthesis would eventually end up in a negative carbon balance. These strategies are illustrated in Fig.1. If $K_{cat(c)}/K_c$ is plotted against $k_{cat(o)}/K_0$, the slope of the lines is equal to $1/s_{rel}$. An enzyme adopting the first strategy will move its relationship to a line of lower slope. The second strategy involves a shift upwards along the same line.
   An examination of the kinetic properties of selected Rubiscos from widely different sources (Table 1) shows that there has been considerable

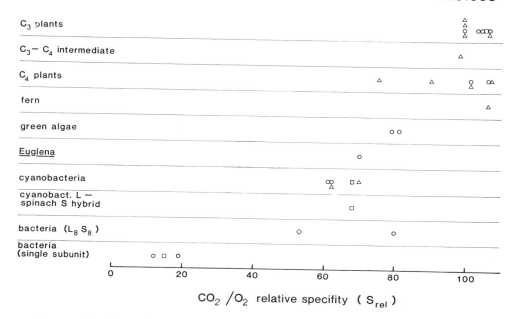

Figure 2. The relative specificity (s_{rel}) scale. Data were taken from: $\circ$, (12); $\triangle$,(13); $\square$, (2); $\diamond$, (15). All values were recalculated assuming a pK_a for the CO_2/HCO_3^- equilibrium of 6.12.

Table 1. Kinetic Properties of Rubisco from Various Sources.

Species	$k_{cat(c)}$[1]	K_c[2]	$\dfrac{k_{cat(c)}}{K_c}$[3]	s_{rel}[4]	$k_{cat(o)}$[5]	$K_{i(o)}$[6]	$\dfrac{k_{cat(c)}}{K_{i(o)}}$[7]	$\dfrac{k_{cat(c)}}{k_{cat(o)}}$	CO_2[8] Pump
Spinach (C_3)	2.9(20)[9]	10(26)	28.6	105(12)	1.1	400(5,12)	0.27	2.7	-
Maize (C_4)	4.3(20)	32(12,25)	13.4	101(12)	0.8	610(5,12)	0.13	5.5	+
Sorghum (C_4) *bicolor*	5.2(20)	30(13,25)	17.2	90(13)	1.1	640(13)	0.19	4.5	+
Amaranthus sp. (C_4)	5.6(20)	16(12,26)	35.0	160(12)	2.1	640(12)	0.33	2.7	+
Chlamydomonas reinhardtii	5.2(20)	29(12)	16.6	79(12)	1.0	480(12)	0.21	5.1	+
Synechococcus sp.	11.4(2)	185(2)	6.2	68(2)	1.2	1300(1)	0.09	9.7	+
Rhodospirillum rubrum	5.9(11)	84(11)	7.0	19(12)	1.9	406(12)	0.47	3.1	?
Chromatium vinosum	6.7(15)	29(15)	23	53(15)	1.25	290(15)	0.43	5.4	?

Footnotes

[1] calculated as s^{-1} per active site at 25°C
[2] µM CO_2 at 25°C, assuming a pKa of 6.12
[3] s^{-1} M^{-1} x 10^{-4}
[4] $S_{rel} = \dfrac{k_{cat(c)}}{K_c} \times \dfrac{K_o}{k_{cat(o)}}$

[5] $k_{cat(o)} = \dfrac{k_{cat(c)} \times K_{i(o)}}{K_c \times S_{rel}}$ (s^{-1})

[6] oxygen inhibition constant for carboxylase reaction
[7] s^{-1} M^{-1} x 10^{-4}
[8] denotes presence or absence of a CO_2 concentrating mechanism
[9] source of data indicated by reference number in brackets

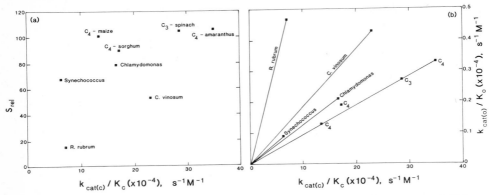

Figure 3. The relationship between (a), $k_{cat(c)}/K_C$ and s_{rel}; and (b) $k_{cat(c)}/K_C$ and $k_{cat(o)}/K_O$, for the Rubisco enzymes detailed in Table 1.

variation in k_{cat} (carboxylase), K_C and the relative specificity (s_{rel}). A more extensive listing of s_{rel} for all species examined so far (Fig.2), indicates that C_3 species, together with some C_4s, have the highest s_{rel} values, being some 5-10 fold higher than the anaerobic, purple non-sulfur bacteria, and 1.5-1.7 fold higher than the cyanobacteria. This trend supports the idea that the s_{rel} of Rubisco has increased in organisms exposed to aerobic, low-CO_2 environments. This increase in s_{rel} has been achieved with a concurrent increase in the $k_{cat(c)}/K_C$ of the enzyme (Fig.3a). The C_3 species, spinach, and surprisingly the C_4(NAD-ME) species, Amaranthus, have the highest values, being some 4-6 fold greater than cyanobacteria and R. rubrum. Thus adaptation of Rubisco to an environment of decreasing $\overline{CO_2/O_2}$ ratio has been through strategy 1 (Fig.1 and Fig.3b). Interestingly, the decrease in s_{rel} has been greatest between the single-subunit enzyme from anaerobes and the aerobic two-subunit Rubiscos. Between the aerobic enzymes, the variation in kinetic properties can be related to the presence or absence of CO_2 concentrating mechanisms. The two-subunit enzyme from the anaerobe, Chromatium vinosum, is interesting in that it has a relatively high $\overline{k_{cat(c)}/K_C}$. Whether this is related to the CO_2 concentration found in its environment remains to be determined.

4.2 CO_2 Concentrating Mechanisms in Aerobic Phototrophs. It has become clear that the majority of photosynthetic, aerobic organisms employ a CO_2-concentrating mechanism to modify the CO_2/O_2 ratio of Rubisco's environment. C_4 higher plants are a prime example, concentrating CO_2 in the bundle sheath (8). However similar effects can be achieved by many aquatic phototrophs (6). Aquatic organisms, ranging from cyanobacteria and green microalgae to both marine and freshwater macrophytes, use an inorganic pumping mechanism to elevate the intracellular CO_2. These pumping mechanisms vary in the level of internal CO_2 which they can sustain. The cyanobacteria seem to be the most effective, elevating CO_2 some 50-1000 fold (6), depending on the external CO_2. Green algae and C_4 plants may achieve a 10-50 fold accumulation (6, 8).

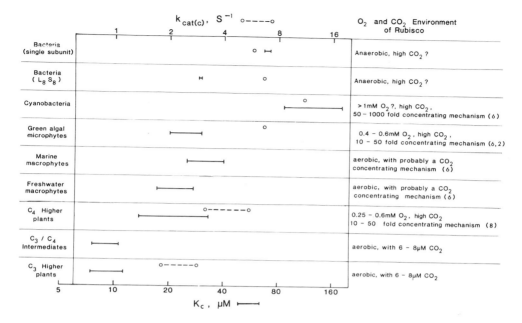

Figure 4. The variation in K_C and $k_{cat(c)}$ for Rubisco from different
groups of photosynthetic organisms. The aerobic organisms
are arranged in approximate order according to their
abilities to concentrate CO_2. Data for each group are from
the following references: C_3 plants, (5,12,13,20,25,26);
C_3/C_4 intermediates, (13,25); C_4 plants, (5,12,13,20,25,26);
freshwater macrophytes, (26); marine macrophytes, (19,26);
green algal microphytes, (12,24); cyanobacteria, (1,4,12,13);
anaerobic bacteria, (12,13,14). All data are expressed for
$25\,^{\circ}C$.

These CO_2 concentrating mechanisms effectively suppress the oxygenase
activity of Rubisco and lead to a dramatic reduction in all
physiological and biochemical manifestations of photorespiration (3,6).
Indeed, development of these suppression systems may be the most common
response to the decreasing CO_2/O_2 ratio; not the evolution of a more
efficient enzyme. Table 1 indicates that, of the Rubisco types that
have been extensively characterised, the C_3 type may be the only variant
which is adapted to the extremes of the present day atmosphere.
    Among the aerobic phototrophs, there is a general relationship between
the kinetic properties of Rubisco and the effectiveness of the CO_2-
concentrating mechanism which supports its activity. Figure 4 shows
that K_C increases some 10-20 fold going from C_3 plants to cyanobacteria.
This is consistent with cyanobacteria having the greatest ability to
concentrate CO_2. Between these extremes lie C_4 plants, aquatic
macrophytes and green microalgae, with K_Cs some 2-4 fold higher than
C_3 plants. All these intermediate groups may have a similar ability to
concentrate CO_2. The $k_{cat(c)}$ of the enzyme also increases as K_C
increases, but to a lesser extent. Consequently, the $k_{cat(c)}/K_C$
decreases over this same range.

4.3 Oxygen Elevation in Pumping Systems. Anaerobic bacteria experience high CO_2 environments accompanied by the absence of O_2. Consequently Rubisco, in this situation, could afford to optimise its carboxylation throughout regardless of what effect this might have on the oxygenase activity. Thus the low s_{rel} of the R. rubrum enzyme would be of no consequence to its physiological performance. This might also be said of aerobes with concentrate CO_2. However, it is not clear whether all oxygenase activity is suppressed in these organisms. The CO_2 concentrating mechanism must operate across a barrier of relatively low gas permeability and, in O_2-evolving organisms, this leads to an elevation in the internal O_2 concentration. The degree of this elevation ranges from being insignificant in the NADP-ME C_4 species lacking photosystem 2 in the bundle sheath, to a 2-3 fold elevation in other C_4s and green microalgae and probably greater than 5-fold elevation in the cyanobacteria (see Table 1 and Fig.4). This elevation of O_2 may influence the selection of a Rubisco which is optimal for each species. Cyanobacteria, although concentrating CO_2 to the greatest extent, probably have to cope with internal O_2 pressures of greater than one atmosphere. It is significant that these organisms show s_{rel} values that are only some 30-40% less than C_3 plants and 4-fold higher than R. rubrum. The retention of high s_{rel} values in most of the C_4 plants examined may also reflect some pressure to deal with elevated O_2.

5. FACTORS INFLUENCING THE ADOPTION OF A CO_2-CONCENTRATING MECHANISM.

Organisms which have chosen to deal with the decreasing CO_2/O_2 ratio by implementing a CO_2-concentrating mechanism have probably been influenced by five main environmental factors.

5.1 The Degree of CO_2 Limitation. CO_2 diffuses some four orders of magnitude more slowly in water than in air. Therefore, in aquatic environments, the concentration of CO_2 available by passive diffusion to Rubisco in an actively photosynthesising cell may be less than $1\mu M$ (6,18). Rather than deal with this problem by evolving a Rubisco with a very high $k_{cat(c)}/K_c$ ratio, most aquatic organisms have chosen to adopt a concentrating mechanism which seems to utilize inorganic carbon in both HCO_3^- and CO_2 forms. This may be an ecological strategy allowing them to exploit low CO_2 environments. Indeed, cyanobacteria, which are the most effective aquatic concentrators, dominate water bodies when inorganic carbon is depleted and the pH is high.

5.2 The Light Regime. Obviously the pumping mechanisms require photosynthetic energy and, if C_4 plants are a guide, the quantum requirements may be some 30-40% higher than in non-pumping C_3 plants. Thus, if light energy is a limiting resource, then a CO_2-pump may not be profitable. Photorespiration is also a costly process, however, and the relative costs of the two processes must be balanced in differing circumstances.

5.3 Temperature. The s_{rel} of Rubisco decreases with temperature (14) and, because of this, the potential for oxygenase activity and photorespiration increases with temperature. Thus pumping mechanisms will be most beneficial at higher temperatures where their energy costs are less than those of photorespiration. The evolution of C_4 plants in warm regions is a prime example of this. Similarly, it is interesting that cyanobacteria generally prefer warm environments.

5.4 <u>Nitrogen Limitation.</u> As a pumping mechanism allows Rubisco to operate near its V_{max} and the inherent $k_{cat(c)}$ of enzymes from pumping organisms is higher, then less Rubisco protein is required per unit of photosynthesis. In addition, since photorespiration is suppressed, loss of nitrogen from the photorespiratory pathway does not occur and investment of protein in photorespiratory components is not necessary. Consequently, a pumping mechanism makes the organism more efficient in its use of limiting nitrogen supplies. Nitrogen is a particularly scarce resource in many aquatic environments and, indeed, it has been found that low nitrogen can in fact induce the CO_2-concentrating mechanism in <u>Chlorella</u> species (7).

5.5 <u>Water-Use Efficiency.</u> In terrestrial environments, carbon can only be acquired by opening stomata and allowing water to be lost. Thus, in dry environments, it is an advantage to acquire carbon with the least expenditure of water. The CO_2-concentrating mechanism of C_4 plants allows them to photosynthesize at a lower intercellular CO_2, and therefore with a lower stomatal conductance, than C_3 plants, making them more efficient in their use of water.

6. TRUE PHOTORESPIRATION IS LARGELY CONFINED TO C_3 PLANTS.

The phenomenon of photorespiration and the associated metabolism of the photosynthetic carbon oxidation cycle is predominantly confined to C_3 terrestrial plants. The majority of aquatic phototrophs appear to lack significant photorespiration (6). Although many aquatic microorganisms have been shown to excrete large amounts of glycolate, this generally only happens when they have been adapted to high-CO_2 conditions and lack the pumping mechanism (23). It seems reasonable to argue that, in the natural environment, these organisms maintain high internal CO_2 concentrations which suppress oxygenase activity. This lack of photorespiratory phenomena extends to many aquatic algae and macrophytes and, in general, those which appear to exhibit positive signs of oxygenase activity are those that occur naturally in regions of relatively good CO_2 supply (6,18).

7. WHY HAVE C_3 PLANTS CHOSEN PASSIVE CO_2 ACQUISITION?

It is not clear why C_3 plants have evolved to maintain their chloroplast environment in passive equilibrium with air. We imagine that it relates to the temperature and the levels of CO_2 available, since these parameters affect the relative costs of photorespiration as opposed to inorganic carbon pumping. Obviously some terrestrial plants have employed the C_4 strategy to increase their performance in regions of high temperature and limiting water supply. However, there is little evidence for a concentrating mechanism which functions at the level of the individual cell or chloroplast. Perhaps some aspect of the mechanism of the pump in cyanobacteria and green microalgae prevents its operation in leaves but, in the absence of detailed knowledge of such mechanisms, we cannot guess what it might be.

Given that C_3 Rubisco functions without the aid of an ancillary mechanism for increasing the CO_2/O_2 ratio, is it perfectly adapted to its difficult environment in the sense that no further increase in s_{rel} is possible without deleterious consequences to other kinetic parameters? The apparent uniformity of the kinetic properties of the C_3 Rubiscos might indicate this. However, comparative studies of the C_3 enzymes have not

been extensive. Furthermore, the resolving power of existing techniques
is not sufficient to detect small differences. Certainly there is no
evidence that a compromise has been struck between s_{rel} and catalytic
efficiency as a carboxylase, since those Rubiscos with the highest s_{rel}
values also have the highest k_{cat}/K_c ratios. Perhaps further increases
in both these parameters might be possible.

REFERENCES.

1   Andrews, T.J. and Abel, K.M. (1981) J. Biol. Chem. 256, 8445-8451
2   Andrews, T.J. and Lorimer, G.H. (1985) J. Biol. Chem. 260, 4632-4636
3   Andrews, T.J. and Lorimer, G.H. (1986) In The Biochemistry of Plants:
    A Comprehensive Treatise. Vol.14: Photosynthesis (Hatch, M.D. and
    Boardman, N.K., eds.), Academic Press Inc, Orlando, Florida, in press
4   Badger, M.R. (1980) Arch. Biochem. Biophys. 201, 247-254
5   Badger, M.R., Andrews, T.J. and Osmond, C.B. (1974) In Proc. of the
    3rd Int. Photosynthesis Cong. (Avron, M., ed.), pp.1421-1429,
    Elsevier, Amsterdam
6   Badger, M.R. (1986) In The Biochemistry of Plants: A Comprehensive
    Treatise. Vol.14: Photosynthesis. (Hatch, M.D. and Boardman, N.K.,
    eds.), Academic Press Inc, Orlando, Florida, in press
7   Beardall, J., Griffiths, H. and Raven, J.A. (1982) J. Expt. Bot. 33,
    729-737
8   Berry, J.A. and Farquhar, G.D. (1978) In Proc. 4th Int. Congr.
    Photosynthesis (Hall, D., Coombs, J. and Goodwin, T., eds.) pp.119-131
    Biochem. Soc., London
9   Broda, E. (1975) "The Evolution of Bioenergetic Processes", Pergamon,
    Oxford
10 Colman, B. and Cook, C.M. (1985). In Inorganic Carbon Uptake By
    Aquatic Photosynthetic Organisms. (Lucas, W.J. and Berry, J.A., eds.),
    the American Society of Plant Physiologists, pp.97-110
11 Gutteridge, S., Sigal, I., Thomas, B., Arentzen, R., Cordova, A. and
    Lorimer, G. (1984) EMBO J. 3, 2737-2743
12 Jordan, D.B. and Ogren, W.L. (1983). Nature 291, 513-515
13 Jordan, D.B. and Ogren, W.L. (1983). Arch. Biochem. Biophys. 227,
    425-433
14 Jordan, D.B. and Ogren, W.L. (1984). Planta 161, 308-313
15 Jordan, D.B. and Chollet, R. (1985). Arch. Biochem. Biophys. 236,
    487-496
16 Lorimer, G.H. (1981). Ann. Rev. Plant Physiol. 32, 349-383
17 Nargang, F., McIntosh, L. and Sommerville, C. (1984). Mol. Gen.
    Genetics 193, 220-224.
18 Raven, J.A., Osborne, B.A. and Johnston, A.M. (1985) Plant Cell Env.
    8, 417-425
19 Robison, P.D., Whitman, W.B., Wadill, F., Riggs, A.F. and Tabita, F.R.
    (1980) Biochemistry 21, 4848-4853.
20 Seemann, J.R., Badger, M.R. and Berry, J.A. (1984) Plant Physiol. 74,
    791-794
21 Spalding, M.H. and Portis, A.R. (1985) Planta 164, 308-320
22 Talling, J.F. (1985) In Inorganic Carbon Uptake By Aquatic
    Photosynthetic Organisms (Lucas, W.J. and Berry, J.A., eds.), the
    American Society of Plant Physiologists, pp.403-420.

23 Tolbert, N.E., Husic, H.D., Husic, D.W., Moroney, J.V. and Wilson, B.J. (1985) In Inorganic Carbon Uptake By Aquatic Photosynthetic Organisms (Lucas, W.J. and Berry, J.A., eds.), the American Society of Plant Physiologists, pp.211-227

24 Tsuzuki, M., Miyachi, M. and Berry, J.A. (1985) In Inorganic Carbon Uptake By Aquatic Photosynthetic Organisms (Lucas, W.J. and Berry, J.A., eds.), the American Society of Plant Physiologists, pp.53-66

25 Yeoh, H., Badger, M.R. and Watson, L. (1980) Plant Physiol. 66, 1110-1112

26 Yeoh, H., Badger, M.R. and Watson, L. (1980) Plant Physiol. 67, 1151-1155

PROPERTIES OF RUBISCO IN RELATION TO THE CONTROL OF PHOTORESPIRATION

KEYS, A.J., PARRY, M.A.J., PHILLIPS, A.L., HALL, N.P. and KENDALL, A.C.
ROTHAMSTED EXPERIMENTAL STATION, HARPENDEN, HERTS. AL5 2JQ, U.K.

1. INTRODUCTION

Ribulose 1,5-bisphosphate carboxylase (EC 4.1.1.39, Rubisco) catalyses both the carboxylation and oxygenation of ribulose 1,5-bisphosphate (RuBP) with CO_2 and O_2 acting as competitive alternative substrates. The oxygenation of the RuBP results in the formation of phosphoglycollate and initiates a metabolic pathway that releases CO_2 during a step in which glycine is converted to serine. The exchange of CO_2 and O_2 between plants and the atmosphere resulting from this pathway is, by one definition, photorespiration. In C_4 plants, and in certain algae and cyanobacteria, photorespiration is controlled by the ability of the organisms to concentrate CO_2 in the cell compartment containing the Rubisco and thereby overcoming the inhibitory effect of oxygen and decreasing the extent of the competitive reaction of the RuBP with oxygen. The concentrating mechanisms require energy to drive them and, in C_4 plants, a more elaborate leaf structure. Possibly these added requirements prevent such mechanisms from having a selective advantage in most land plants adapted to cooler climates.

In C_3 higher plants, Rubisco normally operates below the concentration of CO_2 at equilibrium with the ambient atmosphere because of restricted diffusion of the gas into the leaf to replace that removed by photosynthesis. A recent report [1] claiming that a concentrating mechanism for CO_2 operates in C_3 plants at cool temperatures awaits tests in other laboratories. More generally rates of photosynthesis are near to being consistent with the amount of active Rubisco in the leaf in a manner predictable from properties of the enzyme observed *in vitro* and the measured conductance for CO_2 into the leaf. Likewise photorespiration rates are near to those predicted from the oxygenase activity of the Rubisco present in the leaf and the concentrations of CO_2 and O_2 calculated to be present from the measured values of the gaseous conductances. Descrepancies between predicted and measured rates of exchange of the two gases with the atmosphere are often because of the technical difficulties of the measurements, the existence of other respiratory processes than photorespiration as defined above, and internal recycling of the gaseous substrates. These same difficulties make estimation of fluxes of carbon through the photorespiratory pathway by the use of isotopic tracers subject to uncertainties. Thus, evidence for control of photorespiration in C_3 plants *in vivo* is difficult to obtain. It is, theoretically, much easier to study factors that control the oxygenase activity of Rubisco *in vitro* and to consider their likely significance in the leaf. Many substances are known that affect both carboxylase and oxygenase activities proportionally and therefore both photosynthesis and photorespiration; only changes in the divalent metal ion activating the carboxylase have been reported to make large differences to the specificity factor [2,3].

Biggens, J. (ed.), Progress in Photosynthesis Research, Vol. III. ISBN 90 247 3452 5
© *1987 Martinus Nijhoff Publishers, Dordrecht. Printed in the Netherlands.*

In a separate report, we describe progress in the selection and study of photorespiratory mutants of barley plants. These provide a very valuable means for studying photorespiratory metabolism, identifying any function it may have, and identfying the problems that arise with control beyond the initial reaction - the oxygenation of RuBP.

In what follows, factors affecting the oxygenase and carboxylase activities of purified Rubisco will be discussed with special reference to the carboxylase from wheat leaves. Some preliminary measurements will also be presented concerning the specificity factor of enzyme activated in the presence of low concentrations of CO_2 by inorganic orthophosphate (Pi); sites activated in these conditions seem to have exceptional activity.

2. FACTORS AFFECTING CARBOXYLASE AND OXYGENASE ACTIVITY OF RUBISCO
2.1. Slow and fast activating forms of wheat Rubisco.
Like all Rubiscos, the enzyme from wheat is absolutely dependent on activation by reaction with CO_2 in the presence of Mg^{2+} to reach its highest activity. During the purification of Rubisco from wheat leaves the enzyme changes almost entirely into a form that activates only very slowly when incubated with Mg^{2+} and CO_2 at room temperature [4,5]. This change is associated with prolonged incubations at low temperature after the removal of the activating cofactors, the Mg^{2+} and CO_2, and other ionic stabilizers from the protein. Our practice has been to store purified Rubisco as a freeze-dried powder in which it is largely in this slow activating form (E_S). When dissolved in buffer containing 20 mol $m^{-3} Mg^{2+}$ and 84 mmol $m^{-3} CO_2$, the enzyme activates slowly to a maximum reached after about 5 h at 25°C. Under these conditions the rate of activation can be increased by heating so that at 45°C maximum activation is achieved in some 20 min (Fig. 1). Removal of the Mg^{2+} and CO_2 from the activated enzyme results

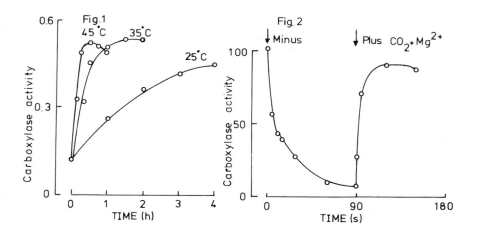

FIGURE 1. Activation of purified wheat Rubisco in the presence of Mg^{2+} (20 mol m^{-3}) and CO_2 (84 mmol m^{-3}) at pH8.2.
FIGURE 2. Effect of removal of Mg^{2+} and CO_2 (time 0) activated as in Fig. 1 and readdition (time 90s). Adapted from reference [5].

in loss of activity. However, if the two cofactors are added back to
the solution activity returns to the maximum in some 2 min so that we
recognize the presence of a rapidly activating form of the enzyme
(E_r) similar to the enzyme often described from spinach (Fig. 2).
We can recognize the formation of E_r in solution of E_s in CO_2-free
buffer without Mg^{2+} by assay after a short incubation (3 min) with
Mg^{2+} and CO_2 for either carboxylase or oxygenase activity. In these
conditions the formation of E_r is favoured by increasing the
temperature or especially by the presence of various known effectors
of the enzyme, NADPH, PGA, inorganic orthophosphate, inorganic
pyrophosphate and sulphate (Fig. 3). The conversion of E_s to E_r is
accompanied by changes in circular dichroism consistent with
significant conformational changes in the protein molecule [6]. It
is not known whether the slow and rapidly activating forms of
carboxylase have significance in the leaf in the control of
photosynthesis and photorespiration. The effects on the
interconversion of the two forms by the metabolites mentioned makes
such an involvement seem likely.

2.2. Activation by carbon dioxide.
Experiments with wheat leaves using extreme conditions to change the
CO_2 concentration in the leaf show that the activity of carboxylase
in vivo is controlled by the external concentration of CO_2 (Fig. 4).

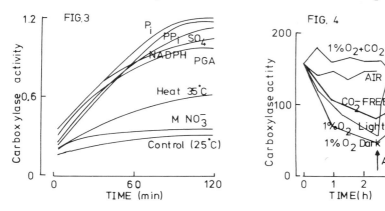

FIGURE 3. Activation of purified wheat Rubisco in the absence of Mg^{2+} and
CO_2. Requirement for heat or effectors. Adapted from [6].
FIGURE 4. Loss of carboxylase activity in leaves in the absence of CO_2.
Adapted from [7,8].

Such effects of CO_2 concentration are undoubtedly modulated by the
concentration of Mg^{2+} in the chloroplast stroma and changes in this
caused by light intensity [9,10]. This explains the rapid recovery
of activity when CO_2 is restored in the light compared to in the dark.

2.3. Activity modulation by effectors.
Two separate effects have to be understood [5]. Firstly there is an
effect on the state of activation of the enzyme and secondly a direct
inhibitory effect on catalysis competitive with RuBP.

2.3.1. <u>Effects on activation state</u>: Effects on activation are studied by adding effectors to a concentrated solution of the enzyme. Samples of this mixture are then diluted into reaction mixtures not containing the effector to measure the activity. The extent of dilution is chosen to avoid any significant direct inhibition. Fig.3 shows that certain effectors increase the conversion of E_S to E_r in the absence of Mg^{2+} and CO_2. The same substances also increase the activation of the enzyme very considerably when Mg^{2+} and CO_2 are present at concentrations below those necessary to produce maximum activation. We believe that the effectors act, not by increasing reaction of the enzyme with activating CO_2 [11], but by an independent mechanism which alters the catalytic function of sites already containing activating CO_2 [12]. At low CO_2 concentrations effectors make sites that are in a 'superactive' state. This conclusion is based on data such as shown in Table 1 in which a high specific activity is reached in the presence of Pi when less than 2 of the eight active sites on the enzyme contain bound activating CO_2 ($^ACO_2$). The activity achieved in this way is indeed greater than can be achieved by activating with CO_2 and Mg^{2+} at saturating concentrations.

TABLE 1. Effect of inorganic orthophosphate on the specific carboxylase activity and on the amount of bound activating CO_2 ($^ACO_2$) for wheat Rubisco. All solutions contained 100 mol m^{-3} Bicine pH8.2, 260 mmol m^{-3} O_2 and 20 mol m^{-3} $MgCl_2$ with the additions shown.

Conditions during activation			
$^{14}CO_2$ (mmol m^{-3})	Pi (mol m^{-3})	Specific Activity (−in 84 mmol m^{-3} $^{14}CO_2$) (μmol min^{-1} mg^{-1})	Bound $^ACO_2$* (mol mol^{-1} holoenzyme)
0	10	0.05	−
8.4	0	0.20	1.28
8.4	10	1.00	1.60
84.0	0	0.60	5.28
84.0	10	1.20	5.60

*measured after reaction with carboxyarabinitol bisphosphate.

Parry (unpublished) has shown that the superactive sites have a greater affinity for RuBP. Therefore, although oxygenase activity was affected as well as carboxylase, it seemed worth making an assessment of the specificity factor for effector activated enzyme and enzyme activated with saturating CO_2 and Mg^{2+}. Results of a preliminary experiment are shown in Table 2. The superactive sites created at low concentrations of CO_2 by the presence of Pi appear to have a higher specificity factor when measured after dilution of the effector. The results need to be confirmed, preferably by a different analytical method. However, differences are of the same order of magnitude already established between carboxylases from wheat and tobacco. On the other hand, similar measurements on phosphate activated wheat carboxylase made at higher bicarbonate concentrations failed to show a difference in specificity factor [8]. With Pi added to the assay mixture the specificity factor was lowered.

TABLE 2. Specificity factors for wheat carboxylase activated (40 min at 40°C) in the presence of Pi at low CO_2 (superactive) or at high CO_2 without Pi. Carboxylase was measured by $^{14}CO_2$ fixation and oxygenase by the sensitized oxygen electrode. Reaction mixtures contained excess carbonic anhydrase. All solutions contained 100 mol m^{-3} Bicine, (pH8.2), 260 mmol m^{-3} O_2 and 20 mol m^{-3} $MgCl_2$ with the additions shown below.

Additions for assay		Additions for activation	
CO_2 (mmol m^{-3})	Pi (mol m^{-3})	8.4mmol m$^{-3}CO_2$ 10 mol m^{-3} Pi	84mmol m$^{-3}CO_2$
		Specificity factor	
8.4/9.2**	0	90.7 ± 3.7	85.8 ± 0.2***
"	0	90.3 ± 0.5***	77.3 ± 1.2***
"	10	84.9 ± 6.1	77.3 ± 1.9
16.7/17.5	0	92.9 ± 4.0	82.2 ± 0.9
"	10	76.6 ± 30	76.9 ± 8.2

*   For enzyme activated in 8.4 mmol m$^{-3}CO_2$.
** For enzyme activated in 84 mol m^{-3} CO_2.
*** Plus 192 mmol m^{-3} x added to the activated anzymes >10 min before assay. x is the natural night-time inhibitor of Rubisco from potatoes (Gutteridge *et al*. submitted to Nature). The amount of inhibitor gave 7.5% inhibition in the presence of Pi and 27.1% inhibition in the absence of Pi - the aim was to block 2 out of the eight active sites.

2.3.2. Competitive inhibition: Effectors such as inorganic orthophosphate are also competitive inhibitors of carboxylase and oxygenase with respect to RuBP. For orthophosphate the inhibition of catalysis offsets the stimulation because of the activating effect at about 10 mol m^{-3} of Pi [13]. This is probably above the concentration of Pi in the chloroplast stroma but this region of the chloroplast contains significant amounts of other effectors particularly 3-PGA, FBP and SBP. The inhibitory effects are additive for the combination of 3-PGA and FBP [14] but more work is necessary to establish the combined effect of the many effectors known to be present in the chloroplast stroma at varying concentrations depending on external conditions.

2.3.3. The natural inhibitor formed in darkness in some species: Gutteridge *et al*. (unpublished) purified a tight-binding inhibitor of Rubisco formed at night in the leaves of *Solanum tuberosum* L. Because of the evidence of negative cooperativity between catalytic sites on the holoenzyme it seemed worth inhibiting a few sites with this substance to see whether this would affect the specificity of catalysis of the remaining sites. The results, included in Table 2, suggest that when added to the enzyme during activation with CO_2 and Mg^{2+} alone the inhibitor decreases the specificity factor. The presence of Pi competitively decreases the effects of the inhibitor and enzyme activated with Pi at low CO_2 retains a high specificity factor even in the presence of the inhibitor.

3. CONCLUSIONS

Rubisco from wheat plants has properties suggesting several possible regulatory mechanisms including a temperature dependent conformational change, sensitivity to CO_2 and Mg^{2+} and sensitivity to various phosphorylated chloroplast components. The major effects of such regulation is on both carboxylation (photosynthesis) and oxygenation (photorespiration). Preliminary experiments show however that changes in specificity factor and hence differential regulation of photosynthesis and photorespiration should not be ruled out in C_3 land plants. The difficulties of measuring both carboxylation and oxygenation with sufficiently high accuracy to provide a ratio that is reproducible and entirely reliable still requires that such results are viewed with caution.

REFERENCES
 1 Lehnherr, B, Mächler, F. and Nösberger, J. (1985) J. exp. Bot. 36, 1835-1841
 2 Wildner, G.F. and Henkel, J. (1978) FEBS Lett. 91, 99-103.
 3 Christeller, J.T. and Laing, W.A. (1979) Biochem. J. 183, 747-750.
 4 Mächler, F., Cornelius, M.J. and Keys, A.J. (1980) J. exp. Bot. 31, 7-14.
 5 Gutteridge, S., Parry, M.A.J. and Schmidt, C.N.G. (1982) Eur. J. Biochem. 126, 597-602.
 6 Schmidt, C.N.G., Gutteridge, S., Parry, M.A.J. and Keys, A.J. (1984) Biochem. J. 220, 781-785.
 7 Boyle, F.A.B. and Keys,m A.J. (1982) Photosyn. Res. 3, 105-111.
 8 Boyle, F.A.B. and Keys,m A.J. (1986) Photosyn. Res. (in press).
 9 Portis, A.R. and Heldt, H.W. (1976) Biochim. Biophys. Acta. 449, 434-446.
10 Portis, A.R., Chon, C.J., Mosback, A. and Heldt, H.W. (1977) Biochim. Biophys. Acta. 461, 313-325.
11 McCurry, S.D., Pierce, J., Tolbert, N.E. and Orme-Johnson, W.H. (1981) J. biol. Chem. 256, 6623-6628.
12 Parry, M.A.J., Schmidt, C.N.G., Cornelius, M.J., Keys, A.J., Millard, B.N. and Gutteridge, S. (1985) J. exp. Bot. 36, 1396-1404.
13 Bhagwat, A.S. (1981) Pl. Sci. Lett. 23, 197-206.
14 Boyle, F.A.B. (1984) Ph.D. Thesis, Univ. London.

QUANTITATIVE SEROLOGICAL STUDIES ON RuBP CARBOXYLASE/OXYGENASE OF TOBACCO
MUTANTS EXHIBITING DIFFERENT PHOTOSYNTHETIC AND PHOTORESPIRATORY ACTIVITIES

Alfons Radunz and Georg H. Schmid
Universität Bielefeld, Lehrstuhl Zellphysiologie, D-4800 Bielefeld 1, FRG

INTRODUCTION

In earlier publications Schmid et al. (1-5) have shown that several
mutants of *Nicotiana tabacum*, who showed substantially higher rates of
photosynthesis, also exhibited considerably higher rates of photorespiration
when compared to the wild type *N. tabacum var. John William's Broadleaf*.
The process of photorespiration is due to the properties of the bifunctional stroma enzyme RuBP carboxylase/oxygenase and depends on external factors such as light and temperature. The carboxylating function of the enzyme is favoured by high CO_2 partial pressures and the oxygenase function
is favoured by high O_2 partial pressures. In this context the question
arises whether the higher rates of photorespiration in mutant plants are
due to molecularly altered RuBP carboxylase/oxygenase or whether the rates
are simply due to higher enzyme concentrations. According to the literature
the enzyme is supposed to make up for 50% of the soluble leaf proteins
(6,7). These data refer to spinach. This would imply that half of the
stroma proteins of chloroplasts is RuBP carboxylase/oxygenase. In the present paper we compare the amount of RuBP carboxylase/oxygenase present by
means of a monospecific antiserum to this enzyme (8-10) prepared of the
wild type *N. tabacum var. John William's Broadleaf* with that of the discussed tobacco mutants, using rocket immuno electrophoresis in agarose gel.

RESULTS

The amount of RuBP carboxylase/oxygenase present in mutants of *Nicoti-*
ana tabacum was compared using the method of rocket immuno electrophoresis
(9,10) with the antiserum to RuBP carboxylase/oxygenase of the wild type
JWB (Fig. 1). The length of the precipitation bands shows in dependence on
the enzyme concentration direct proportionality. The analysis shows, that
the enzyme concentrations in chloroplasts of the different mutants varies
considerably, if the amount of enzyme is referred to the protein portion of
the chloroplasts (Table 1). In all green phenotypes of the mutant series
RuBP carboxylase/oxygenase just as in the wild type makes up for 2-4% of
the chloroplast proteins. This implies that this enzyme occurs only up to
8% in the chloroplast stroma proteins since half of the chloroplast proteins are localized in the stroma, whereas the other half builds up the
thylakoid membrane. On the other hand, it appears that the enzyme content
in chloroplasts of the yellow-green and yellow phenotypes of mutants as
N. tabacum Su/su, N.t. Su/su var. Aurea, Consolation and *Xanthi* is 3 to 6
fold higher. Hence, the stroma proteins of these yellow-green and yellow
phenotypes consist up to 1/3 of RuBP carboxylase/oxygenase.
If in the mutants the higher enzyme concentration is compared to the higher
photosynthetic rates or to the higher photorespiratory activities it seems
as if a dependence of photosynthetic rates on the enzyme concentration
existed. It appears that yellow-green and yellow mutants as *Su/su, Su/su*
var. Aurea and yellow *Consolation* which exhibit a 5 to 10 fold higher rate
of photosynthesis in comparison to the green phenotypes, contain a 4 to 6
fold higher enzyme concentration. However, photorespiration is not dependent on

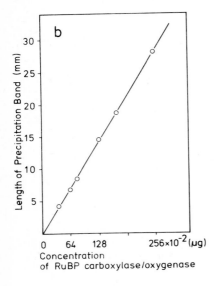

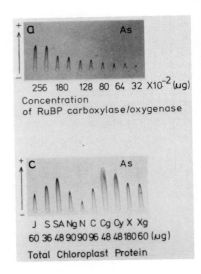

Fig. 1: Quantitative estimation of RuBP carboxylase/oxygenase by rocket immuno electrophoresis.

a. Rocket immuno electrophoresis of RuBP carboxylase/oxygenase in different concentrations in agarose gel, containing 1 per cent of enzyme antiserum

b. Calibration curve of the quantitative estimation. Ordinate: length of the antigen-antibody-precipitation bands. Abscissa: enzyme concentration

c. Rocket immuno electrophoresis in order to estimate the amount of RuBP carboxylase/oxygenase in *N. tabacum var. JWB* and some *N. tabacum* mutants

Antiserum: As, 1 per cent enzyme antiserum in agarose gel

Antigen: Chloroplast preparations of

J	*N.tabacum var.JWB*,green	Cy	*N.tabacum var.Consolation*,yellow	
S	*Su/su*,yellow-green	N	*NC 95*,green	
SA	*Su/su,Aurea*,yellow	Ng	*NC 95*,yellow-green	
C	*Consolation*,green	X	*Xanthi*,green	
Cg	" yellow-green	Xg	*Xanthi*,yellow-green	

a higher RuBP carboxylase/oxygenase content, since some mutants as *N. tabacum var. Consolation* green and var. *Consolation* yellow as well as *Su/su* which exhibit high rates of photorespiration contain particularly low amounts of enzyme (*Consolation* green) as well as particularly high amounts of enzyme (*Su/su*)(Table 1).

The mutant *NC 95* seems to be an exception. Chloroplasts of yellow-green leaf patches contain only twice the enzyme concentration of chloroplasts of green leaf patches. The structure of chloroplasts of yellow leaf areas differs from that of green leaf areas by a lack of grana stacking. The lamellar structure consists of only single stranded isolated thylakoids which exhibit only photosystem I activity (1).

A comparison of our values with those described in the literature namely that RuBP carboxylase/oxygenase (fraction I) would make up for 50% of all

III.9.**619**

Table I: Comparison of the RuBP carboxylase/oxygenase content in chloroplasts of green, yellow-green and yellow phenotypes of different *Nicotiana tabacum* mutant series, which are characterized by different rates of photosynthesis and photorespiration

	Photosynthesis and Photorespiration in comparison to the wild type *JWB*	Chlorophyll content in per cent of proteins	RuBP carb.oxy. in per cent of total chloroplast proteins	Harvesting date of the plants
Nicotiana tabacum var. JWB		10.4	3.8	July 1985
N.t. var. Su/su yellow-green	high rate of photorespiration / high rate of photosynthesis at high light intensities (2)	6.4 / 7.2	10.2 / 14.0	Sept. 1985 / March 1986
N.t. Su/su var. Aurea	high rate of photosynthesis at very high light-intensities (2)	2.4 / 2.1	14.7 / 11.3	Sept. 1985 / Jan. 1985
N.t. var. Consolation, green	high rate of photorespiration (3)	23.2	2.5	May 1985
yellow-green		6.2 / 8.1	10.0 / 14.1	Sept. 1985 / March 1985
yellow	high rate of photorespiration, high rate of photosynthesis (3)	1.7 / 2.3	10.4 / 16.1	Sept. 1985 / Jan. 1986
N.t. var. NC 95, green		26.2	2.7	May 1985
yellow-green	yellow leaf areas of variegated leaf containing only photosystem I (1)	3.5	2.5 / 4.8	Nov. 1985 / March 1986
N.t. var. Xanthi, green		12.5	1.8	Sept. 1985
yellow-green		1.3	5.3	Sept. 1985

soluble leaf proteins in spinach (6,7), is problematic, as it appears unclear in what ratio the soluble proteins are distributed in cell organells and in the cytosol. Our investigations show that the RuBP carboxylase/oxygenase content in spinach chloroplasts reaches the level of 18-20% of the total chloroplast proteins. This was determined by means of an antiserum to the enzyme of *N. tabacum JWB* as well as with an antiserum to the purchased spinach enzyme from Sigma. Moreover, it appears that the RuBP carboxylase/oxygenase content is not constant during the vegetation period, and depends on the developmental state of the plants, on external factors such as light and nutrition and on the season. Thus, our results show that the RuBP carboxylase/oxygenase content of yellow-green and yellow phenotypes of the *N. tabacum* mutants is 40 to 60% higher during the spring months in comparison to the content during the autum months, despite the fact that the plants were grown under identical external conditions in a growth chamber.

REFERENCES
1. Homann, P.H. and Schmid, G.H. (1967) Plant Physiol. 42, 1619-1622
2. Okabe, K., Schmid, G.H. and Straub, J. (1977) Plant Physiol. 60,150-156
3. Schmid, G.H., Bader, K.P., Gerster, R., Triantaphylides, C. and André,M.
   (1981) Z. Naturforsch. 36c, 662-671
4. Ishii, R. and Schmid, G.H. (1982) Z. Naturforsch. 37c, 93-101
5. Ishii, R. and Schmid, G.H. (1983) Plant and Cell Physiol. 24(8),1525-1533
6. Akazawa, T. (1979) in: Enzyclopedia of Plant Physiology, New Series,
   Vol. 6, Photosynthesis II, Photosynthetic Carbon Metabolism and Related
   Processes. ed. by M. Gibbs and E. Latzko, Springer-Verlag, Berlin
   pp. 208-225
7. Kung, S.D. and Marsho, T.V. (1976) Nature 259, 325-326
8. Radunz, A. (1978) Z. Naturforsch. 33c, 731-734
9. Radunz, A., Schmid, G.H., Bertrand, M. and Dujardin, E. (1985) in:
   Proceedings of Intern. Meeting of Regulation of Chloroplast Differentiation, G. Akoyunoglou ed., Rhodos, Greece
10. Dujardin, E., Bertrand, M., Radunz, A. and Schmid, G.H. J. of Plant
   Physiol., in preparation

PHOTOSYNTHETIC CHARACTERISTICS OF HAPLOID TOBACCO PLANTLETS SELECTED FOR RESISTANCE TO OXYGEN STRESS

ISRAEL ZELITCH, DEPT. OF BIOCHEMISTRY AND GENETICS, THE CONNECTICUT AGRICULTURAL EXPERIMENT STATION, P.O. BOX 1106, NEW HAVEN, CT 06504, U.S.A.

New Screening System

Photorespiration increases greatly with increasing O_2 concentration and thereby decreases net photosynthesis. Selecting plants for superior growth because of their resistance to "oxygen stress" (1) might therefore provide a direct method of obtaining mutants of C_3 species with decreased photorespiration and increased net photosynthesis. Previous work has shown that 60% O_2 caused lethality in tobacco seedlings or haploid plantlets in several weeks (2, 3). Elevated O_2 levels may cause deleterious effects unrelated to photorespiration, but stress resulting from excess photorespiration should be reversed at high CO_2 levels (1% or more), and growth should then be the same as in 21% O_2 and high CO_2.

Accordingly, growth experiments were carried out with tobacco seedlings (var. Havana Seed) in constant environments. Their growth was measured using the mean net assimilation rate, in units of mg dry wt/dm^2 leaf area·day, as a conservative measure of photosynthetic efficiency under different O_2 and CO_2 concentrations. The toxic effects of 60% O_2 and low CO_2 were not reversed by 1% CO_2 under these conditions, indicating that some of the toxicity of 60% O_2 is caused by non-photorespiratory mechanisms.

However, the inhibition of growth of tobacco seedlings at 42% O_2 and low CO_2 was completely reversed by high CO_2 (Table 1). In both experiments growth was the same in 21% O_2/1% CO_2 as in 42% O_2/1-3% CO_2. At 42% O_2/low CO_2, growth was always less than in 21% O_2/low CO_2. The results suggest that greatly increased photorespiration is the primary effect of growth inhibition at 42% O_2/low CO_2. These conditions were therefore used as the basis for mass screening of haploid tobacco plantlets for resistance to oxygen stress.

The screening was done with small (about 5 mm) haploid tobacco plantlets derived from anther cultures (4, 5). These were grown in air for one week in light on polyurethane pads (6) in petri plates containing liquid medium without sucrose until they were green. The plantlets were then transferred to illuminated plexiglas chambers that were flushed continuously with 42% O_2/160 μl CO_2/l for two weeks. Of 2,714 haploid plantlets examined, 26 (0.96%) were classified as O_2-resistant as they retained their green appearance and grew while the others yellowed and some died. Resistant and O_2-sensitive (wild type) plantlets from the same plates were recovered in 21% O_2/1% CO_2, transferred to rooting medium (4), and after further development were grown in a commercial potting mixture in a light chamber under controlled conditions. Later, photosynthesis was determined

Biggens, J. (ed.), Progress in Photosynthesis Research, Vol. III. ISBN 90 247 3452 5
© *1987 Martinus Nijhoff Publishers, Dordrecht. Printed in the Netherlands.*

TABLE 1. Comparison of Elevated O_2 and Reduced CO_2 Concentration on the Mean Net Assimilation Rate of Tobacco Seedlings Grown in a Constant Environment. Continuous light 120 μE $m^{-2}s^{-1}$, 29°. Numbers followed by the same letter are not significantly different ($\underline{P}>0.01$) by Duncan's Multiple Range Test.

	NAR
	mg dry wt/$dm^2 \cdot day$
Expt. 1. Growth 5 days	
21% O_2/420 μl CO_2/l	66.8 a
21% O_2/1% CO_2	117.0 b
42% O_2/175 μl CO_2/l	15.7 c
42% O_2/3% CO_2	98.5 b
Expt. 2. Growth 7 days	
21% O_2/160 μl CO_2/l	29.9 a
21% O_2/1% CO_2	72.1 b
42% O_2/160 μl CO_2/l	18.3 c
42% O_2/1% CO_2	79.3 b

on whole plants or leaf discs.

Photosynthesis Measurements

Whole potted plants were placed in a plexiglas chamber in light and flushed continuously with a gas mixture of known composition until steady-state photosynthetic rates were established. This was determined by briefly closing off the chamber and determination of CO_2 depletion rate by removing samples periodically with a syringe and assaying their CO_2 content (7). The CO_2 concentration of the flushing gas was adjusted to about 380 μl CO_2/l so that the average concentration during the CO_2 depletion assay was about 350 μl CO_2/l. A similar system was used for determining photosynthetic rates on leaf discs (1.6 cm diameter) floated on water. At least five determinations of CO_2 assimilation rate were made before the flushing gas was changed and the steady-state photosynthesis rate was determined again under the new conditions. The sequence followed was 21% O_2, 42% O_2, and again 21% O_2. The second set of determinations at 21% O_2 established that the changes in photosynthesis at 1% and 42% O_2 were completely reversible.

Of the 26 O_2-resistant haploid plantlets obtained by screening in 42% O_2/160 μl CO_2/l, five showed photosynthetic characteristics different from wild type. The results on four of these plants, which were obtained on at least two different sets of experiments conducted within a two-week period, are summarized in Table 2. With increasing O_2 concentration the O_2-resistant plants show a progressive increase in net photosynthesis relative to the wild type plantlets. Plantlet 42-12, for example, displayed a 6% increase in photosynthesis compared to a wild type plantlet at 1% O_2, a 12% increase at 21% O_2 ($\underline{P}<0.01$), and a 28% increase at 42% O_2 ($\underline{P}<0.01$). The enhanced photosynthesis is observed most clearly at 42% O_2 and the percent inhibition at 42% compared to 21% O_2 is generally less than the inhibition at 21% compared to 1% O_2, while these values are nearly the same for wild-type plantlets. Net photosynthesis of O_2-

TABLE 2. Effect of O_2 Concentration on Net Photosynthesis of O_2-Resistant Haploid Tobacco Plantlets. Light 300 μE $m^{-2}s^{-1}$, 350 μl CO_2/l, 30°. Values in parentheses are percent differences in mean net photosynthesis of O_2-resistant plantlets (OR) compared to wild types (WT); *, $\underline{P}<0.05$; **, $\underline{P}<0.01$; [a], diploid plant; [b], leaf discs (1.6 cm).

Plantlets Compared	No. of Expts.	O_2 Concentration 1%	21%	42%	% Inhibition 21% vs 1% O_2	42% vs 21% O_2
		means, mg $CO_2/dm^2 \cdot h$				
42-12 (OR)	3	16.1(+6)	10.8(+13)**	7.8(+28)**	33	28
42-13 (WT)		15.2	9.6	6.1	37	36
42-16 (OR) [a]	2	13.2(-1)	8.5(+12)*	4.7(+12)*	36	45
42-17 (WT)		13.3	7.6	4.2	43	45
42-41 (OR)	2	14.9(-7)**	9.5(-8)**	6.8(0)	36	28
42-42 (WT)		16.1	10.3	6.8	36	34
42-1 (OR) [b]	3	15.3(-12)**	9.0(-11)**	7.1(+1)	41	21
42-2 (WT) [b]		17.4	10.1	7.0	42	31

resistant plantlets at 21% O_2 was sometimes greater than wild type and sometimes less, suggesting that to express higher rates per leaf area O_2-resistant photosynthesis must be present in a suitable genetic background.

Biochemical and genetic studies on the mechanisms of O_2-resistant photosynthesis will require fertile diploid plants. Midvein cells of older haploid leaves are diploid, and it is possible to regenerate diploid plants using midvein culture (4). Utilizing this method, we regenerated 898 putative diploid plants from haploid plants with O_2-resistant photosynthesis, but only 15 of these (1.7%) displayed O_2-resistant growth in 42% $O_2/160$ μl CO_2/l. The low frequency of O_2-resistant plantlets regenerated from midveins of O_2-resistant haploids suggests that this phenotype reflects an epigenetic phenomenon that is not transmitted efficiently through the dedifferentiation and shoot morphogenesis stages of midvein culture. Attempts are underway to produce fertile diploid shoots on O_2-resistant haploids using colchicine treatment of floral buds, in order to examine sexual transmission of the O_2-resistant phenotype.

Of 13 O_2-resistant plants tested for O_2-resistant photosynthesis after midvein culture, one displayed characteristics vastly different from a wild type plant obtained from midvein culture. Table 3 shows the photosynthetic characteristics of this plant, 42-19A. It showed superior rates of CO_2 exchange at all O_2 levels, and net photosynthesis increased progressively relative to wild type as the O_2 concentration was raised to 42%. The results are similar to those in Table 2 but more striking.

Discussion

The idea that plants with regulated photorespiration might be

TABLE 3. Effect of O_2 Concentration on Net Photosynthesis of an O_2-Resistant Plant Obtained from Leaf Midvein Culture of a Haploid Plant. Light 300 μE $m^{-2}s^{-1}$, 350 μl CO_2/l, 30°. Values in parentheses are percent differences in mean net photosynthesis of the O_2-resistant plant (OR) compared to the wild type (WT); **, $P<0.01$

| | | | | % Inhibition | |
| | O_2 Concentration | | | 21% vs | 42% vs |
Plants Compared	1%	21%	42%	1% O_2	21% O_2
	means, mg $CO_2/dm^2 \cdot h$				
42-19A (OR)	12.5(+14)**	7.5(+25)**	5.1(+46)**	40	32
42-19B (WT)	11.0	6.0	3.5	45	42

selected by obtaining O_2-resistant photosynthesis (1) has recently been supported by strong evidence that the stoichiometry of photorespiratory CO_2 released per glycolate carbon metabolized is not fixed at the 25% value usually given ((8); see Hanson and Peterson these volumes). With increasing temperature and O_2 levels the stoichiometry can greatly exceed 25% (Table 4). It is reasonable to suppose, therefore, that selection for O_2-resistant growth under appropriate conditions may identify mutations that diminish photorespiration by virtue of lower stoichiometry. The characteristics of O_2-resistant photosynthesis shown in Tables 2 and 3 are consistent with such a mechanism.

TABLE 4. Effect of O_2 and CO_2 Concentration on the Ratio of Photorespiration to Net Photosynthesis and the Stoichiometry of Photorespiration in Inverted Leaf Discs at 32° (based on (8)).

| O_2 Conc. | CO_2 Conc. | Ratio | Photorespiration / Net Photosynthesis | Photorespiration Stoichiometry | | |
				Expected	Excess	Total
%	$\mu l/l$		%	% glycolate C metabolized to CO_2		
21	340		28	25	13	38
42	340		100	25	30	55

Acknowledgments. I wish to thank Jean Pillo for technical assistance.
References
1 Zelitch, I. (1982) BioScience 32, 796-802
2 Heichel, G.H. (1973) Plant Physiol. 51S, 42
3 Zelitch, I. (1984) in Advances in Photosynthesis Research (Sybesma, C., ed.), Vol. 3, pp. 811-816, Martinus Nijhoff/Dr. W. Junk, The Hague
4 Kasperbauer, M.J. and Collins, G.B. (1972) Crop Sci. 12, 98-101
5 Anagnostakis, S.L. (1974) Planta 115, 281-283
6 McHale, N.A. (1985) Plant Physiol. 77, 240-242
7 Peterson, R.B. and Zelitch, I. (1982) Plant Physiol. 70, 677-685
8 Hanson, K.R. and Peterson, R.B. (1986) Arch. Biochem. Biophys. 246, 332-346

THE ISOLATION AND CHARACTERISATION OF PHOTORESPIRATORY MUTANTS OF BARLEY
AND PEA

BLACKWELL, R.D., MURRAY, A.J.S. and LEA, P.J. Department of Biological
Sciences, University of Lancaster LA1 4YQ, U.K.

1. INTRODUCTION
    Mutants of Arabidopsis unable to carry out photorespiration were first
isolated by Somerville and Ogren in 1979 [1]. Since that time the
Rothamsted group have isolated five similar mutants in barley plus lines
deficient in glutamine synthetase and glutamate synthase [2 and Kendall et
al., this volume]. In December 1985, a programme isolating mutants of pea
and barley only viable in 0.7% CO_2 was instituted at Lancaster and to date
fourteen have been characterised.

2. RESULTS AND DISCUSSION
2.1 Glutamine synthetase:
    Four barley mutants are deficient in chloroplast glutamine synthetase
but contain normal levels of the cytoplasm isoenzyme, when exposed to air
the leaves produce ammonia at high rates. By plotting the log of the rate
of ammonia production against the glutamine synthetase activity and
extrapolating to zero glutamine synthetase, a maximum rate of ammonia
production of 140 μ mol h^{-1} g^{-1} fresh weight was obtained. This value
should be equivalent to the rate of photorespiration and was 41% of the
rate of CO_2 fixation in air.

2.2 Ferredoxin dependent glutamate synthase:
    Four barley mutants accumulated ammonia in air but had higher than
normal levels of glutamine synthetase. These mutants were shown to contain
less than 5% ferredoxin dependent glutamate synthase activity and
accumulated high levels of glutamine. Two additional mutants accumulated
glutamine but had normal levels of ammonia and glutamate synthase activity.
Further work is required to establish whether these mutants have impaired
2-oxoglutarate transport into the chloroplast.

2.3 Serine : glyoxylate aminotransferase:
    One mutant plant of barley (designated LaPr 85/84) that accumulated
high levels of serine on transfer to air was further characterised.

TABLE 1. Aminotransferase activity of cv. Maris Mink (M.M.) and LaPr 85/84
         leaf extracts, expressed in nmol hr^{-1}.

Amino acid Donor	Keto Acid Acceptor					
	Glyoxylate		Hydroxypyruvate		Pyruvate	
	M.M.	85/84	M.M.	85/84	M.M.	85/84
Glutamate	841	826	<30	<30	2560	2430
Alanine	924	918	340	<30	--	--
Serine	806	<30	--	--	886	<30
Asparagine	420	<30	611	<30	745	<30
Glycine	--	--	180	<30	<30	<30

In LaPr 85/84, activities of serine : glyoxylate aminotranferase were negligible (Table 1), activities of glutamate : glyoxylate, glutamate : pyruvate and alanine : glyoxylate aminotransferases were normal. In the mutant all other aminotransferase activities were below the limits of detection of the assay used. It has been suggested by Joy and his colleagues [3,4] that asparagine plays a major role in photorespiratory nitrogen cycling and can act as a donor for glycine synthesis via the enzyme serine : glyoxylate aminotransferase. This hypothesis is strengthened by the substrate specificity shown in Table 1 and clearly the asparagine metabolism of LaPr 85/84 would be worthy of further study.

Serine : glyoxylate aminotransferase isolated from leaves of wild type barley (cv. Maris Mink) can be separated into two major peaks of activity, plus a third variable minor peak, by DEAE – Sephacel chromatography (Fig.1). After chromatography of a similar extract isolated from LaPr 85/84, none of the peaks of activity could be detected. Marker activities of two other enzymes involved in photorespiration, hydroxypyruvate reductase and glycerate kinase were similar in both wild type and mutant leaves (Fig.1).

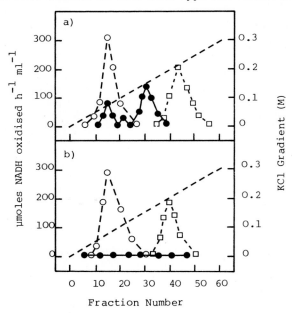

Fig.1. Serine : glyoxylate aminotransferase (●——●), Hydroxypyruvate reductase (○– –○) and glycerate kinase (□··□) activities after DEAE-Sephacel chromatography for (a) cv. Maris Mink (b) LaPr 85/84.

Leaves of LaPr 85/84 exhibited the same rates of CO_2 fixation as wild type leaves when placed in 1% O_2, 340 µl l^{-1} CO_2, but the rate of the mutant fell to 30% of that of the wild type upon transfer to air. Some recovery was observed on transfer back to 1% O_2 (see Fig.2).

As glutamate : glyoxylate aminotransferase is present at normal levels it would be expected that at least half the glyoxylate would be metabolised through glycine and accumulate as serine. However, [1-^{14}C]-Glyoxylate was metabolised to [^{14}C]O_2 at twice the rate in the mutant leaves when compared

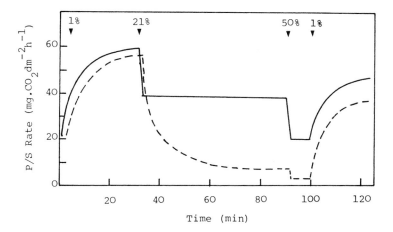

Fig.2. The rates of CO_2 assimilation of barley, cv. Maris Mink (——) and
LaPr 85/84 (— — —). Changes of oxygen concentration (▼) against a background
of 340 μl l^{-1} CO_2 and nitrogen are marked.

to that of the wild type. Amino acid analysis of leaves of LaPr 85/84
exposed to air carried out by Professor K.W.Joy of the University of
Carleton,Canada indicated that the soluble nitrogen accumulated in serine
and that the levels of glutamate, alanine, glutamine, glycine and aspartate
were much lower, (Fig.3).

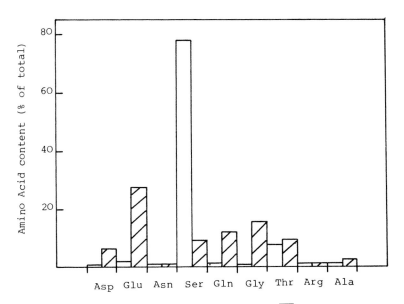

Fig.3. Amino acid accumulation of LaPr 85/84 (☐) and cv. Maris Mink (▨),
after 20 hours exposure to air.

These results suggest that glyoxylate is not converted to glycine but is oxidised directly to CO_2 and formate. Such a pathway was originally proposed for photorespiratory CO_2 release and is now only considered to occur in the absence of amino donors [5,6]. The genetic basis of the mutation is presently being studied.

2.4 Glycine : serine conversion :
      Two mutant plants of barley were identified as containing high concentrations of glycine and low concentrations of serine and glutamate. One plant was later shown to be infertile (as has happened on a previous occasion [2]), but the second (designated LaPr 85/55) was self fertilised and produced M_3 seed. One quarter of the progeny were albino, but the remaining seedlings were green and all those tested accumulated glycine.
      Leaves of LaPr 85/55 converted [^{14}C]-glycine to [^{14}C]O_2 at only 20% of the rate of cv. Maris Mink. Ammonia evolution when leaves of LaPr 85/55 were treated with methionine sulphoximine in the air was only 1% of the wild type control. Activities of serine hydroxymethyltransferase as measured by a modified assay of Taylor and Weissbach [7] were normal in extracts of the leaves of LaPr 85/55, suggesting that the mutation is in the glycine decarboxylase complex [8].

2.5 Pea mutations : On a somewhat reduced scale a number of pea mutants have been selected. In one plant (designated LaPr 85/33), ferredoxin dependent glutamate synthase activity was not detectable and the only soluble amino acid present after exposure to air was glutamine. This mutation will be of interest in examining the nitrogen fixing capacity of root nodules with reduced plant glutamate synthase.

3. CONCLUSIONS
      At the present time we have available 312 barley and 65 pea plants that are unable to grow in normal air but are viable at 0.7% CO_2. So far the biochemical basis of the mutations has only been identified in a very small number of cases. We have established that the inability to grow in air is not due to a deficiency of catalase, glutamate : glyoxylate aminotransferase, phosphoglycollate phosphatase, hydroxypyruvate reductase or glycerate kinase. It would appear that there are other key enzymes and processes other than those involved in photorespiration that are required for normal growth at 0.03% CO_2.

REFERENCES
1 Somerville, C.R. (1984) in Oxford Surveys of Plant Molecular and Cell
   Biology (Miflin, B.J., ed.), Vol.1, pp.103-131, Oxford University Press
2 Kendall, A.C., Hall, N.P., Keys, A.J., Lea, P.J., Turner, J.C. and
   Wallsgrove, R.M. (1986) in Biological Control of Photosynthesis (Marcelle,
   R. et al., eds.), pp.257-265, Martinus Nijhoff Publishers, Dordrecht
3 Ireland, R.J. and Joy, K.W. (1983) Arch.Biochem.Biophys. 223, 291-296
4 Ta, T.C. and Joy, K.W. (1985)Can.J.Bot. 63, 881-884
5 Grodzinski, B. (1978) Planta 144, 31-37
6 Walton, N.J. and Butt, V.S. (1981) Planta 153, 232-237
7 Taylor, R.T. and Weissbach, H. (1968) Anal.Biochem. 13, 80-84
8 Walker, J.L. and Oliver, D.J. (1986) J.Biol.Chem. 261, 2214-2221

ACKNOWLEDGEMENTS
      This research was supported by the AFRC. P.J.L. is grateful to his ex-colleagues at Rothamsted and to Bill Blackledge at Lancaster for help and guidance.

BARLEY PHOTORESPIRATION MUTANTS

KENDALL, A.C., BRIGHT, S.W.J., HALL, N.P., KEYS, A.J., LEA, P.J.*, TURNER, J.C. and WALLSGROVE, R.M.

BIOCHEMISTRY DEPARTMENT, ROTHAMSTED EXPERIMENTAL STATION, HARPENDEN, HERTS. AL5 2JQ, U.K.
*Present address DEPARTMENT OF APPLIED BIOLOGY, UNIVERSITY OF LANCASTER, LANCASTER, LA1 4YQ, U.K.

1. INTRODUCTION
     Photorespiration, the light- and O_2-dependent release of CO_2 from the leaves of certain species, is regarded as the inevitable consequence of the oxygenase activity of ribulose bisphosphate carboxylase/oxygenase, Rubisco [1]. Some authors have suggested that, inevitable or not, the process may have a useful function [2]. It is an apparently wasteful process: 25% or more of fixed CO_2 may be lost and energy is consumed [3]. The metabolic routes involved in the process have long been controversial, particularly regarding the source of photorespiratory CO_2 and the reassimilation of photorespiratory NH_3 [4]. Mutants of *Arabidopsis thaliana* with defects at seven steps in photorespiratory metabolism provided evidence to settle some of these controversies [5]. We have isolated a series of barley (*Hordeum vulgare* L.) mutants, with lesions at additional sites, which provide further insights into photorespiratory metabolism.

2. PROCEDURE
     Materials and methods: Methods for mutant selection, gas-exchange analysis, measurement of ammonia and enzyme assays have all been described in [6].

3. RESULTS AND DISCUSSION
     From more than 120,000 M_2 plants derived from azide-treated barley (cv. Maris Mink) we have obtained about 50 air-sensitive lines, and identified lesions at seven different sites in the photorespiratory pathway (Fig. 1). Deficiencies in the following enzymes were found: catalase [7], 6 lines; ferredoxin-dependent glutamate synthase, Fd-GOGAT [8], 6 lines; chloroplast 2-oxoglutarate uptake [9], 3 lines; phosphoglycolate phosphatase [10], 2 lines; glycine-serine conversion [11], 1 line, not fully characterised (the plants have never produced ears); chloroplast glutamine synthetase, GS_2 [6,12]; low activity of ribulose bisphosphate carboxylase [13]. In all these cases (except that of the infertile glycine/serine mutant) segregation ratios indicate the presence of single nuclear gene mutations. For different lines having the same lesion all mutants so far examined have been allelic.

     Work continues on the identification of other air-sensitive lines. Preliminary results on one of these showed it had only half the wild-type activity of chloroplastic and cytoplasmic superoxide dismutase. A mutant deficient in serine:glyoxylate aminotransferase activity has recently been isolated (see P.J. Lea *et al.*, this meeting).

Biggens, J. (ed.), Progress in Photosynthesis Research, Vol. III. ISBN 90 247 3452 5
© *1987 Martinus Nijhoff Publishers, Dordrecht. Printed in the Netherlands.*

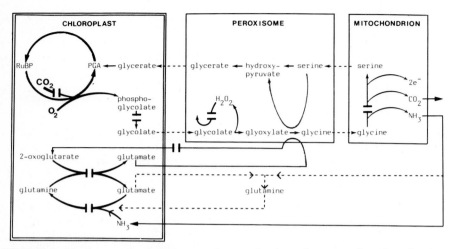

FIGURE 1. Carbon and nitrogen pathways during photorespiration in C_3 plants. Positions of known barley mutations are marked

The phosphoglycolate phosphatase mutant has been used to demonstrate the existence of isoforms of this enzyme [10], confirming earlier work in *Phaseolus vulgaris* [14]. Fig. 2 shows the light response of CO_2 fixation in this mutant under non-photorespiratory conditions. At irradiances above 200 µmol quanta $m^{-2}s^{-1}$ photosynthesis was strongly inhibited: this effect was not reported for the *Arabidopsis* mutant [15], presumably because of the relatively low irradiance (300 µmol quanta $m^{-2}s^{-1}$) used in the measurements. Phosphoglycolate is a potent inhibitor of triose phosphate isomerase *in vitro* [16]. Our results therefore indicate the observed inhibition of photosynthesis is due to the accumulation of phosphoglycolate caused by transient decreases in growth CO_2 concentrations, as we suggested in [10], or that phosphoglycolate production can be sustained even at 2.5% O_2 and 1000 µl l^{-1} CO_2.

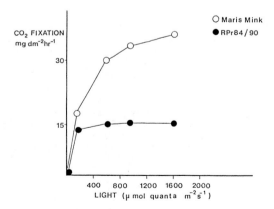

FIGURE 2. Light response of CO_2 fixation in phosphoglycolate phosphatase mutant (2.5% O_2, 1000 µl l^{-1} CO_2).

We have previously reported preliminary results with the Rubisco-deficient barley mutant [6,13]. The mutant had a low photosynthetic rate (about 35% of wild-type) under non-photorespiratory conditions. Rubisco protein was 20-30% of total soluble protein (cf. 50-60% in wild-type). Rubisco activity was 25-30% that of Maris Mink, both immediately after extraction and after HCO_3^- activation: this difference was much less pronounced in older leaves, suggesting impairment of Rubisco synthesis in the mutant.

Mutants lacking GS_2 and Fd-GOGAT have provided conclusive evidence for a photorespiratory nitrogen cycle as proposed in [17] and for the role of chloroplast GS rather than cytoplasmic GS in ammonia reassimilation. Fig. 3 shows the effects of these mutations on CO_2 fixation and NH_3 accumulation, together with that of a double mutant selected from the progeny of a GS_2 x GOGAT mutant cross. The rapid decreases in photosynthesis are not a direct consequence of increased ammonia accumulation and uncoupling of photophosphorylation: similar conclusions have been drawn from work using the GS inhibitor methionine sulphoximine [18]. Under non-photorespiratory conditions all these plants thrive, indicating that the major (and essential) role of GS_2 and Fd-GOGAT is the reassimilation of photorespiratory NH_3.

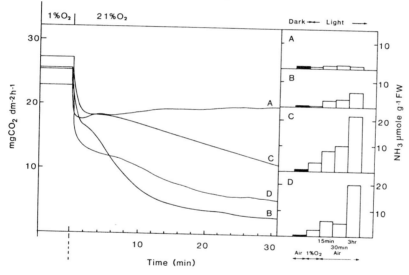

FIGURE 3. CO_2-fixation rate and ammonia accumulation in detached leaves. Leaf sections were illuminated (1000μmol quanta $m^{-2}s^{-1}$) for 1 hour in 1% O_2, 350μl l^{-1} CO_2, and then the gas stream was switched to 21% O_2, 350μl l^{-1} CO_2. At intervals leaf sections were removed and frozen in liquid N_2 prior to ammonia determination.

A – Maris Mink

B – GOGAT mutant

C – GS_2 mutant

D – GS x GOGAT (double mutant)

4. CONCLUSIONS

A range of photorespiration mutants has been isolated in barley, with
some notable differences from that found in *Arabidopsis*. Genetic analysis
has shown that many of the enzymes involved in photorespiration are
encoded by single nuclear genes, whose function is dispensable in
non-photorespiratory conditions. Such mutants may be used both to clarify
controversial aspects of metabolism and to provide starting material for
further investigations into photorespiratory carbon and nitrogen
metabolism and photosynthesis.

REFERENCES
 1 Miziorko, H.M. and Lorimer, G.H. (1983) Ann. Rev. Biochem. 52, 507-535.
 2 Keys, A.J., Bird, I.F. and Cornelius, M.J. (1982) in Chemical
   Manipulation of Crop Growth and Development (McClaren, J.S., ed.), pp.
   39-53, Butterworths, London.
 3 Keys, A.J. (1983) Pestic. Sci. 19, 313-316.
 4 Singh, P., Kumar, A.P., Abrol, Y.P. and Naik, M.S. (1985) Physiol.
   Plant. 66, 169-176.
 5 Somerville, C.R. and Ogren, W.L. (1982) TIBS 7, 171-174.
 6 Kendall, A.C., Hall, N.P., Keys, A.J., Lea, P.J., Turner, J.C. and
   Wallsgrove, R.M. (1986) in Biological Control of Photosynthesis
   (Marcelle, R., Clijsters, H and Van Poucke, M., eds.), pp. 257-265,
   Martinus Nijhoff, Dordrecht.
 7 Kendall, A.C., Keys, A.J., Turner, J.C., Lea, P.J. and Miflin, B.J.
   (1983) Planta 159, 505-511.
 8 Kendall, A.C., Wallsgrove, R.M., Hall, N.P.,Turner, J.C. and Lea, P.J.
   (1986) Planta 168, in press.
 9 Wallsgrove, R.M., Kendall, A.C., Hall, N.P., Turner, J.C. and Lea, P.J.
   (1986) Planta 168, in press.
10 Hall, N.P., Kendall, A.C., Lea, P.J., Turner, J.C. and Wallsgrove, R.M.
   (1986) Photosynth. Res., in press.
11 Lea, P.J., Kendall, A.C., Keys, A.J. and Turner, J.C. (1984) Plant
   Physiol. 75S, 881.
12 Wallsgrove, R.M., Kendall, A.C., Turner, J.C. and Hall, N.P. (1986)
   Plant Physiol. 80S, 283.
13 Hall, N.P., Kendall, A.C., Turner, J.C., Wallsgrove, R.M. and Keys,
   A.J. (1986) Plant Physiol. 80S, 281.
14 Verin-Vergeau, C., Baldy, P. and Cavalié, G. (1979) Phytochem. 18,
   1279-1282.
15 Somerville, C.R. and Ogren, W.L. (1979) Nature 280, 833-836.
16 Anderson, L.E. (1971) Biochim. Biophys. Acta. 235, 237-244.
17 Keys, A.J., Bird, I.F., Cornelius, M.J., Lea, P.J., Wallsgrove, R.M.
   and Miflin, B.J. (1978) Nature 275, 741-743.
18 Ikeda, M., Ogren, W.L. and Hageman, R.H. (1984) Plant & Cell Physiol.
   25, 447-452.

INHERITANCE OF CARBON ISOTOPE DISCRIMINATION IN <u>FLAVERIA</u>

DENNIS KEEFE AND LAURENS METS, DEPARTMENT OF MOLECULAR
GENETICS AND CELL BIOLOGY, UNIVERSITY OF CHICAGO, CHICAGO, IL
60637, USA.

INTRODUCTION

The carbon isotope composition of vascular plants is often used as a primary criterion for distingushing between plants with the C_3 and C_4 pathways of photosynthetic carbon fixation. In general, C_3 plants discriminate more against ^{13}C than C_4 plants. The carbon isotope composition, expressed on the standard $\delta\ ^{13}C$ scale, of C_3 plants is typically -28 o/oo while, C_4 plants generally have $\delta\ ^{13}C$ values of -14 o/oo (1). We are interested in determining the genetic control of the difference in isotope composition between C_3 and C_4 plants. Specifically, we are interested in determining whether there are any uniparental effects influencing the trait, characterization of the dominance relationship of alleles at relevant loci and obtaining an estimate of the number of independently segregating factors influencing the $\delta\ ^{13}C$ value. To address these questions we have made hybrids between C_3 and C_4 individuals from two closely related species within the genus <u>Flaveria</u>. Hybridizations were made between <u>F. brownii</u> (C_4) and <u>F. floridana</u> (C_3) (2,3 and 4). To determine if there are any uniparental effects influencing the $\delta\ ^{13}C$ value, reciprocal F_1 hybridizations were made. To obtain an estimate of the number of genetic factors affecting the $\delta\ ^{13}C$ value in the C_4 parent we have made backcross hybridizations between one of the F_1 hybrids and <u>F. brownii</u>.

MATERIALS AND METHODS

All plants were grown on open benches in the greenhouse facilities at the University of Chicago. <u>F. brownii</u> clone B-6 and <u>F. floridana</u> clone #15 were collected from the field in 1982. Clones of these two individuals and all F_1 hybrids have been maintained by propagation of terminal shoot cuttings.

For hybridizations the parental plants were induced to flower by growth under10 hour photoperiods. To prevent contamination from stray pollen the inflorescences of both the maternal and paternal parents were enclosed in bags constructed of "no-see-um" netting and velcro strips. Hybridizations were performed by applying pollen to

Biggens, J. (ed.), Progress in Photosynthesis Research, Vol. III. ISBN 90 247 3452 5
© *1987 Martinus Nijhoff Publishers, Dordrecht. Printed in the Netherlands.*

emasculated maternal inflorescences. Pollinated infloresescenes were allowed to mature, their seeds collected, dried, and germinated. Reciprocal F_1 hybrids were made between F. brownii clone B-6 and F. floridana clone #15. F_1 clone #535 is the product of a cross in which F. floridana #15 was the maternal parent. Reciprocal backcross hybridizations were made between F_1 #535 and F. brownii B-6.

The hybrid nature of all F_1 hybrids were verified by either zymogram or DNA restriction fragment polymorphism analysis. All F_1 hybrids analyzed in this manner were of hybrid origin (data not shown). Our success in making F_1 hybrids gives us confidence that the backcross hybridizations are also successful.

Carbon isotope ratios were determined in triplicate by mass spectrometry on samples from fully expanded leaf tissue.

RESULTS AND DISCUSSION

The $\delta\,^{13}C$ values of the two parental lines F. brownii B-6 (-16.1 o/oo) and F. floridana #15 (-26.6 o/oo) agree with previously reported values for these species (5). The mean $\delta\,^{13}C$ values of the parental lines differs significantly, suggesting a strong genetic control of the difference between these two species. The mean $\delta\,^{13}C$ value of the reciprocal F_1 hybrids are intermediate between their parents and do not differ significantly from each other (-21.8 o/oo and -21.9 o/oo). This result suggests that there are no major uniparental factors influencing the $\delta\,^{13}C$ value and that most likely the alleles at relevant loci have equal and additive effects. The reciprocal backcross hybrids $\delta\,^{13}C$ values (-18.9 o/oo and -19.9 o/oo) are again intermediate between their parents (F. brownii B-6, -16.1 o/oo and F_1 #535, -22.0 o/oo) and do not differ significantly from each other. This result further supports the conclusion that there are no uniparental factors influencing the $\delta\,^{13}C$ value and that the alleles at relevant loci have equal and additive effects.

Our results contrast with previous studies on the inheritance of δ ^{13}C values. Studies in Atriplex involving unidirectional crosses between A. rosea (C_4) and A. patula (C_3) in which the C_4 plant was the maternal parent resulted in F_1 hybrids with $\delta\,^{13}C$ values similar to the C_3 parent (6). This result suggests that there is dominance of "C_3 like" alleles which cause greater isotope discrimination. Unfortunately, this conclusion could not be tested directly because the reciprocal cross could not be made due to technical difficulties. In another study, reciprocal F_1 hybrids were made between Flaveria palmeri (C_4) and F.

<u>anomala</u> ("C_3-C_4 intermediate") (5).   The δ ^{13}C values of these hybrids were similar to their maternal parent, suggesting a strong maternal influence on the δ ^{13}C value.   These results must be interpreted with caution because no independent test was performed to verify the hybrid nature of these F_1 hybrids.

The lack of a bimodal distribution of the δ ^{13}C values suggests that no single genetic factor can account for the difference in δ ^{13}C value between $\underline{F}$. <u>brownii</u> and $\underline{F}$. <u>floridana</u>.   We are interested in obtaining an estimate of the minimum number of genetic factors contributing to this difference in δ ^{13}C phenotypes between these two species.   An estimate can be made from the magnitude of the genetic variance in the backcross generation (7).   We assume that $\underline{F}$. <u>brownii</u> B-6 is homozygous at relevant loci and that the alleles at these loci have equal and additive effects.   In addition, we assume that the alleles from $\underline{F}$. <u>brownii</u> act to decrease while those from $\underline{F}$. <u>floridana</u> act to increase isotope discrimination.   We also assume disomic segregation during meiosis in the hybrids.

Given these assumptions, there is a direct relationship between the genetic differences between $\underline{F}$. <u>brownii</u> B-6 and $\underline{F}$. <u>floridana</u> #15, and the phenotypic variance of the δ ^{13}C values observed in the backcross generation.   We have used two related methods for estimating the minimum number of independently segregating genetic factors, N_E.

The first method is designed to give an estimate of N_E and is modified from Lande (7).   We can obtain an estimate of the genetic variance (σ_s^2) by subtracting an estimate of the non-genetic sources of variance from the variance observed in the backcross generation.

$$\sigma_s^2 = \sigma_B^2 - (\sigma_{P1}^2 + \sigma_{F1}^2)^2 \qquad\qquad [1]$$

where, σ_{P1}^2, σ_{F1}^2 and σ_B^2 are the observed variances of $\underline{F}$. <u>brownii</u> B-6 $(\sigma_{P1}^2 = 0.42)$, F_1 #535 $(\sigma_{F1}^2 = 0.82)$ and the backcross generation $(\sigma_B^2 = 2.67)$ (7).   By substitution we obtain an estimate of 4.1 for σ_s^2.   An estimate of N_E can then be obtained from:

$$N_E = \frac{(\mu_{P1} - \mu_{P2})^2}{8\sigma_s^2} \qquad\qquad [2]$$

where, μ_{P1} and μ_{P2} are estimates of the mean $\delta^{13}C$ values of the original parents F. brownii B-6 (-16.1 o/oo)and F. floridana #15 (-26.6 o/oo). We obtain an estimate for N_E of 3.4 with a standard error of 1.5.

The second method for estimating N_E makes a prediction of the expected variance in the backcross generation ($\sigma_{B\ exp}$) from a given value of N_E. The 95% confidence interval of the expected variance in the backcross generation is then compared to the observed variance. This method allows for the elimination of a minimum number of independently segregating genetic factors that can account for the observed variance. The expected variance in the backcross generation can be obtained from:

$$\sigma_{B\ exp} = \sigma_{F1} + \frac{(\mu_{F1} - \mu_{P1})^2}{4N_E} \qquad [3]$$

where, μ_{P1} and μ_{F1} are estimates of the mean $\delta^{13}C$ value of F. brownii B-6 (-16.1 o/oo) and F_1 #535 (-22.0 o/oo). By substituting values for N_E we obtain predictions of $\sigma_{B\ exp}$. Comparing the 95% confidence interval of these expected values with the observed variance in the backcross generation we conclude with statistical confidence that more than two independently segregating genetic factors are required to account for the observed variance of $\delta^{13}C$ values in the backcross generation.

Our results allow the following conclusions about the genetic control of the $\delta^{13}C$ value: 1) there are no strong uniparental factors influencing the trait, 2) the alleles at relevant loci have equal and additive effects, and 3) there are at least 3 independently segregating genetic factors contributing to the difference in $\delta^{13}C$ between Flaveria brownii and F. floridana.

REFERENCES
1. O'Leary, M. H. (1981) Phytochem. 20:553-567.
2. Powell, A. M. (1978) Ann. Mo. Bot. Gard. 65:590-636.
3. Bauwe, H. (1984) Biochem. Physiol. Pflanz. (1984) 179:253-268.
4. Reed, J. E. and R. Chollet (1985) Planta 165:439-445.
5. Smith, B. N. and A. M. Powell (1984) Naturwiss. 71:217-218.
6. Boyton, J. E., M. A. Nobs and J. A. Berry (1970) Carn. Inst. Wash. Yearb. 70:507-511.
7. Lande, R. (1981) Genetics 99:544-553.

PHOTOSYNTHETIC CHARACTERISTICS OF RECIPROCAL F_1 HYBRIDS
BETWEEN C_3-C_4 INTERMEDIATE AND C_4 <u>FLAVERIA</u> SPECIES

Shu-Hua Cheng[1], Vincent, R. Franceschi[1], Dennis Keefe[2], Laurens
J. Mets[2], and Maurice S. B. Ku[1]

[1]Dept. of Botany, Washington State Univ., Pullamn, WA 99163
[2]Dept. of Molecular and Cell Biology, Univ. of Chicago,
Chicago, Chicago, IL 60637

1. INTRODUCTION
    Artificial hybrids have been made between plants of
different photosynthetic groups in an attempt to reduce
photorespiration by the introduction of C_4 traits inot C_3
<u>Atriplex</u> (1). The genus <u>Flaveria</u>, which contains C_3, C_4, and
many C_3-C_4 intermediate species (2), is an excellent candidate
for generating different photosynthetic hybrids due to close
phylogenetic relationships among these species. In this study
reciprocal hybrids between <u>F. floridana</u>, a C_3-C_4 intermediate,
and <u>F. brownii</u>, a C_4 plant, were generated by artificial
hybridization and their photosynthetic features examined. The
reciprocal hybrids allow us to begin evaluating the relative
contribution of the cytoplasmic and nuclear genomes to the
expression of C_4 characteristics and the control of
photorespiration.

2. MATERIALS AND METHODS

    <u>Plant Material</u>. Plants of <u>Flaveria brownii</u> A. M. Powell
(clone B6), <u>F. floridana</u> J. R. Johnson (clone #15) as well as
their reciprocal F_1 hybrids <u>F. brownii</u> x <u>F. floridana</u> (clone #
436) and <u>F. floridana</u> x <u>F. brownii</u> (clone #577) were grown in
a growth chamber under conditions as described earlier (3).
Details on the nature of these hybrids plants will be given
elsewhere. Young expanded leaves from the third node were used
throughout all experiments.
    <u>Leaf Anatomy</u>. Tissues were fixed in 2% paraformaldehyde and
2.5% glutaraldehyde in 50 mM PIPES, pH 7.2, and postfixed in 1%
O_sO_4 in 25 mM cacodylate, pH 7.2. Fixed tissues were embedded
in Spurr's resin, cut at 1 μm thickness and stained with
Stevenel Blue.
    <u>Gas Exchange Measurements</u>. The CO_2 compensation point (τ)
was determined at 30°C and at various O_2 levels, using methods
described by (3).
    <u>Enzyme Assays</u>. Assays of C_4 key enzymes were similar to
those of (4).
    <u>Gel Electrophoresis</u>. SDS treated protein samples were
prepared from whole leaf extracts, and SDS-PAGE was performed
according to (5). Proteins were stained with Coomasie
brilliant blue R-250.
    $^{14}CO_2$ <u>Labelling</u>. Procedures for $^{14}CO_2$ labelling and
identification of ^{14}C-products were described in (4).

Biggens, J. (ed.), Progress in Photosynthesis Research, Vol. III. ISBN 90 247 3452 5
© *1987 Martinus Nijhoff Publishers, Dordrecht. Printed in the Netherlands.*

3. RESULTS

Light microscopic examination of leaf cross sections showed that both F_1 hybrids exhibit a distinct Kranz-like leaf anatomy , which is more differentiated than that of the intermediate parent (Fig. 1). Bundle sheath cells of both hybrids contain numerous organelles located in a centripetal position. However, of the two hybrids, F. floridana x F. brownii appeared to have more and larger chloroplasts in the bunlde sheath cells.

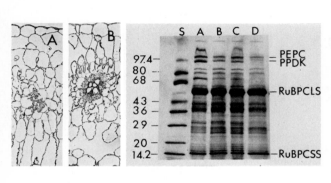

Left & Middle: Fig. 1 leaf cross sections of F_1 hybrids. A, F. brownii x F.floridana :B, F. floridana x F. brownii. Right: Fig. 2 SDS-PAGE of whole leaf soluble protein of various Flaveria species. S, protein standard; A,F.brownii; B, F. brownii x F. floridana; C. F. floridana x F.brownii ; D, F. floridana

Leaves of the reciprocal F_1 hybrids exhibited C_4-like Γ, ranging from 1.4 to 3.2 μl/l as (O_2) increased from 2 to 51% (Table I). There is no apparent difference between the reciprocal hybrids in their Γ at various O_2 levels. The O_2 insensitivity of Γ in the hybrids is similar to that of the C_4 parent, F. brownii, whereas the C_3-C_4 intermediate parent, F. floridana showed higher Γ which responded considerably more to changes in (O_2) . These results indicate that the apparent photorespiration in the F_1 hybrids is greatly reduced to a level similar to that in the C_4 parent.

Table I. Photosynthetic CO_2 compensation points at various (O_2) in F. brownii, F. floridana, and their reciprocal F_1 hybrids.

species	CO$_2$ compensation point		
	2% O$_2$	21% O$_2$	51% O$_2$
	μl/l		
F. brownii (C$_4$)	0.9	1.1	2.7
F. floridana (C$_3$-C$_4$)	4.1	8.3	-
F. brownii x F. floridana (F$_1$)	1.5	1.9	3.2
F. floridana x F. brownii (F$_1$)	1.4	1.6	3.0

Relative to the intermediate parent, both F_1 hybrids possess elevated activities of most key cycle enzymes, except pyruvate, Pi dikinase (Table II). Activities of pyruvate, Pi dikinase, NADP-malate dehydrogenase, and NADP-malic enzyme were similar in the reciprocal hybrids, but the activities of PEP carboxylase, aspartate and alanine aminotransferases were 30-60 % higher in F. floridana x F. brownii than in F. brownii x F. floridana. The analysis of leaf soluble protein by SDS-PAGE, however, revealed no significant differences in protein composition between the hybrids (Fig. 2). The polypeptides of PEP carboxylase and pyruvate,Pi dikinase did stain more intensely in the F_1 hybrids than in the intermediate parent.

The $^{14}CO_2$ pulse experiments showed that after 8 s of $^{14}CO_2$ assimilation, 66% of the total radioactivity incorporated by F. brownii x F. floridana was found in the C_4 acids malate and aspartate, with label being equally distributed between each (Fig. 3A). In F. floridana x F. brownii about 85% of the initial total ^{14}C-label appeared in C_4 acids, with aspartate being predominantly labelled (Fig. 3B). Under similar condition previous studies showed that 77% and 50-55% of the total ^{14}C-label were found in C_4 acids in the C_4 and intermediate parents (6,7), respectively. Therefore, the two hybrids have a higher overall capacity for C_4 photosynthesis than the intermediate parent. Very little label appeared in glycine plus serine in both hybrids, suggesting a further reduction of photorespiration. The turnover of C_4 acids during the $^{12}CO_2$ chase period was slower in both F_1 hybrids than in the C_4 parent. Furthermore, F. brownii x F. floridana showed a faster turnover rate of C_4 acids than F. floridana x F. brownii. Despite little apparent whole leaf photorespiration in both F_1 hybrids, the amount of ^{14}C-label in serine and glycine increased during the chase, and these metabolites became major end products after a 3 min chase. These results suggest that some photorespiratory activities exist in these hybrids, but the photorespired CO_2 must be refixed very efficiently.

Table II. Activities of key C_4 photosynthetic enzymes in F. brownii, F. floridana, and the reciprocal F_1 hybrids.

species	PEPC	NADP-MDH	NADP-ME	PPDK	ASP-AT	ALA-AT
			μmol/mg Chl·h			
F. brownii	588	576	761	217	827	804
F. floridana	202	110	103	44	639	301
F. brownii x F. floridana	385	565	276	62	705	466
F. floridana x F. brownii	620	526	275	45	923	721

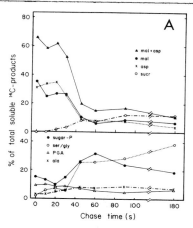
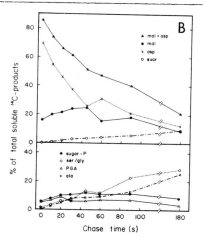

Fig. 3 % distribution of soluble ^{14}C-products. A, F. brownii x
F. floridana; B, F. floridana x F. brownii.

4. DISCUSSION

The results in this study demonstrate that genetic breeding
can produce plants with lower photorespiration, even though the
transfer of carbon from the C_4 to the C_3 cycle in the hybrids
is less efficient than in the C_4 parent. Several factors may
account for this slower transfer of carbon in the F_1 hybrids:
a lack of strict compartmentation of the photosynthetic enzymes
thus causing a futile C_4 cycle; different active pool sizes or
diffusion gradients of C_4 acids; and/or an increase in refixation of photorespiratory CO_2. No strict compartmentation of
PEP carboxylase and RuBP carboxylase was found in either the C_4
or the intermediate parents (6, 8, 9). Also NADP-malic enzyme
is present in mesophyll cells of F. brownii (6). Notably, the
carbon isotope ratios of the hybrids are between those of the
C_4 and intermediate parents (Keefe & Mets, unpublished data),
which tends to support that the hybrids lack a strict compartmentation of photosynthetic enzymes and possess a somewhat
"leaky" C_4 system, relative to the C_4 parent.
   The results show no significant difference between the reciprocal hybrids in leaf anatomy and apparent photorespiration,
suggesting no major cytoplasmic effect on these parameters of
C_4 photosynthesis. However, the low activities of pyruvate,Pi
dikinase in both hybrids, which are similar to those measured
for the intermediate parent, are inconsistent with the enhanced
synthesis of this enzyme in the hybrids as indicated by the degree of protein staining. One should note that the activity of
this enzyme is regulated by a protein factor (10), whose inheritance remains unknown. The reciprocal hybrids show distinct
differences in carbon partitioning between malate and aspartate
and in carbon transfer from the C_4 cycle to the C_3 cycle. The
biochemical and genetic bases for these differences is unknown.
Because the parent plants may be heterozygous for genetic
factors controlling these biochemistry features,there may be
genetic differences among the F_1 plants. Also, maternal genetic
factors may be responsible. Analysis of carbon metabolism in
additional F_1 hybrid plants will allow us to distinguish
between these two possibilities.

5. REFERENCES

1 Björkman, O. (1976) in CO$_2$ Metabolism and Plant Pro ductivity (Burris
   , R.H. & Black, C.C. eds.) Univ. Park Press, Boltimore,   pp287-309
2 Powell, A.M. (1978) Ann. Missouri Bot. Gard. 65:590-636
3 Rumpho, M.E., Ku, S.B., Cheng, S.-H. and Edwards, G.E. (1984) Plant
   Physiol. 75:993-996
4 Moore, B.d., Cheng, S.-H., Edwards, G.E. (1986) Plant Cell Physiol.
   in press
5 Chua, N.A. (1980) in Methods Enzymol. 69:590-636
6 Cheng, S.-H., Moore, B.d. and Ku, M.S.B.(1986) Plant Physiol. 81:S55
7 Monson, R.H., Moore, B.d.,Ku, M.S.B.& Edwards,G.E. Planta, in press
8 Bauwe, H. (1984) Biochem, Physiol. Pflanzen 179:253-268
9 Reed, J.E. and Chollet, R. (1985) Planta 165:439-445
10 Edwards, G.E., Nakamoto, H., Burnell, J.N. and Hatch, M.D. (1985)
   Ann. Rev. Plant Physiol. 36:255-286

EVIDENCE FOR INEFFICIENT C_4-CYCLE FUNCTION IN C_3-C_4 INTERMEDIATE FLAVERIA SPECIES: IMPLICATIONS FOR C_4 EVOLUTION

RUSSELL K. MONSON, DEPARTMENT OF EPO BIOLOGY, CAMPUS BOX 334, UNIVERSITY OF COLORADO, BOULDER, COLORADO, USA 80309

1. INTRODUCTION

Approximately twenty species of plants have been characterized as exhibiting photosynthetic and photorespiratory traits intermediate between the more typical C_3 and C_4 plants (1). In the genus Flaveria (Asteraceae) at least 30% of the species exhibit C_3-C_4 intermediate traits. In many of the Flaveria intermediates atmospheric CO_2 is assimilated through co-functioning C_3 and C_4 pathways (2,3). Two questions that arise concerning these intermediate traits are 1) how efficiently do the C_3 and C_4 pathways co-function in assimilating atmospheric CO_2 and, 2) what are the adaptive advantages of possessing these intermediate traits? In this paper I address these two questions, as well as their importance to a broader question of why C_4 photosynthesis has evolved in vascular plants.

2. MATERIALS AND METHODS

2.1 Plant Material

Plants were grown in growth chambers and greenhouses as previously described (2,3).

2.2 Pulse-chase studies and gas exchange measurements

Procedures for the pulse-chase experiments and gas exchange studies have also been described in detail in a previous report (3). Leaf nitrogen contents were measured using an elemental analyzer (Carlo Erba, 1106). The leaves were dried at 60°C for 72 h, ground in a mortar, and approx. 2 mg were packed into a tin combustion boat. The boats were then loaded into the elemental analyzer and the contents assayed.

3. RESULTS AND DISCUSSION

When leaves of the C_3-C_4 intermediate species were exposed to $^{14}CO_2$ for a 15-s pulse a considerable portion of the assimilated label was recovered as the 4-carbon products malate and aspartate (Table 1). In all of the intermediates examined, between 71 and 85% of the ^{14}C-label recovered in malate and aspartate was present in the C-4 position (B. Moore, personal communication). Although 32% of the ^{14}C-label was recovered in malate + aspartate following a 15-s pulse in the C_3 species, F. cronquistii, subsequent degradation of the C_4-acids revealed that only 50-60% of this label was in the C-4 position. This suggests that much of the label recovered in these products is the result of secondary metabolic processes, not atmospheric CO_2 fixation (B. Moore, personal communication). Following a 15-s pulse with $^{14}CO_2$ and a 3-min chase with $^{12}CO_2$, much less of the label was recovered as malate + aspartate in the intermediate species, F. floridana, F. ramosissima, and F. pubescens (Table 1). A much larger percentage was recovered in sucrose following the chase period. These results are consistent with the assimilation of atmospheric CO_2 through parallel C_3 and C_4 pathways. The C_3 pathway presumably also functions in series

Biggens, J. (ed.), Progress in Photosynthesis Research, Vol. III. ISBN 90 247 3452 5

with the C_4 pathway, providing a path for ^{14}C to move from malate and aspartate to sucrose during the chase period. In the intermediate species, F. linearis and F. anomala, there was no significant decrease in the amount of ^{14}C-label recovered in malate + aspartate after the 15-s pulse + 3-min chase, relative to the 15-s pulse alone. This could be due to inefficient transfer of carbon from the C_4-acid pools to the C_3-cycle, although other factors such as recycling photorespired $^{14}CO_2$ into malate and aspartate, and conversion of ^{14}C-PGA to ^{14}C-PEP and ultimately ^{14}C-malate could also contribute. Reductions in the % of label recovered in malate + aspartate following a 3-min chase period were greatest for the C_4 species, and lowest for the C_3-C_4 intermediate species. The fact that the % of ^{14}C in malate + aspartate decreased less rapidly during the 3-min chase period in the intermediate species is suggestive of less efficient carbon transfer from the C_4-cycle to the C_3-cycle.

TABLE 1. Pulse-chase characteristics of Flaveria leaves. Values in ( ) represent total ^{14}C recovered in units of CPM/cm^2 leaf area ($X10^{-5}$). The latter values are presented to demonstrate that changes in the distribution of ^{14}C are due primarily to a redistribution rather than differential loss of label. Data are based on Monson et al. (3).

Species	^{14}C-mal + ^{14}C-asp $\frac{}{\text{total } ^{14}C\text{-products}}$ X100		^{14}C-sucrose $\frac{}{\text{total } ^{14}C\text{-products}}$ X100	
	15-s pulse	15-s pulse + 3-min chase	15-s pulse	15-s pulse + 3-min chase
F. palmeri (C_4)	67.2 (1.49)	11.4 (1.23)	N.D.	60.2
F. cronquistii (C_3)	32.0 (1.51)	23.0 (1.48)	2.7	40.1
F. anomala (C_3-C_4)	35.3 (1.62)	34.9 (1.59)	0.9	8.6
F. floridana (C_3-C_4)	50.0 (0.78)	20.1 (0.73)	0.7	13.1
F. linearis (C_3-C_4)	23.4 (1.60)	41.1 (1.61)	1.5	6.2
F. pubescens (C_3-C_4)	31.1 (1.48)	15.2 (1.46)	1.6	26.3
F. ramosissima (C_3-C_4)	46.8 (1.61)	10.5 (1.58)	0.7	28.8

N.D. = not detectable

The efficiency of C_3- and C_4-cycle co-function was further examined through a series of quantum requirement studies. Quantum requirements for CO_2 uptake were higher for the C_3-C_4 intermediate species, relative to the C_3 species, F. cronquistii, when analyzed in 2% O_2. This can be explained if the C_3 cycle in the intermediates is accompanied by a parallel C_4 cycle that uses two additional ATP's to assimilate each molecule of atmospheric CO_2. Quantum requirements for the intermediate species in 2% O_2 were not as high as for the C_4 species F. trinervia. This can be explained on the basis that of the assimilated atmospheric CO_2 in the intermediate species, only a portion requires 5 mol of ATP per mole of CO_2 (typical of C_4 fixation), while the remainder requires only 3, or slightly more, moles of ATP per mole of CO_2 fixed, depending on how much residual photorespiration is occurring in the presence of 2% O_2. This lower ATP requirement in the intermediate species in 2% O_2, results in lower quantum requirements, relative to fully-expressed C_4 species. In 21% O_2, quantum requirements were similar for the

C_4 species, the C_3 species, and several of the C_3-C_4 intermediate species (F. linearis, F. anomala, and F. ramosissima). This similarity is presumably due to the offsetting demands on the available ATP by the C_3, C_4, and photo-respiratory cycles, and the ameloriation of photorespiration by C_4-cycle activity (see 3 for further explanation). In the intermediate species, F. pubescens and F. floridana, quantum requirements were higher in 21% O_2, relative to the C_3 or C_4 species. This might be due to futile cycling of CO_2 through inefficient C_4 cycles, using ATP but not resulting in efficient transfer of CO_2 from the C_4 cycle to the C_3 cycle. Incomplete compartmenta-tion of the C_3 and C_4 cycles between mesophyll and bundle-sheath cells in the intermediates might contribute to futile cycles (see 4,5). When taken together the pulse-chase and quantum requirement studies suggest the pre-sence of co-functioning C_3 and C_4 cycles in the C_3-C_4 intermediate Flaverias but at a relatively inefficient level.

TABLE 2. Quantum requirement for CO_2 uptake in Flaveria species. Leaf temp-erature was 30°C, atmospheric CO_2 partial pressure was 34 Pa. Values repre-sent the mean $\pm$ S.E. (n=3-7). Data are based on Monson et al. (3).

Species	Quantum requirement (mol quanta/mol CO_2)	
	2% O_2	21% O_2
F. trinervia (C_4)	18.9 $\pm$ 0.3	19.6 $\pm$ 0.8
F. cronquistii (C_3)	13.9 $\pm$ 0.4	18.9 $\pm$ 0.3
F. anomala (C_3-C_4)	14.3 $\pm$ 0.4	19.6 $\pm$ 0.7
F. floridana (C_3-C_4)	15.4 $\pm$ 0.2	21.7 $\pm$ 0.0
F. linearis (C_3-C_4)	14.5 $\pm$ 0.4	20.0 $\pm$ 0.4
F. pubescens (C_3-C_4)	15.9 $\pm$ 0.5	22.7 $\pm$ 0.0
F. ramosissima (C_3-C_4)	15.9 $\pm$ 0.5	19.2 $\pm$ 0.3

Measurements of instantaneous water- and nitrogen-use efficiencies in leaves of the intermediate species revealed some significant differences, relevant to the C_3 and C_4 species (Table 3). In F. pubescens and F. ramo-sissima NUE's were higher than the C_3 species, but WUE's were lower. In F. floridana, NUE's were similar to the C_3 species, but WUE's were slightly lower. Given that all three species can assimilate between 40-50% of their carbon through a C_4 pathway, at this point their is no consistent pattern of such biochemical intermediacy resulting in greater instantaneous resource use efficiency. Caution must be exercised, however, in interpreting the significance of these results, since they were obtained with plants grown under optimal conditions. Studies are in progress to examine these resource-use efficiencies in water- and nitrogen-stressed plants.

The presence of inefficient C_3 and C_4 cycles in the C_3-C_4 intermediate species raises interesting questions concerning the evolution of C_4 photo-synthesis in the genus Flaveria. Evidence that the C_3-C_4 intermediate species represent the products of an evolutionary progression from fully-expressed C_3 plants to the intermediate state can be derived from several independent lines (see 6). In explaining the existence of C_3-C_4 intermed-iates, the two principal alternatives to directional evolution from C_3 to C_3-C_4 to C_4, are 1) reverse evolution from C_4 towards C_3 and, 2) hybridiza-tion between C_3 and C_4 ancestors. The case of reverse evolution seems unlikely since all of the fully-expressed C_4 species in <u>Flaveria</u> (with the possible exception of <u>F. brownii</u>) are self-compatible, annuals. Both of these traits are considered to be derived from self-incompatible, perennial ancestors in the angiosperms (7,8). Most of the C_3-C_4 intermediate <u>Flaveria</u> species are self-incompatible, perennial species (6). This evolution of breeding-system and life-cycle traits seems to parallel a progression from C_3-C_4 intermediates towards C_4 plants in <u>Flaveria</u>, not the reverse. The case of hybridization orgin of C_3-C_4 intermediates seems unlikely in <u>Flaveria</u>, given the substantial barriers to hybridization in the F_1 and F_2 generations (6), and the recent observation that PEP carboxylase, the principal C_4 carboxylating enzyme is C_3-like, not C_4-like (9), as might be expected if the C_4-traits in the intermediates originated from hybridization events.

TABLE 3. Water- and nitrogen-use efficiencies of three C_3-C_4 intermediates, a C_3 species, and a C_4 species. Water-use efficiency (WUE) is defined as photosynthesis rate/transpiration rate. Nitrogen-use efficiency (NUE) is defined as photosynthesis rate/unit leaf nitrogen. Ci/Ca is the ratio of intercellular (Ci) and atmospheric (Ca) CO_2 partial pressures. Leaf temperature was $30°C$, photon flux density was $1800 \ \mu mol \ m^{-2} \ s^{-1}$, atmospheric CO_2 partial pressure was $330 \ \mu bar$, and the leaf-to-air water vapor concentration gradient was 8-$10 \ mmol/mol$. Values are mean $\pm$ S.E. ($n=4$-8).

Species	Photosynthetic rate ($\mu mol \ m^{-2} \ s^{-1}$)	Ci/Ca	WUE (umol/mmol)	NUE ($\mu mol \ CO_2 \ mol^{-1}$ N s^{-1})
F. <u>trinervia</u> (C_4)	34.8 ± 2.7	$0.53 \pm .03$	11.2 ± 0.7	286 ± 47
F. <u>cronquistii</u> (C_3)	23.7 ± 1.4	$0.67 \pm .02$	8.1 ± 0.2	121 ± 13
F. <u>ramosissima</u> (C_3-C_4)	30.1 ± 3.7	$0.77 \pm .01$	4.5 ± 1.0	366 ± 55
F. <u>floridana</u> (C_3-C_4)	23.9 ± 0.8	$0.74 \pm .02$	7.4 ± 0.2	122 ± 6
F. <u>pubescens</u> (C_3-C_4)	23.4 ± 1.0	$0.75 \pm .01$	6.8 ± 0.3	234 ± 15

Assuming that the C_3-C_4 intermediate <u>Flaveria</u> species represent the products of evolution, it is not apparent what is driving the increased expression of C_4 traits. It does not appear to be improved water- or nitrogen-use efficiencies, since species such as <u>F. floridana</u> exhibit C_3-like WUE and NUE values, yet considerable C_4-cycle activity. These results suggest that we should examine alternatives other than improved resource-use efficiency as the driving force in the initial stages of C_4 evolution.

Acknowledgements
    The author expresses his sincere thanks to B. Moore for conducting the electrophoresis and autoradiography of the pulse-chase products, and G.E. Edwards and M. Ku for allowing the author to use their laboratory and plants for the pulse-chase studies. Supported by NSF grant BSR-8407488.

REFERENCES
1 Monson, R.K., Edwards, G.E. and Ku, M.S.B. (1984) BioScience 34:563-574
2 Ku, M.S.B. et al. (1983) Plant Physiol. 71:944-948
3 Monson, R.K., Moore, B., Ku, M., Edwards, G. (1986) Planta (in press)
4 Bauwe, H. (1984) Biochem. Physiol. Pflanz. 179:253-268
5 Reed, J.E. and Chollet, R. (1985) Planta 165:439-445
6 Powell, A.M. (1978) Ann. Mis. Bot. Gard. 65:590-636
7 Jeffrey, E.C. (1917) The Anatomy of Woody Plants, Univ. of Chicago Press
8 Stebbins, G.L. (1957) Amer. Nat. 91:337-354
9 Adams, C.A., Leung, F., Sun, S.S.M. (1986) Planta 167:218-225

PARTITIONING AND METABOLISM OF PHOTORESPIRATORY INTERMEDIATES.

GRODZINSKI, B., M. MADORE*, R.A. SHINGLES AND L. WOODROW,
Department of Horticultural Science, University of Guelph, Guelph, Ontario
Canada, N1G 2W1.

Many external as well as internal factors control the partitioning of metabolites from source leaves to developing sinks. Initially, however all carbon in the plant and all energy used for the growth and development are derived through photosynthesis. With respect to net assimilation of CO_2, photorespiration (ie. the release of CO_2 in the light) is frequently viewed by breeders as an antagonistic process possibly even wasteful (1,2,3,4,5,6). However, in terms of maintaining whole plant growth in the natural environment photorespiration may have several homeostatic roles (7,8,9) which provide clear advantages to a plant facing adverse conditions. With the view to a better understanding of photorespiration and the role(s) it might preform, our studies have been focussed on the identification of the 'end products' of photorespiration at the cellular level and the relationship of their metabolism to the growth and development of the whole plant. As indicated in Fig. 1, the products of the glycolate pathway which actually leave the cell are CO_2, the transport sugars (primarily sucrose), and amino acids (eg. glycine, serine, glutamate, glutamine).

Most scientists agree that photorespiratory CO_2 release is accounted for by the oxidative decarboxylation of glycolate. Several reviews (1,5,10) outline the manner in which glycolate is metabolised intracellularly. Essentially phosphoglycolate is synthesized by ribulose bisphosphate carboxylase oxygenase (RUBISCO) in the chloroplast. Phosphoglycolate is converted to glycolate within the chloroplast, but the further metabolism of glycolate through glyoxylate, glycine, serine, hydroxypyruvate, and glycerate occurs in a series of linked reactions in at least three other cellular compartments, the peroxisome, the mitochondria and the cytoplasm. Overall two glycolate molecules are cycled intracellurlarly to generate CO_2 (from the carboxyl carbon) and a troise which can be used for the synthesis of hexoses or to replenish the triose phosphate pools of the chloroplast. The origin of the photorespired CO_2 is thought to be glycine, since two glycines are converted to serine, NH_3, and CO_2 in a ratio of 1:1:1 via a tightly linked reaction sequence in the mitochondria (1,2,3,4,10,11,12,13). Many studies attempt to equate, quantitatively, the flux of carbon and nitrogen through glycine and serine in mitochondria to photorespiratory CO_2 release . However, we stress that there are sufficient observations which force us to re-evaluate these assumptions. Of foremost concern is the inherent difficulty involved in quantifying photorespiratory CO_2 release, since photorespiration occurs while the tissue is actively fixing CO_2 during photosynthesis (5,8,14,15). Few studies actually report measurement of intact whole leaf photorespiration rates which can be correlated with analysis of photorespiratory intermediates or the metabolism studied in vitro. For example, the maximum rate of glycine decarboxylase in mitochondria is less

*Present address; Department of Botany and Plant Sciences, University of California, Riverside, California, 92521, U.S.A.

Biggens, J. (ed.), Progress in Photosynthesis Research, Vol. III. ISBN 90 247 3452 5

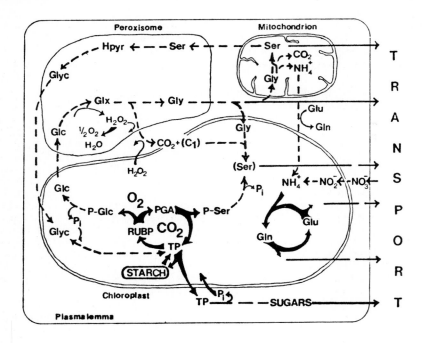

Figure 1. Possible routes for synthesis of sugars and amino acids exported from a typical C_3 leaf cell during photorespration. Glc, glycolate; Gln, glutamine; Glu, glutamate; Glx, glyoxylate; Glyc, glycerate; Hpyr, hydroxypyruvate; P-ser, phosphoserine; Ser, serine; TP, triose phosphate. After Madore and Grodzinski (8).

than or just barely the same order of magnitude as CO_2 release from whole leaf (16,17). Our data suggest that glycine is <u>not</u> the sole source of photorespired CO_2. Glyoxylate as well as glycine is a potential source of CO_2 and C-1 units (1,5,15,17,18,19,20). As shown in Figure 1 once glycolate is formed there are several possible routes for its further metabolism both <u>intracellularly</u> (17,19,20) and <u>intercellularly</u> (8,9,21). For example, as illustrated below serine can be formed from glycine and glyoxylate (ie. formate) <u>without</u> the breakdown of a glycine molecule which results in the release and recycling of NH_3 (17,19,20). Biochemical studies of the glycolate pathway over the last two decades have stressed the 'tightness' of <u>intracellular</u> C and N recycling (ie. chloroplast to peroxisome; peroxisome to mitochondria; mitochondria to peroxisome; peroxisome to chloroplast). We are the first group to demonstrate that under photorespiratory conditions, key photorespiratory intermediates, particularly nitrogenous intermediates such as glycine and serine, are also exported from the leaves and hence not available for recycling inside the cells where they are synthesized (8,9,21,22).

I. Intracellular Glycolate Metabolism

A) Glyoxylate Breakdown versus Conversion to Glycine. Our early studies with isolated peroxisomes (15,18,19,20) showed that normally 90% of the glycolate oxidised to glyoxylate is aminated to produce glycine. However, when the supply of nitrogen donors for glycine synthesis is low or when the temperature is raised, proportionally more H_2O_2 reacts with glyoxylate to produce formate (C-1 unit) and CO_2 (15,19). We proposed therefore that peroxisomal glyoxylate metabolism represents a key branch point in the glycolate pathway. Using peroxisomes isolated from leaf protoplasts and purified on Percoll gradients other workers (12,13) have argued that normally over 95% of the glycolate is converted to glycine. These results are essentially similar to our early observations. Preliminary results with peroxisomes isolated from leaf protoplasts incubated at 35° C and in the presence of glucose oxidase show the pattern of glyoxylate decarboxylation reported previously (15,19). The central role of peroxisomes in glycolate metabolism originally forwarded by Tolbert and his coworkers (10) as the site of H_2O_2 formation and break down to prevent glyoxylate breakdown has been questioned recently by the discovery of a barley mutant which grows only on high CO_2, has only 10% of the wild type catalase activity (4). It is not clear why serine synthesis is dramatically altered in this mutant or how glycolate is metabolised. In some eukaryotic algae glycolate conversion to glyoxylate occurs via a mitochondrial glycolate dehydrogenase not producing H_2O_2 (10,23,24) and glyoxylate is broken down by a $NADPH_2$ -Mn dependent H_2O_2 reaction in the chloroplast similar to the process observed for glyoxylate breakdown in higher plants (19,26,27). Interestingly, the barley glycolate oxidising enzyme reacts with L-lactate and not D-lactate but the protein extracted from leaves and algae have some similarity in electrophoretic mobility (24). Our conclusion that glyoxylate breakdown or conversion to glycine represents a key branch point in the glycolate pathway is supported by studies on Arabidopsis mutants (1,2,3) which are unable to breakdown glycine (**gly D**) and convert the glycine to serine (**stm**). These mutants are still capable of generating photorespired CO_2 in the absence of a surplus of nitrogen donors for glycine synthesis (1,2,3).

B) Photorespiratory Serine Synthesis. Tolbert has argued against the view that the glycolate pathway can be regulated if glyoxylate breakdown with H_2O_2 occurs since formate generated from the methylene carbon would also breakdown to CO_2 readily (10). We have found no evidence that formate oxidation to CO_2 is so dramatic. Although the methylene carbon of glycolate can give rise to photorespired CO_2 (5,20,26), it seems more likely that formate (C-1 units) generated from glycolate or glyoxylate are scavenged (20,28). Studies with Pisum and Arabidopsis leaves show that the enzymes necessary for C-1 metabolism and serine synthesis from glycine and formate are present in both mitochondria and chloroplasts (Table 1). Chloroplasts lack the ability to breakdown glycine to provide the C-1 unit (Table 1), and serine synthesis is dependent on the addition of formate (17) or glyoxylate (Table 2). By comparison in mitochondria formation of an active C-1 unit from formate is not preferred , supporting the view that mitochondrial glycine to serine conversion is "tightly" linked. As pointed out by others this "tightly" linked reaction sequence may also be important in regulation of mitochondrial metabolism in the light (1,5,10,16). Interestingly, the two Arabidopsis mutants (**glyD** and **stm**) lack the ability

Table 1.   Activities of several key enzymes of glycine and serine conversion and C-1 metabolism in mitochondria and chloroplasts from Pisum sativum L. and three lines of Arabidopsis thaliana, a wild type (WT), a glycine decarboxylase mutant (gly D), and a serine hydroxymethyltransferase mutant (stm).   Organelles were isolated from pea and Arabidopsis as described elsewhere (2,17).   Glycine decarboxylase (GLD), ^{14}C-serine synthesis were assayed using [1-^{14}C] glycine as described in (17); serine hydroxymethyltransferase (Ser-OH-MeT); formyltetrahydrofolate synthetase (FTHFS) according to (29), dihydrofolate reductase (DHFS) according to (30) and 5,10 methylene tetrahydrofolate dehydrogenase (5,10-MeTHFD) according to (31). Data represent the mean of 6 determination. Standard deviations were less than 10% of the means.

Enzyme	Enzyme Activity (nmoles/mg protein/min)							
	Pea	Mitochondria			Pea	Chloroplasts		
		Arabidopsis				Arabidopsis		
		WT	stm	glyD		WT	stm	glyD
GLD	60	61	40	5	4	3	2	5
FTHFS	31	46	21	16	68	75	82	41
DHFS	90	82	32	30	133	56	28	30
5,10-MeTHFD	84	40	20	8	72	78	60	29
Ser-OH-MeT	123	83	28	38	80	92	20	38
Serine synthesis	56	73	31	8	60	48	16	25

Table 2.   ^{14}C-Distribution in Serine in Purified Active Pea Chloroplasts. The assay mixture routinely contained 330mM sorbitol, 50mM Hepes-KOH (pH 7.6), 2mM EDTA, 1mM/MgCl$_2$, 100 ul chloroplast (0.1 - 0.2 mg chlorophyl). Following 15 minutes incubation in the dark or light with 10mM sodium bicarbonate ^{14}C-serine formed from ^{14}C-labeled glycine (gly), glyoxylate (glx), or formate (for), was determined (17,32).

| | ^{14}C-SERINE (n moles) | ^{14}C-DISTRIBUTION IN SERINE % of label in | |
		C-1 & C-2	C-3
Dark			
[1-^{14}C]gly + for	990 ± 57	99	1
[2-^{14}C]gly + for	1040 ± 42	88	12
gly + ^{14}C-for	910 ± 55	7	93
gly + [2-^{14}C] glx	30 ± 15	ND	ND
Light			
[1-^{14}C]gly + for	725 ± 26	98	2
[2-^{14}C]gly + for	845 ± 45	96	4
gly + ^{14}C-for	775 ± 30	31	69
gly + [2-^{14}C]glx	286 ± 24	35	65

to produce serine from glycine in mitochondria (1,2,3; Table 1) and from glycine and formate in chloroplasts (Table 1). Activities of several enzymes of C-1 metabolism in both organelles seem to be lower than in the wild type Arabidopsis, consistent with the original view of Somerville and Ogren that C-1 metabolism during photorespiration is disrupted and its major function in leaves may be in maintaining flow of carbon between glycine and serine (1,2,3). However, the data do not definitely support the view that serine synthesis occurs only in mitochondria. The mutations are recessive, involve nuclear genes and appear to involve C-1 metabolism in both mitochondria and chloroplasts (Table 1).

Studies with photosynthetically active intact pea chloroplasts indicate that serine synthesis from formate and glycine occurs and is stimulated by $NADPH_2$ and ATP. Furthermore, in the light in the presence of $NADPH_2$, glyoxylate is readly broken down to generate C-1 units which in the presence of glycine form serine. As indicated in Table 2 one can distinguish between serine formed in mitochondria solely from glycine, and serine produced in chloroplasts from glycine and formate (or glyoxylate). For example, in the chloroplast reaction glycine is not broken down and NH_3 cannot be detected. The serine produced from $[2-^{14}C]$ glycine is labeled primarily in the 2 position, whereas in the mitochondrial reaction serine is labeled in the 2 and 3 positions. In chloroplasts, the serine generated from $[2-^{14}C]$ glyoxylate and glycine is labeled heavily in the 3-C position paralleling results with $[^{14}C]$ formate and glycine. Some randomization of ^{14}C-distribution in serine occurs in the light because formate can be oxidised to CO_2 which presumably is reassimilated. Taken together the ^{14}C distribution data support the view that two molecules of glycolate can be converted to serine via a condensation reaction in the chloroplast without N recycling (Fig. 1).

The energy requirements for serine synthesis from photorespiratory intermediates in chloroplast are difficult to define. Similar to other studies (5,25,26) glyoxylate breakdown in pea chloroplasts requires light and is stimulated by $NADPH_2$ (Shingles, Ph.D thesis 1986). Several key steps in C-1 metabolism require $NADPH_2$ or ATP (33). Light, $NADPH_2$ and ATP enhance incorporation of label from ^{14}C-formate into serine, but during bicarbonate fixation a 10-20% reduction in photorespiratory serine synthesis is observed (Table 2). More significantly, however, CO_2 fixation and serine synthesis proceed at high rates simultaneously in our pea chloroplast system. Addition of glycine and formate to illuminated intact chloroplasts reduces photosynthesis (ie. bicarbonate dependent-O_2 evolution and $^{14}CO_2$ fixation) by only 10%. Rates of 90-100 umoles CO_2 fixed/mg Chl/h are routinely observed and the ratio of O_2 evolved:CO_2 fixed remains 1:1 (Gilmour and Grodzinski, unpublished). These observations agree with an earlier study (34) that at pH 7.6, formate inhibition of chloroplast photosynthesis due to acidification of the stroma is minimal. Glyoxylate, by comparison reduces both photosynthesis and serine synthesis from photorespiratory intermediates in chloroplasts (Table 2). Maximal rates of serine synthesis from formate and glycine are about 20-25 umoles/mg Chl/h. By comparison mitochondrial serine synthesis in peas is about double the chloroplast rate. In vivo photorespiratory CO_2 release from attached young pea leaves is over 90 umoles CO_2/mg Chl/h, which underscores the difficulty in equating these processes.

II. Intercellular Movement of Leaf Products During Photorespiration

The primary function of the chloroplast is the fixation of CO_2 and the export of triose phosphates (TP - see Fig. 1). Chloroplasts do not make transport sugars but they do control the transport of carbohydrates. They either store C in the form of starch or export it as TP. As pointed out above, the classical view of photorespiration stresses the intracellular nature of the glycolate pathway. In order to assess whether the glycolate pathway via serine and trioses to sugars is complete in C_3 plants we fed [^{14}C]-labeled glycolate pathway intermediates (glycine, glyoxylate, and serine) to isolated, photosynthetically active mesophyll cells, leaf discs and attached leaves of Cucumis sativus, Salvia splendens, Apium graveolens, and Helianthus annuus . In addition to starch and amino acids, the transport sugars characteristic of each species are produced. The transport sugars are not labelled in the dark from a variety of photorespiratory intermediates which is consistent with the view that the photorespiratory intermediates (probably glycerate) re-enter the chloroplast (22,35). Cucumber and Salvia were used in our studies because they characteristically produce sugars of the raffinose series (eg. stachyose) for export. The presence of these sugars is an indicator of the capacity of a leaf to export carbohydrate. In cucumber, stachyose was labeled from [^{14}C]-serine in leaf discs but not in isolated leaf cells. These studies reflect an important distinction between the methodologies used to investigate photorespiratory C and N metabolism. Although useful information can be obtained using purified organelles or isolated cells in vitro these studies do not reflect the activity of intact leaf tissue, particularly with respect to the synthesis of transport sugars. In addition, the role of sinks (ie. growing regions) in intact plants needs to be considered more fully. Studies with Salvia plants show that ^{14}C-labeled intermedites can be fed via the xylem through a petiole flap to the source leaves (22). The supplied [^{14}C]-labeled photorespiratory intermediates are converted in the fed leaves into a variety of compounds which can be identified by sampling the leaves and subjecting the tissue extracts to chromatographic analysis. We have been able to show by a series of girdling, and petiole feeding experiements that the Salvia leaves convert the fed intermediates into specific transport sugars (stachyose) and amino acids such as glycine and serine in the leaf before being exported back via the petiole. Increasing the O_2 concentration of the feed leaf environment (eg. 2%, 21%, 50%) enhances photorespiratory activity and results in more of the exported ^{14}C being partitioned into amino acids. Studies with intact attached leaves maintained under steady state conditions at varying CO_2 and O_2 levels partition more newly fixed ^{14}C into amino acids (ie. glycine, serine, glutamine, glutamate) which appear in the transport stream. These studies support earlier pulse chase experiments (8,9) which indicate that in addition to being recycled into the Calvin cycle, intermediates of photorespiration may also serve as transport compounds.

Experiments with short lived radioisotopes have also been informative (21). For a short period (eg. 1 hour) after pulse labeling with $^{11}CO_2$ the rates of movement of ^{11}C-photoassimilates are essentially similar (3-4cm/min) under photorespiratory (21 and 50% O_2) and non-photorespiratory (2% O_2) conditions. These data suggest that in the short term loading and transport occur normally. Using ^{13}N-labeled ammonia gas of high specific activity we have been able to feed and label the leaf directly without relying upon N movement through the xylem or from roots. Consistent with the activity of the GS/GOGAT cycle in leaves of <u>Helianthus</u> and <u>Lupinus</u> ^{13}N-labeled products (ie. amino acids) were transported in the phloem at rates similar to the ^{11}C-labeled photoassimilates at 2,21 and 50% O_2 (21).

Other studies with ^{15}N-labeled amino acids demonstrate that a number of amino acids can participate in photorespiratory metabolism and that photorespiratory N-metabolism may not be a closed cyclic process (36). It is not clear how much N actually enters or leaves cells as a consequence of photorespiration. However, the amount may not be as important in understanding plant growth and development as the "quality" (ie. the nature) of the specific intermediates being synthesised and mobilized during photorespiration.

III Correlating Control of Photorespiration and Growth

The flux of C and N through the photorespiratory pathway is altered by exposing leaves to different analysis conditions (eg. CO_2 , O_2 temperature, water stress, nutrient supply). Photorespiration is frequently rationalised as a mechanism evolved to dissipate excess light energy. Attempts to eliminate or reduce glycolate metabolism by either chemicals or mutation ironically show the importance of branch points in the pathway (Fig. 1) and underscore the view that without the ability to recycle 1 and 2 carbon compounds within the leaves the plants lose an adaptive strength to survive under stress conditions .

Under high CO_2 plant growth and development are usually enhanced (6,9,39). In the short term exposure to a "squirt of CO_2 " favors carboxylation rather than oxygenase activity and the proportion of newly fixed C flowing through the glycolate pathway both <u>intracellularly</u> and <u>intercellularly</u> is diminished. Under high CO_2 total CO_2 fixation increases and more C is partitioned into CHO's (eg. transport sugars and starch) and <u>proportionally</u> less into amino acids ,particularly glycine and serine which are markers of reduced flow through the glycolate pathway. Growth and acclimation to high CO_2 over a longer period (days) can alter nutrient balance (ie nitrogen balance) in the leaves and accelerates turnover of key proteins such carbonic anhydrase and RUBISCO, which helps explain why leaves of CO_2 enriched plants sometimes appear to be have low photosynthetic rates when measured at ambient conditions (37). The reduced flux of C and N through the glycolate pathway may also be important in regulating the synthesis of specific growth hormones, such as C_2H_4, a by-product of sulfur amino acid breakdown (38). Interestingly, it is the commercially used levels of CO_2 enrichment in greenhouses (300 to 2,000 ppm) which have the greatest effect on glycolate metabolism and C_2H_4 production (39).

REFERENCES

1   Artus, N., Somerville, S. and Somerville, C.R. (1986) CRC Critical
    Reviews in Plant Science 4, 121-147
2   Somerville, C.R. and Ogren, W.L. Biochem J. 202,373-380
3   Somerville, C.R. and Ogren, W.L. Plant Physiol 67,666-671
4   Kendall, A.C. et al (1983) Planta 159, 505-511
5   Zelitch, I. (1975) Ann Rev Biochem 44, 123-145
6   Porter,M. and Grodzinski, B. (1985) Hort. Rev. 7, 345-398

7   Grodzinski, B. (1984) Plant Physiol. 74, 871-876
8   Madore, M. and Grodzinski, B. (1984) Plant Physiol. 76, 782-786
9   Madore, M. and Grodzinski, B. (1985) J. Plant Physiol. 121, 59
10 Tolbert, N.E. (1983) In Current Topics in Plant Biochemistry and Physiology (Randall, D., Bleviks, D. Gaud, R., Larson, Eds) Vol 1, 63-77, University of Columbia
11 Keys, A. J. et al (1978) Nature 275, 741-743
12 Chang, C.C. and Huang, A.H.C. (1981) Plant Physiol, 67, 1003-1008
13 Schmitt, M.R., Chestnut, M.H. and Edwards, G.E. (1984) Pl. Sci. Lett 33, 115-118
14 Ludwig, L.J. and Canvin, D.T. (1971) Can J. Bot. 49, 1299-1313
15 Grodzinski, B. and Butt, V.S. (1977) Planta 133, 261-268
16 Day, D.A. and Wiskich, J. T. (1981) Plant Physiol. 68, 425-429
17 Shingles, R.M., Woodrow, L. and Grodzinski, B. (1984) Plant Physiol 74, 705-710
18 Grodzinski, B. and Butt, V.S. (1976) Planta 128, 225-231
19 Grodzinski, B. (1978) Planta 144, 31-37
20 Grodzinski, B. (1979) Plant Physiol. 63, 289-293
21 Grodzinski, B. Jahnke, S. and Thompson, R.G. (1974) J. Expl. Bot. 35, 678-690
22 Madore, M and Grodzinski, B (1985) Plant Physiol. 77S:510
23 Yakota, A. and Kitaoka, S. (1982) Plant Physiol. 70, 760-4
24 Grodzinski, B. and Colman, B. (1972) Phytochem. 11, 1281-1285
25 Yakota, A., Kawabata, A., and Kitaoka, S. (1983) Plant Physiol 71, 722-776
26 Elstner, E.F. and Heupel, A., (1973) Biochim Biophys Acta 325,182-188
27 Zelitch, I., (1972) Arch Biochem Biophys 150, 698-707
28 Woodrow, L., and Grodzinski, B. (1981) in Regulation of Carbon Metabolism (Akoyunoglou G. Ed.) 551-559. Balaban. Pa
29 Gunlack, B. G., Neal, G.E. and Williams, D.C. (1968) Biochem Pharm. 17, 484-487
30 Hiatt, A.J. (1965) Plant Physiol 48, 189-193.
31 Wong, and Cossins Can J. Biochem 44, 1400-1403
32 Brem, S., Ruffner, H.P., Rast, D.M. (1983) Plant Physiol 73, 579-581
33 Cossins, E.A., (1980) In Biochemistry of Plants (Stumpf and Conn E. Eds.), Vol 2 329-336 Academic Press NY
34 Enser, U. and Heber, U. (1980) Biochem Biophys Acta 592, 577-591
35 Howitz,K.T. and McCarty, R.E. (1986) Plant Physiol. 80,390-395
36 Ta, T.C., Joy, K.W. and R.J. Ireland (1985) Plant Physiol.78, 334-337
37 Porter, M. and Grodzinski, B. (1984) Plant Physiol 74, 413-416
38 Grodzinski, B. (1984) Plant Physiol 74, 871-876.
39 Woodrow, L. and Grodzinski, B. (1986) Proceedings of the VII Int. Congress on Photosynthesis ---

ACKNOWLEDGEMENTS   The authors would like to thank Dr. C. R. Somerville for making available seeds of the Arabidopsis thaliana mutant and wild type lines; Mrs. L. Spilsbury for technical assistance and financial support from NSERC.

PHOTOSYNTHETIC GAS EXCHANGE, PHOTOASSIMILATE PARTITIONING, AND DEVELOPMENT IN TOMATO UNDER CO_2 ENRICHMENT.

LORNA WOODROW[1] and BERNARD GRODZINSKI. Dept. of Horticultural Science, University of Guelph, Guelph, Ontario N1G 2W1 Canada. [1] Agriculture Canada

1. INTRODUCTION

Interest in CO_2 enrichment has focussed primarily on its application in horticultural greenhouse production (1). CO_2 may be added to the greenhouse atmosphere to maintain a level of approximately 330ul/L or enriched to higher levels (eg. 700-1500 $\mu l/L$) to further stimulate growth. Elevated CO_2 levels are also experienced by crops in some outdoor production situations. Recently Wallis and Grodzinski (2) reported that sweet pepper plants grown inside polyethylene tunnels on mulched soil experienced CO_2 levels exceeding 3000 $\mu l/L$ due to soil respiration and restricted gas exchange. Potential global increases in CO_2 concentration have also stimulated interest in plant responses to altered CO_2 levels (3). Growth and development of plants is associated with photosynthetic activity and partitioning of carbon (4) as well as with the action and integration of plant growth regulators (5). In this study determinate tomato plants were grown as transplant stock at CO_2 concentrations of 300, 1000, and 3000 $\mu l/L$. Their growth and photosynthetic metabolism was studied during the period of early vegetative development. Leaf ethylene metabolism was examined as it relates to leaf photosynthetic activity and potential growth regulation.

2. MATERIALS AND METHODS

2.1. Plant Material

Lycopersicon esculentum cv. TH318 (Thompson Seed Co.) grown in 1x1x1 meter open flow chambers contained in a greenhouse. The CO_2 levels in the chambers were maintained at 300, 1000, or 3000 $\mu l/L$ by metering CO_2 from cylinders into the airlines.

2.2. Procedures

Groups of 6 plants per chamber (2 chambers per CO_2 level) were harvested for assessment of plant growth. Metabolite levels were determined on the 4th leaf from the apex after extraction with 80% ethanol using the following assays: anthrone (total soluble carbohydrate); arsenomolybdate (reducing sugars); ninhydrin (α-amino nitrogen); amyloglucosidase digestion and anthrone (starch). Photosynthetic gas exchange was assessed on the fourth expanded leaf from the apex using an open flow system. Ethylene release rates by leaf discs were determined in closed flasks. CO_2 levels in the flasks were controlled by enclosing centrewells containing buffered bicarbonate solutions. Ethylene was assayed by gas chromatography.

Biggens, J. (ed.), Progress in Photosynthesis Research, Vol. III. ISBN 90 247 3452 5
© *1987 Martinus Nijhoff Publishers, Dordrecht. Printed in the Netherlands.*

3. RESULTS AND DISCUSSION

Increased plant size in response to growth at elevated CO_2 levels is well documented (1,6). In this cultivar the leaf area and shoot height of the CO_2 enriched and ambient grown plants were similar however dry matter accumulation in leaves, stems, and roots was greater in enriched plants. The total dry weight of plants at 1000 $\mu l/L$ was 64% higher than that of plants grown at 300 $\mu l/L$ (data not shown). The specific leaf weight (SLW) of the ambient plants was 2.45 mg/cm^2 versus 4.17 mg/cm^2 in leaf tissue grown at 1000 $\mu l/L$ CO_2. Higher SLW is an indicator of starch accumulation (7,8) and starch levels in enriched leaves were more than double those in ambient grown leaves. Total soluble carbohydrate was increased from 1.14 umoles glucose equivalent/cm^2 at 300 ul/L to 2.18 umoles glucose eq./cm^2 at 1000 $\mu l/L$ in agreement with similar results obtained by Madsen (9). Amino acid content, as indicated by α-amino nitrogen values, was not altered by the growth condition, however the increased soluble carbohydrate levels result in reduced amino acid to sugar ratio under CO_2 enrichment. Madore and Grodzinski (7) demonstrated that amino acid to sugar ratios were reduced in cucumber under CO_2 enrichment and that amino acid synthesis from newly fixed ^{14}C and export was depressed under CO_2 enrichment while the synthesis of transport sugars was enhanced.

Table I. Net photosynthesis, transpiration and dark respiration in tomato leaf tissue grown at 300, 1000, or 3000 ul/L CO_2. Measurements were conducted at 25°C, 40% R.H., and 700 μmoles quanta/m^2/s. The entire plant was held at the assay CO_2 concentration for the duration of the assay while the leaf under study was contained in a gas exchange cuvette. Numbers in parentheses represent the value above expressed as a percentage of the photosynthesis rate of the 300 $\mu l/L$ CO_2 plants assayed at 300 $\mu l/L$. Photorespiration is expressed as a percent inhibition of net photosynthesis by O_2.

[CO_2] ($\mu l/L$)		Net Pn (μmoles CO_2/m^2/s)		Pr (%)	Trans ($mg H_2O$/m^2/s)	Resp (μmoles CO_2/m^2/s)
		$2\%O_2$	$21\%O_2$			
Growth	Assay					
300	300	13.2±1.4 (144)	9.14±1.19 (100)	44	36.0±1.7	−0.38±0.03
1000	300	11.6±0.5 (127)	8.76±2.41 (96)	32	35.6±2.9	−0.34±0.01
1000	1000	16.4±1.7 (179)	16.4±1.9 (179)	0	23.9±3.8	----
3000	3000	14.1±0.9	18.7±1.6	−25	23.1±2.6	−0.92±0.02

Table I outlines the photosynthetic activity of leaf tissue grown at 300, 1000, and 3000 μl/L CO_2. Net photosynthesis at 21% O_2 increased with CO_2 concentration and transpiration rate is decreased. Plants grown and assayed at 300 μl/L demonstrated a 44% inhibition by 21% whereas plants grown and assayed at 1000 μlL exhibited no detectable inhibition. When assayed at 300 μl/L, plants grown at 1000 μl/L had photosynthetic rates similar to plants grown at 300 μl/L indicating that they are not at a disadvantage as a result of acclimation to high CO_2. In fact O_2 inhibition was only 32% suggesting some benefit from growth at higher CO_2. Inconsistency in the literature with respect to transfer from high to low CO_2 may be due to the selection of experimental material. Porter and Grodzinski (10) have suggested that older leaf tissue would be less active after transfer from high to low CO_2 than younger tissue due to loss of CO_2 fixing protein and reduced turnover. Plants grown and assayed at 3000 ul/L demonstrated a 25% inhibition of CO_2 uptake by 2% O_2 in comparison with 21%O_2. Recently Sharkey **et al.** (11) have suggested that Rubisco activation and the rate of RuBP usage becomes sensitive to low O_2 concentrations when CO_2 assimilation rates are maximal and the capacity for sucrose and starch synthesis nears saturation. The inhibition of CO_2 assimilation by low O_2 at 3000 μl/L may reflect a saturation of the C utilization machanisms. This phenomenon may explain why the advantages of CO_2 enrichment diminish as CO_2 concentration is raised to higher and higher levels.

Table II. Endogenous and ACC-dependent C_2H_4 release from leaf tissue grown at 300, 1000, or 3000 μl/L CO_2. Error bars represent 1 S.E. of the mean.

[CO_2] (ul/L)		Endogenous C_2H_4 release (pmoles/cm^2/33h)	ACC-dependent C_2H_4 release (pmoles/cm^2/h)
Growth	Assay		
300	300	58.0± 9.9	40.2± 2.1
	dark	42.0± 5.5	62.8±11.9
1000	300	57.8± 4.5	34.3± 3.0
	1000	122.4±15.0	48.7± 5.5
	dark	41.3± 3.9	56.4± 2.6
3000	300	61.4±20.9	40.9± 7.0
	3000	186.3±16.6	86.2± 6.3
	dark	45.5± 6.9	78.2 ±12.8

As pointed out above the balance of metabolites (sugars, amino acids) is influenced by CO_2 enrichment. The amounts of specific amino acids are altered by CO_2 levels (7) and in tomato leaves the pools of the amino acid ACC (1-aminocyclopropane-1-carboxylic acid), the immediate precursor of ethylene, is increased under CO_2 enrichment (data not shown). The release of C_2H_4 from tomato leaf tissue is also enhanced by CO_2 enrichment (Table II). Increased C_2H_4 release by ambient grown tissue in response to short-term increases in CO_2 concentration has been reported for many species (12,13,14). In this study leaf tissue supplied with exogenous ACC released C_2H_4 readily in response to CO_2 levels in the incubation flasks. Ethylene release from ACC in the dark indicates that the 'ethylene forming enzyme' system is functional as similar levels in all tissues and does not represent the limiting factor to ethylene release. Ethylene release from endogenous ACC pools illustrates that under high CO_2 growth regimes C_2H_4 release from leaf tissue is increased. However it is not known whether long-term alteration of the CO_2 environment may exert some of its effect on plant development through growth regulators such as ethylene. Changes in metabolite pools and the flux of carbon and nitrogen through photosynthetic and photorespiratory pathways (7) under CO_2 enrichment as they relate to CO_2 mediated effects on plant growth are currently under study.

REFERENCES

1   Porter, M.A. and Grodzinski, B. (1985) Hortic. Rev. 7, 345-398
2   Wallis, M.O. and Grodzinski, B. (1986) HortScience. 21(3), 829
3   Rosenberg, N.J. (1981) Climatic Change. 3, 265-279
4   Gifford, R.M., Thorne, J.H., Hitz, W.D. and Giaquinta, R.T. (1984) Science. 225, 801-808
5   Trewavas, A. (1981) Plant, Cell and Environment. 4, 203-228
6   Kimball, B.A. (1983) Agron. J. 75, 770-778
7   Madore, M.A. and Grodzinski, B. (1985) J. Plant Physiol. 121, 59-71
8   Ehret, D. L. and Jolliffe, P.A. (1985) Can. J. Bot. 63, 2026-2030
9   Madsen, E. (1968) Physiol. Plant. 21, 168-175
10 Porter, M.A. and Grodzinski, B. (1984) Pl. Physiol. 74, 413-416
11 Sharkey, T.D., Seemann, J.R. and Berry, J.A. (1986) Pl. Physiol. 81, 788-791
12 Grodzinski, B., Boesel, I. and Horton, R.F. (1982) J. Exp. Bot. 33, 344-354
13 Kao, C.H. and Yang, S.F. (1982) Planta. 155, 261-266
14 Dhawan, K.R., Bassi, P.K. and Spencer, M.S. (1981) Pl. Physiol. 68, 831-834

ACKNOWLEDGEMENTS

This work was supported by an NSERC grant to B.G., and was conducted while L.W. was on educational leave from Agriculture Canada.

OXYGEN EXCHANGES IN MARINE MACROALGAE

François BRECHIGNAC, Christophe RANGER, Marcel ANDRE, Alain DAGUENET, Daniel MASSIMINO. Laboratoire d'Agrophysiologie, Institut de Recherche Fondamentale, Département de Biologie, CEN Cadarache, 13108, SAINT-PAUL-LEZ-DURANCE, France.

I. INTRODUCTION

Whether photorespiration occurs in marine algae is still very controversial for two main reasons. First, the methods used for its detection are often indirect (O_2 sensitivity of photosynthesis), or known to underestimate the process ($^{14}CO_2$ method). In addition, the nature of the external carbon form acquired for photosynthesis (CO_2 and/or HCO_3^-) is still thought to be quite variable among marine species, and is strongly dependent upon the experimental conditions. Since 1) the HCO_3^- concentration in seawater is about 200 times greater than the free CO_2 concentration, 2) its uptake might drive putative internal DIC accumulation, and 3) the O_2 and CO_2 substrates interact on Rubisco in a competitive manner, it appears essential to know which DIC form is absorbed in order to understand the photorespiratory phenomenon in these organisms.

We present data supporting exogenous HCO_3^- uptake in four marine macroalgae representing three different taxa, with reduced rates of photorespiration measured by $^{18}O_2$ mass spectrometry. In accordance with the influence of O_2 concentration previously studied (1), further investigation of photorespiration in *Chondrus crispus*, through the influence of temperature and various inhibitors on O_2 exchange, indicates that O_2 photoreduction (Mehler-type reactions) may be the major O_2 photoconsuming process. This feature is discussed in relation with the energy requirements for HCO_3^- uptake and/or internal DIC accumulation.

2. MATERIALS AND METHODS

Rates of O_2 exchange were measured on *Enteromorpha compressa* (Chlorophyta, Ulotrichales), *Cystoseria mediterranea* (Pheophyta, Fucales), *Hypnea musciformis* and *Chondrus crispus* (Rhodophyta, Gigartinales), which were grown in the laboratory. The experimental device has been previously described (2,3). It featured a computerized system which continuously monitored the quantitative regulation of CO_2 in an assimilation chamber, while recording the $^{16}O_2$ and $^{18}O_2$ concentrations by mass spectrometry, as well as an exchange column which allowed for the continuous measurement of the free CO_2 concentration during photosynthetic activity by infrared gas analysis (4).

Samples of algae ranging from 10g up to 30g FW were introduced in the assimilation chamber containing 1,5 l of air, and 3 l of N and P-enriched seawater which was accurately thermostatted (20 ± 0.1°C, except for *C. crispus* : 17 ± 0.1°C). The 0_2 concentration was regulated at 20.6 ± 2%, and 250 µmol photon.$m^{-2}.s^{-1}$ were provided with a 18 h light / 6 h dark photoperiod.

In experiments involving inhibitors, the CO_2 concentration was regulated in the gas phase at 80, 330 and 960 µl.l^{-1}, corresponding to free CO_2 concentrations in seawater which were respectively limiting (below the $K_{\frac{1}{2}}$ (CO_2)), just saturating and highly saturating for photosynthesis.

The optimal concentrations of INH, KCN, DCMU and SHAM used were : 2.5 mM, 400 µM, 5 µM and 2.5 mM, respectively.

Biggens, J. (ed.), Progress in Photosynthesis Research, Vol. III. ISBN 90 247 3452 5

3. RESULTS AND DISCUSSION

Figure I shows the influence of the free CO_2 concentration on the steady-state O_2 exchange rates of *E. compressa* (a), *C. mediterranea* (b), *H. musciformis* (c). The general features observed for the four macroalgae studied are summarized in Table I. The major carboxylating enzyme found in the Rhodophyta and the Chlorophyta is Rubisco (5). An additional enzyme is present in the Pheophyta, PEP carboxykinase, but its affinity for CO_2 is much lower than that of the PEP carboxylase found in C_4 plants. Being well below the known affinities for CO_2 of these carboxylating enzymes, the high apparent affinity of photosynthesis for free CO_2 ($K\frac{1}{2}$ $(CO_2) \leqslant 3$ µM) is due to the additional uptake of exogenous HCO_3^-. It has been recently demonstrated from a kinetic analysis that exogenous HCO_3^- uptake from seawater supports more than 90% of the whole photosynthetic rate in *C. crispus* (6).

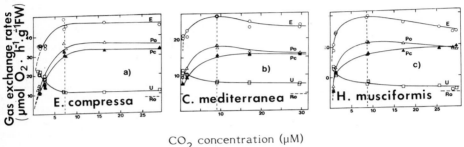

FIGURE I. Influence of CO_2 concentration on the photosynthetic gas exchanges of three marine macroalgae. Po,Pc, net photosynthesis as measured by O_2 and CO_2, respectively ; U, O_2 photoconsumption ; Ro, dark O_2 uptake ; E=Po+U, gross O_2 photosynthesis. All measurements were performed in steady-state with pH, CO_2 and TDIC concentrations being constant.

In contrast to C_3 terrestrial plants, the rate of O_2 photoconsumption did not exceed 45% of net O_2 evolution in natural conditions (10 µM CO_2), a behavior in agreement with the putative internal DIC accumulation brought about by the HCO_3^- uptake which may reduce the RuBP oxygenase activity. O_2 was unable to replace CO_2 as an electron acceptor in conditions of limiting CO_2 : $[U(1µM)-U(10µM)]/U(10µM) \leqslant 46\%$, and an O_2 uptake distinct from RuBP oxygenase activity was still present in saturating CO_2 (Ro/U$\leqslant$79%). This questions wether RuBP oxygenase activity contributes the major amount of O_2 photoconsumption in these marine algae.

	Rhodophyta		Pheophyta	Chlorophyta
	Chondrus c.	*Hypnea m.*	*Cystoseira m.*	*Enteromorpha c.*
$K\frac{1}{2}$ (CO_2) (µM)	1.0	1.3	0.5	3.0
U(10µM)/Po(10µM).100	38%	38%	45%	31%
[U(1µM)–U(10µM)]/U(10µM).100	30%	46%	32%	45%
Ro(27µM)/U(27µM).100	50%	79%	60%	78%

TABLE I. Photosynthetic and photorespiratory characteristics of 4 marine algae.

If different processes consume O_2 in the light, they may have different temperature sensitivities which might allow one to discriminate between them. The influence of temperature on the O_2 exchange of *C. crispus* is presented in Figure 2. The optimal temperature for net O_2 evolution at various CO_2 concentrations was constant at 17°C. For O_2 photoconsumption, in contrast, the optimal temperature raised with increasing CO_2 suggesting the involvement of (at least) two different O_2 uptake processes in the light, whose relative amounts were determined by the CO_2 concentration. Table 2 summarizes the temperature sensitivities (Q_{10}) of O_2 exchange in *C. crispus*. The Q_{10} of net O_2 evolution and dark O_2 uptake were clearly distinct, and unaffected by the CO_2 concentration. In contrast, the Q_{10} of O_2 photoconsumption increased with CO_2, indicating the preponderance of an O_2 photoconsuming process which alternatively would have a temperature sensitivity close to that of RuBP oxygenase activity at limiting CO_2, and close to that of mitochondrial oxidations at saturating CO_2.

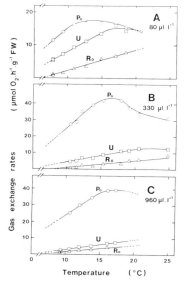

FIGURE 2. Influence of temperature on O_2 exchange rates in *C. crispus*. Each temperature tested was applied for at least one whole photoperiod.

	Temperature sensitivity (Q_{10})		
[CO_2] $(\mu l.l^{-1})$	80	330	960
Net O_2 evolution	– 2.13	2.19	2.12
O_2 photoconsump.	2.39 2.43	3.07	3.12
Dark O_2 uptake	3.08 3.38	4.64	3.34

TABLE 2. Temperature sensitivity of O_2 gas exchange in *C. crispus*. The Q_{10} were calculated from the early linear part of the gas exchange response according to the formula : $Q_{10}=(10\alpha+v)/v$, where α is the slope, and v the rate of gas exchange at 10°C.

O_2 photoreduction and mitochondrial oxidations (1,7,8) have already been suggested to sustain part of the total O_2 photoconsumption. This hypothesis has been further examined by testing various inhibitors on O_2 uptake in the light. The most illustrative results are shown in Table 3, along with a typical experiment (Figure 3).

Rates of O_2 photoconsumption ($\mu mol\ O_2.h^{-1}.g^{-1}$ FW $\pm$ SD)		
[CO_2] $(\mu l.l^{-1})$	80	960
Control	10.5 $\pm$ 1.9 (7)	5.7 $\pm$ 0.9 (8)
DCMU-resistant	3.0 $\pm$ 0.6 (9)	
DCMU-sensitive	5.9 $\pm$ 0.2 (3)	2.7 $\pm$ 0.6 (6)
KCN+SHAM-resistant	2.0 $\pm$ 0.2 (2)	2.5 $\pm$ 0.1 (2)
KCN+SHAM-sensitive	7.5 $\pm$ 0.4 (2)	3.0 $\pm$ 0.7 (3)
INH-sensitive	0	

TABLE 3. Influence of various inhibitors on the rate of O_2 photoconsumption in *C. crispus*.

(in parentheses : number of replicates).

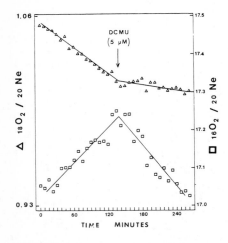

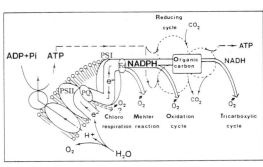

FIGURE 3. Typical inhibition by DCMU of O_2 uptake and O_2 evolution in the light.

FIGURE 4. Schematic representation of the O_2 fluxes during photosynthesis.

These results suggest a partition of the whole O_2 uptake into : 1) 50% Melher reaction and 50% mitochondrial oxidations at saturating CO_2, 2) 35% mitochondrial oxidations, from 0 to 45% RuBP oxygenase activity, and from 20% to 65% Melher reaction at limiting CO_2. Since INH was ineffective and because O_2 photoreduction may be CO_2-sensitive to some extent the Mehler reaction might be the major component of O_2 uptake in the light (closer to 65% than to 20%).

Any O_2 consuming process in the light may be considered as consuming energy since it brings back to a low level (that of water) the high redox potentials generated during the primary photosynthetic reactions (Figure 4). However the bioenergetic consequences of each process are not identical. The survival of a plant will depend upon its ability to balance its photochemistry (energy production) and its biochemistry (energy consumption).

The terrestrial environment imposes a strong carbon constraint which is due to the low atmospheric CO_2 concentration. Consequently, the available light energy is often more than can be usefully consumed through the PCR cycle. Photorespiration (POR cycle), which is an important feature of terrestrial C_3 plants, consumes ATP and NADPH with a stoichiometry close to that of the PCR cycle. Hence, it appears to be the most useful energy-dissipating mechanism for these plants.

In seawater, photosynthesis is already saturated at the natural TDIC concentration, and photorespiration is reduced, both due to the ability of using exogenous HCO_3^-. Nevertheless, some additional energy is required for the operation of the HCO_3^- uptake process and/or the internal DIC accumulation. In addition to mitochondrial oxidations, the Mehler reaction, based on pseudocyclic photophosphorylation, may participate in the generation of this extra ATP.

REFERENCES

1. Bréchignac F. and André M. (1985) Plant Physiol 78;545-550
2. Bréchignac F. et al. (1983) Physiol Vég. 21;665-676
3. Bréchignac F. and André M. (1984) Plant Physiol 75;919-923
4. Bréchignac F. and André M. (1985) Plant Physiol 78;551-555
5. Kremer B. (1981) Oceanogr Mar Biol Ann Rev 19;41-94
6. Bréchignac F. et al. (1986) Plant Physiol 80;1059-1062
7. Glidewell S.M. and Raven J.A. (1976) J Exp Bot 27;200-204
8. Radmer R.J. and Kok B. (1976) Plant PHysiol 58;336-340

INTERREACTION BETWEEN PHOTORESPIRATION AND PHOTOSYNTHESIS IN C_3 PLANTS

Gao Yu-zhu, Wang Zhong and Xiong Fu-sheng

Lab of Crop Photosynthesis, Jiangsu Agricultural College
Yangzhou, Jiangsu Province. The People's Republic of China.

INTRODUCTION

Photorespiration, the O_2-and light-dependent release of CO_2 in C_3 plants, is a curious process which inseparable from photosynthesis. Study on the relationship between the two process, which has not been known well, is important to understand the function of photorespiration and its effects on efficience of photosynthesis in crop plants.

Materials and Methods

The experiments were carried out with C_3 plants: rice, wheat, soybean and spanich in the whole leaves, leaf disc, mesophyll cells and chloroplasts.

Photosynthesis rate in leave was measured by IRGA and Oxygen electrode. CO_2 burst and O_2 inhibition($21\%O_2$) on photosynthesis were measured as photorespiration rate in leave.

The level of glycolate in mesophyll cells and the activity of glycolate oxygenase were assayed by the method of Nishimure(1). RuBPCase/Oase activity and the level of RuBP in leaves were determined with ^{14}C isotope tracing and oxygen electrode.

Results and Discussion

1. The parallel change in the rate of photosynthesis and photorespiration.

There is a parallel change in photosynthesis and photorespiration which was demonstrated by the results we obtained in the treatments of C_3 plants with various factors: temperature, light intensity, water stress, CO_2 and O_2 concentration and so on. As shown in Fig. 1, the rate of photosynthesis and photorespiration increased parallelly with increase

Biggens, J. (ed.), Progress in Photosynthesis Research, Vol. III. ISBN 90 247 3452 5
© *1987 Martinus Nijhoff Publishers, Dordrecht. Printed in the Netherlands.*

of light intensity. From table 1 we can
see, the main reaction spectrum for photorespiration was almost same to that
for photosynthesis, which appeared at
400-480nm and 600-700nm. When photosyn_
thesis was suppressed by DCMU, the rate
of photorespiration shown a corresponding decrease (Fig. 2). In generally,
photorespiration, in most cases, responded to environmental factors parallel
to that of photosynthesis.

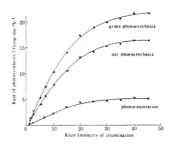

Fig. 1 Effect of light intensity on photosynthesis and
photorespiration.

Table 1   Effects of various light qualities on the rate of
photosynthesis and photorespiration.

Colors of mixed light and main range of wave length		Transmitted light relative energy	Gross rate of photosynthe_sis per energy unit	The rate of photorespiration per energy unit
white	400-760	100.0	15.3	4.0
red	630-760	32.7	19.9	4.9
red	600-760	40.7	22.7	5.5
orange	550-760	59.1	19.2	4.5
yellow	500-760	71.4	15.1	3.4
green	500-580	10.6	15.1	2.6
blue	400-570	21.0	17.1	3.1
violet	400-460	6.4	22.0	6.2
purple	400-430	13.3	27.3	6.7

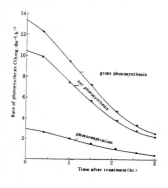

Fig. 2 Effect
of DCMU on ph-
otosynthesis
and photores-
piration in
wheat leaves.

Measurement of the
rate of CO_2 burst and
photosynthesis in diff-
rent light intensity
shown the rate of photo-
respiration was mainly
determined by the rate
of photosynthesis,
which was a support of

Loromer's hypothesis(2).

2. The functional relationship in the parallel change of photosynthesis and photorespiration.

It was found that though there were great difference in the rate of photosynthesis and photorespiration among cultivars and species, the ratio of photorespiration to photosynthesis kept usually stable at a range of 0.25-0.30, which could be fitted by the regression equation as follows: Y=a+0.2551X. Where Y is the rate of photorespiration, X the rate of photosynthesis in 2%O_2, This may be another representation of the Rubis-CO kinetics(3).

3. The relationship between photorespiration and photosynthetic energy metabolism.

When photorespiration was inhibited by HPMS(photorespiratory inhibitor), the level of ATP in the treated mesophyll cells was higher than that in the control(table 2); whereas inhibiting photophosphorylation by K_3AsO_4 and NH_4Cl could result in reduction of photorespiration(Fig. 3).

Table 2 Effect of HPMS on the level of ATP in soybean mesophyll cells

Addition	ATP concentration (nmoles·mgChl^{-1})		Relitive (%)	
HPMS(mM)				
1.0	22.5	19.6*	130.5	122.9*
2.0	17.7	15.2*	103.5	95.6*
5.0	15.6	13.9*	91.2	87.4*
10.0	13.0	13.1*	76.3	76.4*
control	17.1	15.9*	100.0	100.0

*: the results of another assay

In addition, we found, in the HCO_3^--free system, when photosynthetic energy metabolism was suppressed, the rate of carbon fixed going to glycolate decreased markedly which indicated photorespiration, to some extent, was regulated by photosynthetic energy metabolism.

4. The regulating role of photorespiration in photosynthesis.

Under CO_2-free air, with alternatively turning on - off light, we observed a considerable activity of CO_2 burst in wheat leaves, which

could maintained more than
one hour(Fig. 4), from
which we suggested that
during CO_2 burst, some
long-term products of
photosynthetic carbon
fixation were mobiled to
" C_3 cycle " and kept it
running normally.

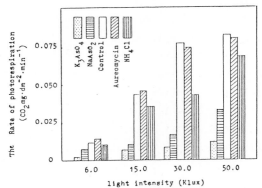

Fig. 3 Effects of photochemical reactor
on the rate of photorespiration under
CO_2-free air with various light intensities.

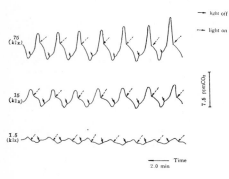

We also found the photoinhibition was more severer when C_3 plants
with HPMS especially under the conditions of strong light and low or
free CO_2 concentration.

From the results above, we conclude that photorespiration is necessary for C_3 plants to keep " C_3
cycle " running normally at lower
CO_2 concentration to reduce the inhibition of strong light and oxyg-
en on photosynthesis and to shorten
" the lag phase of photosynthesis ".

Fig. 4 Effect of light intensity on continuous CO_2 burst of
wheat leave in CO_2-free air.

References

1   Nishimure, M., Douce, K., and Akazawa, T., (1985) Plant physiol.
    78, 343-346
2   Andrews, T.J.,and Lorimer, G.H.,(1978) FEBS Letters 90, 1-9
3   Ogren, W.L. and Bowes, G. (1971) Nature 230, 159-160

MEASUREMENT OF PHOTOSYNTHESIS AND PHOTORESPIRATION IN THE FIELD

D. W. LAWLOR and A. T. YOUNG

PHYSIOLOGY & ENVIRONMENTAL PHYSICS DEPARTMENT, ROTHAMSTED EXPERIMENTAL STATION, HARPENDEN, HERTS. AL5 2JQ, U.K.

1.   INTRODUCTION
     Photorespiration (R_1) by leaves of C3 plants consume products of gross photosynthesis (P_g) and releases CO_2 thus decreasing net photosynthesis (P_n):- $P_n = P_g - R_1$. The ratio of R_1 to assimilation depends on the characteristics of ribulose bisphosphate carboxylase-oxygenase and the partial pressures of CO_2 and O_2 at the enzyme site, and upon temperature (1,2). Other factors may also increase R_1 with respect to P_n: water stress increases it because stomatal closure decreases the internal CO_2/O_2 ratio, and because a larger proportion of CO_2 is evolved by 'dark' respiration as assimilation decreases (3,4). Large applications of nitrogen fertilizers also increase R_1/P_n of flag leaves of field-grown winter wheat (5,6). However, the effects of nitrate in these experiments may have been confounded with water stress; well-fertilized crops produce more leaf area, transpire more and suffer greater water stress than poorly fertilized crops. However, different rates of nitrate supply to well-watered wheat in controlled environments do not alter R_1/P_n at constant temperature (7). Therefore, it is unclear how R_1/P_n is altered by water stress and nitrogen in the field, where it has rarely been measured, despite the importance of R_1 to crop production. We have measured P_g and P_n and hence estimated R_1, on crops grown in the field with two rates of nitrogen fertilizer combined with irrigation or drought.

2.   MATERIALS AND METHODS
2.1.   Crop growth.
       Winter wheat (cv. Avalon) was sown in September 1984 on nitrate-depleted soil. Plots were fertilized with either 300 or 0 kg N ha⁻¹ (H and L respectively) in February 1985. From March the crop was protected from rain with a mobile shelter (8) and plots of both N treatments were either droughted or irrigated weekly (D and I respectively).

2.2.   Measurements of CO_2 exchange.
       P_g and P_n were measured by Ludwig and Canvin's method (9) on fully emerged flag leaves in June and July 1985 during warm weather when R_1 would have been large in relation to P_n, and differences in leaf water potential substantial. Intact leaves still attached to the plant were enclosed in a leaf chamber (Fig. 1) and allowed to equilibrate in a steady flow of air (34 Pa CO_2, 21 kPa O_2, 79 kPa N_2). ¹⁴CO_2 was then supplied in an identical air mixture at specific activity of 670 kBq μmol CO_2⁻¹ and P_g was determined from the depletion of ¹⁴CO_2 within one minute of supplying the labelled gas. ¹²CO_2 concentration was measured throughout the sequence by IRGA

Biggens, J. (ed.), Progress in Photosynthesis Research, Vol. III. ISBN 90 247 3452 5
© 1987 Martinus Nijhoff Publishers, Dordrecht. Printed in the Netherlands.

(PLC chamber and LCA portable IRGA, The Analytical Development Co. Ltd., Hoddesdon, Herts., U.K.). Gas was supplied to the chamber from pressurized bottles at constant flow in excess of that required for the IRGA. ^{14}C was measured using a 15 cm^3 ionization chamber and electrometer (Varian Associates, Reading, U.K.). A switching system (10) controlled the gas supply and timing sequence to the leaf chamber, ionization chamber and IRGA, to achieve rapid, steady state measurements of R_l at near ambient conditions. Stomatal conductance (g) was calculated from transpiration and leaf temperature (T_l); T_l and water vapour pressure of the chamber air were not controlled and measurements of P_g, P_n, g and substomatal CO_2 concentration (C_i) were made at different times under different conditions.

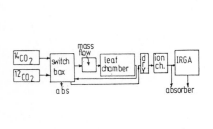

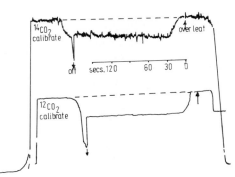

Fig. 1: Schematic of system for measurement of gas exchange of leaves under steady-state photosynthesis.

Fig. 2: Depletion of $^{12}CO_2$ and $^{14}CO_2$ in air passed over a leaf in the field.

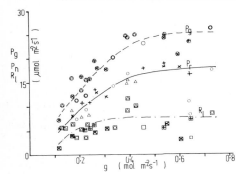

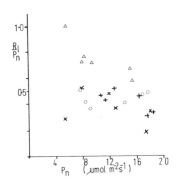

Fig. 3: P_g, P_n and R_l as a function of stomatal conductance for leaves grown with high (H) or low (L) nitrate, with (I) or without (D) irrigation (HI +; HD ▲; LI O; LD ✗; P_g symbols are in circles, R_l symbols are in squares).

Fig. 4: Ratio of photorespiration to net photosynthesis (R_l/P_n) in relation to net photosynthesis; treatments as described in Fig. 3.

2.3.   Water status
        The leaf water potential (Ψ) was measured, using a pressure
        chamber, on similar material to that used for gas exchange, during
        the measurements.

3.   RESULTS AND DISCUSSION
        The time-course of $^{12}CO_2$ and $^{14}CO_2$ content of air passing over a
wheat leaf (Fig. 2) showed greatest depletion of $^{14}CO_2$ after 30 s, decreasing as assimilated ^{14}C was released by photorespiration.

        Most of the measurements were made at light intensities exceeding
1000 μmol quanta (photosynthetically active radiation) m^{-2} s^{-1} and at leaf
temperatures between 26 and 32°C. P_n varied more with g than with light
with leaves of different ages and different water potentials. P_g, P_n and
R_l increased with increasing g up to 0.4 μmol m^{-2} s^{-1} and were constant at
greater g (Fig. 3); P_g reached a maximum of 25 and P_n 18 μmol CO_2 m^{-2} s^{-1}
giving a maximum R_l of about 7 μmol CO_2 m^{-2} s^{-1} i.e. an R_l/P_n ratio of 39%.
Assimilation in HI, LI and LD treatments was similar but in HD, P_g, P_n and
R_l were smaller.   However, the R_l/P_n ratio increased (Fig. 4) at low P_n in
the HD treatment but not others; this is related to the greater water
stress (lower Ψ) measured in HD (Fig. 5).

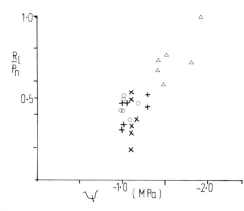 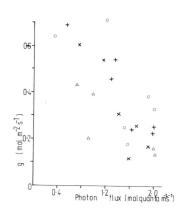

Fig. 5: Ratio of photorespiration   Fig. 6:   Stomatal conductance (g)
        to net photosynthesis   as a function of photon
        (R_l/P_n) as a function of   flux.
        leaf water potential (Ψ).

Droughted crops with deficient N had similar Ψ to irrigated crops (−1 to
−1.3 MPa) whereas Ψ of the HD treatment approached −2 MPa.   There was a
strong positive correlation between g and relative humidity (which was not
controlled) in the chamber.   Leaves were in the chamber for some time
before $^{14}CO_2$ exchange was measured and differences in humidity may have
influenced   and g.   This may explain the stomatal closure at large photon
flux (Fig. 6). R_l/P_n was not correlated with C_i (Fig. 7) suggesting that g
did not alter photorespiration relative to assimilation via changes in the
CO_2/O_2 ratio; rather water stress decreased P_n by changing metabolism,

perhaps by increasing dark respiratory CO_2 production (3,4,11). Nitrate supply did not affect R_l/P_n or g with similar ψ despite the large differences in crop growth and in the size and protein content of leaves.

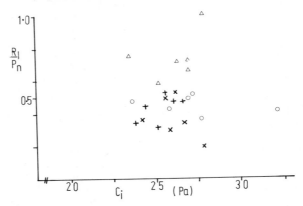

Fig. 7: Ratio of photorespiration to net photosynthesis (R_l/P_n) in relation to calculated sub-stomatal CO_2 partial pressure (C_i) for flag leaves from treatments described in Fig. 3.

REFERENCES

1 Laing, W.A., Ogren, W.L. and Hageman, R.H. (1974) Plant Physiol. 54, 678-685.
2 Lorimer, G.H. (1981) Ann. Rev. Plant Physiol. 32, 349-383.
3 Lawlor, D.W. (1976) Photosynthetica 10, 378-387.
4 Lawlor, D.W. and Fock, H. (1975) Planta (Berl.) 126, 247-258.
5 Thomas, S.M. (1976) in Ann. Rep. Rothamsted Exp. Stn. pp. 35-36.
6 Thomas, S.M. (1977) in Ann. Rep. Rothamsted Exp. Stn. p. 37.
7 Lawlor, D.W., Boyle, F.A., Young, A.T., Keys, A.J. and Kendall, A.C. (1986) In preparation.
8 Legg, B.J., Day, W., Brown, N.J. and Smith, G.J. (1978) J. agric. Sci. 91, 321-336.
9 Ludwig, L.J. and Canvin, D.T. (1971) Can. J. Bot. 49, 1299-1313.
10 Lawlor, D.W., Mahon, J.D. and Fock, H. (1977) Photosynthetica 11, 322-326.
11 Lawlor, D.W. (1983) in Effects of stress on photosynthesis (Marcelle, R., Clijsters, H. and van Pousse, M. eds.) Dr. W. Junk, Publ., The Hague, pp. 33-44.

REGULATING ROLE OF GA_3 ON PHOTOSYNTHETIC AND PHOTORESPIRATORY CARBON METABOLISM IN BARLEY LEAVES

L.P. Popova, T.D. Tsonev, S.G. Vaklinova

Institute of Plant Physiology, Bulgarian Academy of Sciences, Sofia, Bulgaria

Many of the investigators of plant growth regulators share the conception that about 50% of endogenous gibberellins is located within the chloroplasts (1) or chloroplast membranes (2). This suggests that gibberellins are taking an important part in the regulation of the main function of the chloroplasts—the process of photosynthesis. There is scanty and contradictory information published about the effect of gibberellins on the photosynthetic CO_2 fixation and on the activity of the photosynthetic enzymes. The prevalent opinion is that the treatment of the plants with GA_3 results in increasing photosynthesis, enhances ultrastructure morphogenesis of plastides and increases activity of ribulose-1,5-bisphosphate carboxylase (RuBPCase),(3,4). Contrary results were obtained by Huber and Sankhla in their experiments with Pennisetum, a plant with C_4-type of photosynthesis (5). According to De-Shieng Tsai and Arteca when GA_3 is introduced through the roots the response of the plants is not correlating with their belonging to mono-or-dicotyledonous C_3 or C_4 type plants, but is just a species feature (6).

Israelstam and Hiscox have pointed out that the rise in the net photosynthesis, as induced by GA_3, may be viewed as the result of both increased CO_2 fixation and of decreased intensity of photorespiration (7). This would be interpreted that GA_3 is participating in the regulation of CO_2 assimilation and CO_2 evolution as well as the energy balance of the cells.

In the present work the regulatory role of GA_3 in photosynthetic and photorespiratory carbon metabolism is investigated as well as the photosynthesis/photorespiration-ratio and the activity of the enzymes catalyzing the primary reactions of both processes.

MATERIALS AND METHODS

Investigations have been carried out with barley (Hordeum vulgare var, alfa) cultivated in a growth chamber under white fluorescent lamps (35 WM^{-2}) with 12 H light and dark periods. Day/night temperatures were 25/20 C, relative humidity was about 50%. The germinating seeds were cultivated for 7 days on water or water solutions from the required GA_3 concentrations ($10^{-6}-10^{-4}$ M GA_3).

Photosynthetic rates were measured using leaf slices by the method of Rathnam and Chollet (8). The slices were allowed to fix $^{14}CO_2$ for 10 min and the reactions were killed by adding boiling 80% ethanol. Tissues were subsequently extracted eight times with boiling ethanol of the same concentration. The amount of fixed $^{14}CO_2$ was determined using a Packard-TriCarb liquid scintillation counter. The radioactive products of photosynthesis were analyzed by the method of two dimensional paper chromatography and autoradiography.

Biggens, J. (ed.), Progress in Photosynthesis Research, Vol. III. ISBN 90 247 3452 5
© *1987 Martinus Nijhoff Publishers, Dordrecht. Printed in the Netherlands.*

The activity of the enzymes was measured in a leaf homogenate after blending 1 gr leaf tissue with 5 ml buffer containing: 0.33 m sorbitol, 0.05 m hepes-NaOH, 0.001 m $MgCl_2$, 0.002 m KNO_3, 0.002 m EDTA, 0.001 m $MnCL_2$, 0.0005 m K_2HPO_4, 0.02 m NaCl and 0.2 m Na-isoascorbate, pH 7.6. The homogenate was filtered through four layers of cheesecloth and centrifuged at 20,000 g for 15 min.

Activity of RuBP carboxylase was measured radiometrically in a reaction mixture containing in a final volume of 1 ml in μM: $MgCl_2$-10.0, DTT-1.0, $NaHCO_3$-20.0 (containing $10 \mu C/^{14}C$, specific radioactivity 10.36 $\mu C/\mu M$), RuBP-2.0 and enzyme extract (0.30-0.35 mg pr. in each sample), pH 7.8 T^o-25° C.

The reaction mixture for PEP carboxylase contained in 1 ml in μM: $MgCl_2$-20.0, NADH-0.4, $NaHCO_3$-20.0 (containing $10 \mu C^{14}C$, SP. Radioactivity 10.36 $\mu C/M$), PEP-3.0 and enzyme extract (0.3-0.4 mg protein), pH 8.0, T^o-30° C. Reaction time for the both reactions was 5 min.

Activity from the O_2 uptake in a reaction mixture containing: 50.0 mM tris buffer, pH 9.3, 10.0 mM $MgCl_2$, 5 mM DTT, 0.3 mM RuBP and enzyme extract.

Activity of glycolate oxidase was measured by the method of Kolesnikov (9). The amount of glyoxylic acid was assayed spectrophotometrically at 530 nm.

Rate of photorespiration was determined by the method of Catsky and Ticha (10), where photorespiration is estimated as the increase in net CO_2 exchange rate with ambient CO_2 between 20% O_2 and 1% O_2. A closed system was used which included an infrared gas-analyzer (Infralyt 4, GRD) and a paramagnetic O_2 analyzer (Permolyt 2, GRD). The light intensity was 200 W M^{-2}, the leaf temperature was maintained at 25° C.

RESULTS AND DISCUSSION

GA_3 stimulated photosynthesis by 28.3% at 10^{-6}M, 50% at 10^{-5} M and 35.8% at 10^{-4} M (Table 1).

TABLE 1: EFFECT OF GA3 ON THE DISTRIUBTION OF $^{14}CO_2$ IN THE PHOTOSYNTHETIC PRODUCTS IN BARLEY LEAVES
Photosynthetic fixation of CO_2 was 17.10^{-5} dpm g.fr. w.$^{-1}$ for controls. The degree of stimulation by GA_3 was as follows: 28.3% at 10^{-6} M GA_3, 50.3% at 10^{-5} M GA_3 and 35, 8% at 10^{-4} M GA_3. Values are averages from analysis of three separate experiments.

COMPOUNDS	DISTRIBUTION OF ^{14}C			
	CONTROL	10^{-6}M GA_3	10^{-5}M GA_3	10^{-4}M GA_3
	% OF THE TOTAL ^{14}C			
SUGARPHOSPHATE ESTERS	13.9	14.4	18.1	16.9
AMINO ACIDS	32.8	28.2	25.3	25.3
ORGANIC ACIDS	13.2	10.3	9.4	9.7
SUGARS	33.1	41.6	38.4	38.6
UNKNOWN	7.8	6.0	9.1	9.9

Analysis of the products of photosynthesis showed that GA_3 treated plants incorporated more intensive $^{14}CO_2$ than the control plants into the fractions of sugarphosphates and soluble sugars and relatively weaker into the fractions of organic acids and aminoacids.

Among the individual compounds the most significant differences were observed in the amount of 3 PGA, glycine, serine, glycolic acid, sucrose and malic acid (Table 2).

TABLE 2: EFFECT OF GA_3 ON DISTRIBUTION OF $^{14}CO_2$ IN THE MAIN PHOTOSYNTHETIC PRODUCTS IN BARLEY LEAVES
Details are described in the legend to Table 1.

COMPOUNDS	DISTRIBUTION OF ^{14}C			
	CONTROL	10^{-6}M GA_3	10^{-5}M GA_3	10^{-4}M GA_3
	% OF TOTAL CO_2			
URIDINEDIPHOSPHATE GLUCOSE	1.2	0.9	1.7	1.5
HEXOSEMONOPHOSPHATE ESTERS	4.9	2.7	4.5	5.3
HEXOSEDIPHOSPHATE ESTERS	4.0	5.9	6.1	5.5
3-P-GLYCERIC ACID	3.8	4.9	5.8	4.6
ALANINE	16.1	13.8	13.3	12.8
ASPARTIC ACID	4.5	5.7	6.1	4.7
GLYCINE+SERINE	12.2	8.7	5.9	7.8
MALIC ACID	7.8	6.1	5.0	5.0
CITRIC ACID	1.7	1.9	1.6	2.0
GLYCOLIC ACID	3.7	2.3	2.8	2.7
MALTOSE	5.2	5.2	4.5	5.4
SUCROSE	17.7	23.0	26.7	24.0
GLUCOSE	6.9	8.4	4.8	6.5
FRUCTOSE	3.3	5.0	2.4	2.7
UNKNOWN	7.8	6.0	9.1	9.9

In the GA_3 treated plants the relative incorporation of $^{14}CO_2$ into the products of the Calvin cycle: 3 PGA and hexosediphosphate esters was from 20 to 52% higher than those into the control plants for 3 PGA and for the hexosediphosphate esters—from 37 to 50%. A distinct inhibition was noticed in the share of label carbon incorporation into the metabolites of the photorespiratory carbon metabolism: glycolic acid, glycine and serine in GA_3 treated plants. There is almost 25% decrease in the labeling of glycolic acid. At the time this hormone sharply reduces the relative incorporation of ^{14}C into glycine and serine. Analysis of the soluble sugars showed that the amount of ^{14}C incorporated into sucrose was highest.

The results of the present investigation provide grounds for the prediction that GA_3 acts on certain units of the carbon photorespiratory metabolism, especially on both the functions of RuBPCO, which results in a change in the ratio of reductive and oxidative actions of photosynthesis and photorespiration. This assumption has made it necessary to determine the intensity of photorespiration and CO_2 fixation and the activity of the enzymes participating in the initial reactions and also the value of CO_2 compensation point.

TABLE 3: EFFECT OF GA_3 ON THE ACTIVITY OF RUBP CARBOXYLASE-OXYGENASE, PEP
CARBOXYLASE AND GLYCOLATE OXIDASE IN BARLEY
Activity for RuBP Case was $0.68 \mu M \ CO_2 min^{-1} \ mg.pr^{-1}$ for the
controls, PEPCase it was $0.13 \mu M \ CO_2 \ min^{-1} mg.pr.^{-1}$, for RuBP
oxygenase it was $0.27 \mu M \ O_2 min^{-1} mg.pr.^{-1}$ and for glycolate oxydase
it was $0.19 \ mg.glyoxilic \ acid \ min^{-1} mg.pr.^{-1}$.

TREATMENT	RuBP CARBOXYLASE	RuBP OXYGENASE	PEP CARBOXYLASE	GLYCOLATE OXIDASE
	ACTIVITY, % OF CONTROL			
H_2O(CONTROL)	100.0	100.0	100.0	100.0
10^{-6}M GA_3	132.3±0.6***	102.8±6.2	82.6±6.0	94.8±0.8*
10^{-5}M GA_3	144.0±6.1*	102.3±4.4	88.3±4.1	80.6±4.1*
10^{-4}M GA_3	123.5±3.1	112.8±3.8	83.3±2.2*	96.5±6.6

The data presented in Table 3 showed that the activity of RuBP Case
was greater in plants treated with GA_3. The degree of activation depends
on the GA_3 concentration. Greatest effect was found at 10^{-5} M GA_3.
GA_3 had no effect on the oxygenase function of RuBPCO but RuBP
carboxylase/oxygenase ratio was higher in GA_3 treated plants at all
investigated concentrations. PEP carboxylase activity was slightly
inhibited by GA_3. RuBP carboxylase/PEP carboxylase ratio increased in
GA_3 treated plants. Activity of glycolate oxydase was inhibited after
GA_3 treatment. The greatest effect was found at 10^{-5} M GA_3.
The results concerning the effect of GA_3 on the rate of
photorespiration, mitochondrial respiration and the value of CO_2
compensation point are presented in Table 4 and Fig. 1.

TABLE 4: EFFECT OF GA_3 ON THE RATE OF MITOCHONDRIAL RESPIRATION (RD),
PHOTORESPIRATION (RL) AND ON THE CO_2-COMPENSATION POINT ( ).

TREATMENT	/MGCO$_2$DM^{-2}H^{-1}/ RD	/MG CO$_2$DM^{-2}H^{-1}/ RL	/ Γ PPM /
H_2O(CONTROL)	1.21	3.85	61.8
10^{-6}M GA_3	1.45	3.93	56.6
10^{-5}M GA_3	0.70	2.32	44.5
10^{-4}M GA_3	1.60	3.58	60.6

FIG. 1: RESPONSES OF NET CO$_2$ FLUX TO AMBIENT CO$_2$ CONCENTRATION IN BARLEY
SEEDLINGS
Experimental data are marked with "■" for 1% O$_2$ and with "+" for 20%
O$_2$. The dashed line represents regression line for 1% O$_2$, and the
solid line for 20% O$_2$.

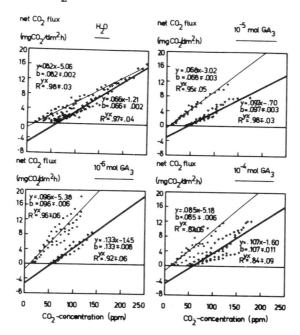

Fig. 1 shows the dependence of the net CO$_2$ exchange rate on the
ambient concentration of CO$_2$ in the controls and the plants cultivated on
10^{-6} M GA$_3$–10^{-4} M GA$_3$, recorded at 20% O$_2$ and 1% O$_2$ for determining the
photorespiration rate. The regression lines of the type Y=a, bx have
been determined. The respiration rates are determined by extrapollating
the dependence of the net CO$_2$ photosynthetic rate to zero CO$_2$
concentration (Fig. 1). Since for the regression lines Y$_x$=0=a, then at
1% O$_2$a=Rd, and at 20% O$_2$a=Rd + Rl, where Rl is the rate of
photorespiration. In an anologous manner at Y=0 estimates CO$_2$
compensation point (Table 4).
      10^{-5}M GA$_3$ decreased about 40% the rate of photorespiration,
mitochondrial respiration and the value of CO$_2$ compensation point. There
are no analogous changes at 10^{-6}M and 10^{-4}M GA$_3$.
      On the basis of the results it may be concluded that GA$_3$ exerts an
influence on the flow of carbon into the photosynthetic products of
barley. Compared with the controls, the leaves of seedlings grown in the
presence of GA$_3$ incorporated a larger proportion of ^{14}CO$_2$ (from 22% to
50%) into the products of the Calvin cycle: 3 PGA and hexosediphosphate
esters, while a lower proportion of ^{14}CO$_2$ was found in the products of
the photorespiratory carbon metabolism: glycolic acid, glycine and
serine. From the fraction of soluble sugars the amount of ^{14}C
incorporated into sucrose was from 30 to 50% higher in the controls than
in GA$_3$ treated plants.

These results are consistent with the stimulating effect of GA_3 on the rate of photosynthetic CO_2 fixation and on the activity of RuBP carboxylase. On the other hand with GA_3 treatment the activity of glycolate oxidase was decreased, the ratio in RuBP carboxylase/oxygenase activities was increased and less $^{14}CO_2$ was incorporated into the products of the photorespiratory carbon metabolism.

The results provide no categorical explanation of the mechanism of GA_3 action. On the basis of our experiments we concluded that GA_3 could regulate the ratio between photosynthesis and photorespiration and this effect depends on its concentration.

It should also be pointed out, last but not least, that the exogenous introduction of phytohormones in the plants disrupts their endogenic balance and the ratio in the levels between the individual phytohormones.

The regulating role of GA_3 on the rate of photosynthesis and photorespiration could be explained by two possibilities: 1) regulation of the synthesis of some chloroplast proteins and 2) an indirect effect mediated by the effect of GA_3 on stomatal opening: increase in the partial pressure of CO_2 and change in the CO_2/O_2 ratio in the chloroplasts.

REFERENCES
1 Loveys, B.R. (1972) Physiol. Plantarum, 40,6.
2 Browning, G. and Saunders, P.F. (1977) Nature, 265, 375.
3 Wellburn, F.A.M., Wellburn A.R., Stoddarrt, J.L., Treharne, K.J. (1973) Planta. III, 337.
4 Popova, L.P., Dimitrova, O.D., Vaklinova, S.G. (1982) Compt. Rend. Acad. Bulg. Sci., 35,6.797.
5 Huber, W. and Sankhla, N. (1985) Z. fur Pflanzenphysiol, 71,5,275.
6 De-Shieng Tsai and Arteca, R.N. (1985) Photosynthesis Research, 6, 147.
7 Israelstam, G.F. and Hiscox, I.D. (1980) Z. fur Pflanzenphysiol., 100, S, 69.
8 Rathnam, C.K.M. and Chollet, R. (1980) Plant Physiol., 65, 489.
9 Kilesnikov, P.A. (1962) Biochimia, 27, 193 (in Russian).
10 Catsky, G. and Ticha, I. (1975) Biol. Plantarum, 17, 405.

THE REGULATION OF SUCROSE SYNTHESIS IN LEAVES

HANS W. HELDT AND MARK STITT
UNIVERSITY OF GÖTTINGEN
INSTITUT FÜR BIOCHEMIE DER PFLANZE, UNTERE KARSPÜLE 2,
3400 GÖTTINGEN, F. R. G.

INTRODUCTION
Photosynthesis uses the energy of light to produce triosephosphates (trioseP) from carbon dioxide, water and phosphate (Fig. 1). In a typical leaf, e.g. from spinach, this trioseP is mainly utilized to synthesize sucrose for transport to other parts of the plant and for storage in the vacuole, and to lesser extent, to synthesize starch for storage in the chloroplasts. As CO_2 fixation is a cyclic process, 5/6 of the trioseP formed during CO_2 fixation have to remain in the cycle for the regeneration of the CO_2 acceptor ribulose 1.5 bisphosphate and only the reminder of 1/6 is available for utilization (Fig. 2). Any withdrawal of trioseP beyond this limit would result in a collapse of the CO_2 fixation cycle. On the other hand, CO_2 fixation is depending on the recycling of Pi released by sucrose and starch synthesis. To ensure a high rate of photosynthesis it seems therefore of vital importance that sucrose synthesis

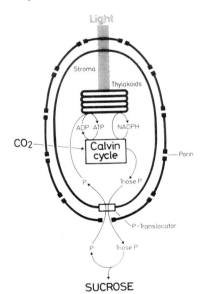

FIGURE 1. Schematic representation of photosynthesis

Abbreviations: PGA, 3-phosphoglycerate; DHAP, dihydroxyacetonephosphate; Fru6P, fructose 6 phosphate; Fru1.6P2, fructose 1.6 bisphosphate; Fru2.6P2, fructose 2.6 bisphosphate; Glc6P, glucose6phosphat.

Biggens, J. (ed.), Progress in Photosynthesis Research, Vol. III. ISBN 90 247 3452 5
© *1987 Martinus Nijhoff Publishers, Dordrecht. Printed in the Netherlands.*

is adjusted in such a way that the withdrawal of trioseP does not exceed the limit set by the CO_2 fixation cycle, and that from this surplus as much as possible is utilized. The mechanism of such a source control of sucrose synthesis will be shown in the following. Furthermore, it will be shown how the synthesis of sucrose might be adjusted to the sink to prevent osmotic damage by excessive accumulation of sucrose in a leaf cell. Finally, additional mechanisms for the regulation of sucrose synthesis will be discussed, which appear to be of importance for a diurnal programming of the partioning of photosynthate between sucrose and starch.

RESULTS AND DISCUSSION
Regulatory sites of sucrose synthesis
Candidates for regulatory sites are those processes which are displaced from equilibrium. Although a certain kinetic limitation appears to be exerted by the phosphate translocator catalyzing the export of trioseP to the cytosol in exchange with Pi (1), the major limiting steps of sucrose synthesis appear to be the reactions catalyzed by the cytosolic fructose 1.6 bisphosphatase, sucroseP synthase and sucrose phosphatase, as indicated from the large negative free energy changes listed in Table 1. The negative ΔG of sucroseP synthase, in contrast to its low ΔG_o (3) may be due to the fact that the activity of sucrose phosphatase is in large excess over the activity of sucroseP synthase (4).

Regulation of cytosolic fructose 1.6 bisphosphatase
As the first irreversible step in the transformation of trioseP into sucrose, the cytosolic fructose 1.6 bisphosphatase reaction, plays an important regulatory role in functioning as an entry valve for the withdrawal of trioseP from the CO_2 fixation cycle. The regulator substance Fru2.6P2 has a decisive function

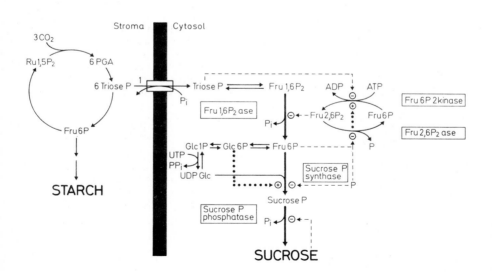

FIGURE 2.   Metabolic pathway of sucrose synthesis

TABLE 1. Free energy changes of cytosolic reactions involved in the synthesis of sucrose. ΔG values have been calculated from cytosolic metabolite concentrations in spinach leaves (for details see ref.1 and 2).

	ΔG° KJ/mol	ΔG
Aldolase	−23	0.8
Fructose 1.6 bisphosphatase	−17	−21
Phosphoglucoisomerase	− 3	0.1
Phosphoglucomutase	7	2
UDPGlc pyrophosphorylase	3	0.4
Sucrosephosphate synthase	− 1.2	− 8
Sucrose phosphatase	−14	−11

in this regulation (5). Micromolar concentrations of Fru2.6P2, as occurring in the cytosol of plant leaf cells, lead to a very strong decrease of the affinity of the cytosolic fructose 1.6 bisphosphatase towards its substrate Fru1.6P2 (Table 2). The level of the regulator substance Fru2.6P2 is due to a control of the rates of synthesis and degradation by metabolite levels. Because of a very rapid turnover, the response of the Fru2.6P2 level upon changes of metabolic conditions is very rapid. In barley leaves, for example, illumination causes a doubling of the Fru2.6P2 level within about a minute (Fig. 3).

The synthesis of Fru2.6P2 catalyzed by Fru6P,2kinase is stimulated by Fru6P and Pi and is inhibited by PGA and DHAP, whereas the hydrolysis of Fru2.6P2 catalyzed by Fru2.6P2ase is inhibited by Fru6P and Pi (7,8). Thus an increase of trioseP results in a lowering of the Fru2.6P2 level and hence to an increase of the affinity of the cytosolic fructose 1.6 bisphosphatase towards its substrate Fru1.6P2. Due to aldolase equilibrium an increase of the trioseP also results in an increase of the concentration of Fru1.6P2. The concurrence of an increase of substrate concentration and an increase of substrate affinity results in a sigmoidal dependence of fructose 1.6 bisphosphatase activity on the trioseP level. This is shown from a functional model for fructose 1.6 bisphosphatase

TABLE 2. Comparison of substrate affinity of cytosolic fructose 1.6 bisphosphatase in leaves from spinach and maize. Data from ref. (6).

Assay conditions	$Fru1.6P_2$ needed for half maximal activity (µM)	
	Spinach	Maize Mesophyll
V_{max}	3	20
+ 1 µM $Fru2.6P_2$	55	500
+ 10 µM $Fru2.6P_2$	250	3500

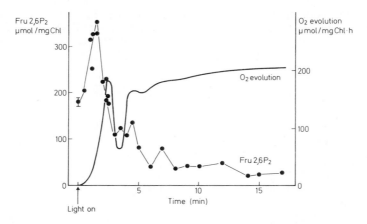

FIGURE 3. O₂ evolution and Fru2.6P2 levels during induction in barley leaves. Data from ref. (2).

activity based on cellular conditions (Fig. 4A). It illustrates the functioning of the entry valve of sucrose synthesis. For withdrawal of fixed carbon for sucrose synthesis a certain threshold level of trioseP has to be exceeded, ensuring that a minimal trioseP level required for an efficient functioning of the CO_2 fixation cycle is maintained. An increase of trioseP beyond this threshold results in a dramatic rise of fructose 1.6 bisphosphatase activity, making all the surplus of trioseP avaiable for sucrose synthesis.

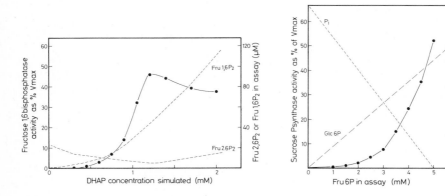

FIGURE 4. Simulation of the response of the cytosolic fructose 1.6 bisphosphatase (A) and sucroseP synthase (B) to increasing levels of trioseP and hexoseP in situ. The simulated enzyme activity is shown as a solid line, the concentration of substrate and regulators are shown in dotted lines. For details see ref. (9,10).

TABLE 3.   Comparison of estimated Fru2.6P$_2$ and DHAP concentrations in leaves from spinach and maize (light, air).

	Spinach (cytosol)	Maize (mesophyll cells)
Fru2.6P$_2$	11 µM (11)	3–30 µM (6)
DHAP	0.3–0.8 mM (1)	11 mM (12)

The threshold principle for the regulation of cytosolic Fru2.6P2ase activity by trioseP levels is supported from recent findings with maize leaves. Here the enzymes of sucrose synthesis, including fructose 1.6 bisphosphatase, and the regulatory substance Fru2.6P2 are largely if not exclusively located in the mesophyll cells. Since maize mesophyll cells supply the bundle sheath cells with redox equivalents by diffusion driven trioseP-PGA shuttle (12), the trioseP levels in the mesophyll cells are about one order of magnitude higher than in C$_3$ plants like spinach (Table 3). On the other hand, the levels of the regulator substance Fru2.6P2, and the relative inhibitory effect of this substance on the cytosolic fructose 1.6 bisphosphatase in maize are similar to spinach. It may be noted, however, that in maize the substrate concentration required for half maximal saturation of fructose 1.6 bisphosphatase is about one order of magnitude higher than in spinach. These findings indicate that in maize the regulation of sucrose synthesis at the site of cytosolic fructose 1.6 bisphosphatase occurs after the same threshold principle as in spinach, with the difference that in maize according to the metabolic demands the threshold level for the withdrawal of trioseP is set about one order of magnitude higher than in spinach.

Regulation of sucrose phosphate synthase
Also sucroseP synthase is a strongly regulated enzyme. This is clearly demonstrated from results of subcellular metabolite analysis in spinach leaves, where the cytosolic levels of Fru6P and UDPGlc do not differ largely under conditions for maximal rates of sucrose synthesis in the light and minimal rates of sucrose synthesis (from starch) in the dark (Table 4). SucroseP synthase has been shown to be allosterically activated by glucose6P and inhibited by Pi (13). As shown in Table 1, the activator Glc6P is in equilibrium with Fru6P, which is

TABLE 4. Cytosolic metabolite levels in spinach leaves. Data from ref (1).

	Light 40 min	Dark 2 h
	nmol/mg Chl	
UDP Glc	28	38
Fru6P	27	30

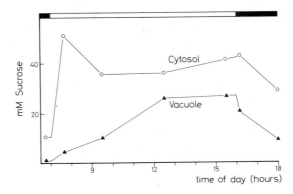

FIGURE 5. Estimated concentrations of sucrose in the vacuole and extra-
vacuolar compartments of spinach leaves. Data from ref. (1).

the substrate of sucroseP synthase. Due to the phosphoglucose isomerase
equilibrium a change of the concentration of the substrate Fru6P results in an
about threefold change of the concentration of the activator. Since the sum of
Pi and phosphorylated intermediates in the cytosol may be regarded as about
constant, the increase of the activator Glc6P will be paralleled by a decrease
of the inhibitory Pi level and vice versa. In a model experiment simulating
these conditions the activity of sucroseP synthase reveals a sigmoidal
dependence on the Fru6P concentration (Fig. 4B). Apparently, also with this
enzyme, the substrate level has to exceed a threshold value for sucroseP
synthesis to occur. Moreover, the regulatory effect of hexosephosphates and Pi
on sucroseP synthase and (via the formation of Fru2.6P2) on fructose 1.6
bisphosphatase makes it possible that the activities of the two enzymes of
sucrose synthesis are adjusted to each other.

Regulation of sucrose phosphatase
In a variety of tissues, sucrose phosphatase was shown to be competitively
inhibited by sucrose to various extent. In spinach, for instance, 100 mM sucrose
resulted in 49% inhibition of the enzyme (4). In view to the high activity of
sucrose phosphatase present, it is presently a matter of debate whether this
observed inhibition bears a physiological relevance. In spinach leaves, during
sucrose synthesis cytosolic sucrose concentrations in the range of 50 mM are
observed (Fig. 5). When spinach leaf discs are floated in a medium containing
100 mM sucrose, a large increase of Glc6P and Fru2.6P2 is found (Table 5) which
might be explained in terms of a feedback control of sucrose on sucrose
phosphatase, resulting in an increase of Fru6P, which by increasing the
regulatory substance Fru2.6P2 inhibits fructose 1.6 bisphosphatase. In this way,
sucrose synthesis might be adjusted to the demands of the sink. An alternative
mechanism of such a feedback control could be that by excessive increase of
sucrose levels in the cell sucrose is split again, yielding hexoses which by
hexokinase activity are being transformed to hexosemonophosphates (C.Foyer
commun.). In fact, leaf discs floated on glucose medium show an even higher
increase of Glc6P and Fru2.6P2 levels as those floated on sucrose (Table 5).
Further experiments are required to elucidate the contribution of sucrose
phosphatase in the regulation of the overall process of sucrose synthesis.

TABLE 5. Effect of sugars on $Fru2.6P_2$ levels in spinach leaf discs (11).

Preincubation with 100 mM sugar	$Fru2.6P_2$	Glc6P
	nmol/mg chl	
Sorbitol (control)	0.20	59
Sucrose	0.35	93
Glucose	0.49	127

Diurnal regulation of sucrose synthesis

In spinach leaves, one observes during the first two hours after the onset of daylight only a low rate of starch synthesis. During the day the rate of starch accumulation gradually increases (Fig. 6) concurrently with a gradual increase of the levels of hexosemonophosphates, trioseP (10), and Fru2.6P2 (Fig. 7) and also of the leaf content of sucrose (Fig. 6). Subcellular analysis revealed that this increase of leaf sucrose content is due to a deposition of sucrose in the large vacuolar space and does not concur with an increase of the cytosolic sucrose level (see Fig. 5). The increased levels of Fru2.6P2 during enhanced starch synthesis made it probable that starch synthesis was enhanced by an inhibition of sucrose synthesis, but this could not be explained by a simple feedback control of sucrose synthesis discussed in the preceeding. A subsequent search for alternative mechanisms for the regulation of sucrose synthesis revealed that diurnal changes of enzyme activities were involved. In desalted leaf extracts obtained at various times of the day one finds in the course of the day a gradual increase of the activity of Fru6P,2kinase and a decrease of Fru2.6P2ase activity (14). The resultant increase of the Fru6P,kinase/Fru2.6P2ase ratio explains the gradual increase of Fru2.6P2 during of the day. It seems probable that these changes in enzyme activities are due to a modification of the enzyme protein. In other tissues, Fru6P,2kinase and Fru2.6P2ase, the activities of both being due to the same protein, are known to be regulated by cAMP dependent protein phosphorylation. In liver, phosphorylation results in a stimulation of Fru2.6P2ase and an inhibition of Fru6P,2kinase (15) whereas in

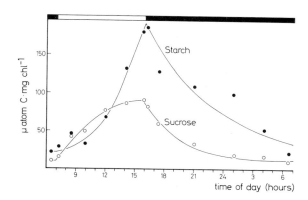

FIGURE 6. Diurnal changes of starch and sucrose content in spinach leaves. Data from ref (1).

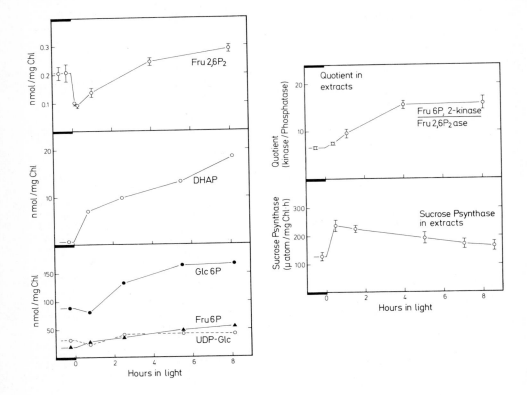

FIGURE 7.   Changes of cytosolic metabolites and of extracted activities of
Fru6P,2kinase, Fru2.6P2ase and sucroseP synthase in spinach leaves
during the day. The enzyme extracts were separated from low
molecular compounds such as metabolites and assayed under standard
conditions. For details see ref. (10).

yeast the two enzyme activities are changed by phosphorylation in the opposite
direction (16,17). It remains to be elucidated whether a regulation by protein
kinases, e.g. like in yeast, might also explain the observed diurnal changes of
activities in plant cells.

Also the activity of sucroseP synthase changes in a diurnal manner (Fig. 7).
Upon illumination there is first a rapid rise in activity, with a gradual
decline afterwards. Diurnal changes in the activities of this enzyme have been
earlier shown to occur (18,19). More detailed investigations showed that the
changes in activity shown in Fig. 7 are actually caused by a change of the
kinetic properties of the enzyme. As shown in Fig. 8, sucroseP synthase
extracted from leaves kept in the dark is much more sensitive towards inhibition
by Pi than the enzyme extracted from leaves illuminated for 15 min. Prolonged
illumination of leaves increased the sensitivity of the enzyme towards Pi. The
mechanism leading to this enzyme modification remain to be investigated.

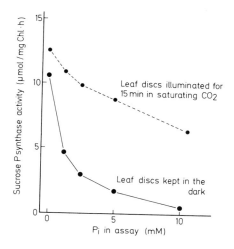

FIGURE 8.   Changes in the sensitivity to Pi of sucroseP synthase extracted from spinach leaf discs which had been kept in the dark and in the light.

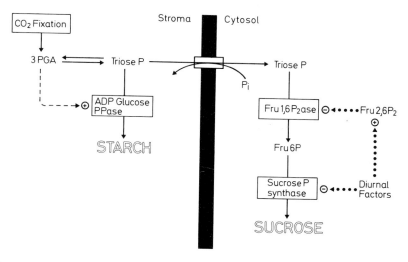

FIGURE 9. Schematic representation of the diurnal regulation of sucrose and starch synthesis.

The apparent modification of enzymes controlling sucrose synthesis may play a role in programming the partitioning between sucrose and starch independently from a source-sink regulation of sucrose synthesis. Fig. 9 summarizes a sequence of events. The increased level of Fru6P as the result of a decline of sucroseP synthase activity together with an increase of Fru6P,2kinase/Fru2.6P2ase activity leads to an increase of Fru2.6P2 level, which in turn will lower the flux of carbon through the cytosolic fructose 1.6 bisphosphatase, and hence to an accumulation of trioseP in the leaf, as can be observed. Although the regulation of starch synthesis is not fully understood at present, an increased stromal PGA/Pi ratio as a consequence of trioseP accumulation will increase the activity of ADPglucose pyrophosphorylase (20) and thus be a major factor in the stimulation of starch synthesis. Diurnal regulation of enzyme activities may therefore be a means to ensure that the starch deposits of the chloroplasts are filled during the day in order to meet the metabolic demands of the plant cell in the following might.

Acknowledgement
This work was supported by the Deutsche Forschungsgemeinschaft.
We thank Bernd Raufeisen for the preparation of the figures.

REFERENCES
1 Gerhardt, R., Stitt, M. and Heldt, H.W. Plant Physiol. submitted
2 Stitt, M., Huber, S. and Kerr, P. in The Biochemistry of Plants (Hatch, M.D. and Boardman, K. eds.), Vol. 8, Academic Press, New York (in print)
3 Barber, G. (1985) Plant Physiol. 79, 1127-1128
4 Hawker, J.S. and Smith, G.M. (1984) Phytochemistry 23, 245-249
5 Cseke, C., Balogh, A., Wong, J.H., Buchanan, B.B., Stitt, M., Herzog, B. and Heldt, H.W. (1984) Trends Biochem. Sci. 9, 533-535
6 Stitt, M. and Heldt, H.W. (1985) Planta 164, 179-188
7 Cseke, C., Stitt, M., Balogh, A. and Buchanan, B.B. (1983) FEBS Letts. 162, 103-106
8 Stitt, M., Cseke, C. and Buchanan, B.B. (1984) Eur.J.Biochem. 143, 89-93
9 Herzog, B., Stitt, M. and Heldt, H.W. (1984) Plant Physiol. 75, 561-565
10 Stitt, M., Gerhardt, R., Wilke, I. and Heldt, H.W. (1986) Physiol. Plant. (in press)
11 Stitt, M., Kürzel, B. and Heldt, H.W. (1984) Plant Physiol. 75, 559-560
12 Stitt, M. and Heldt, H.W. (1985) Biochim. Biophys. Acta 808, 400-414
13 Doehlert, D.C. and Huber, S.C. (1983) Plant Physiol. 73, 989-994
14 Stitt, M., Mieskes, G., Söling, H.-D., Große, H. and Heldt, H.W. (1986) Z. Naturforschung 41c, 291-296
15 Hers, H.-G., Hue, L. and van Schaftingen, E. (1982) Trends Biochem. Sci.7, 329-331
16 Francois, J., van Schaftingen, E. and Hers, H.-G. (1985) Eur. J. Biochem. 145, 187-193
17 Yamashoji, S. and Hess, B. (1985) FEBS Letts 178, 253-256
18 Huber, S.C. and Kerr, P.S. (1984) Plant Physiol. 74, 445-447
19 Sicher, R.C. and Kremer, D.F. (1985) Plant Physiol. 79, 695-698
20 Preiss, J., Robinson, N., Spilatro, S. and McNamara, K. (1985) in Regulation of Carbon Partitioning in Photosynthetic Tissue (Heath R.L. and Preiss J., eds.), pp.1-26, Waverley Press, Baltimore

LIMITATION OF PHOTOSYNTHESIS BY SUCROSE SYNTHESIS

MARK STITT
UNIVERSITY OF GÖTTINGEN
INSTITUT FÜR BIOCHEMIE DER PFLANZE, UNTERE KARSPÜLE 2
3400 GÖTTINGEN, F. R. G.

During photosynthesis, triose P are exported from the chloroplast and converted to sucrose in the cytosol, releasing P_i which returns to the chloroplast to support further photosynthesis. Thus, sucrose synthesis and CO_2 fixation are mutually interdependent processes which will have to be coordinated if high rates of photosynthesis are to be maintained [1]. If sucrose synthesis were to be too slow, metabolites would accumulate and CO_2 fixation would be inhibited by a depletion of P_i. On the other hand, if sucrose synthesis were to be too fast, stromal metabolites would be depleted and the regeneration of $Ru1,5P_2$ would be inhibited. This problem can be seen in isolated chloroplasts, where the supply of P_i outside the chloroplast has to be carefully adjusted to attain the maximum rate of photosynthesis in any given conditions. How is an analogous control achieved <u>in vivo</u>, and is the optimal answer always found?

FRU2,6P_2 AND SPS AS RATE CONTROLLERS OF SUCROSE SYNTHESIS

The cytosolic FBPase and SPS are key control points in sucrose synthesis [1,2]. As discussed in the article by Heldt and Stitt in this volume [2], both enzymes are regulated via changes of metabolites and P_i, but are also subject to a higher level of control. For the cytosolic FBPase, this involves changes in the concentration of $Fru2,6P_2$, which inhibits the enzyme and also decreases its affinity for the substrate, while increasing its sensitivity to inhibition by AMP. The high level of control for SPS involves activation, <u>via</u> a mechanism which has not yet been identified, but probably involves protein modification [1,3]. This leads to increased activity and, in particular, to an alteration in its kinetic properties as the substrate affinity increases [3] while the sensitivity to inhibition by P_i decreases [2].

The alterations of the $Fru2,6P_2$ level and the activation of SPS occur progressively as the rate of photosynthesis increases, allowing sucrose synthesis to be switched on as it is required. As shown in Table 1 and 2, the changes of $Fru2,6P_2$ and SPS activation follow the light response curve of photosynthesis, and are shifted by factors which alter the light-saturation response, including temperature (see Tables) and CO_2 (not shown). Thus, as photosynthesis increases, the available activity of FBPase (modulated by $Fru2,6P_2$) and of SPS (via activation) rises, allowing sucrose to be synthesised more rapidly. As both lead to increased substrate affinity [1,2,3] this allows sucrose synthesis to be activated and P_i to be recycled more rapidly even though the cytosolic metabolite levels remain almost unchanged [4,5]. In other words, a gradual activation of SPS and decline of $Fru2,6P_2$ not only allows them to act as rate controllers of sucrose synthesis over a wide range of rates of photosynthesis, but also allows remarkably constant metabolic conditions to be maintained in the cytosol as the fluxes change.

Biggens, J. (ed.), Progress in Photosynthesis Research, Vol. III. ISBN 90 247 3452 5
© *1987 Martinus Nijhoff Publishers, Dordrecht. Printed in the Netherlands.*

TABLE 1. Fru2,6P$_2$ levels and the light response of photosynthesis in spinach leaves. Measured after 10 minutes illumination with saturating CO_2.

Light Intensity	Photosynthesis 15 °C	30 °C	Fru2,6P$_2$ 15 °C	30 °C
(μmol.m^{-2}.s^{-1})	(μmol.mg Chl^{-1}.h^{-1})		(pmol.mg Chl^{-1})	
Dark	− 9	−21	247	243
275	90	140	192	240
404	141	–	156	–
570	178	259	137	188
1100	188	412	143	161
1500	–	453	–	126

TABLE 2. SPS activation and the light response curve of photosynthesis in barley leaves. Measured after 10 minutes illumination with saturating CO_2.

Light Intensity	Photosynthesis 15 °C	30 °C	SPS activity 15 °C	30 °C
(μmol.m^{-2}.s^{-1})	(μmol.mg Chl^{-1}.h^{-1})			
Dark	− 9	−27	4.0	4.5
40	22	− 3	6.0	4.6
78	47	30	7.1	4.9
144	94	113	7.7	5.7
275	111	217	8.2	6.8
570	127	359	8.6	8.5
1100	140	444	9.0	8.9

In principle, these mechanisms provide a way of adjusting sucrose synthesis to process the triose P made available by photosynthesis, without this impinging on the ability of the chloroplast to carry out photosynthesis. However, we now need to assess whether the regulation always operates effectively in vivo. To do this will require developing ways of recognising when photosynthesis is limited by an inbalance between sucrose synthesis and the chloroplast. The identification of Fru2,6P$_2$ levels and SPS activation as rate controllers for sucrose synthesis provides a way of approaching this question, as they can be measured to provide an estimate of the available capacity for sucrose synthesis at a given time point. If sucrose synthesis is exerting a significant limitation on photosynthesis, then changes of Fru2,6P$_2$, and SPS activation should impact directly on the rate of photosynthesis.

SLOW TRANSIENTS OF PHOTOSYNTHESIS CAN BE RELATED TO SUCROSE
SYNTHESIS

In Fig. 1, barley leaves were illuminated in high light and CO_2, and then
quenched in N_2 and extracted to measure enzyme activation and metabolites.
Sucrose synthesis is activated slowly, as revealed by the adjustment of
$Fru2,6P_2$ and of SPS which requires about 10 minutes. In contrast, the Calvin
cycle enzymes activate rapidly, and within 2 minutes there is a rapid build up
of metabolite pools and a high rate of photosynthesis is achieved. However,
these high rates of photosynthesis cannot be maintained, because sucrose
synthesis has not yet been activated. As P_i is depleted, the ATP/ADP ratio
collapses, leading to a massive accumulation of PGA (equivalent to 20 mM in
the stroma) and an inhibition of O_2 evolution. Subsequently, $Fru2,6P_2$ and
SPS adjust, sucrose synthesis is activated, the ATP/ADP ratio and PGA level
recover partially, and the rate of photosynthesis rises again. During this
recovery there is also a partial deactivation of Rubisco and (not shown) an
adjustment of the redox and energy quenching, showing that the chloroplast
metabolism may also be adjusting to the lowered supply of P_i. Nevertheless,
the activation of sucrose synthesis will be playing a decisive role by allowing
P_i to be recycled more rapidly.

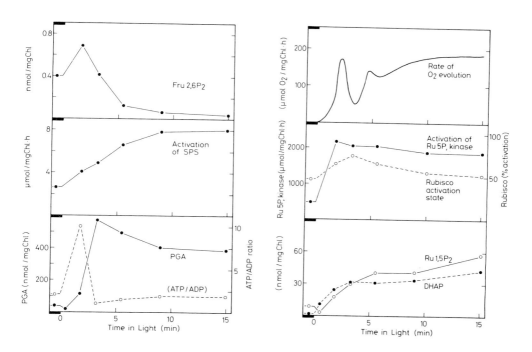

Figure 1. Simultaneous measurement of metabolites and enzyme activation
following illumination of barley leaves. The leaves were held in low
light, then darkened for 5 minutes and reilluminated at 20 °C with
570 $\mu mol.m^{-2}.s^{-1}$ light and saturating CO_2.

OSCILLATIONS OCCUR WHEN SUCROSE SYNTHESIS IS LIMITING

After 10 minutes illumination in Fig. 1, there is a steady rate of O_2 evolution, but the relatively low ATP/ADP ratios and high PGA suggest P_i may still be limiting. This interpretation can be tested by imposing a short interuption in low light or darkness (6), during which electron transport and CO_2 fixation are stopped rapidly but sucrose synthesis continues, consuming the preexistent pools of phosphorylated intermediates and regenerating P_i (5). Upon reillumination, this higher P_i supports a transient stimulation of photosynthesis, which is associated with higher ATP/ADP ratios, decreased PGA, and increased $Ru1,5P_2$ (6), showing how both the electron transport and the Calvin cycle are being stimulated. However, rather than merely returning back to the original rate of photosynthesis as the higher P_i is consumed, oscillations are induced, with the rate of O_2 evolution even being driven below the steady rate, as illustrated in Fig. 2. This experiment reveals how $Fru2,6P_2$ is changing rapidly, and inversely, to the rate of photosynthesis. These oscillations of $Fru2,6P_2$ will be altering the recycling of P_i from the cytosol, and could be contributing to the oscillations of photosynthesis. High $Fru2,6P_2$ will decrease the release of P_i during sucrose synthesis, restricting the electron transport, as revealed by the low triose P/PGA ratio, and inhibiting photosynthesis. Conversely, as $Fru2,6P_2$ decreases, more P_i can be recycled to the chloroplast allowing rapid electron transport, PGA reduction and photosynthesis.

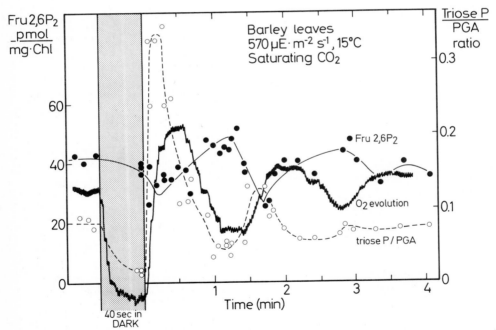

Figure 2. Oscillations of O_2 evolution, $Fru2,6P_2$ and PGA in barley leaves. Leaves were illuminated at 15 °C (saturating CO_2, 570 $\mu mol.m^{-2}.s^{-1}$ light) for 10 minutes, darkened for 40 seconds and reilluminated.

The changes of $Fru2,6P_2$ could result from an interaction between metabolites, the enzymes synthesising and degrading $Fru2,6P_2$, and the cytosolic FBPase. When high $Fru2,6P_2$ inhibits the cytosolic FBPase, Fru6P is depleted as SPS continues to convert hexose P to sucrose. As the hexose P pool declines SPS will also be inhibited, and recycling of P_i decreases and PGA accumulates. Such changes would inhibit synthesis of $Fru2,6P_2$, and stimulate its hydrolysis (see 1), leading to a decrease of $Fru2,6P_2$. This, in turn, reactivates the cytosolic FBPase, regenerating Fru6P and P_i and allowing PGA to be reduced. These changes of metabolites may then stimulate synthesis and inhibit hydrolysis of $Fru2,6P_2$ (1), driving its concentration upwards and reinitiating the cycle. Readjustments in the stroma and thylakoids may also be contributing to the oscillations. Nevertheless, these results support the idea that the oscillations are closely linked with a limitation of photosynthesis by sucrose synthesis. They may even be indicating the relaxation time of $Fru2,6P_2$ in the cytosol. In an analogous way, the gradual activation of photosynthesis during the biphasic induction of Fig. 1 may be closely linked with the gradual activation of SPS which sets an overall maximum for the rate of sucrose synthesis.

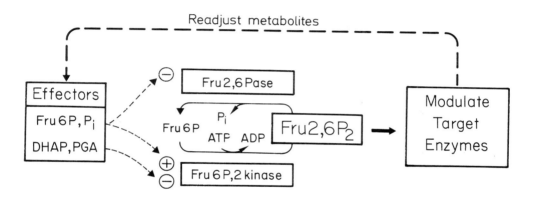

Figure 3. A possible time-loop to explain the oscillations of $Fru2,6P_2$.

ALTERING THE RELATION BETWEEN CO_2 FIXATION AND SUCROSE SYNTHESIS

If oscillations are diagnostic of a limitation in sucrose synthesis, it should be possible to explain their appearance or disappearance in terms of a changing relation between reactions in the chloroplast, and the synthesis of sucrose. For example, oscillations emerge as CO_2 or light are raised (7,8,9), which pushes the rates of CO_2 fixation and electron transport upwards until a ceiling is met in the reactions leading to sucrose (6). Oscillations also emerge when O_2 is lowered to decrease photorespiration (9), which releases more electron transport and Calvin cycle capacity for net CO_2 fixation, imposing an additional load on sucrose synthesis. Alternatively, species with high rates of starch synthesis may show smaller oscillations, due to the internal cycling of P_i in the chloroplast acting as a dampener (7,8).

However, some conditions which modify oscillatory behaviour cannot be explained in terms of a shift in the 'loading' of the chloroplastic and cytosolic reactions. One such case is the appearance of oscillations when O_2 is lowered in the presence of 5% CO_2, when photorespiration is unlikely (7). Indeed, O_2 even stimulates net CO_2 fixation at high light and CO_2 (10). O_2 may be affecting the poising of electron transport (e.g. by decreasing pseudocyclic electron transport flow) and modifying regulatory interactions between the thylakoid reactions. Redox poising outside the chloroplast will also be essential if a high cytosolic adenylate status is to be maintained to allow rapid sucrose synthesis (1). In either case, it seems that O_2 modifies the susceptibility of photosynthesis to limitation by sucrose synthesis.

Lower temperature also leads to oscillations appearing at lower light intensity (6) and CO_2 (9), as well as the appearance of 'O_2-insensitivity', which is also taken as indicating a limitation by sucrose synthesis (see 9). These changes match the shift in the light response curve of photosynthesis at lower temperature (see above) (6). Thus, oscillations appear in all cases as photosynthesis becomes light saturated. But why do the light response curve and oscillations show this temperature dependence?

TEMPERATURE-DEPENDENT ADJUSTMENT OF SUCROSE SYNTHESIS
The simplest answer would be that sucrose synthesis is selectively restricted at lower temperatures, an idea which is supported by Table 3, showing how metabolites accumulate at lower temperatures. Further studies showed that SPS has a fairly high Q_{10} of about 3 (not shown) and that the cytosolic FBPase shows temperature-dependent changes in its properties (Table 4). The sensitivity to inhibition by Fru2,6P_2 and AMP rises markedly at lower temperatures. As Fru2,6P_2 and AMP inhibit by lowering substrate affinity, increasingly high levels of substrate are needed for activity at lower temperatures i.e. the 'threshold' (1,2) for activating sucrose synthesis is raised. This will predispose a leaf to limitation by sucrose synthesis at low temperatures, especially when the potential rate of photosynthesis has been pushed upwards by using high light and CO_2.

At low light or CO_2, this temperature response of sucrose synthesis may be adaptive, as higher metabolite concentrations in the stroma counteract the decline in the rate of enzyme catalysis at lower temperatures. In this way, fluxes around the Calvin cycle are maintained even at lower temperatures. In Table 5, near constant rates of photosynthesis are attained between 8-35 °C, due to far higher metabolite pools being maintained in limiting light at the lower temperatures.

It is interesting that the relatively poor performance in low light above 30 °C is associated with low levels of metabolites (Tables 3,5). Previous explanations for the inhibition of photosynthesis at moderately high temperature and low light or CO_2 levels were in terms of high respiration, or an increased oxygenase/carboxylase ratio. These results suggest an additional factor, namely that at higher temperatures the cytosolic FBPase is too active, due to the action of AMP and Fru2,6P_2 being weakened. This imposes an additional restriction on photosynthesis by draining triose P from the stroma too rapidly and depleting the Calvin cycle pools.

TABLE 3. Metabolite levels at different temperatures in spinach leaves. Measured after 10 minutes illumination with saturating CO_2.

	Temperature °C	Light intensity ($\mu mol.m^{-2}.s^{-1}$)				
		Dark	270	570	1100	1500
Photosynthesis ($\mu mol\ O_2.mg\ Chl^{-1}.h^{-1}$)	15	-10	90	178	188	177
	30	-21	140	259	412	453
Metabolites ($nmol.mg\ Chl^{-1}$) PGA	15	77	105	105	138	178
	30	59	80	104	104	53
Triose P	15	2	23	28	41	38
	30	4	18	19	23	20
Hexose P	15	119	162	197	194	196
	30	60	137	129	159	136

TABLE 4. Inhibition of spinach leaf cytosolic FBPase by AMP and Fru2,6P_2: temperature-dependence. Assayed at pH 7.5, 4 mM $MgCl_2$.

Temperature	Concentration for 50% inhibition with 50 μM AMP		FBP concentrations for half maximal activity with 2 μM Fru2,6P_2 and 100 μM AMP
	Fru2,6P_2	AMP	
(°C)	(μM)		
5	0.44	83	1000
10	0.59	120	793
15	0.75	250	469
20	0.91	470	324
27	1.15	1020	151
30	1.51	1630	95

TABLE 5. Temperature-dependence of photosynthesis and metabolite levels at low light intensities in spinach leaves. Measured after 12 minutes illumination (200 $\mu mol.m^{-2}.s^{-1}$) in saturating CO_2.

Temperature	O_2 evolution	Metabolites			
		Rul,5P_2	PGA	Triose P	Hexose P
(°C)	($\mu mol.mg\ Chl^{-1}.h^{-1}$)	($nmol.mg\ Chl^{-1}$)			
8	43	82	186	23	271
14	56	72	100	21	217
20	63	69	75	19	210
28	75	43	54	18	162
35	65	40	35	15	140

DIAGNOSIS OF LIMITATIONS
Sucrose synthesis can be adjusted to the performance of the chloroplast via allosteric control, time-dependent adjustment of $Fru2,6P_2$ and SPS activation, and temperature-dependent adjustment of the enzyme activity. Nevertheless, it is possible to impose conditions under which these regulatory mechanisms are no longer able to maintain the requisite balance between the cytosolic and chloroplastic reactions. If conditions are created which are particularly favourable for CO_2 fixation, photosynthesis may become limited by the rate of sucrose synthesis. Oscillations in O_2 evolution (and fluorescence) occur, which may be related to rapid changes of $Fru2,6P_2$, and slower transients follow the activation of SPS. Additional indications for the limitation by sucrose synthesis are provided by low ATP/ADP ratios, an accumulation of PGA, and a deactivation of Rubisco, all of which are also typical for isolated chloroplasts in low P_i (1). Conversely, in conditions which selectively favour sucrose synthesis, an additional restriction of photosynthesis can arise because metabolites are drained too rapidly from the stroma. In these conditions photosynthesis decreases, there is a general decline of metabolites, including $Ru1,5P_2$, and the PGA/triose P ratio is low.

These conclusions are based on experiments carried out in simplified conditions, in which the occurence and consequence of a limitation by sucrose synthesis could be more easily analysed. They still need to be extended to more natural conditions, where many factors may interact to determine the rate of photosynthesis. Such studies could provide a widened framework for considering how 'sink' control of photosynthesis may operate, and what aspects of photosynthesis may be involved during acclimatisation or adaptation to varying environmental conditions.

Acknowledgement
This work was supported by the Deutsche Forschungsgemeinschaft, and was carried in close collaboration with Hans W. Heldt.

REFERENCES
1 Stitt, M., Huber, S. C. and Kerr, P. in The Biochemistry of Plants (Hatch, M. D. and Boardman, K., eds.), Vol. 8, Academic Press, London (in press)
2 Heldt, H. W. and Stitt, M. in Proceedings of the VIIth International Congress on Photosynthesis, August 10-15, 1986, Brown University, USA (Biggins, J., ed.), Nijhoff Publishers, Dordrecht
3 Sicher, R. C. and Kremer, D. F. (1985) Plant Physiol. 79, 695-698
4 Stitt, M., Wirtz, W. and Heldt, H. W. (1980) Biochim. Biophys. Acta 593, 85-102
5 Stitt, M., Wirtz, W. and Heldt, H. W. (1983) Plant Physiol. 72, 767-774
6 Stitt, M. (1986) Plant Physiol. (in press)
7 Sivak, M. N. and Walker, D. A. (1986) New Phytol. 102, 499-512
8 Laisk, A. and Walker, D. A. (1986) Proc. R. Soc. Lond. B. 227, 281-302
9 Sharkey, T. D., Stitt, M., Heineke, D., Gerhardt, R., Raschke, K. and Heldt, H. W. (1986) Plant Physiol. (in press)
10 Vill, J., Laisk, A., Oja, V. and Parnik, T. (1977) Photosynthetics (Prague) 11, 251-259

STARCH SYNTHESIS AND ITS REGULATION

JACK PREISS, MATTHEW MORELL, MARK BLOOM, VICKI L. KNOWLES AND TSAN PIAO LIN, DEPARTMENT OF BIOCHEMISTRY, MICHIGAN STATE UNIVERSITY, EAST LANSING, MI 48824

INTRODUCTION

A certain amount of the carbon dioxide that is converted to 3-P-glycerate in plants during photosynthesis is utilized for formation of the reserve polysaccharides present in the leaf starch granule. However, the fraction of carbon accumulated as starch in different plants varies. Species such as barley, spinach and wheat accumulate proportionately more sucrose than starch whereas the leaves of peanuts, soybean and tobacco accumulate more starch than sucrose. It is believed that an important factor in the partitioning of carbon between sucrose and starch is the level of sucrose-P synthase (1,2). The levels of sucrose-P synthase in the leaf cell correlates fairly well with the amount of $^{14}CO_2$ incorporated into sucrose and inversely correlates with the amount of $^{14}CO_2$ fixed into the starch fraction.

Other major factors involved in the partitioning of carbon are the levels of various metabolites, triose phosphate (triose-P), 3-phosphoglycerate (3-PGA) and inorganic phosphate (Pi). They are the regulatory effectors of an enzyme involved in the biosynthesis of starch, ADPglucose pyrophosphorylase (3), in modulating fructose 2,6-bisphosphate levels (4), as well as being the transport ligands of the chloroplast membrane Pi/triose phosphate/3-PGA translocator (5). Fructose 2,6-bisphosphate (6) inhibits the cytoplasmic fructose 1,6-bisphosphatase and thus, sucrose biosynthesis.

Pi is an inhibitor and 3-PGA is the activator of the enzyme catalyzing the synthesis of ADPglucose (ADPG), pyrophosphorylase (3). Pi is an activator of the enzyme catalyzing the synthesis of fructose 2,6-P_2, phosphofructo-2-kinase while 3-PGA and dihydroxyacetone-P are inhibitors (4). Pi is an inhibitor of the fructose 2,6-P_2 bisphosphatase (4). Thus, it is easy to understand how the cytoplasmic and chloroplastic levels of triose-P, 3-PGA and Pi would regulate both starch and sucrose synthesis. The changing levels of these metabolites would undoubtedly participate in the regulation of carbon partitioning.

It was of interest to understand more of the structure-function relationship of the effector sites of the plant leaf ADPglucose pyrophosphorylase since these effector sites are prevalent in proteins involved in the synthesis of starch as well as in sucrose and the binding domains of these binding sites may have great similarity. Pyridoxal-P has been previously used as a site specific probe for the Pi/Triose-P/3-PGA translocator (5,7) and for the bacterial ADPG pyrophosphorylase activator site (8). Results presented here show that it may also be a useful site-specific probe for the spinach leaf ADPglucose pyrophosphorylase.

Biggens, J. (ed.), Progress in Photosynthesis Research, Vol. III. ISBN 90 247 3452 5
© *1987 Martinus Nijhoff Publishers, Dordrecht. Printed in the Netherlands.*

RESULTS

Structural Studies of the Spinach Leaf ADPG pyrophosphorylase. The spinach leaf ADPglucose pyrophosphorylase had been previously purified to homogeneity and was shown to have a native molecular weight of 206,000 (9). The enzyme appears to have two putative subunits of different molecular weights originally and erroneously reported to be 44,000 and 48,000, respectively. However, further studies indicated that these molecular weights are 51,000 and 54,000. Studies of all the previously purified bacterial enzymes and plant ADPglucose pyrophosphorylases purified from non-photosynthetic tissue, potato tuber and maize endosperm (Plaxton and Preiss, unpublished results) indicated that these enzymes had only one subunit with molecular weights ranging from 48 to 55 kilodaltons (3,8). Antibody prepared against the purified spinach leaf enzyme used in Western blotting experiments have shown only one band with the purified maize endosperm enzyme or extracts while results with leaf extracts of rice, wheat, and maize show a reaction with two protein bands (10).

It is quite possible that the appearance of two subunits of the ADPG pyrophosphorylase seen in leaf extracts is due to the lower molecular weight form being derived from the higher molecular weight form via protease action. It is also possible that the two subunits are products from two different genes. Therefore, studies were directed towards determining the nature of the relationship of the two different subunits.

Purification of the Spinach Leaf ADPG Pyrophosphorylase. The previously reported purification procedure of the spinach leaf ADPG pyrophosphorylase resulted in very low yields of the enzyme (9). Thus, it was decided to modify the procedure to obtain higher yield of relatively pure enzyme. The original purification procedure was followed through the DEAE-cellulose chromatography step (9). The DEAE fraction was then subjected to Fast Protein Liquid Chromatography (FPLC) using the Pharmacia Mono Q HR column.

The enzyme was purified about 6-fold with an 85% yield giving it a specific activity of 41.7 µmol of ATP formed per min per mg of protein. The Mono Q HR fraction was then further purified by hydrophobic chromatography using an ethyl-agarose column. The enzyme was retained by the column in the presence of 0.9 to 1.0 M phosphate buffer, pH 7.0, but is immediately desorbed when the eluting solvent is reduced in phosphate concentration to 0.2 M.

This gave a further purification of 2.5-fold with about a 72% recovery at this step. A preparation with an overall recovery of 31% from the initial extract with a specific activity of 105 µmol/min/mg was obtained. About 22 mg of protein were obtained from 8 kg of spinach leaves.

Figure 1 shows the electrophoretic pattern of the various fractions obtained during the purification. One can note that the major protein in the initial extracts is clearly the large subunit, 55 KD, of ribulose bisphosphate carboxylase (Rubisco). However, this was essentially separated from the ADPG pyrophosphorylase via the FPLC step. What is seen in the final C-2 fraction (lane 7) are the two subunits of the ADPG pyrophosphorylase. The enzyme is estimated to be about 95% pure with a slight contamination of peptides of MW of 62 and 64 KD.

Immunoblotting of ADPglucose Pyrophosphorylase. Fig. 2 shows the immunoblotting reaction of antibody with ADPglucose pyrophosphorylase and with Rubisco large subunit. A good reaction is seen with 0.05 µg

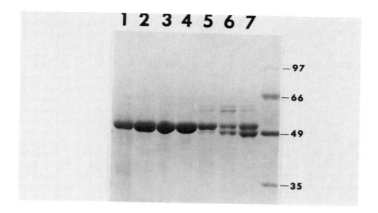

Fig. 1. SDS electrophoresis of spinach ADPGlucose pyrophosphorylase.
Aliquots from various stages of the purification were electrophoresed on
a 10% SDS polyacrylamide gel and stained with Coomassie blue. Lane 1,
initial homogenate (50 µg),; lane 2, heat step (50 µg); lane 3, ammonium
sulfate fraction (50 µg); lane 4, PVP step (50 µg); lane 5, DEAE-cellulose peak (20 µg); lane 6, Mono Q peak (13 µg); lane 7, ethyl agarose
peak (20 µg). Standards: 5 µg each of phosphorylase B (97 KD), BSA (66
KD), E. coli ADPglucose pyrophosphorylase (49 KD) and lactate dehydrogenase (35 KD).

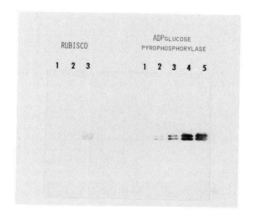

Fig. 2. Western blotting of Rubisco and ADPglucose pyrophosphorylase.
Varying amounts of purified Rubisco (obtained from Sigma) and ADPglucose
pyrophosphorylase from spinach were electrophoresed on a 10% SDS polyacrylamide gel. The gel was electroblotted to nitrocellulose paper and
reacted with sera containing antibodies directed against spinach ADP-
glucose pyrophosphorylase. The immunoreactive proteins were visualized
using an alkaline phosphatase conjugated secondary antibody (11,12).
Rubisco: lane 1, 0.1 µg; lane 2, 1 µg; lane 3, 10 µg. ADPglucose pyrophosphorylase: lane 1, 0.01 µg; lane 2, 0.05 µg; lane 3, 0.1 µg; lane 4,
0.5 µg; lane 5, 1 µg.

and a quite potent reaction is seen with 0.1 µg. Some reaction, however, is barely seen with Rubisco at 1 µg and at 10 µg a perceptible amount of reaction is seen with the Rubisco obtained from Sigma.

It is quite possible that a trace amount of Rubisco was present in the purified ADPG pyrophosphorylase preparation which elicited antibody towards Rubisco or there is some slight immunological relationship between the two enzymes. It is important to note that precaution should be taken in using the ADPglucose pyrophosphorylase antibody in crude extracts because of the great abundance of Rubisco.

Isoelectrofocusing experiments indicate that the pI value for the 54 KD subunit is higher than the 51 KD subunit. Evidence was also obtained to indicate that the presence of two subunits in the enzyme preparation was not due to proteolysis of the larger subunit when the crude extract was prepared (Fig. 3). Preparation of the leaf extract in the presence of 9.5 M urea and detergent still yielded the 54- and 51 KD protein bands after electrophoresis on a two-dimensional gel system according to O'Farrell (13). As expected from the results seen in Figure 2, the third immunospot is Rubisco due to its presence in great excess in leaf extracts. If post-translational modification or proteolysis has occurred to generate the smaller subunit from the large subunit, the process must have occurred in vivo.

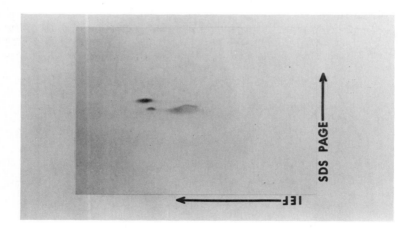

Fig. 3. Detection of ADPglucose pyrophosphorylase in crude extracts of spinach. One leaf (~0.5 g) was excised from a four week old spinach plant. The tissue was immediately homogenized in 1 ml of buffer containing 9.5 M urea, 2.5% (w/v) NP-40, 2% (v/v) Pharmalytes pH 3-10 and 5% (v/v) β-mercaptoethanol using a mortar and pestle. The homogenate was centrifuged at 12,000 xg for 5 minutes and the pellet discarded. An 8 µl (120 µg) aliquot of the supernatant was electrophoresed in a two dimensional gel system by the method of O'Farrell. The gel was electroblotted to nitrocellulose paper and reacted with affinity purified antibody to spinach ADPglucose pyrophosphorylase. Immunoreactive proteins were visualized using an alkaline phosphtase conjugated secondary antibody.

Sequence Studies. The subunits have been separated by FPLC fraction-ation in 8 M urea on the Mono Q anion exchange column. As seen in Fig. 4, N-terminal sequences of the two subunits have been obtained and the sequences are found to be different.

54 kilo-   [1]Ser - Val - Thr - Ala - Asp - Asn - Ala - Ser - Glu - Thr[10]
dalton   Lys - Val - Arg - Glu - Ile - Gly - Gln - Glu - Lys -( )[20]
fraction   (Ser)-(Ser)

51 kilo-   [1]Val - Ser - Asp - Ser - Gln - Asn - Ser - Gln -(Asp)-(Gly)[10]
dalton   (Leu)-(Asp)-(Pro)-(Glu)
fraction

Fig. 4. Amino acid sequences of the spinach leaf ADPglucose pyrophos-phorylase subunits. The amino acids placed in parentheses indicates that identification is only tentative. Residue 20 of the 54 kilodalton subunit was not identified.

Activation of ADPG pyrophosphorylase by Pyridoxal-P (PLP). Figure 5 shows the activation of the ADPglucose pyrophosphorylase by PLP and by 3-PGA. Whereas 15 µM 3-PGA gives 50% of maximal activity ($A_{0.5}$) the $A_{0.5}$ value for PLP is about 6 to 8 µmolar. However, 3-PGA gives about a 25-fold stimulation whereas PLP only gives about a 6-fold activation. Moreover, the maximal velocity observed with optimal concentrations of PLP is only 20% that observed with 3-PGA as the activator. Similar results have been obtained with ADPglucose pyrophosphorylase from Arabidopsis thaliana and from maize endosperm (Plaxton and Preiss, unpublished results). PLP does not appreciably affect the K_m of the

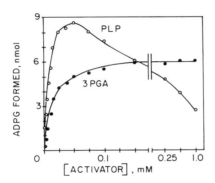

Fig. 5. Activation of ADPG synthesis by PLP and 3-PGA. The reaction mixtures contained 20 µmol of Bicine buffer, pH 8.0, 1 µmol of $MgCl_2$, 0.1 µmol of $[^{14}C]$-glucose-1-P (1105 cpm per nmol), 0.2 µmol of ATP, 50 µg of bovine serum albumin, 0.1 unit of inorganic pyrophosphatase vary-ing concentrations of activator (3-PGA or PLP) as indicated in the figure and enzyme in a volume of 0.2 ml and incubated for 10 minutes at 37°. Eight times more enzyme was used to determine the PLP activation curve.

substrates. However, 3-PGA does significantly reduce the K_m for ATP. PLP can also reverse Pi inhibition as does 3-PGA. In the absence of activator, the enzyme is inhibited 50% by 45 µM Pi. In the presence of 100 µM PLP, however, 50% inhibition requires 230 µM Pi. PLP because of its lower stimulation can inhibit the activation seen with 3-PGA. In the presence of 1 mM 3-PGA; 0.36 mM PLP gives 50% inhibition while in the presence of 0.5 mM 3-PGA one observes 50% inhibition with 0.26 mM PLP.

All these results suggest that PLP does bind at or near the 3-PGA activator site. Since PLP can be irreversibly covalently bound to the protein by reduction with NaBH₄ it was of interest to determine the properties of the reduced phosphopyridoxylated enzyme. If it indeed bound to the activator site and placed the enzyme in the active conformation, one would expect the modified enzyme to be more active in the absence of activator. The results are seen in Table 1.

Table 1. Effect of reductive phosphopyridoxylation on the kinetics of spinach leaf ADPglucose pyrophosphorylase. The reaction mixtures contained in a volume of 1 ml, 0.1 M Bicine buffer, pH 8.0, 50 µM PLP, 0.3 mM NaBH₄, 0.25 mg enzyme and various ligands as indicated in Table 1 at room temperature for 30 min. Controls lacking PLP or NaBH₄ were also run.

Reaction Conditions Additions/Omissions	ADPG Synthesized		+3PGA/-3PGA
	-3PGA	+3PGA	
	µmol/ml		
No PLP, No NaBH₄	23.3	719	30.9
No PLP	16.7	522	31.3
None	83	417	5.0
ADPG, 1 mM; MgCl₂, 6 mM	104	601	5.8
3PGA, 1 mM	58	570	9.9
Pi, 1 mM	33	576	17.4

Reductive phosphopyridoxylation causes the enzyme to be more active in the absence of activator. The activity ratio of ADPG synthesized in presence and in absence of 3PGA is lowered. There is no change in the ratio when either NaBH₄ or PLP is omitted. The presence of the substrate ADPG plus MgCl₂ seems to protect the activity slightly but does not affect the reduction of the +3PGA/-3PGA activity ratio. However, the presence of 3PGA and Pi inhibit the increase in unactivated activity and decrease the extent of +3PGA/-3PGA activity ratio reduction.

The phosphopyridoxylated modified enzyme is quite insensitive to Pi inhibition (Fig. 6). Whereas the untreated enzyme is 50% inhibited by 42 µM Pi only 35% inhibition is noted for the chemically modified enzyme at 0.2 mM Pi. Both Pi and 3PGA when present during the reductive phosphopyridoxylation prevent to a great extent the desensitization of the enzyme to Pi inhibition. These results are also in keeping with the view that PLP binds to the activator site and places the enzyme in a conformation more resistant to Pi inhibition.

These results suggest that the PLP is indeed being covalently bound to the activator site and that 3PGA and Pi prevent to some extent the reductive covalent binding of PLP. Recent experiments utilizing [³H]-PLP show incorporation of labeled PLP into both subunits of ADPG pyrophosphorylase (unpublished). The results suggest at present that

more PLP is bound to the higher molecular weight subunit, but nevertheless both subunits bind PLP and the binding can be inhibited by the
presence of 3PGA.

The binding of labeled PLP to the two subunits of ADPG pyrophosphorylase will enable us to determine the sequence of the activator
binding site of both subunits and whether they are the same. It will
enable us to compare the activator binding site sequence with other
ADPglucose pyrophosphorylases having different activator specificities
(8).

It is of interest to note that pyridoxal-P binds and inhibits the
transport activity of the Pi/triose-P/3PGA translocator (5,13). The
specificity of the effector molecules of the ADPglucose pyrophosphorylase are similar to the metabolites acted upon by the translocator. It
would be of interest to determine and compare the sequences of these
proteins that are involved in carbon partitioning. Since Pi, 3PGA, and
DHAP affect P-fructokinase 2, another enzyme involved in partitioning of
carbon it would certainly be of interest to learn the nature of its
effector binding sites and if there is similarity in sequence. It is
quite possible that in the evolution of the regulatory mechanisms
involved in carbon partitioning, sucrose and starch synthesis that the
DNA portion specifiying peptide sequences for the triose-P, Pi and 3PGA
binding site domains were replicated and incorporated into various structural genes utilized in the above processes.

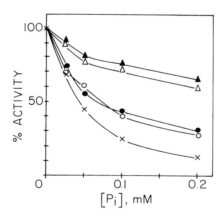

Fig. 6. Inhibition of ADPG pyrophosphorylase chemically modified with
PLP by Pi. The reaction mixture is described in Fig. 4 except that no
activator is present. The various enzymes are, untreated enzyme, X-X;
enzyme reacted with PLP + NaBH₄, Δ-Δ; enzyme reacted with PLP + NaBH₄ in
presence of ADPG plus MgCl₂, ▲-▲; enzyme reacted with PLP + NaBH₄ in
presence of 3PGA, o-o; enzyme reacted with PLP + NaBH₄ in presence of
Pi, ●-●. The chemical modification reactions are described in Table 1.

ACKNOWLEDGEMENTS
     The authors wish to thank Dr. Young Moo Lee for assistance in
obtaining the sequence data, and Ms. Judy Huang for help with the
O'Farrell gel analysis. This research is supported by research grants
from NSF (DMB10088), NIH (AI22835) and the McKnight Foundation (Photosynthesis Training Grant).

REFERENCES
1. Huber, S.C. and Israel, D.W. (1982) Plant Physiol. 69, 691-696.
2. Huber, S.C. (1983) Plant Physiol. 71, 818-821.
3. Preiss, J. (1982) Ann. Rev. Plant Physiol. 33, 431-459.
4. Stitt, M., Cseke, C. and Buchanan, B.B. (1984) Eur. J. Biochem. 143,
     89-93.
5. Fliege, R., Flugge, U.-I., Werdan, K. and Heldt, H.W. (1978)
     Biochim. Biophys. Acta 502, 232-247.
6. Cseke, C., Weeden, N.F., Buchanan, B.B. and Uyeda, K. (1982) Proc.
     Natl. Acad. Sci. USA 79, 4322-4326.
7. Flügge, U.F and Heldt, H.W. (1984) TIBS 9, 530-533.
8. Preiss, J. (1984) Ann. Rev. Microbiol. 38, 419-458.
9. Copeland, L. and Preiss, J. (1981) Plant Physiol. 68, 996-1001.
10. Krishman, B.H., Reeves, C.D. and Okita, T.W. (1986) Plant Physiol.
     81, 642-645.
11. Burnette, W.N. (1981) Anal. Biochem. 112, 195-203.
12. Somerville, C.R., McIntosh, L., Fitcher, J., and Gurevity, M. {1986)
     Meth. Enzymol. 118, 419-433.
13. O'Farrell, P.H. (1975) J. Biol. Chem. 250, 4007-4021.

REGULATION OF STARCH SYNTHESIS IN ZEA MAYS LEAVES.

STEVEN R. SPILATRO AND JACK PREISS, DEPARTMENT OF BIOCHEMISTRY, MICHIGAN STATE UNIVERSITY, EAST LANSING, MI 48824

INTRODUCTION

Synthesis of starch in leaves of Zea mays, an NADP-ME type C_4 plant, is normally restricted to the bundle sheath cells (1). The absence of starch accumulation in maize mesophyll may be related to the distribution of starch metabolizing enzymes between the bundle sheath and mesophyll cells, although this has not been convincingly established (1-3). The stromal enzyme, ADPglucose pyrophosphorylase catalyzes the formation of ADPglucose, the glucosyl donor for starch synthase, and is believed to be the principal site of regulation for starch biosynthesis. This enzyme from all plants studied to date is activated by 3-phosphoglycerate (3-PGA) and inhibited by orthophosphate (Pi) (4). The rate of starch synthesis is believed to be coupled to the rate of photosynthesis and to the rate of photosynthate export from the cell through the interaction of this enzyme and the stromal ratio of 3-PGA/Pi. Activity of this enzyme from whole leaf extracts of sorghum, also a NADH-ME type C_4 plant, is also allosterically regulated by 3-PGA and Pi (5). However, significance of the kinetic properties of bundle sheath and mesophyll specific ADPglucose pyrophosphorylases to localization of starch synthesis in leaves of maize or other C_4 plants has not been previously investigated.

MATERIALS AND METHODS

Protoplasts were prepared from maize leaves 12 days post-germination by cellulase digestion using 2% "Cellulysin" in a buffer solution containing 20 mM Mes (pH 5.5), 0.5 M sorbitol and 5 mM $MgCl_2$. Protoplasts were filtered from the crude digest through an 80 μM screen, and purified using a discontinuous dextran T-40 gradient (6) in 5 mM Hepes (pH 7.5), 0.5 M sorbitol and 5 mM $MgCl_2$. Bundle sheath strands in the crude cellulase digest were freed of protoplasts by initally washing the undigested material in the above buffer containing 0.2 M sorbitol, then allowed to settle through a 800 μM screen. The purified mesophyll protoplasts and bundle sheath strands were either immediately homogenized for enzyme localization studies, or stored at -80°C and used later for ADPglucose pyrophosphorylase purification.

For purification of ADPglucose pyrophosphorylase, the cell fraction homogenates were precipitated by addition of 50% polyethyleneglycol (PEG) solution to obtain fractions of 10% to 20% PEG for mesophyll, and 12% to 18% for bundle sheath. These precipitates were redissolved in a solution containing 30 mM Tris (pH 8.0), 5 mM $MgCl_2$ and 0.5 mM EDTA, and chromatographed with a Pharmacia Fast Protein Liquid Chromatography (FPLC) instrument on a Mono-Q anion exchange column preequilibrated in the above buffer. The column was eluted with a linear 0 to 500 mM NaCl gradient at 10 mM min^{-1} and 1 ml min^{-1} flow rate. The fractions containing activity were pooled and concentrated with an Amicon ultrafiltration apparatus, and used for analysis of enzyme kinetic properties.

ADPglucose pyrophosphorylase was assayed in the pyrophosphorylase direction for enzyme localization studies and during enzyme purification. In this assay the rate of ^{32}P incorporation into ATP is measured during pyrophosphorylysis of ADPglucose with ^{32}PPi (7). Activity was assayed in the ADPglucose synthesis direction for all studies of enzyme catalytic and allosteric properties (8). In this assay, enzyme activity is measured by following the rate of incorporation of ^{14}C into ADPglucose from ^{14}C-glucose-1-P.

RESULTS AND DISCUSSION

Previous studies are in conflict concerning the localization and activities of starch metabolizing enzymes in maize leaves (1-3), and are limited in documentation of cell purity and recovery of enzyme activities. In this study, mesophyll and bundle sheath cells were obtained with less than 1% cross contamination, as judged from the activities of the marker enzymes, RuBP carboxylase and PEP carboxylase. Enzyme activities generally were equal to or greater than previously reported values. On the basis of specific activity, at least 75% of starch synthase, and branching enzyme, and 95% of ADPglucose pyrophosphorylase were in the bundle sheath. In contrast, enzymes of starch degradation, starch phosphorylase and amylase, were more evenly distributed between the cell types and even showed slightly higher activities in the mesophyll. Correlation of a high ratio of starch biosynthetic activity to degradative activity and starch accumulation (1) in the bundle sheath, as compared to a lower ratio and the absence of starch in the mesophyll, suggests that the levels of these enzymes may be partially responsible for the amount of starch accumulating in these cells.

The role of the catalytic and allosteric properties of mesophyll and bundle sheath specific ADPglucose pyrophosphorylases in the regulation of starch synthesis in maize was investigated using enzymes partially purified from the isolated cell types. Purifications of 74-fold and 53-fold with 49% and 25% recoveries were obtained for the bundle sheath and mesophyll ADPglucose pyrophosphorylase, respectively.

The pH optimum of the bundle sheath ADPglucose pyrophosphorylase extended roughly from pH 7.0 to 7.8, whereas the mesophyll enzyme showed a somewhat higher optimum, at pH 7.8 to 8.2. Substrate saturation kinetics were hyperbolic for glucose-1-P and ATP, but highly sigmodial (n=3.1 to 3.2) for $MgCl_2$. Mg^{++} was an essential cofactor for both enzymes. The apparent affinities were, for glucose-1-P, 0.02 and 0.03 mM; for ATP 0.06 and 0.07 mM; and for Mg^+, 1.5 and 1.8 mM for the mesophyll and bundle sheath enzymes, respectively.

Greatest activation of ADPglucose pyrophosphorylase activity was observed with 3-phosphoglycerate, which produced 22-fold and 8-fold activations of the bundle sheath and mesophyll enzymes, respectively. Activation to a lesser extent was observed for a number of other metabolites, including 2,3-bisphosphoglycerate, phosphoenolpyruvate, fructose-6-P, fructose-1,6-bis-P and ribulose-1,5-bis-P. For all these compounds, the bundle sheath enzyme was activated to a greater extent than was the mesophyll enzyme.

TABLE 1. Allosteric parameters for ADPglucose pyrophosphorylase partially purified from bundle sheath and mesophyll cells. Kinetic experiments were carried out measuring enzyme activity in the ADPglucose synthesis direction. Vmax was the experimental value. $A_{0.5}$ and $I_{0.5}$, the concentrations giving half-maximum activation and inhibition, respectively, were determined from Hill plots. Standard assay metabolite concentrations were 0.5 mM glucose-1-P, 1 mM ATP and 5 mM $MgCl_2$ in Hepes buffer at pH 7.4.

Metabolite	[3PGA]	[Pi]	Mesophyll		Bundle Sheath	
			$A_{0.5}(n)$	$I_{0.5}(n)$	$A_{0.5}(n)$	$I_{0.5}(n)$
	mM	mM	mM	mM	mM	mM
3-P-glycerate		0	0.21(0.8)		0.16(1.0)	
		0.2	0.94(2.6)		0.74(1.9)	
		0.4	1.9 (2.8)		1.3 (2.6)	
P_i	0.1			0.04(1.1)		0.05(1.6)
	2.0			0.36(1.7)		0.5 (3.0)
	10			1.0 (2.0)		1.6 (3.1)

Orthophosphate was a powerful inhibitor of both enzymes (Table 1). Activation by 3-PGA was hyperbolic in the absence of Pi, but became increasingly sigmoidal as the Pi concentration was increased. The concentration of 3-PGA giving half-maximum activation, the $A_{0.5}$, was slightly lower for the bundle sheath enzyme in the absence and presence of Pi. Pi decreased the sensitivity of both enzymes to 3-PGA activation. The concentration of Pi giving half-maximum inhibition, the $I_{0.5}$, was lower for the mesophyll enzyme. The presence of 3-PGA modulated the sensitivity of the enzymes to Pi inhibition and increased the sigmoidicity of the curves. The $I_{0.5}$ increased 25-fold and 32-fold for the mesophyll and bundle sheath enzymes, respectively, as the 3-PGA concentration was increased from 0.1 to 10 mM.

The in vivo activity of the maize leaf ADPglucose pyrophosphorylases would appear to be regulated by the stromal 3-PGA/Pi ratio, as has been postulated for this enzyme in other plants (4). The ratios of 3-PGA/Pi required for enzyme activation in maize leaves are significantly higher than what has been previously reported for spinach leaves (8). This may relate to the presence of high levels of 3-PGA in maize leaves (possibly as high as 15 mM in the bundle sheath; 9). High sensitivity to Pi and sigmoidal activation kinetics may allow these enzymes to be regulated at ratios of 3-PGA/Pi of physiologically significance in maize leaves. Starch synthesis in the mesophyll may be limited by the levels of the biosynthetic enzymes present, as well as the allosteric properties of the ADPglucose pyrophosphorylase. Lower sensitivity to 3-PGA activation and higher sensitivity to Pi inhibition of the mesophyll ADPglucose pyrophosphorylase would constrain starch synthesis in the mesophyll as compared to the bundle sheath. Diversification of allosteric properties of the maize leaf ADPglucose pyrophosphorylases indicates a necessity for regulation of their activities under different stromal environments, and not complete repression of starch synthesis in the mesophyll. In fact, starch synthesis in the mesophyll can be induced in plants grown under continuous light (1). In the presence of favorable 3-PGA/Pi ratios starch synthesis in both cell types might be expected in situ.

REFERENCES
1. Downton, W.J.S. and Hawker, J. (1973) Phytochemistry 12, 1551-1556.
2. Echeverria, E. and Boyer, C. (1986) Amer. J. Bot 73, 167-171.
3. Huber, W., DeFekete, A. and Ziegler, H. (1969) Planta 87, 360-364.
4. Preiss, J. (1982) Ann. Rev. Plant Physiol. 33, 431-454.
5. Sanwal, G.G., Greenberg, E., Hardie, J., Cameron, E.C., and J. Preiss (1968) Plant Physiol. 43, 417-427.
6. Moore, B., Ku, M., and Edwards, G. (1984) Plant Sci. Lett. 35, 127-138.
7. Shen, L. and Preiss, J. (1964) Biochem. Biophys. Res. Commun. 17, 424-429.
8. Ghosh, H.P. and Preiss, J. (1966) J. Biol. Chem. 241, 4491-4504.
9. Stitt, M. and Heldt, H.W. (1985) Planta 164, 179-188.

STARCH SYNTHESIS IN ISOLATED <u>CODIUM</u> <u>FRAGILE</u> CHLOROPLASTS IN THE PRESENCE
OF 10mMPi, PGA and G6P.

MICHAEL L. WILLIAMS AND ANDREW H. COBB

Department of Life Sciences, Trent Polytechnic, School of Science,
Nottingham, NG11 8NS, UK.

1. INTRODUCTION
    The chloroplast starch content of the siphonaceous marine alga <u>Codium</u>
    <u>fragile</u> (Suringar) Hariot has previously been reported to be inversely
    proportional to the concentration of stromal inorganic phosphate (Pi),
    and has been considered to reflect the gross regulation of ADP glucose
    pyrophosphorylase activity within the chloroplast (1). This enzyme is
    inhibited by high stromal Pi/PGA ratios (2,3,4), and although its
    presence in <u>C.fragile</u> chloroplasts has yet to be verified, it is
    conserved over a wide range of photosynthetic organisms (5). Experi-
    ments described in this paper attempt to provide supporting evidence
    for the regulation of ADP glucose pyrophosphorylase within <u>C.fragile</u>
    chloroplasts by investigating the effect of exogenous Pi and PGA on
    starch synthesis by intact isolated chloroplasts. Furthermore, as
    <u>C.fragile</u> chloroplast envelope membranes possess an additional Pi
    translocator facilitating the export of G6P in exchange with Pi (6),
    the effect of exogenous G6P on chloroplast starch synthesis is also
    described.

2. MATERIALS AND METHODS
    <u>C.fragile</u> fronds were sampled between September 1983 and November 1984
    from Bembridge, Isle of Wight (UK) and maintained as described
    previously (7). The effect of $10mol.m^{-3}Pi$, PGA and G6P on HCO_3-depen-
    dent O_2 evolution at $200\mu M.m^{-2}s^{-1}$(photosynthetic photon flux density,
    PPFD) was determined for chloroplasts isolated from fronds sampled in
    November 1984 (8). Net starch synthesis at 200 and $250\mu M.m^{-2}s^{-1}$(PPFD)
    in the absence of Pi and $NaHCO_3$ was determined over 3 hours for chloro-
    plasts isolated from fronds sampled in September 1983, November 1983
    and April 1984 using αamylase, amyloglucosidase and Hexokinase/Glucose-
    6-phosphate dehydrogenase coupled-enzyme reactions (7), whilst the
    effect of $10mol.m^{-3}Pi$, PGA and G6P on net starch synthesis after 3
    hours at $200\mu M.m^{-2}s^{-1}$(PPFD) was similarly determined for chloroplasts
    isolated from fronds sampled in November 1984 (8).

3. RESULTS AND DISCUSSION
    Figure 1 illustrates the effect of $10mol.m^{-3}Pi$, PGA and G6P on HCO_3-
    dependent O_2 evolution at $200\mu M.m^{-2}s^{-1}$(PPFD) by <u>C.fragile</u> chloroplasts
    isolated from fronds sampled in November 1984. Photosynthesis (PS)
    was reduced by approx. 20% in the presence of $10mol.m^{-3}Pi$ throughout
    the 15 minute incubation period. This may not primarily be a result
    of a depletion of stromal metabolites via the Pi-translocators, as
    previously reported Pi-inhibition of <u>C.fragile</u> chloroplast $^{14}CO_2$
    fixation was not accompanied by any marked stimulation of photosynthate
    release (9). Indeed, Pi may exert a more direct influence on PS by
    interacting with stromal enzymes of the PCR cycle (10). The PGA-

Biggens, J. (ed.), Progress in Photosynthesis Research, Vol. III. ISBN 90 247 3452 5

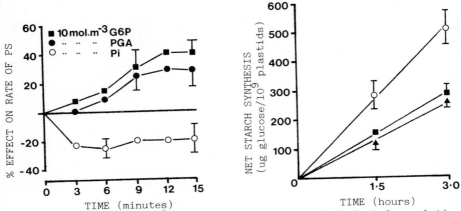

FIG. 1: Effect of 10mol.m^{-3}G6P, PGA and Pi on HCO$_3^-$-dependent O$_2$ evolution at 200μmol.m^{-2}s^{-1}(PPFD) (each point = $\bar{x}$ ± S.E of 9 separate determinations taken from 3 isolates).

FIG. 2: Net starch synthesis at 200 to 250μmol.m^{-2}s^{-1}(PPFD) (Sept. 83:■, Nov. 83:▲, April 84:○, each point = $\bar{x}$ ± S.E of 3 to 8 separate isolates).

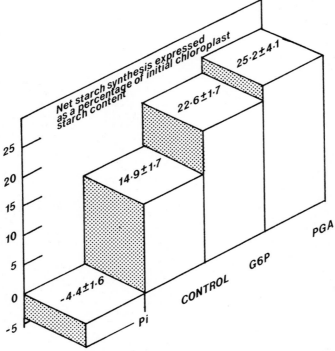

FIG. 3: Effect of 10mol.m^{-3}G6P, PGA and Pi on the net amount of starch synthesised after 3hrs at 200μmol.m^{-2}s^{-1}(PPFD) (values = $\bar{x}$ ± S.E of 12 to 36 determinations taken from 3 to 9 chloroplast isolates).

induced stimulation of O_2 evolution at $200\mu M.m^{-2}s^{-1}$(PPFD) saturated after 9 minutes, being maintained at a value of approx. 25% greater than the control for the remainder of the incubation period. This was probably as a result of both the removal of PCR cycle turnover from the constraints of CO_2/HCO_3^- supply (11) and a reduction in the level of stromal Pi inhibitory on PS (1) via the operation of the Pi/PGA translocator. The G6P-induced stimulation of O_2 evolution at $200\mu M.m^{-2}s^{-1}$(PPFD) was similar to that observed with $10mol.m^{-3}$PGA and was probably due to both a reduction in stromal Pi via the operation of the Pi/G6P translocator and a conversion of G6P within the chloroplast to PCR cycle metabolites (8).

Figure 2 illustrates the marked and consistent linearity of net starch synthesis at 200 and $250\mu M.m^{-2}s^{-1}$(PPFD) for C.fragile chloroplasts isolated from tissue harvested in September, 1983, November, 1983 and April, 1984. Such linearity further illustrates the stability of siphonaceous chloroplasts in vitro (12), in which high rates of net starch synthesis were maintained for up to 3 hours with no added phosphate or $NaHCO_3$ in the incubation medium. This observation has been more specifically addressed in a related paper (13). In contrast, Figure 3 illustrates the effect of exogenous $10mol.m^{-3}$Pi, PGA and G6P on the net amount of starch synthesised over 3 hours at $200\mu M.m^{-2}s^{-1}$ (PPFD) by C.fragile chloroplasts isolated from tissue harvested in November, 1984. PGA markedly stimulated net starch synthesis whilst Pi promoted net starch degradation. Such results are in accordance with the view that ADP glucose pyrophosphorylase, the key enzyme in the regulation of starch synthesis within higher plant chloroplasts, is present in C.fragile. This would be expected as ADP glucose pyrophosphorylase is conserved over a wide range of photosynthetic organisms. It has been isolated and characterised from Chlorella pyrenoidosa, C.vulgaris, Scenedesmus obliquus and Chlamydomonas reinhardii (14) and its presence in the blue-green bacterium Synechococcus 6301 (Anacystis nidulans) has also been recorded (15). In all cases, the regulation of ADP glucose pyrophosphorylase would appear to resemble that from higher plants with respect to PGA activation and Pi inhibition (5). Thus the PGA-induced stimulation of net starch synthesis in C.fragile chloroplasts is most probably a result of allosteric regulation of this enzyme. However, other factors must also be considered. PGA may stimulate PS (Figure 1) and the related removal of PCR cycle turnover from CO_2 availability would be important in the starch metabolism experiments where no $NaHCO_3$ was included in the bathing medium. This may account in part for the observed stimulation of net starch synthesis by G6P which is not a known allosteric activator of ADP glucose pyrophosphorylase. Net starch synthesis in isolated C.fragile chloroplasts at $200\mu M.m^{-2}s^{-1}$(PPFD) is light-saturated and limited by CO_2 availability (7,8). Thus, incoming G6P would be directly channelled into starch synthesis and as a consequence, more PCR products would be available for the regenerative phase of CO_2 reduction.

That $10mol.m^{-3}$Pi not only caused inhibition of net starch synthesis but also net starch breakdown, may be explained by both the influx of Pi inhibiting ADP glucose pyrophosphorylase and the subsequent increase in stromal Pi promoting starch degradation in the light. This supports previous observations for starch-loaded S.oleracea chloroplasts where starch degradation was found to be inhibited at low external Pi concentrations (16).

REFERENCES
1 Cobb, A.H. and Williams, M.L. (1986). Proc. VII. Int. Congr. Photosynthesis.
2 Ghosh, H.P. and Preiss, J. (1966). J. Biol. Chem. 241, 4491-4504.
3 Kaiser, W.M. and Bassham, J.A. (1979a). Plant Physiol. 63, 105-108.
4 Kaiser, W.M. and Bassham, J.A. (1979b). Plant Physiol. 63, 109-113.
5 Preiss, J. and Levi, C. (1979) in Encyclopaedia of Plant Physiology, Vol. VI. (Gibbs, M. and Levi, E., eds), pp. 282-312, Springer, Berlin.
6 Rutter, J.C. and Cobb, A.H. (1983b). New Phytol. 95, 559-568.
7 Williams, M.L. and Cobb, A.H. (1985). New Phytol. 101, 79-88.
8 Williams, M.L. and Cobb, A.H. (1987b). New Phytol. (submitted).
9 Rutter, J.C. and Cobb, A.H. (1983a). New Phytol. 95, 549-557.
10 Cobb, A.H. and Rott, J. (1978). New Phytol. 81, 527-541.
11 Walker, D.A., Cockburn, W. and Baldry, C.W. (1967). Nature, 216, 597-599.
12 Grant, B.R. and Borowitzka, M.A. (1984). Bot. Rev. 50, 267-305.
13 Williams, M.L. and Cobb, A.H. (1987a). New Phytol. (submitted).
14 Sanwal, G.G. and Preiss, J. (1967). Arch. Biochem. Biophys. 119, 454-469.
15 Levi, C. and Preiss, J. (1976). Plant Physiol. 53, 753-756.
16 Stitt, M. and Heldt, H.W. (1981). Biochem. Biophys. Acta. 638, 1-11.

ACKNOWLEDGEMENTS
The authors thank the Science and Engineering Research Council of the UK for financial support.

RELATIONSHIP BETWEEN CHLOROPLAST STARCH, PHOSPHATE CONTENT AND SEASON IN
THE INTERTIDAL ALGAE CODIUM FRAGILE (SURINGAR) HARIOT.

ANDREW H. COBB AND MICHAEL L. WILLIAMS

Department of Life Sciences, Trent Polytechnic, School of Science,
Nottingham, NG11 8NS, UK.

1. INTRODUCTION
    C.fragile chloroplasts possess reserves of inorganic phosphate (Pi) in
    the form of long chain polyphosphate (poly Pi) which vary in both
    quantity and type throughout the life cycle of this alga (1). Acid-
    soluble poly Pi is considered a more readily mobilised form of storage
    poly Pi which may be converted in some algal species to a less readily
    mobilised alkali-soluble form (2). Seasonal fluctuations in the
    C.fragile chloroplast alkali:acid-soluble poly Pi ratio have previously
    been proposed as an adaptation to a fluctuating Pi environment, where
    a high ratio is considered to represent a mobilisation of alkali-
    soluble poly Pi (1). This paper provides additional evidence for
    seasonal variations in C.fragile chloroplast Pi content and underlines
    their influence on starch metabolism by this alga.

2. MATERIALS AND METHODS
    C.fragile fronds were sampled between November 1983 and November 1984
    from Bembridge, Isle of Wight (UK) and maintained as described
    previously (3). Chloroplasts were isolated from C.fragile frond tips
    and purified by gel-filtration (4) and chloroplast starch content
    determined using αamylase, amyloglucosidase and Hexokinase/Glucose-6-
    phosphate dehydrogenase coupled-enzyme reactions (3). Chloroplast Pi
    and poly Pi content was determined by the spectrophotometric
    measurement of molybdate complexes (5) and HCO_3^--dependent O_2 evolution
    measured at $80\mu M.m^{-2}s^{-1}$ (photosynthetic photon flux density, PPFD) for
    chloroplasts isolated from fronds sampled in April 1984 and
    November 1984 (3).

3. RESULTS AND DISCUSSION
    A uniform increase in chloroplast starch content occurred in C.fragile
    fronds sampled from November 1983 (vegetative tissue) to July 1984
    (reproductive tissue) and is illustrated in Figure 1. By the summer
    months chloroplasts were found to contain up to 135% more starch than
    those isolated from tissue sampled in November 1983. Conversely,
    chloroplasts isolated from fronds sampled in November 1984 were found
    to contain markedly less starch than the summer harvest, representing
    a relative decrease in chloroplast starch content from 4.8 to 3.0mg
    glucose/10^9 plastids (cf July 1984). As starch synthesis in isolated
    C.fragile chloroplasts is a function of PCR cycle turnover (3), the
    uniform increase in chloroplast starch content from November 1983 to
    July 1984 may merely reflect the seasonal increase in incident
    irradiance and tidal amplitude within the intertidal zone environment.
    Indeed, starch levels on a frond tissue basis have previously been
    reported to be proportional to irradiance which decreases with the
    depth of the water column (6). However, the life cycle of C.fragile

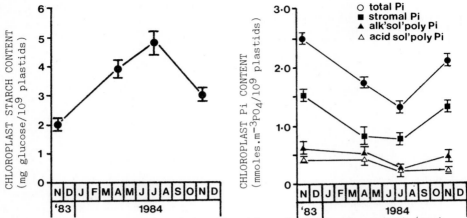

FIG. 1: Seasonal changes in <u>C.fragile</u> chloroplast starch content (each point = $\bar{x}$ ± S.E of 15 to 81 separate determinations taken from 9 to 27 chloroplast isolates).

FIG. 2: Seasonal changes in <u>C.fragile</u> chloroplast Pi content (each point = $\bar{x}$ ± S.E of 18 to 36 separate determinations taken from 6 to 12 chloroplast isolates.

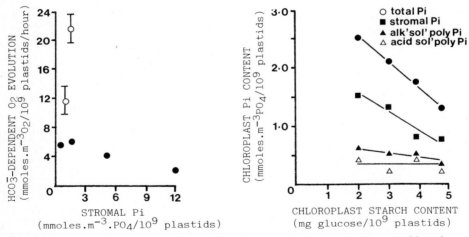

FIG. 3: Effect of stromal Pi on HCO_3^--dependent O_2 evolution at 80μmol. $m^{-2}s^{-1}$(PPFD) for isolated <u>C.fragile</u> chloroplasts (o, data from this study, each point = $\bar{x}$ ± S.E of 3 to 4 separate chloroplast isolates;●, data from Rutter, 1982).

FIG. 4: Effect of chloroplast Pi on chloroplast starch content (each point = mean of 15 to 81 separate determinations taken from 6 to 27 chloroplast isolates).

is such that in the winter months carbon fixation, pigment content and chloroplast size are maximal (1). During this period of optimal vegetative growth, products of photosynthesis are primarily channelled into new cell wall synthesis and the proportion of photosynthate removed to such "sinks" would be expected to influence the finite levels of chloroplast starch observed.

In contrast, levels of stromal Pi and alkali-soluble poly Pi decreased from November 1983 to July 1984 by 50% and 59% respectively, as illustrated in Figure 2, returning to 86% and 91% of their original values by the following November. The chloroplast acid-soluble poly Pi fraction revealed an overall 47% decrease throughout the sampling period. A summation of the seasonal variation in the various chloroplast Pi fractions is illustrated in the total Pi curve (Figure 2), representing an overall decrease in chloroplast Pi status from approx. $2.5 mmol.m^{-3}.PO_4/10^9$ plastids in early vegetative fronds (November 1983) to approx. $1.3 mmol.m^{-3}.PO_4/10^9$ plastids in reproductive fronds (July 1984). Such a decrease in chloroplast Pi status may reflect the proposed gradual depletion of Pi in the intertidal zone due to an increased competition for nutrients from neighbouring algal species (1). Indeed, the observed depletion of the alkali-soluble poly Pi fraction during the summer (Figure 2) supports the role of poly Pi as a chloroplast reserve of Pi, being mobilised under conditions of limiting Pi (1).

As photosynthesis is a Pi-consuming process (7), the observed seasonal variation in C.fragile chloroplast Pi status would be expected to influence finite rates of photosynthesis and hence starch synthesis. Figure 3 illustrates an apparent relationship between stromal Pi and photosynthesis for isolated C.fragile chloroplasts where increasing concentrations of stromal Pi were inhibitory on the rate of O_2 evolution observed at $80\mu M.m^{-2}s^{-1}$(PPFD). To obtain a wider range of stromal Pi values, data from a previous study was also included (5). The observed inhibition of photosynthesis by stromal Pi may reflect an interaction with stromal enzymes of the PCR cycle. Pi concentrations above $2.5 mol.m^{-3}$ have been reported to inhibit FBPase, SBPase and RuBisCO in envelope-free stromal extracts of Pisum sativum (8), whilst a marked inhibition of C.fragile frond tip chloroplast RuBisCO activity has been reported for Pi concentrations of 1 to $5 mol.m^{-3}$ (9).

Figure 4 illustrates the relationship obtained between the various Pi fractions present in isolated C.fragile chloroplasts and chloroplast starch content. A marked linear relationship exists between total Pi and chloroplast starch content of which the major contributor is the stromal Pi fraction. These results support previous observations on isolated Spinacea oleracea chloroplasts in which starch synthesis decreased linearly with respect to increasing concentrations of stromal Pi (10). Although the observed relationship in Figure 4 may be explained in part on the basis of a stromal Pi inhibition of photosynthesis, the decrease in C.fragile chloroplast starch content by increasing concentrations of stromal Pi may also reflect the presence and gross regulation of ADP glucose pyrophosphorylase, the major regulatory enzyme of chloroplast starch synthesis. This enzyme is inhibited by high stromal Pi/PGA ratios (10,11,12,13) and would seem to be conserved over a wide range of photosynthetic organisms (14). This point is more specifically addressed in an accompanying paper (15). The apparent relationship between chloroplast starch content and the alkali-soluble poly Pi fraction may be somewhat overstated in that the total poly Pi fraction of C.fragile chloroplasts increases linearly with respect to stromal Pi (M.L. Williams, unpublished). However, as chloroplast starch content is not necessarily a measure of starch synthesis at the time of chloroplast isolation but a reflection of previous environmental and physiological conditions, the possibility of an interaction between chloroplast starch synthesis and poly Pi content can not be ignored.

REFERENCES
1   Benson, E.E., Rutter, J.C. and Cobb, A.H. (1983). New Phytol. 95, 569-580.
2   Miyachi, S., Kanai, R., Mihara, S., Miyachi, S. and Aoki, S. (1964). Biochim. et Biophys. Acta. 93, 625-634.
3   Williams, M.L. and Cobb, A.H. (1985). New Phytol. 101, 79-88.
4   Cobb, A.H. (1977). Protoplasma, 92, 137-146.
5   Rutter, J.C. (1982). A study of the translocational properties of the chloroplasts of the alga Codium fragile. PhD Thesis, Trent Polytechnic.
6   Wassman, E.R. and Ramus, J. (1973). Marine Biol. 21, 289-297.
7   Walker, D.A. (1976). In Encyclopaedia of Plant Physiology. Vol III, (Pirson, A. and Zimmerman, M., eds), pp. 85-136, Springer, Berlin.
8   Furbank, R.T. and Lilley, R.M.cC. (1980). Biochim. et Biophys. Acta. 592, 65-75.
9   Cobb, A.H. and Rott, J. (1978). New Phytol. 81, 527-541.
10 Heldt, H.W., Chon, C.J., Maronade, D., Herold, A., Stankovic, Z.S., Walker, D.A., Kraminer, A., Kirk, M.R. and Heber, U. (1977). Plant Physiol. 59, 1146-1155.
11 Ghosh, H.P. and Preiss, J. (1966). J. Biol. Chem. 241, 4491-4504.
12 Kaiser, W.M. and Bassham, J.A. (1979a). Plant Physiol. 63, 105-108.
13 Kaiser, W.M. and Bassham, J.A. (1979b). Plant Physiol. 63, 109-113.
14 Preiss, J. and Levi, C. (1979). In Encyclopaedia of Plant Physiology, Vol VI, (Gibbs M. and Levi, E., eds), pp. 282-312, Springer, Berlin.
15 Williams, M.L. and Cobb, A.H. (1986). Proc. VII. Int. Congr. Photosynthesis.

ACKNOWLEDGEMENTS
The authors acknowledge the Science and Engineering Research Council of the UK for financial support.

INTRACELLULAR CARBON PARTITIONING IN GROWING CHLAMYDOMONAS REINHARDII

UWE KLEIN, BOTANICAL INSTITUTE, UNIVERSITY OF BONN, KIRSCHALLEE 1,
5300 BONN 1, FEDERAL REPUBLIC OF GERMANY

1. INTRODUCTION

In leaves of higher plants newly fixed carbon is partitioned into starch and/or sucrose. Intracellular partitioning is assumed to be regulated by the concentration of orthophosphate and triose phosphates inside and outside the plastid [1]. The key enzymes of starch and sucrose metabolism are affected by these metabolites directly or indirectly via fructose-2,6-bisphosphate [1,2]. In addition this rather complex regulation seems to follow a diurnal rhythm in intact plants [3]. In our investigations on metabolic compartmentation in unicellular green algae we also looked for the metabolism of starch. In this report we describe the partitioning of carbon into starch in C. reinhardii.

2. MATERIAL AND METHODS

2.1. Organism

Experiments were done with Chlamydomonas reinhardii, strain 11-32/b from the "Sammlung für Algenkulturen at Göttingen (FRG)". The algae were grown autotrophically in 300 ml glass flasks on a mineral medium at 34°C and synchronized by 12:12 hours light/dark cycles (light fluence rate about 320 W m^{-2}).

2.2. Analytical methods

Starch was determined enzymatically as glucose after hydrolysis of an insoluble fraction from broken cells with amyloglucosidase. Sucrose was extracted from the cells with 5% (v/v) perchloric acid and determined as glucose after hydrolysis with β-fructosidase. Light-dependent incorporation of CO_2 into cell material was measured by incubating the cells with labelled $NaH^{14}CO_3$ (0.04 μCi μmol^{-1}) in the light (15 min) and counting the acid-stable radioactivity with a liquid scintillation counter.

3. RESULTS AND DISCUSSION

3.1. Starch and sucrose

Fig. 1.a. shows the cellular level of starch and sucrose in C. reinhardii during the cell cycle. Starch varies according to a distinct reproducible pattern which is particularly remarkable because of the decrease in starch in the middle of the light period. Sucrose remains at such a low level that it can be regarded as an insignificant metabolite in our cultures. The course of starch accumulation during synchronized growth is so pronounced that it can be followed by electron microscopy (Fig. 2.). It is concluded that newly fixed carbon is partitioned into starch and growth metabolism. This conclusion is supported by experiments where inhibition of growth (which can be induced by withdrawal of nitrogen) results in an increase of starch accumulation (Fig. 1.b.).

Biggens, J. (ed.), Progress in Photosynthesis Research, Vol. III. ISBN 90 247 3452 5

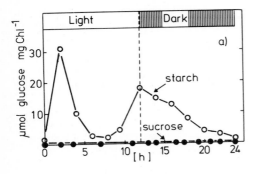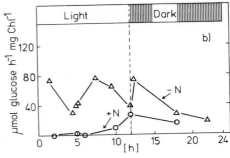

Fig. 1. a) The level of starch and sucrose during the cell cycle in synchronized cells of C. reinhardii.

   b) Rate of starch accumulation in growing (+N) and non-growing (-N) C. reinhardii. Cells were taken from the synchronized cultures at the times indicated, diluted to about 2×10^6 cells per ml with fresh medium (with and without nitrogen), and grown for an hour at full light intensity. Starch was determined before and after this period.

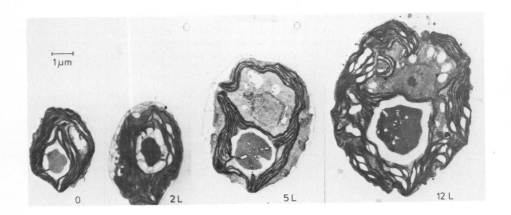

Fig. 2. Electron micrographs of synchronized cells of C. reinhardii taken at different times during the light period. Starch around the pyrenoid and between the thylakoids appears as white ellipsoid granules.

3.2. CO_2 fixation

Since the varying level of starch coud be due to a changing rate of CO_2 fixation this parameter was checked over the cell cycle (Fig. 3.). Directly at the beginning of the light period CO_2 fixation per mg Chl is low but very soon it reaches a high rate which is kept constant for the remainder of the light period. Therefore a variation in CO_2 fixation is not responsible for the pattern of starch accumulation.

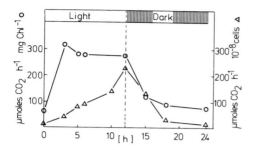

Fig. 3. Rate of CO_2 fixation in synchronized C. reinhardii. Cells were taken from synchronized cultures at the times indicated and incorporation of CO_2 determined as described briefly under "Material and Methods".

3.3. Incorporation of CO_2 into starch

Starch breakdown during the light period - as from the second to sixth hour - could be due to a decrease in starch synthesis, an increase in starch breakdown, or both. Therefore it was investigated whether CO_2 is incorporated into starch at a period when a net decrease of starch is seen. Table 1. shows data for cells from the fourth hour (starch decrease) and tenth hour (starch accumulation) in the light. In both cases about 20% of the incorporated CO_2 are partitioned into starch. This means that synthesis and degradation of starch occur simultaneously in the light. The results also favor the conclusion that in the light not only starch synthesis but also its degradation is regulated.

TABLE 1. Incorporation of CO_2 into starch by synchronized cells of C. reinhardii at 4 and 10 hours in the light. Cells were taken from the cultures, resuspended in fresh medium (+ $NaH^{14}CO_3$), and illuminated for 20 min. Starch was prepared, hydrolysed, and the radioactivity counted as described briefly in the "Material and Methods" section.

Time	μg Chl/ml	Total CO_2 fixed μmol/h · mg Chl	CO_2 in starch μmol/h · mg Chl	% CO_2 in starch
4 L	68.8	275	62.9	22.8
10 L	189.4	275	50.8	18.5

3.4. Cell cycle or diurnal rhythm

Since the pattern of starch accumulation (Fig.1., Fig. 2.) was dependent on growth in light-dark cycles it may be controlled by a

diurnal rhythm or by the cell cycle. To distinguish between these possibilities we compared the starch accumulation of synchronized and de-
synchronized cells, both growing in the same light-dark regime (Fig. 4.).
In contrast to synchronized cells the rate of starch accumulation in de-
synchronized cells does not vary much. This indicates that partitioning of
carbon is mainly linked to the cell cycle in growing C. reinhardii.

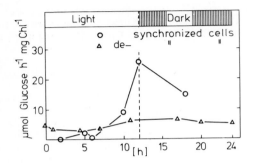

Fig. 4. Rate of starch accumulation in synchronized
and de-synchronized
cells of C. reinhardii.
Samples were taken and
treated as described
under Fig. 1.b..

4. CONCLUSIONS

Growing cells of the unicellular green alga C. reinhardii partition
fixed carbon into starch and growth metabolism but not into sucrose.
The level of starch is regulated by simultaneous synthesis and breakdown and seems to be controlled by cell cycle-dependent processes.

REFERENCES
1 Preiss, J. (1984) Trends Biochem. Sci. 9, 24-27
2 Stitt, M., Kürzel, B. and Heldt, H.W. (1984) Plant Physiol. 75, 554-560
3 Huber, S.C., Rufty, T.W. and Kerr, P.S. (1984) Plant Physiol. 75,
  1080-1084

DIURNAL CHANGES IN CARBON PARTITIONING IN LEAVES

Steven C. Huber, Willy Kalt-Torres, Hideaki Usuda[1] and Mark Bickett

United States Department of Agriculture, Agricultural Research Service, and Departments of Crop Science and Botany, North Carolina State University, Raleigh, NC 27695 (U.S.A), and Laboratory of Chemistry, Faculty of Medicine, Teikyo University, Tokyo, Japan[1]

Sucrose is the primary transport carbohydrate in higher plants such as maize, soybean and spinach, and the regulation of sucrose formation appears to be one of the primary factors in the control of photosynthate partitioning in leaves. Identification of diurnal changes in carbon partitioning in leaves has important implications for whole plant source-sink interactions, and provides a system to probe the biochemical regulation of leaf photosynthetic metabolism.

At the present time, two enzymes of the cytosolic sucrose formation pathway are thought to be key control points. The key enzymes are cytosolic fructose 1,6-bisphosphatase (FBPase) and sucrose phosphate synthase (SPS). Metabolic fine control of the pathway is exerted via regulation of FBPase activity by fructose 2,6-bisphosphate (F26BP), which is a potent inhibitor, and sensitizes the enzyme to control by other metabolites as well (1,2). Fine control of the pathway also involves regulation of SPS activity by glucose-6-P (G6P) and inorganic phosphate (Pi) (3). In addition, coarse control of sucrose formation is exerted by changes in the maximum activity of SPS (4,5). It appears that metabolic fine and coarse control are coordinated to bring about effective regulation of the rate of sucrose formation throughout the diurnal cycle. However, important differences exist among species in the nature of diurnal changes in carbon partitioning in leaves; these differences appear to reflect inherent differences in the underlying biochemical control mechanisms.

Diurnal export patterns. It is generally recognized that C-4 plants, such as maize, have a higher capacity to assimilate and export carbon compared with C-3 plants, such as soybean. This is readily apparent in Fig. 1, which shows the diurnal profiles of assimilate export in greenhouse grown maize and soybean plants. In addition to differences in maximum rates, the diurnal profiles for the two species were different; in maize, export rate generally paralleled changes in photosynthetic rate and light intensity (data not shown), whereas in soybean, export rate was highest in the morning and declined progressively throughout the photoperiod. Interestingly, export rate at night was similar for the two species. In general, spinach behaved similarly to soybean in terms of changes in export rate during the photoperiod (Fig. 2). As previously observed by Stitt et al. (6), sucrose accumulated in spinach leaves throughout the photoperiod, but the rate of accumulation was highest in the morning hours (Fig. 2A). Starch, in contrast, accumulated after an initial lag; the rate of starch accumulation was highest in the afternoon hours (Fig. 2A). Consequently, total accumulation of carbohydrate in spinach leaves remained constant throughout the day (Fig. 2B). Because

Biggens, J. (ed.), Progress in Photosynthesis Research, Vol. III. ISBN 90 247 3452 5
© 1987 Martinus Nijhoff Publishers, Dordrecht. Printed in the Netherlands.

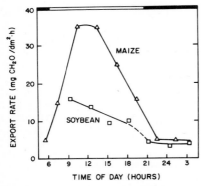

Figure 1. Diurnal changes in assimilate export rate from mature maize and soybean leaves. Plants were grown in the greenhouse. Measurements were made as described previously (5).

Figure 2. Changes in photosynthetic carbon metabolism in attached spinach leaves during the photoperiod. A) accumulation of sucrose and starch in leaves; B) carbon exchange rate (CER), carbohydrate accumulation rate, and assimilate export rate.

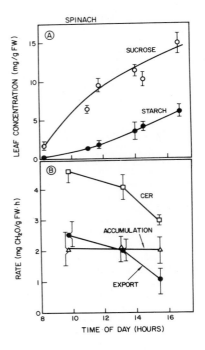

net carbon exchange rate (CER) decreased with time of day, mass carbon export rate also decreased with time (about 50%) (Fig. 2B). Among these three species, two different patterns are observed: in soybean and spinach, export rate is highest in the morning and tends to decrease with time of day, whereas in maize, maximum export rate occurs at midday, coincident with maximum photosynthetic rate.

Coordinate control of sucrose formation. Although two different patterns are observed for diurnal changes in assimilate export rate, the biochemical determinants of carbon flow to sucrose appear to be similar in spinach, soybean and maize. In all three species, assimilate export rate is correlated positively with SPS activity and correlated negatively with leaf F26BP concentration. For example, in maize leaves, SPS is light activated (7) and a relatively high irradiance is required for maximum activation. Consequently, greatest SPS activity is observed at midday (Fig. 3), coincident with maximum assimilate export (Fig. 1). Changes in maize leaf F26BP concentration were essential reciprocal to changes in SPS activity. F26BP concentration decreased about 5-fold early in the photoperiod, remained relatively low and constant during midday, and then increased in the afternoon (Fig. 3). Similar diurnal fluctuations have been documented with barley (8). In contrast, spinach leaf F26BP concentration increased throughout the photoperiod (Fig. 4B and ref. 9), and changes in the dark are fundamentally different from those observed with maize. In addition, spinach leaf SPS activity was highest at the beginning of the photoperiod, and decreased throughout the day (Fig. 4B).

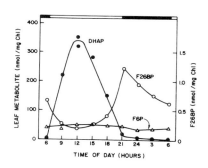

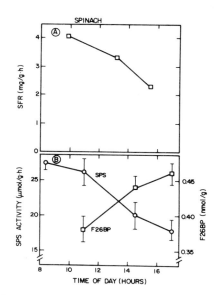

Figure 3. Diurnal changes in maize leaf SPS activity and F26BP concentration.

Figure 4. Changes in photosynthetic carbon metabolism in attached spinach leaves during the photoperiod. A) sucrose formation rate (SFR), which takes into account accumulation and export of sucrose; B) SPS activity and F26BP concentration.

Changes in SPS activity generally paralleled changes in sucrose formation rate in spinach leaves (Fig. 4A and B).

In all three species, leaf F26BP concentration and SPS activity were inversely related (Fig. 5). These results suggest that coarse control (changes in SPS activity) and fine control (changes in F26BP concentration) of the sucrose formation pathway are coordinated during the photoperiod. For example, the reduction in SPS activity during the day in spinach and soybean, coupled with increased F26BP concentration, would

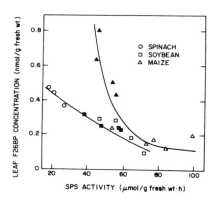

Figure 5. Relationship between SPS activity and F26BP concentration in leaves of spinach, soybean and maize. Open and closed symbols correspond to measurements made during the light and dark periods, respectively.

Table 1. Effect of illumination (20 min) of maize leaves in air or nitrogen atmospheres on metabolite levels and enzyme activation.

DHAP, dihydroxyacetone phosphate; NADP-MDH, NADP malate dehydrogenase

Treatment	DHAP (nmol/mg Chl)	NADP-MDH	SPS (µmol/mg Chl·h)
Dark	<5	<10	68
Light (air)	184	206	140
Light (N_2)	113	873	92

reduce the flux of carbon into sucrose and presumably accounts for the observed reduction in assimilate export rate as the day proceeds. The coordination between SPS activity and F26BP concentration suggests that these two parameters may be controlled by a common mechanism, or that a causal relationship exists between them.

Regulation of SPS Activity. The activity of SPS in soybean leaves fluctuates diurnally (10) and the oscillations persisted in either continuous light or darkness (11). These results indicated that, at least in soybean, SPS activity was controlled by an endogenous rhythm.

It is not yet known whether the diurnal fluctuation in SPS activity in spinach leaves also reflects control by the endogenous clock mechanism. Further, the biochemical basis for the rhythm in SPS activity is also unknown at present. However, differences in SPS activity (at different times of the diurnal cycle) persist during purification, but are not associated with significant changes in kinetic or regulatory properties of the enzyme (Kerr, P.S. and Huber, S.C., unpublished). Thus, the rhythm in activity may be due to changes in the amount of SPS protein, or to a covalent modification that affects only maximum velocity. In contrast, SPS activity in maize leaves can be rapidly modulated by light-dark transitions. The changes in SPS activity with light-dark transitions occur relatively rapidly (within 30 min) and appear to be associated with changes in kinetic properties (7). It appears that the reduction in SPS activity with darkening of leaves can be attributed to a decrease in maximum velocity, and also a decreased affinity for UDP-glucose. Rhythmic fluctuations in maize leaf SPS activity do not occur under constant environmental conditions; in extended darkness, activity declined progressively to essentially undetectable levels after 24 h (Kalt-Torres, W. and Huber, S.C., unpublished), whereas in continuous light, activity remains constant for up to 48 h (A. Lecharny and S.C. Huber, unpublished). Thus, light modulation does not appear to be superimposed on an endogenous rhythm in enzyme activity in maize.

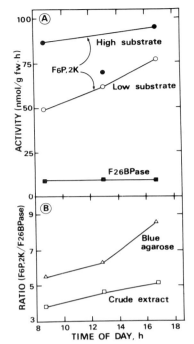

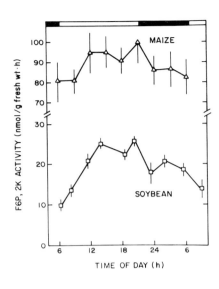

Figure 6. Changes in the activities of F6P,2-kinase and F26BPase in
spinach leaves during the photoperiod. A) After partial purification on
blue agarose; B) activity ratio in desalted crude extracts and after blue
agarose chromatography. High substrate reaction mixtures contained 5 mM
F6P, 5 mM phosphate, 15 mM G6P, 2 mM ATP, 4 mM P-creatine and 4 units/ml
creatine kinase. Low substrate reaction mixtures contained 10-fold lower
concentrations of F6P, phosphate and G6P. All assays contained 50 mM
Tris-HCl (pH 8.0) and 5 mM $MgCl_2$. F26BPase reaction mixtures contained
0.3 µM F26BP in 50 mM Tris (pH 8.0).

Figure 7. Diurnal changes in F6P,2-kinase activity in leaves of maize
and soybean.

It is logical to ask whether sulfhydryl redox changes may be involved in
light modulation of maize leaf SPS activity. It is clear that reductant
(e.g., dithiothreitol) is necessary to maintain maize leaf SPS activity
after extraction, but there is no evidence that sulfhydryl redox changes
are involved in the light modulation process. However, it does appear
that light activation in vivo may be modulated, to some extent at least,
by metabolites. As shown in Table 1, illumination of maize leaves in a
nitrogen atmosphere significantly reduced leaf concentration of DHAP,
relative to leaves illuminated in air. This would be expected, as the
carbon dioxide-free atmosphere would preclude photosynthetic carbon
fixation. As reported earlier (12), illumination in nitrogen results in
a dramatic increase in activation of NADP-malate dehydrogenase (Table 1),
which reflects the fact that light modulation of this enzyme is mediated

by disulfide reduction (13). Importantly, light activation of maize SPS was significantly greater in air relative to nitrogen (Table 1). This result indicates that light activation of SPS may be modulated by one or more metabolites which fluctuate with photosynthetic rate. A likely candidate is DHAP (cf. Figs. 3 and 8), although attempts to activate "dark" SPS in vitro by incubation with DHAP have been unsuccessful.

Regulation of leaf F26BP concentration. The concentration of F26BP in leaves will be a function of the rate of synthesis (by F6P,2-kinase) and degradation (by F26BPase). Both the kinase and phosphatase are known to be highly regulated by metabolic effectors, in particular Pi, F6P and DHAP (14,15). An additional factor may be changes in activity of the kinase and/or phosphatase, which is recognized to be an important level of control in animal systems (16). At the present time, diurnal changes in F6P, 2-kinase activity have been identified in leaves of some, but not all species. Preliminary experiments with spinach leaves compared enzyme activities in desalted crude leaf extracts and after partial purification using blue agarose chromatography. As shown in Fig. 6A, F6P,2-kinase activity (after partial purification) increased during the day, and the changes in activity were more pronounced when assays were conducted using limiting substrate concentrations. F26BPase activity, in contrast, remained relatively constant. Consequently, the kinase/phosphatase activity ratio increased during the photoperiod. The ratio increase was apparent with both crude extracts and partially purified samples; however, activity ratios were significantly higher in the latter case, which reflects removal of "nonspecific" phosphatase (contributing to apparent F26BPase activity) by the blue agarose step. Similar changes in F6P,2-kinase activity have been observed in soybean; activity increased throughout the day and decreased slowly at night (Fig. 7). In contrast, F6P,2-kinase activity in maize leaves showed some fluctuations, but the differences were not statistically significant (Fig. 7). Thus, in spinach and soybean, F6P,2-kinase activity increased during the day concurrent with increased leaf concentrations of F26BP; in maize, kinase activity remained constant while F26BP fluctuated in a distinct pattern from that observed with the dicotyledenous species. Recently, Stitt et al. (17) also reported diurnal changes in F6P,2-kinase and F26BPase in

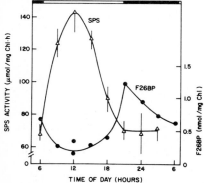

spinach leaves, and concluded that this level of control may reinforce the metabolic regulation of these enzymes.

In maize, metabolic regulation may be the primary, if not only, mechanism for regulation of F26BP concentration. As shown in Fig. 8, F26BP concentration tended to vary inversely with leaf DHAP concentration. F6P, which is one of the substrates of the 2-kinase, remained essentially constant throughout the diurnal cycle (Fig. 8). DHAP antagonized the Pi activation of F6P,2K (15), which may explain the relationship observed.

Figure 8. Diurnal Changes metabolites in maize leaves in concentration of certain

Regulation of F6P,2-kinase by protein factors. It appears that in at
least in some plant species, regulation of leaf F26BP concentration may
be a complex function of allosteric control and enzyme (F6P,2-kinase)
activity changes. A similar situation exists in animals; phosphorylation
of animal F6P,2-kinase/F26BPase (hormone-mediated) inhibits the kinase
activity and increases the phosphatase activity. At the present time,
there is no consistent, positive evidence to suggest that the higher
plant enzyme is regulated by phosphorylation. However, experiments of
this type generally employ the catalytic subunit of cAMP-dependent
protein kinase; it is quite possible that a specific plant protein kinase
is required to achieve phosphorylation in vitro. Alternatively, other
mechanisms may be involved to achieve changes in 2-kinase activity. In
early experiments, it was clear that desalted soybean leaf extracts
contained considerably lower activities of F6P,2-kinase compared with
spinach. Interestingly, the soybean extract could inhibit the activity
present in a spinach leaf extract. The apparent inhibition could not be
ascribed to product degradation, because under the conditions employed,
F26BP was not metabolized by the soybean extract. In addition, ATP
hydrolysis can be ruled out because inclusion of a ATP-regenerating
system had no effect. It is also important to note that the extracts
were desalted, which eliminates the possibility of a low molecular weight
metabolite or inhibitor. The "factor" was inactivated by heating at 80°C
for 3 min, which suggests that it may be proteinaceous in nature.

The inhibitor was partially purified from a soybean leaf extract by
polyethylene glycol-8000 fractionation (5-12%) and mixed with partially
purified spinach F6P,2-kinase/F26BPase. Increasing amounts of the
soybean inhibitor inhibited F6P,2-kinase activity but not F26BPase
activity (data not shown). Thus, some specificity in action of the
inhibitor is apparent. Conceivably, regulation of F6P,2-kinase activity
in soybean leaves (and possibly spinach) may involve changes in the
amount of inhibitor or interaction with the inhibitor in situ.

Conclusions. Pronounced diurnal fluctuations in photosynthetic carbon
metabolism, and the flux of carbon into sucrose, occur in leaves of
spinach, soybean, and maize. However, the diurnal patterns observed are
quite different for the three species. Flux of carbon into sucrose in
maize leaves appears to be closely coordinated with the rate of photosynthesis. As the rate of carbon fixation increases, the concentration of
leaf metabolites (in particular DHAP) will also increase. This appears
to be a major factor regulating the concentration of F26BP in maize
leaves. In addition, the activation state of maize leaf SPS appears to
be either directly or indirectly modulated by metabolic intermediates.
Thus, carbon availability may be the coordinating factor in maize.

In contrast, the diurnal patterns observed, and the underlying
mechanisms, appear to be different in soybean and spinach. In these two
species, SPS activity is not light modulated, and at least in soybean,
has been shown to fluctuate as a result of an endogenous rhythm. As SPS
activity decreases during the day, leaf F26BP concentration may increase
as a result of two factors. First, it is likely that the concentration
of F6P will increase as a result of decreased SPS activity. Because the
affinity of F6P,2-kinase for F6P is rather low (i.e., the Km (F6P) is in

the millimolar range), an increase in F6P would be expected to result in increased F26BP synthesis. In addition, F6P is an inhibitor of F26BPase, which would slow the rate of degradation. Second, the activity of F6P,2-kinase increases as well. A protein factor has been identified that inhibits F6P,2-kinase activity, but has no effect on F26BPase activity. The putative F6P,2-kinase inhibitor may play a role in diurnal regulation of F26BP concentration in certain species.

Overall, the results indicate that regulation of the sucrose formation pathway involves coordinated control of the activities of SPS and cyto-solic FBPase. However, it is becoming apparent that species have adapted different strategies and biochemical mechanisms to regulate the flow of carbon into sucrose.

ACKNOWLEDGEMENTS
Cooperative investigations of the U.S. Dept. of Agriculture, Agricultural Research Service, and the North Carolina Agricultural Research Service, Raleigh, NC 27695-7631. Paper number 10635 of the Journal Series of the N.C. Agricultural Research Service, Raleigh, North Carolina 27695-7601 (USA).

References
1 Herzog, B., Stitt, M. and Heldt, H.W. (1984) Plant Physiol. 75, 561-565
2 Stitt, M. and Heldt, H.W. (1985) Plant Physiol. 79, 599-608
3 Doehlert, D.C. and Huber, S.C. (1983) Plant Physiol. 73, 989-994
4 Rufty, T.W., Jr. and Huber, S.C. (1983) Plant Physiol. 72, 474-480
5 Huber, S.C., Rufty, T.W., Jr. and Kerr, P.S. (1985) Physiol. Plant. 64, 81-87
6 Stitt, M., Kurzel, B. and Heldt, H.W. (1984) Plant Physiol. 75, 554-560
7 Sicher, R.C. and Kremer, D.F. (1985) Plant Physiol. 79, 695-698
8 Sicher, R.C., Kremer, D.F. and Harris, W.G. (1986) Plant Physiol., in press
9 Stitt, M., Gerhardt, R., Kurzel, B. and Heldt, H.W. (1983) Plant Physiol. 72, 1139-1141
10 Rufty, T.W., Jr., Kerr, P.S. and Huber, S.C. (1983) Plant Physiol. 73, 428-433
11 Kerr, P.S., Rufty, T.W., Jr. and Huber, S.C. (1985) Plant Physiol. 77, 275-280
12 Nakamota, H and Edwards, G.E. (1983) Aust. J. Plant Physiol., 10, 279-289
13 Buchanan, B.B. (1980) Annu. Rev. Plant Physiol. 31, 341-374
14 Cseke, C. and Buchanan, B.B. (1983) FEBS Lett. 155, 139-142
15 Stitt, M., Cseke, C and Buchanan, B.B. (1984) Eur. J. Biochem. 143, 89-93
16 Hers, H.G. and Van Schaftingen, E. (1982) Biochem. J. 206, 1-12
17 Stitt, M., Mieskes, G, Soling, H.-D, Grosse, H. and Heldt, H.W. (1986) Z. Naturforschnung. 41c, 291-296

CHARACTERISTICS OF SOYBEAN COTYLEDON PHOTOSYNTHESIS

CHRISTOPHER S. BROWN AND STEVEN C. HUBER   USDA-ARS and Departments of
Crop Science and Botany, N.C. State University, Raleigh, NC, 27695-7631

1. INTRODUCTION

The presence of soybean cotyledons during early seedling growth is
necessary for maximum final yield (1). Past work demonstrated that
stored reserves are mobilized from the cotyledons to support early
seedling growth (2,3,4). The majority of stored reserves are mobilized
within 8-9 days after imbibition (C.S. Brown and S.C. Huber,
unpublished), but the cotyledons persist for several days thereafter.
The reason for cotyledon persistence is at present unclear. Since
cotyledons are epigeal and synthesize chlorophyll upon emergence from
the soil (5), it is possible that the cotyledons contribute carbon to
the growing seedling via photosynthesis. Although cotyledon photosynthesis has been noted previously (5,6), it was ascribed little importance. This apparent anomaly is not dealt with adequately in the literature. Therefore, the objectives of the present study were to: 1)
assess the contribution of cotyledon photosynthesis to total carbon
exported from the cotyledon during seedling growth and; 2) characterize
diurnal changes in photosynthetic rate, lipid and starch mobilization,
and sucrose phosphate synthase activity in cotyledons of different ages.

2. MATERIALS AND METHODS

Soybean (Glycine max L. [Merr] cv. Ransom II) seeds were sown in a
peat moss:sand:vermiculite;perlite medium with commercial slow release
fertilizer (13-6-6) added at the recommended rate. Seedlings were grown
in a controlled environment chamber at a constant $26^\circ C$ with a 12h photoperiod and a light intensity of 450 $uE \cdot m^{-2} \cdot s^{-1}$.

Carbon exchange rate (CER) was measured as CO_2 depletion in a
closed system using a LiCor Model 6000 Portable Photosynthesis System.
Measurements were taken on pairs of intact cotyledons attached to the
plant, a single unifoliate leaf or the central leaflet of the first
trifoliolate leaf.

Immediately after CER measurements, the same pair of cotyledons was
harvested, frozen in liquid nitrogen and stored at $-80^\circ C$. Cotyledon
extracts were prepared by homogenizing the frozen tissue with a
Brinkmann Polytron in 10 volumes of cold buffer containing 50 mM HEPES-
NaOH (pH 7.5), 5mM MgC12, 1 mM EDTA, 2.5 mM DTT, 0.5% (w/v) BSA and 2.0%
(w/v) PEG (MW 20,000). The homogenate was centrifuged at 31,000 x g for
20 minutes and decanted through layers of cheesecloth.

Sucrose phosphate synthase (SPS) activity was assayed in desalted
extracts as described previously (7).

Additional cotyledons were harvested for determination of carbon
reserves. Lipid was quantitated by extracting 100 mg of the lyophilized
powder with petroleum ether (boiling point $37^\circ-55^\circ C$) at $43^\circ C$, and de-
canting the supernatant solution into a pre-weighed vessel. The residue remaining in the vessel after petroleum ether evaporation was the
non-polar lipid material. The pellet was saved for starch determination
as described previously (8).

Biggens, J. (ed.), Progress in Photosynthesis Research, Vol. III. ISBN 90 247 3452 5
© *1987 Martinus Nijhoff Publishers, Dordrecht. Printed in the Netherlands.*

3. RESULTS AND DISCUSSION

3.1 Cotyledon Photosynthesis. The reserves of soybean cotyledons were depleted continuously from imbibition (2,3,4) but net photosynthesis did not begin until some time after emergence from the soil (Fig. 1B). However, carbon fixed by photosynthesis supplied over 75% of the exported carbon from the cotyledon (data not shown) compared to less than 25% from stored reserves It was apparent that photosynthesis provided three times as much carbon than did mobilization of reserves.

Unifoliate and trifoliolate leaves developed net photosynthetic capacity at 8 and 13 days after planting (DAP), respectively (Fig.1). After 10 DAP, the contribution of the cotyledon to whole seedling photosynthesis on a per organ basis was less that 5% (Fig. 1A) even though the highest cotyledon CER was not attained until 14

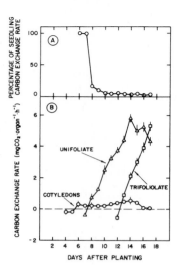

FIGURE 1. A) Soybean cotyledon carbon exchange rate as a percent of total seedling carbon exchange rate and B) absolute carbon exchange rates of cotyledons, unifoliate, and trifoliolate leaves.

DAP. Although cotyledons contributed such a small amount to whole seedling CER, their importance to parameters such as final yield (1) has been established. Since 75% of the carbon exported from the cotyledon was provided by current photosynthesis, this process is probably essential to optimal seedling development and maximum plant productivity.

3.2 Diurnal Changes. Soybean cotyledons exhibited different diurnal patterns of CER at different stages of seedling growth (Fig. 2). At 7 DAP, cotyledon CER was highest early in the photoperiod, declined throughout the mid-portion of the day, then increased near the end of the photoperiod. In contrast, at 14 DAP cotyledons maintained a more constant CER until late in the photoperiod when CER declined somewhat. The diurnal pattern at 14 DAP is similar to that of mature leaves (9,10) and to patterns noted for unifoliate and first trifoliolate leaves (C.S. Brown and S.C. Huber, unpublished). The reason for the substantial mid-day decline in CER in cotyledons at 7 DAP remains unclear at present, but the fact that such distinct diurnal patterns exist for the same parameter at different cotyledon ages suggests that the function of this tissue may change drastically with age.

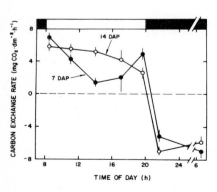

FIGURE 2. Diurnal pattern of soybean cotyledon carbon exchange rate at 7 and 14 days after planting.

At 14 DAP, when cotyledon photosynthetic rates were highest (Fig. 1) and reserve mobilization rates were diminished (Table 1), there was a' marked diurnal pattern in SPS activity, with two peaks of activity, one early and one late in the photoperiod (Fig. 3). This pattern was similar to that observed in mature soybean leaves (9). The activity of SPS was higher at 7 DAP than at 14 DAP, and the diurnal pattern of activity was markedly different. Activity was low during most of the photoperiod but increased late in the day. By comparison, during the dark period, activity was higher and apparently constant. The appearance of diurnal fluctuation in SPS activity similar to leaves appeared to be co-incident with higher CER and lower reserve mobilization rates.

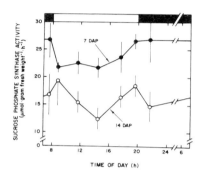

FIGURE 3. Diurnal pattern of sucrose phosphate synthase activity in soybean cotyledons at 7 and 14 days after planting.

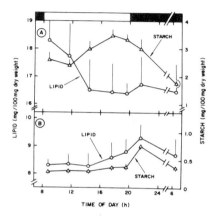

FIGURE 4. Diurnal patterns of lipid and starch concentration in soybean cotyledons at A) 7 days after planting and B) 14 days after planting.

The rate of lipid mobilization and starch accumulation exhibited diurnal fluctuations at 7 DAP, but changes at 14 DAP were less divergent (Fig. 4). Lipid loss from the cotyledon occurred only during the first half of the photoperiod at 7 DAP, and starch accumulated to a small extent during the latter part of the photoperiod and was mobilized at night. At 14 DAP, there was some lipid and starch accumulation throughout the day, with mobilization at night, but these were small amounts. Lower lipid content and a generally slower rate of reserve mobilization may have contributed to lack of a distinct pattern of lipid mobilization at 14 DAP. Low starch accumulation may indicate that most of the concurrently fixed carbon was exported from the cotyledon, presumably as sucrose.

SPS activity was present in 20 fold excess of available carbon at 7 DAP (Table 1), but at 14 DAP, SPS activity and CER were nearly equal, perhaps indicating that SPS activity in older cotyledons was becoming a limiting factor to sucrose formation, similar to what occurs in soybean leaves (11,12).

TABLE 1. Carbon available for sucrose formation and SPS activity in soybean cotyledons

DAP[a]	Reserve mobilization[b]	CER[c]	SPS[d] Activity
	(mg gram dry weight^{-1} h^{-1})		
7	2.0	1.6	77.6
14	0.2	12.9	14.9

[a] Days after planting.
[b] Calculated as the sum of raffinose, stachyose, lipid and soluble protein. Prior to summation, lipid values were multiplied by 0.7 to account for the 70% efficiency of lipid to sucrose conversion (13) and soluble protein was multiplied by 0.43 and 0.96 because only Glu, Asp, Gly, Ala and Ser, comprising 43% of the total amino acid content (14) are potentially gluconeogenic (15) and the molar ratio of sucrose to amino acid is 0.96.
[c] Carbon exchange rate.   [d] Sucrose phosphate synthase.

4. CONCLUSIONS

4.1 Cotyledon photosynthesis was found to be more important to carbon supply for the seedling than previously thought, providing over 75% of the total carbon exported from the cotyledon during the functional life of this organ. However, when unifoliate and trifoliolate leaves became photosynthetic, cotyledon photosynthesis was a minor component to whole seedling photosynthesis.

4.2 CER and SPS activity developed diurnal patterns similar to leaves developed as the cotyledons aged.

4.3 Diurnal patterns of lipid and starch mobilization/accumulation became less evident as the cotyledons aged.

4.4 At 7 DAP, SPS activity was present in cotyledons in excess of carbon available for sucrose formation, but was nearly equal to CER at 14 DAP when the cotyledons were more "leaf-like."

REFERENCES

1 Weber, C.R. and Caldwell, B.E. (1966) Crop Sci. 6, 25-27
2 Pazur, J.H. Shadaksharaswamy, M. and Meidell, G.E. (1962) Arch. Biochem. 99, 78-85
3 Beevers, H. (1980) in The Biochemistry of Plants (Stumpf, P. and Conn, E.E., eds), vol 4, pp. 117-130, Academic Press, New York
4 McCalister, D.F. and Krober, O.A. (1951) Plant Physiol. 26, 525-538
5 Abrahamsen, M. and Mayer, A.M. (1967) Physiol. Plant. 20, 1-5
6 Lane, H.C. and Hesketh, J.D. (1977) Amer. J. Bot. 64(6), 786-790
7 Huber, S.C. (1981) Z. Pflanzenphysiol. 102, 443-450
8 Brown, C.S. Young, E. and Pharr D.M. (1985) J. Amer. Soc. Hort. Sci. 110 (5), 696-701
9 Rufty, T.W. Jr., Kerr, P.S. and Huber, S.C. (1983) Plant Physiol. 73, 428-433
10 Kerr, P.S. and Huber, S.C. (1986) Planta, submitted
11 Huber, S.C. and Israel, D.W. (1982) Plant Physiol. 69, 691-696
12 Huber, S.C. (1983) Plant Physiol. 71, 818-821
13 Beevers, H. (1957) Biochem. J. 66, 23-24
14 Sosulski, F.W. and Holt, N.W. (1980) Can. J. Plant Sci. 60, 1327-1331
15 Stewart, C.R. and Beevers, H. (1967) Plant Physiol. 42, 1587-1595

ACKNOWLEDGEMENTS

Cooperative investigations of the United States Department of Agriculture, Agricultural Research Service and the North Carolina Agricultural Research Service, Raleigh, NC 27695-7631. Paper 10626 of the Journal Series of the North Carolina Agricultural Research Service, 27695-7601.

FRUCTOSE-2,6-BISPHOSPHATE: A REGULATORY METABOLITE DIRECTING CARBON FLOW
BETWEEN SUCROSE AND STARCH.

CSABA CSEKE*, FRASER D. MACDONALD, QUN CHOU, AND BOB B. BUCHANAN, MOLECULAR
PLANT BIOLOGY, UNIVERSITY OF CALIFORNIA, BERKELEY, CA 94720, U.S.A. and *DOW
CHEMICAL RESEARCH CENTER, P.O. BOX 9002, WALNUT CREEK, CA 94598

1.   INTRODUCTION
      Evidence accumulated in the past few years suggests that fructose-2,
6-bisphosphate (Fru-2,6-P2), a regulatory metabolite localized in the
cytosol of plant cells, can play a central role in the coordination of
chloroplast and cytosolic carbohydrate metabolism of leaves (1). In the
chloroplast the principal and ultimate regulator of carbohydrate metabolism
is light. Light, absorbed by chlorophyll, is converted to different
regulatory signals, which modulate selected enzymes in the chloroplast.
Modification of sulphhydryl status of selected enzymes via the ferredoxin/
thioredoxin system and modulation of enzymes by light induced changes in the
concentration of certain ions and metabolites are the major light driven
regulatory mechanisms (2). (Fig. 1.)

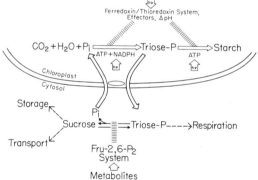

Figure 1. Relation of carbon
processing in the cytosol to
photosynthetic carbon dioxide
assimilation in chloroplasts.

There are several lines of evidence suggesting that chloroplast and
cytosolic carbohydrate metabolism is coordinated and that chloroplasts can
influence metabolic events in the cytosol and vice-versa (2). The biochemical mechanism underlying the communication between these compartments is not
yet fully understood. Light actuated regulatory mechanisms, however, do not
seem to be involved directly in this process. It has been suggested that
triose phosphates such as dihydroxyacetone phosphate (DHAP) and 3-phos-
phoglycerate (PGA) not only serve as a source of energy and carbon in the
cytosol but also, together with inorganic phosphate (Pi), function as
regulatory messengers coordinating chloroplast and cytosolic carbohydrate
metabolism (3,4). Changes in triose phosphate concentration can indeed
communicate significant information, since carbon fixed in photosynthesis exits the chloroplast mainly in the form of DHAP and PGA via a
special carrier protein, the phosphate translocator.
      Research in recent years has produced evidence suggesting that $Fru-2,6-P_2$
concentration in vivo can respond to changes in the concentration of the
above messenger metabolites (DHAP, PGA and Pi) thereby it can serve as a

Biggens, J. (ed.), Progress in Photosynthesis Research, Vol. III. ISBN 90 247 3452 5

regulatory link between carbohydrate metabolism in the chloroplast and the cytosol. Besides utilizing the earlier identified signals, i.e. triose phosphate and Pi concentration, this regulatory system indirectly senses other signals of the carbon and energy status of the cell, such as sucrose concentration (1).

Fru-2,6-P_2 fulfills its function by strongly affecting two cytosolic enzymes. It activates pyrophosphate: fructose-6-phosphate, 1-phosphotransferase (PFP), an enzyme catalyzing the phosphorylation of fructose-6-phosphate (Fru-6-P) by pyrophosphate (PPi), a reversible reaction. The physiological function of PFP is still debated. Originally a glycolytic function was suggested for the enzyme, however it is not clear yet whether PPi is present in sufficient concentration in the cytosol to support the reaction. Alternatively it has been proposed that PFP operates in the reverse direction supplying PPi for UDP-glucose pyrophosphorylase in the sucrose syntheses mediated breakdown of sucrose (2,5). Fru-2,6-P_2 also inhibits cytosolic fructose-1,6-bisphosphatase (Fru-1,6-P_2ase) the enzyme catalyzing the first irreversible step in the pathway of sucrose synthesis from triose phosphates. Changes in Fru-2,6-P2 concentration will reciprocally affect sucrose synthesis and breakdown, high Fru-2,6-P_2 concentration inhibiting and low concentration facilitating sucrose synthesis.

The enzyme activities responsible for the synthesis and hydrolysis of Fru-2,6-P_2, fructose-6-phosphate, 2 kinase (Fru-6-P,2K) and fructose-2, 6-bisphosphate phosphatase (Fru-2,6-P_2ase), respectively, are regulated by certain key metabolites of the glycolytic/gluconeogenetic pathway of the cytosol including the triose phosphates transported by the phosphate translocator (Fig 2.) (1). Fructose-6-phosphate (Fru-6-P), the accumulation of

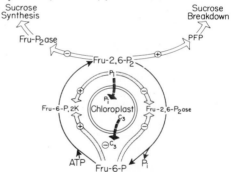

Figure 2. Role of chloroplasts and effector metabolites in Fru-2,6-P_2-linked control of cytosolic sucrose transformations in spinach leaves.

which signals high levels of sucrose in the cytosol (2), and Pi activate Fru-6-P,2K and inhibits Fru-2,6-P_2ase (1). DHAP and PGA, inhibit Fru-6-P,2K (1). As a result Fru-2,6-P2 concentration decreases when triose phosphates accumulate in the cytosol during rapid photosynthesis. Fru-2,6-P_2 concentration will increase, however, when sucrose accumulates in the cytosol signaling a decrease in demand for photosynthate by nonphotosynthetic tissues. The response of Fru-6-P,2K and Fru-2,6-P_2ase activities to these metabolites changes the level of Fru-2,6-P_2 in the cytosol so that sucrose synthesis is favored when the rate of photosynthesis is high, and the sucrose synthetic pathway is blocked when the rate of photosynthesis declines or the demand for carbon decreases in sink tissues.

Fru-2,6-P_2 may also act at points of carbohydrate metabolism other than PFP and Fru-1,6-P_2ase. A Fru-2,6-P_2 activated UDP-glucose phosphorylase has been detected in potato tubers (6), and 6-phosphogluconate dehydrogenase from castor beans was reported to be inhibited by Fru-2,6-P_2 (7).

Fru-6-P,2K and Fru-2,6-P$_2$ase the enzymes catalyzing the synthesis and hydrolysis of Fru-2,6-P$_2$ have been partially purified from spinach leaves. Based on the fact that the two activities cochromatographed throughout the procedure, it was suggested that, in plant tissues as in animal systems, Fru-6-P , 2K and Fru-2,6-P$_2$ase reside on the same protein (1).

In this communication we report that, using a new, medium pressure chromatography system (FPLC), we have been able to separate Fru-6-P,2K and Fru-2,6-P$_2$ase from spinach leaves. We also detected a second activity peak of both enzymes, which, however, is probably a product of the proteolytic degradation of the major enzyme forms.

2. PROCEDURE
2.1 Materials and Methods
     2.1.1 Plant material: Spinach (Spinacea oleracea Var. Hipack was grown in a hydroponic culture in a greenhouse.
     2.1.2 Enzyme purification: Fru-6-P,2K and Fru-2,6-P$_2$ase were purified from spinach leaves. Leaves (20 g) were washed in ice cold distilled water and homogenized in 40 ml of Buffer A (20 mM Tris-HCl pH 7.8,10% glycerol, 0.1% 2-mercaptoethanol, supplemented with 0.5 mM PMSF (phenylmethyl sulphonyl fluoride),20 mM sodium diethyl-dithio-carbamate, 2 mM E-amino n-caproic acid, 2 mM benzamidine-HCl and 1.5% (w/v) insoluble PVP. The homogenate was filtered through four layers of cheesecloth and the filtrate was centrifuged for 20 min at 30,000 xg. The supernatant fraction was applied to a 1.6x15 cm DEAE (Whatman DE52) column, that had been equilibrated with Buffer A. The column was washed with 35 ml of Buffer A and then eluted with 150 ml of a linear, 0 - 0.4 M KCl, gradient prepared in Buffer A. Fractions showing the highest Fru-6-P,2K, and Fru-2,6-P$_2$ase activities were pooled and concentrated by dialysis against 50% glycerol in Buffer A. The enzyme was further purified on a Mono Q (Pharmacia) column using a Pharmacia FPLC system. The enzyme sample was applied to the Mono Q column in Buffer A and eluted with a linear, 0 - 0.4 M KCl, gradient (25 ml) in Buffer A. Fru-6-P,2K and Fru-2,6-P$_2$ase activities obtained from this chromatography were used in the experiments below. The preparation which was completed in 24 h was essentially free of phosphofructokinase, Fru-1,6-P$_2$ase and PFP activities.
     2.1.3 Enzyme assays: Fru-6-P2K and Fru-2,6-P$_2$ase activities were measured using the bioassay described by Cseke and Buchanan (8) and Stitt, et al. (9). For the measurement of Fru-2,6-P$_2$ase activity a spectrophotometric method (10) was also used. The reaction mixture for Fru-6-P,2K assay contained 100 mM Tris-HCl pH 7.3, 5 mM MgCl$_2$, 5 mM Fru-6-P and 1 mM ATP. When the concentration of one of the substrates was varied the other substrate was kept at the concentration indicated above. The reaction mixture for Fru-2,6-P$_2$ase assay contained 100 mM Tris-HCl pH 7.3, 5 mM MgCl$_2$ and Fru-2,6-P$_2$, as indicated in the text. For the spectrophotometric assay of Fru-2,6-P$_2$ase the above reaction mixture was supplemented with 0.1 unit of phosphoglucose isomerase,   0.2 unit of glucose-6-phosphate dehydrogenase and 0.5 mM NADP. The reaction was started by the addition of the enzyme preparation.

3. RESULTS
3.1 Fru-6-P,2K and Fru-2,6-P$_2$ase enzyme forms in spinach leaves. Figure 3 shows the elution profile of Fru-6-P,2K from spinach leaves chromatographed on a Mono Q column. The chromatography profile of the control sample prepared

Figure 3. Activity profile
of spinach leaf Fru-6-P,2K
prepared with and without
protease inhibitors.

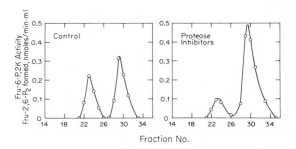

Fraction No.

in the absence of protease inhibitors contains two peaks of Fru-6-P,2K
activity of approximately the same size. Including protease inhibitors in
the extraction medium dramatically reduces the size of peak one (low salt),
suggesting, that only the second peak (high salt) is the physiological form
of the enzyme and the first one is its proteolytic degradation product.
Fru-2,6-P$_2$ase activity followed with the bioassay showed a distribution
pattern and response to the presence of protease inhibitors similar to
Fru-6-P,2K in these experiments.

3.2 Effect of Pi on the Fru-6-P,2K and Fru-2,6-P$_2$ase activities. The two
Fru-6-P,2K peaks show a differential response to Pi. As Figure 4 demonstrates the first peak is more strongly stimulated by Pi than the major form
in the second peak.

Figure 4. Effect of Pi on the
activity of the minor and major
form of spinach leaf Fru-6-P,2K.

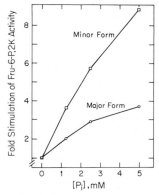

Throughout the Pi concentration range tested (1 - 5 mM) the minor peak
showed a stimulation twice as high as the major form. The Fru-2,6-P$_2$ase
activities associated with the peaks show a similar differential response to
Pi. The minor peak showed only 30% inhibition by 10 mM Pi, when assayed in
the presence of 5 uM Fru-2,6-P$_2$, compared to a 70% inhibition of the major
peak of Fru-2,6-P$_2$ase, under similar conditions.

3.3 Separation of Fru-6-P,2K and Fru-2,6-P$_2$ase activities. As seen in
Figure 5 the majority, when chromatographed on a Mono Q (ion exchange
column), Fru-6-P,2K and Fru-2,6-P$_2$ase activities can be clearly separated

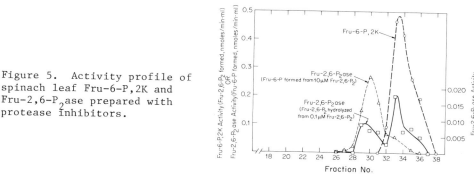

Figure 5. Activity profile of spinach leaf Fru-6-P,2K and Fru-2,6-P$_2$ase prepared with protease inhibitors.

from each other. The main Fru-2,6-P$_2$ase activity peak has negilible Fru-6-P,2K associated with it. This enzyme form is probably a "pure" phosphatase. The Fru-6-P,2K, however contains a significant Fru-2,6-P$_2$ase activity as well, suggesting that this may be a bifunctional enzyme with a particularly high Fru-6-P,2K to Fru-2,6-P$_2$ase ratio.

4. DISCUSSION

We have shown that two Fru-6-P,2K forms can be isolated from spinach leaves, which show different responses to Pi. The two peaks can be separated by ion exchange chromatography. The first peak, however, seems to be the product of proteolytic degradation, since addition of protease inhibitors to the isolation medium decreases its size at least 6-fold. Previous reports have only accounted for one peak of Fru-6-P,2K and Fru-2,6-P$_2$ase activities (1). Since protease inhibitors were not used in these earlier studies it is likely that a mixture of the intact and degraded enzymes were isolated. This may have resulted in errors in the determination of the kinetic properties of the enzyme. The major distinction between the intact and degraded forms of Fru-6-P,2K and Fru-2,6-P$_2$ase however appears to be a quantitative difference in the effect of Pi, which does not significantly alter the conclusions on the regulatory properties of the enzymes. No variation in the effects of DHAP and PGA were observed (results not shown).

Using a medium pressure ion exchange chromatography system we have been able to resolve Fru-6-P,2K and Fru-2,6-P$_2$ase. Earlier reports, using less purified preparations, demonstrated only a single combined peak of Fru-6-P,2K/ Fru-2,6-P$_2$ase. Our present finding points to a significant difference between these enzymes in spinach and liver cells. In the latter the two enzyme activities reside on the same protein. Stitt et al.(11) and Stitt et al.   (12) using crude or partially purified preparations have obtained evidence suggesting covalent modification of Fru-6-P,2K and Fru-2,6-P$_2$ase in plant tissues. The present results, especially the effect of proteases on the chromatographic and also kinetic properties of Fru-6-P,2K and Fru-2,6-P$_2$ase,indicate that results obtained with unpurified preparations should be treated with special caution.

Although, as recently reported for heart and certain other mammalian organisms (13), plant Fru-6-P,2K and Fru-2,6-P$_2$ase seem to be distinct proteins their previously outlined regulatory role in carbohydrate metabolism is still valid. Certain details of that picture need to be modified, however, especially the kinetic properties of the individual enzymes.

(This work was supported in part by a grant from the National Science Foundation.)

REFERENCES
1 Cseke, C., Balogh, A., Wong, J.H., Buchanan, B.B., Stitt, M., Herzog, B. and Heldt, H.W. (1984) Trends Biochem. Sci. 9, 533-535
2 Cseke, C. and Buchanan, B.B. (1986) Biochim. Biophys. Acta
3 Leegood, R.C., Walker, D.A. and Foyer, C. (1985) in Topics in Photosynthesis (Barber, J. and Baker, N.R., eds.), Vol. 6, pp.189-258, Elsevier, Amsterdam
4 Preiss, J. (1984) Trends Biochem. Sci. 9, 24-27
5 apRees, T., Green, J.H. and Wilson, P.M. (1985) Biochem J. 227, 299-304
6 Gibson, D.M. and Shine, W.E. (1983) Proc. Natl. Acad. Sci. USA 80, 2491-2494
7 Miernyk, J.A., MacDougall, P.S. and Dennis, D.T. (1984) Plant Physiol. 76,1093-1094
8 Cseke, C. and Buchanan, B.B. (1983) FEBS Lett. 155, 139-142
9 Stitt, M., Cseke, C. and Buchanan B.B. (1985) Physiol. Veg. 23, 819-827
10 Cseke, C., Stitt, M., Balogh, A. and Buchanan, B.B. (1983) FEBS Lett. 162, 103-106
11 Stitt, M., Cseke, C. and Buchanan, B.B. (1986) Plant Physiol. 80, 246-248
12 Stitt, M., Mieskes, G., Soling, H.D., Grosse, H. and Heldt, H.W. (1986) Z. Naturforsch. 41c, 291-296
13 El-Maghrabi, M.R., Correira, J.J., Heil, P.J., Pate, T.M., Cobb, C.E. and Pilkis, S.J. (1986) Proc Natl. Acad. Sci. USA 83, 5005-5009

CARBOHYDRATE STORAGE STRATEGY, F-2,6-P$_2$, AND CAM

HOLTUM, J.A.M., FAHRENDORF, T., NEUHAUS, H.E., MUKHERJEE, U., LATZKO, E.
Botanisches Inst, Schloßgarten 3, 4400 Münster, Germany (BRD)

1. INTRODUCTION

Twice every 24 h CAM cells effect massive net fluxes of carbon between two carbon storage pools, vacuolar malic acid and storage carbohydrate (CHO) that may be chloroplastic or extra-chloroplastic.

In the light, malate is decarboxylated to produce CO_2 and pyruvate or PEP, most of which is converted to CHO via gluconeogenesis (1). As the enzymes which convert PEP to 3-PGA are cytoplasmic, it has been postulated that for every mole of malate lost at least 0.89 moles of 3-PGA are produced in the cytoplasm (2). In starch-storers, this 3-PGA carbon enters the chloroplast; implying that cytoplasmic sugar production is not favoured and that the chloroplasts are carbon-importing organelles during deacidification. In the extra-chloroplastic CHO storers, the chloroplasts are net carbon exporting organelles and chloroplast carbon uptake is not favoured.

In a study of how CAM cells partition carbon targeted for storage CHO between the chloroplastic and extra-chloroplastic compartments, we measured F-2,6-P$_2$ fluctuations in a chloroplast starch storer Bryophyllum tubiflorum and in an extra-chloroplast sugar storer Ananas comosus (3). We have correlated these changes with the activities of cytoplasmic FBPase, PPi F-6-P phosphotransferase (PFP) and PFK, with the kinetic properties of PFP, and with the ability of CAM chloroplasts to synthesise and exchange PEP.

2. F-2,6-P$_2$ IN BRYOPHYLLUM AND ANANAS DURING DEACIDIFICATION

In both species, the F-2,6-P$_2$ levels were at their daily minimum during deacidification (Fig 1). The daily minimum in Bryophyllum was only 20 pmol g^{-1} fr wt whereas that for Ananas was 800 pmol g^{-1} fr wt. Models of F-2,6-P$_2$ mediated regulation predict that high, or increasing, F-2,6-P$_2$ levels characterise tissues in which chloroplastic CHO formation is favoured and reflect low cytoplasmic 3-PGA-plus-TP levels (4). Low, or decreasing, F-2,6-P$_2$ levels characterise tissues in which cytoplasmic CHO formation is favoured and are indicative of higher cytoplasmic 3-PGA-plus-TP levels.

Biggens, J. (ed.), Progress in Photosynthesis Research, Vol. III. ISBN 90 247 3452 5
© *1987 Martinus Nijhoff Publishers, Dordrecht. Printed in the Netherlands.*

FIGURE 1.   Levels of F-2,6-P$_2$,   malate and the major storage CHOs in Ananas
(left) and Bryophyllum (right) during the light period.

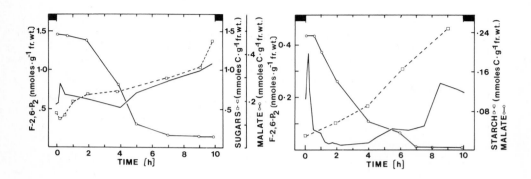

Figure 1 presents an apparent paradox. The chloroplast starch former Bryophyllum has low F-2,6-P$_2$ levels during deacidification but in   Ananas   the
levels are high; both opposite to our initial expectations.

3. THE CYTOSOLIC CAPACITY FOR F-1,6-P$_2$ - F-6-P INTERCONVERSION

In contrast to the starch formers Bryophyllum and Mesembryanthemum,   Ananas
has a high capacity for cytoplasmic F-6-P production (Table 1). Most of the
F-6-P formation capacity in Ananas is due to PFP activity.   This enzyme   is
stimulated by F-2,6-P$_2$(5).

Table 1. Activities of		PFP	FBPase		PFK	MALATE
enzymes which consume			CYTO	CHLOR		LOSS
F-1,6-P$_2$ and F-6-P.	Ananas	600	45	10	20	25
Rates are nkat mg^{-1}	Bryophyllum	17	6	12	17	25
chl	Mesembryanthemum	2	5	7	17	15

In Bryophyllum, the low capacity for F-6-P formation approximates the rates
of malate loss. The low F-2,6-P$_2$ levels, which reflect internal concentrations of around 3 μM, might be sufficient to suppress FBPase activity.
Another possibility is that the cytoplasmic FBPase may have a relatively
low affinity for F-1,6-P$_2$, a substrate normally present in low concentrations. If so,   3-PGA gradients could develop between the cytoplasm and the
chloroplasts despite the presence of cytosolic sugar-producing machinery

which would also compete for the 3-PGA carbon. Such a FBPase is present in maize mesophyll cells (6). High cytoplasmic 3-PGA-plus-TP levels would explain the low $F-2,6-P_2$ levels shown in Figure 1, since the enzyme which synthesises $F-2,6-P_2$, F-6-P 2-kinase, is inhibited by 3-PGA and TP.

4. PPi FRUCTOSE-6-PHOSPHATE 1-PHOSPHOTRANSFERASE (PFP) FROM HOYA CARNOSA

In the preceeding section we suggested a role for PFP in Ananas (see 3). In PFP from the CAM plant Hoya carnosa, the Vmax of the glycolytic reaction is stimulated >70-fold by $F-2,6-P_2$ but the gluconeogenic reaction, which exhibits a lower Vmax, is unaffected (Table 2). $F-2,6-P_2$ however, increases the affinity of PFP for $F-1,6-P_2$ (Table 3). As the cytoplasmic $F-1,6-P_2$ concentration is low in vivo and the Pi:PPi ratio is thought to be high, it is possible that this $F-2,6-P_2$ effect has physiological significance.

Table 2. Effect of $F-2,6-P_2$ on the Vmax of the forward and reverse PFP reactions. Rates are nkats.

	($F-2,6-P_2$) µM	
	0	4
$F-1,6-P_2 \rightarrow F-6-P$	6.5	6.3
$F-6-P \rightarrow F-1,6-P_2$	0.3	22.0

Table 3. Effect of $F-2,6-P_2$ on the Km ($F-1,6-P_2$) of Hoya PFP. Assays contained 0.5 mM Pi.

	($F-2,6-P_2$) µM		
	0	0.1	4
Km ($F-1,6-P$) µM	47	9	7

5. TRANSPORT OF PEP ACROSS THE CHLOROPLAST MEMBRANE

In NADP malic enzyme CAM plants, it has been suggested that pyruvate enters the chloroplast and is transformed to PEP which is exported (2,7). Chloroplasts from CAM Mesembryanthemum possess pyruvate Pi dikinase and can form PEP in the light when presented with pyruvate externally. PEP efflux and formation are stimulated by Pi, 3-PGA and DHAP. The chloroplasts can accumulate PEP in the dark; the Km PEP at 4 °C is 0.22 mM. PEP uptake is inhibited by Pi (Ki=0.19 mM), DHAP and 3-PGA (Ki=0.14 mM), but is unaffected by pyruvate or malate.

The data indicate that CAM Mesembryanthemum chloroplasts possess a Pi translocator with a significant affinity for PEP. The exchange of chloroplastic PEP for cytosplasmic 3-PGA, particularly when cytoplasmic levels of 3-PGA are elevated, is possible. It may be relevent that the known chloroplast CHO storers are NADP malic enzyme plants which possess pyruvate Pi dikinase, whereas those known to be extra-chloroplastic CHO formers are PEP carboxykinase species which lack, or have little, dikinase (7).

6. SYNTHESIS AND POSTULATES

Bryophyllum and Ananas exhibit different carbon-storage strategies during deacidification. In Ananas, 3-PGA produced in the cytoplasm is probably converted to storage sugars via PFP rather than FBPase. Although the high $F-2,6-P_2$ levels present during this period indicate low cytoplasmic concentrations of 3-PGA-plus-TP, and thus $F-1,6-P_2$, it is likely that the affinity of PFP for $F-1,6-P_2$ will be extremely high. Low cytosolic 3-PGA-plus-TP levels also encourage the export, rather than the import, of chloroplast carbon via the Pi translocator. The fate of the PPi produced in the cytoplasm during PFP-mediated sugar synthesis is unclear, particularly as it appears that the activities of PPiase are low in the cytoplasm (3).

In Bryophyllum, 3-PGA produced during decarboxylation is imported into the chloroplast; probably through a Pi translocator that is also involved in the export of PEP from the chloroplasts. The low $F-2,6-P_2$ levels present during deacidification indicate high cytoplasmic 3-PGA-plus-TP levels. Elevated levels of these compounds would promote the import of carbon by the chloroplasts. The low rates of cytoplasmic sugar production probably occur because the capacity for F-6-P formation is low and, secondly, it is postulated that the cytosolic FBPase may belong to the $F-1,6-P_2$-insensitive type recently reported to be present in maize. The relative importance of the cytosolic FBPase and PFP in sugar production during the periods when Bryophyllum produces sugars for export, and the relative importance of hexose transport across the chloroplast membranes, remain to be resolved.

1. Holtum, J.A.M. and Osmond, C.B. (1981) Aust J Plant Physiol 8, 31-44
2. Edwards, G., Foster, J. and Winter, K. (1982) in Crassulacean Acid Metabolism (Ting, I., Gibbs, M. eds.) pp.92-111, Am. Soc. Plant Physiol., MD
3. Black, C., Carnal, N. and Kenyon, W. (1982) in Crassulacean Acid Metabolism (Ting, I.P., Gibbs, M. eds.) pp.51-68, Am. Soc. Plant Physiol., MD.
4. Stitt, M., Huber, S. and Kerr, P. (1987) in Biochemistry of Plants, 2nd edn, Vol. 8 (Hatch, M. and Boardman, K. eds.), Academic Press, New York
5. Carnal, N.W. and Black, C.C. (1983) Biochem Biophys Res Commun 86, 20-26
6. Stitt, M. and Heldt, H.W. (1985) Planta 164, 179-188
7. Osmond, C. and Holtum, J. (1981) in Biochemistry of Plants, Vol 8 (Hatch M.D. and Boardman, N.K. eds.), pp.283-328, Academic Press, New York

PHYSIOLOGICAL FUNCTION AND PHYSICAL CHARACTERISTICS OF THE CHLOROPLAST PHOSPHATE TRANSLOCATOR

ULF-INGO FLÜGGE, UNIVERSITY OF GÖTTINGEN
INSTITUT FÜR BIOCHEMIE DER PFLANZE, UNTERE KARSPÜLE 2,
3400 GÖTTINGEN, F.R.G.

INTRODUCTION
The chloroplast envelope has two distinct membranes: an inner and an outer membrane. The outer membrane is unspecifically permeable for substances with molecular weights up to 10 kD. This is due to the presence of a pore forming protein, named chloroplast porin, with a pore diameter of 2.5 - 3.0 nm which is the largest diameter of all porins pores known so far (1). The inner envelope membrane is the functional barrier between the chloroplasts and the cytosol. It is the site of numerous metabolite translocators which, apparently, are involved in coordinating the metabolism in the stromal and the cytosolic compartment.

RESULTS AND DISCUSSION
The function of the phosphate translocator during CO_2 fixation
The phosphate-triose phosphate-3-phosphoglycerate translocator, in short named phosphate translocator can be regarded as one of the main transport functions of the inner envelope membrane (2). It mediates the export of fixed carbon from the stroma to the cytosol in the form of triose phosphate. The export of triose phosphate (triose P) can proceed by counterexchange with inorganic phosphate (P_i) or with 3-phosphoglycerate (3-PGA), whereby fixed carbon is delivered to the cytosol for sucrose biosynthesis. The P_i formed during sucrose synthesis is shuttled back into the chloroplast in exchange for more triose P. In the chloroplast, P_i is used for the formation of ATP catalyzed by the thylakoid

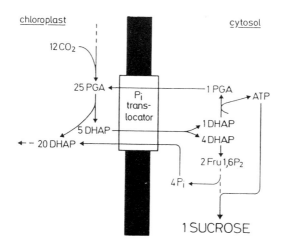

FIGURE 1.   Schematic representation of the metabolic role of the phosphate translocator during CO_2 fixation and sucrose biosynthesis. DHAP, triose phosphate; 3-PGA, 3-phosphoglycerate; P_i, inorganic phosphate

ATP synthase. It is important that the rate of triose P export is well balanced with the rate of CO_2 fixation; only 1/6 of the triose P formed can be removed from the chloroplasts for sucrose synthesis whereas 5/6 must remain with the Calvin cycle for the regeneration of the CO_2 acceptor ribulose 1,5 bisphosphate. Under photorespiratory conditions, however, even less than 1/6 of the newly formed triose P will be available for transport. Any removal of triose P beyond this limit will result in a depletion of the Calvin cycle intermediates and an inhibition of the CO_2 fixation cycle.

In addition, triose P can be also exported in exchange with 3-PGA. This shuttle provides the cytosol with the products of the light reaction, ATP and reducing equivalents. One molecule of ATP is needed for every four molecules of exported triose P which are converted to sucrose. Thus, for the complete pathway of sucrose synthesis, five molecules of triose P have to be removed from the Calvin cycle. Four of these are exchanged with P_i and incorporated directly into sucrose and the other is required to generate ATP needed for sucrose synthesis (Fig. 1).

The functioning of the triose P/P_i and the triose P/3-PGA shuttles implies a preferential export of triose phosphate into the cytosol, even though the level of 3-PGA in the chloroplasts during CO_2 fixation was found to be much higher than that of triose phosphate. In intact chloroplasts it has been actually observed that illumination results in a restriction of 3-PGA export whereas in the dark, 3-PGA was found to be equally well transported in either directions (3). This suggested that the influence of light on the transport is related to the H^+ gradient across the envelope, resulting from the light-driven proton transport into the thylakoids. Since the substrates of the phosphate translocator are transported preferentially as twice negatively charged anions and 3-PGA in contrast to P_i and triose P carries at physiological pH three negative charges, an export of 3-PGA from the chloroplasts in exchange with P_i implies an indirect proton transfer across the envelope. As the pH in the cytosol is more acidic than in the stroma, the export of 3-PGA in the light is directed against the H^+ gradient and should therefore be restricted. On the other hand, the import of 3-PGA into chloroplasts is directed with the H^+ gradient and should be enhanced. The influence of ΔpH on the function of the phosphate translocator can be examined more precisely by measuring its activity in a reconstituted system under defined conditions. As shown below, in addition to the direct effect of pH on the changes of the K_M values, the transmembrane ΔpH exhibits pronouced effects on the translocator activity.

TABLE I. ΔpH induced changes of the kinetic properties of the phosphate translocator and its influence on the fluxes of triose phosphate and 3-phosphoglycerate. The calculations are based on substrate concentrations measured in intact leaves after 40 min illumination (4). The triose P/3-PGA ratio for the divalent anions are 0.52 for the stroma and 0.30 for the cytosol, respectively.

	Translocator activity	
	without ΔpH	with ΔpH
Export triose P/3-PGA	0.71	1.63
Import triose P/3-PGA	0.32	0.15
Export/Import	2.2	10.9

Recently, Gerhardt at al presented data for subcellular metabolite levels in spinach leaves measured by nonaqueous density gradient centrifugation (4). After 40 min illumination 10.7 mM 3-PGA and 0.64 mM triose P were found in the stroma whereas the cytosolic concentrations were 3.3 mM 3-PGA and 0.30 mM triose P, respectively. Assuming pH values in the stroma of about 7.9-8.0 and in the cytosol of 7.4-7.5, these concentrations can be converted into that of the individual twice negatively charged anions. With a simple Michaelis-Menten characteristic for the transport and kinetic constants of the phosphate translocator as measured in (5), the ratios for the import and export rates of triose P/3-PGA transport can be calculated (Table 1). It can be seen that the ratios for triose P/PGA import and export reflect roughly the concentration ratios in both the stromal and the cytosolic compartment. Consequently, 3-PGA is the preferred metabolite imported into and exported out of the chloropoasts. This is the opposite of the measured fluxes during photosynthesis. However, using the same calculation but including the effects of an H^+gradient on the kinetic constants of the phosphate translocator as measured in the reconstituted system (6) it can be predicted that indeed triose P rather than 3-PGA will be preferentially exported and the import of 3-PGA is enhanced (Table 1). This demonstrates how the effect of an H^+gradient on the kinetic constants of the phosphate translocator results in a considerable change in the ratio of the substrates transported.

The phosphate translocator as link between chloroplastic and cytosolic metabolism during sucrose synthesis

Under our experimental growth conditions intact spinach leaves exhibit a photosynthesis rate of about 100 μmol/mg chl x h and a rate of sucrose formation accounting for roughly 50-60% of this value (M. Stitt, personal communication). Assuming that triose P exported into the cytosol is only used for sucrose synthesis, the rate of triose P export is about 40% of the total CO_2 fixation rate as shown in Fig. 1. Thus, the effective rate of triose P export has to be in the range of 20-25 μmol/mg chl x h. Since the V(max)(20°C) of the phosphate translocator was calculated to be 200-250 μmol/mg chl x h (5) it can be concluded that about 10% of the total translocator activity has to be provided for the net export of triose P. It can be also seen from Fig. 1 that 4/5 of this triose P export activity has to be balanced by an exchange with P_i which is released during sucrose synthesis and reimported in order to sustain photosynthesis.

The kinetic constants of the phosphate translocator under conditions of an imposed ΔpH i.e. under photosynthetic conditions can now be used in combination with values for the subcellular concentrations of triose P and 3-PGA (4) for the calculation of the percentage of the total translocator activity involved in both export and import of P_i, triose P and 3-PGA (Fig. 2). Since reliable measurements of the subcellular concentrations of P_i are still not available the extent of translocator activity involved in the transport of the three substrates was calculated for different concentrations of the free divalent phosphate anion in both compartments.

Fig. 2 (left) shows the percentage of the translocator activity involved in the export of metabolites from the chloroplast into the cytosol at increasing concentrations of stromal P_i^{2-}. It becomes obvious that the stromal P_i concentration has to be adjusted to rather low values (below 1mM of free P_i^{2-}) to ensure a preferential export of triose P. At higher stromal P_i concentration, the (counter-productive) export of inorganic phosphate becomes predominant.

At a given stromal P_i concentration the fraction of the total translocator activity involved in triose P export can be defined. As 80% of the exported triose P will be exchanged for P_i (see Fig. 1), this allows us to predict the

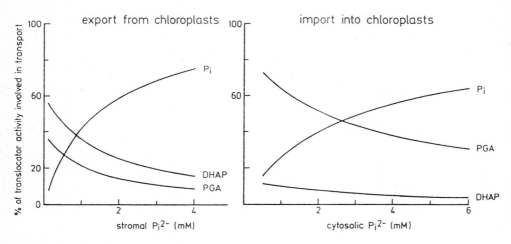

FIGURE 2.   Dependence of the percentage of translocator activity involved in export and import of P_i, triose phosphate (DHAP) and 3-phosphoglycerate (3-PGA) on stromal and cytosolic phosphate concentrations during CO_2 fixation. Values for the subcellular concentrations of DHAP and 3-PGA after 40 min illumination were taken from ref. 4.

rate of P_i import as shown in Fig. 2 (right). In turn, this rate of P_i import (Fig. 2, right) can be used to predict the cytosolic P_i concentration. Comparison between the values obtained for P_i import and triose P export (Fig. 2) indicates that a higher cytosolic/stromal ratio is required for net P_i import into the chloroplasts. Increasing concentrations of the cytosolic P_i lead to an increase of P_i import with an concomitant decrease of 3-PGA import. For the same conditions, the percentage of translocator activity involved in triose P import is affected only to a small extent and does not exceed 10% even at low cytosolic P_i.

The individual percentage of the translocator activity involved in transport of the other metabolites at the given P_i concentration in both compartments can now be estimated allowing the calculation of net fluxes in either direction. The results are shown in Fig. 3. As already mentioned about 10% of the total translocator activity has to be involved in net export of triose P and net import of P_i to sustain efficient formation of sucrose during photosynthesis. This is apparently achieved at a stromal P_i concentration of about 0.5 mM. It must be emphasized that due to the incertainties inherent with the calculations this value is not an absolute number but nevertheless indicates that a low stromal P_i concentration is essential. Fig. 3 shows that at 0.5 mM stromal P_i the cytosolic value would have to be 3-4 times higher to sustain photosynthesis. The rather low stromal value seems to contradict measurements where the stromal plus cytosolic P_i were found to be 20-30 mM in the dark (7,8). However, direct measurements of the intracellular P_i distribution during photosynthesis have not been successful. In addition, part of the P_i might not be freely accessible but may exist in a bound state. On the other hand, the stromal P_i concentration has to be kept high enough to sustain ATP formation during photosynthesis. Measurements of the K_M (P_i) value for the ATP synthase revealed maximal values of about 0.5 mM (9). Thus, a stromal P_i concentration in this range seems to be sufficient to maintain an optimal photosynthesis rate.

Import of P_i is accompanied by the import of 3-PGA and both fluxes are

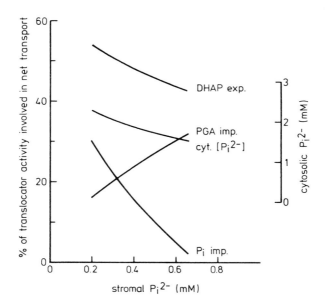

FIGURE 3.   Dependence of the percentage of translocator activity involved in
            net transport of metabolites on the stromal phosphate concentration.

counterbalanced by the export of triose P accounting for only about 45% of the
total translocator activity. Obviously, a great portion of the translocator
activity is taken up in catalyzing non-productive homologous exchange modes.
This result can be expected since both compartments contain triose P, P_i and
3-PGA as well competing for transport in both directions.
As the cytosolic P_i concentration decreases, the export of triose P from the
chloroplasts will be decreased and that of 3-PGA import will be increased (Fig.
3). Under such conditions, sucrose synthesis is lowered and the remaining
surplus of triose P in the stroma can be now converted in the chloroplast to
starch. However, when the photosynthesis rate increases more triose P can be
converted into sucrose. The higher withdrawal of triose P from the chloroplasts
has to be counterbalanced by import of cytosolic P_i. This can be achieved by
decreasing the stromal P_i concentration and by increasing that in the cytosol.
This increased transmembrane P_i gradient would lead to a stimulation of triose
P export from the chloroplast thus enabling a higher sucrose synthesis rate. The
presence of regulatory systems involved in sucrose biosynthesis (fructose 2,6
bisphosphate and modification of sucrose phosphate synthase) will in principle
allow fluxes into sucrose synthesis to be independent of cytosolic P_i
concentration.
It can also be seen from Fig. 3 that at 0.5 mM stromal P_i the amount of the
translocator catalyzing a triose P/3-PGA shuttle is 2-3 times higher than that
of catalyzing a triose P/P_i exchange. This indicates, that during
photosynthesis the export of ATP and reducing equivalents may have the priority
over the export of triose P for the synthesis of sucrose. The implications of
this conclusion remain unclear.
It is also obvious from the results shown in Fig. 2 and 3, that although the
total activity of the phosphate translocator exceeds the rate of photosynthesis,
only a minor portion of the translocator activity is involved in net influx of
P_i and in net efflux of triose P. This indicates that the phosphate

translocator can exert a kinetic limitation during sucrose synthesis <u>in vivo</u> as suggested also by Gerhardt et al (4).

Characterization of the isolated phosphate translocator protein
The membrane protein catalyzing the function of the phosphate translocator has been identified as a major component of the inner envelope membrane with a molecular weight of 29 kD as determined by SDS gel electrophoresis. The protein has been isolated in the nonionic detergent Triton X-100 and reconstituted into liposomes in a functional state, allowing the determination of the kinetic properties of the phosphate translocator under well defined experimental conditions (see ref. 2 and ref. therein).
The estimation of the hydrodynamic parameters of the isolated translocator protein revealed an apparent molecular weight for the protein–detergent micelle of 178 kD. The relative high amount of detergent bound in the complex (180 mol/mol of protein) indicates that the translocator exhibits an extensive membrane contact area as is expected for an integral membrane protein. The molecular weight of the protein moiety of the complex was found to be 61 kD. Since the monomeric form of the translocator exhibits a molecular weight of 29 kD, it appears that the phosphate translocator exists as a dimer made up of two identical subunits (10).
Assuming a prolate ellipsoidal shape of the translocator protein, the axes of the ellipsoid were calculated to be 6.59 nm and 1.59 nm, respectively. It is suggested from these data that the phosphate translocator can span the membrane thereby connecting the stromal and the cytosolic compartment. The amount of protein protruding from both sides of the membrane seems to be sufficient to render the translocator accessible to its substrates. It is supposed that the dimeric translocator exists as a fixed carrier in the membrane exhibiting a C_2 rotational axis perpendicular to the membrane plane as shown for the mitochondrial ATP/ADP transport system (11). This transmembrane arrangement allows the formation of a hydrophilic channel between the two subunits providing the translocation path for the substrate. The gating of this pore is triggered by substrate binding in such a way that the binding site is exposed to the one or the other side of the membrane in turn. This implies that the translocator exhibits only one substrate binding site per dimer.

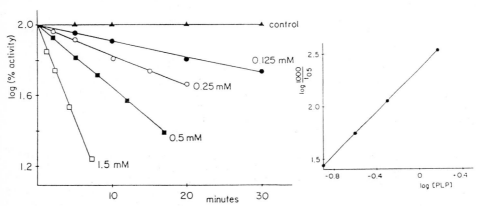

FIGURE 4. Time course of inhibition of phosphate transport by varying pyridoxal 5'-phosphate (PLP) concentrations. Phosphate transport was measured by silicone layer filtering centrifugation. Inset, determination of the order of the inactivation reaction with respect to PLP. Data from ref. 12.

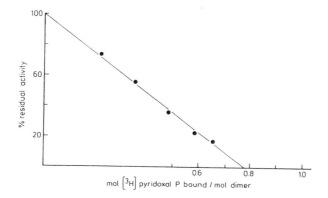

FIGURE 5.   Correlation between the inhibition of phosphate transport and the incorporation of [³H] PLP into the phosphate translocator by varying the PLP concentration (0.05-0.25 mM). Experiment performed by A. Gross.

To verify this hypothesis, a rather specific inhibitor of the phosphate translocator, pyridoxal 5'-phosphate (PLP), has been employed. This inhibitor reacts with ε-amino groups of lysines forming a Schiff base and has originally been used for the identification of the phosphate translocator protein (12). Fig. 4 shows the time course of inhibition of P_i transport by increasing concentrations of PLP. When the logarithm of the reciprocal values of the

TABLE II.   Some characteristics of the chloroplast phosphate translocator

Translocation activity	0.25	μmol/min mg protein (20°C)
Molecular activity (20°C)		
intact chloroplast	5000	min^{-1}
reconstituted system	300	min^{-1}
Translocator density of the envelope	2.5	$pmol/cm^2$
Flux rate per membrane area	7.3	$nmol/min\ cm^2$
Sedimentation coefficient	3.50×10^{-13}	(sec)
Partial specific volume	0.860	cm^3/g
Stokes' radius	6.32	nm
Triton X-100 binding	1.90	g/g protein
Molecular weight		
Monomer (SDS-PAGE)	29,000	
Translocator-detergent micelle	177,500	
Protein moiety	61,000	
Axial ratio of the prolate ellipsoid	3.38	
Semimajor axis	6.59	nm
Semiminor axis	1.95	nm
Number of substrate binding sites/dimer	0.8	

half-time of inhibition ($t_{0.5}$) is plotted as a function of the logarithm of the inhibitor concentration; a straight line is obtained with a slope equal to 1 (Fig. 4, inset). This indicates that one molecule of PLP is required per substrate binding site for the inactivation of the phosphate translocator protein. Furthermore, it reveals the presence of a lysine residue at the substrate binding site.

The evaluation of the number of substrate binding sites present can be accomplished by determining the amount of radioactive labeled inhibitor incorporated into the protein. Unfortunately, this does not seem possible if [^{3}H] NaBH$_4$ is employed to irreversibly reduce the Schiff base formed between PLP and the lysine group present in the active center. Therefore radioactive labeled [^{3}H] PLP (specific activity, 40 Ci/mol) has been synthesized and used for inhibition and incorporation studies. The correlation between the inhibition of transport and the incorporation of the [^{3}H] PLP label into the translocator protein shows that 100% inhibition requires the binding of about 0.8 mol PLP/mol of the dimeric phosphate translocator protein (Fig. 5).

Obviously, in its functional state the phosphate translocator exhibits only one substrate binding site per dimer triggering transport of metabolites into the cytosolic or the chloroplastic compartment in turn (Gross and Flügge, unpublished). Table II shows some characteristics of the isolated chloroplast phosphate translocator.

Acknowledgement
This work was supported by the Deutsche Forschungsgemeinschaft

REFERENCES
1 Flügge, U.I. and Benz, R. (1984) FEBS Lett. 169, 85-89
2 Flügge, U.I. and Heldt, H.W. (1984) Trends in Biochem. Sci. 9, 530-533
3 Heldt, H.W., Flügge, U.I. and Fliege R. (1978) in: The Proton and Calcium Pumps (G.F. Azzone et al. eds.) Elsevier/North Holland Biomedical press, pp 105-114
4 Gerhardt, R., Stitt, M. and Heldt, H.W. (1986) Plant Physiol. submitted
5 Fliege, R., Flügge, U.I., Werdan, K. and Heldt, H.W. (1978) Biochem. Biophys. Acta 502, 232-247
6 Flügge, U.I., Gerber, J. and Heldt, H.W. (1983) Biochim. Biophys. Acta 725, 229-237
7 Foyer, C., Walker, D.A., Spencer, C.A. and Mann, B. (1982) Biochem. J. 202, 429-434
8 Stitt, M., Wirtz, W., Gerhardt, R., Heldt, H.W., Spencer, C.A., Walker, D.A. and Foyer, C. (1985) Planta 166, 354-364
9 Afalo, C. and Shavit, N. (1983) FEBS Lett. 154, 175-179
10 Flügge, U.I. (1985) Biochim. Biophys. Acta 815, 299-305
11 Klingenberg, M. (1981) Nature 290, 449-454
12 Flügge, U.I. and Heldt, H.W. (1977) FEBS Lett. 82, 29-33

THE ROLE OF ESSENTIAL SULPHYDRYL GROUPS IN THE FUNCTIONING OF THE PHOSPHATE TRANSLOCATOR IN MAIZE MESOPHYLL CHLOROPLASTS

A.G. THOMPSON, M.A. BRAILSFORD, N. KADERBHAI AND R.B. BEECHEY, DEPARTMENT OF BIOCHEMISTRY, THE UNIVERSITY COLLEGE OF WALES, ABERYSTWYTH, DYFED, WALES, U.K.

1. INTRODUCTION

The accepted role of the phosphate translocator in maize mesophyll chloroplasts (C4) is that it facilitates the exchange of 3-phosphoglyceric acid (3-PGA), inorganic phosphate (Pi), phosphoenolpyruvate (PEP) and dihydroxy acetone phosphate (DHAP)[1].

Here we report on the inhibition of 3-PGA dependent oxygen evolution by a series of impermeable sulphydryl reagents the acid maleimides (AM). The acid maleimides react irreversibly with sulphydryl groups. They have been synthesised with a polymethylene chain of varying lengths (Table 1) and they have proved to be useful probes for obtaining information on the nature and depth of essential sulphydryl groups in membranes[2].

2. RESULTS AND DISCUSSION

Purification and characterisation of the chloroplasts

Intact maize mesophyll chloroplasts were purified by a procedure based on that of Jenkins and Russ[3]. The procedure purified maize chloroplasts in yields of up to 2mg of chlorophyl per preparation. The chloroplasts were routinely more than 85% intact by the ferricyanide exclusion test[4]. They did not exhibit bicarbonate dependent oxygen evolution which suggests that they were free from contamination by bundle sheath chloroplasts.

In the light and in the presence of phosphate (1mM) the rate of 3-PGA (1mM) dependent oxygen evolution was 0.92 μmol/mg of chl. This activity was phosphate dependent and when 3-PGA was added to chloroplasts without phosphate the rate of oxygen evolution was low or it was absent.

The sensitivity of 3-PGA dependent oxygen evolution to the acid maleimides

Addition of AM resulted in a time dependent inhibition of 3-PGA dependent oxygen evolution. A range of acid maleimides varying in the number of methylene groups were inhibitors of 3-PGA dependent oxygen evolution, AM3-AM11 inhibiting by more than 90% at less than 1.5μmoles/mg of chlorophyll (Table 1). The potency of the inhibitors varied with chain length and AM5, AM7 and AM11 were able to inhibit 3-PGA dependent oxygen evolution by more than 90% at less than 0.6 μmol/mg of chl. (Table 1).

Biggens, J. (ed.), Progress in Photosynthesis Research, Vol. III. ISBN 90 247 3452 5
© *1987 Martinus Nijhoff Publishers, Dordrecht. Printed in the Netherlands.*

Compound	Abbreviation	Distance R_1-R_2 $\overset{o}{A}$	Inhibitor Titre (μmoles/mg of chl)
$R_1.CH_2R_2$	AM1	5.9	9.26
$R_1.(CH_2)_2R_2$	AM2	7.1	3.70
$R_1.(CH_2)_3R_2$	AM3	8.3	0.93
$R_1.(CH_2)_4R_2$	AM4	9.6	0.93
$R_1.(CH_2)_5R_2$	AM5	10.8	0.19
$R_1.(CH_2)_7R_2$	AM7	13.2	0.37
$R_1.(CH_2)_{10}R_2$	AM10	17.1	1.48
$R_1.(CH_2)_{11}R_2$	AM11	18.3	0.56

$R_1 = CO.OH$ $R_2 = $ Maleimide

TABLE 1. The ability of a series of acid maleimides of varying chain length to inhibit 3-PGA dependent oxygen evolution by 90% in intact maize mesophyll chloroplasts.

Chloroplasts (27μg of chl.) and AM were added to 1ml of buffer in an illuminated oxgen electrode. After a 4 min. preincubation substrates (1mM 3-PGA and 1mM Pi) were added. The rate of oxygen evolution was measured 1.5 min. after the addition of substrates.

It has been established that the series of acid maleimides have a similar reactivity towards sulphydryl groups[5] and that differential inhibitor activity is not based upon differences in partition coefficients[2]. The acid maleimides do not fold or pass through plant mitochondrial membranes[6] or red blood cell ghosts[7]. All the available data suggests that the maleimide penetrates the membrane to a depth which is limited by the length of the poly-methylene chain and the presence of an ionised carboxyl group. Hence they will react with sulphydryl groups at a restricted depth within the membrane. Whilst the chloroplast outer envelope membrane is freely permeable to small molecules, they will not cross the inner envelope membrane. Thus it is likely that in maize mesophyll chloroplasts the acid maleimides are reacting with one or more essential sulphydryl groups on the phosphate translocator of the inner envelope membrane. This inhibition restricts the import of 3PGA or phosphate, or export of dihydroxy acetone phosphate and results in inhibition of 3-PGA dependent oxygen evolution.

The data presented in Table 1 demonstrate the potency of AM5, AM7 and AM11 as inhibitors of 3-PGA dependent oxygen evolution. It suggests that the transporter contains at least two essential sulphydryl groups one located at around 12A and the second at a minimum depth of 18A. The intermediate activity of AM3, 4 and 10 may represent mobility of the groups within the membrane.

The effect of light and substrates on the inhibition of 3-PGA dependent oxygen evolution by the acid maleimides

A 4 min. preincubation of chloroplasts with 10μM AM7 in the light without substrates resulted in 100% inhibition of 3-PGA dependent oxygen evolution. A 4 min. preincubation in the light with 10μM AM7 and substrates (1mM Pi, 2mM 3-PGA) reduced the inhibition to 68%. The protection by substrates suggests that the maleimide sensitive sites of the C4 phosphate translocator are not totally available after substrate binding. A similar protection of the Digitaria (C4) phosphate translocator from inhibition by the non permeant inhibitor DIDS has been reported[8].

A 4 min. preincubation of chloroplasts with 10μM AM7 in the dark, either with or without substrates also reduced the inhibition of 3-PGA dependent oxygen evolution by 30%. This protection after dark preincubation suggests a light modulation of the translocator. Light has been shown to modulate the export of triose phosphate and the import of 3-PGA in spinach (C3) chloroplasts. Spinach phosphate translocator has been reconstituted into liposomes, and variations in ΔpH across the liposome membrane which may model light activation of the chloroplast, changed the apparent Km for phosphate and triose phosphate, and changed the Vmax for 3-PGA transport[9]. Protection of the maize phosphate transporter from acid maleimide inhibition in the dark suggests that light may exert a control on this translocator through a reaction involving sulphydryl groups. It is tempting to suggest that sulphydryl groups have a role in regulating the activity of the C4 phosphate translocator. The protection of carrier associated dithiols from reaction with alkylating agents by oxidation, phosphorylation, or imposition of a hydrogen ion gradient have been reported in other carrier systems[9]. The data presented here support the proposition[11] that carrier activity might be associated with dithiol-disulphide interactions between different redox centres in a single carrier.

REFERENCES
1. Day, D.A. and Hatch, M.D. (1981) Arch. Biochem. Biophys. Vol 211, 2, 743-749
2. Griffiths, D.G., Partis, M.D., Sharp, R.N., Beechey, R.B. (1981) FEBS Vol 134, 2, 261-263
3. Jenkins, C.L.D. and Russ, V.J. (1984) Plant Sci. Lett. 35, 19-24
4. Lilley, R.Mc.C., Fitzgerald, M.P., Rients, K.G., Walker, D.A. (1975) New Phytol. 75, 1
5. Griffiths, D.G. PhD Thesis (1984) University of Kent at Canterbury, Kent, U.K.
6. Moore, A.L., Proudlove, M.O., Partis, M.D., Beechey, R.B. (1984) Adv. Photosynth. Res. Proc. Int. Congr. Photosynth. 6th, 3, 909-912
7. Moore, R.J. and Beechey, R.B. (1985) Biochem. Soc. Trans. 13, 690-691
8. Rumpho, M.E. and Edwards, G.E. (1985) Plant Physiol. 78, 537-544
9. Flugge, U.I., Gerber, J., Heldt, H.W. (1983) Biochem. Biophys. Acta, 725, 229-237
10. Poolman, B., Konings, W.N., Robillard, G.T. (1983) Eur. J. Biochem. 135, 41-46
11. Robillard, G.T. and Konings, W.N. (1982) Eur. J. Biochem. 127, 597-604

LEAF PHOSPHATE STATUS AND ITS EFFECTS ON PHOTOSYNTHETIC CARBON PARTITIONING AND EXPORT IN SUGAR BEET

I. MADHUSUDANA RAO, JAVIER ABADIA AND NORMAN TERRY
DEPT. PLANT & SOIL BIOLOGY, UNIVERSITY OF CALIFORNIA, BERKELEY, CA 94720, U.S.A.

1. INTRODUCTION

The level of orthophosphate (Pi) in plant tissues may be involved in the regulation of photosynthate partitioning (1-3). In this study, we report how nutritionally-induced changes in leaf phosphate status affect: i) the partitioning of photosynthate between starch, sucrose and glucose, ii) the activities of key enzymes involved in starch and sucrose metabolism; and iii) the rate of carbon export from the leaf blade to the remainder of the plant.

2. MATERIALS AND METHODS

Sugar beets (cv. F58-554H1) were cultured hydroponically in growth chambers at 25°C, 500 μmol quanta PAR m^{-2} s^{-1} photon flux density and a 16 h photoperiod (4). Low P, control and Pi addition treatments are as described in Rao et al.(5). Measurements were made of: i) the export of photosynthate from the leaf blade at saturating irradiance; ii) starch, sucrose and glucose in leaf blades; and iii) activities of enzymes of starch and sucrose metabolism, including ADP-glucose pyrophosphorylase, phosphorylase, beta-amylase, sucrose phosphate synthase, UDP-glucose pyrophosphorylase, and, cytoplasmic fructose-1,6-bisphosphate (FBP) phosphatase.

3. RESULTS

Low P treatment caused a marked reduction in the export of carbon from the leaf blade. The average rate of export of carbon over a 4-h period of illumination at saturating irradiance (2000 μmol quanta m^{-2} s^{-1}) was decreased by 92% in low-P leaves compared to control leaves (Table 1). When the rate of export is expressed as a percentage of the net rate of CO_2 fixation (Table 1), it is evident that only 6% of the carbon fixed by low-P leaves was exported compared to 77% for the control. When low-P plants were supplied with additional Pi, the average rate of carbon export increased from 8 to 40% (of the control) after 24 h (Table 1). Carbon export in low-P leaves increased from 6 to 33% of the net rate of CO_2 fixation after 24 h of increased Pi supply (Table 1).

The results show that 94% of the net carbon fixed was

Biggens, J. (ed.), Progress in Photosynthesis Research, Vol. III. ISBN 90 247 3452 5
© *1987 Martinus Nijhoff Publishers, Dordrecht. Printed in the Netherlands.*

TABLE 1. Rates of photosynthetic carbon export from leaf blades as affected by low-P treatment and by Pi addition to low-P leaves (Pi-resupply).

Parameter	Control	Low-P	Pi-resupply
Export rate (mg CH_2O dm^{-2} h^{-1})	25	1.8	10
Relative export (% of net photosynthesis)	77	6	33

retained in the leaf blades of low-P leaves over a 4-h period of illumination. This increase in leaf carbon contributed to a 30% increase in dry weight/leaf area (6). Furthermore, since the proportion of plant dry matter invested in the leaf blades increased from 37% to 48% in low P leaves after 14 days of growth at low-P (6), it seems likely that low rates of export of carbon (i.e., sucrose) from the leaf blade of low-P leaves was maintained over the long term.

Carbon export did not recover as rapidly as photosynthesis when Pi supply to low-P leaves was increased. After 24 h of increased Pi supply, the leaves were still accumulating 67% of the newly-formed photosynthate while light-saturated photosynthesis had recovered to 85% of the control values. The diminished ability to export carbon may have been due to smaller sink size (especially new leaf growth which was much smaller in P-limited plants) or to a decreased ability of the leaf to export carbon, e.g., at the phloem loading step which may have a high demand for ATP (7).

Low P treatment increased leaf sucrose and glucose much more than it increased starch (Table 2). In darkness, low P treatment increased glucose by 6.5-fold, sucrose by 3.5-fold and starch by 69%. All three carbohydrates decreased during the first 30 min of illumination, then increased. After 6 h, low P treatment had increased glucose 13-fold, sucrose 2-fold and starch by 14%.

Low P treatment had a pronounced effect on certain enzymes involved in the synthesis and breakdown of starch, sucrose and glucose. Low P treatment increased the total extractable activity of ADP-glucose pyrophosphorylase by 75% (Table 3). Starch breakdown enzymes, beta-amylase and starch phosphorylase, were increased to 157 and 250% of the control values, respectively, by low-P treatment. Low P treatment also increased sucrose phosphate synthase 2.3-fold but decreased UDP-glucose pyrophosphorylase by 40%.

TABLE 2. Effect of low-P treatment on leaf carbohydrates (expressed as μmol C g^{-1} dry wt) after 8 h darkness and 6 h light (in growth chamber).

Treatment		Glucose	Sucrose	Starch
Dark:	Control leaf	20.5	46.1	2782
	Low-P leaf	133.9	162.3	4701
Light:	Control leaf	17.3	158.7	3307
	Low-P leaf	224.4	354.4	3760

The increase in the activity of ADPG-pyrophosphorylase with low-P treatment may be due to an increase in the PGA/Pi ratio which could lead to enhanced activity (3).

TABLE 3. Effect of low-P treatment on the extractable activities (nmol cm^{-2} min^{-1}) of certain enzymes of starch and sucrose metabolism. Plants were dark-adapted for 8 h prior to illumination in the growth chamber for 1 h.

Enzyme	Control	Low-P	% Control
ADPG-pyrophosphorylase	0.8	1.4	175
Beta-amylase	3.5	5.5	157
Phosphorylase	1.5	3.1	250
Cytosolic FBPase	1.5	1.9	127
UDPG-pyrophosphorylase	0.5	0.3	60
Sucrose phosphate synthase	12.8	29.5	232

In the Pi translocator hypothesis, it is proposed that low levels of cytosolic Pi should result in reduced rates of export of triose phosphate from the chloroplast which in turn leads to a build-up in Calvin cycle intermediates and therefore increased starch synthesis (1,8). However, in this study, RuBP and PGA levels were much lower in low-P leaves which almost certainly had low cytosolic Pi levels (6). Furthermore, when Pi supply to low-P leaves was increased, a situation would be obtained where there would be a large gradient of Pi across the chloroplast envelope. According to the Pi translocator theory, export of triose phosphate should occur lowering the concentration of Calvin cycle intermediates and increasing the synthesis of sucrose relative to that of starch. However, in this work the levels of Calvin cycle intermediates increased with the enhanced flux of Pi into the leaf (6) while starch levels increased at 2.5 h then decreased at 5 h.

On the other hand, there was evidence from our work that low P treatment increased the synthesis of starch relative to sucrose synthesis. Starch accumulation in the light was slightly greater in low-P leaves than in control leaves while the rate of photosynthesis (and therefore the rate of formation of starch, sucrose and other carbon products) was smaller; thus, it is highly likely that a greater proportion of newly-fixed carbon was being utilized in starch than in sucrose formation (consistent with the Pi translocator hypothesis). The decreased starch/sucrose ratios in low-P leaves, which we observed in our work (see Table 2 for starch, sucrose values), may in large part have been due to the low rate of export of sucrose from the leaf blade.

4. CONCLUSION

Based on the evidence obtained thus far, we propose that the Pi translocator may form part of an adaptive mechanism to maintain an adequate Pi supply to the chloroplast during periods of P stress. Under low P, the Pi translocator will ensure the uptake of 1 Pi for every triose-P exported from the chloroplast. If Pi becomes seriously limiting to growth, the plant may respond by accumulating starch, sucrose, glucose and other non-phosphorylated compounds in leaf blades while reducing the concentration of phosphorylated compounds such as sugar phosphates and adenylates. This would maximize the inorganic phosphate available for photosynthesis under P stress conditions. At the same time, photosynthate is accumulated so that if the external Pi supply is later increased, the reserve carbohydrates will be available for new growth.

REFERENCES

1 Walker, D. A. (1980) In: Physiological Aspects of Crop Productivity, pp. 195-207, "Der Bund" AG, Bern, Switzerland
2 Herold, A. (1980) New Phytol. 86:131-144
3 Preiss, J. (1982) Annu. Rev. Plant Physiol. 33:431-454
4 Terry, N. (1980) Plant Physiol. 65:114-120
5 Rao, I. M., Abadia, J., and Terry, N. (1986) Plant Sci. 44:133-137
6 Rao, I. M., Abadia, J., and Terry, N. (1986) Proc. VIIth International Congress on Photosynthesis, Martinus Nijhoff/Dr. W. Junk Publishers
7 Giaquinta, R. T. (1983) Annu. Rev. Plant Physiol. 34:347-387
8 Heber, U., and Heldt, H. W. (1981) Annu. Rev. Plant Physiol. 32:139-168

THYLAKOID STRUCTURE AND FUNCTION IN RELATION TO LEAF PHOSPHATE STATUS IN SUGAR BEET

JAVIER ABADIA, I. MADHUSUDANA RAO AND NORMAN TERRY
DEPT. PLANT & SOIL BIOLOGY, UNIVERSITY OF CALIFORNIA, BERKELEY, CA 94720, U.S.A.

1. INTRODUCTION

Based largely on in vitro studies, the level of orthophosphate (Pi) in plant tissues is thought to be involved in the regulation of photosynthesis (1-3). The in vivo role of Pi in photosynthetic regulation may be explored by varying Pi levels in leaves nutritionally and then monitoring changes which occur in various aspects of photosynthesis (4-7). The objective of the present work was to investigate the effects of leaf P status on the structure and function of the thylakoid membrane.

2. MATERIALS AND METHODS

Sugarbeet plants (cv. F58-554H1) were grown hydroponically, with two different levels of phosphate in the culture medium, in growth chambers at $25^{\circ}C$ with an irradiance of 500 μmol quanta PAR m^{-2} s^{-1} supplied over a 16 h photoperiod. Low-P and control plants were raised as described in Rao et al. (7). In some experiments, Pi supply to low-P plants was increased by raising the P concentration in the culture medium (7). Measurements were made of thylakoid polypeptides and pigment-protein complexes, atrazine-binding sites, cytochromes f, b_{563} and b_{559}, fluorescence induction of leaf pieces, PSII and PSI fluorescence at room temperature and 77 K, net photosynthesis and quantum yield of attached leaves, PSI-mediated electron flow from ascorbate plus DCIP to MV, and PSII-mediated electron flow from water to DMBQ.

3. RESULTS AND DISCUSSION

Low-P treatment increased the amount of Chl per area by 22% and decreased the Chl a/Chl b ratio from 3.7 to 3.5 (Table 1). Phosphate supply had relatively small effects on the polypeptide composition of thylakoid membranes. Long, denaturing gels (lithium dodecyl-sulfate polyacrylamide gel electrophoresis) showed that there were no major changes in the relative amounts of the numerous polypeptides resolved. Analysis of the pigment-proteins showed that low-P treatment increased the amounts of Chl associated with LHC II and CP1 by about 26-28% (expressed on a per leaf area basis) while the amount of CPa Chl was unchanged (Table 1). Thus, low P

Biggens, J. (ed.), Progress in Photosynthesis Research, Vol. III. ISBN 90 247 3452 5
© *1987 Martinus Nijhoff Publishers, Dordrecht. Printed in the Netherlands.*

treatment induced an increase in Chl specifically
attributable to increases in PS I and antenna complexes, but
not PSII core complexes. Low-P treatment had little effect
on the number of atrazine binding sites per area (Table 1),
suggesting that the number of PS II reaction centers per
area is about the same in low-P and control plants.

TABLE 1. Effect of low-P treatment on thylakoid membrane
composition and function, photosynthesis and growth

Characteristics	Control	Low-P	% control
Chl (nmol cm^{-2})	47.5	58.8	124
Chl a/Chl b	3.7	3.5	
Chl in CP1	8.6	11.1	128
LHCII (nmol cm^{-2})	21.3	26.8	126
CPa	6.9	7.0	101
Atrazine-binding sites (pmol cm^2)	273	294	107
Cytochromes (pmol cm^{-2}):			
Cyt f	71	80	113
Cyt b_{559}	120	120	100
Cyt b_{563}	282	340	121
Acid-soluble P (% dry weight)	0.41	0.09	22
Total dry matter (g $plant^{-1}$)	23.5	5.3	23
Photosynthesis (μmol CO_2 m^{-2} s^{-1})	30.3	20.6	68
Quantum yield (mol CO_2 mol^{-1} quanta absorbed)	0.117	0.109	93
Electron transport (μmol O_2 cm^{-2} h^{-1}):			
PSII	15.7	13.8	88
PSI	40.2	41.3	103

Leaves of low-P plants also contained similar amounts of Cyt
b_{559} per area as the control plants, whereas Cyt f and Cyt
b_{563} increased by 13 and 21% respectively (Table 1). These
data suggest that low-P treatment increases the
concentration of the cyt b/f complex but has no effect on
the amount of cyt b_{559}, which is believed to be associated
with PSII.

Although low-P treatment decreased soluble leaf phosphorus
contents by 78% and total plant dry matter by 77%, the net
rate of photosynthetic CO_2 fixation (measured at normal

ambient CO_2 concentrations) was decreased by only 32% on a per area basis (Table 1). Photosynthetic quantum yield was slightly decreased by P-deficiency (Table 1). Low-P treatment decreased PSII electron transport per area by about 12% while having no significant effect on PSI electron transport (Table 1).

Low P treatment decreased F_v/F_m (Table 2). When Pi supply was increased to the low P plants, F_v/F_m decreased even more (Table 2). The energy distribution between PSII and PSI was investigated by comparing the fluorescence emissions of leaf pieces at 680 (FII) and 730 (FI) nm, respectively. In control plants, the ratio of FI/FII increased by 45% during illumination with blue light (Table 2). In low-P plants 1 min after illumination, FI/FII was larger than in control plants; however, FI/FII increased only 17% in the same time interval so that by 9 min, FI/FII was smaller in the low-P plants than in the controls. After 5 hours of Pi addition to low-P plants, FI/FII increased at both 1 and 9 min after illumination (Table 2). When fluorescence was measured at 77 K, FI/FII increased from dark to light in control plants. In low-P plants however, there was very little increase in FI/FII from dark to light and the ratio in light was smaller than in the controls. However, after Pi supply was increased for 2.5 hours to the low-P plants, FI/FII in the light increased, becoming similar to or larger than the control values.

TABLE 2. Changes in leaf fluorescence at room temperature

Treatment	F_v/F_m *	FI/FII (1 min)	FI/FII (9 min)
Control	0.63	0.69	1.00**
Low-P	0.50	0.78	0.91
5 h after Pi added	0.37	0.86	0.97

*F_v is the variable Chl fluorescence yield while F_m is the maximum Chl fluorescence yield with PSII traps closed.
**The values of FI/FII found in control after 9 min of illumination were arbitrarily taken as 1.00.

These increases in FI/FII values on illumination could be interpreted in terms of a movement of part of the antenna from appressed to unappressed regions of the thylakoid resulting from the phosphorylation of antenna Chl-proteins (8). When the supply of Pi to low-P plants was increased, the rise in FI/FII values on continuous illumination increased to control levels within 2.5 hours in the 77 K fluorescence experiments and within 5 hours in the room temperature fluorescence experiments. In addition, the variable fluorescence yield associated with PSII decreased substantially within 5 hours of P increase. These results suggest that the ability of the thylakoid membrane kinase to phosphorylate polypeptides of LHCII may directly relate to the Pi available to the chloroplast.

There was some evidence from the fluorescence data that low-P treatment modified the distribution of excitational energy between PSII and PSI. The PS II fluorescence induction curves at room temperature showed that the low-P treatment decreased the variable fluorescence component. Brooks (6) reported a similar effect of P deficiency on fluorescence. Room temperature fluorescence data suggests that low-P treatment increased the antenna size of PSI versus that of PSII in that 1 min after the onset of illumination, FI/FII values were greater in the low-P treatment compared to the control. However, the 77 K fluorescence data for dark adapted leaves indicated only a very slight increase in FI/FII with low-P treatment.

4. CONCLUSIONS

In summary, low-P treatment had little or no effects on thylakoid polypeptide composition, photosynthetic quantum yield, PSI electron transport/area, number of PSII reaction centers/area (measured as atrazine-binding sites), Cyt b_{559}/area and CPa Chl/area and had only small effects on Cyt f/area, Chl a/Chl b ratio, PSII electron transport/area, and the ratio of PSII/PSI fluorescence at 77 K. Low-P treatment appeared to modify thylakoid membrane structure by increasing the per area amounts of PSI and light harvesting Chl-protein complexes, and of the cytochromes f and b_{563}, without affecting the PSII components (including CPa, atrazine binding sites and cyt b_{559}). Most of the effects of low P treatment on thylakoid function were relatively mild and easily reversed within a few hours of increasing the supply of Pi to low P plants.

REFERENCES

1 Walker, D. A. (1976) Curr. Top. Cell Regul. 11, 203-241
2 Walker, D. A. (1980) In: Physiological Aspects of Crop Productivity, pp. 195-207, "Der Bund" AG, Bern/Switzerland
3 Sivak, M. N. and Walker, D. A. (1986) New Phytol. 102, 499-512
4 Foyer, C. and Spencer, C. (1986) Planta 167, 369-375
5 Dietz, K-J. and Foyer, C. (1986) Planta 167, 376-381
6 Brooks, A. (1986) Aust. J. Plant Physiol. 13, 221-237
7 Rao, I. M., Abadia, J. and Terry, N. (1986) Plant Sci. 44, 133-137
8 Barber, J. (1983) Photobiochem. Photobiophys. 5, 181-190

INFLUENCE OF PHOSPHATE NUTRITION ON THE RELATIONSHIPS BETWEEN INORGANIC PHOSPHATE, CARBOHYDRATES AND RuBP-CASE ACTIVITY DURING WHEAT FLAG LEAVES SENESCENCE.

FLECK, I.; FRANSI, A.; FLORENSA, I. AND RAFALES, M.
DEPARTAMENT DE FISIOLOGIA VEGETAL. FACULTAT DE BIOLOGIA. UNIVERSITAT DE BARCELONA.

1. INTRODUCTION
    In previous works (1,2) several physiological and biochemical parameters of flag wheat leaves grown in experimental fields under a Mediterranean climate were studied from anthesis through to senescence. Some of the parameters measured were Inorganic phosphate content, carbohydrate content and RuBP-carboxylase activity. Inorganic phosphorus plays a role on the partitioning of assimilates (3,4) and may affect RuBP regeneration and RuBP-Case activity (5,6). Different relationships between the parameters were found depending on the stage of senescence of the leaves. It was notable, that during the first stages of senescence, daily variations in Pi content of the leaves were directly related to variations in the starch/sucrose ratio. (See Fig. 1). To look into this fact another work has been done with wheat plants grown in substrate under environmental conditions. The plants received different phosphorus nutrition at different stages of development.

2. PROCEDURE
2.1. Materials and methods
    2.1.1. Plant material: Triticum aestivum cv Kolibri was sown in pots with substrate under natural environment conditions in the Experimental Fields of the Faculty of Biology of the University of Barcelona (Spain). Each pot contained finally one wheat plant.
    2.1.2. Fertilization (nutrient solutions). Control: The control solutions was the nitrate-type Hewitt´s solution containing 1.33 mM NaH_2PO_4. Phosphorus-deficient solution: During the first week after sowing: Hewitt´s solution containing 0.665 mM NaH_2PO_4 and complemented with Na_2SO_4. From the second week after sowing: Hewitt's solution without NaH_2PO_4 and with 0.665 mM Na_2SO_4.
    2.1.3. Treatments. Control: Control pots received Hewitt's complete solution during the whole study. PO: The PO pots received the phosphorus deficient solution during the whole study. P1: The P1 pots received phosphorus-deficient solution until anthesis of the plant. From this date, Hewitt's complete solution was applied. P2: The P2 pots received the phosphorus-deficient solution until 10 days after anthesis. From this date, Hewitt's complete solution was applied.
    2.1.4. Sampling. Samples of flag leaves were collected at different stages of development at noon and immediately frozen in liquid nitrogen.
    2.1.5. Methods. RuBP-carboxylase activity was measured by spectrophotometric end point titration of formed D-PGA in a 60 sec. assay at 25°C (7). Inorganic phosphate (Pi) content: Freeze-dried leaves were ground and Inorganic phosphate determined by ion chromatography (HPLC) as described elsewhere (2). Carbohy-

Biggens, J. (ed.), Progress in Photosynthesis Research, Vol. III. ISBN 90 247 3452 5
© *1987 Martinus Nijhoff Publishers, Dordrecht. Printed in the Netherlands.*

drate determination: Carbohydrates were extracted from ground material in boiling water for three minutes. Glucose and fructose, invertase sugars and starch fraction were analysed using an enzymatic method (8).

3. RESULTS AND DISCUSSION

3.1. FIGURE 1. Senescence pattern of the parameters RuBP-Case activity, Inorganic phosphate (Pi) and starch/sucrose ratio of wheat flag leaves grown in experimental fields.

FIGURE 2. RuBP-Case activity during senescence of wheat flag leaves grown in pots containing different phosphorus nutrition.

FIGURE 3. Inorganic phosphate (Pi) content during senescence of wheat flag leaves grown in pots containing different phosphorus nutrition.

FIGURE 4. Sucrose content during senescence of wheat flag leaves grown in pots containing different phosphorus nutrition.

FIGURE 5. Starch content during senescence of wheat flag leaves grown in pots containing different phosphorus nutrition.

FIGURE 6. Production values of the wheat ears grown under different phosphorus treatments.

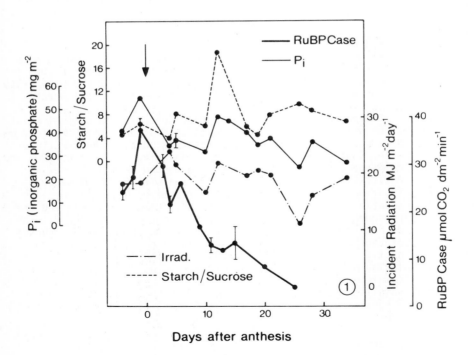

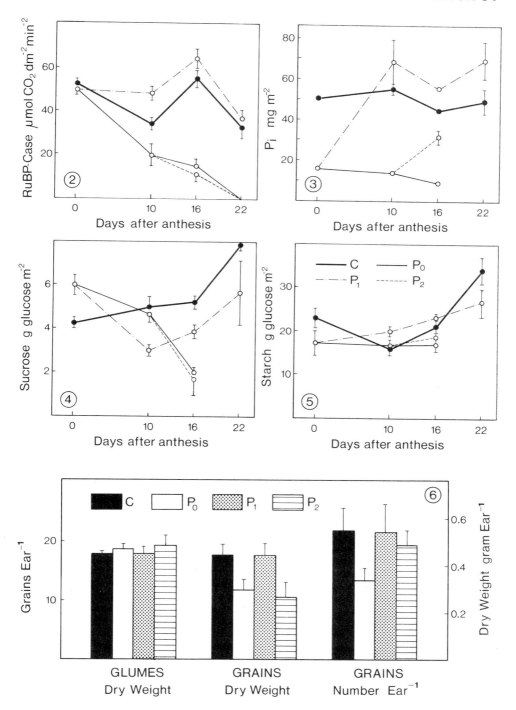

3.2. The effect of the phosphorus nutrition received by the wheat plants clearly affected the Inorganic phosphate (Pi)$_2$ content of the leaves: 50 mg. m^{-2} in the control plants; 12 mg. m^{-2} in phosphorus deficient plants (PO treatment). (Fig. 2). P-deficient flag leaves at anthesis had more sucrose content and less starch content than the control leaves. During the senescence process, PO leaves showed a decrease in in carbohydrate content, whereas control leaves began to accumulate carbohydrates at the third week after anthesis as found in the previous work. (Fig.1,3,4). It is noticeable that starch levels in our flag leaves were very much higher than those of sucrose as found in a previous work (2). The increase in (PGA/Pi) ratio due to high irradiance levels received by the leaves in the fields could account for the high starch synthesis observed (4,9). Several factors may have influenced the relation between starch and sucrose observed in our works e.g. sink demand, age and endogenous control (9,10,11). The activity of RuBP-carboxylase in PO leaves declined rapidly from anthesis whereas in control leaves the decline began two weeks later (Fig.2). The addition of the complete solution to P-deficient plants at anthesis (treatment P1) was very effective. The leaves showed from this moment the same characteristics as the control leaves: the same trends in the carbohydrates patterns during senescence, the same RuBP-Case activity and the same productivity values (Fig. 2,4,5,6). Phosphorus deficient plants receiving complete nutrient solution ten days after anthesis (P2) showed the same values than the permanently phosphorus deficient plants. The results indicate that phosphorus deficient plants are able to recuperate the characteristics of a well fertilized plant if the application of a complete nutrient solution is applied around anthesis. When the complete nutrient solution is applied only ten days later, the plant maintained the characteristics of a phosphorus deficient plant. These events may indicate the initiation of an "irreversible" stage of senescence during the second week after anthesis as pointed out elsewhere (1).

REFERENCES

1 Fleck, I.; Fransi, A. and Vidal, D. (1980) J. Plant Physiol. 123, 327-338
2 Fleck, I.; Fransi, A.; Florensa, I. and Rãfales, M. (1986) J. Plant Physiol. (in press)
3 Heldt, H.W.; Chon, C.J.; Maronde, D.; Herold, A.; Stankovic, Z.S.; Walker, D.A.; Kraminer, A.; Kiak, M.R. and Heber, V. (1977) Plant Physiol. 59, 1146-1155
4 Preiss, J. (1984) TIBBS January 1984
5 Robinson, S.P. and Walker, D.A. (1981) in The Biochemistry of Plants (Hatch, M.D. and Boardman, N.K. eds.), vol. 8, pp. 193-236. Academic Press, New York
6 Parry, M.A.J.; Schmidt, G.; Cornelius, M.; Keys, A.; Millard, B.M. and Gutteridge, S. (1985) J. Exp. Bot. 36, 1396-1404
7 Di Marco, G. and Tricoli, D. (1983) Photosynth. Res. 4, 145-149
8 Azcõn Bieto, J. and Osmond, C.B. (1983) Plant Physiol. 71, 574-581
9 Silvius, J.E.; Chatterton, N.J. and Kremer, D.F. (1978) Plant Physiol. 64, 872-875
10 Claussen, W. and Lenz, F. (1983) Z. Pflanzenphysiol. 109, 459-468
11 Stitt, M.; Wirtz, W. and Heldt, H.W. (1983) Plant Physiol. 72, 767-774

CYANOBACTERIN AND ANALOGS: STRUCTURE AND ACTIVITY
RELATIONSHIPS OF A NATURAL HERBICIDE

Florence K. Gleason, William J. Thoma and Janet L. Carlson.
Gray Freshwater Biological Institute, University of Minnesota,
Navarre, MN 55392 and Dept. of Chemistry, Macalester College,
St. Paul, MN 55108 USA. Minnesota Sea Grant Program,
Contribution # 199.

Cyanobacterin is a secondary metabolite produced by the
cyanobacterium, Scytonema hofmanni UTEX 2349, which inhibits
the growth of other algae. The compound is a diaryl-substi-
tuted γ-lactone (Table 1). Cyanobacterin interrupts photosynthetic electron transport at a site in PS II. Analysis
of PS II activity with various Hill electron acceptors
suggests that the site of action is not identical to the
classical PS II electron transport inhibitors (1).
The chemical synthesis of cyanobacterin has been reported
(2). We have prepared a series of analogs and tested these
for inhibition of electron transport. The objective is to
define which molecular substituents are essential for
activity and to predict the nature of the cyanobacterin
binding site.
We have also used ^{31}P-NMR spectroscopy on intact algal
cells to investigate the in vivo effects of cyanobacterin on
internal pH, nucleotide levels and sugar-phosphate pools.
MATERIALS AND METHODS
Cyanobacterin was extracted from cells of Scytonema
hofmanni (3). Analogs were prepared by a modification of the
reported synthesis (2). Thylakoid membranes were isolated
from Synechococcus sp ATCC 27146, and rates of O_2 evolution
measured with a Clark-type oxygen electrode in the presence
of 2.5 mM $K_3Fe(CN)_6$ (4).
Synechococcus cells for NMR were grown on Bg 11 medium ,
pH 7.6 at 30°C and 670 $\mu E \cdot m^{-2} \cdot s^{-1}$ in a mixture of 5% CO_2 in
air. Cells were harvested in mid exponential growth phase
and resuspended in 200 mM HEPES, pH 7.6; 1 mM EDTA, in Bg 11
without phosphate. The cell density in the NMR tube was
approximately $2 \cdot 10^{11}$ cells/cc.
Cell extracts were prepared from lysates obtained by
sonication. The lysates were extracted with 10% perchloric
acid. The perchloric acid was titrated with KOH to pH 8.0
and the resulting precipitate removed. The pH of the
extract was adjusted to 7.0 and the samples were lyophilized.
The dried samples were resuspended in 80% D_2O buffered
with 50 mM HEPES, pH 7.8, plus 50 mM EDTA.
NMR spectra of cyanobacteria were obtained on a Nicollet
360 MHz wide bore instrument operating at a phosphorus
frequency of 146 MHz. The total time for the accumulation of
each spectrum was 5 min. Illumination of the sample was
provided by a 150 W lamp connected to a fiber optic cable,

Biggens, J. (ed.), Progress in Photosynthesis Research, Vol. III. ISBN 90 247 3452 5

immersed in the sample. The sample was aerated with 5% CO_2 in air. ^{31}P spectra of cell extracts were obtained on a Bruker 275 MHz narrow bore instrument operating at 102.5 MHz.

RESULTS AND DISCUSSION

The effectiveness of cyanobacterin and analogs was determined by monitoring inhibition of the Hill reaction in thylakoid membranes. A comparison of the activities of analogs to the parent compound is shown in Table 1.

TABLE 1. Inhibition of Photosynthetic electron transport by Cyanobacterin and Analogs.

CYANOBACTERIN

COMPOUND	HILL REACTION WITH $K_3Fe(CN)_6$ COMPLETE INHIBITION
1 Cyanobacterin	25 nM
C-8--X	
2 X = H	Not Active
3 X = CH$_3$	Not Active
4 X = Br	60 nM
5	1.5 μM
6	136 nM
7	Not Active
8	Not Active
9	Not Active
10 C-20 = H	143 nM

Apparently a halogenated ring is essential for cyanobacterin
activity. Replacement with a proton (analog 2) or methyl
group yield inactive analogs. A bromo-cyanobacterin (4) has
inhibitory activity similar to that of the natural product.
Retention of the Cl with removal of the methylene dioxy (5)
yields an inhibitor with reduced activity. This implies that
the methylene dioxy group aids in binding. Replacement of
the methylene dioxy with a Cl (6) increases inhibition over
compound 5 suggesting that any hydrophobic substituent at C-9
will yield an active analog. A Cl added para to the
methylene bridge (7) results in an inactive analog. Presumably there is a requirement for an electronegative group meta
to the bridge as in (6) and the parent compound. Removal of
water from the lactone ring results in an inactive anhydro-
derivative (8). This implies that the OH group participates
in binding, most likely by H-bond formation. A lactone is
required as the free-acid (9)has no activity. The methoxyl
group at C-20 can be removed without substantial loss of
activity. From analog studies we conclude that two positions
on the cyanobacterin molecule are crucial for inhibition, an
electronegative group at C-8 on the aromatic ring and an OH
at C-3 in an intact lactone ring.

Using ^{31}P-NMR of Synechococcus cells, we observed intracellular changes on dark to light transition. As shown in
Figure 1, nucleotide levels are low in darkened cells. An
internal pH of 7.2 was calculated from the chemical shift of
the internal phosphorus (Pi) resonance. After illumination
(15 min) there is an increase in both nucleotide levels and
in internal pH of 0.4 units (Fig. 1b). Addition of the
uncoupler, 2,4-dinitrophenol, causes a rapid decline in
nucleotides. The intracellular pH equilibrates with the
medium due to disruption of the cytoplasmic membrane (1c).

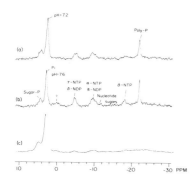

Figure 1. Effect of dark to
light transition and dinitrophenol on ^{31}P spectra of
Synechococcus cells.
(a) Darkened cells. Pi resonance
represents cytoplasmic Pi at pH
7.2.
(b) Illuminated Cells
     1600 $\mu E \cdot m^{-2} \cdot s^{-1}$
(c) Cells in (b) after addition
of 0.3 mM 2,4-dinitrophenol.

Addition of the electron transport inhibitor, DCMU, does not
result in a decline of the nucleotide pools. Partial spectra
shown in Figure 2 illustrate a shift in the sugar-phosphate
metabolites after treatment with 7.2 µM DCMU.

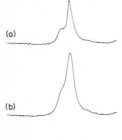

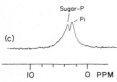

Figure 2. ^{31}P-spectra of sugar-phosphate region in <u>Synechococcus</u> cells.
(a) Illuminated cells. 5.3 ppm, sugar-phosphate pool; 3.0 ppm, Pi.
(b) Spectrum after addition of 7.2 μM DCMU to the cell suspension.
(c) Difference spectrum, light minus DCMU. The increased sugar-phosphate resonance is seen at 3.7 ppm.

^{31}P spectra of cell extracts show an additional resonance at 4.55 ppm after treatment with DCMU. The metabolite which increases was identified as 3-phosphoglyceric acid.

The results of adding 7.8 μM cyanobacterin are shown in Figure 3. The natural product causes no major changes in nucleotide or sugar-phosphate pools. A shift of the Pi resonance indicates a drop in the intracellular pH to 7.3.

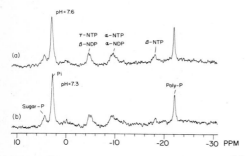

Figure 3. ^{31}P spectra of <u>Synechococcus</u> before and after treatment with cyanobacterin.
(a) Spectrum of illuminated cell suspension.
(b) Spectrum of same cells 15 min after addition of 7.8 μM cyanobacterin.

From the results obtained by ^{31}P-NMR, we conclude that cyanobacterin does not act as an uncoupler of photophosphorylation. Unlike dinitrophenol, it does not disrupt the cytoplasmic membrane. NADP reduction can continue in the presence of cyanobacterin as no increase in phosphoglyceric acid is observed. We propose that cyanobacterin acts by binding to a thylakoid membrane protein which facilitates proton transport.

REFERENCES

1 Gleason, F.K. and Case, D.E. (1986) Plant Physiol. 80,834.
2 Jong, T-T., Williard, P.G. and Porwoll, J.P. (1984) J. Org. Chem. 49, 735-736.
3 Pignatello, J.J., Porwoll, J., Carlson, R.E., Xavier, A., Gleason, F.K. and Wood, J.M. (1983) J. Org. Chem. 48, 4035.
4 Gleason, F.K., Case, D.E., Sipprell, K.D. and Magnuson, T. (1986) Plant Sci., in press.

DETERMINATION OF THE EXCHANGE PARAMETERS OF HERBICIDES ON THE Q_B-PROTEIN OF PHOTOSYSTEM II.

J. DIRK NABER AND JACK J.S. VAN RENSEN.
LABORATORY OF PLANT PHYSIOLOGICAL RESEARCH, AGRICULTURAL UNIVERSITY
WAGENINGEN, GEN. FOULKESWEG 72, 6703 BW WAGENINGEN, THE NETHERLANDS

INTRODUCTION

Many commercially available herbicides block the photosynthetic electron transport of plants, by preventing the oxidation of the primary acceptor Q_A of Photosystem II by the secondary acceptor Q_B. Compounds from different chemical classes, as e.g. DCMU, atrazine and dinoseb, all bind to the same protein. This Q_B-protein, with MW = 32000 dalton, is located on the acceptor side of PS II. It is part of a binding environment for two different plastoquinone molecules, the tightly bound Q_A and the easier exchanged Q_B (1). This Q_B can accept 2 electrons, and then becomes protonated to $Q_B H_2$. Both the oxidized and the fully reduced form are rapidly (50 s^{-1}) exchanging with free plastoquinone molecules from the PQ-pool, but the semi-quinone form is not (2,3).

When a so-called PSII-herbicide is present, this can competitively inhibit binding of a quinone molecule by occupying the same or a closeby located binding environment (4). The reaction center can make one more turnover, under formation of Q_A^-, and then the electron transport is blocked. Most herbicides are exchanged with Q_B on a seconds timescale (5). The kinetic parameters of this exchange are measured in this study. The binding and release rate constants can give information about the working mechanism and the efficiency of the herbicides.

The method for measuring these parameters is based on the flash-induced oxygen evolution patterns of isolated broken chloroplasts. The oxygen release shows a periodicity of four (6). Damping of this pattern is thought to be caused mainly by the occurrence of misses (α) and double hits (β) of reaction centers by the excitation light. The values for α and β are measured in control experiments without herbicides. They are assumed to be determined mainly by the experimental conditions, e.g. light intensity, flash duration and chloroplast concentration.

The exchange parameters are obtained by fitting experimental data to those calculated with a kinetic model. This model is derived from the following equations (7):

$$S_n \cdot Q_A \cdot Q_B + I \underset{E_3}{\overset{E_1}{\rightleftharpoons}} S_n \cdot Q_A \cdot I + Q_B \qquad [1]$$

$$S_n \cdot (Q_A \cdot Q_B)^- + I \underset{E_4}{\overset{E_2}{\rightleftharpoons}} S_n \cdot (Q_A \cdot I)^- + Q_B \qquad [2]$$

In these equations, S_n (where n = 0,1,2,3) represents the redox state of the Oxygen Evolving Complex. In thoroughly dark-adapted thylakoids, nearly 100% of the centers are in the stable S_1-state (8). This explains the maximal O_2-evolution after the third flash. In the presence of slowly exchanging herbicides, with residence times on the Q_B-protein of the same order of magnitude as the duration of the flash train or longer, the oscillation is hardly damped compared to the control. In this case only the amplitude of the signal is diminished. When the herbicide-Q_B exchange is occurring with approximately the same frequency as the flashes are fired, however, the damping of the oscillation is considerably stronger. This is caused by the fact that then reaction centers are blocked for a certain time span, and start making turnovers at the moment the herbicide is displaced by a PQ-molecule. Thus, centers can get out of phase with each other, and produce O_2 at different flashes. By comparing flash patterns obtained with different flash frequencies and herbicide concentrations, the exchange parameters E_1 to E_4 can be calculated.

METHODS

The experiments were carried out using isolated broken thylakoids from 3–4 week old pea (Pisum sativum L.) or Chenopodium album L. seedlings. The isolation medium contained 0.3 M sorbitol, 10 mM NaCl, 5 mM $MgCl_2$ and 50 mM tricine/NaOH pH 7.6. The chloroplasts were broken by incubation during 5 minutes in a medium containing 10 mM NaCl, 5 mM $MgCl_2$ and 10 mM tricine/NaOH pH 7.6.
The measurements were done with a Joliot-type electrode (9). The thylakoids were dark-adapted for at least 1 h after isolation, and transferred to the electrode in near-darkness. They were allowed 5–10 minutes to settle on the platinum surface. For every measurement chloroplasts equivalent to 5 µg chlorophyll were used.
To prevent dilution of added herbicides, these were applied to both the chloroplast sample and the circulating medium. Inhibitor concentrations were chosen around the I_{50}-value. Thus the amplitude of the signal still permits accurate measurements, while at the same time a reasonable damping can be observed.

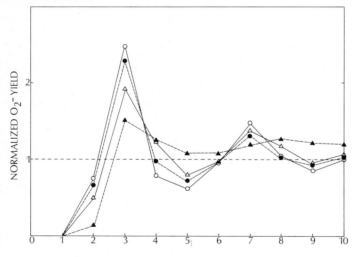

o=4 Hz flash frequency, no inhibitor
△=0.5 Hz flash frequency, no inhibitor
●=4 Hz flash frequency, 0.5 µm atrazine
▲=0.5 Hz flash frequency, 0.5 µm atrazine

Figure 1. Oxygen evolution patterns at different flash frequencies with and without atrazine.

The flash frequencies used were 4, 2, 1 and 0.5 Hz, while the flash duration was estimated to be 8 μs.
Data were stored in a Nicolet digital storage oscilloscope, and processed with the aid of a HP-86 computer. Calculations were carried out on a PDP 11/23 computer.

RESULTS

For most of the experiments chloroplasts isolated from pea seedlings were used. In control experiments α, β and the S_0-fraction were determined. The values for these parameters were $\alpha = 0.13$ and $\beta = 0.08$, while S_0 varied between 0 and 0.10.

The exchange parameters E_1 to E_4 were calculated for different types of herbicides. The values listed are preliminary results, and should be interpreted as an indication about the order of magnitude rather than as absolute values.

Table I. Values for the exchange parameters E_1 to E_4.

	$E_1(\mu M^{-1} \cdot s^{-1})$:	$E_2(\mu M^{-1} \cdot s^{-1})$:	$E_3(s^{-1})$:	$E_4(s^{-1})$:
atrazine *	0.340	0.018	0.071	0.071
atrazine	5.382	0.453	0.100	0.141
DCMU	2.850	0.011	0.135	0.141
dinoseb	0.120	0.005	0.135	0.119
o-phenanthroline	0.141	0.001	0.200	0.238

* This measurement was carried out with Chenopodium chloroplasts

The E_1 and E_2 parameters are dependent on the concentration of free inhibitor. E_3 and E_4, however, are not influenced by the PQ-concentration. This is caused by the fact that E_3 and E_4 are determined rather by the release of the inhibitor from its binding environment than by the binding of a PQ-molecule, since the time for the electron transport from Q_A to PQ is in the range of 1-10 ms. This is much faster than the inhibitor exchange.
The E_1 and E_3 values can be used for an estimation of the I_{50}-concentrations of the applied herbicides. From eqn. (1) it can be derived that

$$d([S_n \cdot Q_A \cdot Q_B])/d(t) = E_3[S_n \cdot Q_A \cdot I] - E_1([S_n \cdot Q_A \cdot Q_B][I]) \qquad [3]$$

In a steady-state situation the concentration of $S_n \cdot Q_A \cdot Q_B$ is constant, so

$$E_3/E_1 = [S_n \cdot Q_A \cdot Q_B][I]/[S_n \cdot Q_A \cdot I] \qquad [4]$$

When 50% of the reaction centers are inhibited, the concentrations of $S_n \cdot Q_A \cdot Q_B$ and $S_n \cdot Q_A \cdot I$ are equal to one another, and thus

$$E_3/E_1 = [I] \qquad [5]$$

where [I] stands for the I_{50}-value in μM.
For the herbicides tested this calculation gives the results listed in Table II.

Table II. Estimation of the I_{50}-values.

	E_3/E_1 (μM):	I_{50}-value (μM):
atrazine	0.5	0.2 - 0.8
DCMU	0.05	0.03 - 0.2
dinoseb	1.15	1.3 - 8
o-phenanthroline	1.4	1.6

CONCLUSIONS

1) The calculated values for the I_{50}-concentrations of the herbicides used are in good agreement with the values found in literature. This is an indication that the measurements give useful results, even though the fit between the experimental and theoretical oxygen evolution patterns was not yet too well in some cases.

2) The values calculated for the E_1-parameters are 20-200 fold higher than those for E_2. This means that the herbicides can bind much easier to the oxidized protein-Q_B complex then when this is in the semi-reduced form. This is in agreement with the hypothesis that Q_B^- has a much higher affinity for its binding environment than Q_B and Q_BH_2.

3) The values for E_3 and E_4 are nearly equal for the tested inhibitors. This is an indication that the displacement of the herbicides from their binding environment by a PQ-molecule is hardly influenced by the redox state of the protein-Q_B complex. The fact that the values are more or less the same for different substances might mean that the herbicides share a common binding site, or that the rate of the displacement is determined by other factors than the inhibitor structure alone.

4) The described method for measuring kinetic parameters may prove to be a useful tool to get more insight in the interaction of herbicides with their binding environment. The planned research with optical isomers from a cyanoacrylate compound, which show a clear difference in their I_{50}-values, may give some more information on the inhibitor-Q_B displacement reactions.

REFERENCES

1 Okamura, M.Y., Debus, R.J., Kleinfeld, D. and Feher, G. (1982) in: Function of Quinones in Energy-Conserving Systems (Trumpower, B.L. ed.) pp. 299-317, Academic Press, New York
2 Velthuys, B.R. (1981) FEBS Lett. 126, 277-281
3 Wraight, C.A. (1981) Isr. J. Chem. 21, 348-354
4 Laasch, H., Pfister, K. and Urbach, W. (1982) Z. Naturforsch. 37c, 620-631
5 Vermaas, W.F.J. (1984) Ph.D. Thesis, Agricultural Univ. Wageningen, The Netherlands
6 Kok, B., Forbush, B. and McGloin, M. (1970) Photochem. Photobiol. 11, 457-475
7 Vermaas, W.F.J., Dohnt, G. and Renger, G. (1984) Biochim. Biophys. Acta 765, 74-83
8 Vermaas, W.F.J., Renger, G. and Dohnt, G. (1984) Biochim. Biophys. Acta 764, 194-202
9 Joliot, P. (1972) Methods Enzymol. 24, 123-134

TERBUTRYN-RESISTANT MUTANTS OF RHODOPSEUDOMONAS VIRIDIS
ARE DOUBLE-MUTATED

I. SINNING and H. MICHEL
MAX-PLANCK-INSTITUT für BIOCHEMIE, D-8033 MARTINSRIED, FRG

ABSTRACT:

The DNA of several independent terbutryn-resistant mutants
of Rhodopseudomonas viridis has been isolated.
Sequence analysis of the L subunit gene of the reaction center
encoding the herbicide binding site shows a double-mutation:
serine L223 is changed into alanine and arginine L217 into
histidine.

INTRODUCTION:

The crystallographic analysis of the reaction center (RC)
crystals from Rhodopseudomonas (Rps.) viridis at 3 Å
resolution provided a complete picture of protein subunits,
pigments and the Q_A-site (1). Since Q_B is lost during
isolation and crystallization, its binding site has been
established by soaking quinones and competitive inhibitors
into the crystals. One of these compounds was the triazine-
herbicide terbutryn.
Triazine-herbicides are known to act as inhibitors of
photosynthetic electron transport in RCs of purple bacteria
as well as in the PSII of higher plants (2). They appear to
block the electron flow between the two RC-bound quinones
Q_A and Q_B by displacement of Q_B.
Sequence homologies and the crystal structure from Rps.
viridis suggest that the core of the RC of PSII resembles
very much the RC of purple bacteria (3). In particular the
D 1 protein (32 kDa-protein or herbicide binding protein)
seems to correspond to the L subunit and the D 2 protein
to the M subunit (see Michel,H. and Deisenhofer,J. ; same
vol.).
To understand the molecular basis of terbutryn resistance,
Rps. viridis is certainly the best suited organism since
details of terbutryn binding have been established by
X-ray crystallography (4).
Towards this goal we have isolated several terbutryn-
resistant mutants of Rps. viridis.
Here we report the results of sequencing the DNA
encoding the protein region which is involved in herbicide
binding.

Biggens, J. (ed.), Progress in Photosynthesis Research, Vol. III. ISBN 90 247 3452 5
© 1987 Martinus Nijhoff Publishers, Dordrecht. Printed in the Netherlands.

MATERIALS AND METHODS:

Spontaneous stable mutants were obtained, in an independent manner, by photosynthetic growth on medium containing 10^{-4}M terbutryn.

Genomic DNA was isolated by standard procedures. DNA was digested using the restriction endonucleases EcoRI and SalI and size fractionated by agarose gel electrophoresis. Fragments of 1.9 kb size were ligated into Bluescribe M13 vector. Screening was done by colony hybridization using the ^{32}P-labelled 1.9 kb EcoRI - SalI fragment of Rps. viridis wild type coding for the L and M subunits of the RC. After isolation of the plasmid from a single positive clone, the purified 1.9 kb fragment was cleaved by PstI and SmaI and size fractionated as above.

The M13mp9 system was used for sequence analysis by the dideoxy-method of Sanger.

All methods were taken from references 3 and 5.

RESULTS AND DISCUSSION:

We found two base pair changes producing the substitution of two amino acids common to all mutants sequenced so far:

$$\begin{array}{ccc} \text{C G T} & & \text{C A T} \\ \text{a r g L217} & \longrightarrow & \text{h i s} \end{array} \quad \text{and} \quad \begin{array}{ccc} \text{T C G} & & \text{G C G} \\ \text{s e r L223} & \longrightarrow & \text{a l a} \end{array}$$

The mutation of serine L223 into alanine could be expected from the mode of herbicide binding (4). There are two hydrogen bonds possible between the protein and terbutryn. The first one is between the polypeptide nitrogen of iso-leucine L224 and the N3 of the s-triazine ring-system, and the second one between the side chain oxygen of serine L223 and the ethylamino nitrogen of terbutryn.

The change of serine L223 into alanine abolishes the second hydrogen bond and, therefore, decreases the binding of terbutryn.

Interestingly it is the same mutation as reported from triazine resistant strains from the green algae Chlamydomonas (6), suggesting that the mode of binding in the RC of purple bacteria is the same as in the PSII of chloroplasts. This confirms the alignment of amino acids from the L and M subunits of Rps. viridis and the D 1 and D 2 proteins of spinach chloroplasts (see Michel,H. and Deisenhofer,J. ; same vol.), where serine L223 is analogous to serine 264 in the D 1 protein.

The second mutation, arginine to histidine, was unexpected, since this residue does not participate to the binding site directly. Simultaneous double-mutation could be excluded on the basis of statistical arguments. We think that the primary mutation is serine to alanine, and the second, arginine to histidine, is somehow able to compensate for a detrimental effect of the first mutation.

Terbutryn resistance can still occur by the mutation of other amino acids which are involved in herbicide binding. In Rhodobacter sphaeroides mutants isoleucine L229 is changed into methionine (Schenck,C.C. ; personal communication) and in Chlamydomonas mutants (6) phenylalanine 255 in the D 1 protein is changed to tyrosine.

We, therefore, expect to find other terbutryn resistant mutants of Rps. viridis in the future.

REFERENCES:

1 Deisenhofer,J., Epp,O., Miki,K., Huber,R. and Michel,H. (1985) Nature, 318, 618-624.
2 Stein,R.R., Castellvi,A., Bogacz,J.P. and Wraight,C.A. (1984) J.Cell Biochem, 24, 243-259.
3 Michel,H., Weyer,K.A., Gruenberg,H., Dunger,I., Oesterhelt, D. and Lottspeich,F. (1986) EMBO J., 5, 1149-1158.
4 Michel,H. Epp,O. and Deisenhofer,J., EMBO J., in press.
5 Maniatis,T., Fritsch,E.F. and Sambrook,J. (1982) Molecular cloning. A Laboratory Manual. Cold Spring Harbor Laboratory Press, NY.
6 Erickson,J.M., Rahire,M., Rochaix,J.D. and Mets,L. (1985) Science, 228, 204-207.

CHARACTERIZATION OF THREE HERBICIDE RESISTANT MUTANTS OF *Rhodopseudomonas sphaeroides* 2.4.1: STRUCTURE - FUNCTION RELATIONSHIP

M. L. PADDOCK, J. C. WILLIAMS, S. H. RONGEY, E. C. ABRESCH, G. FEHER AND M. Y. OKAMURA

DEPARTMENT OF PHYSICS, UNIVERSITY OF CALIFORNIA, SAN DIEGO, LA JOLLA, CALIFORNIA 92093

1. INTRODUCTION

Electron transfer from Q_A^- to Q_B in bacterial reaction centers (RCs) was found to be inhibited by triazine herbicides such as atrazine and terbutryn (1). Triazine resistant mutants of *Rhodopseudomonas sphaeroides* have been isolated by several groups (2-6). The RCs from these mutants display altered electron transfer properties, including decreased inhibition by triazine herbicides. In this study we have sequenced the L and M subunit genes of three herbicide resistant mutants. The changes in the primary structure are correlated with the binding properties of the quinone and of the herbicides terbutryn and o-phenanthroline.

2. MATERIALS AND METHODS

Mutants of the photosynthetic bacterium *R. sphaeroides* were isolated by their ability to grow photosynthetically under increased terbutryn (100 μM) or atrazine (300 μM) concentrations (3). DNA isolation, cloning, and sequencing were performed as described in (7). BamHI - HindIII or PstI fragments were cloned into either pUC8 or pUC118. For sequencing of specific regions, smaller fragments were subcloned into the M13mp18/mp19 vectors, or internal oligonucleotide primers were used.

RCs were solubilized with LDAO and isolated using NH_4SO_4 precipitation and cytochrome c affinity chromatography (8). The electron turnover rate was studied by optically monitoring the rate of cytochrome c oxidation at 550 nm (9) (10 μM cytochrome c, 10 mM PIPES, pH 6.8, 0.025% LDAO, T=20°C) using RCs containing one quinone. The concentration (K_Q) of UQ_0 needed to reach one half of the maximum turnover rate (V_m) was measured for each mutant. Varying amounts of herbicide were added to RCs with 100 μM exogenous UQ_0 to measure the inhibition characteristics of R26 and mutant RCs. The plot of % activity (turnover rate/V_m) versus herbicide concentration was fitted with the relation:

$$\% \text{ activity} = 1 - \frac{1}{1 + K_I/I}$$

where I is the concentration of inhibitor and K_I is the inhibition constant found from this fit. K_Q and K_I are, to a first approximation, the binding constants for the quinone and the inhibitor. The rate constant for electron transfer from Q_A^- to Q_B (k_{AB}) was determined by two different methods. For fast rates ($\tau < 5$ msec), k_{AB} was measured by monitoring the bacteriopheophytin absorbance band at 747nm after a flash (10). This was used to measure k_{AB} for the R26 and the I229M RCs. For slow electron transfer ($\tau > 5$ msec) k_{AB} was estimated from the time required to oxidize a second cytochrome under conditions of continuous illumination with a high concentration of exogenous UQ_{10} to maximize the occupancy of the Q_B site (<0.3 μM RCs, 3 μM UQ_{10}, 10 μM cytochrome c, 10 mM Tris-Cl, pH 8, 0.025% cholate, 0.025% deoxycholate, 0.1 mM EDTA).

Biggens, J. (ed.), Progress in Photosynthesis Research, Vol. III. ISBN 90 247 3452 5
© *1987 Martinus Nijhoff Publishers, Dordrecht. Printed in the Netherlands.*

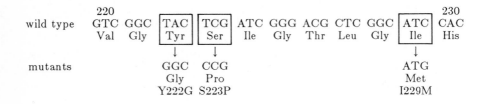

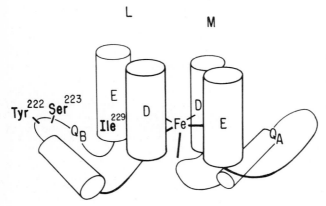

Figure 1. Wild type sequence from residues 220 to 230 and mutations found in the L subunit gene. The schematic representation below shows the binding sites for Q_A and Q_B as found in R. viridis (11). Helices are represented by cylinders. The D and E helices are indicated for the L and M subunits. The sites of the mutations are labeled.

3. RESULTS AND DISCUSSION

3.1. Sequences

The L and M subunit genes of the **I229M** mutant were completely sequenced. Only one nucleotide was found changed (Figure 1), resulting in a methionine (Met) at position 229 of the L subunit in place of the native isoleucine (Ile). This mutation was also found by Schenck, *et al.* (5) and Gilbert, *et al.* (6). The residue is located at the amino terminal end of the fifth (E) transmembrane helix.

About 40% of the L and M subunit genes from **S223P** were sequenced with one nucleotide change resulting in the replacement of serine (Ser) at position 223 of the L subunit with a proline (Pro). This residue is located in a connecting loop between the fourth (D) and fifth (E) transmembrane helices of the L subunit (Figure 1).

The region from the histidine at position 190 of the L subunit to the histidine at position 230 was sequenced for the **Y222G** mutant. Two nucleotide changes were determined resulting in a single amino acid change, a glycine (Gly) at position 222 in place of the native tyrosine (Tyr). This residue is located next to the serine in the same connecting loop between the D and E helices of the L subunit.

The residues that were changed in the mutants are conserved in the wild type strains of *R. sphaeroides* (7), *R. capsulata* (12), and *R. viridis* (13), indicating their importance for efficient electron transfer.

Table 1. Turnover Rates, Binding Constants, and Electron Transfer Rates

strain	V_m (cyt/RC/sec)	K_Q (mM)	k_{AB} (sec^{-1})	$K_I (\mu M)$ Ter	O-phen	mutation
R26	200	0.09	6×10^3	0.1	30	—
I229M	250	1.1	2.5×10^3	10	350	Ile$^{229} \rightarrow$ Met
S223P^a	>14	>1.6	10	>300	130	Ser$^{223} \rightarrow$ Pro
Y222G	60	0.7	50	>300	300	Tyr$^{222} \rightarrow$ Gly

^{a} *The herbicide binding data were obtained from RCs to which an equimolar amount of exogenous* UQ_{10} *was added.*

3.2. Quinone Binding Studies

The **S223P** mutation (Ser$^{223} \rightarrow$ Pro) had the most dramatic effect upon UQ_0 binding (see the large value of K_Q in Table 1). Also the rate constant k_{AB} was found to be greatly reduced. This suggests that Ser^{223} is important for proper action of Q_B and may be hydrogen bonded to the quinone. Hirschberg, *et al.* (14) have proposed that Ser^{264} of the D1 protein is hydrogen bonded to Q_B in photosystem II. These two serines may be functionally equivalent (see Section 3.4).

The **Y222G** mutation (Tyr$^{222} \rightarrow$ Gly) also shows a reduced k_{AB}, but its effect on UQ_0 binding is not as dramatic as with S223P (compare the K_Q values). Of the three mutants Y222G shows the greatest affinity for UQ_0. The maximum rate (V_m) is lower than that of either R26 or I229M, which is attributed to the slower k_{AB} rate constant. The tyrosine appears to be important for electron transfer between Q_A^- and Q_B but less important for Q_B binding. It may be involved in protonation after Q_B accepts an electron.

The **I229M** mutation (Ile$^{229} \rightarrow$ Met) affects UQ_0 binding with little effect on k_{AB} or V_m. This mutation may hinder the binding of Q_B due to the larger size of methionine. It affects the electron transfer less than the other mutants. This suggests that isoleucine is not as critical for efficient electron transfer as the serine or tyrosine residues.

3.3. Herbicide Binding Studies

The binding of terbutryn was reduced most in the S223P and Y222G mutants (K_I values in Table 1). These results suggest that terbutryn binds close to the serine and tyrosine residues. Both of these residues are located in a pocket (Figure 1) near the side farthest from the Fe [based on the crystal structure of *R. viridis*, (11)]. On the other hand the I229M mutation had a large effect on o-phenanthroline binding. This isoleucine is close to His^{230}, one of the Fe ligands, suggesting that o-phenanthroline binds closer to the Fe side of the pocket (Figure 1). These results are in agreement with recent crystallographic data (15). Since both herbicides are competitive inhibitors of UQ_0 at the Q_B site, it seems reasonable to deduce that the quinone binds in the pocket between Ile^{229} and Ser^{223}. This places the Q_B site at a position in the L subunit that is approximately symmetry related to the position of the Q_A site in the M subunit. This is consistent with the presense of a two fold symmetry axis observed in the structure of *R. viridis* (11) and *R. sphaeroides* (16).

3.4. L and D-1 Homology

Several herbicide resistant mutants have been isolated from photosystem II-containing organisms (reviewed in 17). The mutations responsible for the conferred resistance were all located on the D-1 protein of photosystem II RCs. In the bacterial RCs all mutations were located on the L subunit. This suggests a functional similarity between the L subunit and the D1 protein as has been previously noted (4,13,17). The mutations occurred in homologous regions of the proteins. The serine mutations (Ser^{223} in L and Ser^{264} in D-1) show the most striking similarity. The isoleucine mutation in L may correspond to a leucine mutation in D-1 (Ile^{229} in L vs Leu^{275} in D-1). No corresponding mutation to the tyrosine (Tyr^{222} in L) has yet been observed in D-1.

4. SUMMARY AND CONCLUSIONS

These studies have shown the importance of Ser^{223} to the binding of Q_B and to the efficient electron transfer between Q_A^- and Q_B. They support the suggestion that this serine (and the homologous serine in D-1) is hydrogen bonded to Q_B. They have also implicated Tyr^{222} in electron transfer from Q_A^- to Q_B although the detailed mechanism of its action is not known at present. The functional relatedness of L and D-1 is strengthened by the similarity of the herbicide resistant mutants, especially the serine mutations.

REFERENCES

1. Wraight, C. A. (1981) *Israel J. Chem.* **21**, 348-354.

2. Stein, R. R., Castellvi, A. L., Bogacz, J. P., and Wraight, C. A. (1984) *J. Cell. Biochem.* **24**, 243-259.

3. Okamura, M. Y. (1984) in *Biosynthesis of the Photosynthetic Apparatus: Molecular Biology, Development and Regulation* (Thornber, J. P., Staehelin, L. A., and Hallick, R. B., eds.), pp. 381-390, Alan R. Liss, Inc., New York.

4. Brown, A. E., Gilbert, C. W., Guy, R., and Arntzen, C. J. (1984) *Proc. Natl. Acad. Sci. USA* **81**, 6310-6314.

5. Schenck, C. C., Sistrom, W. R., Bunzow, J. R., Rambousek, E. L., and Capaldi, R. A. (1986) *Biochemistry*, in press.

6. Gilbert, C. W., Williams, J. G. K., Williams, K. A. L., and Arntzen, C. J. (1985) in *Molecular Biology of the Photosynthetic Apparatus*, pp. 67-71, Cold Spring Harbor Laboratory, Cold Spring Harbor.

7. Williams, J. C., Steiner, L. A., Ogden, R. C., Simon, M. I., and Feher, G. (1983) *Proc. Natl. Acad. Sci. USA* **80**, 6505-6509; Williams, J. C., Steiner, L. A., Feher, G. and Simon, M. I. (1984) *Proc. Natl. Acad. Sci. USA* **81**, 7303-7307.

8. Feher, G. and Okamura, M. Y. (1978) in *The Photosynthetic Bacteria* (Clayton, R. K. and Sistrom, W. R., eds.), pp. 349-386, Plenum Press, New York; Brudvig, G. W., Worland, S. T., and Sauer, K. (1983) *Proc. Natl. Acad. Sci. USA* **80**, 683-686.

9. Okamura, M. Y., Debus, R. J., Kleinfeld, D., and Feher, G. (1982) in *Function of Quinones in Energy Conserving Systems* (Trumpower, B. L., ed.), pp. 299-317, Academic Press, New York.

10. Vermeglio, A., and Clayton, R. K. (1977) *Biochim. Biophys. Acta* **461**, 159-165.

11. Deisenhofer, J., Epp, O., Miki, K., Huber, R., and Michel, H. (1985) *Nature* **318**, 618-624.

12. Youvan, D. C., Bylina, E. J., Alberti, M., Begusch, H., and Hearst, J. E. (1984) *Cell* **37**, 949-957.

13. Michel, H., Weyer, K. A., Gruenberg, H., Dunger, I., Oesterhelt, D., and Lottspeich, F. (1986) *EMBO J.* **5**, 1149-1158.

14. Hirschberg, J., Bleecker, A., Kyle, D. J., McIntosh, L., and Arntzen, C. J. (1984) *Z. Naturforsch.* **39c**, 412-420.

15. Michel, H., Epp, O., and Deisenhofer, J. (1986) *EMBO J.*, in press.

16. Allen, J. P., Feher, G., Yeates, T. O., Rees, D. C., Deisenhofer, J., Michel, H., and Huber, R. (1986) *Proc. Natl. Acad. Sci. USA*, in press; Allen, J. P., Feher, G., Yeates, T. O., and Rees, D. C., these proceedings.

17. Trebst, A. (1986) *Z. Naturforsch*, **41c**, 240-245.

ACKNOWLEDGEMENT

This work was supported by a grant from the US Department of Agriculture (grant 82-CRCR-1-1043).

THE PRODUCTION OF MONOSPECIFIC ANTIBODIES TO D1 AND D2 POLYPEPTIDES OF
PHOTOSYSTEM II

P.J. Nixon, [*]T.A. Dyer, J. Barber and [+]C.N. Hunter

AFRC Photosynthesis Research Group, [+]Department of Pure & Applied
Biology, Imperial College of Science &Technology, London, SW7 2BB, U.K.
[*]Plant Breeding Institute, Cambridge, CB2 2LQ, U.K.

1.   INTRODUCTION
      It is now assumed that the D1 polypeptide (also known as the 32
kD herbicide binding protein), and the D2 polypeptide are components of
the PSII core complex (1). Their functions, however, remain uncertain.
The amino acid sequences of these polypeptides for a number of species
have been deduced from the nucleotide sequences of psb A and psb D, the
chloroplast genes encoding D1 and D2, respectively (2,3). It has been
noticed that extensive sequence homology exists between D1 and D2 and
the L and M subunits from the reaction centre of R. viridis and R.
capsulata. This homology and the known role for the L and M subunits
has led to speculation that D1 an D2 are the PSII reaction centre
polypeptides (4). At present this is merely conjecture and to help
understand their possible roles monospecific antibodies to D1 and D2
have been produced. In order to produce specific antibodies it was
imperative to immunise with pure antigen. Since D1 and D2 run as
closely spaced, diffuse bands in SDS-polyacrylamide gels and stain
poorly (5), it is difficult to avoid cross-contamination from plant
extracts. The strategy adopted here involved the production of pure D1
and D2 antigen by expressing the parent genes in the bacterium
Escherichia coli.

2.   MATERIALS AND METHODS
      DNA manipulation was carried out using standard techniques (6).
Cell cultures were induced to express lac Z by addition of isopropyl-β
-D-thiogalactoside (IPTG) to 500 μM. SDS-polyacrylamide gradient gel
electrophoresis (SDS-PAGE) was performed essentially as (7) with
samples solubilised at 24°C. Affinity chromatography of crude
antiserum was performed using an antigen-Sepharose column as in (8).
Blotting of protein from gels onto nitrocellulose paper (0.2 μm pore
size) was performed as in (9) with visualisation of bound antibody as
in (10).   PSII enriched particles (11) and cores (12) were prepared
from Pisum sativum (pea). Lys-C protease treatment of core extracts
were carried out according to Marder et al (these proceedings).

3.   RESULTS
3.1   Isolation of psb D gene from wheat
      A Taq I fragment from the psb D gene from spinach (kindly given
by Dr. J. Gray) was used to probe a wheat chloroplast genomic library.
A 2.6 kb Bam HI/Sst I fragment was isolated and sequenced (S. Hird, R.
Barker, and T. Dyer unpublished). By comparison with previously
published sequences (2) the fragment was shown to contain the psb D
gene.

Biggens, J. (ed.), Progress in Photosynthesis Research, Vol. III. ISBN 90 247 3452 5
© *1987 Martinus Nijhoff Publishers, Dordrecht. Printed in the Netherlands.*

3.2 Expression of the psb D gene

In order to produce substantial amounts of easily purifiable D2, it was decided to express D2 as a fusion protein with β -galactosidase. To achieve this aim, a 1.2 kb fragment of psb D, (encoding 97% of D2), was inserted at the 3' end of the lac Z gene into each of the possible reading frames provided by the pUR series of expression vectors - pUR 290, pUR 291 and pUR 292 (13). To analyse potential fused proteins, cultures of E. coli JM101 carrying each hybrid plasmid were grown under non-inducing and inducing conditions. Total cellular protein was then analysed by SDS-PAGE (Fig. 1). As predicted by examination of the DNA sequence (not shown) only when the psb D gene was cloned into pUR 291, (hybrid plasmid pPND2), was D2 fused to the carboxy terminus of β -galactosidase.

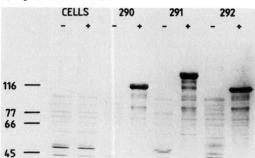

Figure 1. 5-10% gradient SDS-PAGE of total cellular protein produced by uninduced (-) and induced (+) cultures of JM101 containing no plasmid (CELLS) and each of the hybrid plasmids derived from psb D and pUR 290, 291, 292. Molecular weight markers are shown in kD.

3.3 Expression of the psb A gene

Using a similar method to 3.2, a 1.2 kb DNA fragment from Poa annua containing the whole of the psb A gene, (kindly given by Dr. D. Barros), was cloned into the pUR series of vectors. Only the hybrid plasmid pPND1, generated by insertion into pUR 292, produced D1 fused to the carboxy terminus of β -galactosidase (Fig. 2).

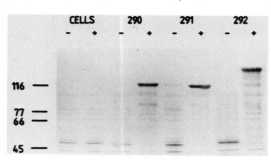

Figure 2. 5-10% gradient SDS-PAGE of total cellular protein produced by uninduced (-) an induced (+) cultures of JM101 containing no plasmid (CELLS) and each of the hybrid plasmids derived from psb A and pUR290,291, 292. Molecular weight markers are shown in kD.

3.4 Production of antibodies

In both instances the fusion protein was excised from polyacrylamide gels and used in a course of injections to raise antibodies in rabbit. Antibodies directed against the D2 - β galactosidase fusion were purified by affinity chromatography. Using an immunoblot procedure, the anti-D2 fusion antibodies cross reacted with a diffuse protein of estimated Mr 31 kDa present in PSII enriched preparations and more specifically PSII core preparations (Fig. 3).

Figure 3.

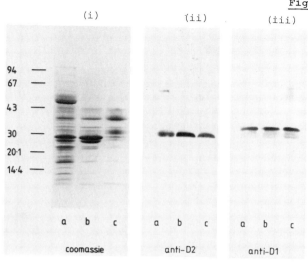

(i)

(ii)

(iii)

94 ——
67 ——
43 ——
30 ——
20·1 ——
14·4 ——

a b c   a b c   a b c

coomassie   anti-D2   anti-D1

(i) 7-17% gradient SDS-PAGE (containing 2M urea) of
  (a) thylakoids (25 μg chlorophyll)
  (b) PSII enriched particles (25 μg chlorophyll)
  (c) PSII cores (15 μg chlorophyll)
Visualisation was by Coomassie staining
(ii) Probed with anti-D2 fusion antibodies
(iii) Probed with anti-D1 fusion antibodies.
Molecular weight markers are shown in kD.

Antiserum raised to the D1 fusion protein cross-reacted predominantly with a 34 kD protein also present in PSII core preparations. Pre-immune sera and antibodies against β-galactosidase showed no cross-reactions with core proteins (not shown).

3.5   Specificity of the antisera
Since D1 and D2 show a large degree of amino acid homology (14) it is possible that antibodies raised to one gene product may crossreact with the other. To investigate this possibility, a PSII core preparation from pea was digested with a lysine specific protease, followed by immunoblotting with either anti-D1 or anti-D2.
From the known sequences of D1 and D2 from pea (2,3) it was expected that D1 would be resistant to digestion unlike D2 which would be sensitive. The results shown in figure (4) show that the anti-D1 serum cross-reacted predominantly with two Lys-C resistant proteins of apparent Mr 34 kD and 30 kD. In contrast the purified anti-D2 antibody cross-reacted with a 31 kD band which was digested to a faster running 29 kD band. These results indicate that the antisera are monospecific for D1 and D2.

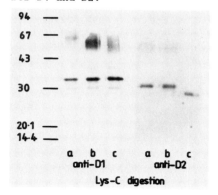

94 ——
67 ——
43 ——
30 ——
20·1 ——
14·4 ——

a b c   a b c
anti-D1   anti-D2

Lys-C digestion

Figure 4.
Immunoblots of Lys-C treated cores (1 ug chlorophyll per lane):
(a) No Lys C
(b) 0.3 μg/ml Lys C
(c) 3.0 μg/ml Lys C
Molecular weight markers are shown in kDa. Digestions were run on 12-25% gradient gels (with 2 M urea) prior to blotting.

4. DISCUSSION

This work provides immunological evidence for the presence of the psb A and psb D gene products in the PSII core complex as isolated by Gounaris and Barber (12).

In the gel system used (with 2M urea present) (7) the psb A gene product runs with apparent Mr 34 kD, virtually coincident with the so called '33 kD extrinsic' protein (15). The psb D gene product migrates more quickly but much more diffusely at about 31 kD. It also appears that a possible breakdown product derived from D1 comigrates with D2 at about 30 kD present in the core preparation. The method used here for producing D1 or D2 as β-galactosidase fusions has several advantages. The hybrid protein is inducible, produced in abundant quantities and of high molecular weight making identification and purification easy.

ACKNOWLEDGEMENTS

We acknowledge financial support from the AFRC and SERC. P.J.N. is a recipient of a SERC studentship.

REFERENCES

1   Tang, X.-S. and Satoh, K. (1985) FEBS Lett. 197, 60-64
2   Zurawski, G., Bohnert, H.J., Whitfeld, P.R. and Bottomley, W. (1982) P.N.A.S. 79, 7699-7703
3   Rasmussen, O.F., Bookjans, G., Stummann, B.M. and Henningsen, K.W. (1984) Plant Molecular Biology 3, 191-199
4   Deisenhofer, J., Epp, O., Miki, K., Huber, R. and Michel, H. (1985) Nature 318, 618-624
5   Chua, N.H. and Gillham, N.W. (1977) J. Cell Biol. 74, 441-452
6   Maniatis, T., Fritsch, E.F. and Sambrook, J. (1982) Molecular Cloning A Laboratory Manual C.S.H.
7   Laemmli, U.K. (1970) Nature 227, 680-685
8   Chua, N.H., Bartlett, S.G. and Weiss, M. (1982) Methods in chloroplast molecular biology, Edelman et al (Eds.) Chapter 85
9   Burnette, W.N. (1981) Anal. Biochem. 112, 195-203
10   Leary, J.J., Brigati, D.J. and Ward, D.C. (1983) PNAS 80, 4045-4049
11   Berthold, D.A., Babcock, G.T. and Yocum, C.J. (1981) FEBS Lett. 134, 231-234
12   Gounaris, K. and Barber, J. (1985) FEBS Lett 188, 68-72
13   Ruether, U. and Mueller-Hill, B. (1983) EMBO Journal 2, (10): 12 91-1794
14   Rochaix, J.-D., Dron, M., Rahire, M. and Malnoe, P. (1984) Plant Molecular Biology 3, 363-370
15   Ghanotakis, D.F. and Yocum, C.F. (1985) Photosynthesis Research 7, 97-114.

COMPARATIVE STUDIES OF ELECTRON TRANSPORT AND ATRAZINE BINDING IN
THYLAKOIDS AND PS II PARTICLES FROM SPINACH

R. Fromme, R. Hagemann and G. Renger, Max-Volmer-Institut für Biophysikalische und Physikalische Chemie, Technische Universität Berlin, Straße des 17. Juni 135, 1000 Berlin 12, FRG

1. INTRODUCTION

Based on different lines of experimental evidence a special protein matrix containing at least two polypeptides (referred to as D-1 and D-2) was inferred to play a key role for the electron transport at the PS II acceptor side and its blockage by a number of herbicides. The lysine free polypeptide D-1 which exhibits remarkable structural homologies and functional analogies with the L-subunit of the reaction center complex of purple bacteria very likely contains the binding site for the secondary plastoquinone Q_B and many PS II herbicides (for recent review see ref. 1). Experiments with a lysine-specific protease led to the conclusion that allosteric modifications affect electron transport and herbicide binding (2). Recently lipids were shown to be of functional relevance for PS II electron transport (3). Accordingly detergent treatment of thylakoids for preparation of PS II particles might also give rise to changes of the structural and functional pattern of PS II. Here this problem is attacked by comparative studies of electron transport and the modifications of atrazine binding by specific proteases and $K_3[Fe(CN)_6]$ in thylakoids and PS II particles, respectively.

2. PROCEDURE
2.1 Materials and Methods

Thylakoids and PS II particles were prepared as described in (4) and (5), resp., with modifications as in (6). Proteolytic treatment was performed in suspensions containing sample material (50 μM chlorophyll) 10 mM NaCl, 5 mM $MgCl_2$, 0.33 M sorbitol, commercially available (Boehringer) proteases of 0.25 activity units and either 50 mM MES/NaOH, pH = 6.0 or 50 mM Tricine/NaOH, pH = 7.6. $[^{14}C]$-atrazine was added either 5 min before starting proteolysis (30 min) or after 30 min proteolytic treatment. $[^{14}C]$-atrazine binding was determined as in (7). The average oxygen yield per flash was measured with a Clark-type electrode as outlined in (8).

3. RESULTS AND DISCUSSION

In order to analyze the origin of the marked dependence of oxygen evolution on the type of the exogenous electron acceptor used in PS II particles for the determination of the rate at saturating continuous light excitation the average oxygen yield per flash was measured as a function of the time t_d between the flashes. The data obtained are presented in fig. 1. If phenyl-p-benzoquinone (Ph-p-BQ) is used as exogenous electron acceptor the average oxygen yield per flash attains its maximum value at ca. 150 ms between the flashes. On the other hand, in the presence of $K_3[Fe(CN)_6]$ instead of Ph-p-BQ a more complex pattern is observed. A smaller fraction of

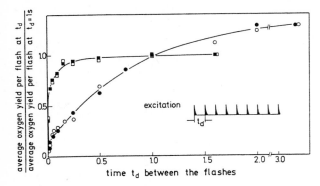

Figure 1 Average oxygen yield per flash as a function of the time t_d between the flashes.
Electron acceptors: 100 μM Ph-p-BQ without (□) or with 10 mM CaCl$_2$ (■), 300 μM K$_3$ [Fe(CN)$_6$] without (○) or with (●) 10 mM CaCl$_2$ other conditions as in Materials and Methods.

ca. 25 % of all PS II centers exhibit a comparatively fast electron transfer to the exogenous acceptor while a large fraction appears to be rather slow. The rate limiting step is probably located at the PS II acceptor side, because the donor side reactions from water are much faster (9). Recently it was found that the well established Ca^{2+}-effect in PS II particles not only refers to the donor side. CaCl$_2$ was shown to modify in a nonspecific way also the acceptor side as reflected by the protection to tryptic attack (10, 11). At 10 mM CaCl$_2$ the p-benzoquinone (BQ) mediated oxygen yield was highly protected to mild trypsin treatment at pH = 6.0 in PS II particles, while thylakoids did not exhibit this protection (10). A more thorough analysis indicates that the CaCl$_2$ induced protection is not completely absent in thylakoids but higher CaCl$_2$ concentrations (> 30 mM) are required to exhibit this effect. The CaCl$_2$ induced structural modification giving rise to the protection to tryptic attack leads to the question whether or not CaCl$_2$ does also affect the rate limiting step of the electron transfer to exogenous acceptors. The data in Fig. 1 however reveal that these reactions are almost invariant to CaCl$_2$. However, we observed a 20 - 30 % increase of the average oxygen yield per flash (data not shown). This result indicates that some centers are completely blocked in the absence of CaCl$_2$. The origin of this phenomenon remains to be clarified.

Another sensitive tool for monitoring structural changes in the microenvironment of polypeptides at the PS II acceptor side is the measuring of PS II herbicide binding. It was found that the affinity of [^{14}C] - atrazine binding was lower in PS II particles than in thylakoids whereas the number of binding sites closely correlates with the number of PS II reaction centers (data not shown). Previous studies with pea thylakoids revealed that atrazine binding becomes markedly reduced after proteolytic attack by trypsin and a lysine-specific protease, whereas an arginine-specific protease was shown to be without effect (2). It was further reported that [^{14}C] -azidoatrazine could also label a 34 kDa polypeptide which is assumed to be involved in donor side reactions (12). Therefore, it appears worthwhile to compare the effect of proteolytic enzymes in PS II particles and thylakoids, because in the former sample both sides of

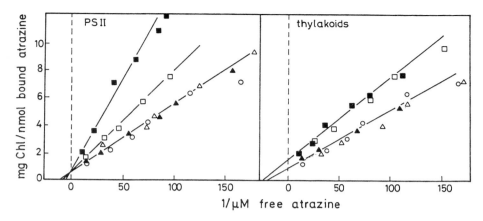

Figure 2 Double reciprocal plot for binding of ^{14}C —atrazine to
thylakoids and PS II membranes.

O control,◻ ■ lysine-specific protease,△▲ arginine-specific
protease, open symbols: inhibitor added prior protease, closed
symbols: inhibitor added after proteolysis.
Other conditions as in Materials and Methods

the membrane are susceptible to an attack, whereas in thylakoids only the
outer surface is exposed. As the effect of trypsin has been presented in
a recent studies (10, 11) here only the action of a lysine-specific and
an arginine-specific protease has been analyzed. The data obtained are
presented in fig. 2. In spinach thylakoids basically the same characteristics were found as in pea thylakoids (2). Prior addition of the inhibitor before starting the proteolysis was without effect. A different
pattern however is observed in PS II particles. The effect of the lysine-
specific protease depends on the presence of $[^{14}C]$ -atrazine during the
proteolytic treatment. This finding was surprising because the incubation
time was rather long and the inhibitor concentration comparatively low
(maximal one molecule per PS II reaction center). For the time beeing
we cannot offer a simple explanation and we will refrain from speculations
about long range allosteric effects and / or conformational changes induced by atrazine binding which slowly relax after inhibitor release. Regardless of the underlying mechanism the data exhibit a remarkable difference between thylakoids and PS II particles. No differences however are
observed for the inability of the arginine-specific protease to affect
atrazine binding.
   Another interesting component at the PS II acceptor side is the high
spin Fe^{2+} which is located between Q_A and Q_B. Recently oxidation of Fe^{2+}
to Fe^{3+} was shown to reduce drastically the DCMU-affinity whereas atrazine
binding remained unaffected (13). The effect of Fe^{2+} oxidation on herbicide binding could provide another sensitive test for possible differences
in the microenvironment of the binding site. Comparative studies of the
influence of $[^{14}C]$ - atrazine binding were performed in thylakoids and
PS II particles. The experimental data depicted in fig. 3 reveal that
$K_3 [Fe(CN)_6]$ decreases the affinity of atrazine binding in PS II particles,
while no change takes place in thylakoids. This finding provides further
evidence for differences at the PS II acceptor side between thylakoids and

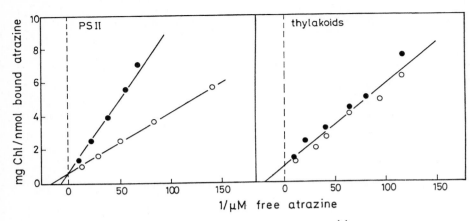

Figure 3 Double reciprocal plot for binding of $[^{14}C]$-atrazine to thylakoids and PS II particles in the absence (o) and presence of
500 µM K_3 $[Fe(CN)_6]$ (●).
Experimental conditions as described in Materials and M

PS II particles. These differences have to be taken into accoun
detailed mechanistic studies in PS II particles.

In future work the possible role of the lipidic surroundings of functional polypeptides in PS II remains to be analyzed in detail.

ACKNOWLEDGEMENTS
The financial support of this work by ERP-Sondervermögen (ERP 2520) is
gratefully acknowledged.

REFERENCES
1 Renger, G. (1986) Physiol. Veg. 24, 509-521
2 Renger, G., Hagemann, R. and Vermaas, W.F.J. (1984), Z. Naturforsch.
  39 c, 362-367
3 Gounaris, K., Whitford, D. and Barber, J. (1983) FEBS Letters 163,
  230-234
4 Winget, G.D., Izawa, S. and Good, N.E. (1965) Biochem. Biophys. Res.
  Commun. 21, 438-443
5 Berthold, D.A., Babcock, G.T. and Yocum, C.F. (1981) FEBS Lett. 134,
  231-234
6 Völker, M., Ono, T., Inoue, Y. and Renger, G. (1985) Biochim. Biophys.
  Acta 806, 25-34
7 Tischer, W. and Strotmann, H. (1977) Biochim. Biophys. Acta 460, 113-125
8 Renger, G. (1972) Biochim. Biophys. Acta 256, 428-439
9 Renger, G. and Weiss, W. (1986) Biochem. Soc. Trans. 14, 17-20
10 Renger, G., Hagemann, R. and Fromme, R. (1986) FEBS Letters 203, 210-214
11 Völker, M., Eckert, H.J. and Renger, G. (1987), These Proceedings
12 Metz, J.G., Bricker, T.M. and Seibert, M. (1985) FEBS Letters 185,
   191-196
13 Wraight, C.A. (1985) Biochim. Biophys. Acta 809, 320-330

ISOLATION AND CHARACTERIZATION OF THE RAPIDLY TURNING OVER PROTEIN OF
PHOTOSYSTEM II

G.F. WILDNER, R. BERZBORN, N. DEDNER, R. DMOCH, C. FIEBIG* AND H.E. MEYER+
BIOCHEMIE DER PFLANZEN* AND PHYSIOL. CHEMIE+, RUHR-UNIVERSITÄT, POSTFACH
10 21 48, D-4630 BOCHUM, FRG

1. INTRODUCTION

The photosystem II complex of Chlamydomonas reinhardii contains a rapidly turning over protein with a molecular weight of 32-35 kDa (1-3). Since
many proteins with similar molecular weights in this range can not be
separated by one-dimensional SDS-gel electrophoresis, it was of special
interest to investigate how new separation techniques, such as HPLC-
reverse phase chromatography, could be applied for these very hydrophobic
proteins.

The purified rapidly turning over protein was used for immunization of
rabbits and the resulting antiserum for immunological analysis of cross
reactivity. The western blots with photosystem II preparations from spinach, trypsinated particles, photosynthetic strains of Chlamydomonas reinhardii and Scenedesmus obliquus, and the blue-green algae Synechococcus
showed a similar protein. It was of special interest also to include in
these studies the reaction center complex of Rhodopseudomonas sphaeroides.

The immunological analysis was extended to the trypsin-digested fragments of the purified rapidly labeled protein in association with protein
analysis (amino acid composition and N-terminus sequencing).

2. PROCEDURE

2.1. Materials and Methods

2.1.1. In Vivo Labeling of the Cells with ^{35}S-Sulfate

Mixotrophic cultures of Chlamydomonas reinhardii were adjusted to
10 mg chlorophyll/300 ml and 30 mg cycloheximide was added prior
to the addition of 35S-sulfate (570 kBq/mg Chl). After one hour of
illumination the incorporation was terminated by the addition of
cold 1M sulfate.

2.1.2. Isolation of the Rapidly Labeled Protein

The cells were broken by sonication and the resulting thylakoids
were isolated by gradient centrifugation. The purified thylakoids
were solubilized according to the method of Chua (4) and the proteins separated on polyacrylamide gels with a linear gradient from
10 to 15% (4). The labeled protein bands were identified by autoradiography and the dominantly labeled protein was electroeluted
(5). The crude preparation of the rapidly labeled protein (RLF) was
subjected to a second gel electrophoresis in presence of 6M urea,
and the procedure repeated.

2.1.3. Purification of the RLP by HPLC-Chromatography

The crude RLF could be further purified by reverse phase chromatography with consecutive linear gradients with different polarity.
The reverse phase chromatography was performed on a Vydac C 18
column (15x0.4 cm) 218 TP 5415 with a linear gradient of the binary

system (A_1: 0.1% TFA in H_2O, B_1: 0.1% TFA in 84% acetonitrile) and a subsequent run with a linear gradient of a binary solvent system (A_2: 0.1% TFA in 60% HCOOH, B_2: 0.1% TFA in 60% HCOOH and 40% n-propanol). The protein peaks were recorded either at 214 or at 280 nm and simultaneously the tryptophan fluorescence was measured (285/340). Three proteins were separated at this final step.

2.1.4. Amino Acid Analysis and Determination of the N-Terminus

The protein samples were hydrolysed for 24, 48 and 72 hours in 6N HCl and the amino acids in the hydrolysates were analysed as phenylthiocarbamyl derivatives on a Sprisorb ODS II 3um column.

The N-terminal amino acid of the RLP and of the trypsinated 20 kDa fragment was performed on a Applied Biosystems model 470 A gas-phase-sequencer with an on-line PTH-Analyser.

2.1.5. Immunological Studies

The purified RLP was injected into rabbits as a paste with Freund's complete adjuvant followed by an intravenous booster injection four weeks later as described (6). The specificity of the antisera was established by the western blot technique of Towbin et al. (7) modified for the use of horse radish peroxidase conjugated second antibody.

3. RESULTS

3.1. Labeling of the Cells

The incorporation of ^{35}S-sulfate into the thylakoids corresponded to 15% of the total radioactivity of 98% incorporated into the cells. The light dependent (6500 lx) uptake was performed in the presence of cycloheximide to limit the number of labeled proteins. The kinetics of the incorporation of 35S into the protein occured with a t1/2 value of 20 minutes.

3.2. Purification of the Rapidly Labeled Protein

The autoradiography of the gel with the solubilized thylakoids showed that in the presence of cycloheximide one major band was labeled at 34 kDa (90% of the total labeled proteins) with minor bands at 30, 43 and 51 kDa. The RLP-band was poorly stainable with coomassie-blue, and therefore, the identification was done by autoradiography. The yield of the 35S incorporation into the RLP-band was equivalent to 0.5 to 1% of the 35S-sulfate. The recovery of the RLP after the step of electroelution was rather high and approximately 85% of the original material.

The electroeluate was used for further purification by HPLC-reverse phase chromatography. The first binary gradient pair eluted unlabeled material from the column (Fig. 1a). The change to the second gradient system resulted in the separation of three protein peaks as seen in Fig. 1b. The first peak with an app. molecular weight of 33 kDa was lysine-rich. This protein was phosphorylated in dark-light transition experiments with 32P-phosphate and intact cells, but not labeled with 35S-sulfate. The second and main peak had most of the radioactivity (97%) of the 35S-labeled proteins and only minute amounts of lysine. The third peak in the elution pattern of the chromatogramm contained a lysine-rich protein and was neither labeled with 35S nor with 32P.

The question was studied whether the RLP is identical to the herbicide binding protein (Q_B-protein). Thylakoids were labeled with 14C-azidoatrazine and the purification procedure was repeated including the step with the HPLC-chromatography. All the radioactivity of the photoaffinity label was detected in the second and major peak of the elution profile. This result suggests that both proteins have similar properties; the question whether this protein peak consists of only one component could be answered by sequencing of the peak material. Unfortunately, the N-terminal amino acid residue was blocked and therefore trypsin digested fragments have to be used for these studies.

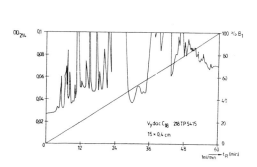
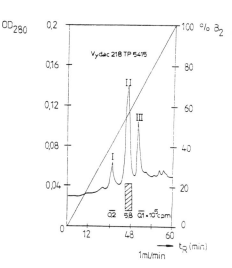

FIGURE 1. HPLC-chromatogramm with the solvent system A_1B_1 (Fig. 1a) and with the solvent system A_2B_2 (Fig. 1b). Thylakoids equivalent of 4.2 mg chlorophyll were used for the preparation yielding 0.24 mg RLP (10^6 cpm). The recovery of the RLP after the HPLC chromatography corresponded to 60%.

3.3. Immunological Studies

The purified RLP was used for the immunization of rabbits and with the resulting antiserum their immunological specificity was studied. The 35S-labeled thylakoids from Chlamydomonas reinhardii were sepa-rated by gel electrophoresis and the results of the western blot analysis with the antiserum against RLP are presented in Table I. The immunological reaction was specific for the 34 kDa protein and the autoradiography of the blot showed identity. The photosynthetic mutant FuD7 (a deletion of the psb A gene, see for details 8) was analysed as described above and no positive reaction could be ob-served. The interpretation of this result is not as easy since the omission of one core-protein of the photosystem II complex (Q_A-pro-tein and Q_B-protein are considered essential elements of this complex, see ref. 9) prevents the stable assembly of the residual proteins and therefore, only transitory pools can be detected.

The antiserum did not only react with thylakoids of Chlamydomonas reinhardii but also with a similar protein from other organisms such as blue-green algae (Synechococcus), green algae (Scenedesmus obli-quus) and with thylakoids and photosystem II preparations of higher plants (results are presented in Table I).

TABLE I. Western Blot Analysis with the Antiserum against RLP

Sample	Blot reaction with protein bands (in kDa)	
	SDS gel	SDS-urea gel
Chlamydomonas reinhardii		
wild-type thylakoids	34	30
crude RLP	67,34	30
HPLC-peak 2 fraction	34	n.d.
trypsinated RLP	12,8	n.d.
Scenedesmus obliquus	34	n.d.
wild-type thylakoids		
spinach thylakoids	34	30
BBY preparation	34	30
PS II (Octylgluc.)	66,34	60,30
PS II trypsinated	(34),12	n.d.
Synechococcus PS II prep.	34	30

It was of special interest to investigate the reaction center complex of the photosynthetic bacteria (Rhodopseudomonas sphaeroides) for immunological cross-reactivity. The established homology between the L-subunit and the Q_B-protein (9) could implicate a similar immunological relationship. However, in the western blot the H-subunit was the reactive species.

ACKNOWLEDGEMENTS
    The mutant FuD7 was provided by Dr. Rochaix (Geneva) and the Scenedesmus cultures by Dr. Bishop (Corvallis). The photosystem II particles from Synechococcus were prepared by Dr. Satoh (Tokyo), the reaction center complex from Rh. sphaeroides by Mrs. Preuße (Bochum) and the different spinach preparations were contributed by Mrs. Depka. These studies were financially supported by DFG (SFB 168-A3).

REFERENCES
1 Chua, N.H. and Gillham, N.W. (1977) J. Cell. Biol. 74, 441-452
2 Delepelaire, P. (1983) Photochem. Photobiophys. 6, 279-292
3 Wettern, M., Owens, J.C. and Ohad, I. (1983) Methods Enzymol. 97, 554-567
4 Chua, N.H. (1980) Methods Enzymol. 69, 434-446
5 Hunkapiller, M.W., Lujan, E., Ostrander, F. and Hood, L.E. (1983) Methods Enzymol. 91, 227-236
6 Berzborn, R.J. (1980) Methods Enzymol. 69, 492-502
7 Towbin, H., Staehelin, T. and Gordon, G. (1979) Proc. Natl. Acad. Sci. USA 76, 4350-4354
8 Bennoun, P., Spierer-Herz, M., Erickson, J., Girard-Bascou, J., Pierre, Y., Delosme, M. and Rochaix, J.-D. (1986) Plant Mol. Biol. 6, 151-160
9 Trebst, A. (1986) Z. Naturforsch. 41c, 240-245

INTERACTION OF THE Q_B PROTEIN OF VARIOUS SPECIES WITH CROSS LINKING REAGENTS AND THEIR USE FOR ITS ISOLATION.

NOAM ADIR, JOSEPH HIRSCHBERG* AND ITZHAK OHAD, DEPARTMENTS OF BIOLOGICAL CHEMISTRY, AND GENETICS*, HEBREW UNIVERSITY, JERUSALEM, 91904 ISRAEL.

INTRODUCTION

The herbicide binding 32 kDa protein of thylakoid membranes functions as the apoprotein of the secondary acceptor Q_B of photosystem II (PS II) (1). Additional roles have been ascribed to this protein including control of assembly of PS II reaction center (RC II) (2) and participation in the formation of RC II (3). Thus one would expect that the Q_B protein should be located close to other PSII complex polypeptides. Attempts to identify polypeptides located in close proximity to the Q_B protein by use of hydrophobic or hydrophilic bifunctional crosslinking reagents, have demonstrated that the Q_B protein of **Chlamydomonas reinhardtii** thylakoids is not cross linked. However the protein cross links if the thylakoids are pretreated with β-D-octylglucoside (4). The amino acid sequence of the Q_B protein is well conserved among various prokariotes, algae and higher plants (1), and thus its organization within PSII should be similar in thylakoids of various species. If this were the case then refractility to cross-linking reagents in its native state should be a general property of the Q_B protein. In the present work we demonstrate that indeed this is the case. Furthermore this property can be exploited for the isolation of the Q_B protein.

RESULTS AND DISCUSSION

1) Effect of cross linking on the Q_B protein of thylakoids from various species.

Similar results to those reported for **Chlamydomonas** (4) were obtained when thylakoids of the cyanophyte **Spirulina platensis** and the C_3 plant **Soybean** were used (Fig. 1). Following cross linking of the thylakoid polypeptides with GA, the Q_B protein retains its mobility as can be seen by coomasie blue stain and by ^{35}S labeling of the thylakoid proteins (Fig. 1). In **Spirulina** thylakoids an additional protein of 95 kDa, which might be identical with the protein involved in the binding of the phycobilisomes to the thylakoid membrane (5) is also not cross linked. Addition of β-D-octylglucoside prior to cross-linking reagents resulted in complete cross-linking of the Q_B protein in both **Spirulina** and **Soybean** thylakoids.

The above results could be explained if one considers that the Q_B protein of **Chlamydomonas** as well as of several higher plants does not contain Lysine and the Arginine residues might not be sufficiently reactive to allow cross-linking in nondenatured membranes. To test this possibility, thylakoids in which the Q_B protein contains Lysine residues obtained from **Anacystis nidulans** (6), a cyanophyte, and **Phalaris paradoxa** a C_3 plant, (Fig. 2, Ben-Yehuda, A., Hirschberg, J., in preparation) were crosslinked with GA with or without pretreatment with β-D-octylglucoside. The results (Fig. 3) demonstrate that the Q_B protein of these thylakoids is cross-linked only after their solubilization with detergent as in thylakoids containing Lysine-less Q_B protein. Similar results were obtained with thylakoids of Maize, a C_4 plant (Fig. 3).

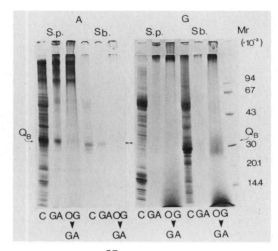

Figure 1. Effect of GA on ^{35}S-labeled thylakoid polypeptides from **Spirulina platensis** (S.P.) and **Soybean** (Sb.). All reactions were performed in the dark at 4°C for 15 min. at a chlorophyll concentration of 200 µg/ml in 100mM phosphate buffer, (pH7.5), including 100mM Nacl and 5mM MgCl$_2$. After cross linking the thylakoids were pelleted and electrophoresed as described in Ref. 4. C, control; GA, glutaraldehyde treated thylakoids; OG → GA, thylakoid membranes pretreated with 2.5% ß-octylglucoside for 60 min. at 25°C in the dark followed by glutaraldehyde cross-linking. G, gel; A, autoradiograph.

```
     220              225             230
1) Thr Ser Ser Leu Ile Arg Glu Thr Thr Glu Asn Glu Ser Ala Asn
2) Thr Ser Ser Leu Ile Arg Glu Thr Thr Glu Asn Glu Ser Ala Asn
     235        238    240                245
1) Glu Gly Tyr |Lys| Phe Gly Glu Glu Gln Gln Thr
2) Glu Gly Tyr |Arg| Phe Gly Glu Glu Gln Gln Thr
```

Figure 2. The amino acid sequence of the Q_B protein hydrophylic segment from residue 220 to 245. 1), the C_3 plant **Phalaris paradoxa**, 2), published "consensus" sequence of **Chlamydomonas** (7) and higher plants (8). Note the change of Arginine 238 to a Lysine in **Phalaris paradoxa**. Arginine 238 has been reported to be a Trypsin cleavage site (9) indicating this segments hydrophilic nature.

2) Isolation of the Q_B protein
The isolation of the Q_B protein from various sources, to be used in biochemical studies or preparation of antibodies is difficult due to its hydrophobicity. Elution of the Q_B band from LDS or SDS PAGE slabs does not ensure its purity since several other polypeptides comigrate with the Q_B protein or very close to it. However following treatment of thylakoids with GA, one could use LDS or SDS PAGE for isolation of the Q_B protein since most of the comigrating polypeptides do not

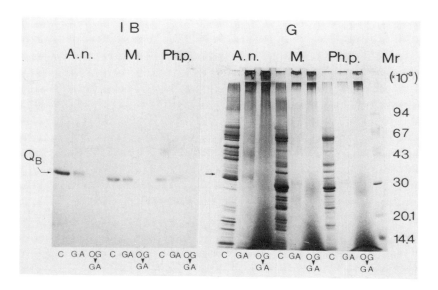

Figure 3. Effect of glutaraldehyde on thylakoid membrane polypeptides of **Anacystis nidulans** (A.n.), **Maize** (M.) and **Phalaris paradoxa** (Ph.p.), treatments as in Fig. 1. The Q_B protein is identified by protein immunoblotting (IB) with anti-Q_B protein antibodies decorated with ^{125}I- donkey anti-rabbit IgG and autoradiography. G, stained gel.

penetrate the 5% stacking gel (Fig. 1,3-4). An alternative way which can further facilitate isolation of the Q_B protein in relatively large amounts was devised in which fully cross linked thylakoids are solubilized in LDS. The solute is centrifuged shortly to remove unsolubilized material. The supernatant is then filtered through an Amicon XM-100A filter membrane (100 kDa cut-off limit) and the filtrate is concentrated using an Amicon Centricon-30 (30 kDa cut-off limit). The results of this isolation procedure are shown in figure 4.

CONCLUSION AND SUMMARY
1) The results presented here and elsewhere (4) demonstrate that the orientation and conformation of the Q_B protein, as well as its immediate surroundings within thylakoids of **Chlamydomonas** are such as to prevent it's Cysteine, Arginine, or free amino terminal from forming cross links with neighboring polypeptide amino acid residues to a distance of at least 1.2 nm.
2) The lipid surroundings of the Q_B protein participate in the organization and maintenance of this conformation. Alteration of the membrane lipid phase by detergent results in Q_B cross-linking.
3) Not only the amino acid sequence of the Q_B protein is preserved among different and distant eukariotic and prokariotic photosynthetic species, but also its orientation and the immediate lipid and protein environment.
4) The refractility of the native Q_B protein to cross linkers including that of Lysine containing Q_B, can be used for its easy isolation and purification.

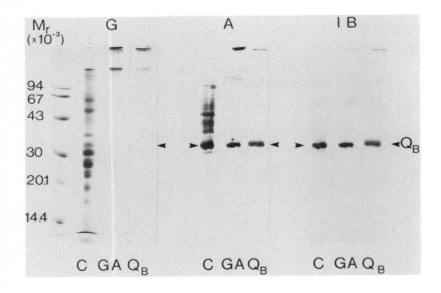

Figure 4. Isolation of the Q_B protein. ^{35}S-labeled thylakoids from **Chlamydomonas reinhardtii** were treated as described in figure 1. The thylakoids were solubilized in 2% LDS, and then centrifuged for 15 min at 10000xg to remove unsolubilized material. The supernatant was diluted to a final concentration of 0.3% LDS, and then filtrated twice through an Amicon XM-100A ultrafiltration membrane under 50 p.s.i. nitrogen. The filtrate was concentrated on an Amicon Centricon-30. The Q_B protein is identified by both its strong radioactive label (A) and by protein immunoblotting (IB) decorated with goat-anti-rabbit alkaline phosphatase conjugate. Some material which does not penetrate into the resolving gel, contains Q_B protein, as can be seen by comparing the upper portion of lanes GA and Q_B (the isolated protein) in the immunoblot (IB) and the autoradiograph (A). G, stained gel.

REFERENCES
1. Kyle, D.J., (1984) Photochem. Photobiol., 41, 107-116.
2. Bennoun, P., Spierer-Herz, M., Erickson, J.M., Girard-Bascou, J., Pierre, Y. and Rochaix, J.D., (1986) Plant Mol. Biol., 6, 151-160.
3. Deisenhofer, J., Epp, O., Miki, K., Huber, R. and Michel, H., (1985) Nature, 318, 618-624.
4. Adir, N. and Ohad, I., (1986) Biochim. Biophys. Acta, 850, 264-274.
5. Kirilovsky, D., Lavintman, N., Ish-Shalom, D. and Ohad, I. (1984) Biochim. Biophys. Acta, 767, 451-459.
6. Golden, S. and Haselkorn, R., (1985) Science, 229, 1104.
7. Erikcson, J.M., Rahire, M. and Rochaix, J.D., (1984) EMBO. J., 3, 2753-2762.
8. Zurawsky, G., Bohnert, H.J., Whitefield, P.R. and Bottomley, W., (1982) Proc. Natl. Acad. Sci. USA, 79, 7099-7703.
9. Marder, J.B., Goloubinoff, P. and Edelman, M., (1984) J. Biol. Chem., 259, 3900-3908.

EVIDENCE FOR MULTIPLE ROLES OF THE 32-34 KDA CHLOROPLAST POLYPEPTIDE OF SCENEDESMUS OBLIQUUS IN REACTIONS OF PHOTOSYSTEM II

NORMAN I. BISHOP
Department of Botany and Plant Pathology, Oregon State University
Corvallis, Oregon, 97331 U. S. A.

1. INTRODUCTION

The 32-34 kDa intrinsic polypeptide of chloroplast membranes has been equated with the rapidly metabolized chloroplast protein as originally discovered by Ellis and coworkers (1). It has also been described as the Q_b-binding protein, the D_1-protein and the herbicide-binding protein (2). The primary role of this protein has been deduced to be the site of binding of a plastoquinone molecule, Q_b, that functions in electron transport on the reducing side of PS-II; DCMU and atrazine compete at the Q_b binding site. This interpretation is supported by the preferential labeling of a 32-34 kDa PS-II core protein by the photoaffinity label, azidoatrazine (3). A comprehensive review of this subject area is presented in (2).

Through analysis of mutant phenotypes of Scenedesmus obliquus, blocked only in water photolysis (4) a 32-34 kDa polypeptide has been identified as an essential component of the oxidizing side of PS-II. Loss of this restricted portion of PS-II in the low fluorescent (LF) mutant, e.g., LF-1, is accompanied by the appearance of a 34-36 kDa polypeptide rather than the normal 32-34 kDa form (4). This approximate 2 kDa differential in apparent M_r is retained in PS-II reaction center core preparations of the WT and mutant LF-1 (5). Revertant strains of LF-1 possessing normal photosynthesis show recovery of the 32-34 kDa polypeptide (4). The observation by Metz, et al. that both the 32-34 and 34-36 kDa forms of this polypeptide are specifically labeled by azido[^{14}C]atrazine (5) questions the identification of the Q_b-protein as a moderator of electron transport only on the reducing side of PS-II. In this paper additional information gained from studies on the light sensitivity of growth, atrazine binding capacities and fluorescence properties of the LF-series of Scenedesmus mutants having either modified forms of or deficient in the 32-34 kDa polypeptide are presented. The data are interpreted to suggest that the rapidly labeled-rapidly metabolized 32-34 kDa chloroplast protein may function in reactions involving both the reducing and oxidizing sides of PS-II.

2. EXPERIMENTAL

For this study the wild-type and several non-photosynthetic mutant phenotypes of the green alga, Scenedesmus obliquus, were used as the experimental organisms. All methods for maintenance, growth, and mutant and revertant induction and recovery were according to procedures previously described (4). Algal chloroplast membranes used for this study were prepared from cells grown heterotrophically at 30C for 48 hrs. Procedures employed for rupturing of the algal cells, the buffers employed and centrifugation regimens were the same as previously described (4).

Biggens, J. (ed.), Progress in Photosynthesis Research, Vol. III. ISBN 90 247 3452 5
© *1987 Martinus Nijhoff Publishers, Dordrecht. Printed in the Netherlands.*

Manganese, atrazine binding, and polypeptide analyses were performed on unfrozen chloroplast membrane fractions obtained by density gradient centrifugation (4).

3. RESULTS AND DISCUSSION

To extent earlier efforts of this laboratory on the role of the intrinsic 32-34 kDa chloroplast membrane polypeptide in reactions of PS-II several new mutant phenotypes of <u>Scenedesmus</u> unable to synthesize normal levels of this protein have been isolated. Where appropriate, revertants of these new strains were also sought. In Tables 1 and 2 several parameters of whole cells and isolated chloroplast membranes of these new strains and their respective revertants are compared to those of the WT and mutant LF-1. The four mutant phenotypes display very low levels of photosynthesis but normal levels of photoreduction (data not shown) and of PS-I cell free reactions. Chloroplast reactions in which water serves as the electron donor are absent in all the mutants. The PS-II catalyzed reduction of dichlorophenol indophenol with diphenyl carbazide as the electron donor is retained near WT levels only in LF-1. Revertants developed from LF-1,. LF-18 and LF-22 show near normal levels of all reactions analyzed (Table I). Data of Table II illustrate that the loss of PS-II activity in mutants LF-18, LF-22 and LF-23 is accompanied by a decrease in chloroplast manganese and cytochrome b-559$_{HP}$ similar to that previously noted for LF-1 (4). The distinguishing features among the mutant strains are the noted inabilities to synthesize normal levels of the 32-34 kDa polypeptide and the apparently related photosensitivity of mixotrophic growth. Separation by LDS-PAGE and subsequent autoradiography of the polypeptides of [^{35}]S-labeled thylakoids of the mutant strains (Figs. 1 & 2) were used to determine the amounts of the 32-34 kDa polypeptide shown in Table 2. Comparable analysis of the available revertant phenotypes (Table 2) confirmed the presence of this polypeptide in all cases.

TABLE 1. Comparison of various whole cell and isolated chloroplast particle photochemical reactions of the WT, select water-side mutant and respective revertants.

Strain	PS*	H2O-DCPIP**	DPC-DCPIP**	DCPIP-MV**
Wild type	1084	220	180	310
LF-1	120	0	168	386
LF-1 RVT-1	1016	206	186	325
LF-18	220	0	28	322
LF-18 RVT-1	1048	212	168	347
LF-22	116	0	25	316
LF-22 RVT	1050	220	175	366
LF-23	90	0	tr	320

*umoles O_2 produced/mg Chl-hr; **umoles DCPIP reduced/mg Chl-hr.

TABLE 2. Manganese, cytochrome b-559(H.P.), relative abundance of "34 kDa" polypeptide and photoinhibition of heterotrophic growth in chloroplast particles and whole cells of WT, select water-side mutants and their respective revertants of Scenedesmus obliquus.

Strain Designation	Chl/Mn [*]	Chl/cytochrome b-559 H.P.[**]	"34 kDa" Protein[a]	Photoinhibition of growth[b]
Wild type	86 (4.6)	524	100	0
LF-1	302 (1.3)	2180	100 [***]	20
LF-1RVT-1	111 (3.6)	590	100	0
LF-18	320 (1.2)	2200	25	68
LF-18RVT-1	81 (4.9)	560	100	0
LF-22	350 (1.1)		0	75
LF-22RVT-1	112 (3.6)	510	85	10
LF-23	360 (1.1)		0	75

Values are molecules chlorophyll/Mn atom (Mn atoms/400 chlorophyll molecules).[*] [**]Mole ratio of chlorophyll/cytochrome b-559$_{HP}$. [a]Estimated abundance (%WT) of the 32-34 kDa polypeptide. [b]Inhibition of growth (%WT) under mixotrophic growth conditions. Growth measured as chlorophyll content per 100 ul of cells. [***]This values represents the level of the 34-36 kDa polypeptide in LF-1.

TABLE 3. Atrazine equilibrium binding constants and sites for chloroplast membranes of WT, mutants LF-1, LF-18, LF-22, LF-23 and revertants.

Strain	Binding Constant (K_b) (nanomolar)	# of Binding Sites (Chlorophyll/Atrazine)
WT	42	432
LF-1	43	468
LF-1 RVT-1	41	455
LF-18	165	2730
LF-18 RVT-1	40	417
LF-22	182	3235
LF-22 RVT-1	49	450
LF-23	168	3900

The observations that the 32-34 and 34-36 kDa polypeptides of WT and LF-1 are preferentially labeled by azidoatrazine(5), and that mutant strains LF-18, LF-22 and LF-23 showed light sensitivity and lacked the 32-34 polypeptide to varying degrees led to an evaluation of the atrazine binding characteristics of these mutants. The results, as summarized in Table 3, reveal that the atrazine binding constant (high affinity) and the number of binding sites are unaltered in LF-1 (and related phenotypes not considered here) but are acutely changed in the other mutants. As was anticipated the atrazine binding kinetics of the revertant phenotypes were normal.

Based on the information presented here and the knowledge that antibodies prepared against the herbicide binding protein of Chlamydomonas and Amaranthus cross-react with the 32-34 kDa polypeptide of Scenedesmus chloroplasts (personal communications, (J. Metz and G. F.Wildner), it is concluded that the 32-34 kDa polypeptide is essential for the integrity of the oxygen evolving enzyme and preferential alteration of it can influence PS-II activities to varying degrees including induction of photolability (6).

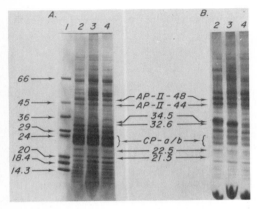

Figure 1. Analysis of purified, ^{35}S labeled chloroplast membrane polypeptides of WT, LF-1 and LF-18 by LDS-PAGE. (A) Slots #2, 3, and 4 contained samples equivalent to 10 ug protein of LF-1, WT and LF-18. Slot #1 contained protein standards of indicated MW. Coomassie Blue stained. (B). Autoradio-graphic pattern of gel slab illustrated in part A.

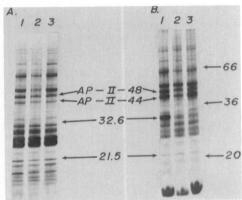

Figure 2. LDS-PAGE patterns of ^{35}S-labeled membrane polypeptides of WT and two low-fluorescent, water-side mutants, LF-22 and LF-23. (A.) Slots 1, 2, and 3 contained samples equivalent to 10 ug protein of WT, LF-2 and LF-23 chloroplast membranes. Proteins stained with Coomassie Blue. (B) Autoradiographic pattern of gel slab shown in part A.

ACKNOWLEDGMENTS
Supported the United States Department of Agricultural (82-CRCR-1-1079).

REFERENCES
1 Ellis, J.R. (1981)   Ann. Rev. Plant Physiol. 32, 111-137.
2 Kyle, D.J. (1985)   Photochem. Photobiol. 41, 107-116.
3 Pfister, K., Steinbach, K. E., Gardner, G. and Arntzen, C. J. (1981)
   Proc. Natl. Acad. Sci. USA 78, 981-985.
4 Bishop, N.I. (1984)   In: Advances in Photosynthesis Research
   (Sybesma, C.,ed.) M. Nijhoff/Dr W.Junk Publ. The Hague. I, pp.321-332.
5 Metz, J.B., Bricker, T.M. and Seibert, M. (1985) FEBS Lett., 185,
   191-196.
6 Kyle, D.J., Ohad, I. and Arntzen, C.J. (1984) Proc. Natl. Acad.
   Sci. USA 81, 4070-4074.

POSTTRANSLATIONAL ACYLATION AND INTRA-THYLAKOID TRANSLOCATION OF SPECIFIC
CHLOROPLAST PROTEINS

FRANKLIN E. CALLAHAN[1], MARVIN EDELMAN[2], and AUTAR K. MATTOO[1]
Plant Hormone Laboratory, USDA/ARS, Beltsville Agricultural Research
Center[1], Beltsville, MD. 20705, USA and Plant Genetics, Weizmann
Institute of Science[2], Rehovot, Israel

1. INTRODUCTION
    Our knowledge of the molecular events that target newly synthesized
components of the photosynthetic complexes to specific loci within the
thylakoids is minimal. The 32,000 dalton (32kDa) herbicide receptor
protein is a well-studied integral component of photosystem II (1-3).
This rapidly-metabolized (4-6) thylakoid protein is a major, light-driven
translation product (7) of the chloroplast genome. It contains
overlapping receptor sites for several classes of herbicides, including
the s-triazines, ureas and phenylureas (8). The 32kDa protein is first
synthesized as a 33,500-34,500 dalton (33.5 kDa) precursor polypeptide
(4,5,9). Both the precursor and mature protein are membrane-associated
(5,9).

    In this report we summarize our data indicating that the 33.5kDa
precursor protein is exclusively associated with the unstacked stromal
lamellae. Processing to the mature 32kDa form occurs in this membrane
fraction as well. A undirectional translocation of mature 32kDa protein
to the stacked granal lamellae takes place after processing.

    In combination with this translocation event we have examined the
possibility that posttranslational modification of the 32kDa protein is
involved in its membranal deployment. Posttranslational modifications
such as glycosylation, methylation, phosphorylation and acylation can have
major effects on the activity, conformation or stability of a protein. In
plants, glycosylation of seed storage proteins (10) and phosphorylation of
chloroplast membrane proteins (11,12) are well known. Reports on protein
acylation (13,14), however, have been mainly confined to viral glyco-
proteins (15) and a number of bacterial lipoproteins (16, 17), viral
transforming proteins (18) and animal membrane proteins (19).

    In the second part of this study we report that palmitic acid
selectively binds to certain chloroplast membrane proteins, including
prominently, the 32kDa protein. Bound palmitic acid is exclusively
associated with the processed 32kDa form, and is readily observed in the
granal membrane fraction. These findings are summarized in a model
depicting the life history of the 32kDa herbicide-receptor protein.

2. METHODS
    Spirodela oligorrhiza was grown phototrophically for 10-15d as axenic
cultures (6) containing 0.5% sucrose. Cultures were transferred to
mineral medium lacking sucrose for 2-3d prior to pulse labeling
experiments. Whole thylakoids were isolated and fractionated into granal

Biggens, J. (ed.), Progress in Photosynthesis Research, Vol. III. ISBN 90 247 3452 5
© *1987 Martinus Nijhoff Publishers, Dordrecht. Printed in the Netherlands.*

and stromal lamellae as previously described (20). Polypeptides were
analyzed on 10-20% gradients by SDS-PAGE (2).

3. RESULTS AND DISCUSSION
    The aquatic higher plant, Spirodela oligorrhiza, rapidly incorporates
nutrient precursors into macromolecules. We radiolabeled such plants with
[35S] methionine for 3 min and subsequently chased the label by
incubation in excess non-radioactive methionine. Membrane fractions were
isolated at 0, 3, 6, 60 and 120 min of chase time and then were analyzed
by SDS-PAGE and fluorography.

    Coomassie stained gels showed distinct polypeptide compositions of the
isolated membrane fractions. Generally, granal lamellae were enriched in
PSII polypeptides, while stromal lamellae displayed increased abundances
of PSI (e.g., CPI) and ATPase (e.g., alpha/beta subunits) proteins.
Electron microscopic visualization further showed single-membrane vesicles
versus double membrane sheets for stromal and granal lamellae isolates,
respectively (in collaboration with W. P. Wergin, USDA, Beltsville, MD).

    Fluorographs revealed that at 0t chase practically all of the
incorporated radioactivity was in the stromal lamellar fraction with the
33.5kDa precursor protein as the prominent band. Processing of the
precursor to the 32kDa product occurred exclusively on the stromal
lamellae during the chase (t0.5 = 3 - 6 min). Coincident with processing
of the 33.5dKa protein on stromal membranes, we observed appearance of the
mature 32kDa protein on granal lamellae. After 120 min chase, virtually
all of the labeled 32K protein was concentrated in the granal fraction.
Interestingly, longer fluorographic exposures suggested that another
protein of ∿30kDa was labeled and translocated in a similar manner, while
other proteins (e.g. alpha/beta subunits of ATPase) remained associated
with the stromal lamellae.

    Similar analyses of the isolated membrane fractions, but from plants
pulse-labeled for 1-3 min with [3H] palmitic acid, revealed the
following: 1) light-dependent [3H]-labeling of two bands which
coelectrophorese with [35S] methionine labeled 32kDa and Coomassie
stained LHCP, respectively; 2) primarily the granal associated-mature
32kDa protein is labeled; 3) the [3H] label extracted from hydrolyzed
32kDa protein comigrates with palmatic acid standard during HPLC (in
collaboration with H. Norman and J. B. St. John, USDA, Beltsville, MD).
Additionally, we found that partial trypsin digestion of granal membranes
from [3H] palmitic acid and [35S] methionine labeled plants
generates equivalent radioactive fragments (T20 and T22, ref. 28) known to
comprise the membrane anchor region of the 32kDa protein.

    Figure 1 is our attempt to incorporate these data with other recent
findings to form a working model for further study of the metabolism of
the 32kDa herbicide-receptor protein. Elucidation of the specific role
that palmitoylation has in the life history of the 32kDa and other
thylakoid proteins is being actively pursued.

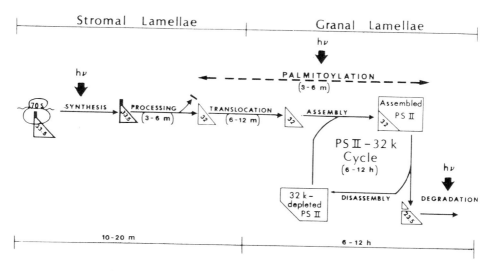

FIGURE 1.   A model depicting course of events in the life of the 32kDa protein.

All the values given in minutes (m) or hours (h) represent estimated t0.5's for different processes for the protein in *Spirodela* cultured at 30 μ mol.m-2.s-1 PAR. Light-dependent synthesis (6,7) of nascent 33.5kDa precursor protein takes place on 70S thylakoid-bound ribosomes (21-23) attached most likely to unstacked stromal lamellae (24). *In vivo* the 33.5kDa protein is associated exclusively with the stromal lamellae following completion of translation (as reported here). At this location, carboxyl-terminal processing (25) occurs with a t0.5 of 3-6 minutes (5) resulting in a membrane-integrated 32kDa protein. Following processing, the 32kDa protein is translocated to the stacked, granal lamellae with a t0.5 of 6-12 minutes (as reported here). In its new location the 32kDa protein is assembeled into a largely 32kDa protein depleted PSII complex. The t0.5 of its life span in the granal lamellae is 6-12 hours (26), considerably less than for other protein components of PSII. This relatively rapid decay of the 32kDa protein in the light (4,6,26) is the driving force behind the PSII-32kDa cycle. Light-intensity dependent degradation (6) of the 32kDa protein occurs in the granal lamellae with production of a 23.5kDa membrane-bound intermediate (27). A different 32kDa protein molecule is now assimilated into the recycled 32kDa protein depleted PSII complex. At some stage(s) between 32kDa protein translocation and degradation, a light-induced, post-translational palmitoylation (as reported here) of the membrane anchor region of the protein occurs.

ACKNOWLEDGEMENTS
    This investigation was supported in part by a U.S.-Israel Binational
Agricultural Research and Development Grant.

REFERENCES
 1. Gardner, G. (1981) Science 211, 937-940.
 2. Mattoo, A.K., Pick, U. Hoffman-Falk, H. and Edelman, M. (1981) Proc.
    Natl. Acad. Sci. USA 78, 1572-1576.
 3. Hirschberg, J., Bleecker, A., Kyle, D.J. McIntosh, L. and Arntzen,
    C.J. (1984) Z. Naturforsch 39C, 412-420.
 4. Ellis, R.J. (1981) Ann. Rev. Plant Physiol 32, 111-137.
 5. Reisfeld, A., Mattoo, A.K. and Edelman, M. (1982) Eur. J. Biochem.
    124, 125-129.
 6. Mattoo, A.K., Hoffman-Falk, H., Marder, J.B. and Edelman, M. (1984)
    Proc. Natl. Acad. Sci. USA 81, 1380-1384.
 7. Fromm, H., Devic, M. Fluhr, R. and Edelman, M. (1985) Embo J. 4,
    291-295.
 8. Trebst, A. (1984) in Biochemical and Physiological Mechanisms of
    Herbicide Action (S.O. Duke, ed.) pp 45-57, Tallahassee, Florida.
 9. Grebanier, A.E. Coen, D.M., Rich, A. and Bogorad, L.J. (1978) J. Cell
    Biol. 78, 734-746.
10. Higgins, T.J.V. (1984) Ann. Rev. Plant Physiol. 35, 191-221.
11. Bennett, J. (1977) Nature 269, 344-346.
12. Staehelin, L.A. and Arntzen, C.J. (1983) J. Cell Biol. 97, 1327-1337.
13. Schmidt, M.F.G. (1982) Trends Biochem. Sci. 7, 322-324.
14. Magee, A.I. and Schlesinger, M.J. (1982) Biochim. Biophys. Acta 694,
    279-289.
15. Schmidt, M.F.G. (1982) Virology 116, 327-338.
16. Braun, V. (1975) Biochim. Biophys. Acta 415, 335-377.
17. Tokunaga, M., Tokunaga, H. and Wu, H.C. (1982) Proc. Natl. Acad. Sci
    USA 79, 2255-2259.
18. Grand, R.J.A., Roberts, C. and Gallimore, P.H. (1985) FEBS Lett 181,
    229-235.
19. Soric, J. and Gordon, J.A. (1985) Science 230, 563-566.
20. Leto, K. J., Bell, E. and McIntosh, L. EMBO J. 4, 1645-1653 (1985).
21. Ellis, R.J. (1977) Biochim. Biophys. Acta 463, 185-215.
22. Minami, E-I and Watanabe, A. (1984) Arch. Biochem. Biophys. 235,
    562-570.
23. Herrin, D. and Michaelis, A. (1985) FEBS Lett. 184, 90-95.
24. Yamamoto, T., Burke, J., Autz, G. and Jagendorf, A.T. (1981) Plant
    Physiol. 67, 940-949.
25. Marder, J.B., Goloubinoff, P. and Edelman, M. (1984) J. Biol. Chem.
    259, 3900-3908.
26. Edelman, M., Mattoo, A.K. and Marder, J.B. in Chloroplast Biogenesis
    (Ellis, R.J., ed.), pp 282-302, Cambridge Univ. Press, London.
27. Greenberg, B.M., Mattoo, A.K., Gaba, V. and Edelman, M. (1986) Plant
    Physiol. S67, 245.
28. Marder, J.B., Mattoo, A.K. and Edelman, M. (1986) Methods Enzymol.
    118, 384-396.

EFFECT OF THE PHOTOSYNTHETIC HERBICIDES DCMU AND ATRAZINE ON
THE GROWTH OF PHOTOAUTOTROPHIC AND PHOTOHETEROTROPHIC CELL SUS-
PENSION CULTURES OF CHENOPODIUM RUBRUM

PAUL ZIEGLER and ANTHONY R. ASHTON, LEHRSTUHL PFLANZENPHYSIO-
LOGIE, UNIVERSITÄT BAYREUTH, 8580 BAYREUTH, F.R.G.

1. INTRODUCTION

A structurally diverse range of herbicides, e. g., phenyl
ureas, triazines and uracils, inhibit photosynthesis by apparently adhering to a plastoquinone binding site on the thylakoid membrane and thereby blocking photosynthetic electron
transport (1). That atrazine neither binds to thylakoids nor
interferes with light-dependent electron transport in triazine
herbicide-resistant weeds (2) strongly indicates that this
inhibition is responsible for the herbicidal action of these
compounds. Some uncertainty still exists, however, as to the
physiological basis of their action (3). Death of the plant
may result from starvation due to obstructed photosynthesis,
from the action of harmful radicals generated as a result of
continued light absorption by a blocked photosynthetic electron
transport system, or from a combination of both these alternatives. Inhibition of the photoheterotrophic growth of plant
cell cultures by photosynthetic herbicides (4) supports the
toxic radical hypothesis (5), but these results were obtained
using concentrations of herbicides much higher than necessary
to inhibit photosynthetic electron transport in vivo and may
thus have revealed non-specific effects of the herbicides. The
present study shows that low concentrations of herbicide sufficient to completely prevent the photoautotrophic growth of a
cell suspension culture of Chenopodium rubrum do not detectably
inhibit the growth of these cells using sucrose as a carbon and
energy source.

2. MATERIAL AND METHODS

The cell suspension culture of Chenopodium rubrum developed by
Hüsemann und Barz (6) has been grown photoautotrophically for
several years under long day conditions as described by (7).
The cultures in two-tiered flasks were exposed to a 2 % CO_2 atmosphere and a light (75 ueinstein/m^2/sec)/dark regime of 16/8
h at 26°-27°. These cells were also grown photoheterotrophically by innoculation into growth medium containing 1 or 2 %
sucrose and maintainance in air under the same light and temperature conditions as for photoautotrophic growth. DCMU and
atrazine, both 99% pure (from Serva), were added to the cultures as solutions in dimethyl sulphoxide, while an equal
amount of dimethyl sulphoxide was added to control cultures.
Photosynthetic O_2 evolution was measured in a Clark-type O_2
electrode at saturating light and CO_2.

Biggens, J. (ed.), Progress in Photosynthesis Research, Vol. III. ISBN 90 247 3452 5
© *1987 Martinus Nijhoff Publishers, Dordrecht. Printed in the Netherlands.*

3. RESULTS AND DISCUSSION

The growth patterns of the cell cultures without herbicides are illustrated in Fig. 1 (solid lines). The photoautotrophic cultures grow exponentially for 11 days (doubling time 2.2 days), after which the growth rate gradually decreases. Chlorophyll and wet weight increase in parallel (up to 1 mg chl/g cells) to an approximately 30-fold extent after 3 weeks. These cultures thus now grow about 3 times faster than the original cell line (8) and contain considerably more chlorophyll (1 mg chl/g cells as compared to 0.2 mg/g). These cells perform light- and CO_2-dependent O_2 evolution at rates of up to 230 µmoles/mg chl/h, with the highest rates being observed at about 12 days. The cells grow even more rapidly under photoheterotrophic conditions as long as sucrose is still present in the medium, with wet weight increasing exponentially for about 13 days (doubling time 2.0 days) and showing an 80-fold accumulation after 20 days. Although chlorophyll accumulation with these cultures shows an initial lag and the cells thus contain only up to 0.5 mg chl/g cells, the photoheterotrophic cells are capable of rates of photosynthesis quite comparable to and in the early growth phase even higher than those observed with photoautotrophic cells.

Light- and CO_2-dependent O_2 evolution of both photoautotrophic and photoheterotrophic culture cells is inhibited within seconds upon addition of DCMU or atrazine. These compounds cause 50% inhibition of photosynthetic O_2 evolution at approximately 0.1 µM and 0.5 µM respectively in the case of photoautotrophic cells and were found to be similarly effective with photoheterotrophic cells. Photoautotrophic growth was completely suppressed by 0.2 µM DCMU (Fig. 1) and took place only very slowly in the presence of 0.5 µM atrazine. Photoheterotrophic growth, however, was not affected by 10-fold higher concentrations of the herbicides (2 µM DCMU: Fig. 1 and 5 µM atrazine) as long as sucrose was still present in the medium. Neither photoautotrophic nor photoheterotrophic cells cultured in the presence of the herbicides were capable of any net photosynthetic O_2 evolution.

Photoautotrophic growth was completely inhibited when the cells were deprived of light or the high CO_2 concentrations required for this type of growth and were thus unable to assimilate carbon. When, however, the possibility of starvation is circumvented by the presence of an alternative carbon source as in the case of the photoheterotrophic cells (which also grow rapidly in the dark), the presence of the herbicides results in no inhibition of growth. Any injurious effects on growth due to the production of toxic radicals resulting from continued light absorbtion by a blocked photosynthetic electron transport system would have been evident during the first two weeks of photoheterotrophic culture in the presence of the herbicides (since these cells were shown to be photosynthetically competent in control cultures). That radical-producing conditions would indeed severely inhibit the growth of even the photoheterotrophic cells is evident from Table 1.

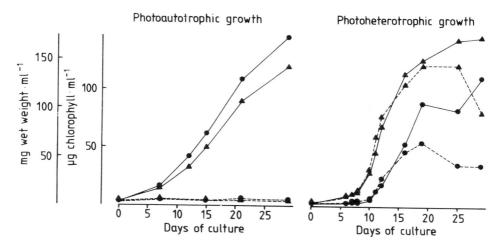

Fig. 1: Effect of DCMU on the growth of C. rubrum cells. Cells were grown either a) photoautotrophically in the absence (———) or presence (-----) of 0.2 μM DCMU or b) photohetero-trophically (2% sucrose) in the absence (———) or presence (-----) of 2 μM DCMU. Chlorophyll per ml culture suspension is represented by (●), wet weight by (▲). In the case of the photoheterotrophic cultures (both with and without herbicide) less than 1% of the original sucrose content was present in the medium after 16 days.

Table 1: Effect of methyl viologen on the growth of C. rubrum cells. Photoautotrophically growing cells were cultured under both photoautotrophic and photoheterotrophic conditions either in the absence or presence of 2 μM methyl viologen and harves-ted after 3 weeks, when the values shown were determined. Imme-diately following innoculation the chlorophyll contents of the culture suspensions were 1.1 μg/ml (equivalent to 1.4 mg wet weight/ml).

Type of growth	Wet weight mg/ml culture	Chlorophyll content	
		μg/ml culture	μg/g cells
Photoheterotrophic			
control	122	42.1	350
2 μM methyl viologen	2.3	0.18	80
Photoautotrophic			
control	86	61.5	715
2 μM methyl viologen	0.8	0.07	87

No growth under either set of culture conditions was observed when 2 µM methyl viologen, which is known to generate radicals upon reduction by Photosystem I (1), was present in the medium. If free radicals (or other toxic substances) are indeed being produced in the light in the presence of the low concentrations of DCMU and atrazine employed, it must be assumed that the photoheterotrophic cells posess the resources to cope with these toxins (e. g. see 9, 10) until, possibly, their source of energy and carbon is exhausted (see Fig. 1: photoheterotrophic growth after 20 days).

4. CONCLUSIONS

Our data are consistent with the conclusion that DCMU and atrazine are inhibitory to growth of C. rubrum cells only in the absence of an additional energy source and that any photooxidative damage alone caused by continued light absorption by a blocked photosynthetic apparatus does not detectably inhibit cell growth. Any deleterious side effects of the herbicides should have been revealed in the growth pattern of the photoheterotrophic cells. The present work thus indicates that the photoautotrophic cell cultures provide a convenient model for the study of the effects of photosynthetic herbicides and that the same cells growing photoheterotrophically provide a guide to the specificity of such inhibitors.

5. REFERENCES

1 Fedtke, C. (1982) Biochemistry and Physiology of Herbicide Action, Springer, Berlin
2 Arntzen, C.J., Pfister, K. and Steinback, K.E. (1982) in: Herbicide Resistance in Plants (LeBaron, H.M. and Gressel, J., eds.), pp. 185-214, Wiley, New York
3 LeBaron, H.M. (1984) in: Biosynthesis of the Photosynthetic Apparatus: Molecular Biology, Development and Regulation (Thornber, J.P., Staehlin, L.A. and Hallick, R.B., eds.), pp. 351-356, Liss, New York
4 Zilkah, S. and Gressel, J. (1978) Pesticide Biochem. Physiol. 9, 334-339
5 Gressel, J. (1979) Z. Naturforschung 34, 905-913
6 Hüsemann, W. and Barz, W. (1977) Physiol. Plant. 40, 77-81
7 Campbell, W. H., Ziegler, P. and Beck, E. (1984) Plant Physiol. 74, 947-950
8 Hüsemann, W., Plohr, A. and Barz, W. (1979) Protoplasma 100, 101-112
9 Halliwell, B. (1978) Prog. Biophys. Molec. Biol. 33, 1-54
10 Badger, M. R. (1985) Ann. Rev. Plant Physiol. 36, 27-53

6. ACKNOWLEDGEMENTS

We thank Dr's Hüsemann and Barz for generously providing the C. rubrum cell cultures. A. R. A. was the recipient of a fellowship of the Alexander von Humboldt Stiftung. Helga Simper provided excellent technical assistance.

BIOCHEMICAL AND MOLECULAR ANALYSES OF HERBICIDE RESISTANT MUTANTS IN
CYANOBACTERIA.

NIR OHAD, IRIS PECKER AND JOSEPH HIRSCHBERG, DEPARTMENT OF GENETICS,
THE HEBREW UNIVERSITY OF JERUSALEM, JERUSALEM 91904, ISRAEL

INTRODUCTION

Many widely used herbicides are inhibitors of photosynthesis. Their
mechanism of action was found, in the case of triazine and urea
herbicides, to involve binding of the herbicide molecule to a
thylakoid-membrane polypeptide of 32000 dalton (the Q_B protein) and
blocking the photosynthetic electron transport at the second stable
electron acceptor of photosystem II (PSII) (1,2). The Q_B protein is
part of PSII complex and it is encoded by the chloroplast gene psbA.
It has been shown previously (3), that a point mutation in psbA,
resulting in a serine to glycine substitution at position 264 of Q_B,
is responsible for the resistance to the herbicide atrazine (a
triazine derivative) in two herbicide resistant mutants of higher
plants. The same serine residue is changed to alanine in herbicide
resistant mutants of Clamydomonas (4) and Cyanobacteria (5). Two other
mutations in psbA confering herbicide resistance, have been identified
in Chlamydomonas (6). In order to study the structure and function of
Q_B, we have isolated and characterized herbicide resistant mutants in
the Cyanobacterium Synechococcus R2 (Anacystis nidulans R2).

RESULTS AND DISCUSSION

1. Isolation of herbicide resistant mutants

Two herbicide resistant mutants -Di1 and D5, have been isolated
following a mutagenic treatment with ethyl methane sulfonate (EMS) of
Synechococcus R_2 cells and selecton on atrazine containing medium.
PSII dependent electron transport was measured in isolated
photosynthetic membranes from wild type (w.t.) and mutant strains,
using H_2O as electron donor and DCPIP as electron acceptor (7). The
concentration of various herbicides that inhibit 50% of the control
rate of electron flow (I_{50} concentration) was determined. Based on

Biggens, J. (ed.), Progress in Photosynthesis Research, Vol. III. ISBN 90 247 3452 5
© *1987 Martinus Nijhoff Publishers, Dordrecht. Printed in the Netherlands.*

these data, the relative resistance of the mutants was calculated (Table 1).

TABLE I. Relative resistance (I_{50} mutant/I_{50} w.t.) of mutants Di1 and D5 to various herbicides.

	mutant	
herbicide	Di1	D5
atrazine	25	500
terbutryne	16	650
metribuzin	5000	2000
metamitron	30	3
diuron	150	300
tebuthiuron	2.5	2.5
ustilan	33	20
Bromacil	300	30

2. Identification, cloning and sequencing of psbA genes

Synechococcus R2 contains three psbA genes (5). Probing synechococcus R2 DNA, digested with Hind III, BamHI and SalI, with an internal fragment of the psbA gene from Amarantus hybridus (8), at different stringencies of hybridizations, revealed that two genes of psbA, copy II and copy III, are more homologous to psbA of higher plants than copy I (data not shown). The 3.5 kb BamHI and 2.7kb BamHI-HindIII fragments, containing two psbA genes, copy I and copy II, respectively, were cloned in the plasmid pBR328.

Nucelotide sequence analysis of two psbA genes - copy I and copy II revealed a high degree of homology in the deduced amino acid sequence of Q_B from Synechococcus R2 and Q_B from chloroplasts of higher plants and algae (see fig. 1). Q_B encoded by copy I is 85% homologous to spinach and 93% homolgous to Q_B of copy II. The homology between Q_B of copy II and Q_B of spinach is 89%.

3. Point mutations in psbA copy I of herbicide resistant mutants

The psbA genes copy I and copy II from mutants Di1 and D5 were sequenced and compared to the wild type genes. Point mutations were found only in copy I of the two mutants (see fig. 2), one mutation was found in Di1 and two mutations in D5.

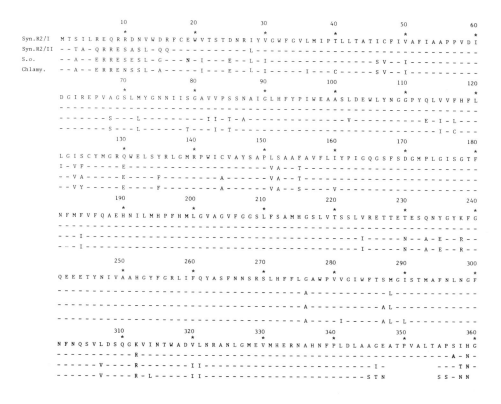

FIGURE 1. Comparison of the deduced amino acid sequences of the Q_B protein from Synechococcus R2 (two psbA genes copy I and copy II) Spinach (S.o.) (9) and Chlamydomonas (Chlamy) (10).

```
w.t.  CAC GGT TAC TTC GGT CGC TTG ATC TTC CAA TAC GCA TCG TTC AAC
      His Gly Tyr Phe Gly Arg Leu Ile Phe Gln Tyr Ala Ser Phe Asn
              255                                 264
Di1   ─────────────────────────────────────GCG───────────
      ─────────────────────────────────────Ala───────────


D5    ────────────TAC─────────────────────GCG───────────
      ────────────Tyr─────────────────────Ala───────────
```

FIGURE 2. Nucleotide sequence and deduced amino acid sequence of a portion of psbA copy I from w.t., Di1 and D5. Numbers indicate position on the Q_B. Differences are only given for the mutants.

CONCLUSIONS

The molecular basis of herbicide resistance in Cyanobacteria is similar to that which is found in higher plants and algae.

Cross resistance to various herbicides as a result of a single or double amino acid substitution in the Q_B polypeptide, demonstrates that these herbicides share the same binding site in Q_B. Correlating mutations in psbA with binding affinity of Q_B to different herbicides would enable the definition of this domain in the polypeptide.

Mutations in only one of the three psbA genes of Synechococcus R2 confer herbicide resistance, suggesting that this gene - copy I, is most strongly expressed.

REFERENCES

1. Pfister, K., Steinback, K.E., Gardner, G. and Arntzen, C.J. (1981). Proc. Natl. Acad. Sci. USA 78, 981-985.

2. Vermaas, W.F.J., Arntzen, C.J., Gu, L.-Q. and Yu, C. -A. (1983). Biochim. Biophys. Acta 723, 266-275.

3. Hirschberg, J., Bleecker, A., Kyle, D.J. McIntosh, L. and Arntzen, (1984). Z. Naturforsch. 39c, 412-420.

4. Erickson, J.M., Rahire, M., Bennoun, P., Delepalaire, P., Diner, B. and Rochaix J.-D. (1984). Proc. Natl. Acad. Sci. USA 81, 3617-3621.

5. Golden, S.S. and Haselkorn, R. (1985). Science 229, 1104-1107.

6. Erickson, J.M. Rahire, M., Rochaix, J.-D. and Mets, L.(1985). Science 228, 204-207.

7. Cahen, D.S., Malkin, S., Shochat, S. and Ohad, I. (1976). Plant Physiol. 60, 845-849.

8. Hirschberg, J. and McIntosh, L. (1983). Science 222, 1346-1349.

9. Zurawski, G., Bohnert, H.J., Whitfeld, P.R. and Bottomley, W. (1982). Proc. Natl. Acad. Sci. USA 79, 7699-7703.

10. Erickson, J., Rahire, M., Rochaix, J.-D. (1984). EMBO J. 3, 2753-2762.

Acknowledgement: This research was supported by a grant from N.C.R.D., Israel and G.S.F., Munchen, Germany.

THE CHLOROPLAST-ENCODED TYPE OF HERBICIDE RESISTANCE IS A RECESSIVE
TRAIT IN CYANOBACTERIA

IRIS PECKER, NIR OHAD AND JOSEPH HIRSCHBERG, DEPARTMENT OF GENETICS,
THE HEBREW UNIVERSITY OF JERUSALEM, JERUSALEM, 91904 ISRAEL

INTRODUCTION

Triazine and urea herbicides inhibit photosynthesis by blocking
photosystem II (PSII) dependent electron transport. The herbicide
binding site was identified as the 32000 dalton Q_B polypeptide of
PSII. The gene psbA, which codes for Q_B, is mapped in the chloroplast
genome of higher plants and algae. A mutation in psbA, leading to a
single amino acid change in Q_B, is responsible for herbicide
resistance in higher plants, Chlamydomonas and Cyanobacteria (1-3).
This is in accordance with the maternal mode of inheritance of
atrazine resistance in higher plants (4). In order to determine the
dominance-recessiveness nature of psbA-encoded herbicide resistance,
we used the Cyanophyte Synechococcus R2 as a model system. PSII
structure as well as the molecular basis of herbicide resistance in
this organism resemble chloroplasts of higher plants (3,5).

RESULTS AND DISCUSSION

Construction of synecochoccus R2 diploid to psbA

The mutant D5 of Synechococcus R2 (Anacystis nidulans R2) is highly
resistant to the herbicide atrazine (6). Two point mutations, that
result in the substitutions of phenylalanine at position 255 of Q_B to
tyrosine and of serine at position 264 to alanine, were identified in
psbA (copy I) from D5. A 3.5Kb DNA fragment from the strain D5
containing this gene, was cloned in the plasmid pBR328. The
recombinant plasmid was designated pAN35D5. Following transformation
(7) of the wild type (w.t.) herbicide susceptible, strain with pAN35D5
DNA, cells were selected for: 1. atrazine resistance, which is a
result of a reciprocal exchange between the plasmid and the
Cyanobacterial chromosome (see fig. 1a), since the plasmid pBR328 is

Biggens, J. (ed.), Progress in Photosynthesis Research, Vol. III. ISBN 90 247 3452 5
© *1987 Martinus Nijhoff Publishers, Dordrecht. Printed in the Netherlands.*

not capable of replicating in Cyanobacteria; or 2. Chloramphenicol
resistance, which is encoded by the CAT gene on the plasmid
vector, as a result of plasmid integration into the Cyanobacterial
chromosome (7) (see fig. 1b).

(a)   **(b)**

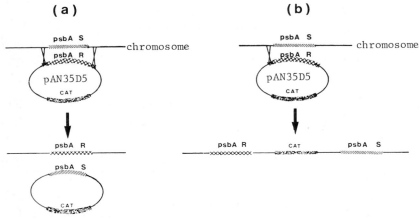

FIGURE 1.   A model for possible recombination events between the
plasmid pAN35D5 and the Cyanobacterial chromosome.

Southern hybridization analysis of DNA isolated from chloramphenicol
resistant transformants reveals the existance of two complete copies
of psbA-copy I (see fig.2) as a result of plasmid integration.

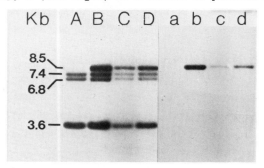

FIGURE 2. Total DNA of Synechococcus R2 w.t. strain (A,a) and three
chloramphenicol resistant transformants (B,b; C,c; D,d) was
digested with the restriction enzyme XhoI, which cleaves
pAN35D5 only once, electrophoresed and blotted. Southern
hybridization was carried out using the cloned psbA-copy I
(A-D) or pBR328 (a-d) as probes.

Phenotypic expression of the partial diploid strains

The partially diploid strains of Synechococcus R2, heterozygous for the herbicide resistant gene, were sensitive to atrazine indicating that atrazine resistance is recessive.

Evidence for true heterozygosity of the partial diploid strain

1. The mutated psbA gene was rescued from the partially heterozygous strain. Total DNA was extracted from a chloramphenicol resistant transformant and cleaved with Hind III. Following self ligation, the DNA was transfected into E. coli. Ampicillin resistant colonies contained a plasmid with a complete sequence of psbA, which was capable of transforming w.t. cells to become herbicide resistant (as in fig. 1a).

2. Chloramphenicol resistant colonies of the herbicide resistant mutant D5, which have been selected following transformation with the w.t. psbA gene cloned in pBR328 (reciprocal to the transformation shown in fig. 1b), turned out to be sensitive to atrazine.

3. Sensitivity of PSII dependent electron transport to atrazine was measured in the w.t., the atrazine resistant transformant (as described in fig. 1a) and the "heterozygote" strain (as described in fig. 1b). Rate of electron flow through PSII was measured in isolated membranes, using H_2O as electron donor and DCPIP as electron acceptor (8).

    I_{50} concentration of atrazine: w.t. $1.2 \times 10^{-7} M$

                               resistant transformant $6.0 \times 10^{-5} M$

                               heterozygote $5.0 \times 10^{-6} M$

This intermediate resistance demonstrates that in the partial diploid, both genes are expressed. The difference between the in vivo behaviour of the heterozygote strain (complete sensitivity) and the in vitro (partial resistance) can be explained by photo induced turn over of the Q_B protein (9), which is prevented by herbicide binding (10) and occurs only in vivo. PSII complexes in the heterozygote are composed of two types of Q_B-susceptible and resistant to atrazine. In the presence of atrazine, the "herbicide resistant" Q_B proteins are degraded, but the w.t. proteins are not. Thus, only atrazine-resistant proteins are replaced up to saturation of all PSII complexes with w.t. Q_B blocked by atrazine.

CONCLUSIONS

The psbA-encoded herbicide resistance is a recessive trait in Cyanobacteria. This nature of herbicide resistance should be considered in developing strategies for genetic engineering of herbicide resistant plants, since introducing the herbicide resistant psbA gene into susceptible plants will not result in herbicide resistance unless expression of the native, w.t. gene is inhibited.

REFRENCES

1. Hirschberg, J. and McIntosh, L. (1983) Science 222, 1346-1349.
2. Erickson, J.M., Rahire, M., Rochaix, J-D. and Mets, L. (1985) Science 228, 204-207.
3. Golden, S.S. and Haselkorn, R. (1985) Science 229, 1104-1107.
4. Darr, S., Souza Machado, V. and Arntzen, C.J. (1981). Biochim. Biophys. Acta 643, 219-228.
5. Ho, K.K. and Krogmann, D.W. (1982) in The Biology Of Cyanobacteria (Carr, N.G. and Whitton, B.A. eds) p.191, University of California Press, Berkley.
6. Ohad, N., Pecker, I. and Hirschberg, J. (1986) in this volume.
7. Williams, J.G.K. and Szalay, A.A. (1983) 24, 37-51.
8. Cahen, D.S., Malkin, S., Shochat, S. and Ohad, I. (1976). Plant Physiol. 60, 845-849.
9. Mattoo, A.K., Hoffman-Falk, H., Marder, J.B. and Edelman, M. (1984). Proc. Natl. Acad. Sci. USA 81, 1380-1384.
10. Kyle, D.J., Ohad, I. and Arntzen, C.J. (1984). Proc. Natl. Acad. Sci. USA 81, 4070-4074.

Acknowledgement: This work was supported by a grant from the N.C.R.D. Israel and G.S.F. München, Germany.

INHIBITOR AND QUINONE BINDING TO CHROMATOPHORES AND REACTION CENTERS FROM THE PHOTOSYNTHETIC BACTERIUM RHODOPSEUDOMONAS SPHAEROIDES

SYLVANA PREUSSE AND WALTER OETTMEIER, LEHRSTUHL BIOCHEMIE DER PFLANZEN, RUHR-UNIVERSITÄT, POSTFACH 10 21 48, D-4630 BOCHUM 1, F. R. G.

1. INTRODUCTION

From all photosystems of living organisms the reaction center of photosynthetic bacteria is so far the most simple one. It consists of only three proteins, known as the L-, M-, and H-subunit. The photosynthetic pigments and the primary and secondary quinone acceptors Q_A and Q_B are closely associated with the L- and M-subunit, whereas the H-subunit seems to stabilize the reaction center core complex (1). Recent amino acid sequence analysis has established a high sequence homology between the L-subunit of the bacterial reaction center and a protein D1 ("Q_B-protein", "herbicide binding protein") of photosystem II of higher plants and algae. Similarily, a high sequence homology also exists between the M-subunit of the bacterial reaction center and a protein D2 from photosystem II. It has been speculated that D1 and D2 constitute the reaction center core complex of photosystem II as do the L- and M-subunit in the bacterial reaction center (2). Therefore, the bacterial reaction center may serve as a simple model system for photosystem II. We have used common photosystem II inhibitors and herbicides to test their inhibitory activity and investigate their binding characteristics in bacterial chromatophores and reaction centers as well.

2. RESULTS AND DISCUSSION

2.1. Inhibition of photosynthetic electron transport by various inhibitors and herbicides in bacterial reaction centers

Inhibition constants of photosystem II herbicides in reaction centers from the carotenoidless mutant Rhodopseudomonas sphaeroides R-26 have been determined by Stein et al. (3). We have assayed several "DCMU-type", phenolic, chromone and 1,4-benzoquinone inhibitors in reaction centers from wild type Rhodopseudomonas sphaeroides. In the anaerobic testing system reduced cytochrome c served as the electron donor and ubiquinone-6 as the electron acceptor. The results are shown in Table 1. We basically agree with Stein et al. (3) that only s-triazines like terbutryn are efficient inhibitors of electron transport in isolated reaction centers. However, contrary to the results by Stein et al. (3), metribuzin was also an efficient inhibitor in our system. With the exception of 2-iodo-4-nitro-6-isobutylphenol, phenolic inhibitors are of only moderate activity or inactive at all (Table 1.). The recently introduced chromone derivative stigmatellin (4) also stands out as a powerful inhibitor. In addition, the four tetrahalogen-substituted 1,4-benzoquinones are also characterized by a high inhibitory activity (Table 1). Their biological activity increases from fluorine to iodine. Other halogenated 1,4-benzoquinones are only weak inhibitors or function as electron acceptors provided their half wave reduction potentials in cyclic voltammetry are below -195 mV.

Biggens, J. (ed.), Progress in Photosynthesis Research, Vol. III. ISBN 90 247 3452 5
© *1987 Martinus Nijhoff Publishers, Dordrecht. Printed in the Netherlands.*

Table 1. pI_{50}-values for Inhibition of Photosynthetic Electron Transport by Various Herbicides and Inhibitiors in Bacterial Reaction Centers

A. "DCMU-type"			i-dinoseb	none
terbutryn	6.25		dinoterb	none
metribuzin	5.22			
DCMU	4.30		C. Chromone type	
phenisopham	none		stigmatellin	6.09
B. phenol type			D. 1,4-benzoquinone type	
2-iodo-4-nitro-6-iso-			tetrafluoro	5.00
butylphenol	4.80		tetrachloro	5.25
picric acid	<4		tetrabromo	5.60
ioxynil	<4		tetraiodo	6.00

Cytochrome c oxidation was followed spectrophotometrically at 550 nm according to (5). The reaction mixture contained in a volume of 1 ml 10 mM Tris, pH 7.5; reduced cytochrome c, 10 uM; ubiquinone-6, 3 uM; and reaction centers, 0.5 uM.

2.2. Binding of radioactively labeled herbicides to bacterial chromatophores
Binding experiments with radioactively labeled herbicides in isolated thylakoids of higher plants or algae are common practice (6). So far, similar experiments have not yet been performed with isolated

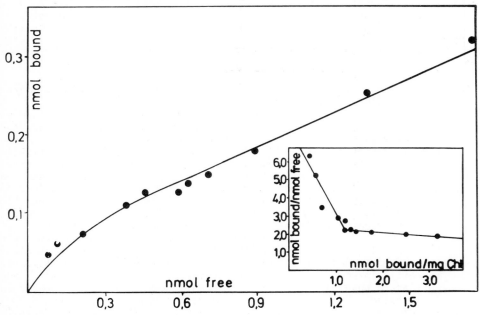

Fig. 1. Binding of $[^{14}C]$terbutryn to isolated bacterial chromatophores. Inset:Scatchard plot of binding data

bacterial chromatophores. Fig. 1 shows the binding characteristics of [^{14}C]terbutryn to isolated bacterial chromatophores. The corresponding half reciprocal Scatchard plot is shown as an inset in Fig. 1. Clearly a two-phasic binding, attributed to a high and low affinity binding can be visualized. In this respect, there exists no difference between chromatophores and thylakoids from higher plants. From the Scatchard plot for the high affinity binding a binding constant K_b=113 nM and a number of binding sites x_t=1.70 nmol/mg Bchl can be calculated. A similar binding curve is obtained for [^{14}C]metribuzin (K_b=90 nM; x_t=0.12). The number of binding sites for metribuzin is much lower as compared to terbutryn, but could be confirmed in independent experiments. Other "DCMU-type" photosystem II herbicides like DCMU itself, atrazine or phenisopham exhibited no high affinity binding at all. Similarily, from all phenolic herbicides tested, high affinity binding was only found for [^{3}H]2-iodo-4-nitro-6-isobutylphenol (K_b=79 nM; x_t=1.88 nmol/mg Bchl).

2.3   Covalent labeling of bacterial reaction centers
For identification of inhibitor binding proteins the photoaffinity labeling technique has proven to be a powerful tool. Azido-atrazine in bacterial reaction centers predominantly binds to the L-subunit (7, 8) and, hence, the L-subunit was identified as the Q_B binding site in accordance with the X-ray cristallographic data (1). An azido-plasto-quinone (9) preferentially also tags the L-subunit, and to a smaller extent in addition the M-subunit (data not shown). We could recently

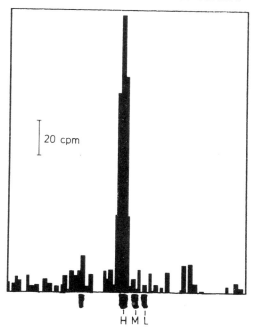

20 cpm

H M L

Fig. 2. Photograph of a SDS polyacrylamide electrophoresis gel (10-15 %) and radioactivity distribution therein of reaction centers from Rhodopseudomonas sphaeroides labeled with [^{14}]bromanil (10 nmol/mg Bchl)

demonstrate that halogen-substituted 1,4-benzoquinones as vinylogous acid halides in an addition/elimination reaction can react with nucleophiles under formation of a covalent linkage (10; see also Dostatni, Masson and Oettmeier, this proceedings). This property allows for identification of a quinone binding site with a suitable radiolabeled 1,4-benzoquinone. As demonstrated in 2.1, tetrabromo-1,4-benzoquinone (bromanil) with a pI_{50}-value of 5.6 is a fairly good inhibitor of electron transport in the bacterial reaction center. A $[^{14}C]$bromanil has been synthesized (10). If bacterial reaction centers are incubated with this compound, radioactivity is almost exclusively found in the H-subunit (Fig. 2). Though the H-subunit is not directly involved in Q_B binding, recent results have suggested a major role for the H-subunit in defining the environment of the Q_B-site (11). This is judged from the fact that the rate of electron transfer from Q_A to Q_B is greatly diminished in reaction centers where the H-subunit has been removed. Furthermore, in reaction centers depleted of the H-subunit the sensitivity to inhibitors of electron transfer like terbutryn is reduced by about two orders of magnitude (11). The labeling of the H-subunit by bromanil further corroborates the importance of the H-subunit for the Q_B binding site and the entire reaction center core complex.

3. ACKNOWLEDGEMENT

This work was supported by Deutsche Forschungsgemeinschaft.

REFERENCES
1 Deisenhofer, J., Epp, O., Miki, K., Huber, R. and Michel, H. (1985) Nature 318, 618-624
2 Trebst, A. (1986) Z. Naturforsch. 41c, 240-245
3 Stein, R. R., Castellvi, A. L., Bogacz, J. P. and Wraight, C. A. (1984) J. Cell. Biochem. 24, 243-259
4 Oettmeier, W., Godde, D., Kunze, B. and Höfle, G. (1985) Biochim. Biophys. Acta 807, 216-219
5 Clayton, R. K., Fleming, H. and Szuts, E. Z. (1972) Biophys. J. 12, 46-63
6 Tischer, W. and Strotmann, H. (1977) Biochim. Biophys. Acta 460, 113-125
7 de Vitry, C. and Diner, B. A. (1984) FEBS Lett. 167, 327-331
8 Brown, A. E., Gilbert, C. W., Guy, R. and Arntzen, C. J. (1984) Proc. Natl. Acad. Sci. USA 81, 6310-6314
9 Oettmeier, W., Masson, K., Soll, H. J., Hurt, E. and Hauska, G. (1982) FEBS Lett. 144, 313-317
10 Oettmeier, W., Masson, K. and Dostatni, R. (1986) submitted for publication
11 Debus, R. J., Feher, G. and Okamura, M. Y. (1985) Biochem. 24, 2488-2500

USE OF BARLEY MUTANTS TO PROBE THE ROLE OF PHOTOSYNTHETIC ELECTRON
TRANSPORT IN THE MODE OF ACTION OF NITRODIPHENYL ETHER HERBICIDES

J R BOWYER[1], P CAMILLERI[2], AND S A LEE[2], [1]DEPARTMENT OF BIOCHEMISTRY,
ROYAL HOLLOWAY AND BEDFORD NEW COLLEGE, EGHAM HILL, EGHAM, SURREY TW20 OEX
AND [2]SITTINGBOURNE RESEARCH CENTRE, SITTINGBOURNE, KENT ME9 8AG, UK.

1. INTRODUCTION

   Nitrodiphenyl ether herbicides cause a rapid light and oxygen-dependent
pigment bleaching and peroxidation of unsaturated membrane lipids in leaves
(1). These effects resemble those of methyl viologen, the action of which
depends on an enhancement of superoxide formation by catalysing electron
transfer from Photosystem I to oxygen (2). However, a number of observations indicate that the underlying mechanism in nitrodiphenyl ether
(NDPE) toxicity is different from that of methyl viologen. In most cases,
NDPEs do not enhance superoxide formation in experiments with isolated
thylakoids (3). In experiments with intact tissue, the photosynthetic
electron transport inhibitor DCMU[1] provides a marked protection from
methyl viologen, but, except in the case of the alga Scenedesmus obliquus,
this is not seen with NDPEs (4,5). Furthermore, chlorodiphenyl ether
analogues of NDPEs which have the same herbicidal characteristics, have
chemical properties which preclude radical formation by Photosystem I (6).

   In order to demonstrate whether Photosystem I is required for NDPE
toxicity, we have used a barley mutant, viridis-zb[63], which totally lacks
Photosystem I, but has normal levels of Photosystem II (7). We have used
another mutant, viridis-zd[69], which has more than 50% of the wild type
activity for Photosystem I, but which is highly deficient in PhotosystemII
(8). We introduce a novel NDPE, 5-[2-chloro-4-trifluoromethyl phenoxy]-2-
nitroacetophenone oxime-o-(acetic acid, methyl ester) referred to below
as DPEI.

2. MATERIALS AND METHODS

2.1 Effects of herbicides on leaf discs

   Leaf discs (11 mm in diameter) were cut from the leaves of tobacco
   (Nicotiana tabacum) avoiding the main vein. Sufficient discs from a
   number of plants were mixed together to obtain a random distribution
   and then used for various experiments. In each experiment, 20 discs
   were submerged in 30 ml water containing the herbicide in a Dreschel
   bottle bubbled with water-saturated O_2 at a rate of 100 ml min^{-1}.
   Illumination was provided by three 20W fluorescent strip lights at
   an intensity of 4.5 Wm^{-2}. The chlorophyll content of the discs was
   estimated by grinding 10 discs in 10 ml of 4:1 v/v acetone: water

---

[1]Abbreviations: DCMU, 3-(3', 4'-dichlorophenyl)-1,1-dimethylurea; NDPE,
p-nitrodiphenyl ether; DPEI, 5-[2-chloro-4-(trifluoromethyl)phenoxy]-2-
nitroacetophenone oxime-o-(acetic acid, methyl ester).

Biggens, J. (ed.), Progress in Photosynthesis Research, Vol. III. ISBN 90 247 3452 5
© 1987 Martinus Nijhoff Publishers, Dordrecht. Printed in the Netherlands.

using a Potter Elvehjem homogeniser and measuring the absorption of the solution at 652 nm after centrifugation (9).

2.2 Hydrocarbon formation in herbicide-treated barley

Barley seedlings grown in vermiculite were treated with herbicides 6 days after germination. Aqueous solutions of herbicide were applied by dipping the leaves in the solution. Seedlings were then kept at 25°C under an illumination of 35 Wm^{-2}. After appropriate periods of illumination, two shoots per treatment were removed, placed in a 100 x 10 mm test tube sealed with a Subaseal and placed under a bright illumination (330 Wm^{-2}) for 4 hours. Sample heating was avoided using a perspex tank of water through which water flowed. A 1 ml sample of the head space was removed using a gas tight syringe and analysed by gas chromatography.

3. RESULTS AND DISCUSSION

3.1 Effects of DPEI on leaf discs

Experiments in which leaf discs were treated with DPEI but maintained in darkness or bubbled with nitrogen confirmed that both light and oxygen were required to elicit chlorophyll bleaching by DPEI. The results in Table 1 below show that pre-treatment of the leaf discs with 10µM DCMU almost completely eliminated the effect of 100µM methyl viologen on chlorophyll degradation, but had little or no effect on the action of DPEI.

TABLE 1. Chlorophyll degradation by DPEI and methyl viologen in tobacco leaf discs: effect of DCMU

Leaf discs were floated on 10µM DCMU (added in dimethylsulphoxide to give 0.1% v/v) solution or on control solution containing only 0.1% v/v dimethylsulphoxide for 28 hours in the dark before being transferred into the respective herbicide solution with or without 10µM DCMU, in the light. DPEI was added from a stock solution in dimethylsulphoxide to give 0.1% v/v final concentration of dimethylsulphoxide.

Treatment	Chlorophyll content (mg chlorophyll. g fresh wt^{-1})
Control	1.41
DPEI (1 x 10^{-4}M)	0.60
Methyl viologen (1 x 10^{-4}M)	0.25
DCMU (1 x 10^{-5}M)	0.94
DPEI + DCMU	0.64
Methyl viologen + DCMU	1.23

These results further confirm that DPEI behaves like other NDPEs in experiments with higher plants and suggest that, at least in higher plants, whole chain photosynthetic electron transport plays a less important role in NDPE-toxicity than in methyl viologen toxicity. In contrast, in Scenedesmus obliquus, we have confirmed using DPEI the observation of Kunert and Boger (5) that NDPE-toxicity is eliminated by DCMU, implying a greater role for photosynthetic electron transport in this organism.

3.2 Effect of DPEI on barley mutants

When 7 day old wild type barley seedlings were treated with methyl viologen or DPEI, visible signs of necrosis were apparent after 24 hours, and the seedlings were severely necrotic after 48 hours. In contrast, under the same conditions, seedlings of the viridis-zb[63] and zd[69] mutants were apparently unaffected by methyl viologen but were severely affected by DPEI. These visible effects were confirmed by an electron microscopy investigation of the ultrastructure of the tissues (not shown). In seedlings treated with DPEI, swollen chloroplasts and vacuoles in the cytoplasm were observed 25 hours after herbicide application. This was followed by rupture of the tonoplast and of chloroplast envelopes and then separation of the cell membrane from the cell wall due to loss of turgor. Unfortunately, these ultra-structural effects give few clues as to the primary target of the herbicide: they may simply reflect those membranes which are most susceptible to lipid peroxidation or to changes in cell turgor.

The herbicidal actions of methyl viologen and NDPEs, although probably resulting from different primary processes, both lead to the peroxid-ation of unsaturated lipids. These peroxides decompose to produce hydrocarbon gases (10) which can be measured to provide a more quantitative assay of herbicidal toxicity. The results of experiments performed using wild type barley and the viridis-zb[63] mutant lacking Photosystem I are shown in Figures 1 and 2.

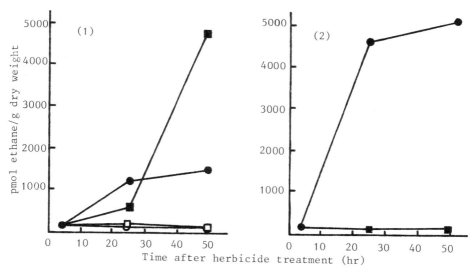

Figure 1. Production of ethane from barley leaves treated with DPEI.(◐) wild type leaves treated with 0.1% v/v dimethylsulphoxide, (●) wild type leaves treated with 10^{-5}M DPEI, (□) vir-zb[63] mutant leaves treated with 0.1% v/v dimethylsulphoxide, (■) vir-zb[63] mutant leaves treated with 10^{-5}M DPEI.

Figure 2. Production of ethane from barley leaves treated with methyl viologen. (●) wild type leaves treated with 2.5 x 10^{-4}M methyl viologen (■) vir-zb[63] mutant leaves treated with 2.5 x 10^{-4}M methyl viologen.

The results appear to confirm the observations made by a visual inspection and electron microscopy, that lipid peroxidation occurs in the wild type but not viridis-zb[63] after methyl viologen treatment, whereas DPEI causes lipid peroxidation in both strains. Similar results were obtained using the viridis-zd[69] mutant (not shown).

The results establish clearly that electron transfer through Photosystem I is not essential for NDPE toxicity in barley. This rules out the hypothesis that DPEI must be reduced to a radical by Photosystem I in order to elicit its herbicidal effects (5). The results also confirm that the presence of Photosystem I is essential for methyl viologen toxicity, because methyl viologen is not reduced at significant rates by Photosystem II. Furthermore, the results show that Photosystem II is not essential for DPEI toxicity, but is required for methyl viologen activity in order to provide reducing equivalents to Photosystem I. The results rule out mechanisms for NDPE action which involve cyclic electron transport around Photosystem I, oxygen reduction by Photosystem I or effects on the activation of Calvin cycle enzymes by Photosystem I (11).

The mode of action of NDPEs remains a mystery. Radical reactions have been implicated in lipid peroxidation, but the results reported here and elsewhere indicate that radical formation by the NDPE itself is not required. Specific structural features required for the toxicity of NDPEs imply that they bind to a receptor protein (C Swithenbank, unpublished results). Binding to this receptor may facilitate a photochemical electron transfer reaction involving an unidentified chromophore, resulting in radical formation. An action spectrum for NDPE toxicity in Chlamydomonas eugametos indicated that light absorbed by both chlorophylls and carotenoids elicited NDPE toxicity, although the lack of protection by DCMU suggested that this light was not needed to drive whole chain photosynthetic electron transport (12).

Acknowledgements: We thank P P G Industries Inc for supplying us with DPEI. We are indebted to Prof von Wettstein and Dr D Simpson for providing seeds of the barley mutants and for their kind hospitality during a visit to the Carlsberg Laboratory in September 1984.

REFERENCES

1 Orr, G.L. and Hess, F.D. (1982) A.C.S. Symp. Ser. 18,131-152
2 Summers, L.A. (1980) 'The Bipyridinium Herbicides' Academic Press, London
3 Lambert, R., Kroneck, P.H. and Boger, P. (1984) Z Naturforsch. 39c, 486-491
4 Ensminger, M.P. and Hess, F.D. (1985) Plant Physiol. 78, 46-50
5 Kunert, K.J. and Boger, P. (1981) Weed Sci. 29, 169-173
6 Ensminger, M.P., Hess, F.D. and Bahr, J.T. (1985) Pestic. Biochem. Physiol. 23, 163-170
7 Nugent, J.H.A., Møller B.L. and Evans, M.C.W. (1980) FEBS Lett. 121, 355-357
8 Machold, D, Simpson, D.J. and Møller, B.L. (1979) Carlsberg Res. Commun. 44, 235-254
9 Arnon, D.I. (1949) Plant Physiol. 24, 1-15
10 Dumelin, E.E. and Tappel, A.L. (1977) Lipids 12, 894-900
11 Wettlaufer, S.H., Alscher, R. and Strick C. (1985) Plant Physiol. 78, 215-220
12 Ensminger, M.P. and Hess, F.D. (1985) Plant Physiol. 77, 503-505

DIFFERENCES IN PHOTOSYNTHESIS BETWEEN A DIURON-RESISTANT, AN ATRAZINE-RE-
SISTANT, A BENTAZON-RESISTANT AND A SUSCEPTIBLE BIOTYPE OF THE GREEN ALGA
Monoraphidium pusillum (PRINTZ).

BENGT LUNDEGARDH, DEPT OF PLANT PHYSIOLOGY, SWEDISH UNIV. OF AGRIC.
SCIENCES, S-750 07 UPPSALA, SWEDEN

1. Introduction
  A diuron-resistant, an atrazine-resistant and a bentazon-resistant biotype of the unicellular green alga *Monoraphidium pusillum* were selected
during an adaptation period of 9 months. Both the diuron-resistant and
the atrazine-resistant biotype seem to be affected in the herbicide
binding protein of the photosystem II complex but not at the same site
(cp. Kyle 1985). However, the mutant resistant against bentazon seems to
be resistant in two ways; one change in the thylakoidmembrande and one
metabolic change.
  The aim of this study is (1) to indicate differences between the biotypes under the growth phase of the cultures and (2) to establish changes
in thylakoidmembrane with luminescence.

2. MATERIALS AND METHODS
  The biotypes were continuously cultured in a sterile nutrient solution
(0.24 mM $CaCl_2$, 0.30 mM $MgSO_4$, 0.22 mM K_2HPO_4, 0.19 mM Na_2CO_3, 23.3 mM
$NaNO_3$, micronutrients and tris pH 7.0) with 8 µM diuron (Karmex 80) to
the diuron-resistant biotype, 15 µM atrazine (Gesaprin 50) to the atrazine-resistant and 2 mM bentazon (Basagran 480) to the bentazon-resistant.
  The oxygen evolution was determined polarographically using a Clarke
type electrode in a thermostated cuvette at 20ºC in red light produced
by a xenon lamp provided with a cut off filter >600 nm. Before the measurements the algae were centrifuged during 3 minutes at 2200xg. The pellet
was suspended in new nutrient solution and adjusted to an absorbance of
1.2 at 550 nm and 1 cm thickness.
  The chlorophyll and the carotenoid content was measured in methanol
extract according to Senger (1970).
  The luminescence study was performed as described by Mellvig and
Tillberg (1984). The algae were suspended in HCO_3^- free medium.

3. RESULTS AND DISCUSSION
  The growth, measured as the change in absorbance at 550 nm (see Fig. 1.),
is very low for both the bentazon-resistant and the atrazine-resistant biotype at the beginning of the growth phase. After 4 days the growth rate
for the bentazon-resistant increases which could be due to detoxification
of bentazon. At the same time the growth rate for the control and the
diuron-resistant biotype decreases.
  Oxygen evolution of the control in red light decreases under the growth
phase from over 300 µmol O_2/mg chlxh to 62 µmol O_2/mg chlxh. The resistant
biotypes have the lowest oxygen evolution compared to the control between
7 and 10 days and the highest after 15 days. This could be due to a CO_2
deficit in the control culture, developed already after a couple of days

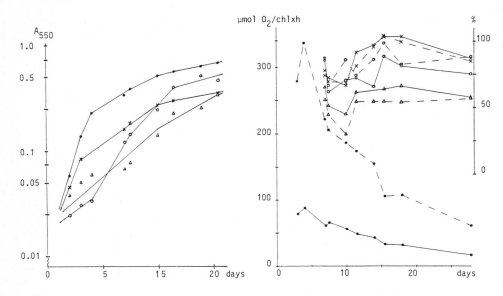

FIGURE 1. The growth of the four biotypes measured as absorbance at 550 nm

FIGURE 2. Oxygen evolution of the control and of the three resistant biotypes, given as percentage of control

✗——✗ , ✗---✗ Diuron-resistant 73 W/m^2 and 320 W/m^2 red light

△——△ , △---△ Atrazine-resistant   "

o——o , o---o Bentazone resistant   "

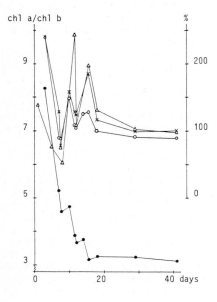

FIGURE 3. Chl a/chl b ratio in the
control and in the three
resistant biotypes, given
as percentage of control

✗——✗ Diuron-resistant

△——△ Atrazine-resistant

o——o Bentazon-resistant

TABLE 1. Early output (during 1 minutes) of luminescence as percentage of total output during 3 minutes 5 s 700 nm light or 30 s white light.

alga	12 days old cells		35 days old cells	
	700 nm	white	700 nm	white
Control	79	82	79	92
Diuron-resistant	89	85	88	79
Atrazine-resistant	94	94	95	94
Bentazon-resistant	84	78	83	75

TABLE 2. Total number of photons emitted per mg chlorophyll during 3 minutes as percentage of the control after 5 s 700 nm light or 30 s white light.

alga	12 days old cells		35 days old cells	
	700 nm	white	700 nm	white
Diuron-resistant	64	82	47	51
Atrazine-resistant	67	103	56	77
Bentazon-resistant	59	89	62	75

as no decrease in the growth capacity was observed during the first 7 days if the cultures were aerated. After 4 weeks the photosynthetic capacity is about 85% of the control for the diuron-resistant and the bentazon-resistant types and 56% for the atrazine-resistant one.

At the beginning of the growth phase both the chl a/chl b ratio and the chlorophyll/carotenoid ratio fluctuate very much in the resistant biotypes, probably due to a heterogenous cell division. However, the fluctuation, especially in the chl a/chl b-ratio, is less in the bentazon-resistant biotype than in the other two. During the first 18 days the three resistant biotypes mostly show a higher chl a/chl b-ratio than the control. Obviously there is a larger difference in the chlorophyll/carotenoid ratio between the four biotypes. The mean value of the atrazine-resistant type is, during the same time, lower than the control while the bentazon-resistant shows a higher mean value.

At the end of the growth phase the fluctuation decreases, and the diuron-resistant and the atrazine-resistant biotype get a mean chl a/chl b ratio close under the value of the control. The chl a/chl b ratio in the bentazon-resistant is much lower. The results may be explained by the fact that the cells of the bentazon-resistant biotype are larger than the those of the other types and have a lower rate of division (Lundegårdh, unpublished).

The luminescence from 12 and 35 days old cells indicates a slow emission of photons in the bentazon-resistant biotype (Table 1. and 2.). This confirms that the bentazon-resistant type has a more drawn-out lifecycle.

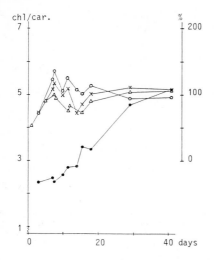

FIGURE 4. Chlorophyll/carotenoid
ratio in the control and
in the three resistant
biotypes, given as percentage of control

×——× Diuron-resistant

△——△ Atrazine-resistant

○——○ Bentazon-resistant

However, this biotype can show both a high growth rate (see Fig. 1.) and
oxygen evolution (see Fig. 2.). The ratio between early and late emission
of photons is independent of cell age and light treatment in the atrazine-resistant biotype. Further, the output of luminescence is very high
during the first minutes. These results further confirm that triazine-resistant biotypes have a change in the Q_B-protein in photosystem II which
leads to an increased backflow of electrons through photosystem II. This
may be due to a change in the redox poising of the Q_A^-/Q_B reaction pair
and/or a change in the affinity of PQ to the Q_B-protein (Pfister and
Arntzen 1969). This stabile luminescence in the atrazine-resistant biotype could explain the low change in oxygen evolution in this biotype
compared to the control. Of the three biotypes the diuron-resistant is
most similar to the control, in growth rate, oxygen evolution and luminescence.

REFERENCES
1 Kyle, D.J. (1985) Photochem. Photobiol. 41, 107-116
2 Senger, H. (1970) Planta 90, 243-266
3 Mellvig, S. and Tillberg, J.-E. (1984) Photochem. Photobiol. 39, 817-822
4 Pfister, K., Arntzen, C.J. (1979). Z. Naturforsch, 34c, 996-1009

FLUORESCENCE   INVESTIGATION   of   DCMU   and   S-TRIAZINE   HERBICIDE
ACTIVITY in CROP and WEED SPECIES

Panneels P.[*] , Van Moer A.[**], Reimer P.[*] , Salis P.[*] , Chouhiat A.[*] ,
Lannoye R.[*] , and Figeys H.[**]

[*] Lab. Physiologie végétale, Av. Paul Heger, 28/ 1050 Bruxelles
[**] Lab. Chimie organique et Chimie organique physique
    Av. F.D. Roosevelt, 50 / 1050 Bruxelles /   Belgium

1. Introduction

DCMU/ triazine type herbicides are known to block the electron
transport between the primary and secondary electron accepting
plastoquinone (refered as Q_a and Q_b respectively) at the
photosystem II (PSII) acceptor side (1, 2). The detection of
herbicide effects in whole plants is greatly eased by using
chlorophyll fluorescence as an intrinsic probe of PSII primary
photochemistry (3). Numerous attempts have been reported to use
fluorescence measurements as a reliable analytical tool for
detecting plants resistant to atrazine, monitoring translocation
and detoxification of herbicides (4).
The present study describes results obtained in barley and in a
weed species with atrazine and a newly synthesized active
molecule using an improved computerized device which allows a
precise measure of the fluorescence rise.

2. Materials and methods

Chlorophyll fluorescence (above 710 nm) of leaf discs (diameter
of 8 mm) was measured with a BRANKER SF10 fluorometer (Branker
Research Ltd, Ottawa). Intensity of the exciting light (LED
MV522c) was 11.2 microEm^{-2}s^{-1} unless stated otherwise. The
signals were digitized on line in real-time by an analogic to
digital (A/N) 8 bits conversion card and stored in a CBM64
microcomputer, using machine-code language to derive the ratios
P/O, I/O, I/P. Seedlings of Solanum nigrum (seeds kindly provided
by Dr. Stryckers, RUG-Gent) were grown in soil in a glass house
for approximately five weeks. Chloroplasts were isolated as
described in (5). Chloroplasts were also isolated from leaves of
10-15 days old barley seedlings (Hordeum vulgare CV Monix) as
previously reported (6). The Hill reaction was measured by oxygen
evolution with a Clark type oxygen electrode. Herbicide solutions
were either introduced directly into the reaction chamber of the
electrode or used for leaf disc infiltration (7).

3. Results and discussion

3.1 The computerized data acquisition system

A cheap 8 bits micro-computer (CBM64) was used as the heart of a data sampling system. It performs, in machine code essentially, all calculations about the collected data and controlls all the digitizing process. We added to it a self-constructed analog to digital (A/D) converter device (using A/D MN574 by Micro-Networks) performing an 8 bit conversion in less than 20 microseconds. This sytem was developped with fluorescence measurements in mind, but could, however, be used with totally different probes to measure any kind of electrical signal because it is software driven. The fluorescence signal collected with the above described fluorometer is connected to the A/D converter. The initiation and the sampling time is controlled by the computer software allowing maximal flexibility to analyse different kinds of fluorescence processes. The range allowed begins at a 100 microseconds sampling intervall to an indefinite time (practically we never sample with more than a 1 second and less than a one millisecond intervall). Some measurements require a 5 minute time sampling and this is also controlled by the computer, freeing up operator's work. The signal is first stored in the available memory (RAM) of the computer. This memory is up to 50 kilobytes big, allowing the processing of even rather small sampling intervalls to be used over extended time. Practically however, using some advanced programming techniques, it has been possible to compress a 5 minute long fluorescence signal with all the precision required in a 4 kilobyte memory space. This data is then processed to be displayed on a graphic video screen with a 320x200 pixels resolution. Various horizontal and vertical magnifications ratios may be selected. Other parameters are calculated (like I/P ratio etc.). Signal can also be normalized and integrated, allowing a beter comparison between different treatments. The data is easily stored on a floppy-disk allowing further treatment of the collected data, after many measurements. Collected data may also be meaned together to produce a so-called mean-curve, which reduces the normal random variations between repetitions. A graphic printout with the associated calculated parameters is also implemented in machine code for quick printing on an EPSON RX-80 printer, and can also be printed in various scales. The CBM VC1526 plotter is also supported.

Future development

The program fits actually on floppy-disk and will be totally transferred in Read Only Memory (ROM), which allows an immediate start-up of the machine, freeing up some more RAM memory. It will also be possible to save the data to a little cassette recorder. Adding to this a battery power supply and a LCD display, the system will be autonomous and very handy, allowing even easy on-field measurements and data storage to be made.

3.2 Fluorescence induction transients in infiltrated leaf discs

When a dark adapted leaf disc is illuminated at room temperature, chlorophyll fluorescence rises immediately to a F_0 level, then we assist to a fast time-dependent increase in fluorescence through an inflexion point I, a dip D to a peak level P. The following decline of fluorescence refelects the light-activation of the electron transport activity of the PSI acceptor side and the development of several fluorescence quenching processes. The first fluorescence rise O-I, reflects the reduction of Q_a in equilibrium with the second acceptor Q_b. The level may be correlated to $Q_a^- Q_b \rightleftharpoons Q_a Q_b^-$ equilibrium which may coexist with a fully oxydized plastoquinone pool. The I-P rise is controlled by the progressive reduction of Q_a and Q_b to the final state $Q_a^- Q_b^-$ as the plastoquinone pool become reduced. The I/P ratio was used to measure the increase of the photochemical rise (O-I) induced by atrazine and HF200e. Linear relationships were obtained between the increase of the I/P ratio, expressed as percent of the control sample, and the log of herbicide concentrations used in the infiltration step.

TABLE 1

Linear regression between the increase of the I/P ratio of treated samples (y) expressed as percent of control, and the log. of herbicide concentrations used for infiltration (x) with n = 10 observations.

	Atrazine	HF200e
Barley	77.7 x + 6.5 r= .95	177.7 x - 149.4 r= .95
Solanum nigrum Susceptible	58.9 x + 36.1 r= .94	103.6 x - 15.1 r= .96
Solanum nigrum Resistant	17.2 x - 13 r= .94	180.1 x - 119 r= .99

The angular coefficients of this effect-concentration linear relation are accurate parameters to quantify the herbicide activity in whole plants and for detecting plant resistance to atrazine. They clearly show the powerfull phytotoxic effect of the experimental herbicide HF200e on the atrazine susceptible and resistant biotype of *Solanum nigrum*.

3.3 Hill reaction

Photoreduction of DCPIP was used to compare the inhibitory effect of atrazine and of the newly synthesized molecule HF200e in triazine-resistant and triazine-susceptible chloroplasts of *Solanum nigrum* and in barley chloroplasts.

The degree of resistance (the R/S value) is defined as the I50 (half-inhibitory concentration) of the resistant biotype divided by I50 of the susceptible biotype for the same compound. The I50 concentration of atrazine and HF200e were in the range of 6.10^{-4} mole.m^{-3} in chloroplasts of susceptible *Solanum nigrum* and in the range of 2.10^{-4} mole.m^{-3} in chloroplasts of barley. Two thousands and three fold higher concentrations were needed respectively for atrazine and HF200e, to cause the same inhibition in the chloroplasts of the resistant biotype. The low degree of resistance to HF200e, a S-triazine molecule, found for the resistant biotype is a very interesting observation which needs more research. It is worth noting that this increased susceptibility was also detected when the fluorescence test was applied to infiltrated leaf-discs. Further research should reveal if choloroplasts of other weed species exhibit the same increased susceptibility to HF200e.

4. Conclusions

This is the first report of a new S-triazine molecule acting on the susceptible biotype of *Solanum nigrum* as well as on the resistant one. The chemical design of this experimental herbicide was based on reliable structure activity relationships, revealed by quantum mechanically calculated electrostatic potential maps and electron densities (8) and on the chemical interpretation of the necessary properties for an active herbicide-quinone binding environment, revealed by conformational analysis of the 32 K-Dalton protein of both susceptible and resistant biotype (J.Ruysschaert, personal communication).

REFERENCES

1 Vanrensen, J.J.S. (1982) Physiol. Plant. 54, 515-521
2 Pfister, K. and Urbach, W. (1983) in Encyclopedia of Plant Physiology (Lange, L. et al., ed.), Vol. 12D, pp. 329-391, Sringerr, Berlin
3 Voss, M., Renger, G., Graber, P. and Kötter, C. (1984) Z. Naturforsch. C 39c (5), 359-361
4 Ducruet, J.M, Gaillardon, P. and Vienot, J. (1984) Z. Naturforsch. C 39c (5), 354-358
5 Chapman, D.J., Defelice, J. and Barber, J. (1985) Planta 166, 280-285
6 Barber, J., Chow, W.S., Scoufflaire, C. and Lannoye, R. (1980) Biochim. Biophys. Acta 591, 92-103
7 Steucek, G., Hill, R. and Norbeck, M. (1985) Carolina Tips 12, 48-49
8 Figeys, H.P., Geerlings, P., Berkmans, D. and Van Alsenoy, C. (1981) J.Chem. Soc. Faraday Trans. II, 77, 721-729

We wish to thank l'Institut pour l'Encouragement de la Recherche Scientifique dans l'Industrie et l'Agriculture (IRSIA) for financial support and prof. Stryckers for supplying us with seeds of triazine susceptible and triazine-resistant biotypes. We also thank Drs. J.M. Ruysschaert and R. Brasseur for the communication of unpublished results.

PYRUVATE PHOSPHATE DIKINASE FROM MAIZE: PURIFICATION, CHARACTERIZATION, ANALYSIS OF INHIBITORS, AND ITS USE FOR BIORATIONAL DESIGN OF HERBICIDES

ARTHUR L. LAWYER, SARITA R. KELLEY, AND JUDITH I. ALLEN
CHEVRON CHEMICAL COMPANY, ORTHO RESEARCH CENTER, P.O. BOX 4010, RICHMOND, CALIFORNIA, USA

1. INTRODUCTION

Pyruvate phosphate dikinase (PPDK) was considered as a potential target for the action of novel herbicides. PPDK was chosen because of its central role in C4 photosynthetic metabolism (a process necessary for the viability of C4-type plants) and the absence of appreciable or essential PPDK activity in C3 plants and mammalian species. The relative uniqueness of PPDK to C4 plants was thought to be an important criteria for enhancing the selectivity of the "C4 herbicides" and, at the same time, minimizing the potential for mammalian toxicity. The spectrum of activity expected from a C4 herbicide was also considered to be beneficial since most grassy weeds are C4 plants while most major crops are C3 plants (the major exceptions being maize, sorgham, and sugar cane). Targeting of C4 enzymes for herbicide action has been explored by other laboratories (C. L. D. Jenkins, M. D. Hatch, personal communication).

Compounds, either designed in-house or purchased commercially, were tested for their activity against PPDK isolated from maize. Tests included determination of both reversible and irreversible inhibition of enzyme activity. The specificity of test compounds toward PPDK was also examined by determining their relative activity against phosphoenolpyruvate carboxylase (PEPCase) from wheat. Results obtained with commercially obtained compounds will be discussed.

2. PROCEDURES

2.1 Purification: (1,2) Laminar tissue, 50 g, from 2.5 to 3 week old maize (Bantam Golden Cross) were homogenized in 200 ml of 100 mM Tris-HCl, pH 8.0, 10 mM $MgSO_4$, 2.0 mM K_2HPO_4, 2.5 mM Na pyruvate, 1 mM EDTA, 10 mM mercaptoethanol, 0.5% ascorbate, 1% PVPP, and 20% glycerol. The homogenate was then filtered through 150 μ nylon mesh and centrifuged for 8 min at 10,500 g and 4 C. The supernatant was stored under N_2 at -20 C and then diluted 15 % with chromatography buffer before placing it directly onto a DEAE-cellulose column (2.5 x 47 cm, Whatman DE-52) equilibrated with chromatography buffer (100 mM Tris-HCl, pH 8.0, 5 mM $MgSO_4$, 2.5 mM Na pyruvate, 1 mM EDTA, 10 mM mercaptoethanol, and 20% glycerol. The column was washed with 300 ml of chromatography buffer at a rate of 2.3 ml min^{-1} followed by a linear gradient (1.1 l) of 0.0 to 0.5 M KCl. Fractions containing PPDK free of PEPCase activity were pooled and concentrated on an Amicon XM-300 membrane. The specific activity of the concentrated fraction was about 2 U mg protein^{-1}, 5-

Biggens, J. (ed.), Progress in Photosynthesis Research, Vol. III. ISBN 90 247 3452 5
© *1987 Martinus Nijhoff Publishers, Dordrecht. Printed in the Netherlands.*

to 10-fold higher than the starting material. The enzyme was stable for over a month when stored at -20 C under N_2.

2.2 <u>PPDK Assay:</u> PPDK activity was followed spectrophotometrically at 340 nm. Typical assays contained 100 mM HEPES buffer, pH 8.0, 5 mM DTT, 10 mM $MgSO_4$, 2.5 mM K_2HPO_4, 50 mM $NaHCO_3$, 5 mM glycine, 5 mM glucose-6-P, 330 μM NADH, 1.25 mM ATP, 2.5 U malate dehydrogenase (porcine heart, Calbiochem), 4 U PEPCase (wheat, Boeheinger-Mannheim), and 0.033 U of PPDK in a final volume of 1.0 ml at 25 C. Background absorbance changes were followed before initiating the assay with the addition of pyruvate. When determining reversible inhibition, test compounds were allowed to incubate with the enzyme for 6 min prior to assaying enzyme activity. Inhibition constants, Ki's, were determined by Dixon analysis (1/v vs. [I]) using pyruvate concentrations around the $Km_{(pyruvate)}$. Ki values were calculated assuming that the inhibition was both competitive and reversible.

2.3 <u>PEPCase Assay:</u> The conditions and concentrations used for measuring PEPCase activity were identical to those described above for PPDK except that only 0.8 U of PEPCase was used and ATP was eliminated.

2.4 <u>Irreversible Inhibition:</u> PPDK, 0.3 U in 90 μl, was mixed with 80 μl assay buffer and incubated at 25 C for 10 min before adding the inhibitor (10 μl) at a final concentration of 5.5 x Ki. After a 10 min incubation, 20 μl portions were removed and diluted 50-fold into the appropriate assays for PPDK or PEPCase activities, as detailed above.

3. RESULTS AND DISCUSSION
The abbreviated procedure for purifying PPDK from maize leaves described above was found to be beneficial in obtaining large quantities of PPDK, adequately separated from contaminating PEPCase activity. The presence of 2.5 mM pyruvate and 20 % glycerol enhanced the stability of PPDK during both purification and storage. Stability of PPDK during storage at -20 C was enhanced significantly when performed anaerobically. About 20 U of enzyme with a specific activity of around 2 was routinely obtained from 50 g of leaf tissue. Significant improvement (2.7-fold) in the activity of PPDK was obtained by substituting HEPES buffer for Tris buffer (both buffers at 100 mM, pH 8.0). Enzyme activity was also enhanced by the presence of 5 mM glycine and/or glucose-6-P.

Examples of the inhibition constants for PPDK activity obtained with commercial obtained compounds are given in Table 1. The constants are calculated assuming the inhibitors are classic reversible competitive inhibitors. The values should be compared to the Km for pyruvate of 220 ± 120 μM (the Km for ATP was 300 ± 20 μM).

Table 2 lists the commercially obtained compounds which were found to be nonspecific inhibitors of PPDK. A compound was considered nonspecific for PPDK when the inhibition constant (Ki) for that compound toward PEPCase activity was as strong or stronger than that observed toward PPDK activity. The test for compound specificity was a necessary control for our assay of PPDK activity since the assay coupled product

Table 1. Specific Inihibitors of Pyruvate Phosphate Dikinase

Compound Name	Inhibition Constant (Ki)
	µM ± std. err.
Ethyl-3-trichloropyruvate	46 ± 9
3-Bromopyruvic acid	290 ± 70
3-Fluoropyruvic acid	330 ± 60
Ethyl-3-bromopyruvate	340 ± 150
2-Phenylglyoxylic acid sodium salt	990 ± 200
Ethyl-4-chloroacetoacetate	1,400 ± 200
2-(Bromomethyl)-acrylic acid	3,900 ± 100
Ethyl-1-piperidine-glyoxylate	41,000 ± 30,000

Table 2. Nonspecific Inhibitors of Pyruvate Phosphate Dikinase

Compound Name	Inhibition Constant (Ki)
	µM ± std. err.
Methyl-2,4-diketo-n-pentanate	1.7 ± 0.4
3-Indolepyruvic acid monohydrate	4.8 ± 0.8
3-Indoleglyoxylic acid	18 ± 2
Ethyl-2,3-dibromoproprionate	11,000 ± 3,000
3-Dimethylpyruvate sodium salt	35,000 ± 14,000

formation by PPDK through PEPCase and malate dehydrogenase, the two
enzymes used in the PEPCase/specificity test. It should be noted that
some of the nonspecific inhibitors listed in Table II were the tightest
binders we observed among the commercially available compounds. However, since the inhibition of PEPCase activity was at least as strong as
that measured in the PPDK assay, the target of these strong inhibitors
was probably not PPDK.

Halogenation at the 3-position of pyruvate maintained and, for some
compounds, even enhanced binding to PPDK. For example, 3-fluoropyru-
vate, 3-bromopyruvate, and even 3-trichloropyruvate bind PPDK as well as

pyruvate (Km = 220 μM). The relatively strong binding by these compounds may have been due to enhancing the electrophilicity of the 3-position. The binding by 3-halogenated pyruvate analogs was not correlated to significant irreversible inhibition of PPDK activity.

Irreversible inhibition of PPDK was observed with 2-(bromomethyl)--acrylic acid (51% for PPDK compared to 10% for PEPCase). The irreversible activity was possibly due to the production of a stable intermediate upon nucleophilic attack by the enzyme at the 3-position. This nucleophilic attack could be performed by the purported cystidyl residue located in that area of the PPDK active site (3) though other residues could be involved (4). In comparison, the next strongest irreversible inhibitor tested was 3-fluoropyruvate (15% on PPDK and 0% on PEPCase). It is important to note that none of the test compounds completely inactivated PPDK. This was true even when the inhibitors were incubated at higher concentrations or incubated for longer time periods.

Of the compounds tested, very few were determined to be specific inhibitors of PPDK activity. The C4-herbicide program targeting PPDK for pesticide action was deferred for several reasons. The lack of specificity in the compounds tested gave us little structural leads to synthesize around. In addition, it was difficult to design inhibitors and potential herbicides analogous to the substrate (pyruvate) and product (PEP) of PPDK due to their small size.

REFERENCES

1 Sugiyama, T. (1973) Biochem. 12, 2862-2868
2 Burnell, J.N. and Hatch, M.D. (1984) Biochem. Biophys. Res. Commun. 111, 65-72
3 Yoshida, H. and Wood, H.G. (1978) J. Biol. Chem. 253, 7650-7655
4 Phillips, N.F.B., Goss, N.H. and Wood, H.G. (1983) Biochem. 22, 2518-2523

INDEX OF NAMES